Basics of the Finite Element Method

Basics of the Finite Element Method

Solid Mechanics, Heat Transfer, and Fluid Mechanics

Paul E. Allaire
University of Virginia

wcb
Wm. C. Brown Publishers
Dubuque, Iowa

Wm. C. Brown *Chairman of the Board*
Mark C. Falb *President and Chief Executive Officer*

wcb

Wm. C. Brown Publishers, College Division

Lawrence E. Cremer *President*
James L. Romig *Vice-President, Product Development*
David A. Corona *Vice-President, Production and Design*
E. F. Jogerst *Vice-President, Cost Analyst*
Bob McLaughlin *National Sales Manager*
Marcia H. Stout *Marketing Manager*
Craig S. Marty *Director of Marketing Research*
Marilyn A. Phelps *Manager of Design*
Eugenia M. Collins *Production Editorial Manager*
Mary M. Heller *Photo Research Manager*

Book Team

David K. Faherty *Editor*
Nova A. Maack *Associate Developmental Editor*
Lisa Bogle *Designer*
Kevin P. Campbell *Production Editor*

Library of Congress Catalog Card Number: 84-71807

ISBN 0-697-00119-9

Printed in the United States of America
10 9 8 7 6 5 4 3 2 1

This book is dedicated to my wife, Jan

Contents

Appendixes 603

Preface

The finite element method is one of the most flexible tools available for solving engineering problems of the kind involved in analyzing the deformation of solids, the transfer of heat, the flow of fluids, or electrical problems. It can be applied to systems with virtually any geometric configuration or boundary conditions. Most textbooks on finite elements have been devoted almost exclusively to structural applications, and only recently have books appeared that stress the general applicability of the method to wide-ranging fields of engineering. Engineers in industry, government, and universities have made the finite element method the basis of computer programs for the analysis of a variety of problems, but because of the narrow scope in which the method has usually been presented, the typical user of such a program is not familiar with either the philosophy or the details of the technique. To this person, the analysis is a black box that somehow transforms input into output.

This text is intended as an introduction to the finite element method that will enable the engineer or scientist who has had no previous exposure to it to understand the formulation and solution procedures that are commonly used. Topics are presented here in the order of increasing difficulty of solving the differential equations that govern the phenomena being analyzed. Thus, the method is first considered in the context of problems in which the relationships between independent and dependent variables can be expressed by ordinary differential equations. The discussion then progresses to scalar partial differential equations and, finally, to vector partial differential equations. After each technique is developed, it is applied to examples of engineering analysis, including heat conduction, heat convection, ideal fluid flow, viscous fluid flow, electrical and pipe networks, axial and lateral deflection of beams, plane stress, plane strain, and other topics.

In each chapter the relevant equations are developed in substantial detail, so that one not familiar with finite elements may follow the complete derivation, starting with the differential equation and associated boundary conditions governing an engineering problem and proceeding with the full finite element solution of that problem. Concepts such as interpolation functions within elements, residual methods, variational methods, methods of assembling global matrices, and the final solution techniques for obtaining nodal values are introduced, both in general and in specific examples. In keeping with the introductory nature of this text, advanced discussions of higher order elements and other esoteric topics have been omitted.

This text is designed for use in the third or fourth year of an undergraduate curriculum in mechanical or civil engineering or in the first year of graduate study. An undergraduate class should be able to cover most—but not all—of the material; Chapter 13, on isoparametric elements, could be left out, for example. For graduate students, the advanced topics, such as residual methods and isoparametric elements, can be stressed. The text has been used in a fourth-year mechanical engineering course and in a first-year graduate course in applied mechanics at the University of Virginia. Most of the problems included in the book have been tested by members of those classes.

Besides its use as a college text, the book describes methods that can be used to solve problems encountered in industrial engineering practice. Many of the examples appearing in the text and in the problems are typical of such real-world situations. The appendixes contain computer programs for the efficient solution of differential equations. These programs incorporate general input and output formats that make them useful in a variety of applications.

Acknowledgements

One thing I have learned about the publication process is that it requires a lot of effort from many people. I have tried to give credit to as many of those people as I can here.

Many faculty members encouraged me to teach the course at the University of Virginia that resulted in my writing this book. These include Dr. Lloyd Barrett, Dr. David Lewis, Dr. Edgar J. Gunter, and Dr. John Thacker. Dr. Furman Barton in the Civil Engineering department has for years patiently answered my questions on finite elements for structures. Faculty in environmental science, engineering physics, chemical engineering, electrical engineering, and biomedical engineering have sent their students to take the course.

Several graduate students helped me with the text and computer programs. Dr. Chip Quietzsch (now graduated) worked on programs FINONE and STRESS. Ms. Rita Schnipke worked with the material on time-transient methods. Dr. Mitchell Rosen (also graduated) helped with comments along the way. Mr. Robert Krout helped with the fluid flow material in chapter 5.

I was also fortunate to have two excellent typists. Mrs. Susan Bailey typed the first draft and some of the constantly needed revisions as I learned how to write a book (they don't teach you that in grad school) while teaching a class from it. Mrs. Sandy Smith typed the major revisions. Both people were tireless in their efforts.

Many of my students were kind enough to proof the manuscript. Mr. Joe Scholander checked nearly the entire first draft. Ms. Amy Gadsden did an outstanding job with the revised text.

At Wm. C. Brown, I must thank Mr. Robert Stern, who was my editor while I was writing this text. When I met Mr. Stern in the late 1970s, I gave him what I thought was an essentially finished manuscript of about 500 typed pages. He obtained numerous reviews, evaluated them, discussed them with me, and recommended changes. What was originally intended to be a first graduate level book eventually became a combined advanced undergraduate/first year graduate level text. Much material was added, including all of chapter 4. The final manuscript was about 1,000 typed pages. Mr. Stern stuck with me all the way.

Other people at Wm. C. Brown who were instrumental in completing the text included associate editor Nova Maack and production editor Kevin Campbell. Mrs. Maack redid all of the numbering schemes for the equations, figures, and tables. Her unfailing good humor made her phone calls a pleasure. Mr. Campbell did an outstanding job of keeping me informed about the production schedule so that I knew whether the index or the proofs for chapter 14 should be done next.

Last, but far from least, come the reviewers. Dr. Rod W. Douglass of the University of Nebraska and Dr. Henry Sneck of Rensselaer Polytechnic Institute had a major hand in shaping the text so that it would be suitable for undergraduates. Their comments and criticisms were very strong influences, even when their views differed from mine. Other skillful reviewers were Dr. Thomas R. Rogge, Iowa State University; Dr. Ronald L. Sack, University of Idaho; and Dr. Kaspar William, University of Colorado-Boulder.

I have enjoyed writing this book. I hope you find it helpful.

Part 1
Introduction, Element Properties, and Assembly

Chapter 1

General Concepts

1.1 Analysis of Complex Engineering Problems

In modern-day engineering design analysis, numerical methods are often required in order to obtain information about potential design changes. Temperatures, fluid velocities, or stresses to be calculated for a particular engineering problem are often complicated by geometry, variable properties, and other difficulties. The numerical method with the most capability for solving a wide range of engineering problems is the finite element method discussed in this text. With proper use, it can make possible the analysis of large-scale engineering systems.

Much of engineering training is concerned with the development of boundary value problems that describe various physical phenomena. In some cases, simple differential equations with simple boundary conditions are the result. These may be solved to obtain the variation of the particular physical characteristics as a function of either space or time or both. Usually this situation does not occur because the physical system is complex, the differential equations are complex, and there is no simple solution.

The major advantage of the finite element method is that it can be used to solve virtually any engineering problem for which a differential equation can be written. When analytic techniques of solving differential equations fail, finite elements or some other numerical method must be used to obtain a solution. Perhaps the major disadvantage of the finite element method is that it is somewhat complex. Even the solution of differential equations describing simple physical systems can be difficult. The reader is asked to bear with the introduction of new concepts in the hope that a significant increase in analytical capability will result.

Generally the complexity of the finite element method for a particular problem will be proportional to the complexity of the differential equation for that particular problem. That is, a simple heat conduction problem resulting in a second-order differential equation will result in a correspondingly simple finite element analysis, while the vibration of a structure with many degrees of freedom, such as an automobile frame, described by many differential equations will result in a quite complicated finite element analysis. Fortunately, the principles learned in understanding the simple analysis can be extended to the more complicated problems without great difficulty.

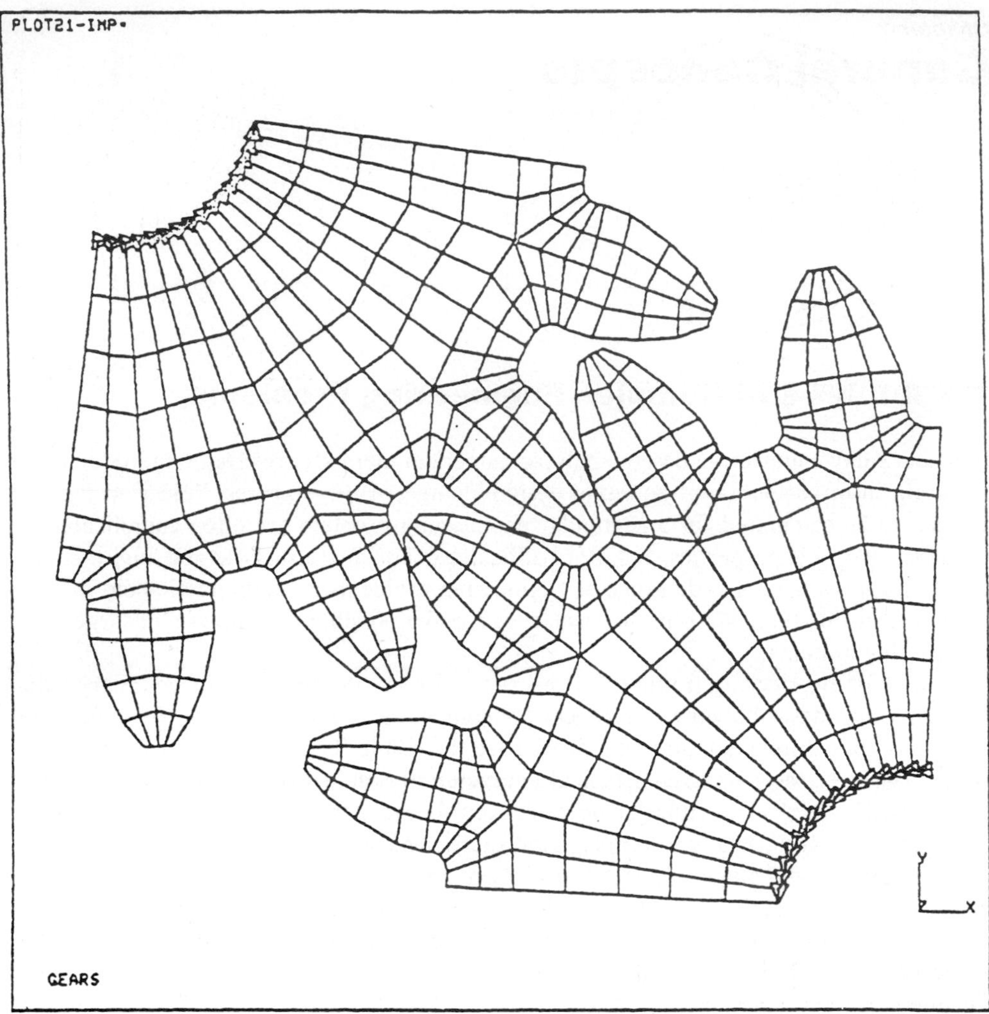

Figure 1.1 Finite element model of two spur gears in contact. Plot from ANSYS computer program (reproduction courtesy of Swanson Analysis Systems, Inc.)

Finite elements are extensively used in stress analysis. Figure 1.1 shows the meshing of two spur gears that were modeled by using the widely used, commercially available computer program ANSYS. (The plots in this section were supplied courtesy of Swanson Analysis Systems, Inc.) In this case, one quarter of each spur gear is modeled. The objective is to determine how the gear teeth deform during contact and what the resulting stresses will be. For analytical purposes, each quadrant is divided into small quadrilateral areas called finite elements.

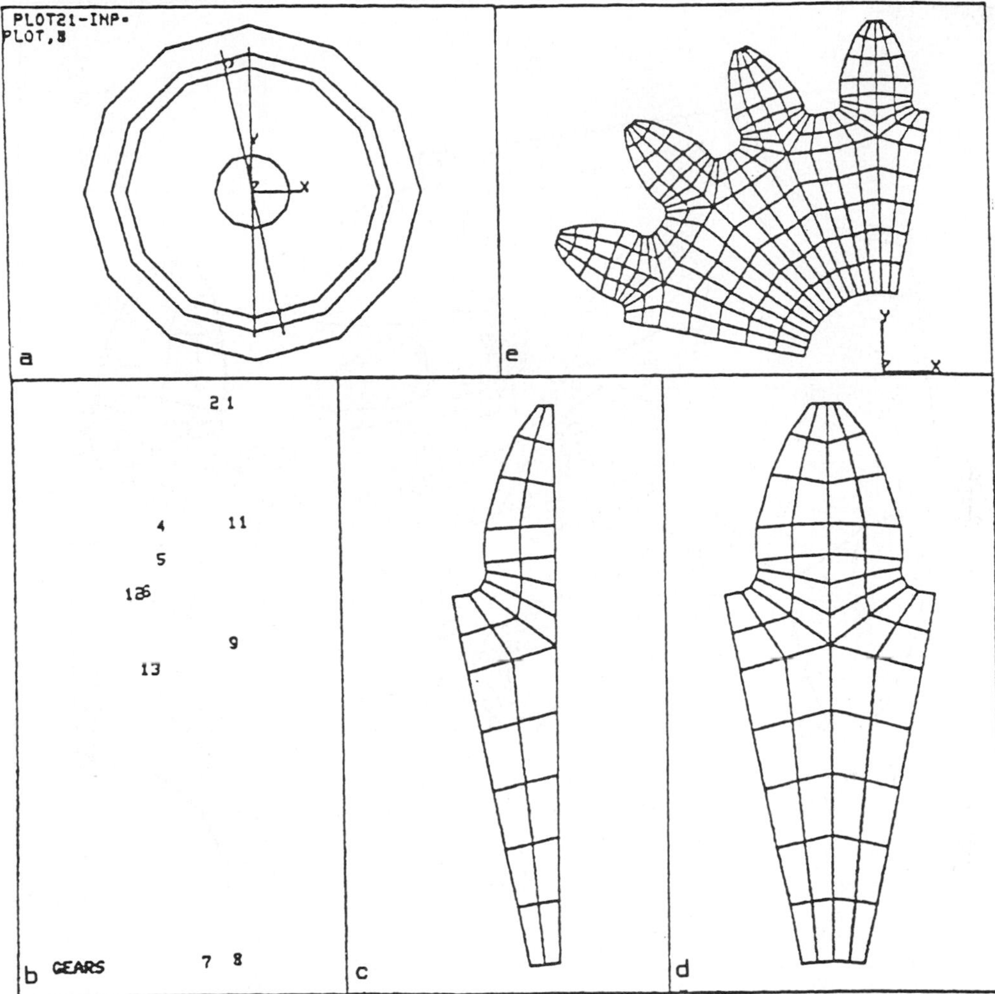

Figure 1.2 Construction of finite element model of spur gear, using ANSYS (reproduction courtesy of Swanson Analysis Systems, Inc.)

A finite element model of the gear is developed according to the steps shown in Figure 1.2. The gear is divided up into teeth, as shown in window (*a*), and then further subdivided into half teeth, as shown in window (*c*). Key points in the half tooth are indicated in window (*b*). Once the elements are determined for this, the coordinates are "folded" over the line of symmetry down the center of the tooth as shown in window (*d*). These results are then rotated about the origin of the *xy* coordinate system, shown in window (*e*), to obtain the finite element model for four teeth in one quarter of the gear.

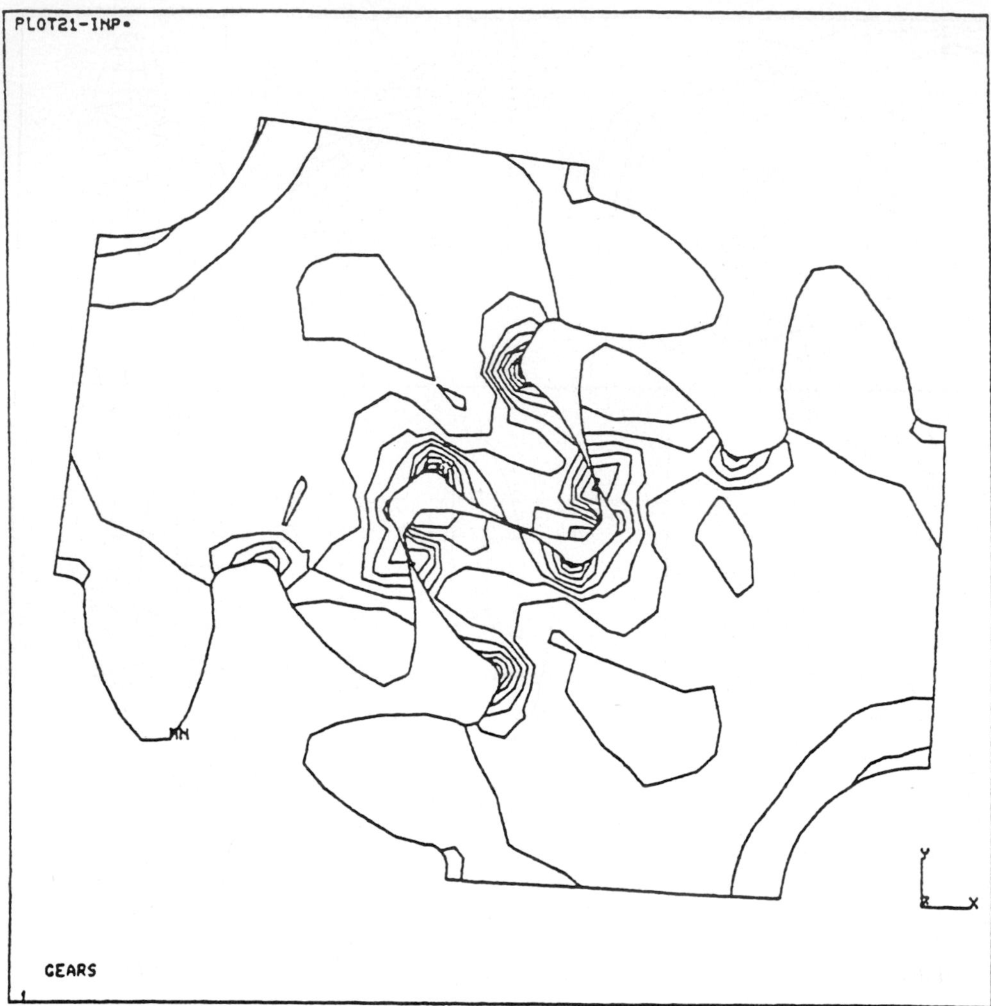

Figure 1.3 Stress contours in gear teeth calculated with ANSYS (reproduction courtesy of Swanson Analysis Systems, Inc.)

Torque is transmitted from one gear to the other. Lines of constant stress at one contact position of the gears are plotted in Figure 1.3. The stress along any of the lines can be obtained from the output of the finite element program used (ANSYS, in this case). A more detailed look at the stresses is shown in Figure 1.4. High stresses occur in the areas at which the teeth come in contact and also around the roots of the teeth. Usually the gear must be designed so that the stresses in these two areas do not exceed some limiting value.

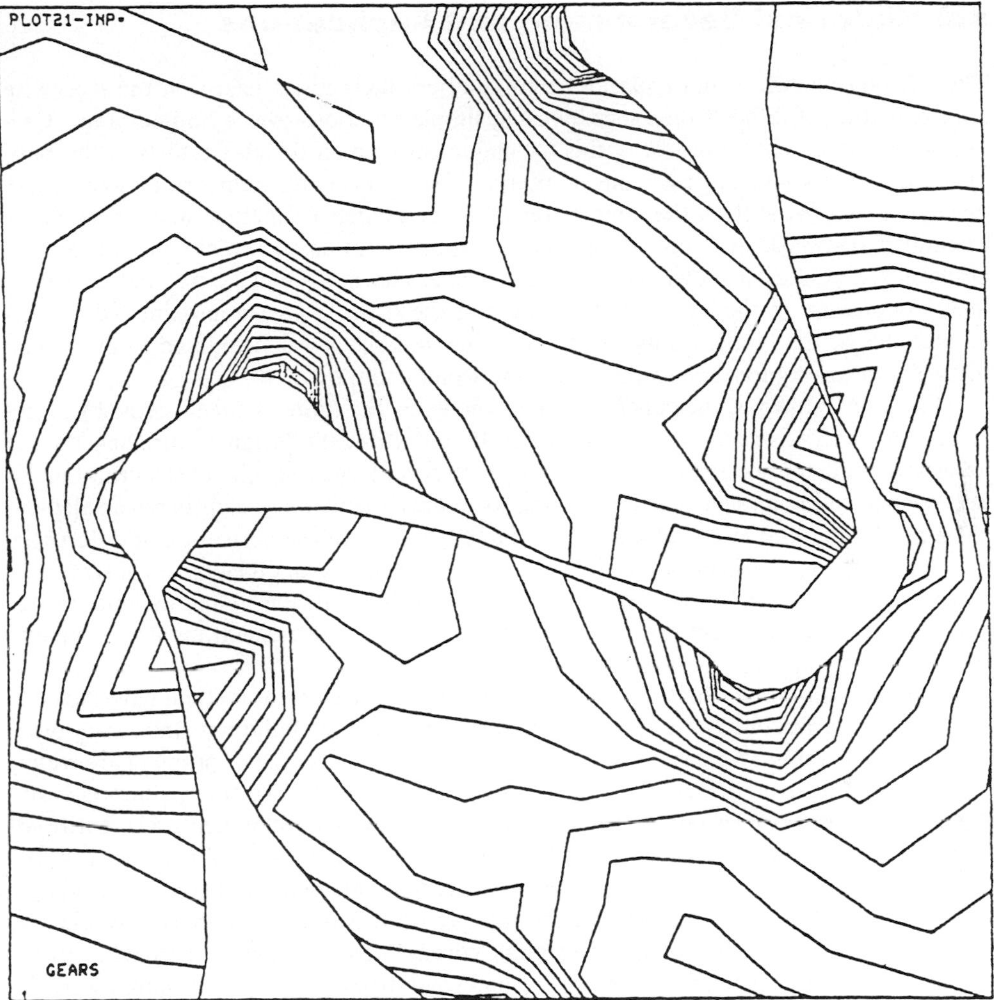

Figure 1.4 Details of stress contours in gear teeth determined with finite element program ANSYS (reproduction courtesy of Swanson Analysis Systems, Inc.)

The steps involved in this type of analysis are discussed in this text. Chapter 2 indicates how the node points should be numbered to minimize solution time on the computer. Chapter 3 presents various element types and their properties. Chapter 4 shows how the finite element equations are developed and assembled. Finally, Chapter 14 indicates how two-dimensional stress of this type is analyzed.

1.2 Historical Development and Applications

The solution of differential equations by numerical methods is based on the piecewise approximation of difficult functions by using simple functions over a limited range. Certainly the oldest written record of this technique is found on the clay tablets of the Babylonians. They knew and used linear interpolation to evaluate numbers between those given in tables. We still use the same method for trigonometric functions and others. Early oriental mathematicians applied geometrical approximation to the circumference of a circle. Straight lines of known length were placed as chords around a circle and their total length added up. In the limit the total length of the straight lines approximated the circumference of the circle. Using this method the number π was evaluated to more than forty significant figures before the rise of the Roman empire.

Early in the 1950s, the general method known as the finite element method was developed to analyze structural frames in aircraft. The stringent design requirements of the aerospace industry necessitated an accurate analysis of large numbers of structural members connected in nonregular fashion. Each structural member was considered an element to be assembled into an entire air frame, thus the name. In the derivation of differential equations for engineering one usually looks at infinitesimally small elements of structure. In the new approach, however, each structural member was considered as finite in size and then modeled in its entirety. Thus, the full name of finite element method came into being.

In recent years the structural applications of the finite element method have included analysis of the Boeing 747 airplane, earthquake analysis of buildings and many other structural problems. Quite general programs to carry out this type of problem are widely available in industrial and government use. One example is the extremely comprehensive NASTRAN program, which required many man-years of development. The entire air frame of the United States Space Shuttle has been analyzed with the aid of this program.

By the early 1960s, it was recognized that the finite element method is more general than had been previously known. During this period it was used to analyze conductive heat transfer and the flow of fluids through porous media. General approaches to the problems of heat transfer, fluid mechanics, and solid mechanics in a continuum were developed. These advancements would have been impossible without the advent of large-scale, high-speed computers to carry out the calculations involved. There is no reason to doubt that much more is to come.

1.3 Outline of the Finite Element Approach

Solving an engineering problem by the finite element approach follows an orderly step-by-step process, as outlined in Table 1.1. The steps are described as follows:

Step 1. Formulate Governing Equations and Boundary Conditions

Usually an engineering problem can be described by a governing differential equation and associated boundary conditions. Once these are determined, the appropriate finite element solution algorithm can be obtained.

Table 1.1
Steps in Finite Element Method

Step	Procedure
1	Formulate governing equations and boundary conditions
2	Divide analysis region into finite elements
3	Select interpolation functions
4	Determine element properties
5	Assemble global equations
6	Solution of global equations
7	Verification of solution

Step 2. Divide Analysis Region Into Elements

The next step is to divide the solution region into appropriately shaped elements. In a typical one-dimensional problem, a rod over which the axial displacement is to be found may be divided up into sections of some desired length. Two-dimensional areas may be divided up into triangles, rectangles, or other appropriately shaped elements. Similarly, three-dimensional regions may be divided up into tetrahedrons, rectangular prisms, or elements of more complicated shapes. Often, different shaped elements are used within the same solution region. For example, when a structure containing both plates and beams is analyzed, it is convenient to choose elements that approximate those particular shapes.

Step 3. Select Interpolation Functions

Within an element the physical variable such as displacement, temperature, pressure, stress, or other variable is to be approximated by a simple function, such as a linear polynomial. Specific points within the elements, such as the corners of a triangular element, are designated as nodal points. If the temperature at each of the three corners of the triangle is determined and a linear interpolation function for the temperature is used within the element, the temperature over the entire element is known. Usually, polynomials are used as interpolation functions because they are easy to differentiate and integrate. The degree of the polynomial chosen is related to the number of nodes in the element and other factors to be discussed in detail later.

Step 4. Determine Element Properties

Each element makes a contribution to the overall region that is a function of element geometry, material properties (such as thermal conductivity or Young's modulus), number of node points, degree of interpolation function, and other variables. Because of the relatively simple nature of the assumed state of temperature, pressure, stress, or other physical variables within the particular element, it is generally easy to determine the element properties.

Step 5. Assemble Global Equations

All of the element properties must be assembled to form a set of algebraic equations for the nodal values of the physical variables. In the case of linear systems described by linear differential equations, the resulting algebraic equations will be linear in form and can be assembled using matrix techniques. For nonlinear systems a set of nonlinear algebraic equations results. The procedures used in Step 4 of determining the element properties should be coordinated with the assembly of global equations so that the nomenclature makes the process easy to follow. Generally the global equations for the entire region have forms similar to those for individual elements except that they contain many more terms because they include all nodes.

Step 6. Solution of Global Equations

If the global equations are linear, many standard techniques are available to solve them. Some of these are direct, such as the Gauss elimination method, while others are iterative. Sets of nonlinear algebraic equations are often very much more difficult to solve, but techniques for these are available in several good textbooks.

Step 7. Verification of Solution

The accuracy of the numerical solution of differential equations must be verified. The exact solution to the problem being analyzed usually is not known. Therefore the accuracy of the approximate solution obtained through the finite element method may not be known. For example, it is not known beforehand how many nodes are enough to insure an accurate solution. This can be checked by increasing the number of nodes and determining whether the solution at the node points changes or not. Generally, if the value of the physical variables at the node points is not significantly changed, the solution is considered to be accurate. When iterative techniques are used, some sort of resubstitution into the original differential equation should be used in order to verify that an accurate solution has been obtained.

1.4 Methods of Formulating Finite Element Equations

The finite element method consists primarily of replacing a set of differential equations in terms of unknown variables with an equivalent but approximate set of algebraic equations where each of the unknown variables is evaluated at a nodal point. Several different approaches may be used in the evaluation of these algebraic equations, and finite element methods are often classified as to the method used. Unfortunately, no one method is suitable for all problems likely to be encountered in engineering today, so several of the methods may have to be examined in order to choose the proper one for a particular problem.

Three methods are discussed here: direct, variational, and residual. While other, less common methods are sometimes used, they will not be discussed here.

A. Direct Method

The direct method was originally used to develop the finite element method for aircraft structures in the early 1950s. Individual structural members, such as bars, were analyzed with techniques similar to those used in analyses of simple trusses and frames. The displacement caused by applied forces was expressed by a set of equations convertible into a stiffness matrix for each of the structural members. Techniques were developed to assemble the element stiffness matrices together to form a large (global) matrix representing the stiffness of the entire structure or airframe.

Direct methods are still employed today in structural analysis. Practically all parameters employed in this approach may be interpreted on physical grounds. Unfortunately, the method is difficult to apply to two- and three-dimensional problems. As these are precisely the cases for which the finite element method is most useful, this limitation is quite severe. The direct method will be discussed only for bars and trusses in Chapter 4 of this text.

B. Variational Method

The variational method involves a quantity called a functional, which is introduced in Chapter 5; methods for deriving it are given in Chapter 6. The functional I may be obtained either from some sort of energy expression (for solid mechanics problems, usually) or from a boundary value problem. The solution T to the problem is approximated by the finite element function T^*: $T^* \cong T$, where T and T^* represent displacement, temperature, or fluid velocity. The approximate solution is defined as the sum of a set of local functions, one for each element:

$$T^* = \sum_{e=1}^{E} T^{(e)}$$

Here E is the total number of elements. Each of the element functions $T^{(e)}$ depends upon the values of T at the node points and some relatively simple function, usually a polynomial, in between. The variational method minimizes the value of the functional I with respect to each of these nodal values.

Advantages of the variational method, which is the most popular of the three methods for solid mechanics problems, include the familiarity of energy techniques in solid mechanics and easy extension to two- and three-dimensional problems. Disadvantages include the lack of a functional for certain classes of problems (for example, those dealing with the flow of viscous fluids) and the difficulty of finding variational methods for other problems, even when they exist. The lack of a functional for some problems means that other methods, such as residual methods, must be mastered in addition to the variational method.

C. Residual Methods

Residual methods are the most general of the three techniques and, also, the most difficult to understand. They are introduced in Chapter 8.

Residual methods usually start with a governing boundary value problem. The differential equation is written so that zero occurs on one side of the equal sign. If the exact solution T could be substituted into the equation, the result would be zero. The exact solution is not known, so some approximation of the exact solution $T^* \cong T$ is employed instead. Substitution of the approximate solution into the differential equation results in an erroneous value r, rather than zero. The error r is then multiplied by a weighting function W, and the product is integrated over the solution region. The result is called the residual R and is set equal to zero. Actually, there is a weighting function W and a residual R for each unknown nodal value, so the result is a global set of algebraic equations.

The major advantage of the residual method is that it can be applied to any problem for which a governing boundary value problem can be written. Once the techniques are learned, the details are relatively straightforward. It is basically a mathematical approach, and often little physical significance can be attached to the various parameters as one proceeds through a solution. The residual method is rapidly gaining in popularity.

Chapter 2

Division into Finite Elements

2.1 Introduction

Engineering problems involve the evaluation of temperatures, pressures, velocities, stresses, or other variables as functions of spatial coordinates (x,y,z) and time (t). The region in space being considered is often not geometrically regular. One step of the finite element method consists of dividing this nonregular region into small regular regions. One-dimensional objects are subdivided into short line segments. Two-dimensional bodies may be divided up into triangles, rectangles, quadrilaterals, or any other appropriate subregions. Finally, tetrahedral elements, rectangular prismatic elements, pie-shaped elements, and many others are employed in three-dimensional problems. A great variety of methods of dividing regions into finite elements have been employed in engineering practice.

It should be noted that there is usually not a single right way, all other ways being wrong. Often, different methods of dividing a region into elements will result in approximately the same computational difficulty and require approximately the same amount of solution time. In some cases, the division of the solution region into elements is an important factor because it requires an engineer's time. If computer costs are not large, then methods that significantly reduce the amount of setup time are very important. In other problems the setup time is slight, and much effort is expended in reducing computational cost. One example is problems that are amenable to automatic mesh generation. The input data may then be kept to a bare minimum and the computer may be programmed to do all of the detailed work of subdividing the solution region.

At the present time, the decision of how to divide up solution regions into elements must be based on engineering judgment. Our experience with finite element analysis has not yet been sufficient to permit definitive guidelines in this area. As individual engineers or groups of engineers gain experience with particular problems, they may be able to do so for that class of problems.

The following sections of this chapter discuss the various types of commonly used finite elements and some of the reasons why those particular elements are appropriate for specific problems. Advantages and disadvantages of various methods of division of regions into finite elements are discussed, as are labeling techniques for elements and nodes.

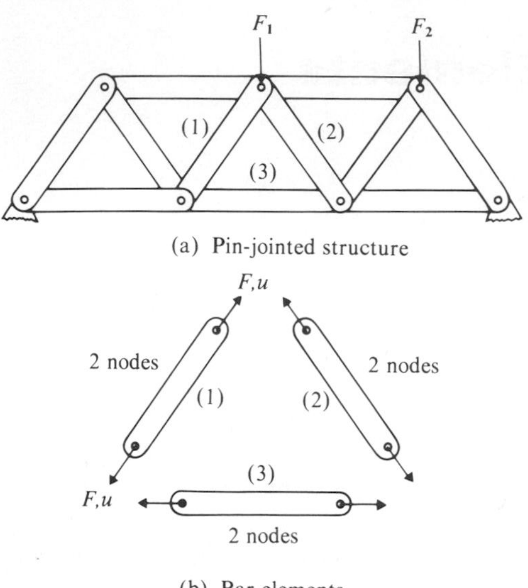

Figure 2.1 Simple structure and bar elements

2.2 Types of Elements

A large variety of finite elements have been used for different problems. In what follows, the simplest elements normally used are presented, along with examples of more advanced elements. No attempt is made here to try to cover all elements that have been used, because this would require more of a catalog than a textbook.

A. One-Dimensional Elements

Figure 2.1(a) shows a pin-jointed structure with applied loads F_1 and F_2. Each bar is a two-force member. That means that the force F and displacement u in the bar are directed along the bar. They are both one-dimensional in the sense that the force, for example, has an unknown magnitude but known direction. Bar elements representing part of the structure are shown in Figure 2.1(b) and are classified as one-dimensional elements. Because this type of bar has a constant cross-sectional area and force acting in it, a single element can be used to model the entire bar.

Many relations in engineering can be expressed as a function of one independent variable such as the coordinate x. Properties such as cross sectional area or modulus of elasticity vary with the position x. The governing differential equations for the problem can be expressed as ordinary differential equations in terms of the independent variable. Finite elements used in solving this type of problem are called one-dimensional elements. Examples of one-dimensional elements are shown in Figure 2.2. The number of node points in an element may vary from two up to any value desired. Raising the number of nodes

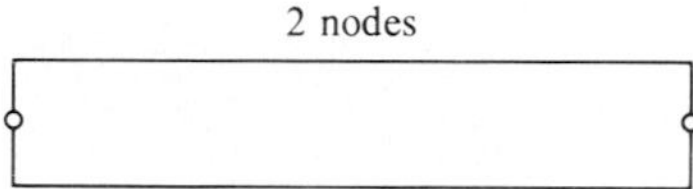

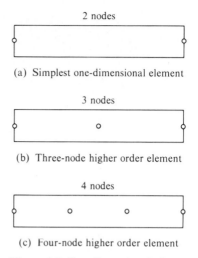

(a) Simplest one-dimensional element

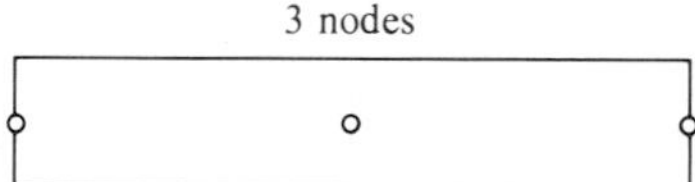

(b) Three-node higher order element

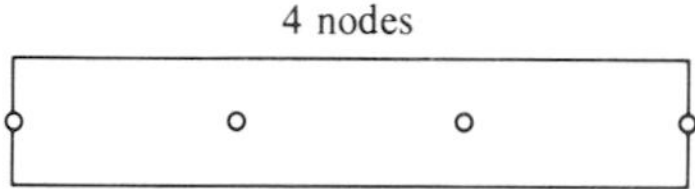

(c) Four-node higher order element

Figure 2.2 One-dimensional elements

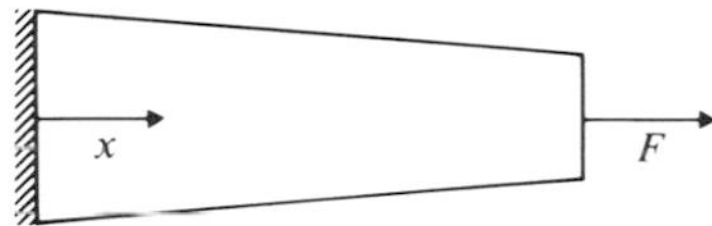

(a) Axial displacement of rod with variable area

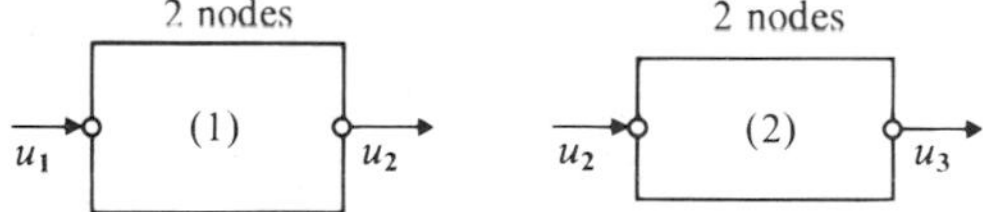

(b) Finite elements (bar elements) with element area approximating actual area

Figure 2.3 Axial displacement of rod

in a given element increases the accuracy of the solution but also the calculational complexity for that element. Elements (b) and (c) (in Figure 2.2) have a polynomial approximation higher than first order so they are called higher order elements.

A nonuniform rod subject to an axial force is shown in Figure 2.3(a). A two-element model of the rod is indicated in Figure 2.3(b). Both elements are the simplest possible, with a node at each end. Breaking the rod up into more than one element leads eventually to an expression for the axial displacement of the rod that is more accurate than one based on a single element could be. The cross sectional area of the bar considered here varies with x, and the cross sectional area of each element can be adjusted to approximate the actual area over the length of the element.

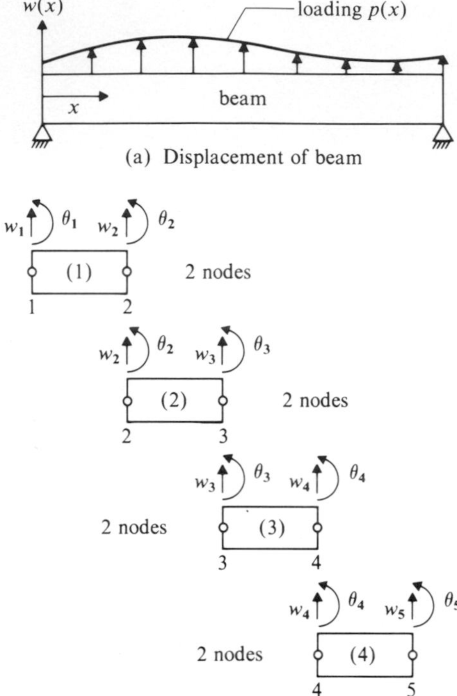

Figure 2.4 Transverse deflection of beam

A beam, shown in Figure 2.4(a), is subject to a load $p(x)$ per unit length and has a resulting displacement of $w(x)$. The differential equation for this problem is

$$\frac{d^2}{dx^2}\left(EI\frac{d^2w}{dx^2}\right) - p = 0$$

where $E(x)$ = Young's modulus, $I(x)$ = moment of inertia, $p(x)$ = transverse load, and $w(x)$ = transverse displacement. The solution with variable E, I, p is not easily found. Division of the beam into four finite elements is shown in Figure 2.4(b); the five nodal values of the displacement w and slope $\theta = dw/dx$ are to be found as part of the solution. These beam elements have only two nodes each (the minimum), but two dependent variables—displacement and slope—are to be found at each node point. Thus each is called a higher (higher than first) order element.

Applications for all of the elements shown here can easily be found. One dimension is sufficient in dealing with problems of heat dissipation in cooling fins, temperature distributions in long cylindrical bodies, and heat transfer in spherically symmetric regions. In fluid mechanics, the straight elements may be used for fluid flow in axially symmetric ducts of slowly varying area, while curved elements may be easily employed for axially symmetric ducts where the center line follows some slowly varying space curve. Structural applications may be easily thought of in two major areas; one is axial stresses or strains

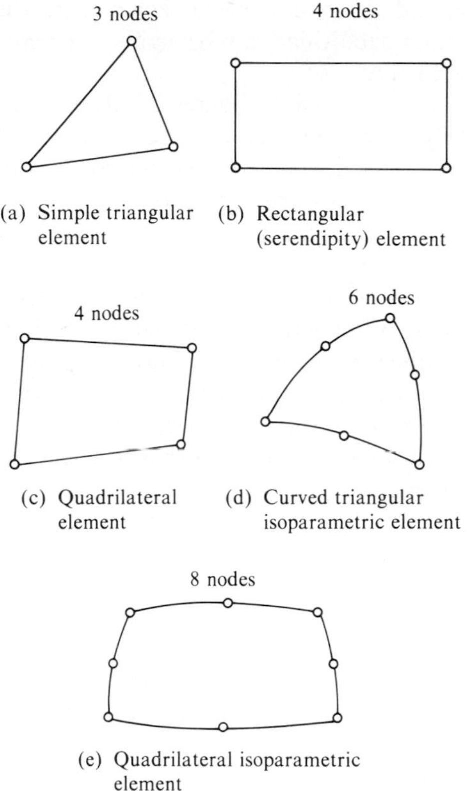

Figure 2.5 Two-dimensional elements

in a two-force member and the other is transverse deflections of a beam. Complex changes in the geometry and properties of the material along the length of the structure may be easily taken into account by varying the element properties.

B. Two-Dimensional Elements

Two-dimensional finite elements are used if the temperature or displacement is a function of two coordinates, say x and y. Such systems are governed by partial differential equations rather than the ordinary differential equations that govern the one-dimensional case. Some two-dimensional elements, both simple and higher order, are shown in Figure 2.5.

The simplest two-dimensional element is the triangular element shown in Figure 2.5(a) with three nodal points, one at each of the corners of the triangle. The temperatures, displacements, or other field variables are evaluated at these three nodal points. One major advantage of triangular elements is that any two-dimensional region can be divided into triangles in relatively simple fashion. Rectangular elements, such as shown in Figure 2.5(b), and other quadrilateral elements have four sides and four nodes, one at each corner. It is apparent that quadrilateral elements can be further subdivided into triangular elements simply by connecting either set of opposite nodal points.

Curved triangular and quadrilateral elements are shown in Figures 2.5(d) and (e). With these, curved sides of regions may be closely approximated. Higher order elements use larger numbers of node points, located at various points on the element boundaries and within the element. The most complex finite elements are used extensively in advanced computer programs. The difficulties in incorporating these elements into computer codes may be significant.

A large number of two-dimensional engineering problems is well known to anyone with a few years' engineering experience or training. Conductive heat transfer, fluid flow in ducts of various shapes, ideal or viscous flow about two-dimensional or axially symmetric bodies, displacement of two-dimensional frames, and plane stress or plane strain are examples. While one-dimensional engineering problems may often be solved by analytical techniques, the finite element method, with its many different shaped elements, is very useful for solving two-dimensional problems.

C. Three-Dimensional Elements

Many engineering problems inherently involve three dimensions. One- or two-dimensional approximations are simply not appropriate. The velocity or other variable is a function of three spatial coordinates, say the three cartesian coordinates x, y, z. The governing differential equations are also, of course, functions of these same three independent coordinates.

Figure 2.6 shows examples of three-dimensional elements. A tetrahedral element in Figure 2.6(a), consisting of four sides and four nodal points placed at the corners of the element, is the simplest. Just as any two-dimensional region can be divided into triangular elements, all three-dimensional regions can be approximated using tetrahedral elements. Rectangular prisms or irregular hexahedrons, shown in Figures 2.6(b) and (c), are often used for three-dimensional problems; these have six sides and eight nodal points. Any six-sided solid element can be further subdivided into tetrahedral elements if necessary, but at least five tetrahedral elements result from this process. Thus, the advantage of the rectangular prism or the irregular hexahedral solid element is that a much smaller number of elements may be used to represent some three-dimensional regions.

Curved higher order elements are shown in Figures 2.6(d) and (e). The primary application for these types of elements is in subdivision of three-dimensional regions that have curved boundaries.

2.3 Methods of Division

A region of space or an interval of time to be analyzed can be divided into finite elements in many different ways. Visual examination of a simple region may suggest a freehand subdivision into a coarse network of elements that can serve as the basis of a satisfactory solution. A region of reasonably regular geometry can be divided into suitable elements

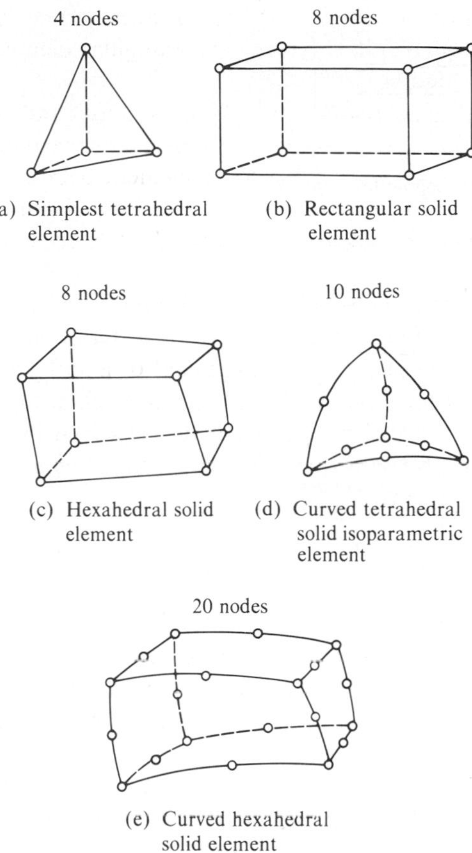

Figure 2.6 Three-dimensional elements

by means of a simple computer subroutine. If the region is highly complex and irregular, its division into finite elements may require the use of a very sophisticated computer program.

A. One-Dimensional Regions

It is generally quite easy to divide one-dimensional regions into finite elements; the process is equivalent to cutting a line into segments according to criteria selected by the analyst. The following treatments of a single example illustrate some of the alternatives.

Figure 2.7(a) shows a cross section of a tapered cooling fin that conducts heat away from a hot wall; the temperature of the fin decreases as the distance from the wall increases (from left to right in the figure). If the interest of the engineer is restricted to the distribution of temperature along the fin, rather than across it, this three-dimensional object can be treated as if it were one-dimensional. Figure 2.7(b) shows a division of the fin into four simple elements of equal length, resulting in five nodal points. Division into a larger number of simple elements would provide a larger number of nodal points and

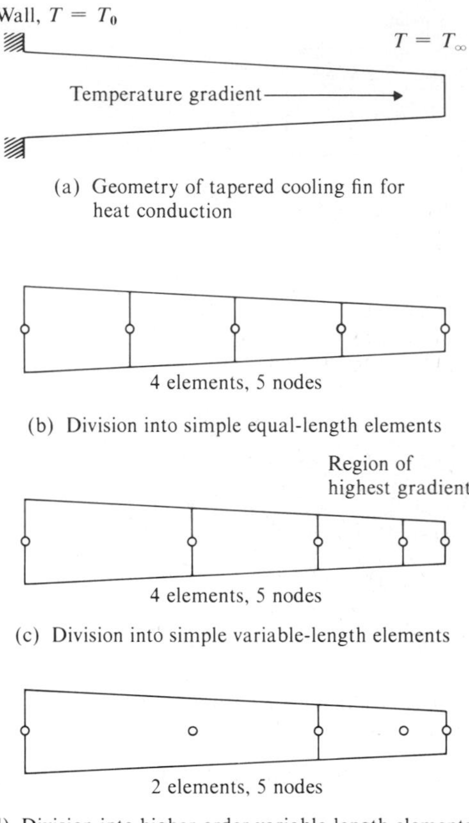

Figure 2.7 Division of one-dimensional cooling fin into elements

lead to more accurate values of the temperature along the fin. Any division of a one-dimensional region into simple elements with nodes at the ends leads to a simple relation. The number of elements E will be equal to the number of nodes N minus one; that is, $E = N-1$.

The accuracy of the solution can also be improved by using an alternative division scheme, in which the number of elements is kept the same but their lengths differ from each other. It is known from analytical solutions that the largest temperature gradient occurs near the small end of a tapered cooling fin. Figure 2.7(c) shows a division into four simple elements with lengths chosen so as to make the nodal points successively closer together toward the tip of the fin. At the beginning of the analysis the temperature gradients are not known, so the density of the nodes cannot be made proportional to the gradient, but an advantageous choice—such as that shown—is easily devised. The basis of such choices may be experience gained in solving analogous problems, as suggested above, or during the solution of the problem at hand. That is, if a naive initial selection of elements and nodal points leads to a solution that indicates a better division, the scheme may be easily revised to concentrate the nodal points in the appropriate area.

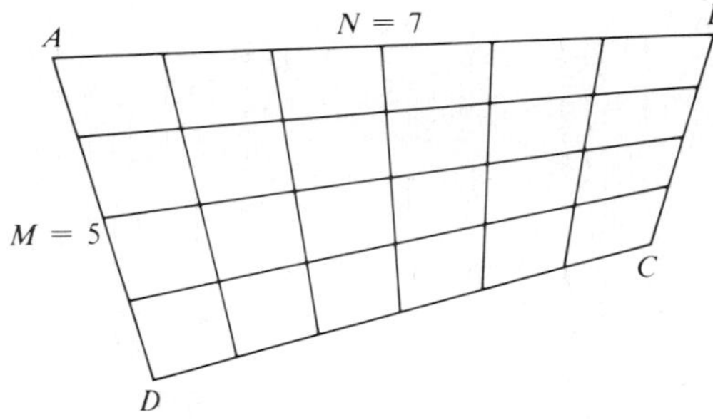

(a) Division of quadrilateral region into quadrilateral mesh; each side may be divided into segments of equal length or variable length

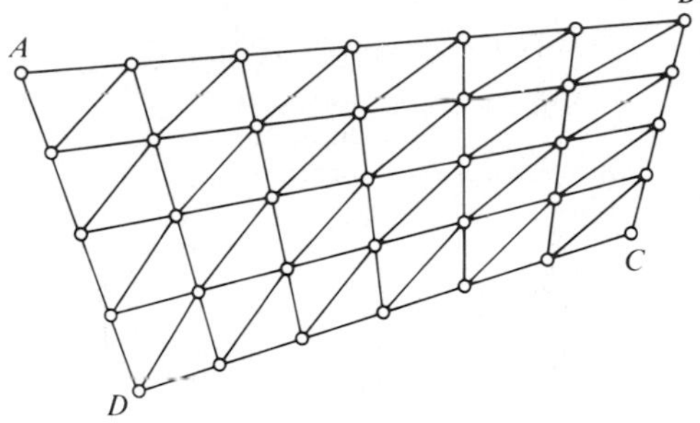

(b) Subdivision of quadrilateral mesh into triangular elements; $E = 2(M - 1)(N - 1) = 48$

Figure 2.8 Two-dimensional quadrilateral region divided into regular mesh of triangular elements

A division into higher order elements, each containing more than two nodes, is shown in Figure 2.7(d). In this case the five nodes have been placed at the same locations as in Figure 2.7(c). Criteria affecting the location of interior nodal points in higher order elements are discussed in later chapters of this book.

B. Two-Dimensional Regions

The finite element approach is fully adaptable to the solution of problems involving two-dimensional regions, regardless of their shape. Whether the boundaries are straight or curved, they can be approximated as closely as desired by the sides of elements of various shapes. Classical finite difference methods are not well suited to the engineering analysis of regions that have irregular boundaries; the irregularities commonly necessitate the imposition of exceptional conditions that must be specially formulated for each new problem. No such difficulties hinder the use of the finite element method.

The solution of various kinds of engineering problems requires the consideration of quadrilateral regions like ABCD in Figure 2.8(a). In this figure, seven nodes have been defined at the ends of, and at equal intervals along, sides AB and CD; five nodes have

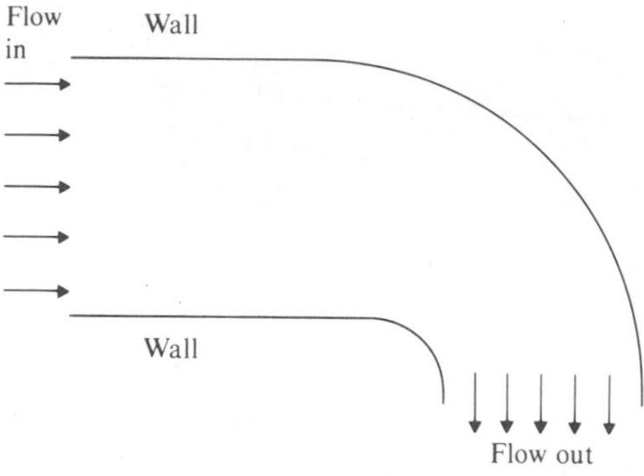

(a) Geometry of contracting duct

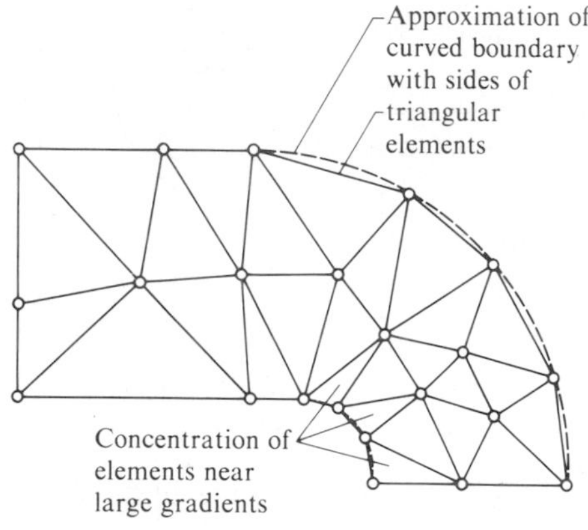

(b) Division into simple triangular finite elements

Figure 2.9 Fluid flow in two-dimensional contracting duct: Division into finite elements

been similarly defined on sides AD and BC. Lines have then been drawn through corresponding pairs of these nodes. These lines divide the region into a set of quadrilateral elements, and the intersections of the lines (the corners of the quadrilaterals) are designated nodes. Often, the calculations are simpler if they are based on triangular elements; a set of such elements can be produced by drawing the diagonal of each quadrilateral, as shown in Figure 2.8(b), without affecting the nodal points. If there is any clue concerning the direction in which the gradients of the variables will be greatest, a more accurate solution will be obtained if the diagonals are drawn in that direction. (The number of triangular elements produced in this way is clearly twice the number of quadrilateral elements formed in the earlier stage.) A further improvement in the accuracy of the solution may result if the mesh is made denser in certain areas; computerized mesh generation schemes simplify such modifications.

Consider the two-dimensional contracting duct that turns through a 90° angle, as shown in Figure 2.9(a). This section must be divided into elements to analyze the flow of fluid through the duct. Figure 2.9(b) shows a set of triangular elements resulting from a

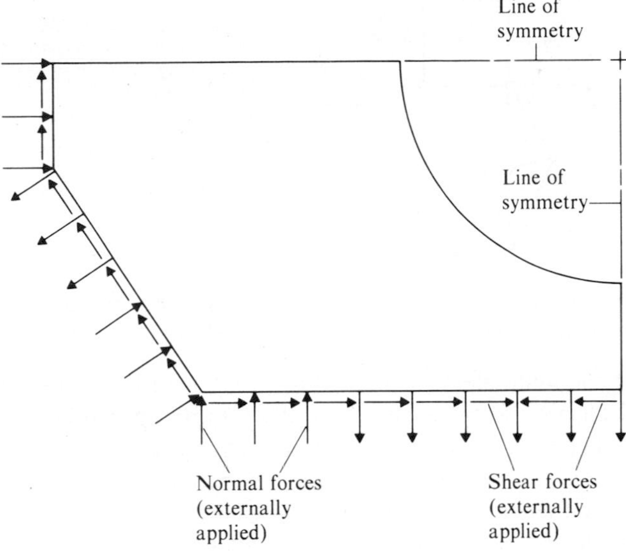

Figure 2.10 Two-dimensional plane stress or strain in plate with center hole (because the plate is symmetrical, only one quadrant is shown)

freehand division in which no particular attempt was made to keep the division scheme regular. In the entrance region, where the sides are straight, the sides of the elements represent the wall exactly. Where the wall curves, the straight sides of the triangular elements approximate the curved boundary between the nodal points, which lie exactly on the boundary. Clearly, if a larger number of node points were used, the approximation would be even more accurate. It is expected that the gradients of pressure, horizontal velocity, or vertical velocity will be the largest near the place where the curvature is largest. Therefore, the elements are made small in this region to obtain greatest accuracy. This concentration of grid points is not easy to bring about when the standard finite difference formulation is used.

The finite element method is similarly useful in the analysis of stresses or strains acting in the plane of a flat plate of irregular shape. Figure 2.10 shows one quadrant of an octagonal plate with a central circular hole; the figure also shows a set of externally applied stresses along the boundary of the plate. In the model drawn in Figure 2.11, the plate has been divided into simple triangular elements, curved isoparametric triangular elements (around the circular arc), and rectangular elements. For clarity in the figures shown here, the plate has been represented by only a small number of elements; usually the stress or displacement must be determined at many more points than are shown, so a much finer grid system would be used.

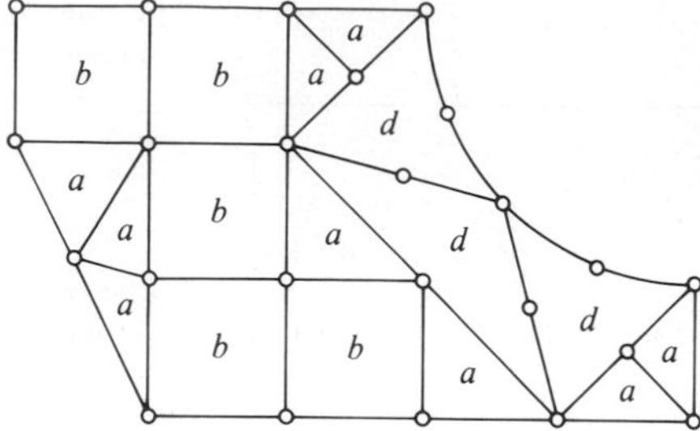

Figure 2.11 Division of plate into elements of types shown in Figure 2.5: *a*, simple triangular element; *b*, rectangular element; *d*, curved triangular isoparametric element

In applying the finite element method to situations like the two just mentioned, an engineer would choose the types and sizes of elements appropriate to each part of the solution region. The selections would then be made part of the input for the computer program to be employed. An automatic mesh generation scheme would then perform the task of subdividing the entire region into finite elements.

Some desirable and undesirable choices of two-dimensional finite elements are illustrated in Figure 2.12. Desirable triangular elements are approximately equilateral; if such elements are being formed by dividing quadrilaterals, the shorter diagonal is therefore preferred. Divisions that leave empty spaces between elements must never be used: these gaps represent a failure to account for the whole solution region and lead to erroneous results, even if the problem is solved by successive approximations in which the sizes of the elements are made smaller and smaller. If very large elements are adjacent to very small ones, numerical difficulties not associated with the original differential equations and boundary conditions may arise.

C. Three-Dimensional Regions

The division of three-dimensional regions is based on the same general principles as those just applied to two-dimensional regions. Usually there are many more elements, and the details may become tedious. Several computer routines have been developed for generating three-dimensional meshes, but they are beyond the scope of this text, which contains no further discussion of such schemes.

2.4 Labeling Techniques

The manner in which nodes are labeled can strongly influence the efficiency of a computer program. The numbering scheme should be chosen to minimize computer storage requirements as well as execution time.

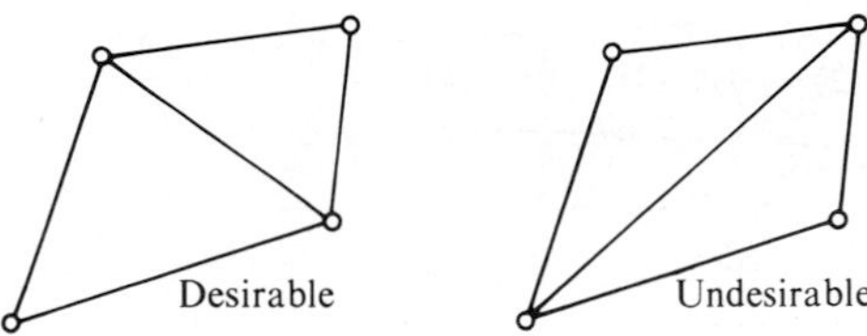

(a) Alternative divisions of quadrilateral into triangles

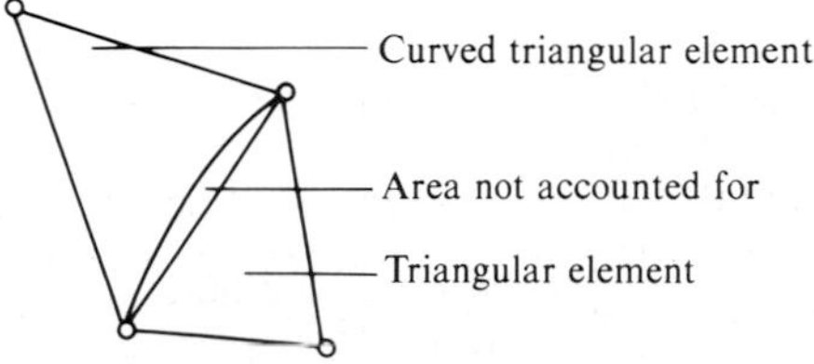

(b) Very undesirable; leaving an area not within either element

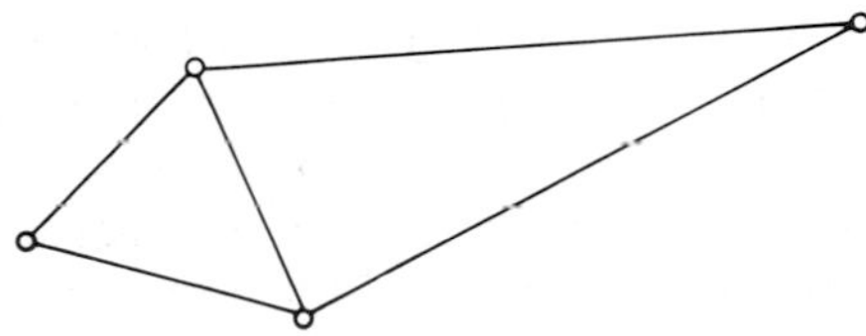

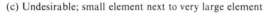

(c) Undesirable; small element next to very large element

Figure 2.12 Desirable and undesirable divisions of quadrilateral into elements

A. Bandwidth

A labeling scheme for a one-dimensional system of finite elements is shown in Figure 2.13. The node points are labeled with plain numbers, while the element numbers are enclosed in parentheses. This is really the only technique necessary for one-dimensional problems.

As shown in later chapters, the set of algebraic equations that would result from an axial displacement analysis in one dimension (for example) would appear as follows:

| NBW |

$$\begin{bmatrix} C & C & 0 & 0 & 0 \\ C & C & C & 0 & 0 \\ 0 & C & C & C & 0 \\ 0 & 0 & C & C & C \\ 0 & 0 & 0 & C & C \end{bmatrix} \begin{Bmatrix} u_1 \\ u_2 \\ u_3 \\ u_4 \\ u_5 \end{Bmatrix} = \begin{Bmatrix} C \\ C \\ C \\ C \\ C \end{Bmatrix} \qquad \textbf{(2.1)}$$

The five unknown nodal displacements u_i ($i = 1$–5) are shown in a column matrix. In the square coefficient matrix on the left and the column matrix on the right, C denotes a

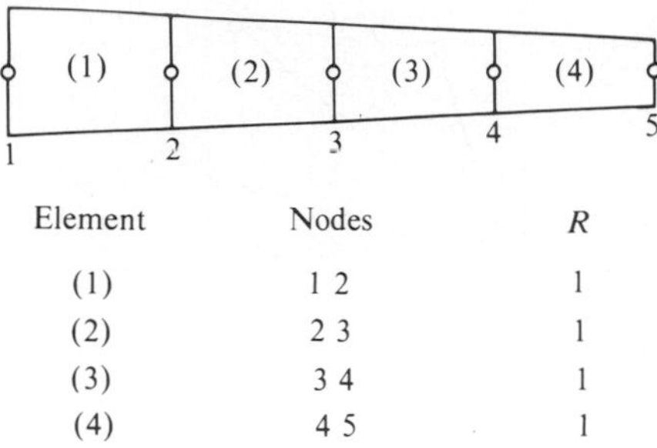

Element	Nodes	R
(1)	1 2	1
(2)	2 3	1
(3)	3 4	1
(4)	4 5	1

Figure 2.13 Labeling of nodes and elements for one-dimensional problem: Element numbers are enclosed in parentheses, node numbers are not

nonzero number. All of the nonzero elements in the coefficient matrix are located on or near the principal diagonal, as shown. This property makes possible a very efficient solution of the five linear algebraic equations in five unknowns through a banded Gauss elimination method, for example.

The bandwidth NBW is given by

$$\text{NBW} = (R + 1)(\text{NDOF}) \tag{2.2}$$

where

R = largest difference between the node numbers in a single element (all elements must be examined to determine the value of R)

NDOF = number of degrees of freedom at each node point

For the one-dimensional system shown in Figure 2.13, substitution of the values $R = 1$ and NDOF = 1 in Equation 2.2 yields a bandwidth, NBW, of 2.

The beam shown in Figure 2.4 has the same numbering scheme as Figure 2.13 (four elements and five node points) with the same value $R = 1$ for each element. However, the number of degrees of freedom is two (NDOF = 2; displacement w and slope θ). Therefore the bandwidth is NBW = (1 + 1) (2) = 4, and the algebraic equations for the nodal values of w and θ will have the form

$$|\leftarrow \text{NBW}=4 \rightarrow|$$

$$\begin{bmatrix} C & C & C & C & 0 & 0 & 0 & 0 & 0 & 0 \\ & C & C & C & C & 0 & 0 & 0 & 0 & 0 \\ & & C & C & C & C & 0 & 0 & 0 & 0 \\ & & & C & C & C & C & 0 & 0 & 0 \\ & & & & C & C & C & C & 0 & 0 \\ & & & & & C & C & C & C & 0 \\ & & & & & & C & C & C & C \\ & & & & & & & C & C & C \\ & \text{(Symmetric)} & & & & & & & C & C \\ & & & & & & & & & C \end{bmatrix} \begin{Bmatrix} w_1 \\ \theta_1 \\ w_2 \\ \theta_2 \\ w_3 \\ \theta_3 \\ w_4 \\ \theta_4 \\ w_5 \\ \theta_5 \end{Bmatrix} = \begin{Bmatrix} C \\ C \\ C \\ C \\ C \\ C \\ C \\ C \\ C \\ C \end{Bmatrix}$$

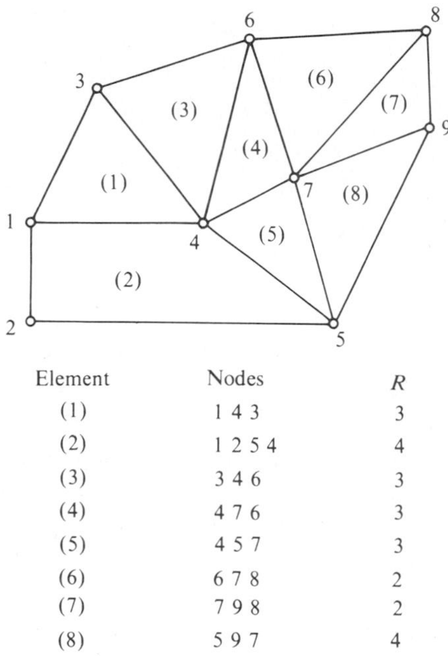

Element	Nodes	R
(1)	1 4 3	3
(2)	1 2 5 4	4
(3)	3 4 6	3
(4)	4 7 6	3
(5)	4 5 7	3
(6)	6 7 8	2
(7)	7 9 8	2
(8)	5 9 7	4

Figure 2.14 Efficient labeling of nodes and elements in two-dimensional problem

B. Minimization of Bandwidth

Both the storage and execution time are reduced by decreasing the bandwidth. For example, consider the storage of a problem with 100 node points (NP = 100), each with one degree of freedom, and a bandwidth of 12. The storage space required for the full symmetric matrix is NP $\times$ NP = 100 $\times$ 100 = 10,000. If only the upper part of the banded portion is stored—not the zeros—the storage requirement is NP $\times$ NBW = 100 $\times$ 12 = 1,200. In two- and three-dimensional problems, a method of labeling the nodes to minimize the bandwidth should be used.

Figure 2.14 shows a two-dimensional region that has been divided into eight elements, seven triangular and one quadrilateral. As before, the nine nodal points are labeled with plain numbers and the eight elements are labeled with numbers in parentheses. The Element/Node Connectivity Data, tabulated in the same figure, include the eight elements and the nodes present in each of those elements. Within each element the nodes are cited in counterclockwise sequence, which is the conventional method of labeling. Counterclockwise labeling produces the proper signs for element properties, which are discussed in Chapter 3.

The right-hand column contains the value of R for each element (the largest difference between the node numbers in that element). The largest value resulting from this particular labeling scheme is $R = 4$. In this example, the bandwidth for a problem with a single degree of freedom would be 5.

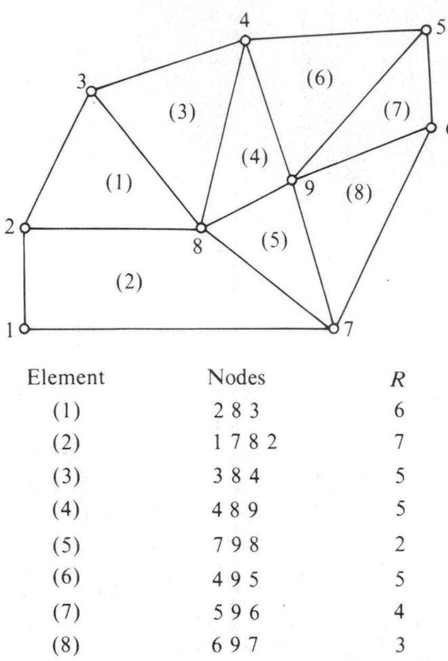

Element	Nodes	R
(1)	2 8 3	6
(2)	1 7 8 2	7
(3)	3 8 4	5
(4)	4 8 9	5
(5)	7 9 8	2
(6)	4 9 5	5
(7)	5 9 6	4
(8)	6 9 7	3

Figure 2.15 Inefficient labeling of nodes and elements in two-dimensional problem

An alternative labeling scheme for the same set of nodes and elements is shown in Figure 2.15. The nodes were labeled by starting in the lower left-hand corner and simply progressing around the exterior of the region and then labeling the last two nodes in the center 8 and 9. The resulting labeling of the nodes in each element and the corresponding value of R is also shown in the figure. As the largest value of R is 7, the bandwidth for this problem would be 8. It can be seen from this analysis that the first labeling scheme yields a smaller bandwidth than the second scheme.

The writer and user of finite element programs must keep in mind the effect of labeling schemes on bandwidth. Usually the best labeling sequence results from traversing the grid in the direction of the smallest number of nodes. This method is illustrated by the first example. As appropriate points occur later in the text, more details will be provided about favorable strategies.

2.5 Problems

2.1 A one-dimensional nozzle (with slowly varying cross section) for studying subsonic flow is shown in Figure 2.16(a). Divide the length into six equal elements with node points at the ends of each element, label the elements (using parentheses), and label the node points. Determine the bandwidth of the resulting algebraic equations.

2.2 Figure 2.16(b) shows a plate in plane stress. Divide the plate into 12 triangular elements, each with node points at the three corners. Label the elements and nodes in some systematic fashion.

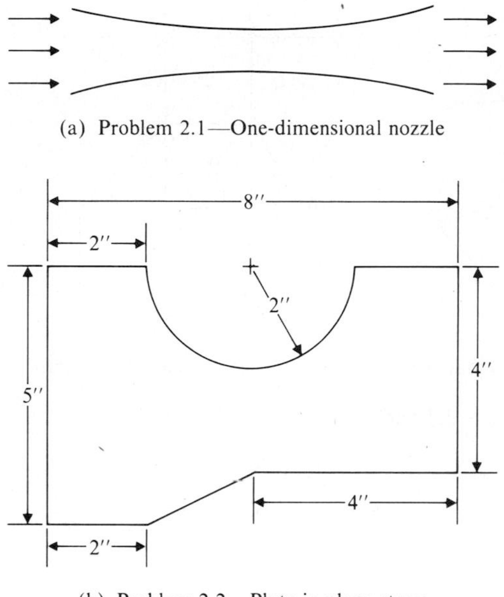

(a) Problem 2.1—One-dimensional nozzle

(b) Problem 2.2—Plate in plane stress

Figure 2.16 Diagrams for Problems 2.1 and 2.2

2.3 A region, shown in Figure 2.17(a), has been divided into elements. Label the nodes to minimize the bandwidth of the resulting system of algebraic equations. Consider only a single degree of freedom at each node.

2.4 The two-dimensional plate, originally shown in Figure 2.10, has been labeled as indicated in Figure 2.17(b). Determine the bandwidth of the resulting system of algebraic equations. Renumber the node points to produce the minimum bandwidth. Assume a single degree of freedom per node.

2.5 Label the elements and node points in the contracting duct (Figure 2.9) to produce a bandwidth of 6 or less.

2.6 Label the elements and node points in the quadrilateral region (Figure 2.8) systematically to produce a minimum bandwidth. Obtain a simple formula for the bandwidth of an M by N quadrilateral mesh divided in the manner shown.

2.7 Centrifugal water pumps present an interesting application for finite element analysis. The bladed circular impeller, rotating with angular velocity ω, imparts a high velocity (kinetic energy) to the water. The region between the rotating impeller and fixed pump casing is called the volute. At the entrance to the volute, the flow slows down and the pressure increases substantially (kinetic energy is converted into potential energy). It spirals outward from the tongue around to the outlet pipe to allow for increased volume of flow. The volute region is to be analyzed to determine the velocities and pressures at various node points.

Figure 2.18 shows a coarse mesh of quadrilateral elements, with a node at each corner, that is to be used for the analysis. Label the nodes and elements to produce a minimum bandwidth. Indicate how you would add elements both radially and circumferentially and number them without radically increasing the bandwidth.

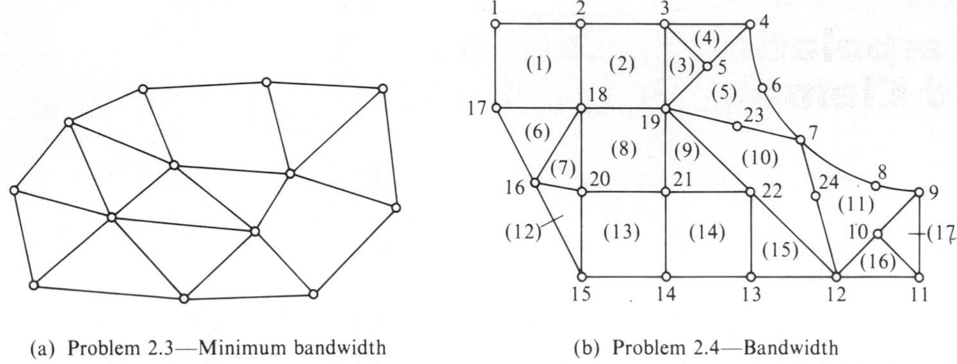

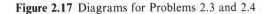

(a) Problem 2.3—Minimum bandwidth

(b) Problem 2.4—Bandwidth

Figure 2.17 Diagrams for Problems 2.3 and 2.4

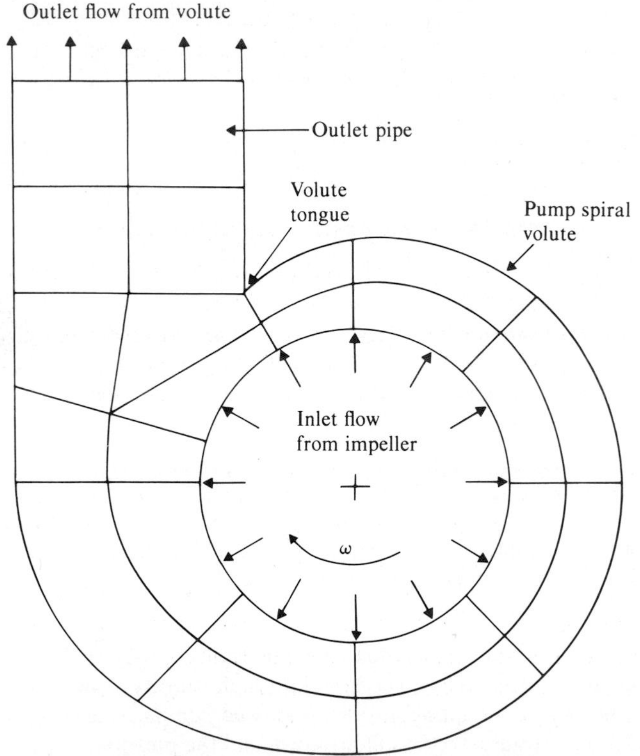

Figure 2.18 Diagram for Problem 2.7 (fluid flow in centrifugal pump volute): Division into quadrilateral elements

Chapter 3

Interpolation Functions and Element Properties

3.1 Approximation Functions

It was seen in Chapter 2 how complex spatial regions can be divided up into simpler elements of various shapes. The same idea is extended to the displacements, temperatures, pressures, velocities, stresses, and other quantities that vary continuously in space or time. Within a given element, the continuous variable is approximated by a simple function, such as a polynomial. The complete function over the entire region is made up of the element polynomials summed together as follows:

$$T = T^{(1)} + T^{(2)} + T^{(3)} + \ldots + T^{(E)} = \sum_{e=1}^{E} T^{(e)}$$

A few approximating functions other than polynomials have been used, but polynomials are convenient because they are easily differentiated and integrated. They are also easily extended from simple one-dimensional problems all the way through complex three-dimensional problems. Only polynomial interpolating functions will be considered in this text.

A. Interpolation Functions

Once the solution region has been divided into elements, the field variable $T(x)$ within each element is approximated by the interpolation function, expressed in terms of the nodal values of the dependent variable and sometimes the nodal values of its derivatives up to a certain order. Choosing a linear interpolation function for $T(x)$ over an element insures that the gradient of the field variable T over the element will be constant. Higher order polynomials clearly have first derivatives that vary within the element.

Two general characteristics are usually required of interpolation functions: compatibility and completeness. The first requirement ensures that the piecewise approximation of the dependent variable be continuous. Any polynomial interpolation function will always be continuous within an element, so this condition actually applies to the interelement boundaries. For a two- or three-dimensional region, this means that $T(x)$ is continuous from one element to the next. Elements whose interpolation functions satisfy this requirement are called *compatible* or conforming elements. To meet the second requirement, the interpolation function must be chosen so that if the dependent variable is a constant in

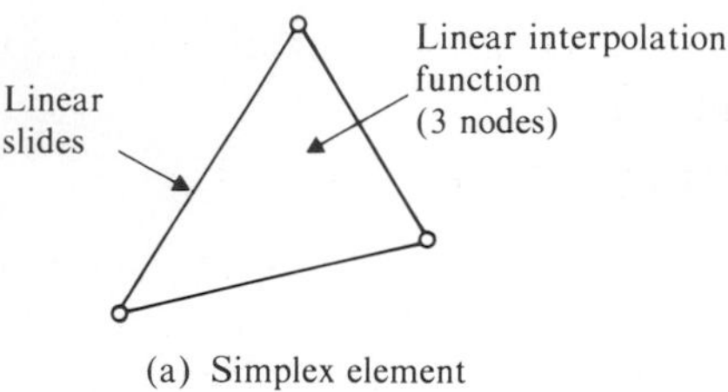

(a) Simplex element

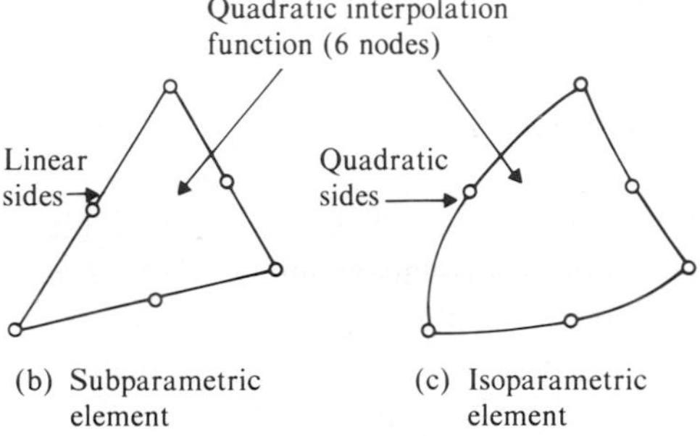

(b) Subparametric element

(c) Isoparametric element

Figure 3.1 Types of two-dimensional elements

the real problem it will also be a constant in the finite element solution. Other problems, such as those in solid mechanics, may require that rigid body displacements (uniform values of displacement over a body) and constant strain states (uniform first derivative values of displacement over the body) are matched by finite element solutions. It can be shown that *completeness* is necessary to insure convergence of the approximate finite element solution to the exact solution (of the appropriate governing differential equation) as the number of elements becomes very large.

Linear interpolation functions and a few higher order polynomials will be considered in this chapter. Linear elements with the simplest possible geometry are called simplex elements. A linear approximating polynomial consists of a constant term plus a linear term in each of the dimensions. Such a function for a three-dimensional element has the form

$$T^{(e)}(x,y,z) = \alpha_1 + \alpha_2 x + \alpha_3 y + \alpha_4 z$$

In such an equation the number of coefficients in the polynomial is equal to the number of spatial coordinates plus one.

B. Element Classification

A convenient classification of elements employs three groups: simplex, subparametric, and isoparametric. *Simplex* elements have linear polynomials, as discussed above. A two-dimensional example is a triangular element with a linear interpolation function of the form $T^{(e)}(x,y) = \alpha_1 + \alpha_2 x + \alpha_3 y$. Figure 3.1 illustrates these examples. *Subparametric* elements are those with higher order interpolation functions and lower order shape functions. An example is a quadratic function of the form

$$T^{(e)}(x,y) = \alpha_1 + \alpha_2 x + \alpha_3 y + \alpha_4 x^2 + \alpha_5 xy + \alpha_5 y^2$$

in an element with straight (linear) sides. The final category is the *isoparametric* element, which is discussed in detail in Chapter 13. In this case, both the interpolation function and the shape of the element sides are of the same order (second order in x and y). *Superparametric* elements are those in which the sides of the element are of higher order than $T^{(e)}(x,y)$. These are not used because of accuracy problems.

3.2 One-Dimensional Simplex Element

A. Element Interpolation Function

Each simple or simplex element, introduced in Section 2.2, has properties (using temperature T as the example field variable)

$$\begin{aligned} L^{(e)} &= \text{element length} \\ T^{(e)} &= \text{element interpolation function for temperature} \end{aligned}$$

where the element is identified by the superscript in parentheses (to differentiate from powers of L or T). As shown in Figure 3.2, the first nodal coordinate is $x = X_i$, and the second is $x = X_j$. Here i denotes the nodal coordinate with smaller x value, j denotes the nodal coordinate with larger x coordinate, and capital letters are employed to indicate nodal properties. The i,j labeling system for the element is called the local labeling system (where only one element is considered).

Nodal field variables are

$$\begin{aligned} T_i &= \text{first nodal temperature} \\ T_j &= \text{second nodal temperature} \end{aligned}$$

The values of T_i and T_j may be determined as part of the finite element solution or—in some cases, such as the area of a cooling rod—be specified by the problem geometry.

The linear polynomial for element e is defined as

$$T^{(e)} = \begin{cases} \alpha_1 + \alpha_2 x, & X_i \leq x \leq X_j \\ 0 & \text{elsewhere} \end{cases} \tag{3.1}$$

where α_1 and α_2 are constants to be determined. Within the element, the interpolation function is linear; outside the element, it is zero. (Often the latter part of the definition is assumed but not specifically stated.) The linear polynomial must pass through both nodal points, leading to the equations $T^{(e)} = T_i$ at $x = X_i$, $T^{(e)} = T_j$ at $x = X_j$. Substituting from Equation (3.1)

$$\begin{aligned} T_i &= \alpha_1 + \alpha_2 X_i \\ T_j &= \alpha_1 + \alpha_2 X_j \end{aligned} \tag{3.2}$$

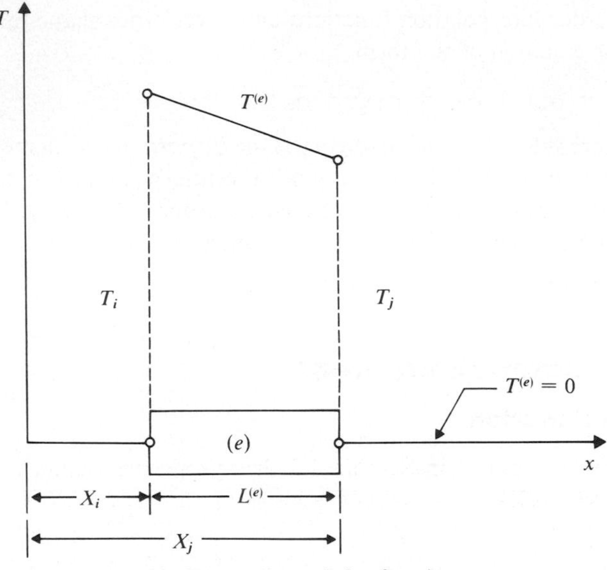

(a) Element interpolation function

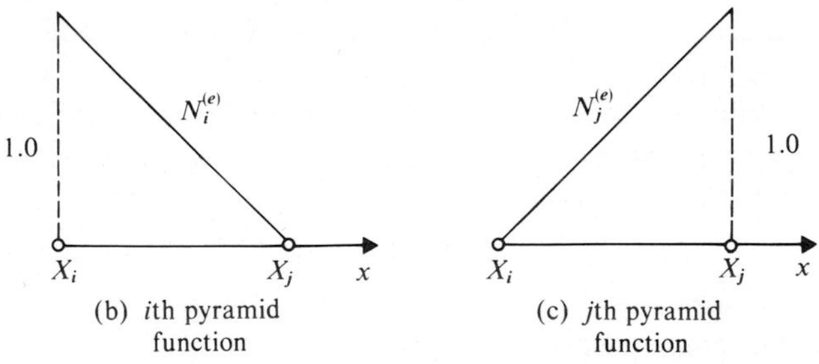

(b) ith pyramid function

(c) jth pyramid function

Figure 3.2 Linear interpolation function for simplex element

and solving for α_1, α_2

$$\alpha_1 = \frac{1}{L^{(e)}}(X_j T_i - X_i T_j)$$

$$\alpha_2 = -\frac{1}{L^{(e)}}(T_i - T_j)$$

$$L^{(e)} = X_j - X_i$$

Resubstituting these expressions into Equation (3.1) yields

$$T^{(e)} = \frac{X_j - x}{L^{(e)}} T_i + \frac{x - X_i}{L^{(e)}} T_j, \quad X_i \leq x \leq X_j \tag{3.3}$$

The element interpolation polynomial is a function of the nodal coordinates, element length, and nodal temperatures as well as a linear function of x.

B. Pyramid Functions

It is useful to write the interpolation function as

$$T^{(e)} = N_i^{(e)} T_i + N_j^{(e)} T_j \tag{3.4}$$

where

$$N_i^{(e)} = \text{first } (i\text{th}) \text{ pyramid function}$$
$$N_j^{(e)} = \text{second } (j\text{th}) \text{ pyramid function}$$

The functions N are given, by comparison with Equation (3.3), as

$$N_i^{(e)} = \frac{X_j - x}{L^{(e)}}$$
$$N_j^{(e)} = \frac{x - X_i}{L^{(e)}} \tag{3.5}$$

As shown in Figure 3.2, N has the value 1.0 at the corresponding node point and 0.0 at the opposite node point. As suggested by the shape, the N functions are called pyramid functions. Similar functions for two- and three-dimensional elements will be defined to have the value unity at one node and zero at the others. It should be noted that the pyramid functions are normally used instead of α_1, α_2.

Example 3.1 *Element Properties for Tapered Rod*

Given A tapered cylindrical cooling rod has cross sectional area S, as shown in Figure 3.3, where $S = 40$ mm² at $x = 50$ mm, and $S = 10$ mm² at $x = 100$ mm, and the area varies linearly between the two points. The symbol S is used rather than A so as not to conflict with an element area later.

Objective Determine the element properties for a simplex element.

Solution The field variable is the area, so the symbol T is replaced by S. Otherwise, all of the equations developed in this section are valid. The local nodal coordinates and element length are $X_i = 50$, $X_j = 100$, $L^{(e)} = 50$ where the units (mm) are omitted for clarity. Nodal areas are $S_i = 40$, $S_j = 10$.

As given in Equation (3.1), the linear interpolation polynomial is

$$S^{(e)} = \begin{cases} \alpha_1 + \alpha_2 x, & 50 \le x \le 100 \\ 0 & \text{elsewhere} \end{cases}$$

At the nodal points, $S^{(e)}$ must equal the nodal values, or $S^{(e)} = 40$ at $x = 50$, and $S^{(e)} = 10$ at $x = 100$. Substituting these values into $S^{(e)}$ yields $40 = \alpha_1 + \alpha_2\,(50)$, $10 = \alpha_1 + \alpha_2\,(100)$, and solving for α_1, α_2; $\alpha_1 = 70$, $\alpha_2 = -0.60$. Thus the interpolation polynomial is finally

$$S^{(e)} = \begin{cases} 70 - 0.60\,x, & 50 \le x \le 100 \\ 0 & \text{elsewhere} \end{cases}$$

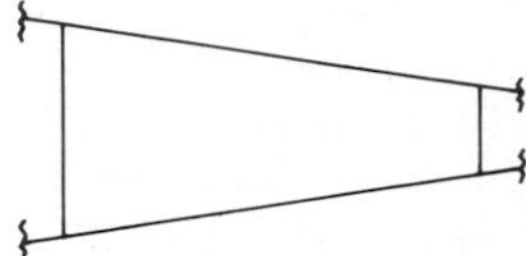

(a) Tapered cylindrical cooling rod

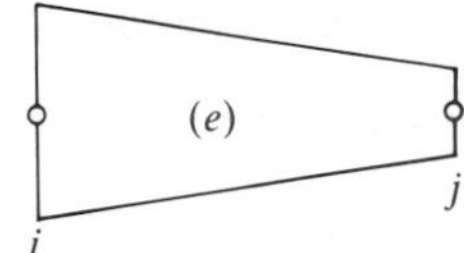

(b) One-dimensional simplex element

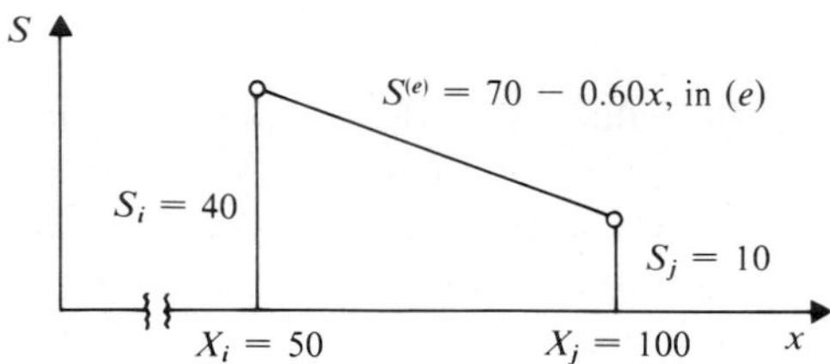

(c) Element linear interpolation function

Figure 3.3 One-dimensional simplex element (Example 3.1—Element properties for tapered rod)

An alternative approach is simply to substitute in Equation (3.3), which gives the same result.

The pyramid functions are also easily obtained as

$$N_i^{(e)} = \frac{100 - x}{50}, \quad N_j^{(e)} = \frac{x - 50}{50}$$

It may be easily seen that the ith pyramid function has the value 1.0 and 0.0 when x is 50 and 100, respectively. Also the jth pyramid function is 0.0 and 1.0 when x is 50 and 100, respectively. In matrix form the interpolation function may be written as

$$S^{(e)} = [N_i^{(e)} \; N_j^{(e)}] \begin{Bmatrix} S_i \\ S_j \end{Bmatrix}$$

where the zero value outside the element has been omitted.

$T^{(e)} = N_i T_i + N_j T_j$

(Interpolation function)

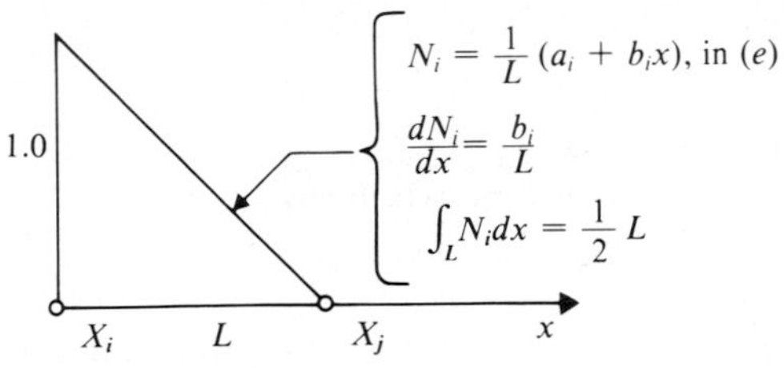

$N_i = \frac{1}{L}(a_i + b_i x)$, in (e)
$\frac{dN_i}{dx} = \frac{b_i}{L}$
$\int_L N_i dx = \frac{1}{2} L$
1.0
X_i
L
X_j
x

(a) ith pyramid function

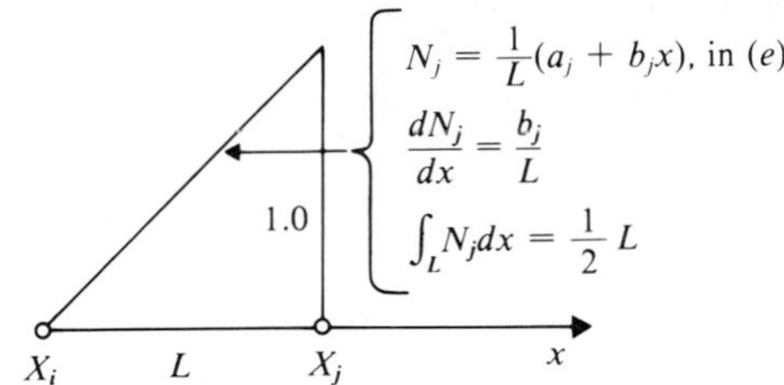

$N_j = \frac{1}{L}(a_j + b_j x)$, in (e)
$\frac{dN_j}{dx} = \frac{b_j}{L}$
1.0
$\int_L N_j dx = \frac{1}{2} L$
X_i
L
X_j
x

(b) jth pyramid function

Figure 3.4 One-dimensional simplex element: Pyramid functions and properties

Another method of formulating the interpolation function is suggested by the form of the pyramid functions. Let N be linear functions

$$N_i = \frac{1}{L}(a_i + b_i x), \quad \begin{cases} a_i = X_j \\ b_i = -1 \end{cases}$$
$$N_j = \frac{1}{L}(a_j + b_j x), \quad \begin{cases} a_j = -X_i \\ b_j = 1 \end{cases} \tag{3.6}$$

from Equation (3.5). Figure 3.4 gives some of the element properties.

C. Element Derivatives and Integrals

Certain derivatives and integrals of the interpolation function are important for later use. If the interpolation function is $T^{(e)} = N_i T_i + N_j T_j$, the derivative is

$$\frac{dT^{(e)}}{dx} = \frac{dN_i}{dx} T_i + \frac{dN_j}{dx} T_j$$

From the definition of N, the x derivatives are

$$\frac{dN_i}{dx} = \frac{1}{L} b_i, \quad \frac{dN_j}{dx} = \frac{1}{L} b_j \tag{3.7}$$

and

$$\frac{dT}{dx} = \frac{1}{L} (b_i T_i + b_j T_j) \tag{3.8}$$

Often these expressions are written in matrix form as follows

$$T^{(e)} = \{N\}\{T\} \tag{3.9}$$

Here the pyramid function row is $\{N\} = \{N_i \quad N_j\}$. The nodal column is

$$\{T\} = \begin{Bmatrix} T_i \\ T_j \end{Bmatrix}$$

Also the element derivative matrix $[D]$ is given by

$$\frac{\partial T^{(e)}}{\partial x} = [D]\ \{T\}$$

where

$$[D] = \begin{bmatrix} \dfrac{\partial N_i}{\partial x} & \dfrac{\partial N_j}{\partial x} \end{bmatrix}$$

and from Equation (3.7)

$$[D] = \frac{1}{L} [b_i \quad b_j] \tag{3.10}$$

These matrix relations will be used extensively later, mainly for two- and three-dimensional problems.

An integral over the element must involve the integrals of the pyramid functions. Because of the triangular shape of the pyramid function, with base L and height 1.0,

$$\int_L N_i \, dx = \int_L N_j \, dx = \frac{1}{2} L$$

Thus

$$\int_L T^{(e)} \, dx = \int_L (N_i T_i + N_j T_j) \, dx$$

with the result

$$\int_L T^{(e)} \, dx = \frac{1}{2} L \, (T_i + T_j) \tag{3.11}$$

It has been shown that the general integration formula is

$$\int_L N_i^{\alpha} N_j^{\beta} \, dx = \frac{\alpha! \beta!}{(\alpha + \beta + 1)!} L \tag{3.12}$$

where α, β are integers. The terms $\alpha!$ are factorials; for example, $4! = (4)(3)(2)(1) = 24$ and $0! = 1$.

Example 3.2 *Simplex Element for Fluid Velocity Component*

Given The horizontal component of fluid velocity is labeled u. A term in the momentum (Navier-Stokes) equation, called the convective inertia, is often important. It is written as $\rho u \dfrac{\partial u}{\partial x}$, ρ = density (constant). Let the element be defined by $X_i = 2$ and $X_j = 5$.

Objective Determine the convective inertia term for a one-dimensional simplex element.

Solution From Equation (3.6), $a_i = X_j = 5$, $b_i = -1$, $a_j = -X_i = -2$, $b_j = 1$, and $L = X_j - X_i = 5 - 2 = 3$. Then

$$N_i = \frac{1}{L}(a_i + b_i x) = \frac{1}{3}(5 - x)$$

$$N_j = \frac{1}{L}(a_j + b_j x) = \frac{1}{3}(-2 + x)$$

The interpolation function is

$$u^{(e)} = \frac{1}{3}(5 - x)\, u_i + \frac{1}{3}(-2 + x)\, u_j$$

$$\frac{du^{(e)}}{dx} = [D]\,\{u\} = \frac{1}{L}[b_i \quad b_j]\begin{Bmatrix} u_i \\ u_j \end{Bmatrix} = \frac{1}{3}(-u_i + u_j)$$

with the final convective inertia term

$$\rho u^{(e)} \frac{du^{(e)}}{dx} = \frac{\rho}{9}[(5 - x)\, u_i + (-2 + x)\, u_j]\,[-u_i + u_j]$$

3.3 Two-Dimensional Simplex Element

A. Interpolation Functions

If the field variable T (which may represent the temperature or other variable) is a function of two coordinates x and y, the simplest or simplex element is a triangle. The sides are straight and a node point is located at each of the corners. Using notation similar to that for the one-dimensional simplex element, the area and interpolation function are denoted

$A^{(e)}$ = element area
$T^{(e)}$ = element interpolation function for temperature

As shown in Figure 3.5, the local labeling system i,j,k for the node points is defined in the counterclockwise direction (or by using the right-hand rule), in agreement with most other authors. The coordinates of the three nodes or corners are

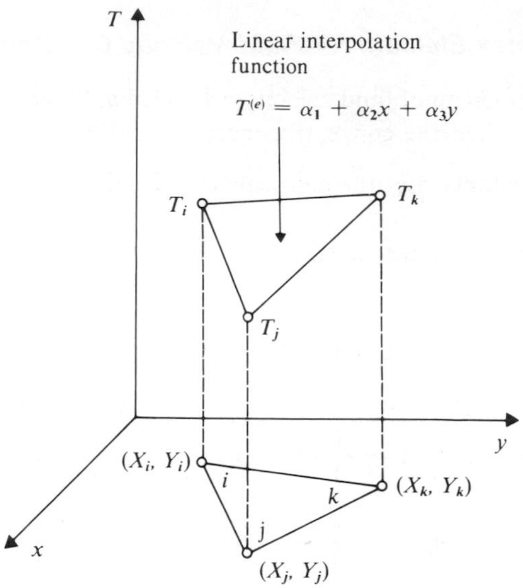

Figure 3.5 Two-dimensional simplex element: Three node points at corners of triangular element

(X_i, Y_i) = first (ith) nodal coordinates
(X_j, Y_j) = second (jth) nodal coordinates
(X_k, Y_k) = third (kth) nodal coordinates

At the node points, the nodal field variables are

T_i = first (ith) nodal temperature
T_j = second (jth) nodal temperature
T_k = third (kth) nodal temperature

The values of T_i, T_j, T_k may be given or may be found as part of the finite element solution.

A linear interpolating polynomial must be of the form

$$T^{(e)} = \begin{cases} \alpha_1 + \alpha_2 x + \alpha_3 y & \text{in } A \\ 0 & \text{elsewhere} \end{cases} \tag{3.13}$$

where the superscript (e) has been dropped from A. From now on, the zero value outside of the element is not explicitly stated but should not be forgotten. At each of the node points, the interpolation function should match the nodal temperatures, or

$$\begin{aligned} T^{(e)} &= T_i, \quad x = X_i, \quad y = Y_i \\ T^{(e)} &= T_j, \quad x = X_j, \quad y = Y_j \\ T^{(e)} &= T_k, \quad x = X_k, \quad y = Y_k \end{aligned}$$

Substituting these three expressions into Equation (3.13) yields the system of equations

$$\begin{aligned} T_i &= \alpha_1 + \alpha_2 X_i + \alpha_3 Y_i \\ T_j &= \alpha_1 + \alpha_2 X_j + \alpha_3 Y_j \\ T_k &= \alpha_1 + \alpha_2 X_k + \alpha_3 Y_k \end{aligned} \tag{3.14}$$

Solving for the values for $\alpha_1, \alpha_2, \alpha_3$ yields

$$\alpha_1 = \frac{1}{2A}[(X_jY_k - X_kY_j)T_i + (X_kY_i - X_iY_k)T_j + (X_iY_j - X_jY_i)T_k]$$

$$\alpha_2 = \frac{1}{2A}[(Y_j - Y_k)T_i + (Y_k - Y_i)T_j + (Y_i - Y_j)T_k]$$

$$\alpha_3 = \frac{1}{2A}[(X_k - X_j)T_i + (X_i - X_k)T_j + (X_j - X_i)T_k]$$

where A, the area of the triangle, is given by the formula

$$A = \frac{1}{2}(X_jY_k - X_kY_j + Y_jX_i - Y_kX_i + X_kY_i - X_jY_i) \tag{3.15}$$

This form of the interpolation function is not particularly useful.

B. Pyramid Functions

Rearrange the function $T^{(e)}$ in the pyramid function form

$$T^{(e)} = N_i\, T_i + N_j\, T_j + N_k\, T_k \tag{3.16}$$

where, from the expressions for $\alpha_1, \alpha_2, \alpha_3$ just obtained,

$$N_i = \frac{1}{2A}(a_i + b_ix + c_iy), \quad \begin{cases} a_i = X_jY_k - X_kY_j \\ b_i = Y_j - Y_k \\ c_i = X_k - X_j \end{cases}$$

$$N_j = \frac{1}{2A}(a_j + b_jx + c_jy), \quad \begin{cases} a_j = X_kY_i - X_iY_k \\ b_j = Y_k - Y_i \\ c_j = X_i - X_k \end{cases} \tag{3.17}$$

$$N_k = \frac{1}{2A}(a_k + b_kx + c_ky), \quad \begin{cases} a_k = X_iY_j - X_jY_i \\ b_k = Y_i - Y_j \\ c_k = X_j - X_i \end{cases}$$

and

$$2A = b_ic_j - b_jc_i \tag{3.18}$$

Then

N_i = first (ith) pyramid function
N_j = second (jth) pyramid function
N_k = third (kth) pyramid function

as shown in Figure 3.6.

As in the one-dimensional pyramid functions, the value of N is unity at the corresponding nodal point and zero at the remaining ones. This is the most commonly used form for the two-dimensional simplex element properties.

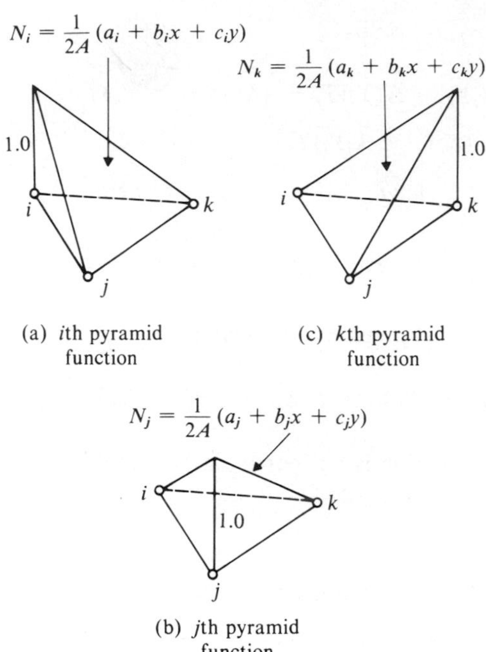

Figure 3.6 Pyramid functions for two-dimensional simplex element

This is written in matrix form as

$$T^{(e)} = \{N\}\ \{T\} \tag{3.19}$$

where

$$\{N\} = \underbrace{\{N_i \quad N_j \quad N_k\}}_{\substack{\text{pyramid function} \\ \text{row } (1\times 3)}}, \quad \{T\} = \underbrace{\begin{Bmatrix} T_i \\ T_j \\ T_k \end{Bmatrix}}_{\substack{\text{nodal column} \\ (3\times 1)}}$$

The pyramid row is of order (1×3) and the nodal column is of order (3×1).

It can also be shown that the constants a,b,c can be written as

$$\begin{bmatrix} a_i & a_j & a_k \\ b_i & b_j & b_k \\ c_i & c_j & c_k \end{bmatrix} = 2A \begin{bmatrix} 1 & X_i & Y_i \\ 1 & X_j & Y_j \\ 1 & X_k & Y_k \end{bmatrix}^{-1} \tag{3.20}$$

$$= \begin{bmatrix} X_jY_k - X_kY_j & X_kY_i - X_iY_k & X_iY_j - X_jY_i \\ Y_j - Y_k & Y_k - Y_i & Y_i - Y_j \\ X_k - X_j & X_i - X_k & X_j - X_i \end{bmatrix}$$

and

$$2A = \begin{bmatrix} 1 & X_i & Y_i \\ 1 & X_j & Y_j \\ 1 & X_k & Y_k \end{bmatrix} \tag{3.21}$$

Section 3.4 shows the derivation of very similar formulas in three-dimensional elements. If the element is labeled in a clockwise manner (rather than counterclockwise), the area formula yields the same numerical value, except that it is negative. The formulation of the actual finite element method for a particular problem may use the most appropriate form developed above.

Example 3.3 *Element Properties in Temperature Field*

Given Three temperatures are given at the nodal points

$$\begin{aligned} T &= 70°\text{C}, \quad (x,y) = (0,0) \text{ mm} \\ T &= 56°\text{C}, \quad (x,y) = (3,1) \text{ mm} \\ T &= 84°\text{C}, \quad (x,y) = (-2,4) \text{ mm} \end{aligned}$$

and the interior temperature is to be approximated by a linear interpolation function. Figure 3.7 indicates the nodal locations and element properties.

Objective Determine the interpolation function for a two-dimensional simplex element.

Solution The nodal temperatures and coordinates are obviously

$$\begin{aligned} T_i &= 70, \quad (X_i, Y_i) = (0, 0) \\ T_j &= 56, \quad (X_j, Y_j) = (3, 1) \\ T_k &= 84, \quad (X_k, Y_k) = (-2, 4) \end{aligned}$$

Employing the pyramid function form of the interpolation function, Equation (3.16),

$$T^{(e)} = N_i\, T_i + N_j\, T_j + N_k\, T_k$$

and, from Equation (3.17),

$$N_\beta = \frac{1}{2A}\,[a_\beta + b_\beta x + c_\beta y] \qquad \beta = i,j,k$$

The constants a,b,c are given by (3.17) as

$$\begin{aligned} a_i &= X_jY_k - X_kY_j = (3)(4) - (-2)(1) = 14 \\ a_j &= X_kY_i - X_iY_k = (-2)(0) - (0)(4) = 0 \\ a_k &= X_iY_j - X_jY_i = (0)(1) - (3)(0) = 0 \\ b_i &= Y_j - Y_k = 1 - 4 = -3 \\ b_j &= Y_k - Y_i = 4 - 0 = 4 \\ b_k &= Y_i - Y_j = 0 - 1 = -1 \\ c_i &= X_k - X_j = (-2) - 3 = -5 \\ c_j &= X_i - X_k = 0 - (-2) = 2 \\ c_k &= X_j - X_i = 3 - 0 = 3 \end{aligned}$$

and, from Equation (3.18),

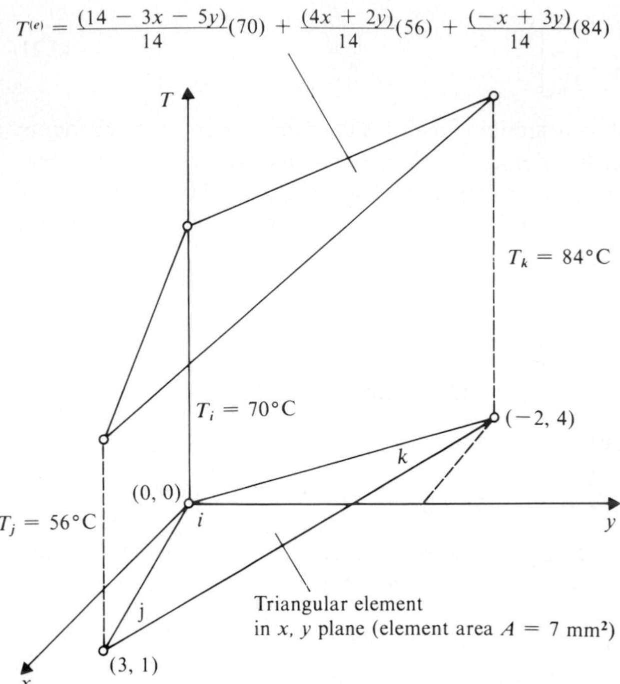

Figure 3.7 Two-dimensional simplex element (Example 3.3—Element properties in temperature field)

$$2A = b_i c_j - b_j c_i = (-3)(2) - (4)(-5) = -6 + 20 = 14$$

The element area is 7.

Substituting for the constants in the pyramid functions yields the interpolation function

$$T^{(e)} = \frac{1}{14}[(14 - 3x - 5y)\ T_i + (4x + 2y)\ T_j + (-x + 3y)\ T_k]$$

In matrix form this is

$$\lfloor N \rfloor = \frac{1}{14}\lfloor 14-3x-5y \qquad 4x+2y \qquad -x+3y \rfloor$$

and $T^{(e)} = \lfloor N \rfloor \{T\}$ where the matrix multiplication gives the above result. Note that the nodal temperatures have not been used so far. Only the nodal coordinates are required, and these are usually chosen by the analyst for convenience. Using matrix form,

$$T^{(e)} = \frac{1}{14}\lfloor 14-3x-5y \qquad 4x+2y \qquad -x+3y \rfloor \begin{Bmatrix} 70 \\ 56 \\ 84 \end{Bmatrix}$$

which multiplies out to

$$T^{(e)} = (14-3x-5y)(5) + (4x+2y)\ (4) + (-x+3y)\ (6)$$

C. Element Derivatives and Integrals

Useful derivatives include both x and y directions.

$$\frac{\partial T^{(e)}}{\partial x} = \frac{\partial N_i}{\partial x} T_i + \frac{\partial N_j}{\partial x} T_j + \frac{\partial N_k}{\partial x} T_k$$

$$\frac{\partial T^{(e)}}{\partial y} = \frac{\partial N_i}{\partial y} T_i + \frac{\partial N_j}{\partial y} T_j + \frac{\partial N_k}{\partial y} T_k$$

From the definitions of N in Equation (3.17),

$$\begin{aligned} \frac{\partial N_\beta}{\partial x} &= \frac{1}{2A} b_\beta, \quad \beta = i,j,k \\ \frac{\partial N_\beta}{\partial y} &= \frac{1}{2A} c_\beta, \quad \beta = i,j,k \end{aligned} \tag{3.22}$$

and the derivatives are given by

$$\begin{aligned} \frac{\partial T^{(e)}}{\partial x} &= \frac{1}{2A} (b_i T_i + b_j T_j + b_k T_k) \\ \frac{\partial T^{(e)}}{\partial y} &= \frac{1}{2A} (c_i T_i + c_j T_j + c_k T_k) \end{aligned} \tag{3.23}$$

This is conveniently written in matrix form as

$$\begin{Bmatrix} \dfrac{\partial T}{\partial x} \\ \dfrac{\partial T}{\partial y} \end{Bmatrix} = [D]\{T\} \tag{3.24}$$

where

$$[D] = \begin{bmatrix} \dfrac{\partial N_i}{\partial x} & \dfrac{\partial N_j}{\partial x} & \dfrac{\partial N_k}{\partial x} \\ \dfrac{\partial N_i}{\partial y} & \dfrac{\partial N_j}{\partial y} & \dfrac{\partial N_k}{\partial y} \end{bmatrix}^{(e)}$$

or, from the above results,

$$[D] = \frac{1}{2A} \begin{bmatrix} b_i & b_j & b_k \\ c_i & c_j & c_k \end{bmatrix}^{(e)} \tag{3.25}$$

This is called the element derivative matrix. It is extensively used in formulating element matrices. For the simplex element it is of order (2,3).

Integrating a pyramid function gives

$$\int_A N_i \, dA = \frac{1}{3} A$$

(from the volume of a pyramid, $Ah/3$, where A is the area of the base and h is the altitude $= 1.0$), with the same results for the other two pyramid functions. Thus the integral of $T^{(e)}$ is

$$\int_A T^{(e)}\, dA = \int_A (N_i T_i + N_j T_j + N_k T_k)\, dA$$

with the result

$$\int_A T^{(e)}\, dA = \frac{1}{3} A\, (T_i + T_j + T_k) \tag{3.26}$$

In summation form,

$$\int_A T^{(e)}\, dA = \frac{1}{3} A \sum_{\beta=1}^{3} T_\beta$$

Another useful formula for integrating pyramid functions over a triangular element is

$$\int_A N_i^\alpha\, N_j^\beta\, N_k^\alpha dA = \frac{\alpha!\beta!\alpha!}{(\alpha + \beta + \gamma + 2)!}\, 2A \tag{3.27}$$

Because much algebra is involved in proving this, no proof will be presented here.

Example 3.4 Heat Flow in Temperature Field

Given The simplex element just analyzed in Example 3.3.

Objective Obtain the heat flow per unit cross sectional area

$$q_x = -k\,\frac{\partial T}{\partial x}, \quad q_y = -k\,\frac{\partial T}{\partial y}$$

in the x and y directions, where $k = 0.25$ W/mm°C.

Solution From Equation (3.22),

$$\begin{aligned} q_x &= -\frac{k}{2A}\,(b_i T_i + b_j T_j + b_k T_k) \\ &= -\frac{0.25}{14}\,[(-3)(70) + (4)(56) + (-1)(84)] \\ &= 1.25 \text{ W/mm}^2 \end{aligned}$$

and

$$\begin{aligned} q_y &= -\frac{k}{2A}\,(c_i T_i + c_j T_j + c_k T_k) \\ &= -\frac{0.25}{14}\,[(-5)(70) + (2)(56) + (3)(84)] \\ &= -0.25 \text{ W/mm}^2 \end{aligned}$$

The element derivative matrix (3.25) is

$$[D] = \frac{1}{2A}\begin{bmatrix} b_i & b_j & b_k \\ c_i & c_j & c_k \end{bmatrix}$$

$$= \frac{1}{14}\begin{bmatrix} -3 & 4 & -1 \\ -5 & 2 & 3 \end{bmatrix}$$

In matrix form once again (3.24),

$$\begin{Bmatrix} q_x \\ q_y \end{Bmatrix} = -k \begin{Bmatrix} \dfrac{\partial T}{\partial x} \\ \dfrac{\partial T}{\partial y} \end{Bmatrix} = -k\ [D]\{T\}$$

$$= -\frac{0.25}{14}\begin{bmatrix} -3 & 4 & -1 \\ -5 & 2 & 3 \end{bmatrix}\begin{Bmatrix} 70 \\ 56 \\ 84 \end{Bmatrix}$$

$$= \begin{Bmatrix} 1.25 \\ -0.25 \end{Bmatrix} \text{W/mm}^2$$

3.4 Three-Dimensional Simplex Element

A. Interpolation Function

The three-dimensional simplex element is a tetrahedron with a node at each of the corners. The element properties (volume and interpolation function) are

$V^{(e)}$ = element volume
$T^{(e)}$ = element interpolation function for temperature

As indicated in Figure 3.8, the local labeling proceeds counterclockwise through i,j,k in one face and ℓ at the remaining node (following the right-hand rule). Simply extending the notation in Section 3.3 for two dimensions, the nodal coordinates are $(X_\beta, Y_\beta, Z_\beta)$ = nodal coordinate, $\beta = i,j,k,\ell$, and the nodal field variables are T_β = nodal temperature, $\beta = i,j,k,\ell$.

The linear interpolation polynomial is of the form

$$T^{(e)} = \alpha_1 + \alpha_2 x + \alpha_3 y + \alpha_4 z \tag{3.28}$$

At each of the nodal points, the interpolation polynomial must equal the nodal temperature. The resulting equations are

$$T_i = \alpha_1 + \alpha_2\, X_i + \alpha_3\, Y_i + \alpha_4\, Z_i$$
$$T_j = \alpha_1 + \alpha_2\, X_j + \alpha_3\, Y_j + \alpha_4\, Z_j$$
$$T_k = \alpha_1 + \alpha_2\, X_k + \alpha_3\, Y_k + \alpha_4\, Z_k$$
$$T_\ell = \alpha_1 + \alpha_2\, X_\ell + \alpha_3\, Y_\ell + \alpha_4\, Z_\ell$$

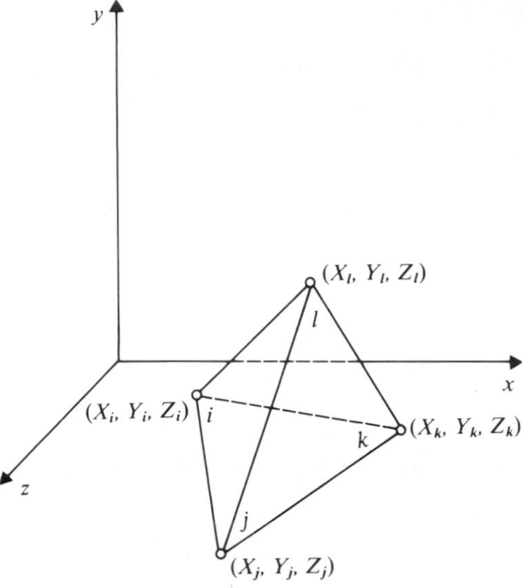

Figure 3.8 Three-dimensional simplex element: Four node points at corners of tetrahedron

A solution for α_1, α_2, α_3, α_4 could be obtained by hand, but a better procedure is to use matrices and let the computer do the work. The four equations become

$$\begin{Bmatrix} T_i \\ T_j \\ T_k \\ T_\ell \end{Bmatrix} = \begin{bmatrix} 1 & X_i & Y_i & Z_i \\ 1 & X_j & Y_j & Z_j \\ 1 & X_k & Y_k & Z_k \\ 1 & X_\ell & Y_\ell & Z_\ell \end{bmatrix} \begin{Bmatrix} \alpha_1 \\ \alpha_2 \\ \alpha_3 \\ \alpha_4 \end{Bmatrix}$$

Taking the inverse of the square matrix, it becomes apparent that

$$\begin{Bmatrix} \alpha_1 \\ \alpha_2 \\ \alpha_3 \\ \alpha_4 \end{Bmatrix} = \begin{bmatrix} 1 & X_i & Y_i & Z_i \\ 1 & X_j & Y_j & Z_j \\ 1 & X_k & Y_k & Z_k \\ 1 & X_\ell & Y_\ell & Z_\ell \end{bmatrix}^{-1} \begin{Bmatrix} T_i \\ T_j \\ T_k \\ T_\ell \end{Bmatrix} \tag{3.29}$$

Rewriting the interpolation polynomial in the matrix form gives

$$T^{(e)} = \{1,x,y,z\} \begin{Bmatrix} \alpha_1 \\ \alpha_2 \\ \alpha_3 \\ \alpha_4 \end{Bmatrix}$$

and substituting for the α values from Equation (3.29) yields

$$T^{(e)} = [1,x,y,z] \begin{bmatrix} 1 & X_i & Y_i & Z_i \\ 1 & X_j & Y_j & Z_j \\ 1 & X_k & Y_k & Z_k \\ 1 & X_\ell & Y_\ell & Z_\ell \end{bmatrix}^{-1} \begin{Bmatrix} T_i \\ T_j \\ T_k \\ T_\ell \end{Bmatrix} \tag{3.30}$$

B. Pyramid Functions

If the pyramid functions are desired, the interpolation function may be written as

$$T^{(e)} = \lfloor N \rfloor \{T\} \tag{3.31}$$

where

$$\lfloor N \rfloor = \lfloor N_i \quad N_j \quad N_k \quad N_\ell \rfloor$$

is the pyramid row, and

$$\{T\} = \begin{Bmatrix} T_i \\ T_j \\ T_k \\ T_\ell \end{Bmatrix}$$

is the nodal column. Here $\lfloor N \rfloor$ is of order (1×4) while $\{T\}$ is of order (4×1). Comparison with Equation (3.30) yields

$$\lfloor N \rfloor = \lfloor 1,x,y,z \rfloor \begin{bmatrix} 1 & X_i & Y_i & Z_i \\ 1 & X_j & Y_j & Z_j \\ 1 & X_k & Y_k & Z_k \\ 1 & X_\ell & Y_\ell & Z_\ell \end{bmatrix}^{-1} \tag{3.32}$$

One last formulation, similar to the two-dimensional case, Equation (3.17), involves the terms a,b,c,d:

$$\lfloor N \rfloor = \frac{1}{6V} \lfloor 1,x,y,z \rfloor \begin{bmatrix} a_i & a_j & a_k & a_\ell \\ b_i & b_j & b_k & b_\ell \\ c_i & c_j & c_k & c_\ell \\ d_i & d_j & d_k & d_\ell \end{bmatrix} \tag{3.33}$$

where

$$\begin{bmatrix} a_i & a_j & a_k & a_\ell \\ b_i & b_j & b_k & b_\ell \\ c_i & c_j & c_k & c_\ell \\ d_i & d_j & d_k & d_\ell \end{bmatrix} = 6V \begin{bmatrix} 1 & X_i & Y_i & Z_i \\ 1 & X_j & Y_j & Z_j \\ 1 & X_k & Y_k & Z_k \\ 1 & X_\ell & Y_\ell & Z_\ell \end{bmatrix}^{-1} \tag{3.34}$$

and the volume of the tetrahedron is given by the determinant

$$V = \frac{1}{6} \det \begin{bmatrix} 1 & X_i & Y_i & Z_i \\ 1 & X_j & Y_j & Z_j \\ 1 & X_k & Y_k & Z_k \\ 1 & X_\ell & Y_\ell & Z_\ell \end{bmatrix} \tag{3.35}$$

C. Element Derivatives and Integrals

It should be apparent from comparison with the one- and two-dimensional cases that the derivatives are

$$\frac{\partial T^{(e)}}{\partial x} = \frac{1}{6V}(b_i\, T_i + b_j\, T_j + b_k\, T_k + b_\ell\, T_\ell)$$

$$\frac{\partial T^{(e)}}{\partial y} = \frac{1}{6V}(c_i\, T_i + c_j\, T_j + c_k\, T_k + c_\ell\, T_\ell) \tag{3.36}$$

$$\frac{\partial T^{(e)}}{\partial z} = \frac{1}{6V}(d_i\, T_i + d_j\, T_j + d_k\, T_k + d_\ell\, T_\ell)$$

The element derivative matrix is

$$\left\{\begin{matrix} \frac{\partial T}{\partial x} \\ \frac{\partial T}{\partial y} \\ \frac{\partial T}{\partial z} \end{matrix}\right\} = [D]\{T\} \tag{3.37}$$

where the 3 × 4 element derivative matrix

$$[D] = \begin{bmatrix} \frac{\partial N_i}{\partial x} & \frac{\partial N_j}{\partial x} & \frac{\partial N_k}{\partial x} & \frac{\partial N_\ell}{\partial x} \\ \frac{\partial N_i}{\partial y} & \frac{\partial N_j}{\partial y} & \frac{\partial N_k}{\partial y} & \frac{\partial N_\ell}{\partial y} \\ \frac{\partial N_i}{\partial z} & \frac{\partial N_j}{\partial z} & \frac{\partial N_k}{\partial z} & \frac{\partial N_\ell}{\partial z} \end{bmatrix}$$

Finally

$$[D] = \frac{1}{6V}\begin{bmatrix} b_i & b_j & b_k & b_\ell \\ c_i & c_j & c_k & c_\ell \\ d_i & d_j & d_k & d_\ell \end{bmatrix} \tag{3.38}$$

The integral of the interpolation function is

$$\int_V T^{(e)}\, dV = \int_V (N_i T_i + N_j T_j + N_k T_k + N_\ell T_\ell) dV$$

or, evaluating,

$$\int_V T^{(e)}\, dV = \frac{1}{4} V (T_i + T_j + T_k + T_\ell) \tag{3.39}$$

In summation form,

$$\int_V T^{(e)}\, dV = \frac{1}{4} V \sum_{\beta=1}^{4} T_\beta$$

In determining the integral above, the general formula was used:

$$\int_V N_i^\alpha N_j^\beta N_k^\gamma N_\ell^\delta \, dV = \frac{\alpha!\,\beta!\,\gamma!\,\delta!\,6V}{(\alpha + \beta + \gamma + \delta + 3)!} \tag{3.40}$$

Example 3.5 *Interpolation Function for Tetrahedron*

Given The coordinate points for a tetrahedral element are (0,0,0), (1,0,3), (2,0,0), (1,2,1). A schematic is shown in Figure 3.9.

Objective Determine the volume and interpolation function for the element.

Solution The volume is given by Equation (3.35) as

$$V = \frac{1}{6}\det\begin{bmatrix} 1 & 0 & 0 & 0 \\ 1 & 1 & 0 & 3 \\ 1 & 2 & 0 & 0 \\ 1 & 1 & 2 & 1 \end{bmatrix}$$

$$= \frac{1}{6}\det\begin{bmatrix} 1 & 0 & 3 \\ 2 & 0 & 0 \\ 1 & 2 & 1 \end{bmatrix}$$

$$= -\frac{2}{6}\det\begin{bmatrix} 0 & 3 \\ 2 & 1 \end{bmatrix} = -\frac{2}{6}(0 - 6)$$

or $V = 2$ (units)^3. Also from Equation (3.34), the inverse of the coordinate matrix (obtained by a simple computer operation) is

$$\begin{bmatrix} 1 & 0 & 0 & 0 \\ 1 & 1 & 0 & 3 \\ 1 & 2 & 0 & 0 \\ 1 & 1 & 2 & 1 \end{bmatrix}^{-1} = \frac{1}{6}\begin{bmatrix} 6 & 0 & 0 & 0 \\ -3 & 0 & 3 & 0 \\ -1 & -1 & -1 & 3 \\ -1 & 2 & -1 & 0 \end{bmatrix}$$

Thus

$$\begin{bmatrix} a_i & a_j & a_k & a_\ell \\ b_i & b_j & b_k & b_\ell \\ c_i & c_j & c_k & c_\ell \\ d_i & d_j & d_k & d_\ell \end{bmatrix} = \frac{6(2)}{6}\begin{bmatrix} 6 & 0 & 0 & 0 \\ -3 & 0 & 3 & 0 \\ -1 & -1 & -1 & 3 \\ -1 & 2 & -1 & 0 \end{bmatrix}$$

The pyramid functions are

$$N_\beta = \frac{1}{6V}(a_\beta + b_\beta x + c_\beta y + d_\beta z), \quad \beta = i,j,k,\ell$$

or

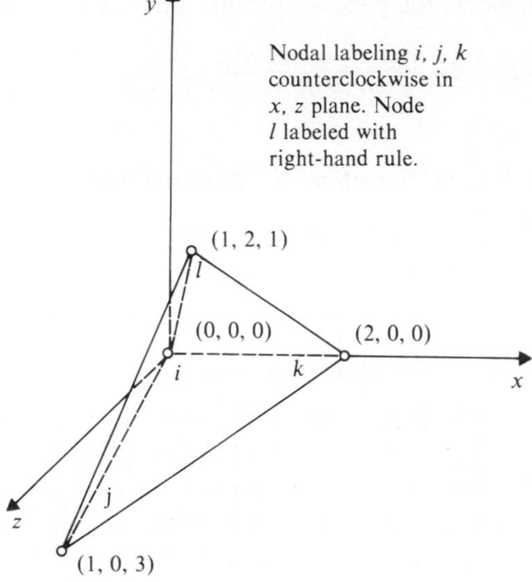

Figure 3.9 Nodal labeling for three-dimensional simplex element (Example 3.5—Interpolation function for tetrahedron)

$$N_i = \frac{1}{6}(6 - 3x - y - z)$$

$$N_j = \frac{1}{6}(-y + 2z)$$

$$N_k = \frac{1}{6}(3x - y - z)$$

$$N_\ell = \frac{1}{2}y$$

Element derivative matrices and other element properties can be easily determined.

3.5 Simplex Elements for Vectors

A. Nodal Labeling

In the previous sections, the field variable was a scalar quantity such as temperature, pressure, velocity potential, or stress function. The concepts developed for the simplex elements in one, two, and three dimensions are easily extended to include vector functions as well. Usually the vectors considered are force-type variables (such as gravitational forces or stresses) or position-type variables (such as displacement, strain, velocity, or acceleration). The vector variables are resolved into components as follows:

$$u = u(x), \text{ one-dimensional vector}$$

$$\left.\begin{aligned} u &= u(x,y) \\ v &= v(x,y) \end{aligned}\right\}, \text{ two-dimensional vector}$$

$$\left.\begin{aligned} u &= u(x,y,z) \\ v &= v(x,y,z) \\ w &= w(x,y,z) \end{aligned}\right\}, \text{ three-dimensional vector}$$

where

$$\begin{aligned} u &= x\text{-component of vector} \\ v &= y\text{-component of vector} \\ w &= z\text{-component of vector} \end{aligned}$$

As the symbols u,v,w are often used to represent displacements in solid mechanics and velocities in fluid mechanics, these appear to be good symbols to use. In each case the vector component is taken as positive in the positive coordinate direction.

A formulation of the computer solution for vector problems does not normally employ the vector components as written above. The same variable, say U, is used for all of the components, but the subscripts are changed. Of course, in one dimension only one component u exists, so the nodal values are simply

$$\begin{aligned} u &= U_i, \quad x = X_i \\ u &= U_j, \quad x = X_j \end{aligned} \tag{3.41}$$

Figure 3.10 shows the vector components. In two dimensions, the nodal equivalences are

$$\begin{aligned} &\left.\begin{aligned} u &= U_{2i-1} \\ v &= U_{2i} \end{aligned}\right\}, x = X_i, \quad y = Y_i \\ &\left.\begin{aligned} u &= U_{2j-1} \\ v &= U_{2j} \end{aligned}\right\}, x = X_j, \quad y = Y_j \\ &\left.\begin{aligned} u &= U_{2k-1} \\ v &= U_{2k} \end{aligned}\right\}, x = X_k, \quad y = Y_k \end{aligned} \tag{3.42}$$

A similar relationship follows for three dimensions.

B. Interpolation Functions

Interpolation functions are defined in the same manner as in the previous sections. For one dimension,

$$u^{(e)} = N_i U_i + N_j U_j$$

or, in matrix form,

$$u^{(e)} = \{N_i \quad N_j\} \begin{Bmatrix} U_i \\ U_j \end{Bmatrix} \tag{3.43}$$

Two-dimensional problems yield

$$\begin{aligned} u &= N_i U_{2i-1} + N_j U_{2j-1} + N_k U_{2k-1} \\ v &= N_i U_{2i} + N_j U_{2j} + N_k U_{2k} \end{aligned}$$

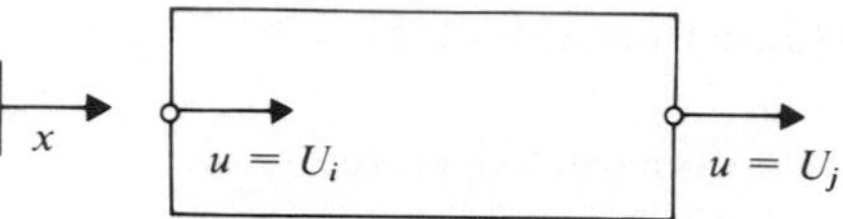

(a) One-dimensional element

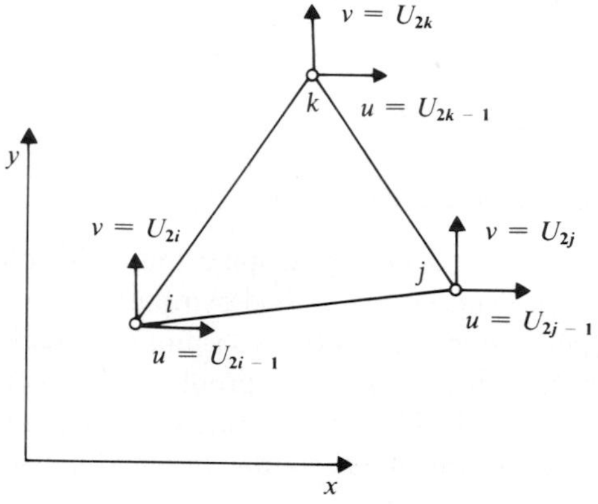

(b) Two-dimensional element

Figure 3.10 Vector variables in simplex elements (one and two dimensions)

and, including all nodal values of U in the element,

$$u = N_i U_{2i-1} + 0\,U_{2i} + N_j U_{2j-1} + 0\,U_{2j} + N_k U_{2k-1} + 0\,U_{2k}$$

$$v = 0\,U_{2j-1} + N_i U_{2i} + 0\,U_{2j-1} + N_j U_{2j} + 0\,U_{2k-1} + N_k U_{2k}$$

In matrix notation, these become

$$\left\{ \begin{matrix} u \\ v \end{matrix} \right\} = [N]\{U\} \tag{3.44}$$

where

$$[N] = \begin{bmatrix} N_i & 0 & N_j & 0 & N_k & 0 \\ 0 & N_i & 0 & N_j & 0 & N_k \end{bmatrix}$$

$$\{U\} = \left\{ \begin{matrix} U_{2i-1} \\ U_{2i} \\ U_{2j-1} \\ U_{2j} \\ U_{2k-1} \\ U_{2k} \end{matrix} \right\}$$

The pyramid functions are the same as defined in Equation (3.17).

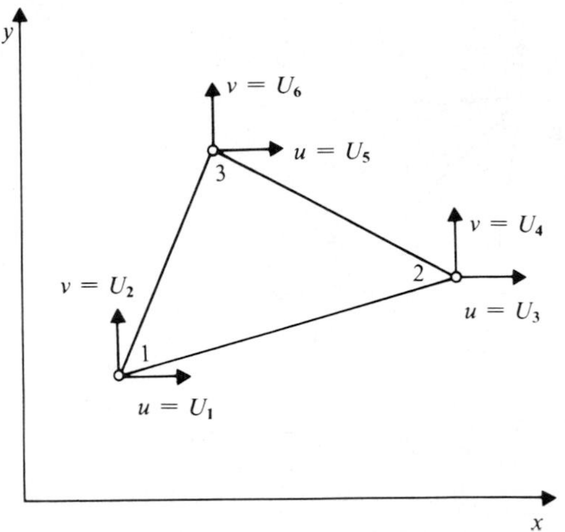

Figure 3.11 Two-dimensional simplex element (Example 3.6—Vector properties)

Example 3.6 *Vector Properties of Two-Dimensional Simplex Element*

Given A two-dimensional simplex element has the nodal coordinates (1,1), (4,2), (2,3). The displacements in two directions are u and v. A schematic is given in Figure 3.11.

Objective Find the element properties in matrix form.

Solution Let the nodes be labeled 1,2,3, where

$$\begin{aligned} X_1 &= 1, \quad Y_1 = 1 \quad (i = 1) \\ X_2 &= 4, \quad Y_2 = 2 \quad (j = 2) \\ X_3 &= 2, \quad Y_3 = 3 \quad (k = 3) \end{aligned}$$

Equations (3.17) and (3.18) yield the interpolation function as

$$N_1 = \frac{1}{5}(8 - x - 2y)$$

$$N_2 = \frac{1}{5}(-1 + 2x - y)$$

$$N_3 = \frac{1}{5}(-2 - x + 3y)$$

and the area of the element as 2.5. The displacement interpolation functions are given by

$$\begin{aligned} u &= N_1 U_1 + N_2 U_3 + N_3 U_5 \\ v &= N_1 U_2 + N_2 U_4 + N_3 U_6 \end{aligned}$$

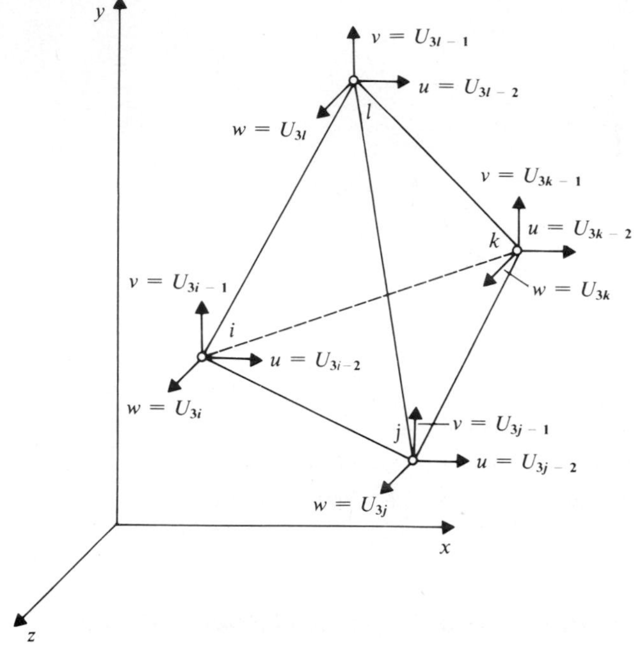

Figure 3.12 Vector variables in three-dimensional simplex elements

Equation (3.44) becomes

$$\begin{Bmatrix} u \\ v \end{Bmatrix} = \begin{bmatrix} N_1 & 0 & N_2 & 0 & N_3 & 0 \\ 0 & N_1 & 0 & N_2 & 0 & N_3 \end{bmatrix} \begin{Bmatrix} U_1 \\ U_2 \\ U_3 \\ U_4 \\ U_5 \\ U_6 \end{Bmatrix}$$

where the values for N_1,N_2,N_3 are those given above.

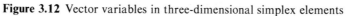

The same procedure gives the three-dimensional formulation, Figure 3.12, as

$$\begin{Bmatrix} u \\ v \\ w \end{Bmatrix} = [N]\{U\} \tag{3.45}$$

where

$$[N] = \begin{bmatrix} N_i & 0 & 0 & N_j & 0 & 0 & N_k & 0 & 0 & N_\ell & 0 & 0 \\ 0 & N_i & 0 & 0 & N_j & 0 & 0 & N_k & 0 & 0 & N_\ell & 0 \\ 0 & 0 & N_i & 0 & 0 & N_j & 0 & 0 & N_k & 0 & 0 & N_\ell \end{bmatrix}$$

and

$$\{U\} = \begin{Bmatrix} U_{3i-2} \\ U_{3i-1} \\ U_{3i} \\ U_{3j-2} \\ U_{3j-1} \\ U_{3j} \\ U_{3k-2} \\ U_{3k-1} \\ U_{3k} \\ U_{3\ell-2} \\ U_{3\ell-1} \\ U_{3\ell} \end{Bmatrix}$$

The pyramid functions are the same as in Equation (3.33). The element derivative matrix is easy to ascertain by comparison with the two-dimensional case.

Other problems, such as the transverse deflection of beams (discussed in Section 3.6), may have more than one degree of freedom per node. Both deflection and slope can be considered degrees of freedom at one node point. More details are given in Chapter 7.

3.6 One-Dimensional Higher Order Elements

Up to now in this chapter, all of the interpolation functions used for elements have been linear in form. The minimum number of node points has been used for each element. One of the advantages of finite element analysis is that higher order polynomials (and more node points) can be employed in elements, with a resulting increase in accuracy. This section will give only an outline of some of the properties of these functions. Additional details are given in later chapters, such as Chapter 7 (cubic beam elements), Chapter 12 (rectangular elements), and Chapter 13 (curved isoparametric elements).

A. Element Classification

Elements may be classified in several different ways, as shown in Table 3.1. Some have already been discussed, such as the dimension of the problem (one, two, three) and the number of degrees of freedom at each node point (one, two, three, etc.). Elements are also classified by the order of interpolation polynomial, shape, and degree of interelement continuity. The order of the interpolation polynomial is determined by the highest complete polynomial (all possible terms present), not by the order of the highest term. The shape classification is relatively obvious. Interelement continuity refers to the continuity of interpolation functions and their derivatives from one element to the next.

An important consideration for element accuracy is the continuity of the interpolation function and its derivatives between elements. When higher order elements use more

Table 3.1
Element Classifications

Dimension	Degrees of Freedom	Order of Polynomial	Shape	Interelement Continuity
One	One	Linear	Straight	Function
Two	Two	Quadratic	Triangular	First derivative
Three	Three	Cubic	Rectangular	Second derivative
	. . . .	Quartic	Tetrahedral	. . . .
		. . . .	. . . .	

node points (at which the field variable is evaluated), shape functions that have zeros at specified points are used extensively. These polynomials are called Lagrange polynomials. Other elements employ more than one degree of freedom at each node point. Usually the derivatives of the interpolation functions are treated as degrees of freedom. Polynomials that have zeros and whose first derivatives have zeros at prescribed points are called Hermite polynomials. When the first derivative is specified at each end of an element, then the derivative will be continuous across interelement boundaries. Generally, this improves solution accuracy.

B. Quadratic One-Dimensional Element

A second-order one-dimensional element, shown in Figure 3.13, has three node points and a Lagrange interpolation function of the form

$$T^{(e)} = \alpha_1 + \alpha_2 x + \alpha_2 x^2 \tag{3.46}$$

At each node point, this relation must hold, giving the three equations

$$\begin{aligned} T_i &= \alpha_1 + \alpha_2 X_i + \alpha_3 X_i^2 \\ T_j &= \alpha_1 + \alpha_2 X_j + \alpha_3 X_j^2 \\ T_k &= \alpha_1 + \alpha_2 X_k + \alpha_3 X_k^2 \end{aligned}$$

These may be solved for α_1, α_2, α_3, if necessary.

For this discussion, it will be assumed that the nodes are equally spaced and that the coordinate system starts at the first node (this causes no loss of element generality, as x can then be thought of as a local coordinate within the element). Now the nodal coordinates are $X_i = 0$, $X_j = \frac{1}{2} L$, $X_k = L$. Using the shape function form for T,

$$T^{(e)} = G_i T_i + G_j T_j + G_k T_k$$

where the symbol G has been used rather than N (which will denote the linear pyramid function). The functions G_i, G_j, G_k are called shape functions rather than pyramid functions when the order of the polynomial is higher than one. The resulting functions are stated here as

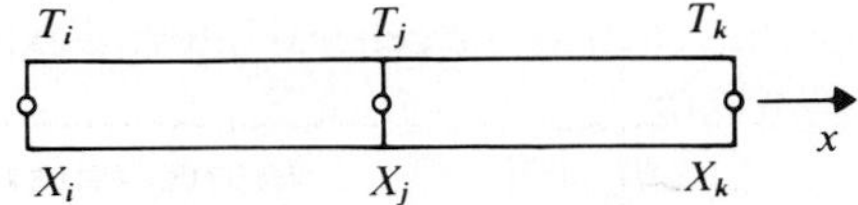

(a) Quadratic element

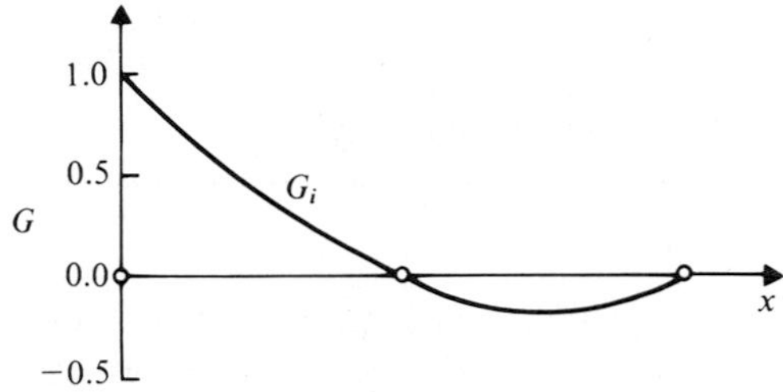

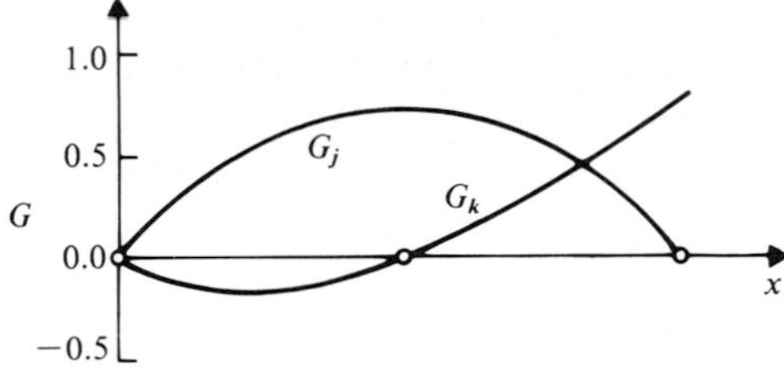

(b) Shape functions

Figure 3.13 Quadratic Lagrange element in one dimension

$$
\begin{aligned}
G_i &= \frac{1}{2L^2}(2x-L)(2x-2L) \\
G_j &= -\frac{1}{L^2}(2x)(2x-2L) \qquad \textbf{(3.47)} \\
G_k &= \frac{1}{2L^2}(2x)(2x-L)
\end{aligned}
$$

Each shape function has the value unity at its corresponding node point and the value zero at the other node points.

It is also possible to express each shape function G in terms of the linear pyramid functions N

$$
\begin{aligned}
G_i &= N_i(2N_i - 1) \\
G_j &= 4N_iN_k \qquad \textbf{(3.48)} \\
G_k &= N_k(2N_k - 1)
\end{aligned}
$$

This allows the use of the pyramid function properties such as the general integral expression (3.12). In matrix form this is $T^{(e)} = \{G\}\{T\}$ where

$$\{G\} = \{G_i \quad G_j \quad G_k\}$$

$$\{T\} = \begin{Bmatrix} T_i \\ T_j \\ T_k \end{Bmatrix}$$

Derivatives for the second-order, one-dimensional element are given by

$$\frac{\partial T^{(e)}}{\partial x} = \frac{\partial G_i}{\partial x} T_i + \frac{\partial G_j}{\partial x} T_j + \frac{\partial G_k}{\partial x} T_k$$

In matrix form this is

$$\frac{\partial T^{(e)}}{\partial x} = [D]\{T\}$$

where

$$[D] = \begin{bmatrix} \dfrac{\partial G_i}{\partial x} & \dfrac{\partial G_j}{\partial x} & \dfrac{\partial G_k}{\partial x} \end{bmatrix}$$

or, evaluating the terms, the 1×3 element derivative matrix is

$$[D] = \frac{1}{L^2}[4x-3L \quad -8x+4L \quad 4x-L] \tag{3.49}$$

Note that $[D]$ is now a function of x rather than a constant, as in the simplex element. While the interpolation functions $T^{(e)}$ are continuous from one element to the next, the derivatives are not continuous, even though the polynomial is second-order.

Example 3.7 *Axial Displacement of Quadratic Element*

Given A rod is subjected to an axial force F. If the rod is divided into a number of quadratic elements, let element 3 be located between the points $x = 3.0$ mm and $x = 7.0$ mm. The axial displacement u at the three nodal points is

$$\{u\} = \begin{Bmatrix} u_i \\ u_j \\ u_k \end{Bmatrix} = \begin{Bmatrix} 0.0045 \\ 0.0053 \\ 0.0060 \end{Bmatrix} \text{ mm}$$

Objective Determine the shape functions and the element strain given by

$$\epsilon^{(e)} = \frac{\partial u^{(e)}}{\partial x}$$

Solution The length of the element is $L^{(e)} = 4.0$ mm. Then the shape functions are (3.47)

$$G_i = \frac{1}{32}(2x-4)(2x-8)$$

$$G_j = -\frac{1}{16}(2x)(2x-8)$$

$$G_k = \frac{1}{32}(2x)\ (2x-4)$$

The strain is

$$\epsilon^{(e)} = \frac{\partial u^{(e)}}{\partial x} = [D]\{u\}$$

$$\epsilon^{(e)} = \frac{1}{16}[4x-12 \quad -8x+16 \quad 4x-4] \begin{Bmatrix} 0.0045 \\ 0.0053 \\ 0.0060 \end{Bmatrix}$$

$$\epsilon^{(e)} = [(1.125x-3.375) + (-2.65x+5.3) + (1.5x-1.5)]\frac{1}{1000}$$

Finally,

$$\epsilon^{(e)} = \frac{(-0.025x + 0.425)}{1000}\frac{\text{mm}}{\text{mm}}$$

C. Cubic Elements

Another element is the cubic one-dimensional element, shown in Figure 3.14, which has one more node point. The interpolation function has the form

$$T^{(e)} = \alpha_1 + \alpha_2 x + \alpha_3 x^2 + \alpha_4 x^3$$

Let the node points be equally spaced and

$$X_i = 0,\ X_j = \frac{1}{3}L,\ X_k = \frac{2}{3}L,\ X_\ell = L$$

If T has the form

$$T^{(e)} = G_i T_i + G_j T_j + G_k T_k + G_\ell T_\ell$$

then the shape functions are

$$G_i = -\frac{1}{6L^3}(3x-L)(3x-2L)(3x-3L)$$

$$G_j = -\frac{1}{2L^3}(3x)(3x-2L)(3x-3L)$$

$$G_k = -\frac{1}{2L^3}(3x)(3x-L)(3x-3L) \qquad \textbf{(3.50)}$$

$$G_\ell = \frac{1}{6L^3}(3x)(3x-L)(3x-2L)$$

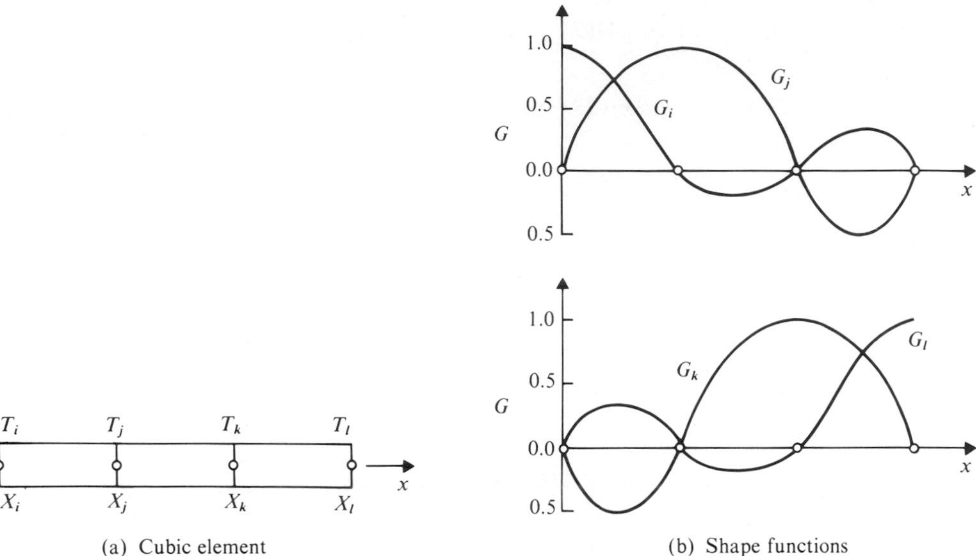

Figure 3.14 Cubic Lagrange element in one dimension

The derivative $\partial T/\partial x$ is now a second-order polynomial. More details are given in Chapter 7.

A cubic one-dimensional element that is widely used, particularly for transverse beam deflections, has two node points with the values of both the interpolation function and its first derivative as degrees of freedom. Let the nodal values of the first derivative (slope) be denoted by θ_i and θ_j, as shown in Figure 3.15(a).

$$\theta_i = \frac{dT}{dx}\bigg|_{x=X_i=0}, \qquad \theta_j = \frac{dT}{dx}\bigg|_{x=X_j=L}$$

If the element interpolation function has the form

$$T^{(e)} = G_{Ti}T_i + G_{\theta i}\theta_i + G_{Tj}T_j + G_{\theta j}\theta_j$$

with the Hermite polynomials giving the shape functions as

$$\begin{aligned} G_{Ti} &= 1 - 3\left(\frac{x}{L}\right)^2 + 2\left(\frac{x}{L}\right)^3 \\ G_{\theta i} &= x\left[\left(\frac{x}{L}\right) - 1\right]^2 \\ G_{Tj} &= \left(\frac{x}{L}\right)^2\left[3 - 2\left(\frac{x}{L}\right)\right] \\ G_{\theta j} &= \frac{x^2}{L}\left[\left(\frac{x}{L}\right) - 1\right] \end{aligned} \qquad \textbf{(3.51)}$$

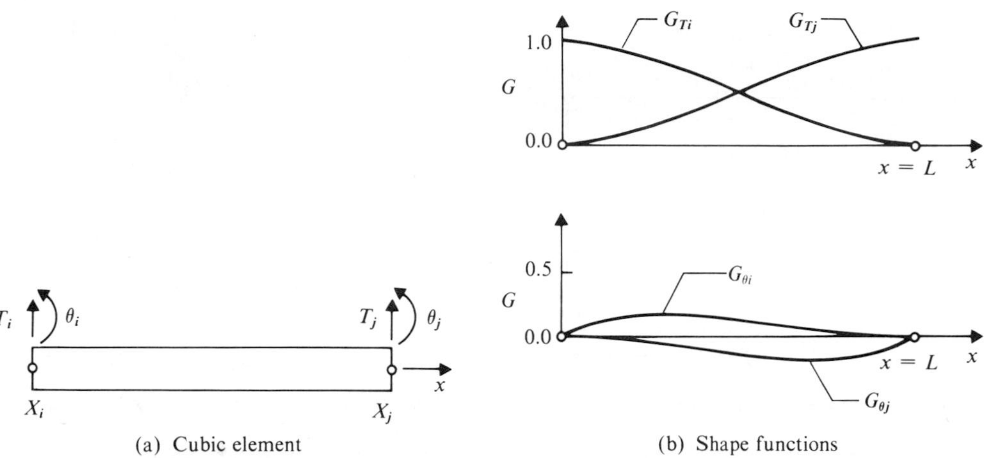

Figure 3.15 Cubic Hermite beam element in one dimension

In terms of the linear pyramid functions, these expressions are

$$\begin{aligned} G_{Ti} &= N_i^2\,(2N_j+1) \\ G_{\theta i} &= N_i^2\,N_jL \\ G_{Tj} &= (2N_i+1)N_j^2 \\ G_{\theta j} &= -N_iN_j^2\,L \end{aligned}$$

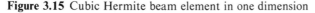

The shape functions are plotted in Figure 3.15(b), which shows that G_{Ti} and G_{Tj} have the normal value of 1.0 at nodes i and j, respectively, with the value zero at the opposite node. The shape functions for θ_i and θ_j are somewhat different. Here the slope of $G_{\theta i}$ has the value 1.0 at node i, while the slope of $G_{\theta j}$ is 1.0 at node j. The numerical value of $G_{\theta i}$ and $G_{\theta j}$ is zero at both node points.

Because the first derivative (slope) is treated as a nodal value, the slope is continuous from element to element. This gives a high level of accuracy to the element for problems in which the slope is physically important, as in beam problems. Chapter 7 discusses this element in much more detail.

3.7 Two-Dimensional Higher Order Elements

Higher order two-dimensional elements are widely used in finite element analysis. This section will treat only elements with simple shapes, such as triangular and rectangular. Curved-sided elements, such as isoparametric elements, which are considered in Chapter 13, require some relatively advanced techniques, which are not appropriate at this stage of the text. All of the two-dimensional elements discussed in this section are classified subparametric.

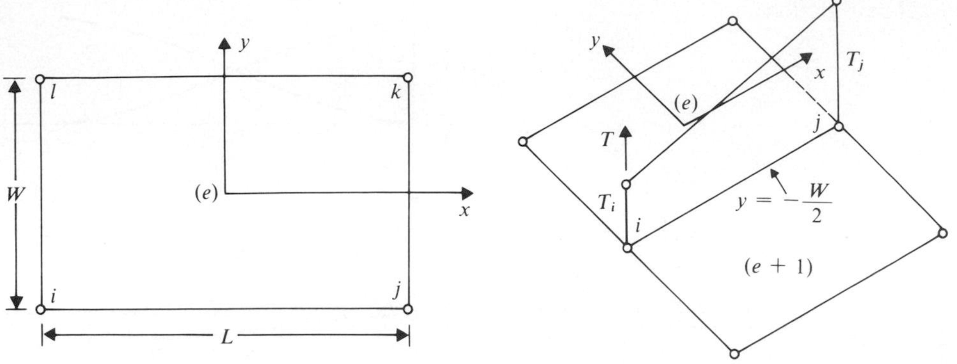

(a) Rectangular (serendipity) element (b) Interpolation function along interelement boundary

Figure 3.16 Element geometry and interelement continuity for rectangular serendipity element

A. Rectangular Element

The simplest higher order element is the rectangular (or square) element. Figure 3.16(a) shows the element of length L and width W. Four nodes, one at each corner, are used. The element interpolation function is assumed to have the form

$$T^{(e)} = \alpha_1 + \alpha_2 x + \alpha_3 y + \alpha_4 xy \tag{3.52}$$

This is called a bilinear function of x and y. Note that two alternative polynomials could have been assumed, $T^{(e)} = \alpha_1 + \alpha_2 x + \alpha_3 y + \alpha_4 x^2$ or $T^{(e)} = \alpha_1 + \alpha_2 x + \alpha_3 y + \alpha_4 y^2$. These have not been used because the bilinear function is linear along each side of the element.

Let the interpolation be written as $T^{(e)} = G_i T_i + G_j T_j + G_k T_k + G_\ell T_\ell$ or

$$T^{(e)} = \{G\}\{T\} \tag{3.53}$$

where

$$\{G\} = \{G_i \quad G_j \quad G_k \quad G_\ell\}$$

$$\{T\} = \begin{Bmatrix} T_i \\ T_j \\ T_k \\ T_\ell \end{Bmatrix}$$

The shape functions are

$$
\begin{aligned}
G_i &= \frac{1}{LW}\left(\frac{L}{2} - x\right)\left(\frac{W}{2} - y\right) \\
G_j &= \frac{1}{LW}\left(\frac{L}{2} + x\right)\left(\frac{W}{2} - y\right) \\
G_k &= \frac{1}{LW}\left(\frac{L}{2} + x\right)\left(\frac{W}{2} + y\right) \\
G_\ell &= \frac{1}{LW}\left(\frac{L}{2} - x\right)\left(\frac{W}{2} + y\right)
\end{aligned} \tag{3.54}
$$

Again local coordinates x,y are used without loss of generality. The shape functions have the value unity at the associated node point and the value zero at the other three node points.

The variation of T along the boundary determines the interelement continuity. Consider any side of the rectangular element, say side 1 (between nodes i and j). Here $y = -W/2$, and the shape functions have the form

$$G_i = \frac{1}{L}\left(\frac{L}{2} - x\right), \quad G_j = \frac{1}{L}\left(\frac{L}{2} + x\right), \quad G_k = 0, \quad G_\ell = 0$$

and the interpolation function becomes

$$T^{(e)} = \frac{1}{L}\left(\frac{L}{2} - x\right)T_i + \frac{1}{L}\left(\frac{L}{2} + x\right)T_j$$

as shown in Figure 3.16(b). Along this side, the relation is linear between nodes i and j. If another rectangular element, labeled $(e + 1)$, is next to the one considered above, the interpolation function is linear along its adjacent boundary so the function must be continuous from one element to the next. When a simplex triangular element is placed next to a rectangular one, interelement continuity is again observed. Thus the element is sometimes called a serendipity element.

The element derivative matrix is given by

$$\begin{Bmatrix} \dfrac{\partial T}{\partial x} \\ \dfrac{\partial T}{\partial y} \end{Bmatrix} = [D]\{T\}$$

where

$$[D] = \begin{bmatrix} \dfrac{\partial G_i}{\partial x} & \dfrac{\partial G_j}{\partial x} & \dfrac{\partial G_k}{\partial x} & \dfrac{\partial G_\ell}{\partial x} \\ \dfrac{\partial G_i}{\partial y} & \dfrac{\partial G_j}{\partial y} & \dfrac{\partial G_k}{\partial y} & \dfrac{\partial G_\ell}{\partial y} \end{bmatrix}$$

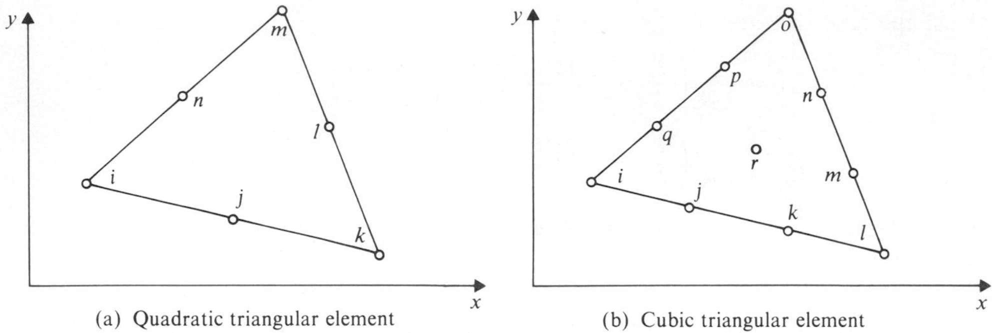

Figure 3.17 Higher order subparametric triangular elements

Evaluating the terms gives the 2 × 4 element derivative matrix

$$[D] = \frac{1}{LW}\begin{bmatrix} -\left(\frac{W}{2}-y\right) & \left(\frac{W}{2}-y\right) & \left(\frac{W}{2}+y\right) & -\left(\frac{W}{2}+y\right) \\ -\left(\frac{L}{2}-x\right) & -\left(\frac{L}{2}+x\right) & \left(\frac{L}{2}+x\right) & \left(\frac{L}{2}-x\right) \end{bmatrix} \tag{3.55}$$

Note that the x derivative is a linear function of y alone, while the y derivative is a linear function of x alone.

B. Quadratic Triangular Element

Another element is the quadratic triangular element, shown in Figure 3.17(a), with three corner nodes and three midside nodes. These additional three nodes allow the interpolation function to have three additional terms (compared to the simplex element). This is a subparametric element because it has a second-order interpolation function but linear sides.

$$T^{(e)} = \alpha_1 + \alpha_2 x + \alpha_3 y + \alpha_4 x^2 + \alpha_5 xy + \alpha_6 y^2$$

Note that all second-order terms (x^2, xy, y^2) are included in the function, so it is called a quadratic element.

Express the interpolation function in the form

$$T^{(e)} = G_i T_i + G_j T_j + G_k T_k + G_\ell T_\ell + G_m T_m + G_n T_n$$

where the G's are the second-order shape functions. Here the second-order shape functions $G(x,y)$ can be expressed in terms of the linear pyramid functions $N(x,y)$ given by Equation (3.17). They are

$$\left.\begin{aligned} G_i &= N_i(2N_i - 1) \\ G_j &= 4N_iN_k \\ G_k &= N_k(2N_k - 1) \\ G_\ell &= 4N_kN_m \\ G_m &= N_m(2N_m - 1) \\ G_n &= 4N_kN_i \end{aligned}\right\} \quad \begin{array}{l} k \text{ replaces } j \text{ and } m \text{ replaces} \\ k \text{ in linear pyramid} \\ \text{functions} \end{array} \qquad \textbf{(3.56)}$$

As usual, the shape functions G have the value unity at the associated node and the value zero at the other five nodes.

C. Cubic Triangular Element

A cubic triangular element is illustrated in Figure 3.17(b). Note that ten node points are required for the complete cubic interpolation function

$$\begin{aligned} T^{(e)} = \alpha_1 + \alpha_2 x + \alpha_3 y + \alpha_4 x^2 + \alpha_5 xy + \alpha_6 y^2 + \alpha_7 x^3 + \alpha_8 x^2 y \\ + \alpha_9 xy^2 + \alpha_{10} y^3 \end{aligned}$$

Nodes are located along the element boundaries, as shown, and the last node is usually placed at the center; this is the Lagrange type of element. A Hermite element can also be developed with the function T and both its first derivatives $\partial T/\partial x$ and $\partial T/\partial y$ at each of three corner nodes. There are nine nodal variables, but it takes ten terms to complete the cubic polynomial. Either another nodal variable must be added or the element is only quadratic in accuracy. Chapter 13 gives more details, including isoparametric elements.

3.8 Problems

3.1 A very wide heat-conducting fin can be considered one-dimensional. Let one end of an element be at $x = 0.0$ mm and the other at $x = 40.0$ mm. The temperature in the element is to be modeled with a linear interpolation function of the form

$$T^{(e)} = \alpha_1 + \alpha_2 x$$

Also calculate the linear interpolation function in pyramid function form

$$T^{(e)} = N_i\, T_i + N_j\, T_j$$

The temperature at node i is 330.00°C and at node j is 77.54°C. If the width of the fin is 160 mm and the thickness is 1.25 mm, calculate the rate of heat flow through the element with the formula

$$q = -kA\frac{dT}{dx}$$

where $k = 0.20$ W/mm-°C.

3.2 A one-dimensional simplex element has its ends located at $x = 20$ and $x = 80$. Calculate the pyramid functions and linear interpolation function for the element.

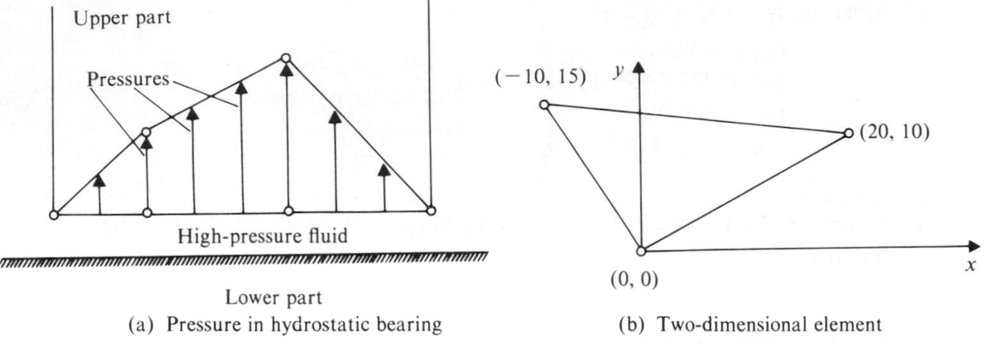

Figure 3.18 Diagrams for Problems 3.3 and 3.5

3.3 The node points and pressures under a very long hydrostatic bearing are given by

X (in)	p (lb/in^2)
0.0	0.0
1.0	50.0
2.5	100.0
4.0	0.0

The width is 30 inches. Calculate the force exerted by the fluid on the upper part of the bearing. A schematic is shown in Figure 3.18(a).

3.4 The one-dimensional tapered cooling rod considered in Example 3.1 has two nodal temperatures, T_i and T_j. Evaluate the term $S\,dT/dx$ for the element.

3.5 A two-dimensional simplex element for a thin heat-conducting plate has nodal coordinates (in millimeters) as follows

Node	X	Y
i	0	0
j	20	10
k	−10	15

Find the element area and interpolation function using the pyramid functions. A diagram is shown in Figure 3.18(b).

3.6 A two-dimensional simplex element has the nodal coordinates

Node	X	Y
18	0.5	1.0
17	1.5	1.0
20	1.0	2.5

Find the element area and interpolation functions using the pyramid functions.

3.7 The nodal temperatures (°C) for the two-dimensional element in Example 3.4 are

Node	T
i	80
j	20
k	130

Determine the heat flow per unit cross sectional area in the x and y directions if $k = 0.50$ W/mm-°C. If the heat loss due to convection out of upper and lower surfaces of the element is given by

$$q_{\text{convection}} = 2h \int_{A^{(e)}} (T - T_\infty)\, dA$$

where $h = 4 \times 10^{-4}$ W/mm²–°C and $T_\infty = 10$°C, calculate the convective heat loss for the element. Here h is the convective heat transfer coefficient and T_∞ is the temperature of the surrounding fluid.

3.8 Evaluate the integrals

$$\int_{L^{(e)}} N_i N_j^2 \, dx =$$

$$\int_{A^{(2)}} N_7 N_{10}^3 N_{11}^2 dA =$$

$$\int_{A^{(e)}} N_j \, dA =$$

$$\int_{V^{(e)}} N_i^2 N_j N_\ell \, dV =$$

3.9 A three-dimensional simplex element has the nodal coordinates given by

Node	X	Y	Z
1	0	0	0
2	2	−3	0
3	4	0	0
4	0	1	2

Determine the volume of the element and the interpolation function constants a,b,c,d.

3.10 The velocity components (u,v) for a two-dimensional simplex element are

Node	u	v
8	u_8	v_8
10	u_{10}	v_{10}
11	u_{11}	v_{11}

Determine labeling subscripts to be used in the vector equation (3.44).

3.11 At three points $x = 35$ m, $x = 45$ m, and $x = 55$ m, a hanging cable has the vertical positions

$$\{w\} = \begin{Bmatrix} 1.73 \\ 1.95 \\ 2.20 \end{Bmatrix}$$

Determine the shape functions for a quadratic element modeling this cable and the slope of the shape function $\theta = dw/dx$ over the element.

3.12 A rectangular element has length 4 mm and width 3 mm. The nodal temperatures are

$$\begin{Bmatrix} T_1 \\ T_2 \\ T_3 \\ T_4 \end{Bmatrix} = \begin{Bmatrix} 80 \\ 76 \\ 68 \\ 73 \end{Bmatrix} \quad (°\text{C})$$

Determine element interpolation functions and the heat flow

$$\begin{Bmatrix} q_x \\ q_y \end{Bmatrix} = -k \begin{Bmatrix} \dfrac{\partial T}{\partial x} \\ \dfrac{\partial T}{\partial y} \end{Bmatrix}$$

where $k = 0.50$ W/mm-°C.

3.13 A quadratic one-dimensional element has length 100 mm. If

$$\begin{aligned} T_1 &= 30°\text{C}, x = 0 \text{ mm} \\ T_2 &= 10°\text{C}, x = 50 \text{ mm} \\ T_3 &= -5°\text{C}, x = 100 \text{ mm} \end{aligned}$$

determine the value of the interpolation function $T^{(e)}(x)$ at the quarter points $x = 25$ mm and $x = 75$ mm.

3.14 A quadratic triangular element has known nodal coordinates for all six node points. Determine an expression for the element derivative matrix $[D]$

$$\begin{Bmatrix} \dfrac{\partial T}{\partial x} \\ \dfrac{\partial T}{\partial y} \end{Bmatrix} = [D]\{T\}$$

in terms of the corner pyramid functions N_i, N_k, N_m and their derivatives with respect to x and y.

Chapter 4

Springs, Networks, Bars, and Trusses—The Direct Method

4.1 Introduction

A. Purpose of Chapter

The purpose of this chapter is to solve some practical engineering problems involving springs, networks, bars, and trusses. Springs are considered as finite elements. (They were not presented in Chapter 3 because they are taken as single units rather than having properties that vary with x.) Pipe and electrical networks are then introduced. Simplex bar elements are shown to have properties very similar to those used to represent springs. Finally, the bar elements are used to analyze plane trusses. Truss problems are now commonly solved by finite element analysis in industry.

An example of a plane bridge truss is shown in Figure 4.1. It is composed of 11 bar members (elements) and 7 pins (node points). It has the externally applied loads as shown. Sections 4.5 and 4.6 illustrate the solution techniques used to find the displacements, stresses, forces, etc.

Some new finite element concepts are also introduced in this chapter—element matrices and assembly. These are shown in Section 4.3, using spring elements and networks (pipe and electrical) as examples. The concepts are then used throughout finite element practice.

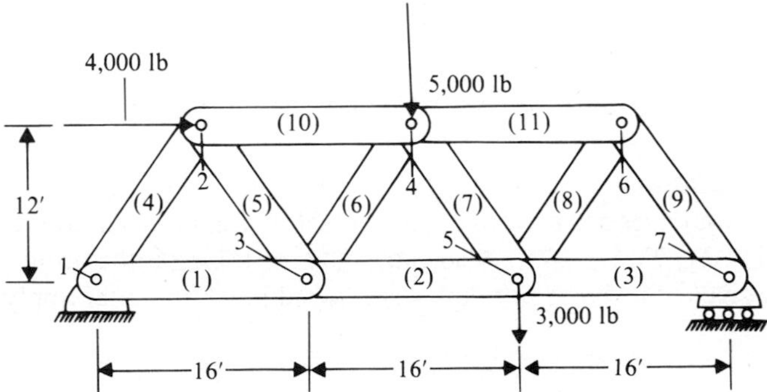

Figure 4.1 Plane bridge truss with 11 elements and 7 node points

Table 4.1
Steps in Finite Element Method

Step	Procedure
1	Formulate governing equations and boundary conditions
2	Divide analysis region into finite elements
3	Select interpolation functions
4	Determine element properties
5	Assemble global equations
6	Solve global equations
7	Verify solution

B. Review of Steps in Finite Element Method

At this point, it would be good to review the steps in the finite element method and how they are treated in this chapter. Table 4.1 repeats the seven steps (originally presented in Section 1.3).

Step 1. Formulate Governing Equations and Boundary Conditions

Several specific engineering problems involving springs and bars are formulated in this chapter.

Step 2. Divide Analysis Region into Elements

Division of the problems in this chapter into elements is quite simple.

Step 3. Select Interpolation Functions

Only simplex elements will be used in this chapter. Higher order elements are discussed in detail starting with Chapter 7 on beam elements.

Step 4. Determine Element Properties

Most of the element properties have been developed in Chapter 3. This chapter uses those results and develops element matrices for springs and bars. The concepts associated with element matrices then are applied to other elements in the chapters that follow. The simplest method—the direct method (as discussed originally in Section 1.4)—is used in this chapter.

Step 5. Assemble Global Equations

This chapter shows how element matrices are used to obtain the set of algebraic equations for the unknown nodal values $T_1, T_2, T_3, \ldots T_N$. Also, a method for taking into account prescribed nodal values of T is given. The basic principles developed in this chapter apply to nearly all finite elements.

Step 6. Solve Global Equations

Global equations are usually solved by some standard technique for simultaneous linear equations, such as Gauss elimination, which is presented in Chapter 11.

Step 7. Verify Solution

This topic is treated in various sections as the opportunity arises.

4.2 Finite Element Model of a Spring

Springs in series are used here as an introduction to finite element modeling. Each spring is represented as an element. The solution for a system of springs illustrates the basic finite element techniques just discussed. The spring problems discussed in this chapter are fairly easily solved by other methods normally studied in the first two solid mechanics courses: statics and strength of materials. Finite elements provide an equally good approach and have the additional advantage of introducing the solution of much more difficult problems. Also, spring elements are used in very advanced finite element programs for cases when springs are found in engineering designs. These may be actual springs or other devices, such as fluid film bearings, that are represented as springs in a rotating machine.

A. Springs in Series

As an introduction to finite element modeling, consider the spring shown in Figure 4.2, which is attached to a wall at the left end. A force F acting at the right end (a node point) produces the displacement u given by the equation $F = ku$, where k is the stiffness of the spring. Note that k is the slope of the F vs. u plot and has the units of force per length. Given F and k, the solution for u is simply $u = F/k$.

A more interesting problem involves a set of E springs in series, shown in Figure 4.3, all being displaced in the same direction (both positive and negative—to the left—displacements are allowed). The springs are the finite elements located between node points. The stiffness of the spring i is $k^{(e)}$ where e varies from 1 to E. The connections between springs are node points, and they are labeled 1,2,3, . . . , N, as shown. Each node point i has a displacement labeled u_i. In this case, the number of nodes is just one more than the number of elements, so $N = E + 1$. Also, each node point i has an externally applied force, which is labeled F_i. When all of the forces are applied, a set of displacements results. Normally at least one of the nodal displacements is specified.

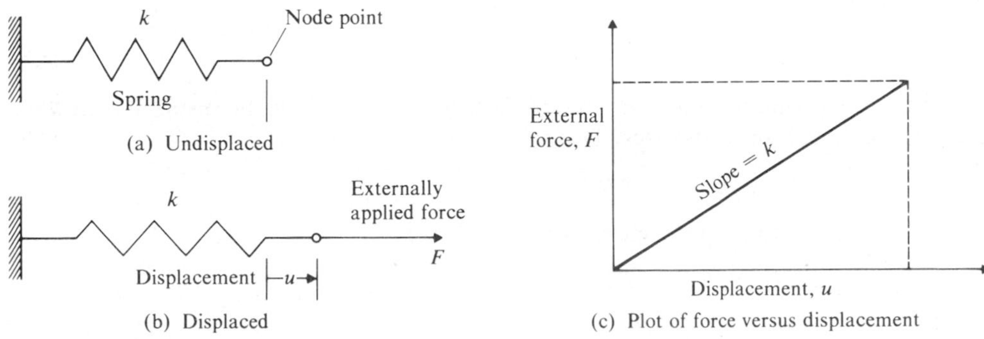

Figure 4.2 Displacement of a spring

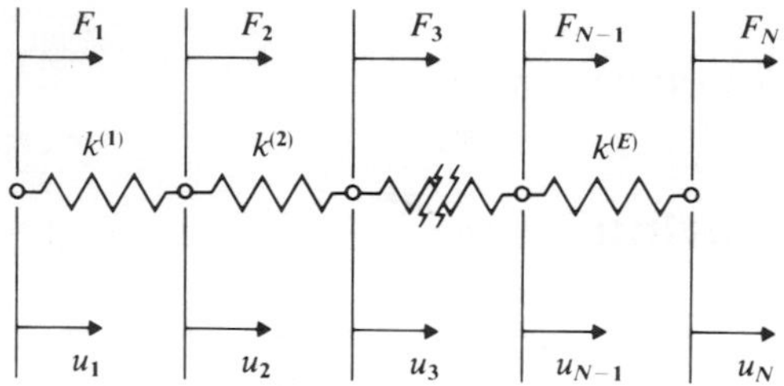

Figure 4.3 Springs in series with externally applied forces

B. Spring Element

Consider a single spring isolated from the series and shown in Figure 4.4. Let $u_i^{(e)}$ and $u_j^{(e)}$ denote the displacements of the left and right node points, respectively, of spring element e. Also, let $R_i^{(e)}$ and $R_j^{(e)}$ be the reaction forces (internal forces acting on each end of the spring) at the left and right ends. Recall that when drawing a free body diagram of a single spring, formerly internal forces are replaced by equivalent external forces R_i and R_j.

The equations for the displacement of the spring element are developed in four steps:

a. Apply the force $R_i^{(e)}$ required to produce a displacement $u_i = 1.0$ while $u_j = 0$.
b. Calculate the force $R_j^{(e)}$ required to keep the element in equilibrium for step a.
c. Apply the force $R_j^{(e)}$ required to produce a displacement $u_j = 1.0$ while $u_i = 0$.
d. Calculate the force $R_i^{(e)}$ required to keep the element in equilibrium for step c.

This is one version of the direct method. It is also sometimes called the influence coefficient method.

Consider the steps in order. For step a, Figure 4.5(b) shows the spring with force R_i and displacement $u_i = 1.0$. Node j has zero displacement. Then

$$R_i^{(e)} = k^{(e)}u_i = k^{(e)} \tag{4.1}$$

Step b sums the forces acting on the element. The sum must be zero to keep the element in equilibrium.

$$\sum F_x = R_i^{(e)} + R_j^{(e)} = 0$$

Solving for $R_j = -R_i$ or

$$R_j^{(e)} = -k^{(e)}u_i = -k^{(e)} \tag{4.2}$$

For step c, the spring has force R_j and displacement $u_j = 1.0$, while displacement u_i is zero. Figure 4.5(c) illustrates this. The relation is

$$R_j^{(e)} = k^{(e)}u_j = k^{(e)} \tag{4.3}$$

Step d sums forces to keep the element in equilibrium.

$$\sum F_x = R_i^{(e)} + R_j^{(e)} = 0$$

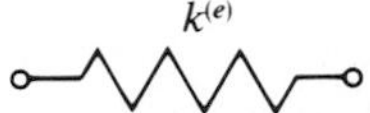

(a) Undisplaced spring element

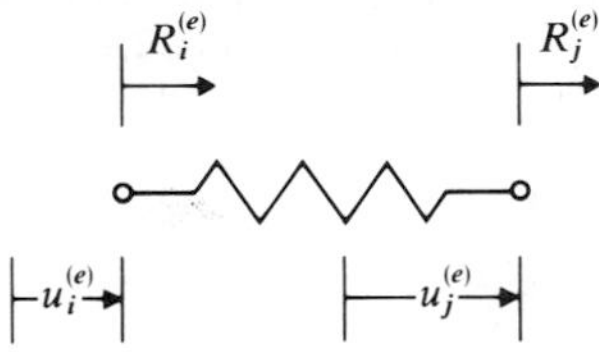

(b) Displaced spring element

Figure 4.4 Nodal displacements and internal reaction forces on spring element

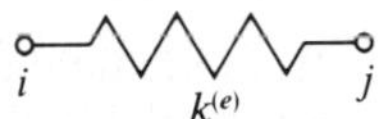

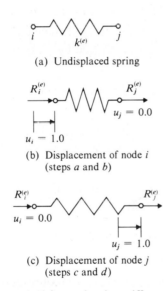

(a) Undisplaced spring

$R_i^{(e)}$ $R_j^{(e)}$

$u_j = 0.0$

$u_i = 1.0$

(b) Displacement of node i
(steps a and b)

(c) Displacement of node j
(steps c and d)

Figure 4.5 Direct method for evaluating stiffness matrix for spring element

Solving again

$$R_i^{(e)} = -k^{(e)}u_j = -k^{(e)} \tag{4.4}$$

which completes the four steps.

When the element is subjected to both nodal displacements, the nodal reaction forces are obtained from Equations (4.1)–(4.4). The forces at node i add together to give

$$R_i^{(e)} = k^{(e)}(u_i - u_j) \tag{4.5}$$

and for node j

$$R_j^{(e)} = k^{(e)}(-u_i + u_j) \tag{4.6}$$

These are the equilibrium equations for the element.

C. Equilibrium Equations for All Springs

The springs are connected at the node points. Each node point has two springs that act on it, except for the end node points, which have only one spring each. In order for each node point to be in equilibrium (not moving), the sum of the reaction forces must be equal to the external force on that node. Summing the reaction forces for each node gives the set of equations

$$\begin{aligned}
&\textit{Node 1} \qquad && \sum R_1 = R_1^{(1)} = F_1 \\
&\textit{Node 2} && \sum R_2 = R_2^{(1)} + R_2^{(2)} = F_2 \\
&\textit{Node 3} && \sum R_3 = R_3^{(2)} + R_3^{(3)} = F_3 \\
&\quad \cdots \\
&\textit{Node N} && \sum R_N = R_N^{(E)} = F_N
\end{aligned} \tag{4.7}$$

These equations essentially follow the method of pins from elementary statics.

The reaction forces on each end of an element are given by (4.5) and (4.6). Writing all of them out yields

$$\begin{aligned}
&\textit{Element 1} \qquad && R_1^{(1)} = k^{(1)}(u_1 - u_2) \\
& && R_2^{(1)} = -k^{(1)}(u_1 - u_2) \\
&\textit{Element 2} && R_2^{(2)} = k^{(2)}(u_2 - u_3) \\
& && R_3^{(2)} = -k^{(2)}(u_2 - u_3) \\
&\textit{Element 3} && R_3^{(3)} = k^{(3)}(u_3 - u_4) \\
& && R_4^{(3)} = -k^{(3)}(u_3 - u_4) \\
&\quad \cdots \\
&\textit{Element E} && R_{N-1}^{(E)} = k^{(E)}(u_{N-1} - u_N) \\
& && R_N^{(E)} = -k^{(E)}(u_{N-1} - u_N)
\end{aligned} \tag{4.8}$$

Substituting the expressions here for the reaction forces in the equilibrium equations (4.7) yields

$$\begin{aligned}
&\textit{Node 1} \qquad && k^{(1)}(u_1 - u_2) = F_1 \\
&\textit{Node 2} && -k^{(1)}(u_1 - u_2) + k^{(2)}(u_2 - u_3) = F_2 \\
&\textit{Node 3} && -k^{(2)}(u_2 - u_3) + k^{(3)}(u_3 - u_4) = F_3 \\
&\quad \cdots \\
&\textit{Node N} && -k^{(E)}(u_{N-1} - u_N) = F_N
\end{aligned} \tag{4.9}$$

This is the set of linear algebraic equations for the nodal displacements u_1, u_2, u_3, . . . , u_N.

In matrix form, the equations become

$$\underbrace{\begin{bmatrix} k^{(1)} & -k^{(1)} & 0 & \cdot & 0 \\ -k^{(1)} & k^{(1)}+k^{(2)} & -k^{(2)} & \cdot & 0 \\ 0 & -k^{(2)} & k^{(2)}+k^{(3)} & \cdot & 0 \\ \cdot & \cdot & \cdot & \cdot & \cdot \\ 0 & 0 & 0 & \cdot & k^{(E)} \end{bmatrix}}_{\text{Global matrix } (N \times N)} \begin{Bmatrix} u_1 \\ u_2 \\ u_3 \\ \cdot \\ u_N \end{Bmatrix} = \underbrace{\begin{Bmatrix} F_1 \\ F_2 \\ F_3 \\ \cdot \\ F_N \end{Bmatrix}}_{\text{Force column } (N \times 1)} \quad \textbf{(4.10)}$$

This is called the global set of equations for the displacements. The large matrix contains the coefficients of the linear set of algebraic equations. It is a function only of the spring stiffnesses, so it is often called the stiffness matrix. In other problems it may be known by other names. For example, in fluid mechanics, it is often called a fluidity matrix. The global matrix is symmetric, as it is for many of the other finite elements. The column on the right side is called the global force column (or vector).

D. Prescribed Nodal Displacements

The set of equations (4.10) has no solution. The physical reason is that one or more nodes must have prescribed displacements to prevent the series of springs from moving (hold it in equilibrium). An example would be a restrained first node: the prescribed value would then be $u_1 = 0$. Another example might be a numerical value prescribed at node 3, denoted by $u_3 = 0.32$ mm. The mathematical reason is the global stiffness matrix in (4.10) is singular unless at least one nodal value is prescribed.

The method for including prescribed nodal values is treated in general fashion in Section 4.3. For the purposes of this section, only the prescribed nodal value $u_1 = 0.0$ is treated here. In row 1 of the global equations (4.10), this boundary condition is assigned by replacing the term in column 2 by zero and replacing the force by zero: $-k^{(1)} = 0.0$, $F_1 = 0.0$. In row 2, the term $-k^{(1)}u_1$ is moved to the right hand side, but since $u_1 = 0.0$ this means no change to the force column. The final result of the two above steps is

$$\begin{bmatrix} k^{(1)} & 0 & 0 & \cdot & 0 \\ 0 & k^{(1)}+k^{(2)} & -k^{(2)} & \cdot & 0 \\ 0 & -k^{(2)} & k^{(2)}+k^{(3)} & \cdot & 0 \\ \cdot & \cdot & \cdot & \cdot & \cdot \\ 0 & 0 & 0 & \cdot & k^{(E)} \end{bmatrix} \begin{Bmatrix} u_1 \\ u_2 \\ u_3 \\ \cdot \\ u_N \end{Bmatrix} = \begin{Bmatrix} 0 \\ F_2 \\ F_3 \\ \cdot \\ F_N \end{Bmatrix} \quad \textbf{(4.11)}$$

Once again the global stiffness matrix is symmetric. The first row clearly gives the relation $k^{(1)}u_1 = 0.0$, which yields the prescribed nodal value of $u_1 = 0.0$.

Example 4.1 *Two Springs in Series*

Given Two springs, with spring constants $k^{(1)} = 30$ N/mm and $k^{(2)} = 20$ N/mm, form a spring system as shown in Figure 4.6(a). The left end is attached to a wall. Forces of the following magnitude are applied to the two remaining node points: $F_2 = -2$ N and $F_3 = 6$ N.

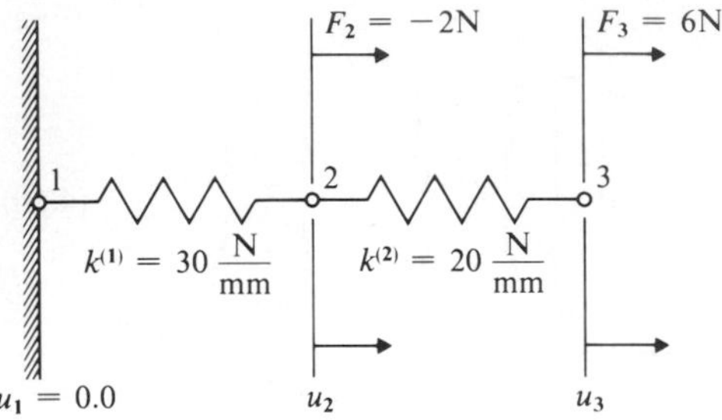

(a) Springs with applied forces

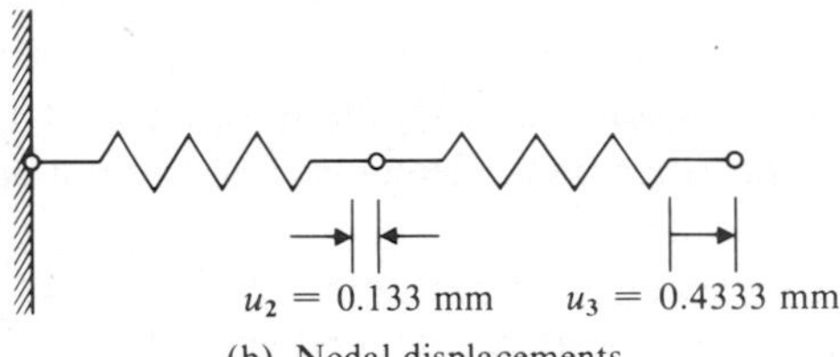

(b) Nodal displacements

Figure 4.6 Aligned springs (Example 4.1—Two springs in series)

Objective Determine the nodal displacements and internal reaction forces.

Solution The equation for the nodal displacements (4.10) reduces to three equations for the three nodal displacements.

$$\begin{bmatrix} k^{(1)} & -k^{(1)} & 0 \\ -k^{(1)} & k^{(1)} + k^{(2)} & -k^{(2)} \\ 0 & -k^{(2)} & k^{(2)} \end{bmatrix} \begin{Bmatrix} u_1 \\ u_2 \\ u_3 \end{Bmatrix} = \begin{Bmatrix} F_1 \\ F_2 \\ F_3 \end{Bmatrix}$$

Substituting the numerical values gives

$$\begin{bmatrix} 30 & -30 & 0 \\ -30 & 30 + 20 & -20 \\ 0 & -20 & 20 \end{bmatrix} \begin{Bmatrix} u_1 \\ u_2 \\ u_3 \end{Bmatrix} = \begin{Bmatrix} F_1 \\ -2 \\ 6 \end{Bmatrix}$$

Note that the force F_1 is not known.

The node at the wall is restrained, so its displacement is zero: $u_1 = 0$. This prescribed nodal displacement is taken into account as just described in (4.11). In this case the -30 is set to zero in row 1, as is F_1. Also the -30 in row 2 is set to zero, with the result

$$\begin{bmatrix} 30 & 0 & 0 \\ 0 & 50 & -20 \\ 0 & -20 & 20 \end{bmatrix} \begin{Bmatrix} u_1 \\ u_2 \\ u_3 \end{Bmatrix} = \begin{Bmatrix} 0 \\ -2 \\ 6 \end{Bmatrix}$$

This is easily solved with the result

$$\begin{Bmatrix} u_1 \\ u_2 \\ u_3 \end{Bmatrix} = \begin{Bmatrix} 0.0 \\ 0.1333 \\ 0.4333 \end{Bmatrix} \quad \text{mm}$$

Figure 4.6(b) shows the results.

The reaction forces may now be found for each element from (4.8). For element 1,

$$R_1^{(1)} = k^{(1)}(u_1 - u_2) = 30[0 - (0.1333)] = -4.0 \text{ N}$$

$$R_2^{(1)} = -k^{(1)}(u_1 - u_2) = 4.0 \text{ N}$$

and for element 2,

$$R_2^{(2)} = k^{(2)}(u_2 - u_3) = 20[(0.1333) - (0.4333)] = -6.0 \text{ N}$$

$$R_3^{(2)} = -k^{(2)}(u_2 - u_3) = 6.0 \text{ N}$$

The first reaction force of -4.0 newtons is the internal force in the wall required to keep the series of springs in equilibrium by balancing the externally applied forces -2 and 6 newtons. Summing forces for node 2 yields

$$\sum R_2 = R_2^{(1)} + R_2^{(2)} = 4 - 6 = -2 = F_2$$

Similarly for node 3

$$\sum R_3 = R_3^{(2)} = 6 = F_3$$

Thus, all of the force relations are satisfied.

4.3 Element Matrices and Assembly— Springs and Networks

The purpose of this section is to introduce three new concepts. These are element matrices, assembly of nodal equations, and prescribed nodal values. First is the arrangement of element properties, such as those of the spring just discussed, into a convenient matrix called an element matrix. Second is the method of using these element matrices to obtain the set of global equations for the nodal displacements. These new concepts can be illustrated using the spring element. Third, the method of including prescribed nodal displacements is presented. Examples are carried out with springs, pipe networks, and electrical networks.

A. Element Matrix

An element matrix has the form

$$[B]^{(e)} = \begin{bmatrix} B_{ii} & B_{ij} \\ B_{ji} & B_{jj} \end{bmatrix}^{(e)} \tag{4.12}$$

It is a square matrix labeled $[B]$ in this text. A matrix of order two by two is shown here but it may be three by three, four by four, and so on for different elements.

For the spring, the element reaction forces and displacements are related by (4.5) and (4.6)

$$R_i^{(e)} = k^{(e)}u_i - k^{(e)}u_j$$

$$R_j^{(e)} = -k^{(e)}u_i + k^{(e)}u_j$$

These are written in matrix form as

$$\underbrace{\begin{Bmatrix} R_i \\ R_j \end{Bmatrix}^{(e)}}_{\substack{\text{Reaction}\\ \text{column}\\ (2\times1)}} = \underbrace{\begin{bmatrix} k & -k \\ -k & k \end{bmatrix}^{(e)}}_{\substack{\text{Element}\\ \text{matrix}\\ (2\times2)}} \underbrace{\begin{Bmatrix} u_i \\ u_j \end{Bmatrix}^{(e)}}_{\substack{\text{Nodal}\\ \text{displacements}\\ (2\times1)}} \tag{4.13}$$

The element matrix contains only the spring stiffness k and is often called the stiffness matrix. It is of order (2×2) for the spring described here.

Using the $[B]$ notation for the element matrix, this is written as

$$\begin{Bmatrix} R_i \\ R_j \end{Bmatrix} = [B]^{(e)} \begin{Bmatrix} u_i \\ u_j \end{Bmatrix} \tag{4.14}$$

where

$$[B]^{(e)} = \begin{bmatrix} k & -k \\ -k & k \end{bmatrix}^{(e)}$$

Then the terms in the element matrix are

$$B_{ii}^{(e)} = k^{(e)}, \qquad B_{ij}^{(e)} = -k^{(e)}$$

$$B_{ji}^{(e)} = -k^{(e)}, \quad B_{jj}^{(e)} = k^{(e)}$$

This can be simplified somewhat by factoring out the constant k.

$$[B]^{(e)} = k^{(e)} \begin{bmatrix} 1 & -1 \\ -1 & 1 \end{bmatrix} \tag{4.15}$$

It can also be easily seen that the matrix is symmetric, as is often true of element matrices.

Writing out element matrices for all of the elements in the series of E springs gives

$$\begin{aligned} \begin{Bmatrix} R_1 \\ R_2 \end{Bmatrix}^{(1)} &= \begin{bmatrix} k & -k \\ -k & k \end{bmatrix}^{(1)} \begin{Bmatrix} u_1 \\ u_2 \end{Bmatrix}^{(1)} \\ \begin{Bmatrix} R_2 \\ R_3 \end{Bmatrix}^{(2)} &= \begin{bmatrix} k & -k \\ -k & k \end{bmatrix}^{(2)} \begin{Bmatrix} u_2 \\ u_3 \end{Bmatrix}^{(2)} \\ \begin{Bmatrix} R_3 \\ R_4 \end{Bmatrix}^{(3)} &= \begin{bmatrix} k & -k \\ -k & k \end{bmatrix}^{(3)} \begin{Bmatrix} u_3 \\ u_4 \end{Bmatrix}^{(3)} \\ &\cdots \\ \begin{Bmatrix} R_{N-1} \\ R_N \end{Bmatrix}^{(E)} &= \begin{bmatrix} k & -k \\ -k & k \end{bmatrix}^{(E)} \begin{Bmatrix} u_{N-1} \\ u_N \end{Bmatrix}^{(E)} \end{aligned} \tag{4.16}$$

This indicates that the element stiffness matrices are

$$
\begin{array}{l}
[B]^{(1)} = \begin{array}{c} \\ 1 \\ 2 \end{array} \begin{array}{c} \begin{array}{cc} 1 & \quad 2 \end{array} \\ \begin{bmatrix} k & -k \\ -k & k \end{bmatrix}^{(1)} \end{array} \\
[B]^{(2)} = \begin{array}{c} \\ 2 \\ 3 \end{array} \begin{array}{c} \begin{array}{cc} 2 & \quad 3 \end{array} \\ \begin{bmatrix} k & -k \\ -k & k \end{bmatrix}^{(2)} \end{array} \\
[B]^{(3)} = \begin{array}{c} \\ 3 \\ 4 \end{array} \begin{array}{c} \begin{array}{cc} 3 & \quad 4 \end{array} \\ \begin{bmatrix} k & -k \\ -k & k \end{bmatrix}^{(3)} \end{array} \\
\cdots \\
[B]^{(E)} = \begin{array}{c} \\ N-1 \\ N \end{array} \begin{array}{c} \begin{array}{cc} N-1 & \quad N \end{array} \\ \begin{bmatrix} k & -k \\ -k & k \end{bmatrix}^{(E)} \end{array}
\end{array}
\tag{4.17}
$$

The associated node numbers for each element matrix are indicated along the left side and top of each matrix. These numbers are obtained from the element/node connectivity data. They permit the assembly of the element matrices. It is easily seen that once one element matrix is derived, as in the relation (4.13), there is no need to write out these relations for all elements as in (4.16).

B. Assembly of Nodal Equations

The solution of nodal unknowns in a finite element analysis results in a set of algebraic equations. These may be called the global equations. If the original problem is a linear one, the algebraic equations are linear also. Good examples are the spring problem just discussed and the elastic bar problem presented in the next section. The method of evaluating all of the terms that form those equations from individual element matrices is called assembly. Of course, once the assembly is completed and prescribed nodal values taken into account, the set of algebraic equations can be solved for the nodal unknowns.

Let the global equations have the general form $[A]\{T\} = \{F\}$ where the N by N global matrix $[A]$ is

$$
[A] = \begin{bmatrix}
a_{11} & a_{12} & a_{13} & \cdot & a_{1N} \\
a_{21} & a_{22} & a_{23} & \cdot & a_{2N} \\
a_{31} & a_{32} & a_{33} & \cdot & a_{3N} \\
\cdot & \cdot & \cdot & \cdot & \cdot \\
a_{N1} & a_{N2} & a_{N3} & \cdot & a_{NN}
\end{bmatrix}
$$

and the nodal column $\{T\}$ and global column (or vector) $\{F\}$ are

$$\{T\} = \begin{Bmatrix} T_1 \\ T_2 \\ T_3 \\ \cdot \\ T_N \end{Bmatrix}, \qquad \{F\} = \begin{Bmatrix} f_1 \\ f_2 \\ f_3 \\ \cdot \\ f_N \end{Bmatrix}$$

The global matrix $[A]$ always contains a large number of zeros (except for very small problems), as discussed in Section 2.4. Usually the nonzero terms are confined to a small bandwidth.

The global matrix $[A]$ for the springs in series is obtained from (4.10)

$$|\leftarrow \text{NBW} = 2 \rightarrow|$$

$$[A] = \begin{array}{c} \\ 1 \\ 2 \\ 3 \\ \cdot \\ N \end{array} \begin{array}{c} \begin{array}{ccccc} 1 & 2 & 3 & \cdot & N \end{array} \\ \begin{bmatrix} k^{(1)} & -k^{(1)} & 0 & \cdot & 0 \\ -k^{(1)} & k^{(1)} + k^{(2)} & -k^{(2)} & \cdot & 0 \\ 0 & -k^{(2)} & k^{(2)} + k^{(3)} & \cdot & 0 \\ \cdot & \cdot & \cdot & \cdot & \cdot \\ 0 & 0 & 0 & \cdot & k^{(E)} \end{bmatrix} \end{array}$$

The bandwidth (NBW) for this matrix is two, and all terms outside of this bandwidth are zero. Also the matrix is symmetric. Node numbers are written along the left side and top for easy reference.

In similar fashion, the global column is obtained from (4.10)

$$\{F\} = \begin{array}{c} 1 \\ 2 \\ 3 \\ \cdot \\ N \end{array} \begin{Bmatrix} f_1 \\ f_2 \\ f_3 \\ \cdot \\ f_N \end{Bmatrix}$$

The global column contains the externally applied forces at the node points. Thus it is often called the force column (or vector). Once again, the node numbers are written along the left side for reference.

It is easily seen that the global matrix $[A]$ is obtained from the element matrices. Figure 4.7 illustrates the procedure for three elements. A term from an element matrix has a row and column number (shown on the left side and top). This term is entered into the location in the global matrix with the same row and column number. All terms from the element matrix are entered. This process is repeated until all of the elements are entered.

This procedure is quite general and is not limited to spring elements; it works for all other elements as well. The computer does all of this bookkeeping work, so an engineer need not do it by hand. Note that the element matrices need not be assembled in consecutive order 1 through E, but may be taken in any order. The assembly procedure is equivalent to satisfying the equilibrium equations for each node point (4.7). It is left to the student to verify this.

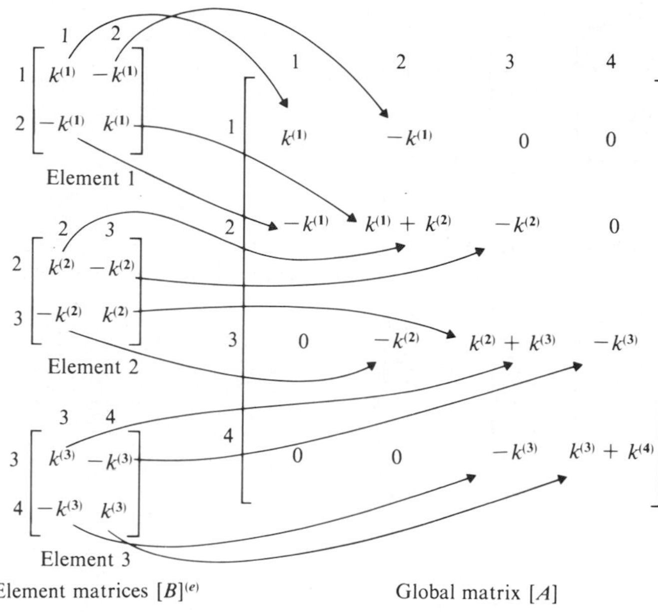

Figure 4.7 Assembly of global matrix for series of springs

C. Prescribed Nodal Values

One method of including the prescribed nodal values in the global equations $[A]$ is discussed here. It involves modifying the $[A]$ matrix to include the given nodal values. This is done without changing the size of the matrix. An alternative method involves renumbering and partitioning the matrices into two parts: unknown and known. Both methods work efficiently, but the second is not discussed.

Consider a general banded matrix $[A]\{T\} = \{F\}$ just assembled. Let the known nodal value be $T_i = \theta_i$. An outline of the steps to be followed is

1. Set all coefficients a_{ij} in row i to zero (except a_{ii}) and replace f_i by $a_{ii}\theta_i$.
2. Add the value $a_{ji}\theta_i$ to the existing value of f_j in row j and then set a_{ji} to zero.

Step 2 is applied to all rows $j = 1, \ldots, N$ where θ_i appears, except row i. The resulting matrix remains symmetric and banded.

As an example, consider the five linear equations with bandwidth 3 shown in Figure 4.8(a). Let the known nodal value be $T_2 = \theta_2$. First consider equation 2, which is $a_{21}T_1 + a_{22}T_2 + a_{23}T_3 + a_{24}T_4 = f_2$. This satisfies the boundary condition if $a'_{21} = 0$, $a'_{23} = 0$, $a'_{24} = 0$, $f'_2 = a_{22}\theta_2$, where the primed numbers indicate new values. Equation 2 then becomes $a_{22}T_2 = a_{22}\theta_2$, which clearly gives back $T_2 = \theta_2$. The new set of equations is shown in Figure 4.8(b). Dotted lines outline the areas of change. Note that it is not desirable to divide by a_{22}: if a_{22} is very large or very small, an ill-conditioned $[A]$ matrix may result.

$$\begin{bmatrix} a_{11} & a_{12} & a_{13} & 0 & 0 \\ a_{21} & a_{22} & a_{23} & a_{24} & 0 \\ a_{31} & a_{32} & a_{33} & a_{34} & a_{35} \\ 0 & a_{42} & a_{43} & a_{44} & a_{45} \\ 0 & 0 & a_{53} & a_{54} & a_{55} \end{bmatrix} \begin{Bmatrix} T_1 \\ T_2 \\ T_3 \\ T_4 \\ T_5 \end{Bmatrix} = \begin{Bmatrix} f_1 \\ f_2 \\ f_3 \\ f_4 \\ f_5 \end{Bmatrix}$$

(a) Original equations

$$\begin{bmatrix} a_{11} & a_{12} & a_{13} & 0 & 0 \\ 0 & a_{22} & 0 & 0 & 0 \\ a_{31} & a_{32} & a_{33} & a_{34} & a_{35} \\ 0 & a_{42} & a_{43} & a_{44} & a_{45} \\ 0 & 0 & a_{53} & a_{54} & a_{55} \end{bmatrix} \begin{Bmatrix} T_1 \\ T_2 \\ T_3 \\ T_4 \\ T_5 \end{Bmatrix} = \begin{Bmatrix} f_1 \\ a_{22}\theta_2 \\ f_3 \\ f_4 \\ f_5 \end{Bmatrix}$$

(b) Equations after Step 1

$$\begin{bmatrix} a_{11} & 0 & a_{13} & 0 & 0 \\ 0 & a_{22} & 0 & 0 & 0 \\ a_{31} & 0 & a_{33} & a_{34} & a_{35} \\ 0 & 0 & a_{43} & a_{44} & a_{45} \\ 0 & 0 & a_{53} & a_{54} & a_{55} \end{bmatrix} \begin{Bmatrix} T_1 \\ T_2 \\ T_3 \\ T_4 \\ T_5 \end{Bmatrix} = \begin{Bmatrix} f_1 - a_{12}\theta_2 \\ a_{22}\theta_2 \\ f_3 - a_{32}\theta_2 \\ f_4 - a_{42}\theta_2 \\ f_5 \end{Bmatrix}$$

(c) Equations after Step 2

Figure 4.8 Modification of global equations for prescribed nodal value

At this point, only row 2 has been changed. However, the other equations also contain T_2, which is known. Equation 1 is $a_{11}T_1 + a_{12}T_2 + a_{13}T_3 = f_1$. The T_2 term can be transferred to the right side and then a_{12} set to zero. New parameters are defined as $f_1' = f_1 - a_{12}\theta_2$, $a_{12}' = 0$. Equations 3 and 4 contain T_2 as well. They may be changed by writing $f_3' = f_3 - a_{32}\theta_2$, $a_{32}' = 0$ and $f_4' = f_4 - a_{42}\theta_2$, $a_{42}' = 0$. The final results are shown in Figure 4.8(c). Dotted lines indicate both the row and column changes.

Several features can be observed. The resulting $[A]$ matrix is symmetric and banded, just as it was originally. Also, the changes only occur a distance of NBW -1 terms away from the a_{22} term. This completes the inclusion of $T_2 = \theta_2$ in the global equations.

Example 4.2 *Four Springs in Series*

Given A set of five node points and four spring elements is shown in Figure 4.9(a). The prescribed nodal displacements are $u_2 = 0.2$ mm and $u_5 = 0.0$ mm, and the externally applied nodal forces are $F_1 = 2$ N, $F_3 = -3$ N, and $F_4 = -1$ N. The springs have

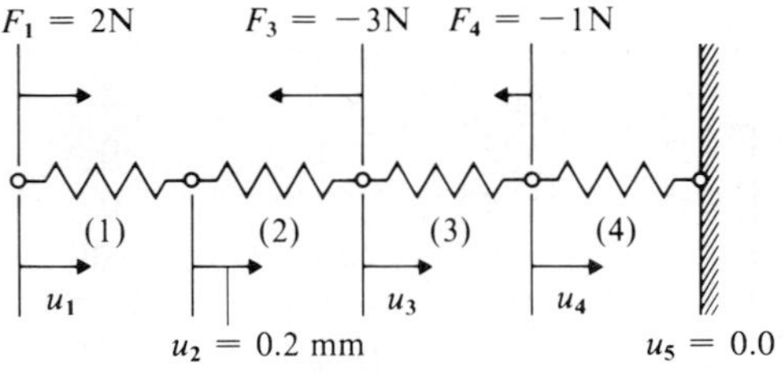

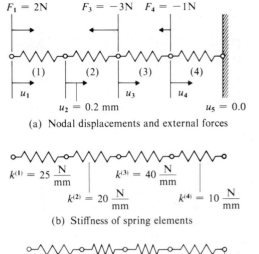

(a) Nodal displacements and external forces

$k^{(1)} = 25 \frac{\text{N}}{\text{mm}}$ $k^{(3)} = 40 \frac{\text{N}}{\text{mm}}$

$k^{(2)} = 20 \frac{\text{N}}{\text{mm}}$ $k^{(4)} = 10 \frac{\text{N}}{\text{mm}}$

(b) Stiffness of spring elements

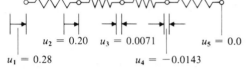

$u_2 = 0.20$ $u_3 = 0.0071$ $u_5 = 0.0$

$u_1 = 0.28$ $u_4 = -0.0143$

(c) Nodal displacements

Figure 4.9 Series of springs with applied forces (Example 4.2—Four springs in series)

stiffnesses $k^{(1)} = 25$ N/mm, $k^{(2)} = 20$ N/mm, $k^{(3)} = 40$ N/mm, and $k^{(4)} = 10$ N/mm. Figure 4.9(b) illustrates the elements.

Objective Determine nodal displacements, using the methods of this section.

Solution Element matrices for all elements are (4.17)

$$[B]^{(1)} = \begin{matrix} \\ 1 \\ 2 \end{matrix} \begin{matrix} \begin{matrix} 1 & \quad 2 \end{matrix} \\ \begin{bmatrix} 25 & -25 \\ -25 & 25 \end{bmatrix}^{(1)} \end{matrix}$$

$$[B]^{(2)} = \begin{matrix} \\ 2 \\ 3 \end{matrix} \begin{matrix} \begin{matrix} 2 & \quad 3 \end{matrix} \\ \begin{bmatrix} 20 & -20 \\ -20 & 20 \end{bmatrix}^{(2)} \end{matrix}$$

$$[B]^{(3)} = \begin{matrix} \\ 3 \\ 4 \end{matrix} \begin{matrix} \begin{matrix} 3 & \quad 4 \end{matrix} \\ \begin{bmatrix} 40 & -40 \\ -40 & 40 \end{bmatrix}^{(3)} \end{matrix}$$

$$[B]^{(4)} = \begin{matrix} \\ 4 \\ 5 \end{matrix} \begin{matrix} \begin{matrix} 4 & \quad 5 \end{matrix} \\ \begin{bmatrix} 10 & -10 \\ -10 & 10 \end{bmatrix}^{(4)} \end{matrix}$$

The assembled global matrix is

$$[A] = \begin{array}{c} \\ 1 \\ 2 \\ 3 \\ 4 \\ 5 \end{array}\begin{array}{c} \begin{array}{ccccc} 1 & 2 & 3 & 4 & 5 \end{array} \\ \begin{bmatrix} 25 & -25 & 0 & 0 & 0 \\ -25 & 25+20 & -20 & 0 & 0 \\ 0 & -20 & 20+40 & -40 & 0 \\ 0 & 0 & -40 & 40+10 & -10 \\ 0 & 0 & 0 & -10 & 10 \end{bmatrix} \end{array}$$

using the assembly techniques in Figure 4.7.

The final set of five linear algebraic equations is obtained by adding the nodal forces to the right hand side

$$\begin{bmatrix} 25 & -25 & 0 & 0 & 0 \\ -25 & 45 & -20 & 0 & 0 \\ 0 & -20 & 60 & -40 & 0 \\ 0 & 0 & -40 & 50 & -10 \\ 0 & 0 & 0 & -10 & 10 \end{bmatrix} \begin{Bmatrix} u_1 \\ u_2 \\ u_3 \\ u_4 \\ u_5 \end{Bmatrix} = \begin{Bmatrix} 2 \\ f_2 \\ -3 \\ -1 \\ f_5 \end{Bmatrix}$$

There are no external nodal forces specified at nodes 2 and 5. Now the prescribed nodal values must be included.

For node 2, step 1 involves row 2. All terms in row 2 (-25 and -20) of the global stiffness matrix $[A]$ except the main diagonal are set to zero and $f_2 = 45\ (0.2) = 9$. Step 2 involves transferring all terms in column 2 of $[A]$ except the main diagonal term to the right side of the equation. Thus the term 2 in the global column (row 2) is replaced by $2 - (-25)\ (0.2) = 7$, while the term -3 in row 3 of $\{F\}$ is replaced by $-3 - (-20)\ (0.2) = 1$. The final result, after taking into account the displacement at node 2, is

$$\begin{bmatrix} 25 & 0 & 0 & 0 & 0 \\ 0 & 45 & 0 & 0 & 0 \\ 0 & 0 & 60 & -40 & 0 \\ 0 & 0 & -40 & 50 & -10 \\ 0 & 0 & 0 & -10 & 10 \end{bmatrix} \begin{Bmatrix} u_1 \\ u_2 \\ u_3 \\ u_4 \\ u_5 \end{Bmatrix} = \begin{Bmatrix} 7 \\ 9 \\ 1 \\ -1 \\ f_5 \end{Bmatrix}$$

The matrix remains symmetric.

For node 5, the first step acts upon row 5. Again, all terms in the row (-10) except the main diagonal are set to zero. Then the right side is $f_5 = 10(0.0) = 0$. The second step transfers all terms in column 5 of $[A]$ except the main diagonal to the right side. The term -1 in the global column (row 4) is replaced by $-1-(-10)(0.0) = -1$. The result is

$$\begin{bmatrix} 25 & 0 & 0 & 0 & 0 \\ 0 & 45 & 0 & 0 & 0 \\ 0 & 0 & 60 & -40 & 0 \\ 0 & 0 & -40 & 50 & 0 \\ 0 & 0 & 0 & 0 & 10 \end{bmatrix} \begin{Bmatrix} u_1 \\ u_2 \\ u_3 \\ u_4 \\ u_5 \end{Bmatrix} = \begin{Bmatrix} 7 \\ 9 \\ 1 \\ -1 \\ 0 \end{Bmatrix}$$

Solving the final set of nodal equations gives

$$\begin{Bmatrix} u_1 \\ u_2 \\ u_3 \\ u_4 \\ u_5 \end{Bmatrix} = \begin{Bmatrix} 0.2800 \\ 0.2000 \\ 0.0071 \\ -0.0143 \\ 0.00 \end{Bmatrix} \quad \text{(mm)}$$

Reaction forces on each of the springs can easily be calculated from these values. Figure 4.9(c) illustrates the nodal displacements.

D. Flow in Pipe Networks

Other problems can be solved using the finite element techniques just discussed. These include many types of "network" problems, such as fluid flow in pipe networks and current flow in electrical networks. A pipe network is chosen as the first example here; an electrical network follows. These networks differ from the aligned springs in that an element may be connected to several adjacent elements.

An incompressible fluid flows through a pipe, as shown in Figure 4.10, with volume rate of flow Q. For laminar flow, the Reynolds number Re must be less than 2,000. This number is defined by the equation

$$\text{Re} = \frac{\rho V D}{\mu} < 2{,}000$$

where ρ = density, V = velocity, D = pipe diameter, and μ = viscosity. The volume rate of flow is related to the pressure gradient dP/dx along the pipe by the formula

$$Q = -\frac{\pi D^4}{128\mu}\frac{dP}{dx}$$

which can be obtained from any undergraduate fluid mechanics text. If the pressure in the pipe drops by the amount ΔP along its length L, the pressure gradient is

$$\frac{dP}{dx} = \frac{\Delta P}{L}$$

The pressure at the ith end of the pipe and the pressure at the jth end of the pipe are thus related by

$$Q = \frac{\pi D^4}{128\mu L}(P_i - P_j) \tag{4.18}$$

This is the key formula for laminar pipe flow. Note that the volume rate of flow is positive if P_i is greater than P_j.

Let the flow in a pipe (element) be denoted by $Q^{(e)}$. The flow at node i of the element is

$$Q_i^{(e)} = \left[\frac{\pi D^4}{128\mu L}P_i - \frac{\pi D^4}{128\mu L}P_j\right]^{(e)}$$

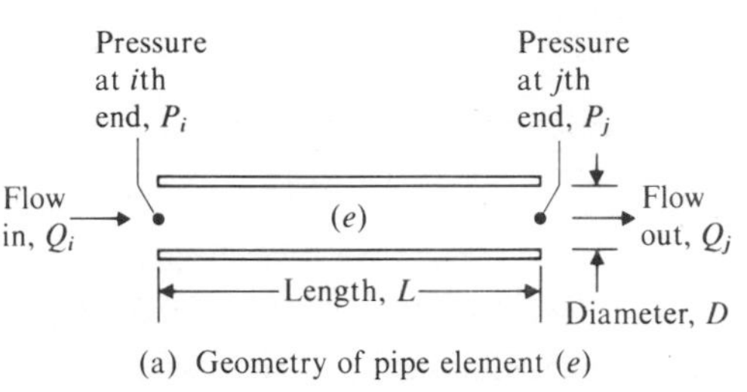

(a) Geometry of pipe element (e)

(b) Network of pipe elements

Figure 4.10 Finite element model of pipe network

assuming flow into the pipe to be positive. Alternatively the flow out of node j is

$$Q_j^{(e)} = \left[-\frac{\pi D^4}{128\mu L} P_i + \frac{\pi D^4}{128\mu L} P_j \right]^{(e)}$$

where the sign change occurs because the flow is out of the pipe. In element matrix form this becomes

$$\begin{Bmatrix} Q_i \\ Q_j \end{Bmatrix}^{(e)} = \begin{bmatrix} \dfrac{\pi D^4}{128\mu L} & -\dfrac{\pi D^4}{128\mu L} \\ -\dfrac{\pi D^4}{128\mu L} & \dfrac{\pi D^4}{128\mu L} \end{bmatrix}^{(e)} \begin{Bmatrix} P_i \\ P_j \end{Bmatrix}^{(e)}$$

This simplifies to

$$\begin{Bmatrix} Q_i \\ Q_j \end{Bmatrix} = \underbrace{\left(\frac{\pi D^4}{128\mu L} \right)^{(e)} \begin{bmatrix} 1 & -1 \\ -1 & 1 \end{bmatrix}^{(e)}}_{\text{Element matrix for incompressible laminar pipe flow}} \begin{Bmatrix} P_i \\ P_j \end{Bmatrix}^{(e)} \qquad \textbf{(4.19)}$$

This has exactly the same form as the spring element matrix, where

$$k^{(e)} = \left(\frac{\pi D^4}{128\mu L} \right)^{(e)}$$

Of course, the units are different.

At any node n, the sum of all of the internal flows into pipes connected to that node must be equal to the externally imposed volume of flow Q'. The nodal flow equations are written out for all of the nodes as

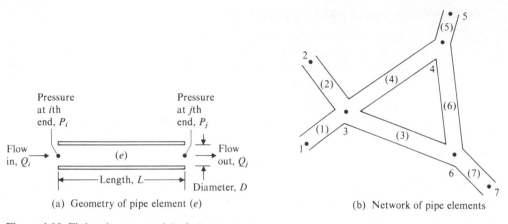

$$\text{Node 1} \quad \sum Q_1 = Q'_1$$

$$\text{Node 2} \quad \sum Q_2 = Q'_2$$

$$\text{Node 3} \quad \sum Q_3 = Q'_3$$

$$\cdots$$

$$\text{Node } N \quad \sum Q_N = Q'_N$$

Here $Q'_1, Q'_2, Q'_3, \ldots, Q'_N$ represent the externally imposed flow rates at the respective node points, and each summation is taken over all pipes connected to that node. This set of relations is the same as that for the summation of forces at a node point, Equation (4.8), with the exception that the external forces are zero. The assembly process is exactly the same.

Example 4.3 *Flow in Three-Pipe Network*

Given A three-pipe network is shown in Figure 4.11. The known pressures at the three exterior nodes are $P_1 = 20 \text{ lbf/in}^2$, $P_3 = 17 \text{ lbf/in}^2$, and $P_4 = 15 \text{ lbf/in}^2$. The fluid viscosity is $\mu = 1.5 \times 10^{-6} \text{ lbf-sec/in}^2$ and density $\rho = 1.93 \text{ slug/ft}^3$.

Objective Determine the pressure at node 2 and the flow rates in the pipes.

Solution For element 1, the coefficient is

$$\left(\frac{\pi D^4}{128\mu L}\right)^{(1)} = \frac{3.14159}{128} \times (2.0 \text{ in})^4 \times \frac{1}{1.5\times10^{-6}} \frac{\text{in}^2}{\text{lbf-sec}} \times \frac{1}{40{,}000 \text{ in}}$$

$$= 6.544 \frac{\text{in}^5}{\text{lbf-sec}}$$

The element matrix is

$$[B]^{(1)} = \begin{matrix} \\ 1 \\ 2 \end{matrix} \begin{matrix} 2 & 2 \\ \begin{bmatrix} 6.544 & -6.544 \\ -6.544 & 6.544 \end{bmatrix}^{(1)} \end{matrix}$$

For element 2, the coefficient is

$$\left(\frac{\pi D^4}{128\mu L}\right)^{(2)} = 0.818 \frac{\text{in}^5}{\text{lbf-sec}}$$

and the element matrix

$$[B]^{(2)} = \begin{matrix} \\ 2 \\ 3 \end{matrix} \begin{matrix} 2 & 3 \\ \begin{bmatrix} 0.818 & -0.818 \\ -0.818 & 0.818 \end{bmatrix}^{(2)} \end{matrix}$$

For element 3,

$$\left(\frac{\pi D^4}{128\mu L}\right)^{(3)} = 0.545 \frac{\text{in}^5}{\text{lbf-sec}}$$

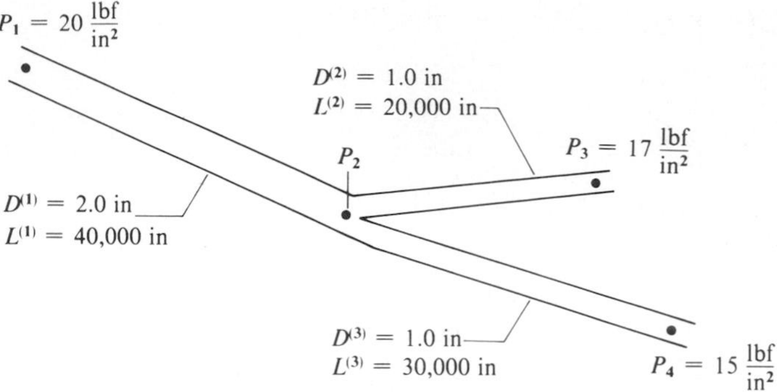

Figure 4.11 Problem parameters (Example 4.3—Flow in three-pipe network)

with an element matrix

$$[B]^{(3)} = \begin{matrix} \\ 2 \\ 4 \end{matrix} \begin{matrix} 2 & 4 \\ \begin{bmatrix} 0.545 & -0.545 \\ -0.545 & 0.545 \end{bmatrix}^{(3)} \end{matrix}$$

Note that the assembly numbers for element 3 (2 and 4 associated with the nodes in that element) differ by more than one. Thus the assembled matrix will have nonzero terms more than one diagonal away from the main diagonal.

Assembling the elements into the global matrix gives

$$\begin{matrix} \\ 1 \\ 2 \\ \\ \\ 3 \\ 4 \end{matrix} \begin{matrix} \quad 1 \qquad\qquad 2 \qquad\qquad 3 \qquad\qquad 4 \\ \begin{bmatrix} 6.544 & -6.544 & 0 & 0 \\ -6.544 & \begin{matrix} 6.544 \\ +0.818 \\ +0.545 \end{matrix} & -0.818 & -0.545 \\ 0 & -0.818 & 0.818 & 0 \\ 0 & -0.545 & 0 & 0.545 \end{bmatrix} \end{matrix} \begin{Bmatrix} P_1 \\ P_2 \\ P_3 \\ P_4 \end{Bmatrix} = \begin{Bmatrix} Q'_1 \\ 0 \\ Q'_3 \\ Q'_4 \end{Bmatrix}$$

The external flow rates Q' at nodes 1, 3, 4 are unknown, but the pressures are known at those nodes. After putting in the prescribed values for pressure, the result is

$$\begin{bmatrix} 6.544 & 0 & 0 & 0 \\ 0 & 7.907 & 0 & 0 \\ 0 & 0 & 0.818 & 0 \\ 0 & 0 & 0 & 0.545 \end{bmatrix} \begin{Bmatrix} P_1 \\ P_2 \\ P_3 \\ P_4 \end{Bmatrix} = \begin{Bmatrix} 130.88 \\ 152.96 \\ 13.91 \\ 8.17 \end{Bmatrix}$$

Solving for the pressures gives

$$\begin{Bmatrix} P_1 \\ P_2 \\ P_3 \\ P_4 \end{Bmatrix} = \begin{Bmatrix} 20 \\ 19.34 \\ 17 \\ 15 \end{Bmatrix} \text{ (lbf/in}^2\text{)}$$

Not much pressure drop occurs in pipe 1. The flow rates through each pipe are

$$\begin{aligned} Q^{(1)} &= 6.544\ (20 - 19.34) = 4.27\ \text{in}^3/\text{sec} \\ Q^{(2)} &= 0.818\ (19.34 - 17) = 1.91\ \text{in}^3/\text{sec} \\ Q^{(3)} &= 0.545\ (19.34 - 15) = 2.36\ \text{in}^3/\text{sec} \end{aligned}$$

Of course the flow through pipes 2 and 3 must match that through pipe 1.

E. Electrical Networks

Electrical networks can be analyzed by using a finite element in a manner very similar to that used for pipe networks. The voltage drop V across a resistor, shown in Figure 4.12(a) is related to the resistance R and current I by the formula $V = RI$. Let the voltages at the ends be V_i and V_j. The current is related to the voltages by the relation

$$I = \frac{1}{R}(V_i - V_j)$$

As with the pipe flow, a decrease in voltage across the resistor from i to j produces a current with positive flow from end i to end j.

A network of resistors is shown in Figure 4.12(b). Let each resistor be considered as an element. The current flow in an element is denoted as $I^{(e)}$. At node i, the current flow for the element is

$$I_i^{(e)} = \left[\frac{1}{R} V_i - \frac{1}{R} V_j\right]^{(e)}$$

For node j, the current flow is

$$I_j^{(e)} = \left[-\frac{1}{R} V_i + \frac{1}{R} V_j\right]^{(e)}$$

The element matrix is thus easily obtained as

$$\begin{Bmatrix} I_i \\ I_j \end{Bmatrix}^{(e)} = \begin{bmatrix} \frac{1}{R} & -\frac{1}{R} \\ -\frac{1}{R} & \frac{1}{R} \end{bmatrix} \begin{Bmatrix} V_i \\ V_j \end{Bmatrix}^{(e)}$$

Factoring out the resistance R yields

$$\begin{Bmatrix} I_i \\ I_j \end{Bmatrix}^{(e)} = \underbrace{\left(\frac{1}{R}\right)^{(e)} \begin{bmatrix} 1 & -1 \\ -1 & 1 \end{bmatrix}^{(e)}}_{\text{Element matrix for electrical network}} \begin{Bmatrix} V_i \\ V_j \end{Bmatrix} \tag{4.20}$$

The equivalent spring constant is

$$k^{(e)} = \frac{1}{R^{(e)}}$$

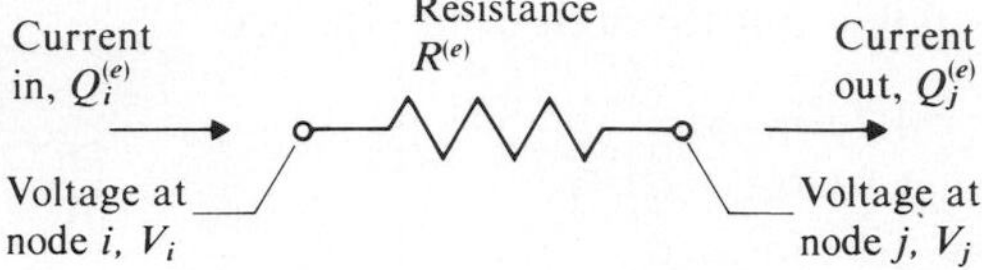

(a) Finite element model of resistor

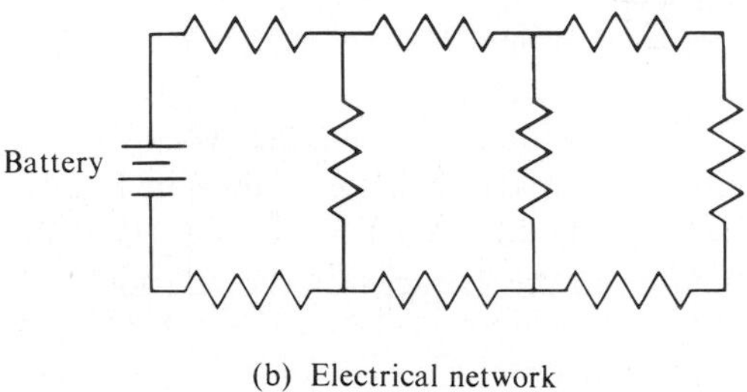

(b) Electrical network

Figure 4.12 Finite element model of electrical network

The sum of all of the internal currents at any node point n must vanish (unless some external current I' is introduced at that node point). The equations describing this are written as

$$\textit{Node 1} \quad \sum I_1 = 0$$

$$\textit{Node 2} \quad \sum I_2 = 0$$

$$\textit{Node 3} \quad \sum I_3 = 0$$

$$\cdots$$

$$\textit{Node N} \quad \sum I_N = 0$$

Each summation is taken over all of the resistors connected to that node. This yields the same assembly process as any finite element modeling.

Example 4.4 *Six Resistor Electrical Network*

Given An electrical network with one 20-volt battery and six resistors is shown in Figure 4.13. Assume that node 1 is grounded (0 volts) and node 2 is at a potential of 20 volts.

Objective Determine the voltages at all of the node points and the currents through all of the resistors.

Solution The element matrices for each element (resistor) are easily determined as

$$\textit{Element 1} \quad \begin{matrix} & 1 & 3 \\ 1 & 0.25 & -0.25 \\ 3 & -0.25 & 0.25 \end{matrix} \; ^{(1)} \text{(mhos)}$$

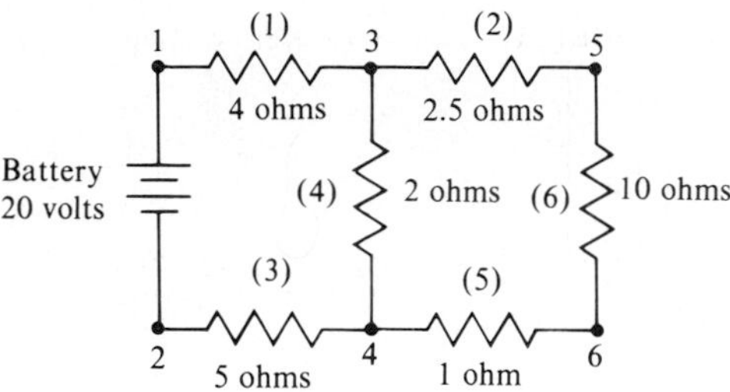

Figure 4.13 Electrical network for Example 4.4

$$\textit{Element 2} \quad \begin{matrix} & 3 & 5 \\ 3 & 0.40 & -0.40 \\ 5 & -0.40 & 0.40 \end{matrix}^{(2)} \quad \text{(mhos)}$$

$$\textit{Element 3} \quad \begin{matrix} & 2 & 4 \\ 2 & 0.20 & -0.20 \\ 4 & -0.20 & 0.20 \end{matrix}^{(3)} \quad \text{(mhos)}$$

$$\textit{Element 4} \quad \begin{matrix} & 3 & 4 \\ 3 & 0.50 & -0.50 \\ 4 & -0.50 & 0.50 \end{matrix}^{(4)} \quad \text{(mhos)}$$

$$\textit{Element 5} \quad \begin{matrix} & 4 & 6 \\ 4 & 1.00 & -1.00 \\ 6 & -1.00 & 1.00 \end{matrix}^{(5)} \quad \text{(mhos)}$$

$$\textit{Element 6} \quad \begin{matrix} & 5 & 6 \\ 5 & 0.10 & -0.10 \\ 6 & -0.10 & 0.10 \end{matrix}^{(6)} \quad \text{(mhos)}$$

The numbers on the left side and above each element matrix indicate the nodes to which the resistors are connected. They also give the location for assembly of the element matrices into the global matrix. Figure 4.14 shows the global matrices just after assembly but before the known voltages are included.

Known voltages are $V_1 = 0$, $V_2 = 20$. After including these in the global matrices, the equations to be solved for the nodal voltages become

$$\begin{bmatrix} 0.25 & 0 & 0 & 0 & 0 & 0 \\ 0 & 0.20 & 0 & 0 & 0 & 0 \\ 0 & 0 & 1.15 & -0.50 & -0.40 & 0 \\ 0 & 0 & -0.50 & 1.70 & 0 & -1.00 \\ 0 & 0 & -0.40 & 0 & 0.50 & -0.10 \\ 0 & 0 & 0 & -1.00 & -0.10 & 1.10 \end{bmatrix} \begin{Bmatrix} V_1 \\ V_2 \\ V_3 \\ V_4 \\ V_5 \\ V_6 \end{Bmatrix} = \begin{Bmatrix} 0 \\ 4.0 \\ 0 \\ 4.0 \\ 0 \\ 0 \end{Bmatrix}$$

This system of equations should be solved by a computer program of some sort. Note that nonzero terms such as -0.40 appear further away from the diagonal than just immediately above and below it (as they do in the spring problems). Here NBW = 3.

$$
\begin{array}{c}
\\ 1 \\ 2 \\ 3 \\ 4 \\ 5 \\ 6
\end{array}
\begin{array}{c}
\begin{array}{cccccc} 1 & 2 & 3 & 4 & 5 & 6 \end{array} \\
\begin{bmatrix}
0.25 & 0 & -0.25 & 0 & 0 & 0 \\
0 & 0.20 & 0 & -0.20 & 0 & 0 \\
-0.25 & 0 & \begin{matrix} 0.25 \\ +0.40 \\ +0.50 \end{matrix} & -0.50 & -0.40 & 0 \\
0 & -0.20 & -0.50 & \begin{matrix} 0.20 \\ +0.50 \\ +1.00 \end{matrix} & 0 & -1.00 \\
0 & 0 & -0.40 & 0 & \begin{matrix} 0.40 \\ +0.10 \end{matrix} & -0.10 \\
0 & 0 & 0 & -1.00 & -0.10 & \begin{matrix} 1.00 \\ +0.10 \end{matrix}
\end{bmatrix}
\end{array}
\begin{Bmatrix} V_1 \\ V_2 \\ V_3 \\ V_4 \\ V_5 \\ V_6 \end{Bmatrix}
=
\begin{Bmatrix} 0 \\ 0 \\ 0 \\ 0 \\ 0 \\ 0 \end{Bmatrix}
$$

Figure 4.14 Global matrices just following assembly (Example 4.4—Six Resistor Electrical Network)

The solution for the nodal voltages is

$$
\begin{Bmatrix} V_1 \\ V_2 \\ V_3 \\ V_4 \\ V_5 \\ V_6 \end{Bmatrix}
=
\begin{Bmatrix} 0 \\ 20 \\ 7.448 \\ 10.691 \\ 8.048 \\ 10.450 \end{Bmatrix}
\qquad \text{(volts)}
$$

Once these are found, the currents in the resistors are easily obtained as

$$
\begin{array}{ll}
I^{(1)} = -1.862 & I^{(4)} = -1.622 \\
I^{(2)} = -0.240 & I^{(5)} = 0.240 \\
I^{(3)} = 1.862 & I^{(6)} = -0.240
\end{array}
\qquad \text{(amps)}
$$

4.4 Simplex Bar Elements

Two related physical systems that are easily modeled with finite elements are axially aligned elastic bars and a rod whose properties vary along its length. The finite element formulation is similar to that for the spring just discussed, except that the axial displacement in either the bars or rod is a function of x. Once again the direct method is used. Section 4.5 treats simple two-dimensional trusses with these simplex bar elements.

A. Aligned Series of Bars

Figure 4.15 shows a series of E bars with N pin joints. The bars are axially aligned so that all forces and displacements are in the axial direction. All bars have pinned joints at the ends, and forces are applied only at the pin joints (nodes). Thus, each bar is a two-force member (the force in the bar is directed along the axis of the bar), as discussed in elementary statics courses.

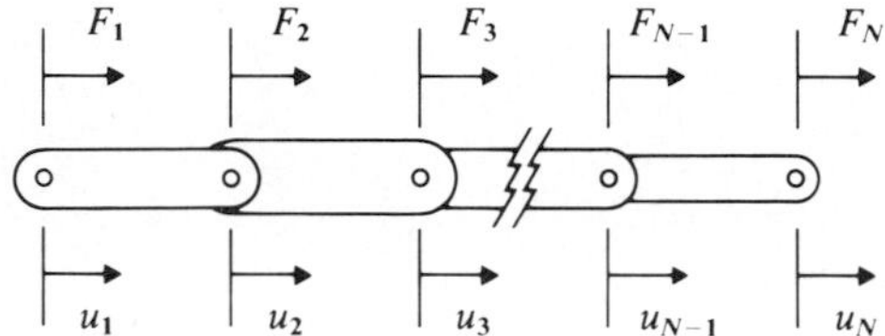

(a) Nodal displacements and external forces on series of bars

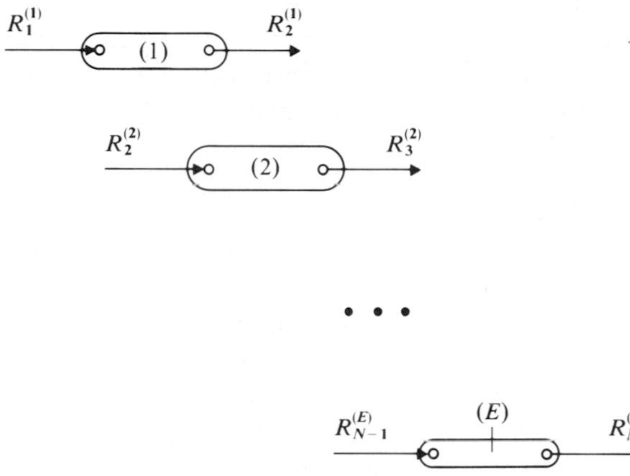

(b) Simplex bar elements and reaction forces

Figure 4.15 Aligned series of bars

Each bar element has the properties $E^{(e)}$ = Young's modulus, $A^{(e)}$ = cross sectional area, and $L^{(e)}$ = length, where Young's modulus E and the cross sectional area A are constant. The properties E, A, and L can be different from element to element. Let the axial deformation of the bar be denoted by $u(x)$. The displacement within the element is evaluated later in this section.

Bar elements are shown in Figure 4.15(b). As in the spring elements, reaction forces act at each pin joint in the element. These forces are assumed positive in the direction indicated by the arrows.

The axial strain or unit elongation $\epsilon(x)$ in a bar subject to axial deformation is given by

$$\epsilon = \frac{du}{dx} \qquad \textbf{(4.21)}$$

For a linearly elastic material, the stress $\sigma(x)$ is related to the strain by Hooke's law

$$\sigma = E\,\epsilon \qquad \textbf{(4.22)}$$

or from the relation for the strain

$$\sigma = E \frac{du}{dx} \tag{4.23}$$

The internal reaction force $R(x)$ in the bar is

$$R = A\sigma = EA \frac{du}{dx} \tag{4.24}$$

Here both the stress σ and reaction force R are assumed positive when acting in the positive x direction (positive to the right) on a free body diagram of the left-hand side of the rod. This gives the standard sign convention that tension in the bar is positive and compression is negative. Determination of the actual element properties from these equations is postponed to the section following the introduction of the axially variable rod.

B. Rod with Axially Variable Properties

A rod of length L with axially variable properties has a finite element model that is very similar to that just discussed for the axially aligned series of bars. Figure 4.16(a) illustrates the rod with properties $E = E(x) =$ Young's modulus, $A = A(x) =$ cross sectional area. These properties are assumed to vary reasonably slowly with x so that the rod deforms in one dimension. If an external force F is applied at some point along the rod (say, the right end, at which $x = L$), an axial displacement $u(x)$ results. The displacement at one point must be specified to keep the rod from moving axially under unbalanced forces.

The rod is divided into any desired number of short lengths as indicated in Figure 4.16(b). They need not be of equal length. Each segment is an element of length $L^{(e)}$. The elements are labeled 1 through E. For simplex elements, node points are placed at the division lines. The nodes are labeled from 1 to N, where $N = E + 1$.

Within each element, the element cross sectional area A and Young's modulus E are assumed constant: $A = A^{(e)}$, $E = E^{(e)}$. Figure 4.16(b) illustrates the constant area for each element. If simplex elements are used, the constant property assumption gives a degree of accuracy consistent with linear interpolation functions. If either property varies with x, the element value may be taken as the average of the nodal values A_i, A_j, or E_i, E_j.

$$A^{(e)} = \frac{1}{2}(A_i + A_j)$$

$$E^{(e)} = \frac{1}{2}(E_i + E_j)$$

This technique is illustrated in Example 4.6.

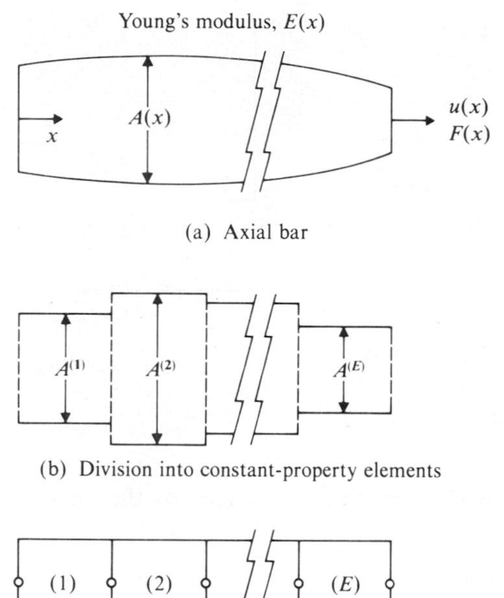

(a) Axial bar

(b) Division into constant-property elements

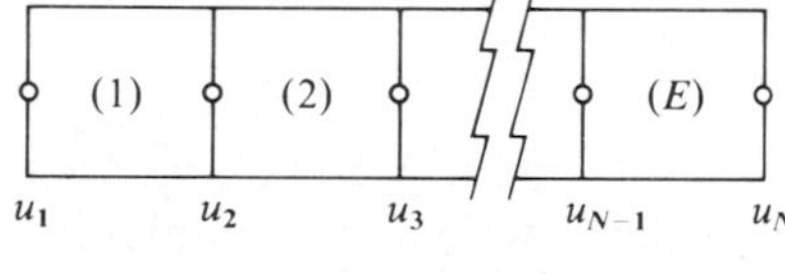

(c) Labeling of nodes and elements

Figure 4.16 Finite element model of axial rod with variable properties

The equations describing the strain, stress, and reaction forces of the rod are given by the same equations as for the bars in series (4.21) through (4.24).

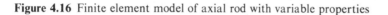

$$\epsilon = \frac{du}{dx}, \quad \sigma = E\,\epsilon, \quad \sigma = E\frac{du}{dx}, \quad R = EA\frac{du}{dx}$$

Of course, the variable Young's modulus $E(x)$ and cross sectional area $A(x)$ are functions of x within these equations.

C. Simplex Bar Element

Both problems, aligned series of bars and axial rod with variable properties, use the same simplex element. Figure 4.17(a) shows the two-node element with properties $E^{(e)}$ = Young's modulus, $A^{(e)}$ = cross sectional area, and $L^{(e)}$ = length of element. Both E and A are assumed constant over the element. The element has nodal displacements $u_i^{(e)}$ and $u_j^{(e)}$ and internal reaction forces $R_i^{(e)}$ and $R_j^{(e)}$.

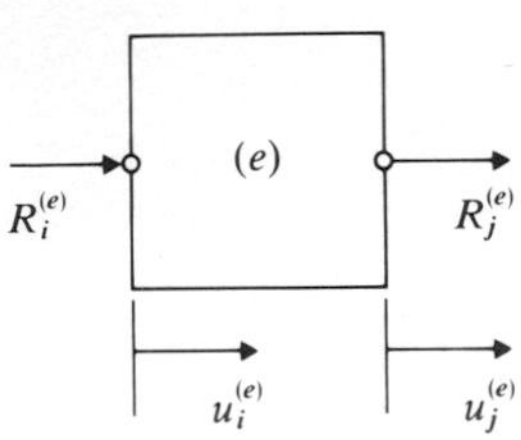

(a) Free body diagram of simplex bar element

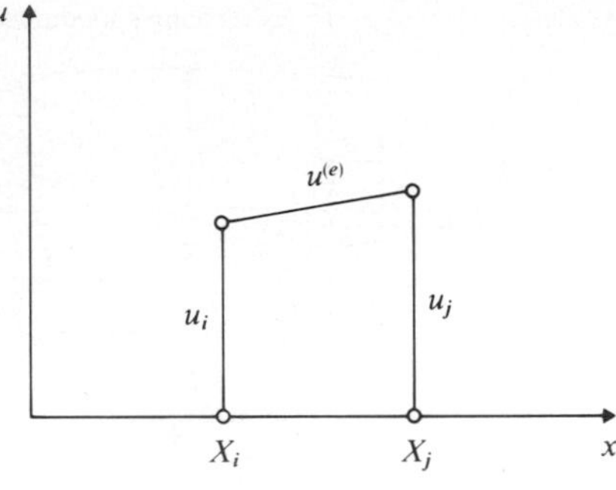

(b) Interpolation function

Figure 4.17 Simplex bar element

The displacement $u(x)$ within the element is given by the linear interpolation function (3.4)

$$u^{(e)} = N_i^{(e)}u_i + N_j^{(e)}u_j$$

where the pyramid functions N_i, N_j are given by (3.6), and u_i, u_j, are the nodal displacements. Here the variable $u(x)$ replaces the function $T(x)$ employed in Section 3.2. In matrix form, the interpolation function is written (3.9)

$$u^{(e)} = \lfloor N \rfloor \{u\}$$

where

$$\lfloor N \rfloor = \lfloor N_i \quad N_j \rfloor , \quad \{u\} = \begin{Bmatrix} u_i \\ u_j \end{Bmatrix}$$

The element strain ϵ is obtained from (4.21) as

$$\epsilon^{(e)} = \frac{du^{(e)}}{dx}$$

This physically represents the change in length

$$(\Delta L)^{(e)} = u_j - u_i$$

divided by the element length

$$\epsilon^{(e)} = \left(\frac{\Delta L}{L}\right)^{(e)}$$

It is a dimensionless measure of the elongation of the bar. Substituting the interpolation function,

$$\epsilon^{(e)} = \frac{\partial N_i}{\partial x} u_i + \frac{\partial N_j}{\partial x} u_j$$

the final result is

$$\epsilon^{(e)} = \frac{b_i}{L} u_i + \frac{b_j}{L} u_j \qquad \textbf{(4.25)}$$

where $b_i = -1$ and $b_j = 1$. In matrix form, the strain is

$$\underset{(1\times1)}{\{\epsilon\}^{(e)}} = \underset{(1\times2)}{[D]^{(e)}} \quad \underset{(2\times1)}{\{u\}^{(e)}} \tag{4.26}$$

where the terms are written out as

$$\epsilon^{(e)} = \underbrace{\begin{bmatrix} \dfrac{-1}{L} & \dfrac{1}{L} \end{bmatrix}^{(e)}}_{\substack{\text{Element}\\ \text{derivative}\\ \text{matrix } [D]}} \begin{Bmatrix} u_i \\ u_j \end{Bmatrix}^{(e)}$$

The element derivative matrix $[D]$ is given by (3.10). For simplex bar elements, the strain ϵ is constant.

Using Hooke's law (4.22), the stress can also be obtained for the element. The expression is $\sigma^{(e)} = E^{(e)}\,\epsilon^{(e)}$ or

$$\sigma^{(e)} = E^{(e)}\left(-\frac{1}{L}\,u_i + \frac{1}{L}\,u_j\right)^{(e)} \tag{4.27}$$

Again in matrix form

$$\{\sigma\}^{(e)} = E^{(e)}\,[D]^{(e)}\{u\}^{(e)} \tag{4.28}$$

where the terms are written out as

$$\sigma^{(e)} = E^{(e)}\begin{bmatrix} -\dfrac{1}{L} & \dfrac{1}{L} \end{bmatrix}\begin{Bmatrix} u_i \\ u_j \end{Bmatrix}$$

Since the strain is constant within the element, the stress σ must be constant as well. Finally the internal reaction force in the element is obtained from (4.24) as

$$R^{(e)} = (AE)^{(e)}\left(-\frac{1}{L}\,u_i + \frac{1}{L}\,u_j\right)^{(e)} \tag{4.29}$$

This may also be written as

$$R^{(e)} = \left(\frac{AE}{L}\right)^{(e)}(\Delta L)^{(e)}$$

In matrix form

$$\underset{(1\times1)}{\{R\}^{(e)}} = (AE)^{(e)}\,\underset{(1\times2)}{[D]^{(e)}} \quad \underset{(2\times1)}{\{u\}^{(e)}} \tag{4.30}$$

where the terms are

$$R^{(e)} = (AE)^{(e)}\begin{bmatrix} -\dfrac{1}{L} & \dfrac{1}{L} \end{bmatrix}^{(e)}\begin{Bmatrix} u_i \\ u_j \end{Bmatrix}^{(e)}$$

The reaction force R within the element is constant.

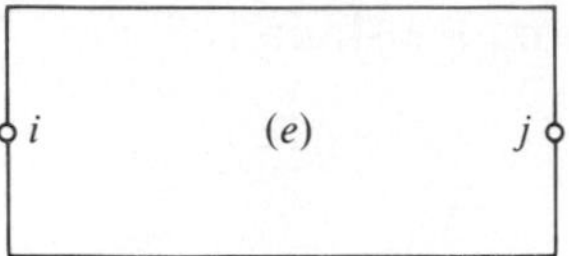

(a) Undisplaced bar element

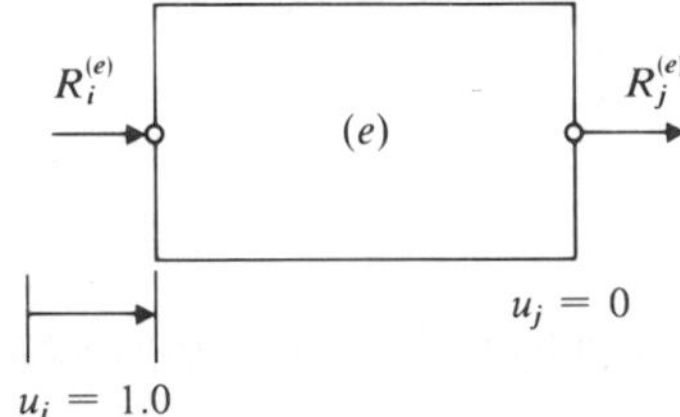

(b) Displacement of node i (steps a and b)

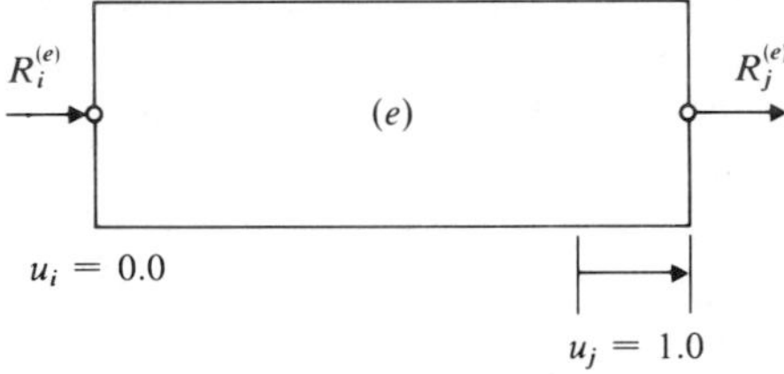

(c) Displacement of node j (steps c and d)

Figure 4.18 Direct method for evaluating element stiffness matrix for bar element

D. Element Stiffness Matrix

An element matrix—the name element stiffness matrix is explained a bit later—is obtained using the same direct method as employed for the spring element in Section 4.2. Steps a through d are repeated here. Step a involves applying a reaction force R_i required to produce a displacement $u_i = 1.0$ while u_j is zero. Step b uses element equilibrium to evaluate R_j. Similarly, step c applies reaction force R_j to produce a displacement $u_j = 1.0$ when $u_i = 0$. Step d determines R_i with element equilibrium again.

Figure 4.18 illustrates the steps. Actually using displacements of unity, for example, $u_i = 1.0$ in step a and $u_j = 1.0$ in step c, works just as well as the algebraic symbols. This produces the terms in the element matrix directly, as seen in what follows. The element matrix has the form (4.14)

$$\begin{Bmatrix} R_i \\ R_j \end{Bmatrix}^{(e)} = \begin{bmatrix} B_{ii} & B_{ij} \\ B_{ji} & B_{jj} \end{bmatrix}^{(e)} \begin{Bmatrix} u_i \\ u_j \end{Bmatrix}^{(e)}$$

or writing out the two equations

$$R_i^{(e)} = B_{ii}^{(e)} u_i^{(e)} + B_{ij}^{(e)} u_j^{(e)}$$

$$R_j^{(e)} = B_{ji}^{(e)} u_i^{(e)} + B_{jj}^{(e)} u_j^{(e)}$$

These give the relations needed.

For steps a and b, let the displacements be $u_i = 1.0$ and $u_j = 0.0$. The two equations are

$$R_i^{(e)} = B_{ii}^{(e)}$$

$$R_j^{(e)} = B_{ji}^{(e)}$$

The reaction force on the left end is

$$R_i^{(e)} = \left(\frac{AE}{L}\right)^{(e)} = B_{ii}^{(e)} \tag{4.31}$$

which gives the one term in the element matrix. Summing the forces to keep the element in equilibrium yields

$$\sum F_x = R_i^{(e)} + R_j^{(e)} = 0$$

with the result

$$R_j^{(e)} = -R_i^{(e)} = -\left(\frac{AE}{L}\right)^{(e)} = B_{ji}^{(e)} \tag{4.32}$$

This completes steps a and b.

Steps c and d have the displacements $u_i = 0.0$ and $u_j = 1.0$. The two equations become

$$R_i^{(e)} - B_{ij}^{(e)}$$

$$R_j^{(e)} = B_{jj}^{(e)}$$

On the right end, the reaction force is

$$R_j^{(e)} = \left(\frac{AE}{L}\right)^{(e)} = B_{jj}^{(e)} \tag{4.33}$$

and the other end gives

$$\sum F_x = R_i^{(e)} + R_j^{(e)} = 0$$

Solving for R_i

$$R_i^{(e)} = -R_j^{(e)} = -\left(\frac{AE}{L}\right)^{(e)} = B_{ij}^{(e)} \tag{4.34}$$

All steps are now complete.

Gathering all of the terms in the element matrix yields

$$[B]^{(e)} = \underbrace{\begin{matrix} & i & j \\ i \\ j \end{matrix}\begin{bmatrix} \dfrac{AE}{L} & -\dfrac{AE}{L} \\ -\dfrac{AE}{L} & \dfrac{AE}{L} \end{bmatrix}^{(e)}}_{\substack{\text{Element stiffness matrix} \\ \text{for bar } (2\times 2)}} \tag{4.35}$$

Note that all of the terms in the element matrix are similar. Factoring the common terms outside gives

$$[B]^{(e)} = \left(\frac{AE}{L}\right)^{(e)} \begin{matrix} & i & j \\ i & \left[\begin{matrix} 1 \\ -1 \end{matrix}\right. & \left.\begin{matrix} -1 \\ 1 \end{matrix}\right]^{(e)} \\ j & & \end{matrix}$$

The matrix is clearly symmetric and of the same form as the element matrix for the spring. The constant term AE/L has units

$$\frac{AE}{L} = \frac{(\text{length})^2}{\text{length}} \frac{\text{force}}{(\text{length})^2} = \frac{\text{force}}{\text{length}}$$

This has units of stiffness, as does the element matrix for the spring. The reactions are

$$\begin{Bmatrix} R_i \\ R_j \end{Bmatrix}^{(e)} = \begin{bmatrix} \dfrac{AE}{L} & -\dfrac{AE}{L} \\ -\dfrac{AE}{L} & \dfrac{AE}{L} \end{bmatrix}^{(e)} \begin{Bmatrix} u_i \\ u_j \end{Bmatrix} \tag{4.36}$$

E. Assembly

Determining the set of N linear algebraic equations for the nodal displacements proceeds exactly as it did for the springs in series, as discussed in Sections 4.2 and 4.3. The sum of the reaction forces must equal the externally applied nodal forces $F_1, F_2, F_3, \ldots,$ F_N. The equations are (4.7)

Node 1 $\quad \sum R_1 = R_1^{(1)} = F_1$

Node 2 $\quad \sum R_2 = R_2^{(1)} + R_2^{(2)} = F_2$

Node 3 $\quad \sum R_3 = R_3^{(2)} + R_3^{(3)} = F_3$

. . .

Node N $\quad \sum R_N = R_N^{(E)} = F_N$

The elements are assembled in the manner shown in Figure 4.7. Prescribed nodal values are also treated as indicated in Figure 4.8.

Example 4.5 Statically Indeterminate Bars

Given Two bars, one made of steel and the other made of aluminum, form an axially aligned set, as shown in Figure 4.19. The set has fixed ends, where the displacement is zero. An axial force of 10 kN is applied at the center pinned joint, as indicated. This problem is statically indeterminate but easily solved with finite elements.

The element properties are as follows. The first bar is made of steel with properties $E^{(1)} = 207$ kN/mm², $A^{(1)} = 25$ mm², and $L^{(1)} = 40$ mm, while the second bar is made of aluminum with properties $E^{(2)} = 69$ kN/mm², $A^{(2)} = 40$ mm², and $L^{(2)} = 60$ mm.

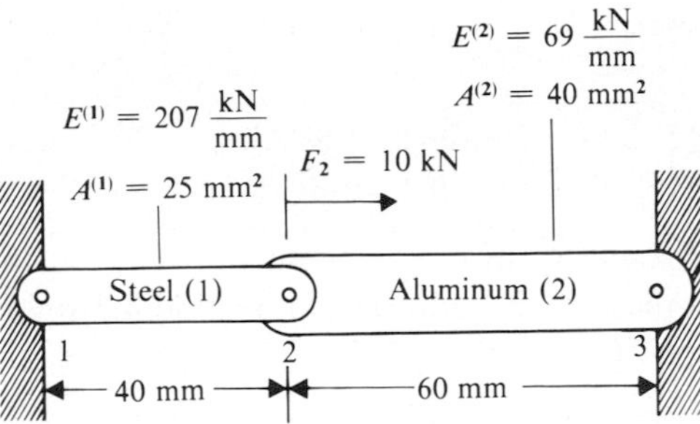

Figure 4.19 Axially aligned bars (Example 4.5—Statically indeterminate bars)

Objective Obtain the displacement of the pinned joint between the bars as well as the axial strain, stress, and internal force in each bar.

Solution In this problem there are three nodes ($N=3$) and two elements ($E=2$). Nominally there are three nodal displacements, but two are specified as zero. Thus there is actually only one unknown in this simple example. In spite of this the global matrix is kept at (3×3) to illustrate the techniques used.

The element matrix for the first element is obtained from (4.35) as

$$[B]^{(1)} = \begin{matrix} & 1 & 2 \\ 1 & \\ 2 & \end{matrix}\begin{bmatrix} \frac{AE}{L} & -\frac{AE}{L} \\ -\frac{AE}{L} & \frac{AE}{L} \end{bmatrix}^{(1)}$$

The stiffness term is

$$\left(\frac{AE}{L}\right)^{(1)} = 25 \text{ mm}^2 \times 207 \frac{\text{kN}}{\text{mm}^2} \times \frac{1}{40 \text{ mm}} = 129 \frac{\text{kN}}{\text{mm}}$$

and the element matrix becomes

$$[B]^{(1)} = \begin{matrix} & 1 & 2 \\ 1 & \\ 2 & \end{matrix}\begin{bmatrix} 129 & -129 \\ -129 & 129 \end{bmatrix}^{(1)} \frac{\text{kN}}{\text{mm}}$$

For element 2, the element matrix is

$$[B]^{(2)} = \begin{matrix} & 2 & 3 \\ 2 & \\ 3 & \end{matrix}\begin{bmatrix} \frac{AE}{L} & -\frac{AE}{L} \\ -\frac{AE}{L} & \frac{AE}{L} \end{bmatrix}^{(2)}$$

with the stiffness

$$\left(\frac{AE}{L}\right)^{(2)} = 40 \text{ mm}^2 \times 69 \frac{\text{kN}}{\text{mm}^2} \times \frac{1}{60 \text{ mm}} = 46 \frac{\text{kN}}{\text{mm}}$$

Finally the second element matrix is

$$[B]^{(2)} = \begin{matrix} & 2 & 3 \\ 2 & 46 & -46 \\ 3 & -46 & 46 \end{matrix}^{(2)} \frac{\text{kN}}{\text{mm}}$$

The steel bar is the stiffer element by a ratio of 2.8 to 1.

Assembling the element matrices as indicated in Section 4.3 yields the global equations for the nodal displacements as

$$\begin{matrix} 1 \\ 2 \\ 3 \end{matrix}\begin{bmatrix} 129 & -129 & 0 \\ -129 & 129 + 46 & -46 \\ 0 & -46 & 46 \end{bmatrix} \begin{Bmatrix} u_1 \\ u_2 \\ u_3 \end{Bmatrix} = \begin{Bmatrix} f_1 \\ 10 \\ f_3 \end{Bmatrix}$$

Only one nodal force appears in the force column. Combining terms gives

$$\begin{bmatrix} 129 & -129 & 0 \\ -129 & 175 & -46 \\ 0 & -46 & 46 \end{bmatrix} \begin{Bmatrix} u_1 \\ u_2 \\ u_3 \end{Bmatrix} = \begin{Bmatrix} 0 \\ 10 \\ 0 \end{Bmatrix}$$

This is the global matrix and column before prescribing nodal values. Here the terms f_1 and f_3 have been replaced by zero as they would be in a computer program to solve the problem (the computer stores numbers, not algebraic symbols). In any case, the replacement of f_1 and f_3 by zero does not affect the final solution.

Prescribing the first nodal displacement $u_1 = 0$, again following Section 4.3, yields

$$\begin{bmatrix} 129 & 0 & 0 \\ 0 & 175 & -46 \\ 0 & -46 & 46 \end{bmatrix} \begin{Bmatrix} u_1 \\ u_2 \\ u_3 \end{Bmatrix} = \begin{Bmatrix} 0 \\ 10 \\ 0 \end{Bmatrix}$$

Taking into account the third nodal displacement $u_3 = 0$ gives

$$\begin{bmatrix} 129 & 0 & 0 \\ 0 & 175 & 0 \\ 0 & 0 & 46 \end{bmatrix} \begin{Bmatrix} u_1 \\ u_2 \\ u_3 \end{Bmatrix} = \begin{Bmatrix} 0 \\ 10 \\ 0 \end{Bmatrix}$$

This is easily solved. The solution is

$$\begin{Bmatrix} u_1 \\ u_2 \\ u_3 \end{Bmatrix} = \begin{Bmatrix} 0.0 \\ 0.0571 \\ 0.0 \end{Bmatrix} \text{ mm}$$

From (4.26), the element strains are

$$\epsilon^{(1)} = \begin{bmatrix} -\frac{1}{L} & \frac{1}{L} \end{bmatrix}^{(1)} \begin{Bmatrix} u_1 \\ u_2 \end{Bmatrix} = \begin{bmatrix} -\frac{1}{40} & \frac{1}{40} \end{bmatrix} \begin{Bmatrix} 0.0 \\ 0.0571 \end{Bmatrix}$$

$$\epsilon^{(1)} = 0.00143 \frac{\text{mm}}{\text{mm}}$$

Table 4.2
Element Properties (Example 4.5: Statically Indeterminate Bars)

Element Property	Element 1 (Steel)	Element 2 (Aluminum)
Change in length, ΔL (mm)	0.0571	−0.0571
Strain, ϵ (mm/mm)	0.00143	−0.00095
Stress, σ (kN/mm)	0.296	−0.066
Nodal reaction force, R_2 (kN)	7.38	2.62

and

$$\epsilon^{(2)} = \begin{bmatrix} -\frac{1}{60} & \frac{1}{60} \end{bmatrix} \begin{Bmatrix} 0.0571 \\ 0.0 \end{Bmatrix}$$

$$\epsilon^{(2)} = -0.00095 \frac{\text{mm}}{\text{mm}}$$

Also, the element stresses are (4.27)

$$\sigma^{(1)} = E^{(1)} \epsilon^{(1)} = 0.296 \frac{\text{kN}}{\text{mm}^2}$$

and

$$\sigma^{(2)} = E^{(2)} \epsilon^{(2)} = -0.066 \frac{\text{kN}}{\text{mm}^2}$$

Finally the reaction forces at node 2 for each element are (4.32)

$$R_2^{(1)} = \left(\frac{AE}{L}\right)^{(1)} u_2 = 7.38 \text{ kN}$$

$$R_2^{(2)} = \left(\frac{AE}{L}\right)^{(2)} u_2 = 2.62 \text{ kN}$$

Summing forces at node 2

$$\sum R_2 = R_2^{(1)} + R_2^{(2)} = F_2$$

$$7.38 + 2.62 = 10$$

The reaction forces match the externally applied force. Table 4.2 lists the element properties for both elements. Most of the external force is opposed by the steel bar with stiffness 129 kN/mm as compared to the aluminum bar with stiffness 46 kN/mm.

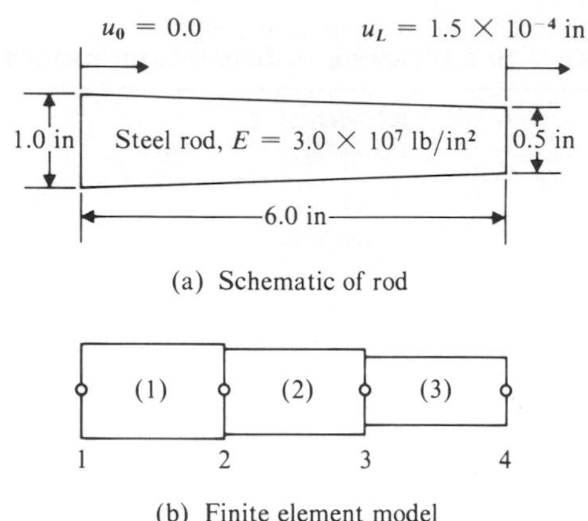

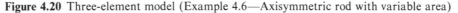

Figure 4.20 Three-element model (Example 4.6—Axisymmetric rod with variable area)

Example 4.6 *Displacement of Axisymmetric Rod*

Given An axisymmetric steel rod has a linear taper such that the diameter is 1.0 in at $x = 0.0$ in and 0.5 in at $x = 6.0$ in. It is observed that the rod has a total displacement of 1.5×10^{-4} in over its length.

Objective Determine the displacement, strain, stress and force in the rod, using three equal-length simplex bar elements as shown in Figure 4.20(a).

Solution For this problem

$$E = 3.0 \times 10^7 \text{ lb/in}$$

$$A = \frac{\pi d^2}{4} = \frac{\pi}{4}(1.0 - 0.0833x)^2 \text{ in}^2$$

$$L = 6.0 \text{ in}$$

with the prescribed displacements $u_0 = 0$, $u_L = 1.5 \times 10^{-4}$ in. Dividing the rod into three bar elements, as shown in Figure 4.20(b), the lengths are $L^{(1)} = 2.0$, $L^{(2)} = 2.0$, $L^{(3)} = 2.0$. Now the values of area at the node points are $A_1 = 0.7854$ in², $A_2 = 0.5454$ in², $A_3 = 0.3491$ in², and $A_4 = 0.1963$ in². The element areas are calculated by taking the average of the two areas of the node (end) points

$$A^{(1)} = \frac{1}{2}(A_1 + A_2) = 0.6654 \text{ in}^2$$

$$A^{(2)} = \frac{1}{2}(A_2 + A_3) = 0.4472 \text{ in}^2$$

$$A^{(3)} = \frac{1}{2}(A_3 + A_4) = 0.2726 \text{ in}^2$$

Now the element matrices can be calculated. The element stiffness matrix for element 1 is

$$[B]^{(1)} = \begin{bmatrix} \dfrac{AE}{L} & -\dfrac{AE}{L} \\ -\dfrac{AE}{L} & \dfrac{AE}{L} \end{bmatrix}^{(1)}$$

or

$$[B]^{(1)} = \begin{matrix} \\ 1 \\ 2 \end{matrix} \begin{matrix} \begin{matrix} 1 & & & & & & 2 \end{matrix} \\ \begin{bmatrix} 0.9981 \times 10^7 & -0.9981 \times 10^7 \\ -0.9981 \times 10^7 & 0.9981 \times 10^7 \end{bmatrix}^{(1)} \end{matrix}$$

For element 2

$$[B]^{(2)} = \begin{bmatrix} \dfrac{AE}{L} & -\dfrac{AE}{L} \\ -\dfrac{AE}{L} & \dfrac{AE}{L} \end{bmatrix}^{(2)}$$

or

$$[B]^{(2)} = \begin{matrix} \\ 2 \\ 3 \end{matrix} \begin{matrix} \begin{matrix} 2 & & & & & & 3 \end{matrix} \\ \begin{bmatrix} 0.6705 \times 10^7 & -0.6705 \times 10^7 \\ -0.6705 \times 10^7 & 0.6705 \times 10^7 \end{bmatrix}^{(2)} \end{matrix}$$

Similarly for element 3

$$[B]^{(3)} = \begin{matrix} \\ 3 \\ 4 \end{matrix} \begin{matrix} \begin{matrix} 3 & & & & & & 4 \end{matrix} \\ \begin{bmatrix} 0.4090 \times 10^7 & -0.4090 \times 10^7 \\ -0.4090 \times 10^7 & 0.4090 \times 10^7 \end{bmatrix}^{(3)} \end{matrix}$$

Using the assembly procedure developed in Section 4.3, the matrix equations are

$$\begin{matrix} \\ 1 \\ 2 \\ \\ 3 \\ \\ 4 \end{matrix} \begin{matrix} \begin{matrix} 1 & 2 & 3 & 4 \end{matrix} \\ \left[\begin{array}{c|c|c|c} 0.9981 \times 10^7 & -0.9981 \times 10^7 & 0.0 & 0.0 \\ \hline -0.9981 \times 10^7 & \begin{matrix} 0.9981 \times 10^7 \\ +0.6705 \times 10^7 \end{matrix} & -0.6705 \times 10^7 & 0.0 \\ \hline 0.0 & -0.6705 \times 10^7 & \begin{matrix} 0.6705 \times 10^7 \\ +0.4090 \times 10^7 \end{matrix} & -0.4090 \times 10^7 \\ \hline 0.0 & 0.0 & -0.4090 \times 10^7 & 0.4090 \times 10^7 \end{array}\right] \end{matrix} \begin{Bmatrix} u_1 \\ u_2 \\ u_3 \\ u_4 \end{Bmatrix}$$

$$= \begin{Bmatrix} 0.0 \\ 0.0 \\ 0.0 \\ 0.0 \end{Bmatrix}$$

No external forces are applied. Applying the prescribed nodal displacements $u_1 = 0.0$ and $u_4 = 1.5 \times 10^{-4}$ gives

$$10^7 \begin{bmatrix} 0.9981 & 0 & 0 & 0 \\ 0 & 1.6686 & -0.6705 & 0 \\ 0 & -0.6705 & 1.0795 & 0 \\ 0 & 0 & 0 & 0.4090 \end{bmatrix} \begin{Bmatrix} u_1 \\ u_2 \\ u_3 \\ u_4 \end{Bmatrix}$$

$$= \begin{Bmatrix} 0 \\ 0 \\ 6.135 \times 10^2 \\ 6.135 \times 10^2 \end{Bmatrix}$$

Here the common factor of 10^7 has been moved outside of the global matrix. The solution is

$$\begin{Bmatrix} u_1 \\ u_2 \\ u_3 \\ u_4 \end{Bmatrix} = \begin{Bmatrix} 0.0 \\ 3.043 \times 10^{-5} \\ 7.573 \times 10^{-5} \\ 1.500 \times 10^{-4} \end{Bmatrix} \quad \text{three-element solution}$$

Figure 4.21 plots the displacement as a function of position.

The differential equation for the axial rod is obtained by noting that the internal reaction force must be a constant since there is no externally applied force over the length of the rod. Then

$$\frac{dR}{dx} = \frac{d}{dx}(A\sigma) = \frac{d}{dx}\left(AE\,\frac{du}{dx}\right) = 0$$

with the boundary conditions $u = u_0$, $x = 0$ and $u = u_L$, $x = L$. After substituting numerical values, this becomes

$$\frac{d}{dx}\left[2.356\,(10^7)\,(1.0 - 0.08333x)^2\,\frac{du}{dx}\right] = 0$$

with $u = 0$, $x = 0$ and $u = 1.5 \times 10^{-4}$, $x = 6.0$. The exact solution evaluated at the node points is

$$\begin{Bmatrix} u_1 \\ u_2 \\ u_3 \\ u_4 \end{Bmatrix} = \begin{Bmatrix} 0.0 \\ 3.000 \times 10^{-5} \\ 7.500 \times 10^{-5} \\ 1.500 \times 10^{-4} \end{Bmatrix} \quad \text{exact solution}$$

It is easily seen that the error at the node points is less than 2%. The exact solution is plotted in Figure 4.21.

The strain in each element is

$$\epsilon^{(e)} = \frac{du^{(e)}}{dx}$$

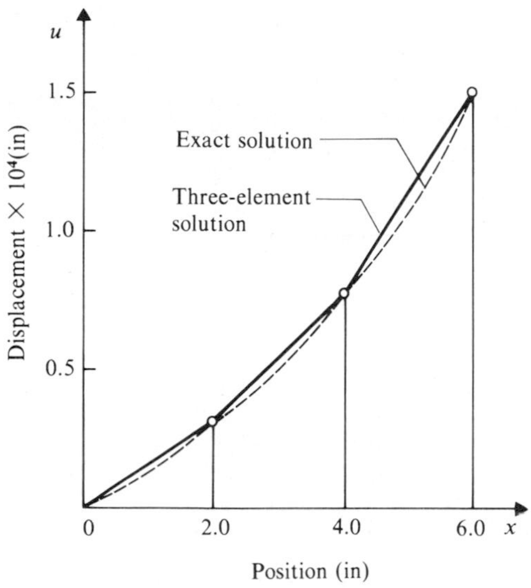

Figure 4.21 Displacement versus position (Example 4.6—Displacement of axisymmetric rod)

or

$$\begin{aligned}\epsilon^{(1)} &= 1.522 \times 10^{-5}\ \text{in/in}\\ \epsilon^{(2)} &= 2.265 \times 10^{-5}\ \text{in/in}\\ \epsilon^{(3)} &= 3.713 \times 10^{-5}\ \text{in/in}\end{aligned}$$

Evaluating the stress, $\sigma^{(e)} = E\epsilon^{(e)}$ or $\sigma^{(1)} = 456.6$ lb/in², $\sigma^{(2)} = 679.5$ lb/in², and $\sigma^{(3)} = 1114.0$ lb/in². Also, the internal force (which should be constant over the length of the bar) is $R^{(e)} = \sigma^{(e)} A^{(e)}$. Using the average area values, $R^{(1)} = 303.8$ lb, $R^{(2)} = 303.9$ lb, and $R^{(3)} = 303.7$ lb. The reaction force in each element varies less than 0.1% along the rod.

4.5 Finite Element Analysis of Trusses

This section extends the bar element to the analysis of plane trusses. Basically a coordinate transformation is used here to develop an element matrix. Plane trusses are relatively simple to analyze in concept and have quite a number of practical applications. Finite elements provide an excellent method to calculate displacements, strains, stresses, and forces in the truss.

A. Finite Element Model of a Truss

Figure 4.22 shows a plane truss. The name "plane truss" arises from the two-dimensional nature of the truss (in the x,y plane). All members are connected by pin joints at the ends. External loads are applied only at the joints of the truss—never in the middle of a member.

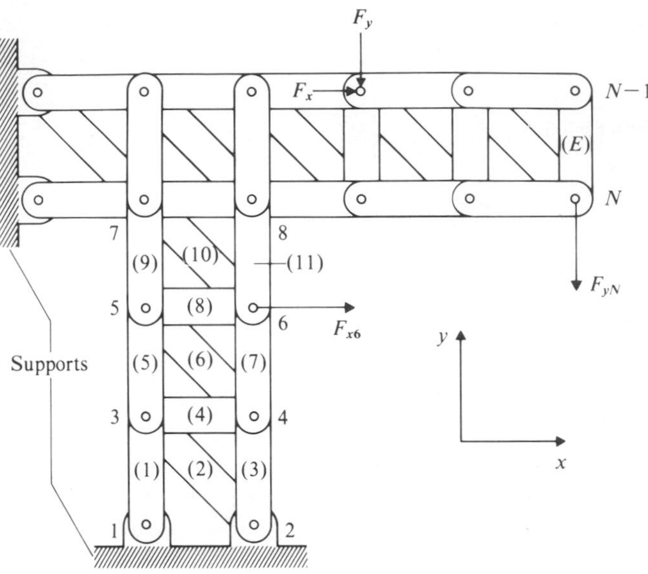

Figure 4.22 Plane truss with externally applied forces

Each member is thus a two-force member, as discussed in the previous section on axially aligned bars. The difference is that the pins (nodes) can move in two directions rather than just axially. Springs attached to the truss can easily be taken into account by the same element analysis developed here for axial bars.

Trusses can be classified as either statically determinate or statically indeterminate. Statically determinate trusses are those for which all of the forces acting on the joints and members can be determined from the static equations of equilibrium. For plane trusses three equations are given: the sums of the forces in the x and y directions equal zero and the moment about a point equals zero. The nodal displacements can be found after this analysis. Three forces act at the supports of the truss to keep it in equilibrium (prevent it from moving). Statically indeterminate trusses have forces acting at the supports such that the three equations of statics alone are not sufficient to analyze it. Displacements must be taken into account. The method of doing this is covered in typical courses on strength of materials. The finite element method described here works for either case without difficulty. It should be noted that a minimum of three nodal displacements should be prescribed to prevent the truss from moving. When these are specified, the support forces are calculated as a part of the finite element analysis, following the determination of nodal displacements.

Solving truss problems with finite elements is actually carried out by a computer program. A teaching program, called STRESS, has been developed for this text. It is described in some detail in Appendix C. The examples shown in the next section were solved with this program. The development of the finite element method in the rest of this section is integrated with program STRESS.

Let the number of pin joints (nodes) be denoted by N. The nodal coordinates of each joint are specified by the user. Nodes should be numbered so as to keep the bandwidth

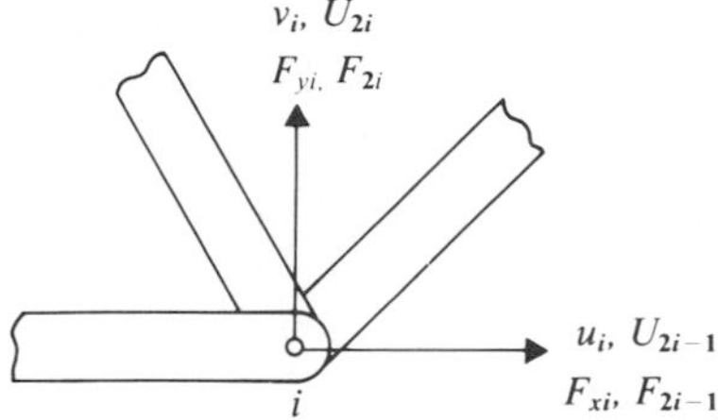

Figure 4.23 Nodal displacements in two dimensions

low, as discussed in Section 2.4. Usually this is easily accomplished for trusses by numbering in the direction of the smallest number of nodes in a systematic fashion.

Nodal displacements occur in both horizontal (x) and vertical (y) directions. These are labeled u and v, with the nodal values at node i denoted as u_i, v_i. These must all be stored in a single column of length $2N$ with terms labeled U, as discussed in Section 3.5 for vector problems. Forces are applied only at the node points (joints). These forces have two components F_{xi}, F_{yi}, as shown in Figure 4.23. These are stored in a single column of forces of length $2N$.

There are E bars (elements) in the truss. For each element, the node/element connectivity data are specified by the user, as was discussed in some detail in Section 2.4. Each element has only two node points, so there are only two entries.

As noted in Section 2.4, the global bandwidth is determined from the element/connectivity data using the formula NBW $= (R+1)$ (NDOF). Here NDOF $= 2$ because each node point has two degrees of freedom u,v. Program STRESS calculates NBW automatically from the element/node connectivity table.

B. Bar Element

A simplex bar element, shown in Figure 4.24, has two node points i, j. It has the properties $E^{(e)}$ = Young's modulus, $A^{(e)}$ = cross sectional area, and $L^{(e)}$ = length of element. As in the previous section, it is assumed in this section that E and A are constant within the element.

Nodal coordinates are $x_i^{(e)}$, $y_i^{(e)}$ and $x_j^{(e)}$, $y_j^{(e)}$. These values are obtained from the nodal coordinate table and element/node connectivity table. The length of the element is then given by the formula

$$L^{(e)} = [(x_j - x_i)^2 + (y_j - y_i)^2]^{1/2} \tag{4.37}$$

Let the angle ϕ denote the angle of the bar relative to the positive x axis. Thus ϕ ranges from 0 to 360 degrees (0 to 2π radians). The sine and cosine of the element angle are given in terms of the nodal coordinates as

$$\begin{aligned} \sin \phi^{(e)} &= \left(\frac{y_j - y_i}{L}\right)^{(e)} \\ \cos \phi^{(e)} &= \left(\frac{x_j - x_i}{L}\right)^{(e)} \end{aligned} \tag{4.38}$$

These relations are used for the element coordinate transformation.

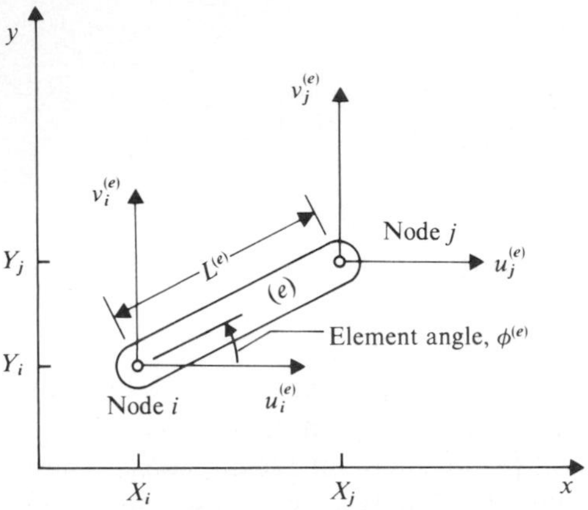

Figure 4.24 Simplex bar element

If ΔL represents the element axial change in length and R represents the reaction force (exerted by the pin on the bar) the relation (4.29) from the previous section becomes

$$R^{(e)} = \left(\frac{AE}{L}\right)^{(e)} (\Delta L)^{(e)} \tag{4.39}$$

This gives the basis for the derivation of an element stiffness matrix.

The four nodal displacements for the element are $u_i^{(e)}$, $v_i^{(e)}$, $u_j^{(e)}$, and $v_j^{(e)}$ as shown in Figure 4.24. For convenience, the element nodal displacements are written in matrix form as

$$\{u\}^{(e)} = \underbrace{\begin{Bmatrix} u_i \\ v_i \\ u_j \\ v_j \end{Bmatrix}^{(e)}}_{\substack{\text{element}\\ \text{nodal}\\ \text{displacements}\\ (4\times 1)}} = \begin{Bmatrix} U_{2i-1} \\ U_{2i} \\ U_{2j-1} \\ U_{2j} \end{Bmatrix}^{(e)} \tag{4.40}$$

Here the numbering of the second column follows the methods of Section 3.5. Similarly, internal reaction forces (applied by the node to the element) for the element are R_{xi}, R_{yi}, R_{xj}, and R_{yj}, or, in matrix form,

$$\{R\}^{(e)} = \underbrace{\begin{Bmatrix} R_{xi} \\ R_{yi} \\ R_{xj} \\ R_{yj} \end{Bmatrix}^{(e)}}_{\substack{\text{element} \\ \text{nodal} \\ \text{reaction} \\ \text{column} \\ (4\times 1)}} = \begin{Bmatrix} R_{2i-1} \\ R_{2i} \\ R_{2j-1} \\ R_{2j} \end{Bmatrix}^{(e)} \tag{4.41}$$

C. Element Stiffness Matrix

For the two-dimensional bar element, the element matrix is of order four by four with the form

$$\begin{Bmatrix} R_{2i-1} \\ R_{2i} \\ R_{2j-1} \\ R_{2j} \end{Bmatrix}^{(e)} = [B]^{(e)} \begin{Bmatrix} U_{2i-1} \\ U_{2i} \\ U_{2j-1} \\ U_{2j} \end{Bmatrix}^{(e)} \tag{4.42}$$

where

$$[B] = \begin{bmatrix} B_{2i-1,2i-1} & B_{2i-1,2i} & B_{2i-1,2j-1} & B_{2i-1,2j} \\ B_{2i,2i-1} & B_{2i,2i} & B_{2i,2j-1} & B_{2i,2j} \\ B_{2j-1,2i-1} & B_{2j-1,2i} & B_{2j-1,2j-1} & B_{2j-1,2j} \\ B_{2j,2i-1} & B_{2j,2i} & B_{2j,2j-1} & B_{2j,2j} \end{bmatrix}$$

element matrix for bar in two dimensions (4×4)

The terms within the element matrix are found in terms of the element properties A, E, L, ϕ alone.

The direct method of evaluating the terms in the element matrix is similar to that used in the previous sections. It proceeds as follows:

a. A displacement (say $u_i = 1.0$) is applied to one node point. The other three nodal displacements are set to zero ($v_i = 0.0$, $u_j = 0.0$, $v_j = 0.0$). The forces at this node (R_{xi}, R_{yi}) required to produce this displacement are calculated.
b. The forces at the other node (R_{xj}, R_{yj}) required to keep the element in equilibrium are calculated.

This process is carried out for all four nodal displacements.

Consider the first case, where the element displacements for step a are

$$\{u\} = \begin{Bmatrix} u_i \\ v_i \\ u_j \\ v_j \end{Bmatrix} = \begin{Bmatrix} 1.0 \\ 0.0 \\ 0.0 \\ 0.0 \end{Bmatrix}$$

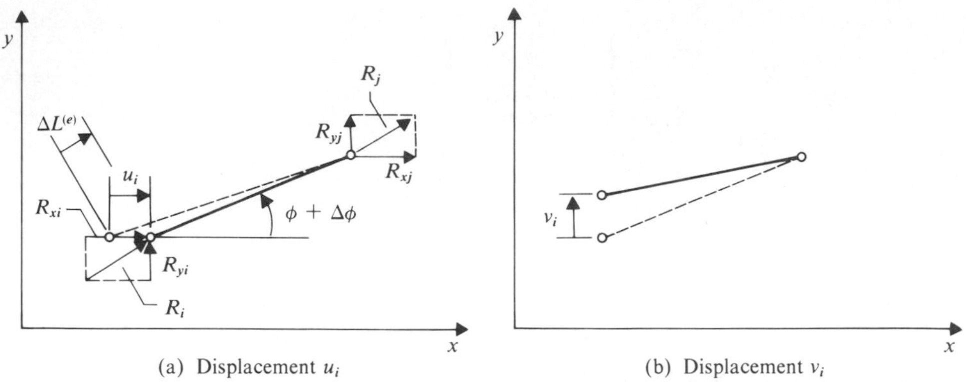

Figure 4.25 Displacements for direct method of calculating element stiffness matrix (displacements u_j and v_j not shown)

Figure 4.25(a) illustrates the geometry. The dashed line indicates the original element position, while the solid line shows the deformed bar. The element matrix reduces to the four equations

$$\begin{aligned} R_{xi} &= B_{2i-1,2i-1}u_i = B_{2i-1,2i-1} \\ R_{yi} &= B_{2i,2i-1}u_i = B_{2i,2i-1} \\ R_{xj} &= B_{2j-1,2i-1}u_i = B_{2j-1,2i-1} \\ R_{yj} &= B_{2j,2i-1}u_i = B_{2j,2i-1} \end{aligned}$$

Thus this first case, $u_i = 1.0$, gives the first column of the element matrix.

The change in length ΔL of the bar due to u_i is given by $\Delta L = u_i \cos(\phi + \Delta\phi)$, where $\Delta\phi$ represents the change in angle of the bar (relative to the x axis). It is assumed that the displacements are very small compared with the length of the bar so $\cos(\phi + \Delta\phi) \cong \cos\phi$, and the change in length can be approximated by $\Delta L \cong u_i \cos\phi \cong \cos\phi$. The axial force R_i required to deform the bar ΔL is

$$R_i = \frac{AE}{L}\Delta L$$

Taking x and y components of this force, as shown in Figure 4.25, yields

$$R_{xi} = \frac{AE}{L}\Delta L \cos\phi$$

$$R_{yi} = \frac{AE}{L}\Delta L \sin\phi$$

Substituting the expression for ΔL from above gives

$$R_{xi} = \frac{AE}{L}\cos^2\phi$$

$$R_{yi} = \frac{AE}{L}\sin\phi\cos\phi$$

These are two of the terms in the first column in the element stiffness matrix.

Now, for step b, the sum of the reaction forces must equal zero. In the x direction, $\sum F_x = R_{xi} + R_{xj} = 0$, and solving,

$$R_{xj} = -R_{xi} = -\frac{AE}{L}\cos^2\phi$$

Similarly in the y direction, $\sum F_y = R_{yi} + R_{yj} = 0$, and solving,

$$R_{yi} = -R_{yi} = -\frac{AE}{L}\sin\phi\cos\phi.$$

These are the other terms in the first column of the element stiffness matrix.

Writing these forces in column form gives

$$\begin{Bmatrix} R_{xi} \\ R_{yi} \\ R_{xj} \\ R_{yj} \end{Bmatrix}^{(e)} = \left(\frac{AE}{L}\right)^{(e)} \begin{Bmatrix} \cos^2\phi \\ \sin\phi\cos\phi \\ -\cos^2\phi \\ -\sin\phi\cos\phi \end{Bmatrix}^{(e)}$$

Repeating these steps for v_i, u_j, v_j yields the full element stiffness matrix. Figure 4.25(b) illustrates a displacement v_i. It is not necessary to go through all of the details. The results are

$$[B]^{(e)} = \left(\frac{AE}{L}\right)^{(e)} \underbrace{\begin{bmatrix} c^2 & sc & -c^2 & -sc \\ sc & s^2 & -sc & -s^2 \\ -c^2 & -sc & c^2 & sc \\ -sc & -s^2 & sc & s^2 \end{bmatrix}}_{\text{element stiffness matrix for bar in two dimensions } (4\times4)} \tag{4.43}$$

where

$$s = \sin\phi = \left(\frac{Y_j - Y_i}{L}\right)^{(e)}$$

$$c = \cos\phi = \left(\frac{X_j - X_i}{L}\right)^{(e)}$$

Note that only the terms s^2, sc, c^2 appear in the element matrix above.

This is the element stiffness matrix for a two-force bar. The terms represent a coordinate transformation from an axial coordinate system to the global x,y coordinate system. This is the element matrix suitable for analysis of trusses.

D. Assembly and Prescribed Nodal Displacements

The set of $2N$ linear algebraic equations for the $2N$ nodal displacements is obtained by summing the reaction forces at each node point and setting the result equal to the externally applied nodal force.

Node 1	$\sum R_{x1} = F_{x1}$ $\sum R_{y1} = F_{y1}$
Node 2	$\sum R_{x2} = F_{x2}$ $\sum R_{y2} = F_{y2}$
Node 3	$\sum R_{x3} = F_{x3}$ $\sum R_{y3} = F_{y3}$
. . .	
Node N	$\sum R_{xN} = F_{xN}$ $\sum R_{yN} = F_{yN}$

This results in an assembly procedure along the lines of Section 4.3, except that the element matrix is four by four instead of two by two. The next example illustrates the technique in some detail.

Any number of nodal values can be prescribed. The method of including these is the same as that in Section 4.3.

4.6 Truss Examples

This section carries out two truss examples to illustrate the finite element method in detail. The first example is a simple three-bar triangular truss with simple supports. The second example is a more complex bridge truss that shows a practical application of the method. Computer program STRESS, described in Appendix C, was used to obtain the results for the second example.

Example 4.7 Triangular Truss

Given A truss composed of three bars—two aluminum and one steel—supports a weight of 0.4 kN. Figure 4.26 shows the geometry. The aluminum bar (1) has the properties $E^{(1)} = 69$ kN/mm², $A^{(1)} = 200$ mm², $L^{(1)} = 260$ mm, and $\phi^{(1)} = 0.0$ degrees. The other aluminum bar (2) has the properties $E^{(2)} = 69$ kN/mm², $A^{(2)} = 200$ mm², $L^{(2)} = 150$ mm, and $\phi^{(2)} = 90.0$ degrees. The steel bar has the properties $E^{(3)} = 207$ kN/mm², $A^{(3)} = 100$ mm², $L^{(3)} = 300$ mm, and $\phi^{(3)} = 30$ degrees. The yield stresses of the two materials are

$$\sigma_{\text{aluminum}} = 0.0375 \, \frac{\text{kN}}{\text{mm}^2}$$

$$\sigma_{\text{steel}} = 0.0586 \, \frac{\text{kN}}{\text{mm}^2}$$

These must not be exceeded.

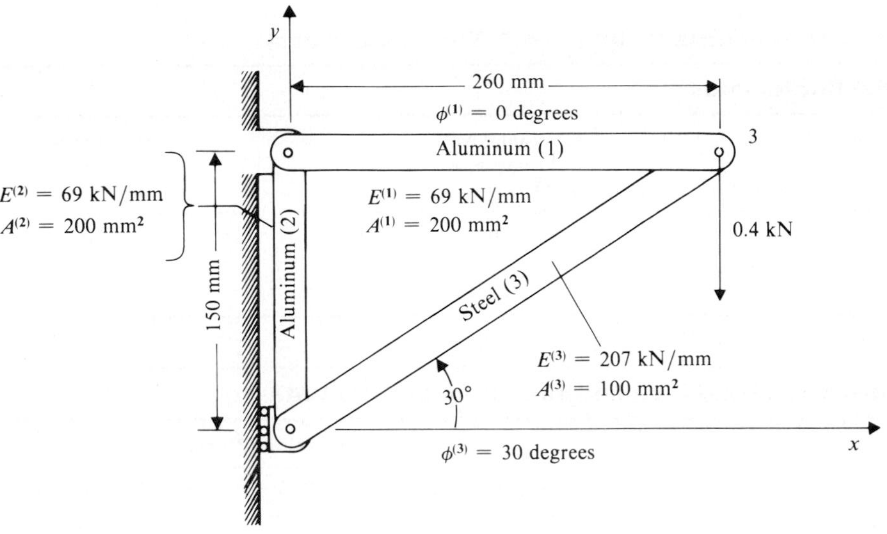

Figure 4.26 Problem geometry (Example 4.7—Triangular truss)

At the wall, joint 1 has no displacement: $u_1 = 0$, $v_1 = 0$. Joint 2 has a roller support, which allows vertical movement but not horizontal: $u_2 = 0$. The problem is statically determinate. Only one prescribed nodal force is given: $F_{y3} = -0.4$ kN. Support forces at nodes 1 and 2 can be found at the end of the solution procedure.

Objective Obtain the displacement of joints 2 and 3 as well as the axial strain, stress, and internal force in each member. Determine if the bars have a design factor of safety of 5 (can support a load of $F_{y3} = -2.0$ kN at node 3).

Solution The number of nodes and elements is NP = 3, NE = 3. Nodal locations are given in Table 4.3(a). The element type and element/node connectivity data are presented in Table 4.3(b). A bar element is labeled as 21 for 2 nodes, with the 1 indicating the linear interpolation function. The nodes k and ℓ are not used here. Element properties E and A are indicated in Table 4.3(c). Prescribed nodal forces and displacements are shown in Tables 4.3(d) and 4.3(e). An exploded view of the elements is presented in Figure 4.27.

For the first element (aluminum), the length is given by (4.37)

$$\begin{aligned} L^{(1)} &= [(X_3 - X_1)^2 + (Y_3 - Y_1)^2]^{1/2} \\ &= [(260 - 0)^2 + (150 - 150)^2]^{1/2} \\ &= 260 \text{ mm} \end{aligned}$$

and the sinusoidal terms are (4.38)

$$s = \sin \phi^{(1)} = \left(\frac{Y_3 - Y_1}{L}\right)^{(1)} = \frac{150 - 150}{260} = 0.0$$

$$c = \cos \phi^{(1)} = \left(\frac{X_3 - X_1}{L}\right)^{(1)} = \frac{260 - 0}{260} = 1.0$$

Table 4.3
Node and Element Data (Example 4.7: Triangular Truss)

(a) Nodal Coordinates

Global Node Number	Horizontal Nodal Coordinate, x (mm)	Vertical Nodal Coordinate, y (mm)
1	0.0	150.0
2	0.0	0.0
3	260.0	150.0

(b) Element Type and Element/Node Connectivity Data

Element Number	Element Type	Global Node Number			
		i	j	k	ℓ
1	21	1	3	—	—
2	21	2	1	—	—
3	21	2	3	—	—

(c) Element Properties

Element Number	Young's Modulus, E (kN/mm^2)	Poisson's Ratio, ν	Cross Sectional Area, A (mm^2)
1	69	—	200
2	69	—	200
3	207	—	100

(d) Prescribed Nodal Forces

Global Node Number	Direction of Force Component ($1=x$, $2=y$)	Magnitude of Applied Force Component, F_x, F_y (kN)
3	2	−0.4

(e) Prescribed Displacements

Global Node Number	Direction of Displacement Component ($1=x$, $2=y$)	Magnitude of Displacement Component, u, v (mm)
1	1	0.0
1	2	0.0
2	1	0.0

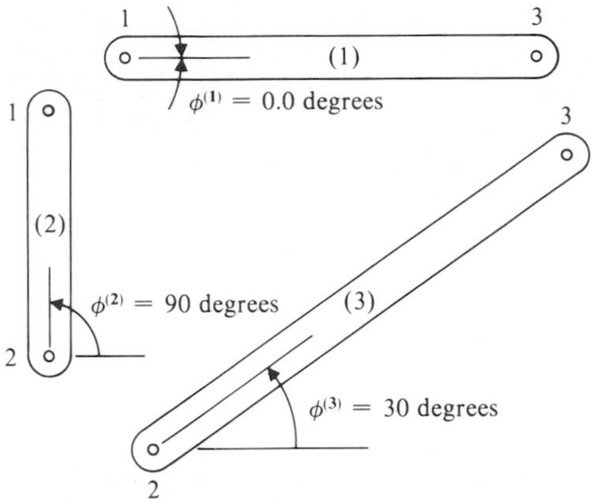

Figure 4.27 Exploded view of elements (Example 4.7—Triangular truss)

The terms in the coordinate transformation matrix (4.43) are $s^2 = 0.0$, $sc = 0.0$, $c^2 = 1.0$. The stiffness term EA/L is (4.39)

$$\left(\frac{EA}{L}\right)^{(1)} = 69\ \frac{\text{kN}}{\text{mm}^2} \times 200\ \text{mm}^2 \times \frac{1}{260\ \text{mm}} = 53.1\ \frac{\text{kN}}{\text{mm}}$$

Now from (4.42) the element matrix has the form

$$\begin{Bmatrix} R_1 \\ R_2 \\ R_5 \\ R_6 \end{Bmatrix}^{(1)} = [B]^{(1)} \begin{Bmatrix} U_1 \\ U_2 \\ U_5 \\ U_6 \end{Bmatrix}^{(1)}$$

where (4.43)

$$[B]^{(1)} = (53.1) \begin{array}{c} \\ 1 \\ 2 \\ 5 \\ 6 \end{array} \begin{array}{c} \begin{array}{cccc} 1 & 2 & 5 & 6 \end{array} \\ \begin{bmatrix} 1.0 & 0.0 & -1.0 & 0.0 \\ 0.0 & 0.0 & 0.0 & 0.0 \\ -1.0 & 0.0 & 1.0 & 0.0 \\ 0.0 & 0.0 & 0.0 & 0.0 \end{bmatrix}^{(1)} \end{array} \frac{\text{kN}}{\text{mm}}$$

This element is aligned with the x axis, as were the elements considered in the previous section. If the rows and columns associated with the x direction (1 and 5) are retained, while those in the y direction (2 and 6) are dropped, the same element matrix (4.35) results, as it should. The element matrix above includes y direction motions as well.

The second element has length $L^{(2)} = 150$ mm, and the s,c terms are

$$s = \sin \phi^{(2)} = \left(\frac{Y_1 - Y_2}{L}\right)^{(2)} = \frac{150 - 0}{150} = 1.0$$

$$c = \cos \phi^{(2)} = \left(\frac{X_1 - X_2}{L}\right)^{(2)} = \frac{0 - 0}{150} = 0.0$$

Then $s^2 = 1.0$, $sc = 0.0$, $c^2 = 0.0$. The stiffness term is

$$\left(\frac{EA}{L}\right)^{(2)} = 69\ \frac{\text{kN}}{\text{mm}^2} \times 200\ \text{mm}^2 \times \frac{1}{150\ \text{mm}} = 92\ \frac{\text{kN}}{\text{mm}}$$

Now the element matrix has the form (with $i = 2, j = 1$)

$$\begin{Bmatrix} R_3 \\ R_4 \\ R_1 \\ R_2 \end{Bmatrix}^{(2)} = [B]^{(2)} \begin{Bmatrix} U_3 \\ U_4 \\ U_1 \\ U_2 \end{Bmatrix}^{(2)}$$

where

$$[B]^{(2)} = (92) \begin{array}{c} \\ 3 \\ 4 \\ 1 \\ 2 \end{array} \begin{array}{c} \begin{array}{cccc} 3 & 4 & 1 & 2 \end{array} \\ \begin{bmatrix} 0.0 & 0.0 & 0.0 & 0.0 \\ 0.0 & 1.0 & 0.0 & -1.0 \\ 0.0 & 0.0 & 0.0 & 0.0 \\ 0.0 & -1.0 & 0.0 & 1.0 \end{bmatrix}^{(2)} \end{array} \frac{\text{kN}}{\text{mm}}$$

Element 2 is aligned with the y axis, so only the terms associated with the y direction (4 and 2) have nonzero entries.

For the third element (steel), the length is

$$\begin{aligned} L^{(3)} &= [(X_3 - X_2)^2 + (Y_3 - Y_2)^2]^{1/2} \\ &= [(260 - 0)^2 + (150 - 0)^2]^{1/2} \\ &= 300\ \text{mm} \end{aligned}$$

and

$$s = \sin \phi^{(3)} = \left(\frac{Y_3 - Y_2}{L}\right)^{(3)} = \frac{150 - 0}{300} = 0.500$$

$$c = \cos \phi^{(3)} = \left(\frac{X_3 - X_2}{L}\right)^{(3)} = \frac{260 - 0}{300} = 0.866$$

Again, the terms in the transformation matrix are $s^2 = 0.250$, $sc = 0.433$, $c^2 = 0.750$. The stiffness term is

$$\left(\frac{EA}{L}\right)^{(3)} = 207\ \frac{\text{kN}}{\text{mm}^2} \times 100\ \text{mm}^2 \times \frac{1}{300\ \text{mm}} = 69.0\ \frac{\text{kN}}{\text{mm}}$$

The element matrix has all nonzero terms of the form

$$\begin{Bmatrix} R_3 \\ R_4 \\ R_5 \\ R_6 \end{Bmatrix}^{(3)} = [B]^{(3)} \begin{Bmatrix} U_3 \\ U_4 \\ U_5 \\ U_6 \end{Bmatrix}^{(3)}$$

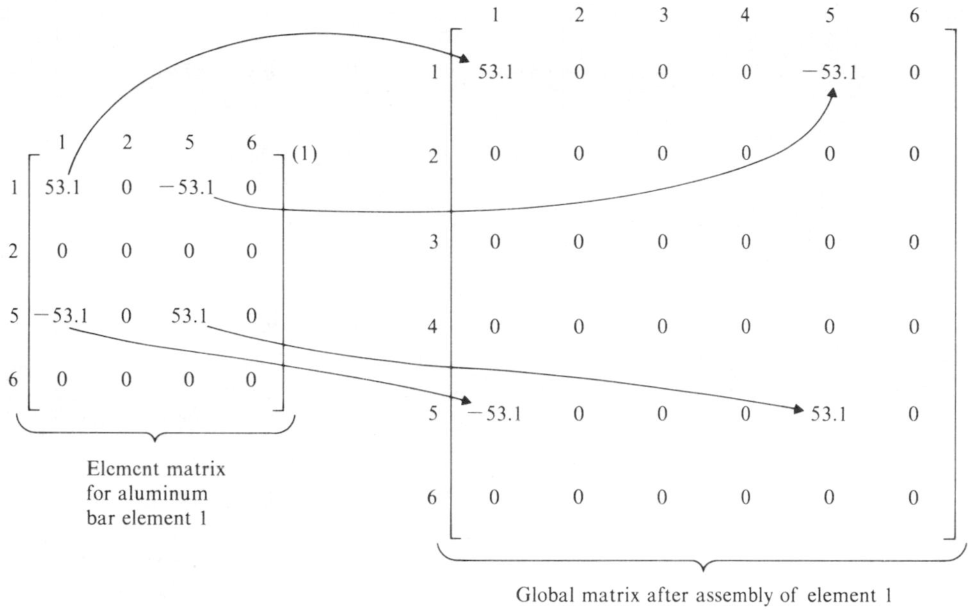

Figure 4.28 Assembly of element 1 (Example 4.7—Triangular truss)

where

$$[B]^{(3)} = (69)\begin{array}{c} \\ 3 \\ 4 \\ 5 \\ 6 \end{array}\overset{\begin{array}{cccc} 3 & \quad 4 & \quad 5 & \quad 6 \end{array}}{\begin{bmatrix} 0.75 & 0.433 & -0.75 & -0.433 \\ 0.433 & 0.25 & -0.433 & -0.25 \\ -0.75 & -0.433 & 0.75 & 0.433 \\ -0.433 & -0.25 & 0.433 & 0.25 \end{bmatrix}}^{(3)} \frac{\text{kN}}{\text{mm}}$$

The element matrices are complete.

Assembling the three elements proceeds nearly as indicated in Section 4.3. Figure 4.28 illustrates the assembly for element 1. The rows and columns (1,2,5,6) corresponding to the node points are added to the global matrix. Arrows indicate the nonzero entries from the element matrix into the global matrix. Similarly Figure 4.29 demonstrates the assembly of element 2.

Figure 4.30 shows the assembly of element 3. All of the terms are nonzero, so they must all be added to the global matrix. Once again, arrows demonstrate the assembly process. Figure 4.31 presents the assembled set of equations with the prescribed force on node 3.

Prescribed nodal displacements must now be included. This is done in the manner used so far (developed in Section 4.3). Figure 4.32 gives the final set of equations.

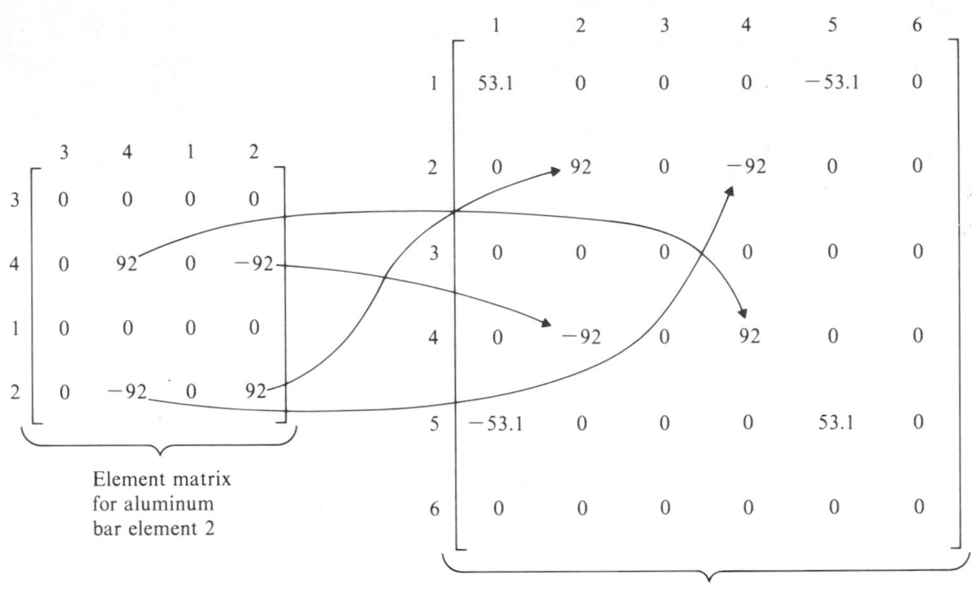

Figure 4.29 Assembly of element 2 (Example 4.7—Triangular truss)

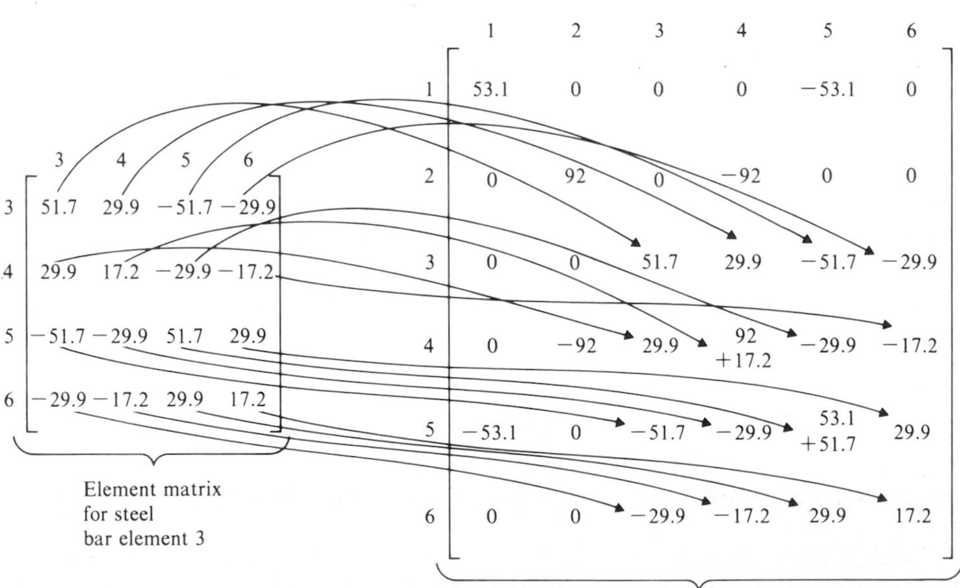

Figure 4.30 Assembly of element 3 (Example 4.7—Triangular truss)

$$\begin{bmatrix} 53.1 & 0 & 0 & 0 & -53.1 & 0 \\ 0 & 92 & 0 & -92 & 0 & 0 \\ 0 & 0 & 51.7 & 29.9 & -51.7 & -29.9 \\ 0 & -92 & 29.9 & 109.2 & -29.9 & -17.2 \\ -53.1 & 0 & -51.7 & -29.9 & 104.8 & 29.9 \\ 0 & 0 & -29.9 & -17.2 & 29.9 & 17.2 \end{bmatrix} \begin{Bmatrix} U_1 \\ U_2 \\ U_3 \\ U_4 \\ U_5 \\ U_6 \end{Bmatrix} = \begin{Bmatrix} 0 \\ 0 \\ 0 \\ 0 \\ 0 \\ -0.4 \end{Bmatrix}$$

Figure 4.31 Nodal equations after assembly (Example 4.7—Triangular truss)

$$\begin{bmatrix} 53.1 & 0 & 0 & 0 & 0 & 0 \\ 0 & 92 & 0 & 0 & 0 & 0 \\ 0 & 0 & 51.7 & 0 & 0 & 0 \\ 0 & 0 & 0 & 109.2 & -29.9 & -17.2 \\ 0 & 0 & 0 & -29.9 & 104.8 & 29.9 \\ 0 & 0 & 0 & -17.2 & 29.9 & 17.2 \end{bmatrix} \begin{Bmatrix} U_1 \\ U_2 \\ U_3 \\ U_4 \\ U_5 \\ U_6 \end{Bmatrix} = \begin{Bmatrix} 0 \\ 0 \\ 0 \\ 0 \\ 0 \\ -0.4 \end{Bmatrix}$$

Figure 4.32 Nodal equations after including prescribed nodal values (Example 4.7—Triangular truss)

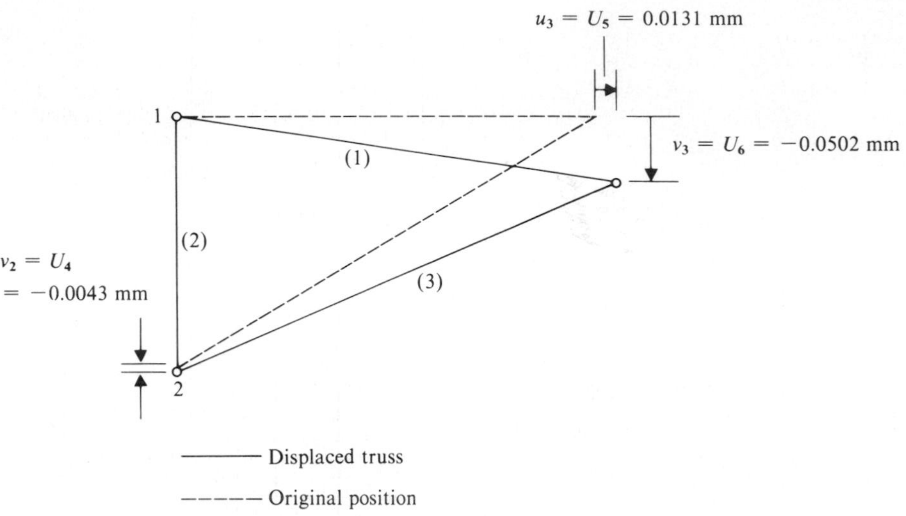

Figure 4.33 Nodal displacements, shown greatly exaggerated (Example 4.7—Triangular truss)

Solution of the final nodal equations is

$$\begin{Bmatrix} U_1 \\ U_2 \\ U_3 \\ U_4 \\ U_5 \\ U_6 \end{Bmatrix} = \begin{Bmatrix} 0 \\ 0 \\ 0 \\ -0.0043 \\ 0.0131 \\ -0.0502 \end{Bmatrix} \text{ mm}$$

As expected, the vertical displacement is downward at node 3. Also, the horizontal displacement is positive (to the right). Figure 4.33 illustrates the results.

For element 1, the change in element length is $\Delta L^{(1)} = 0.0131$ mm and the strain is

$$\epsilon^{(1)} = \left(\frac{\Delta L}{L}\right)^{(1)} = \frac{0.0131 \text{ mm}}{260 \text{ mm}} = 5.02 \times 10^{-5}$$

The element axial stress is

$$\sigma^{(1)} = E^{(1)}\,\epsilon^{(1)} = 69\,\frac{\text{kN}}{\text{mm}^2} \times 5.02 \times 10^{-5} = 0.00347\,\frac{\text{kN}}{\text{mm}^2}$$

and the axial force is

$$R^{(1)} = \sigma^{(1)}\,A^{(1)} = 0.00347\,\frac{\text{kN}}{\text{mm}^2} \times 200 \text{ mm}^2 = 0.693 \text{ kN}$$

Table 4.4
Element Strains and Stresses (Example 4.7: Triangular Truss)

Element Property	Element 1 (Aluminum Bar)	Element 2 (Aluminum Bar)	Element 3 (Steel Bar)
Change in length, ΔL (mm)	0.0131	0.0043	−0.0116
Strain (mm/mm)	5.02×10^{-5}	2.90×10^{-5}	-3.87×10^{-5}
Stress (kN/mm²)	0.00347	0.00200	−0.0080
Element axial force, R (kN)	0.693	0.400	−0.800
Stress with five times load	0.0174	0.0100	−0.040
Yield stress	0.0375	0.0375	0.0586

Comparing the design stress of 0.00347 kN/mm² to the yield stress of 0.0375 kN/mm² indicates a safe value. Even in the case of five times the load, the design stress would be 0.0174 kN/mm², which is well below the yield stress. A summary of all of these figures is given in Table 4.4. The strains, stresses, and forces for element 2 are given in Table 4.4.

The change in length in element 3 is given by

$$\Delta L^{(3)} = u_3 \cos\phi + v_3 \sin\phi - v_2 \sin\phi$$
$$= 0.0131 \cos 30 - 0.0502 \sin 30 + 0.0043 \sin 30$$
$$= -0.0116$$

and the strain is

$$\epsilon^{(3)} = \left(\frac{\Delta L}{L}\right)^{(3)} = \frac{-0.0116 \text{ mm}}{300 \text{ mm}} = -3.87 \times 10^{-5}$$

The element axial stress is

$$\sigma^{(3)} = E^{(3)}\,\epsilon^{(3)} = 207\,\frac{\text{kN}}{\text{mm}^2} \times (-3.87 \times 10^{-5}) = -0.0080\,\frac{\text{kN}}{\text{mm}^2}$$

and the axial force is

$$R^{(3)} = \sigma^{(3)}\,A^{(3)} = -0.0080\,\frac{\text{kN}}{\text{mm}^2} \times 100 \text{ mm}^2 = -0.800 \text{ kN}$$

For element 3, the design stress is −0.0080 kN/mm², and the factor of safety of five gives a stress of −0.040 kN/mm². This is below the yield stress for steel of 0.0586 kN/mm². The steel bar is closer to failure than the aluminum bar.

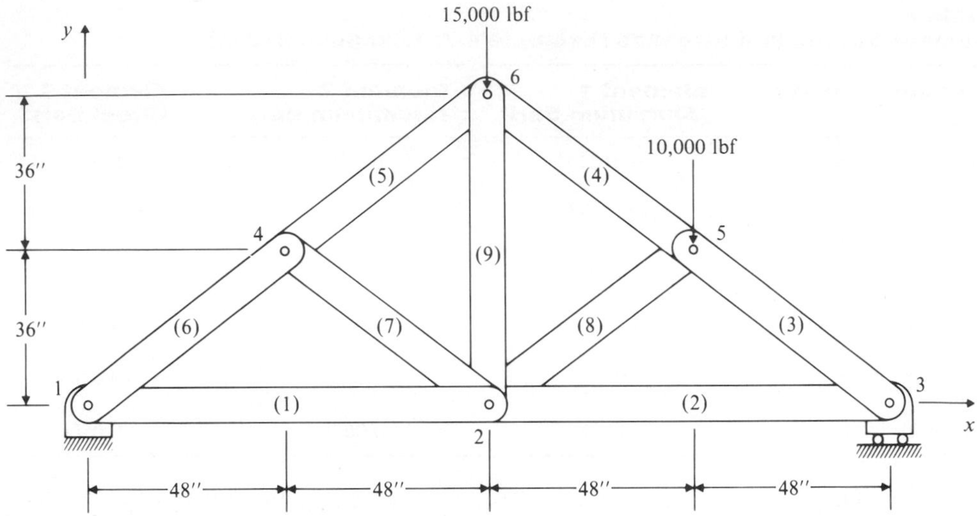

Figure 4.34 Problem diagram (Example 4.8—Bridge truss)

Example 4.8 *Bridge Truss*

Given A simple plane bridge truss consisting of rigid bars connected by pinned joints is shown in Figure 4.34. The origin of the x,y coordinate system is taken at node 1. It is pinned at node 1 and acts on a roller support at node 3. All members are made of aluminum with Young's modulus of

$$E = 10 \times 10^6 \frac{\text{lbf}}{\text{in}^2}$$

The outer members all have a cross sectional area of $A = 4\ \text{in}^2$ while all of the interior members have a cross sectional area of $A = 2\ \text{in}^2$. Two forces are applied as shown: $F_{y6} = -15{,}000$ lbf and $F_{y5} = -10{,}000$ lbf. Both are taken as vertically downward for this example but other forces could be prescribed in the horizontal direction as well.

Two cases are to be treated in this example.

1. Right end roller supported (statically determinate)
2. Right end pinned (statically indeterminate)

The displacement at the left end, node 1, is prescribed as $u_1 = 0.0$, $v_1 = 0.0$ for both of the above cases. For case 1, the right end is on a roller support, as shown in Figure 4.34. The prescribed displacement is $v_3 = 0.0$. These three specified nodal displacements would allow the support and all internal forces to be calculated from the three equations of static equilibrium. Actually STRESS does it, so there is no need to do it by hand. For case 2, the right end is pinned so both the horizontal and vertical displacements are set to zero: $u_3 = 0.0$, $v_3 = 0.0$. The four boundary conditions overconstrain the truss, making it statically indeterminate.

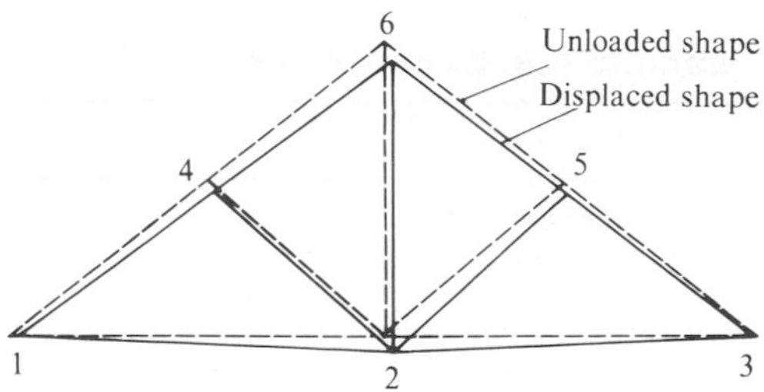

Figure 4.35 Displaced bridge shape for Case 1 (roller right end) (Example 4.8—Bridge truss)

Objective Analyze this bridge truss, using simplex bar elements with program STRESS. Obtain the displacements at each node. Also determine the axial strains and stresses in each bar.

Solution The program input, described in Appendix C, starts with Number of nodes, NP = 6; Number of elements, NE = 9; Plane truss indicator, NAN = 3. Figure 4.34 illustrates the nodal and element numbering. The nodal coordinates shown in Table 4.5(a) are input. Element type and element/node connectivity data are given in Table 4.5(b). The calculated bandwidth, NBW, is 10 for this problem.

Element properties must be input next. The coefficient parameter indicates that the element properties are not all the same, so ICO = 2. Table 4.5(c) gives the Young's modulus and cross sectional area for each element. Note that the length of the element is calculated automatically by the program from the nodal coordinates.

The number of prescribed nodal forces, NPF, is 2 for this problem. The nodal forces are indicated in Table 4.5(d) including the node number, direction of force, and force component in that direction. For case 1 with the roller right end, the number of prescribed nodal displacements, NPD, is 3. Table 4.5(e) shows the values used for the node number, direction, and prescribed displacement component.

Finally a suitable output parameter NOUT is chosen. All of the data has been input to the program. Element matrices and output, if desired by the user, are calculated by program STRESS. Each element is assembled into the global stiffness matrix. Prescribed nodal forces and displacements are taken into account as described earlier in this chapter. The final equations are solved and nodal displacements calculated.

The nodal displacements are given in Table 4.6. Figure 4.35 plots the unloaded (undeformed) shape as dashed lines and the displaced shape as solid lines. As expected, the truss moved horizontally as well as vertically, even though the load was only applied vertically.

Element strains and stresses are automatically calculated by program STRESS. Table 4.7 gives the results. The axial force is easily obtained by multiplying the stress by the bar cross sectional area. Numerical values are given in Table 4.7 as well. Figure 4.36 illustrates the forces on the truss drawing, where tension is designated (T) and compression is designated (C). Support forces may be obtained by summing the element forces at nodes 1 and 3.

Table 4.5
Node and Element Data (Example 4.8: Bridge Truss)

(a) Nodal Coordinates

Global Node Number	Horizontal Nodal Coordinate, x (inches)	Vertical Nodal Coordinate, y (inches)
1	0.0	0.0
2	96.0	0.0
3	192.0	0.0
4	48.0	36.0
5	144.0	36.0
6	96.0	72.0

(b) Element Type and Element/Node Connectivity

Element Number	Element Type	Global Node Number			
		i	j	k	ℓ
1	21	1	2	—	—
2	21	2	3	—	—
3	21	3	5	—	—
4	21	5	6	—	—
5	21	6	4	—	—
6	21	4	1	—	—
7	21	4	2	—	—
8	21	5	2	—	—
9	21	6	2	—	—

(c) Element Properties

Element Number	Young's Modulus, E (lbf/in^2 $\times$ 10^7)	Poisson's Ratio, ν	Cross Sectional Area, A (in^2)
1	1.0	—	4.0
2	1.0	—	4.0
3	1.0	—	4.0
4	1.0	—	4.0
5	1.0	—	4.0
6	1.0	—	4.0
7	1.0	—	2.0
8	1.0	—	2.0
9	1.0	—	2.0

Table 4.5 *Continued*

(d) Prescribed Nodal Forces

Global Node Number	Direction of Force Component ($1 = x$, $2 = y$)	Magnitude of Applied Force Components, F_x, F_y (lbf)
5	2	−10,000
6	2	−15,000

(e) Prescribed Nodal Displacements for Case 1 (Roller Right End)

Global Node Number	Direction of Displacement Component ($1 = x$, $2 = y$)	Magnitude of Displacement Component, u, v (in)
1	1	0.0
1	2	0.0
3	2	0.0

Table 4.6
Nodal Displacements for Case 1 (Roller Right End)
(Example 4.8: Bridge Truss)

Node	Horizontal Displacement, u (inches)	Vertical Displacement, v (inches)
1	0.0	0.0
2	0.03200	− 0.16508
3	−0.08000	0.0
4	0.06228	− 0.12471
5	0.00191	− 0.16663
6	0.04781	− 0.14708

In Case 2, the right-end roller support is replaced by a pinned end. No schematic figure of this is shown. The prescribed nodal displacements are given in Table 4.8. This is the replacement for Table 4.5(e) and is the only input change that needs to be made for program STRESS.

Nodal displacements are once again output by STRESS. The results are given in Table 4.9(a). The displaced shape is indicated in Figure 4.37 by the solid lines. It is quite different from Case 1, primarily at nodes 2 and 5, which are shifted to the left. Similarly the strains, stresses, and axial forces are presented in Table 4.10. The axial forces along each bar change only along the two bars between the supports. All the rest are unchanged.

Table 4.7
Element Strains, Stresses, and Axial Forces for Case 1 (Example 4.8: Bridge Truss)

Element	Element Strain, ϵ (in/in $\times$ 10^4)	Element Stress, σ (lbf/in²)	Element Axial Force, R (lbf)
1	3.333	3333	13,333
2	5.000	5000	20,000
3	−6.250	−6250	−25,000
4	−4.167	−4167	−16,667
5	−4.167	−4167	−16,667
6	−4.167	−4167	−16,667
7	0.0	0.0	0.0
8	−4.167	−4167	− 8,333
9	2.500	2500	5,000

Table 4.8
Prescribed Nodal Displacements for Case 2 (Pinned Right End) (Example 4.8: Bridge Truss)

Global Node Number	Global Direction of Displacement (1=x, 2=y)	Component of Prescribed Nodal Displacement (Inches)
1	1	0.0
1	2	0.0
3	1	0.0
3	2	0.0

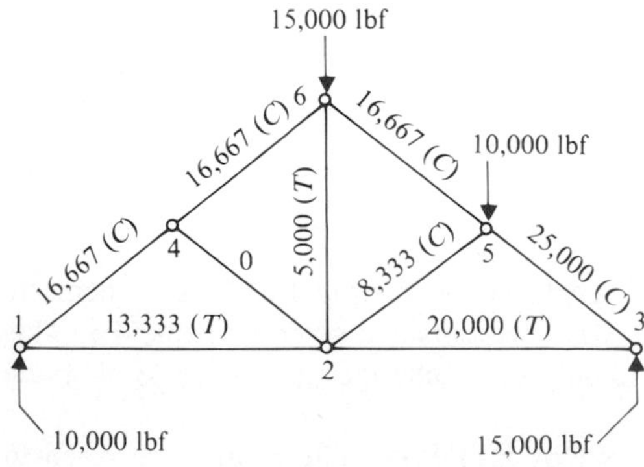

Figure 4.36 Axial forces in bar elements for Case 1 (Example 4.8—Bridge truss)

Table 4.9
Nodal Displacements for Case 2 (Pinned Right End)
(Example 4.8: Bridge Truss)

Node	Horizontal Displacement, u (Inches)	Vertical Displacement, v (Inches)
1	0.0	0.0
2	− 0.00800	− 0.11175
3	0.0	0.0
4	0.02228	− 0.07137
5	− 0.03809	− 0.11329
6	0.00781	− 0.09375

Table 4.10
Element Strains, Stresses, and Forces for Case 2
(Example 4.8: Bridge Truss)

Element	Element Strain, (in/in)	Element Stress, (lbf/in²)	Element Axial Force, R (lbf)
1	−0.833	−833	3,333
2	0.833	833	3,333
3	−6.250	−6250	−25,000
4	−4.167	−4167	−16,667
5	−4.167	−4167	−16,667
6	−4.167	−4167	−16,667
7	0.0	0.0	0.0
8	−4.167	−4167	− 8,333
9	2.500	2500	5,000

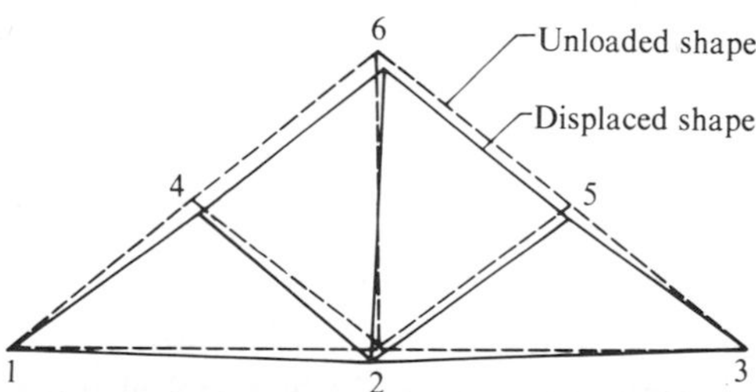

Figure 4.37 Displaced bridge shape for Case 2 (pinned right end) (Example 4.8—Bridge truss)

In summary, the finite element method is used in this example to solve two different cases with little difficulty. All of the methods developed in this chapter were used (actually incorporated into program STRESS). Any plane truss problem can be solved in a similar manner.

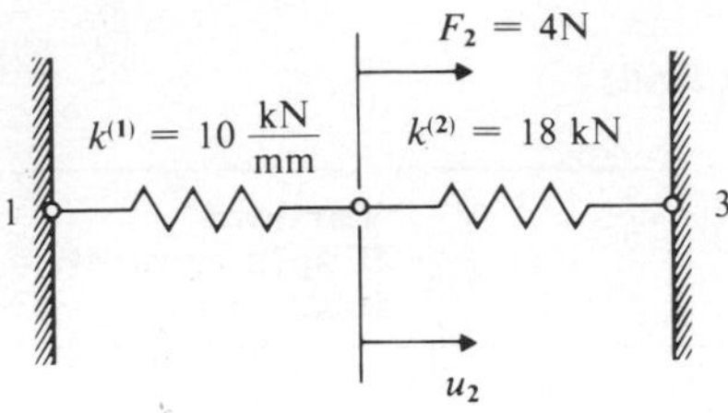

Figure 4.38 Diagram for Problem 4.1

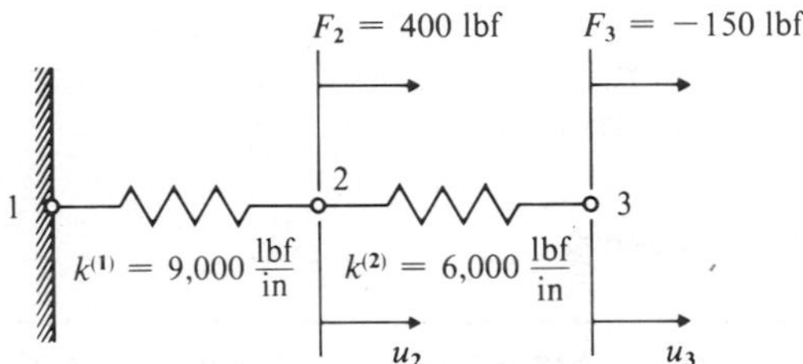

Figure 4.39 Springs for Problem 4.2

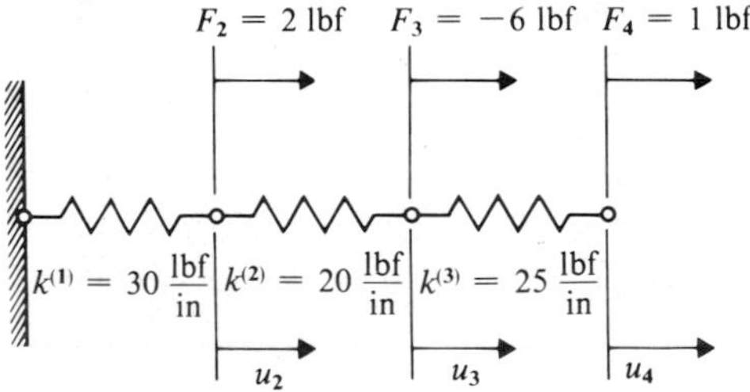

Figure 4.40 Diagram for Problem 4.3

4.7 Problems

A. Springs and Networks

4.1 Two springs, shown in Figure 4.38, are fixed at both ends. A force of 4 N is applied at the center node. Determine the displacement of the center node, using the finite element method.

4.2 Figure 4.39 shows two springs in series. Determine the displacements of the two nodes, using the element matrix and assembly techniques.

4.3 A series of springs and applied forces is illustrated in Figure 4.40. Determine the displacement of nodes 2,3,4, using the element matrix and assembly technique.

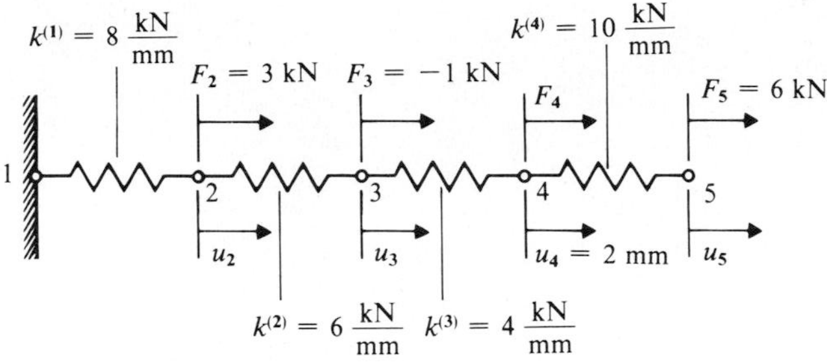

Figure 4.41 Springs for Problem 4.4

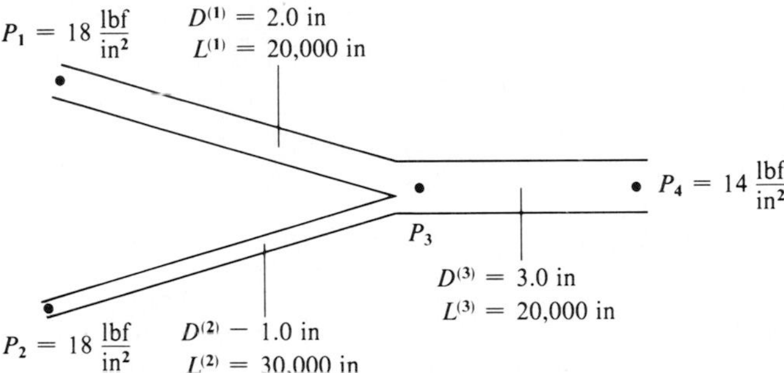

Figure 4.42 Pipe network for Problem 4.5

4.4 Four springs are shown in Figure 4.41. Three forces are applied as indicated. Two displacements are prescribed. Determine the displacements of nodes 2,3,5 and the force on node 4, using the element matrix and assembly method.

4.5 An incompressible fluid flows through a pipe network shown in Figure 4.42. The fluid has viscosity and density $\mu = 2.0 \times 10^{-6}$ lbf-sec/in², $\rho = 1.93$ slugs/ft³. Determine the pressure at node 3 and the volume rate of flow through each pipe. Also find the Reynolds number for each pipe to determine whether the flow is laminar.

4.6 A simple electrical network is indicated in Figure 4.43 with a battery and three resistors. Assume that node 1 is grounded (zero volts). Determine the voltage at node 2 and the current through each resistor, using the element matrix and assembly procedure.

4.7 Flow occurs in the eight-pipe network shown in Figure 4.44. The water in the pipe network has the viscosity $\mu = 1.5 \times 10^{-6}$ lbf-sec/in² and density $\rho = 1.93$ slugs/ft³. Determine the pressures at each of the four unknown node points and the volume flow rate in each pipe. (The resulting equations must be solved with a computer routine.) Verify that the pipe with the largest flow rate has a laminar flow.

4.8 An electrical network is shown in Figure 4.45. Node 2 is grounded (0 volts); node 1 is at 50 volts; node 5 is grounded; node 6 is at 80 volts. Find the nodal voltages and the current through each resistor. (Only two unknown nodes are present, so this problem is easily solved by hand.)

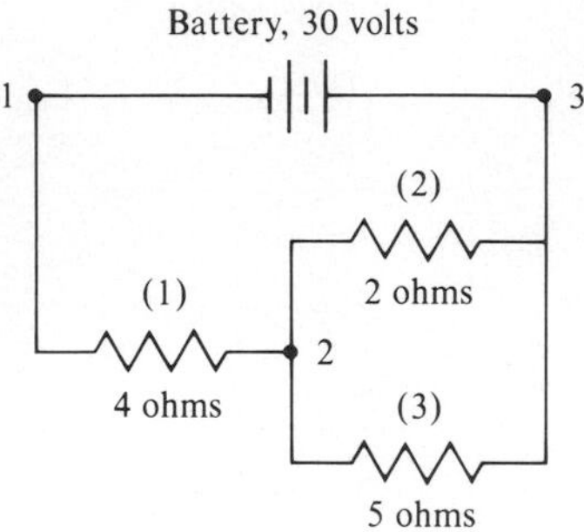

Figure 4.43 Electrical network for Problem 4.6

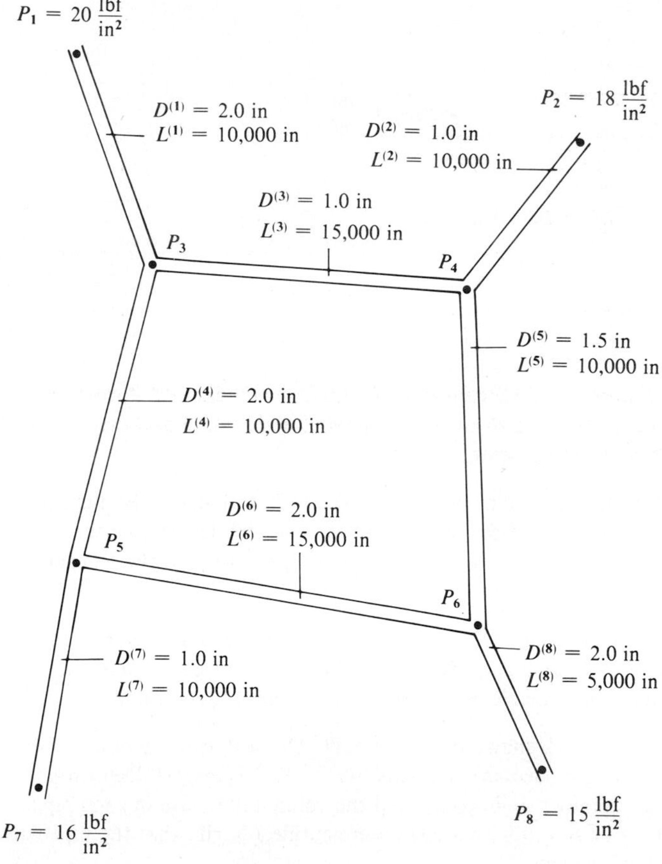

Figure 4.44 Pipe network for Problem 4.7

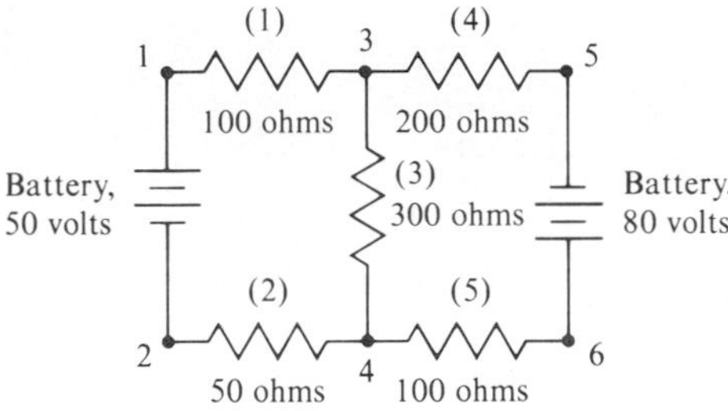

Figure 4.45 Electrical network for Problem 4.8

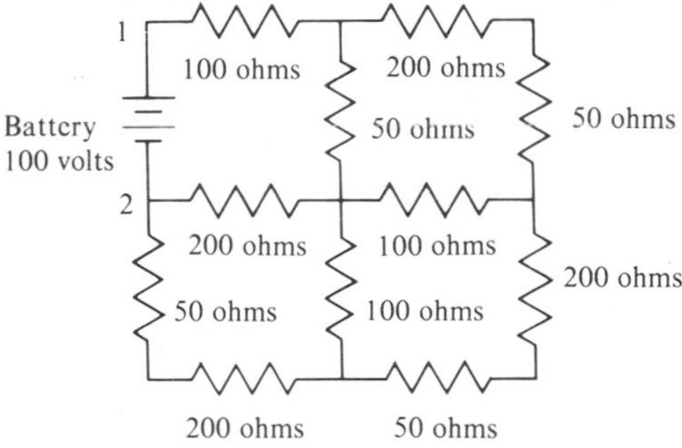

Figure 4.46 Electrical network for Problem 4.9

4.9 Figure 4.46 shows a large electrical network. Node 2 is grounded (0 volts) and node 1 is at 100 volts. Determine the nodal voltages and current through each resistor. (The resulting equations must be solved with a computer routine.)

4.10 Make up a problem of your own choice involving aligned springs. Include at least two unknown nodes.

4.11 Make up a problem of your own choice involving a pipe or electrical network. Include at least two unknown node points.

B. Aligned Bars

4.12 A bronze bar and a steel bar are shown in Figure 4.47. Calculate the displacements, strains, and stresses in the two bars with the finite element method.

4.13 A concrete support with variable cross sectional area is indicated in Figure 4.48. Use a three-element model to solve for the displacement and stress in the support at the locations 9 inches and 18 inches from the left end. Do not use program STRESS.

4.14 Do Problem 4.13 with $E = 3.0 \times 10^7$ lbf/in^2 (steel).

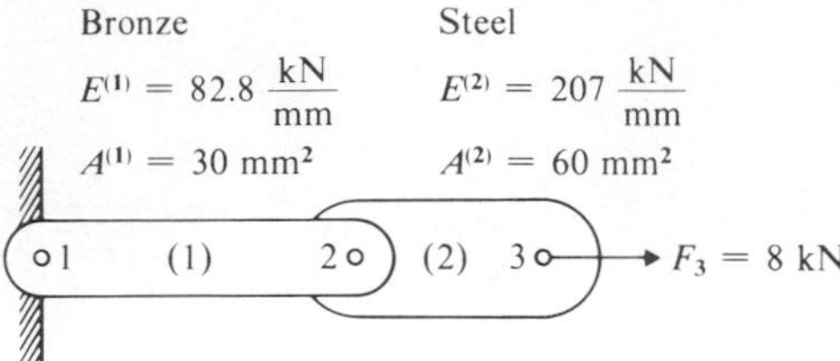

Figure 4.47 Diagram for Problem 4.12

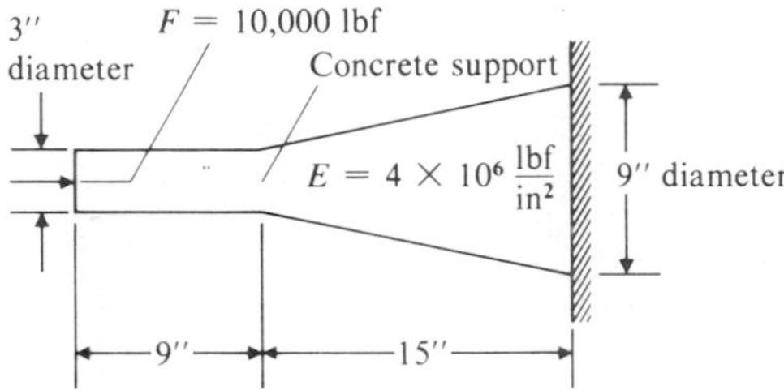

Figure 4.48 Diagram for Problem 4.13

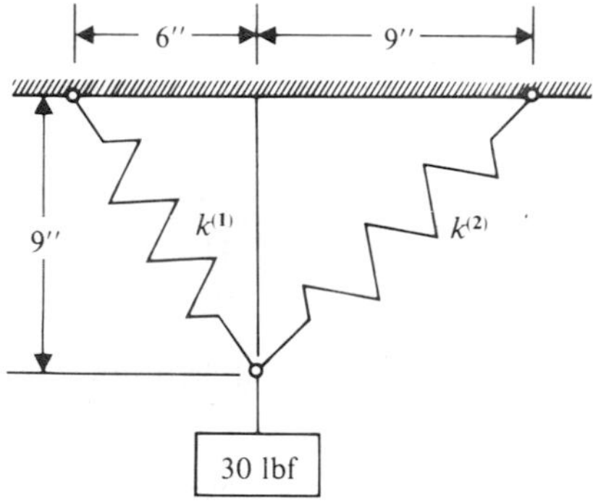

Figure 4.49 Diagram for Problem 4.15

C. Trusses

4.15 A set of two springs is shown in Figure 4.49 in their initial undeformed position. The springs have stiffnesses $k^{(1)} = 140$ lbf/in, $k^{(2)} = 200$ lbf/in. Model the springs as equivalent bars in a truss. Determine the displacement of the lower node point after a weight of 30 lbf is applied. Use the finite element method to obtain the solution. Solve by hand to become more familiar with the method.

4.16 Three bars are subject to a force of 12,000 lbf, as shown in Figure 4.50. All three bars are made of aluminum; Young's modulus and cross sectional area are $E = 10 \times 10^6$ lbf/in², $A = 6.5$ in². Use the finite element method to obtain the stresses in the three bars. Carry out the analysis by hand.

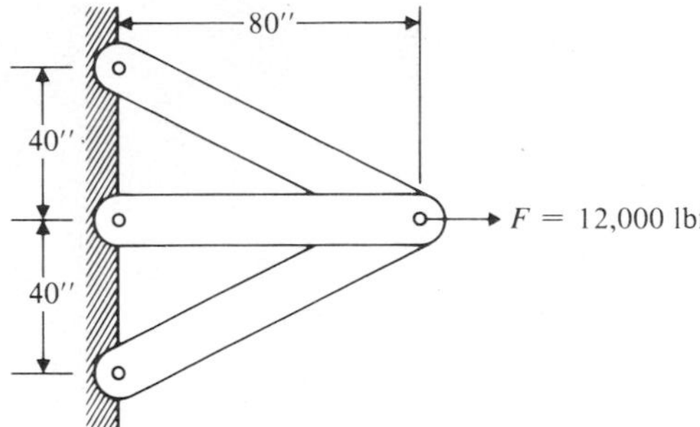

Figure 4.50 Diagram for Problem 4.16

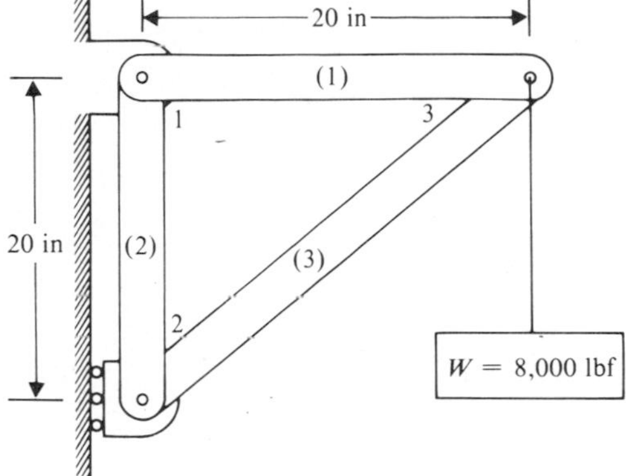

Figure 4.51 Diagram for Problem 4.17

4.17 A three-bar truss supports an 8,000-lbf load, as shown in Figure 4.51. The properties for the bars are

$$\left.\begin{aligned} E^{(1)} &= 30.0 \times 10^6 \text{ lbf/in}^2 \\ A^{(1)} &= 3.0 \text{ in}^2 \end{aligned}\right\} \quad \text{Steel}$$

$$\left.\begin{aligned} E^{(2)} &= 4.0 \times 10^6 \text{ lbf/in}^2 \\ A^{(2)} &= 4.0 \text{ in}^2 \end{aligned}\right\} \quad \text{Concrete}$$

$$\left.\begin{aligned} E^{(3)} &= 4.0 \times 10^6 \text{ lbf/in} \\ A^{(3)} &= 4.0 \text{ in}^2 \end{aligned}\right\} \quad \text{Concrete}$$

Use the finite element method to solve for the nodal displacements, strain in each bar, and stress in each bar. Carry out the solution by hand to become familiar with each step.

4.18 Figure 4.52 illustrates a truss subjected to two loads, as shown. All five bars are made of steel with the properties $E = 207$ kN/mm², $A = 100$ mm². Use computer program STRESS to solve for the nodal displacements. Also determine the strain, stress, and axial force in each element.

4.19 Solve Problem 4.18 with the roller support at the left end replaced by a pinned support.

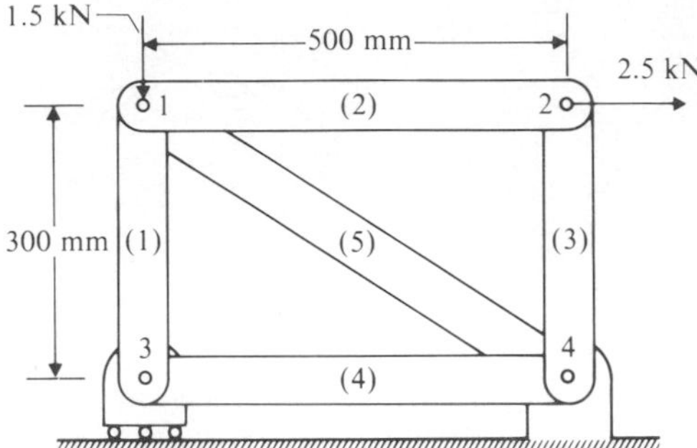

Figure 4.52 Diagram for Problem 4.18

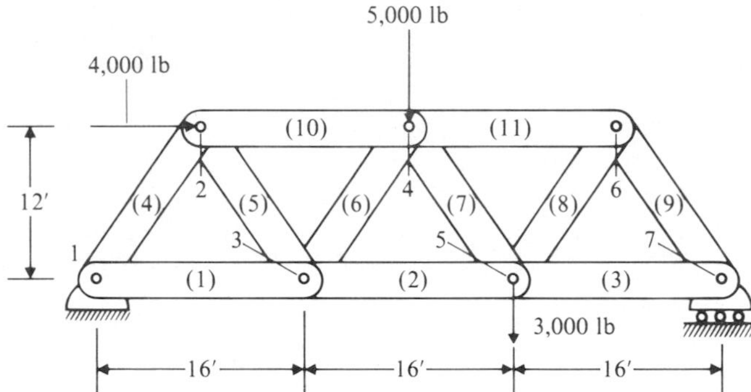

Figure 4.53 Plane bridge truss for Problem 4.20

4.20 A plane truss bridge, shown in Figure 4.53, has three loads applied to it. The five horizontal members all have the properties $E = 30 \times 10^6$ lbf/in², $A = 6.0$ in², while the six connecting members have the properties $E = 30 \times 10^6$ lbf/in², $A = 3.0$ in². Model the bridge truss using program STRESS. Determine the nodal displacements and axial stresses in each member. Sketch the bridge, indicating stress on each bar.

4.21 An overhanging roof truss is shown in Figure 4.54. Each of the truss members are made of steel with the properties $E = 30 \times 10^6$ lbf/in², $A = 3.0$ in². Model the roof truss using program STRESS. A distributed snow load of 2,000 lbf/ft is to be applied to the roof truss. Assume that one half of the total load on each of the three upper bars is applied at each end of that bar. Sketch the truss and the resulting displacements of the node points. Show the stresses in the bars.

4.22 Re-solve Problem 4.21 but move the joint (node) 5 vertically upward one foot. What is the change in stress in the bars near node 5?

4.23 A K bridge truss is shown in Figure 4.55. Each bar in the truss has the properties $E = 30 \times 10^6$ lbf/in², $A = 4.0$ in². Model the bridge using program STRESS. Use symmetry about the center line of the bridge if you can. Sketch the resulting displacements of the node points and indicate the stress in each bar.

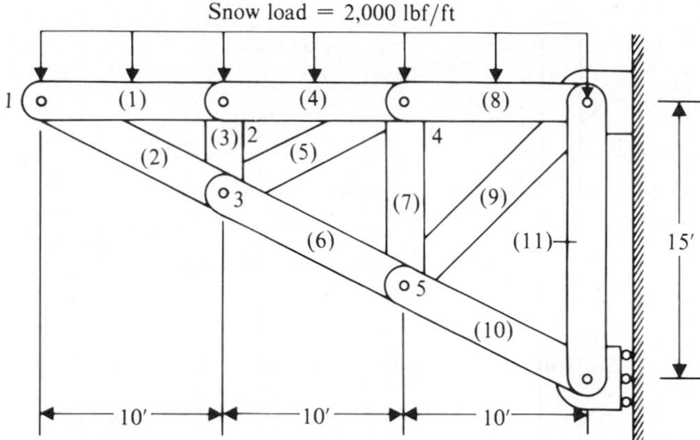

Figure 4.54 Roof truss for Problem 4.21

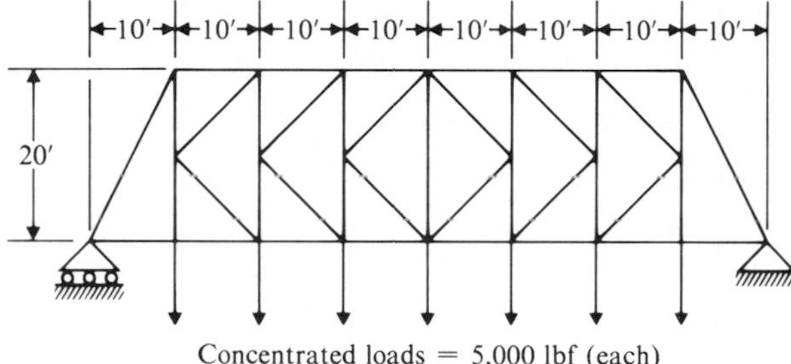

Figure 4.55 K bridge truss for Problem 4.23

4.24 A counterweighted crane for lifting building materials of weight W is shown in Figure 4.56. All bars are made of steel with properties $E = 207$ kN/mm², $A = 200$ mm². The steel cable has the same Young's modulus and cross sectional area $A = 100$ mm². The cable is prestressed to provide a stiffness equivalent to a bar of the same E and A. The counterweight is 1.2 kN and the lifted weight is $W_L = 0.75$ kN. Treat the counterweight as one half of the 1.2 kN applied at each node. Use program STRESS to determine the stress in each bar and the cable. Indicate the maximum stress. Sketch the crane, indicating the stress in each member.

4.25 Solve Problem 4.24, but increase the weight W_L until the maximum stress occurs somewhere in the crane.

$$\sigma_{\mathbf{max}} = 0.0586 \text{ kN/mm}^2$$

This determines the peak allowable load (including the safety factor).

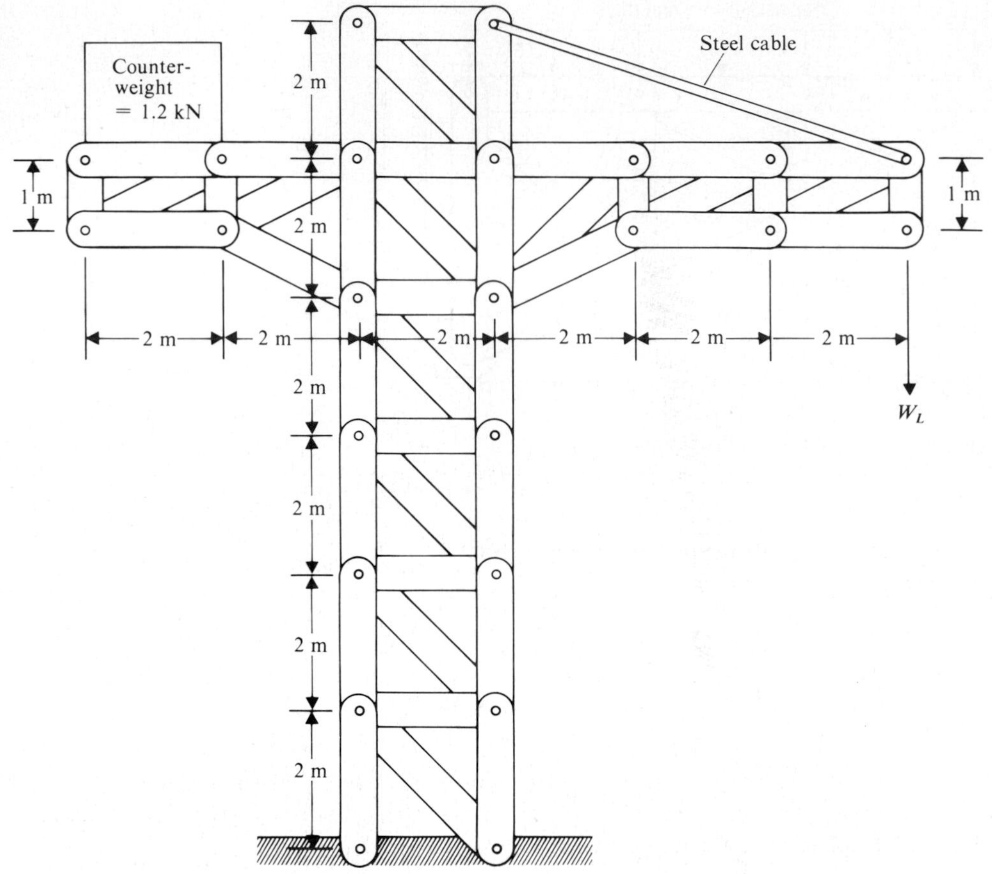

Figure 4.56 Diagram of counterweighted crane truss for Problems 4.24 and 4.25

Part 2
Ordinary Differential Equations

Chapter 5

Solid Mechanics, Heat Transfer, Fluid Mechanics—The Variational Method

5.1 Introduction

The direct method was used in Chapter 4 to obtain element matrices for springs, bars, and trusses. This method becomes very difficult to use for more advanced problems; therefore it is not used for much more in this text. As indicated in Chapter 1, the variational method is widely used. This chapter introduces the variational method as employed in finite element analysis.

A. Functional

The variational method is concerned with the variations of a quantity called a functional I. The functional is not dependent upon position or coordinate direction within a physical problem. In fact, the name functional is used to distinguish it from properties such as displacement or temperature, which are functions of position. In many engineering applications, the functional corresponds to a physical concept. Solid mechanics problems are usually based upon the potential energy, which has a single value for the entire structure.

Generally the objective of finding the solution to engineering problems using a variational method is to find a stationary value of I. As in elementary calculus, a stationary point can be any one of three possibilities: a minimum, an inflection point, or a maximum. Figure 5.1 illustrates the possible stationary points of the functional I. All of the topics in this text produce a minimum value of I.

B. First and Second Variations

Let δI represent a small change of I away from a minimum value. This is called the variation of the functional. A stationary value of I is expressed in equation form as

$$\delta I = 0 \tag{5.1}$$

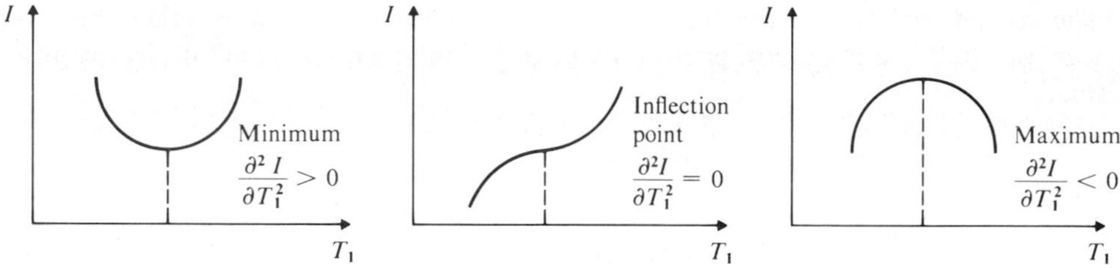

Figure 5.1 Minimum, maximum, and inflection stationary points

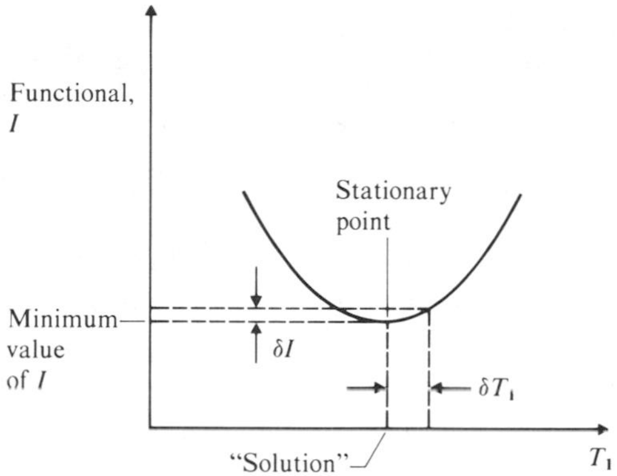

Figure 5.2 Variations of T_1 and I

In finite element analysis, only an approximate solution is obtained to any problem. Thus the minimum of I is only approximately attained. Consider a simple problem with one node point and one nodal unknown T_1. This minimum of the functional is determined by the relation

$$\delta I = \frac{\partial I}{\partial T_1} \delta T_1 = 0 \tag{5.2}$$

Here δT_1 is a small variation of the nodal value (away from that value which produces the smallest value of I). Figure 5.2 illustrates the variations of the unknown nodal value and functional. The actual minimum is given by the equation

$$\frac{\partial I}{\partial T_1} = 0 \tag{5.3}$$

which produces an approximate solution to the problem. This is the minimization equation for the finite element formulation.

The second variation of the functional can be used to determine whether the stationary value of I is a minimum or maximum. It is similar to the second derivative test in calculus.

$$\frac{\partial^2 I}{\partial T_1^2} > 0 \text{ (minimum)}$$

$$\frac{\partial^2 I}{\partial T_1^2} = 0 \text{ (inflection point)}$$

$$\frac{\partial^2 I}{\partial T_1^2} < 0 \text{ (maximum)}$$

Often this test can be applied to engineering problems.

If there are N unknown nodal values, denoted by $T_1, T_2, \ldots, T_N$, the variation of the functional is given by

$$\delta I = \frac{\partial I}{\partial T_1}\delta T_1 + \frac{\partial I}{\partial T_2}\delta T_2 + \frac{\partial I}{\partial T_3}\delta T_3 + \ldots + \frac{\partial I}{\partial T_N}\delta T_N = 0 \qquad \textbf{(5.4)}$$

Each of the nodal variations $\delta T_1, \delta T_2, \delta T_3, \ldots, \delta T_N$ is a small change for that particular nodal value while the others are held constant. The above relation, which is satisfied if

$$\begin{aligned}
&\textit{Node 1} && \frac{\partial I}{\partial T_1} = 0 \\
&\textit{Node 2} && \frac{\partial I}{\partial T_2} = 0 \\
&\textit{Node 3} && \frac{\partial I}{\partial T_3} = 0 \\
&\ldots \\
&\textit{Node N} && \frac{\partial I}{\partial T_N} = 0
\end{aligned} \qquad \textbf{(5.5)}$$

gives a set of nodal solutions $T_1, T_2, \ldots, T_N$. These are the N minimization equations for the N unknown nodal values.

The stationary value of I obtained by the minimization equations may or may not be a minimum. This can be determined by applying the second derivative test at each node point. For a minimum to occur at all node points

$$\frac{\partial^2 I}{\partial T_1^2} > 0$$

$$\frac{\partial^2 I}{\partial T_2^2} > 0$$

$$\ldots$$

$$\frac{\partial^2 I}{\partial T_N^2} > 0$$

Usually this is satisfied for physical problems found in engineering applications. Alternatively all of the second derivatives could be less than zero, giving a set of maximization equations. If some are positive and some are negative, the result is a set of multidimensional inflection points. They are called "saddle points," which have a minimum with respect to one value and a maximum with respect to another.

Example 5.1 *Stationary Value of a Functional*

Given Consider the functional $I = 4T_1^2 + 2T_2^2 - 3T_1 - T_2$, which is a function of both nodal values T_1 and T_2.

Objective Obtain the values of T_1 and T_2 that make I a stationary value. Determine whether this value is a minimum or a maximum.

Solution The first variation δI is given by (5.4)

$$\delta I = \frac{\partial I}{\partial T_1}\delta T_1 + \frac{\partial I}{\partial T_2}\delta T_2 = 0$$

which is satisfied if

$$\frac{\partial I}{\partial T_1} = 0$$

$$\frac{\partial I}{\partial T_2} = 0$$

Substituting for the functional I yields

$$\frac{\partial I}{\partial T_1} = \frac{\partial}{\partial T_1}(4T_1^2 + 2T_2^2 - 3T_1 - T_2) = 0$$

$$\frac{\partial I}{\partial T_2} = \frac{\partial}{\partial T_2}(4T_1^2 + 2T_2^2 - 3T_1 - T_2) = 0$$

and evaluating the derivatives gives $8\,T_1 - 3 = 0$, $4\,T_2 - 1 = 0$. Solving for T_1 and T_2 gives $T_1 = 0.375$, $T_2 = 0.25$. These give the stationary value of I.

The second variation is

$$\frac{\partial^2 I}{\partial T_1^2} = 8, \quad \frac{\partial^2 I}{\partial T_2^2} = 2$$

Both are positive, so the stationary value is a minimum.

5.2 Solid Mechanics—Spring Examples

The general discussion of functionals presented in the preceding section is rather mathematical in nature. This section gets more specific for solid mechanics problems. Springs are used as examples.

A. Solid Mechanics Problems

In solid mechanics problems, the functional involved is the total potential energy I. The total potential energy is due to two sources: the internal strain energy U due to elastic deformation and the potential W of the external load to do work. The functional is

$$I = U + W \tag{5.6}$$

where I = total potential energy, U = internal strain energy due to deformation, and W = potential of the external force to do work. The physical significance of U and W is discussed in some detail.

The governing principle is the principle of minimum potential energy. It can be shown in general that the total potential energy of an elastically deformed body in equilibrium is stationary. Further, it can be shown that the energy is always a minimum for stable (not moving) bodies.

Finite element analysis is carried out for structures that are already deformed. Only linearly elastic materials are considered in this text. The internal strain energy U is the work that could be recovered if the structure were allowed to return to its undeformed position. Since the structure is linearly elastic, U is simply the negative of the work put in originally.

A simple spring, such as discussed in Chapter 4, has a displacement at one end denoted by u. The recoverable work is the integral of the internal reaction force R over the distance moved.

$$U = \int_0^u R du \tag{5.7}$$

This has the units of force times length, or work. If the spring is linear with stiffness k, $R = ku$, and the internal strain energy is

$$U = \int_0^u kudu = \frac{1}{2}ku^2 \tag{5.8}$$

This is discussed in much more detail in the examples.

If the spring endpoints have displacements u_i and u_j, the actual deformation of the spring is $u = -u_i + u_j$, and the internal strain energy becomes

$$U = \frac{1}{2}k(-u_i + u_j)^2 \tag{5.9}$$

Figure 5.3 shows a series of E springs with displacements $u_1, u_2, u_3, \ldots, u_N$. The internal strain energy for all of the springs is

$$U = \sum_{e=1}^{E} \left[\frac{1}{2}k(-u_i + u_j)^2\right]^{(e)} \tag{5.10}$$

This is the elastic energy stored in the system.

Now consider the potential W of an externally applied force F to do work. The structure is assumed to be deformed. The external load does work if the structure is displaced

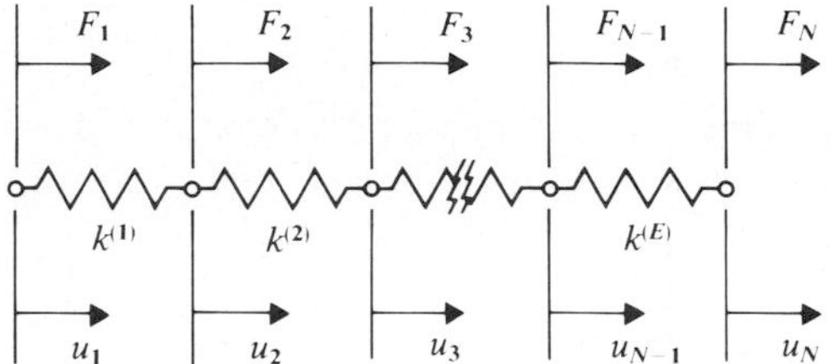

Figure 5.3 Springs in series with externally applied forces

to its original undeformed position. During this displacement, the external force remains constant. The potential W is the capacity of the force F to perform work through deformation u.

$$W = -Fu \tag{5.11}$$

If external concentrated loads $F_1, F_2, F_3, \ldots, F_N$ are applied at the node points on the structure, the potential W is the capacity of each force F_i to perform work through deformation u_i

$$W = -\sum_{i=1}^{N} F_i u_i \tag{5.12}$$

where $u_1, u_2, u_3, \ldots, u_N$ are the nodal displacements.

Combining the internal strain energy and potential of the external force gives $I = U + W$, or for a single spring and force

$$I = \frac{1}{2}ku^2 - Fu \tag{5.13}$$

With a series of E springs and N externally applied forces, it is

$$I = \sum_{e=1}^{E} \left[\frac{1}{2}k(-u_i + u_j)^2\right]^{(e)} - \sum_{i=1}^{N} F_i u_i \tag{5.14}$$

The minimization equations are given by (5.5).

B. Examples Using Springs

Two examples employing springs are given now to illustrate the methods just presented. The first shows a spring/mass system and gives specific numerical values. The second example solves a problem involving four springs in series.

Example 5.2 Spring/Mass System

Given A horizontal spring of stiffness $k = 500$ lbf/in is fixed to a wall at the left end and attached to a spring over a pulley at the other end. Figure 5.4(a) illustrates the problem. The spring supports a mass m with a weight of 100 lbf hanging vertically.

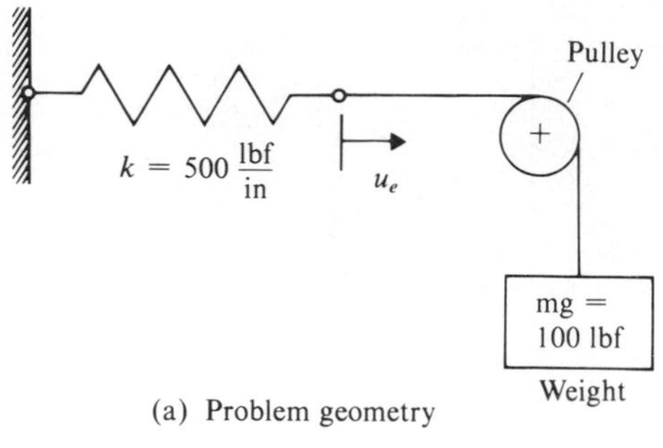

(a) Problem geometry

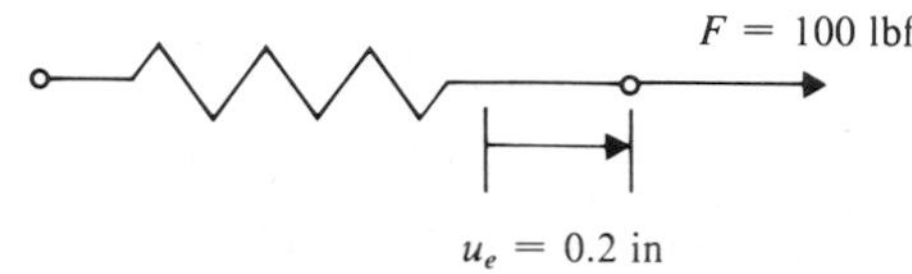

(b) Equilibrium displacement

Figure 5.4 Displacement of spring (Example 5.2—Spring/mass system)

Objective Evaluate the total potential energy (functional) for the spring/mass system and employ it to find the equilibrium displacement of the spring.

Solution The externally applied force F is the weight supported by the string: $F = mg = 100$ lbf. The pulley only changes the direction of the force. It is easy to calculate the equilibrium displacement of the spring as

$$u_e = \frac{F}{k} = 100 \text{ lbf} \times \frac{\text{in}}{500 \text{ lbf}} = 0.20 \text{ in}$$

Figure 5.4(b) illustrates the displaced spring. While this equilibrium displacement u_e is simple to obtain for this example, it will be assumed that u_e is unknown. Its value will be found by the variational method.

The internal strain energy U in the deformed spring is

$$U = \frac{1}{2}ku^2$$

Substituting the numerical value gives

$$U = \frac{1}{2} \times 500\, u^2 = 250\, u^2$$

Figure 5.5 shows a plot of the strain energy versus displacement u_e.

In this case, the work W is done against gravity. The gravitational potential is taken relative to some reference height. For convenience here, this height is chosen as the undeformed height of the 100 lbf weight. The potential work that can be done by the weight is $W = -mg\, u$. Note that the work that can be done by the weight decreases as the weight goes lower. Thus W is negative for this example. The resulting potential is $W = -100\, u$. This is also plotted in Figure 5.5.

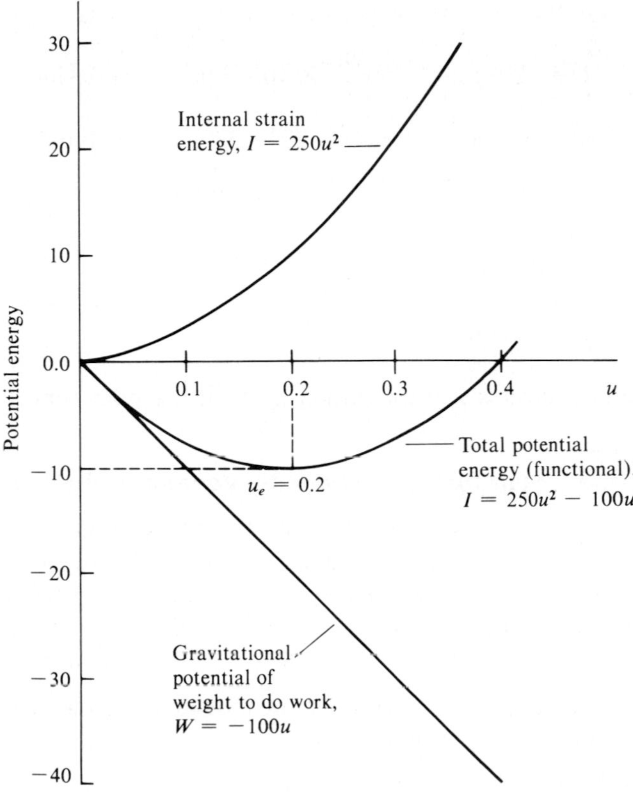

Figure 5.5 Potential energy (Example 5.2—Spring/mass system)

Combining these two gives the total potential energy I, $I = U + W = 250\ u^2 - 100u$. The functional is plotted in Figure 5.5, where the minimum point is obvious. The first variation is

$$\delta I = \frac{\partial I}{\partial u}\,\delta u = 0$$

or

$$\frac{\partial I}{\partial u} = 0$$

Evaluating

$$\frac{\partial I}{\partial u} = 500\ u - 100 = 0$$

and solving for u at equilibrium yields

$$u_e = \frac{500}{100} = 0.2 \text{ in}$$

Clearly the equilibrium displacement, u_e, is the same as calculated earlier.

At equilibrium the terms in the potential energy are

$$U = 250\, u_e^2 = 250\, \frac{\text{lbf}}{\text{in}} \times (0.2)^2\ \text{in}^2 = 10\ \text{lbf-in}$$

$$W = -100\, u_e = -100\ \text{lbf} \times 0.2\ \text{in} = -20\ \text{lbf-in}$$

$$I = U + W = 10 - 20 = -10\ \text{lbf-in}$$

All have units of work.

The second derivative is

$$\frac{\partial^2 I}{\partial u^2} = 500$$

The positive value indicates that the stationary point is a minimum.

Example 5.3 Four Springs in Series—Variational Method

Given This problem is the same as Example 4.2. The variational method is used this time, whereas the direct method was used in Section 4.3. Figure 5.6 shows the geometry. The prescribed nodal displacements are $u_2 = 0.2$ mm and $u_5 = 0.0$ mm, and the externally applied forces are $F_1 = 2$ N, $F_3 = -3$ N, and $F_4 = -1$ N. The springs have stiffnesses $k^{(1)} = 25$ N/mm, $k^{(2)} = 20$ N/mm, $k^{(3)} = 40$ N/mm, $k^{(4)} = 10$ N/mm.

Objective Determine the functional I for the problem, its first variation, and the displacements that make it a minimum.

Solution The functional I is given by (5.14) as

$$\begin{aligned} I = {} & \frac{1}{2}k^{(1)}(-u_1 + u_2)^2 + \frac{1}{2}k^{(2)}(-u_2 + u_3)^2 \\ & + \frac{1}{2}k^{(3)}(-u_3 + u_4)^2 + \frac{1}{2}k^{(4)}(-u_4 + u_5)^2 \\ & - F_1u_1 - F_2u_2 - F_3u_3 - F_4u_4 - F_5u_5 \end{aligned}$$

The first four terms represent the internal strain energy in the four springs, while the last five terms represent the potential of the external forces to do work. Substituting the numerical values gives

$$\begin{aligned} I = {} & \frac{1}{2}(25)(-u_1 + u_2)^2 + \frac{1}{2}(20)(-u_2 + u_3)^2 \\ & + \frac{1}{2}(40)(-u_3 + u_4)^2 + \frac{1}{2}(10)(-u_4 + u_5)^2 \\ & - (2)\, u_1 - F_2(0.2) - (-3)\, u_3 - (-1)\, u_4 - F_5\,(0.0) \end{aligned}$$

Here the forces F_2 and F_5 are carried along but do not affect the result.

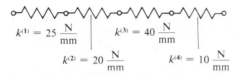

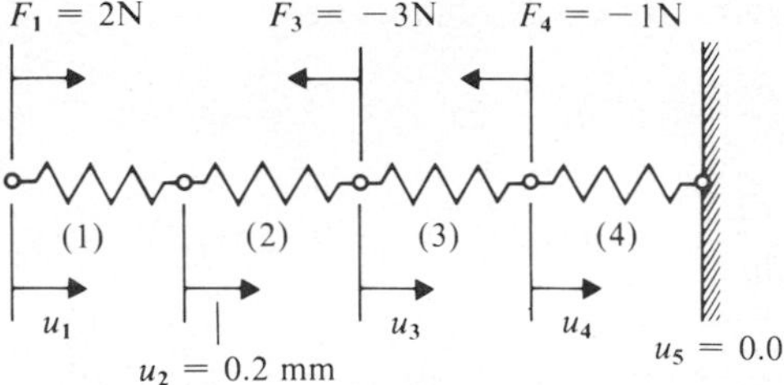

(a) Nodal displacements and external forces

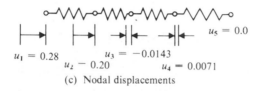

$k^{(1)} = 25 \dfrac{\text{N}}{\text{mm}}$ $k^{(3)} = 40 \dfrac{\text{N}}{\text{mm}}$

$k^{(2)} = 20 \dfrac{\text{N}}{\text{mm}}$ $k^{(4)} = 10 \dfrac{\text{N}}{\text{mm}}$

(b) Stiffness of spring elements

$u_5 = 0.0$

$u_1 = 0.28$ $u_3 = -0.0143$

$u_2 = 0.20$ $u_4 = 0.0071$

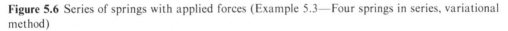

(c) Nodal displacements

Figure 5.6 Series of springs with applied forces (Example 5.3—Four springs in series, variational method)

In this problem, there are only three unknown nodal displacements: u_1, u_3, and u_4. The first variation (5.5) is

$$\delta I = \frac{\partial I}{\partial u_1}\delta u_1 + \frac{\partial I}{\partial u_3}\delta u_3 + \frac{\partial I}{\partial u_4}\delta u_4 = 0$$

Both u_2 and u_5 are known, so these are not varied. The first variation is zero if the relations

Node 1 $\quad \dfrac{\partial I}{\partial u_1} = 0$

Node 3 $\quad \dfrac{\partial I}{\partial u_3} = 0$

Node 4 $\quad \dfrac{\partial I}{\partial u_4} = 0$

are satisfied. Each of these three equations is a minimization equation associated with a particular node.

The minimization equations are easily obtained as

$$\text{Node 1} \quad \frac{\partial I}{\partial u_1} = 25\, u_1 - 25\, u_2 - 2 = 0$$

$$\text{Node 2} \quad \frac{\partial I}{\partial u_3} = -20\, u_2 + 20\, u_3 + 40\, u_3 - 40\, u_4 + 3 = 0$$

$$\text{Node 3} \quad \frac{\partial I}{\partial u_4} = -40\, u_3 + 40\, u_4 + 10\, u_4 - 10\, u_5 + 1 = 0$$

The resulting set of equations is expressed in matrix form as

$$\begin{bmatrix} 25 & -25 & 0 & 0 & 0 \\ 0 & -20 & 60 & -40 & 0 \\ 0 & 0 & -40 & 50 & -10 \end{bmatrix} \begin{Bmatrix} u_1 \\ u_3 \\ u_4 \end{Bmatrix} = \begin{Bmatrix} 2 \\ -3 \\ -1 \end{Bmatrix}$$

This is the same set of linear algebraic equations as found in Example 4.2, except for the prescribed nodal values at nodes 2 and 5. The final results for the nodal displacements are the same also.

5.3 Solid Mechanics—Bar Elements

In most solid mechanics problems the displacement or deformation is a function of position. This makes the problem a little harder. The expression for internal strain energy equation must be evaluated somewhat differently. An axially deformed bar is used as the example in this section. So far, only concentrated forces have been considered. This section adds axially distributed forces.

Again the total potential energy is given by $I = U + W$. These terms U and W must now be evaluated for a bar element.

A. Internal Strain Energy

Consider the bar shown in Figure 5.7(a) and divided into elements in Figure 5.7(b). The recoverable work at a position x in the bar is the integral of the internal stress σ (force per unit area) over the strain (displacement per unit length):

$$\int_0^{\epsilon} \sigma d\epsilon = \text{internal strain energy per unit volume} \tag{5.15}$$

Clearly this has the units of force-length/(length)^3, or work per unit volume. The total internal strain energy is then

$$U = \int_V \left[\int_0^{\epsilon} \sigma d\epsilon \right] dV = \text{internal strain energy} \tag{5.16}$$

where the second integral is taken over the full volume of the body.

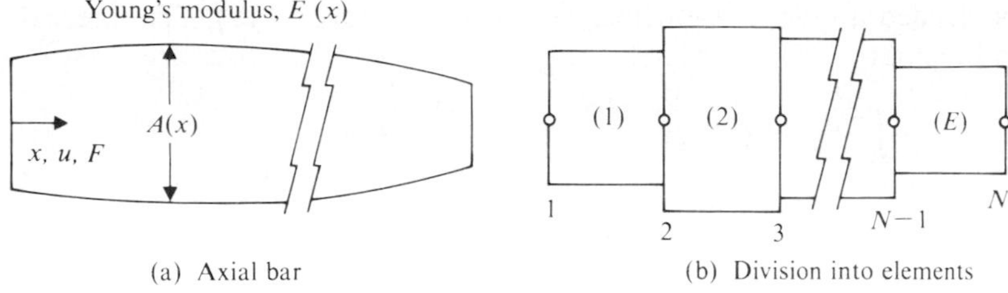

Figure 5.7 Axial bar

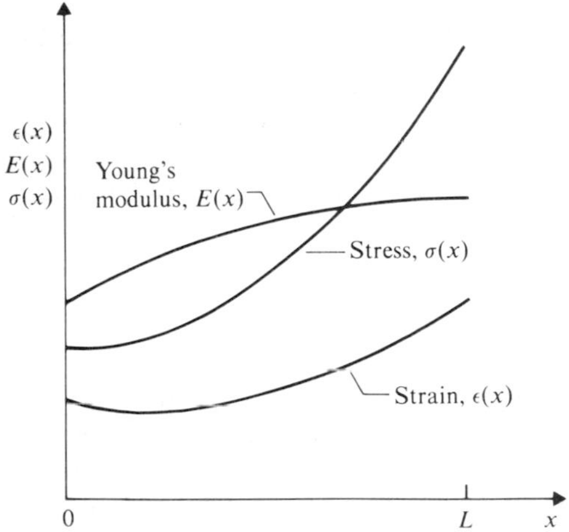

Figure 5.8 Strain, Young's modulus, and stress versus position in axial bar

Generally the axial displacement is a function of x, $u = u(x)$, and the strain is also a function of x, $\epsilon = \epsilon(x)$. Figure 5.8 illustrates the variation of the strain. The Young's modulus E can be a given function of x as well. Hooke's law gives the axial stress $\sigma(x)$ as $\sigma(x) = E(x)\epsilon(x)$ for a linearly elastic bar. Both $E(x)$ and $\sigma(x)$ are shown in Figure 5.8.

At any position x along the bar, the strain energy per unit volume (5.15) becomes

$$\int_0^\epsilon \sigma d\epsilon = \int_0^\epsilon E\epsilon d\epsilon = \frac{1}{2}E\,\epsilon^2 = \frac{1}{2}E(x)\,[\epsilon(x)]^2$$

Also the differential volume dV is $dV = A\,dx$, where A is the cross sectional area. The total internal strain energy

$$U = \frac{1}{2}\int_0^L (EA\epsilon^2)\,dx \tag{5.17}$$

where the integration is now performed over the length of the bar.

Let the bar be divided up into E elements, as shown in Figure 5.7(b). The internal strain energy then becomes

$$U = \frac{1}{2}\int_{x_1}^{x_2} (EA\epsilon^2)^{(1)}\,dx + \frac{1}{2}\int_{x_2}^{x_3} (EA\epsilon^2)^{(2)}\,dx + \ldots + \frac{1}{2}\int_{x_{N-1}}^{x_N} (EA\epsilon^2)^{(E)}\,dx$$

where the integrations are performed over each element. This is also written as the sum of element strain energies,

$$U = \sum_{e=1}^{E} U^{(e)} \tag{5.18}$$

where

$$U^{(e)} = \frac{1}{2}\int_{L^{(e)}} (EA\epsilon^2)^{(E)}\,dx$$

The integral is over the length of the element.

B. Potential Energy Due to External Forces

If a set of concentrated external axial forces $F_1, F_2, F_3, \ldots, F_N$ are applied to the bar, the potential of these forces to do work is

$$W_C = -\sum_{i=1}^{N} F_i u_i \tag{5.19}$$

where $u_1, u_2, u_3, \ldots, u_N$ represent the displacements at the points of application. Here W_C denotes the potential due to the concentrated loads.

The bar is subject to an axial force per unit length. It is denoted as $p(x)$. This might be either a surface force on the outside of the bar or a body force acting on the mass of the bar. Figure 5.9 illustrates these. An example of the surface force might be a shear stress (force per unit area) multiplied by the circumference of the bar to give the distributed force $p(x)$ per unit length. An example of the body force is the weight per unit length of the bar. Usually this acts only when the bar is vertically oriented. The potential of the external force per unit length $p(x)$ is

$$W_D = -\int_0^L pu\,dx \tag{5.20}$$

Here W_D denotes the potential due to the distributed load. This expression may also be broken up into element integrals of the form

$$W_D = -\sum_{e=1}^{E} \int_{L^{(e)}} (pu)^{(e)}\,dx \tag{5.21}$$

The total potential energy can now be formed.

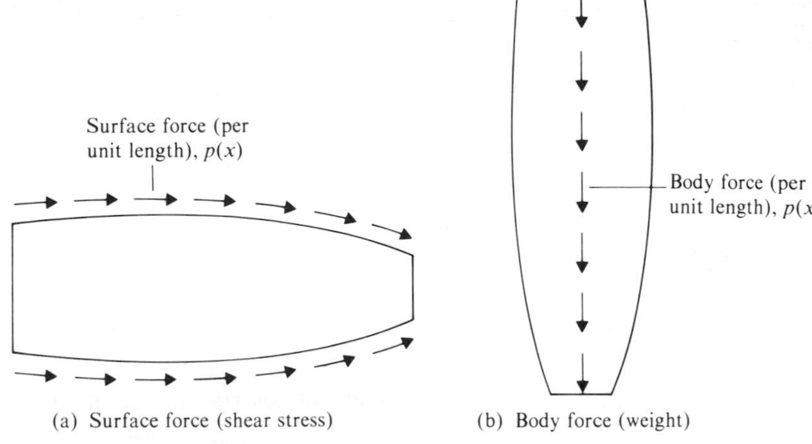

(a) Surface force (shear stress) (b) Body force (weight)

Figure 5.9 Distributed forces on bar

C. Total Potential Energy

The total potential energy is (5.6)

$$I - U + W_C + W_D$$

From the results just obtained, the expression for I becomes

$$\begin{aligned} I = \sum_{e=1}^{E} \frac{1}{2} \int_{L^{(e)}} (EA\epsilon^2)^{(e)}\, dx - \sum_{i=1}^{N} F_i u_i \\ - \sum_{e=1}^{E} \int_{L^{(e)}} (pu)^{(e)}\, dx \end{aligned} \tag{5.22}$$

This is now in a form suitable for the development of an element matrix.

D. Element Matrix and Column

As in Section 4.4, take an element with constant properties $E^{(e)}$ = Young's modulus, $A^{(e)}$ = cross sectional area, $p^{(e)}$ = distributed force as shown in Figure 5.10. The axial displacement $u(x)$ within the element is given by

$$u^{(e)} = N_i^{(e)} u_i + N_j^{(e)} u_j$$

and the element strain is

$$\epsilon^{(e)} = \frac{1}{L^{(e)}} (b_i u_i + b_j u_j)^{(e)}$$

This is a constant-strain element.

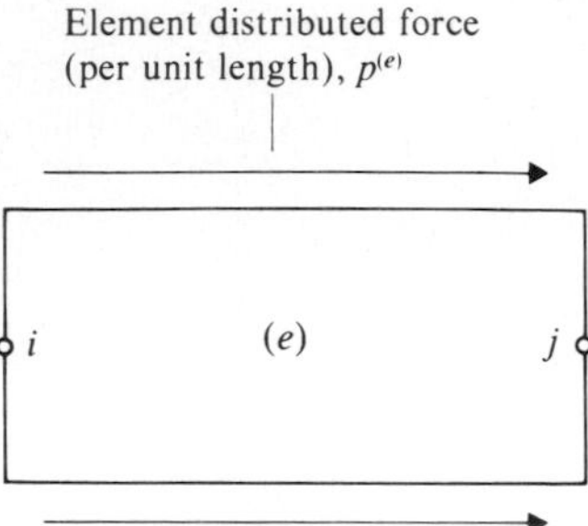

Figure 5.10 Bar element with distributed force

The element matrix and column are given by the element contribution to the minimization of I.

$$\left\{\begin{matrix} \dfrac{\partial I}{\partial u_i} \\ \dfrac{\partial I}{\partial u_j} \end{matrix}\right\}^{(e)} = [B]^{(e)} \left\{\begin{matrix} u_i \\ u_j \end{matrix}\right\}^{(e)} - \{C\}^{(e)}$$

Here $[B]$ is the same element matrix as used in Section 4.4 and $\{C\}$ is an element column. Writing out the element contribution to the minimization for node i gives

$$\text{Node } i \quad \frac{\partial I^{(e)}}{\partial u_i} = \frac{\partial U^{(e)}}{\partial u_i} + \frac{\partial W_C^{(e)}}{\partial u_i} + \frac{\partial W_D^{(e)}}{\partial u_i}$$

and the element contribution to the minimization for node j yields

$$\text{Node } j \quad \frac{\partial I^{(e)}}{\partial u_j} = \frac{\partial U^{(e)}}{\partial u_j} + \frac{\partial W_C^{(e)}}{\partial u_j} + \frac{\partial W_D^{(e)}}{\partial u_j}$$

The individual terms can now be evaluated. As in the spring element, the internal strain energy gives the element matrix, while the external force terms determine the element column.

The element strain energy is

$$U^{(e)} = \frac{1}{2} \int_{L^{(e)}} \left[\frac{EA}{L^2} (b_i u_i + b_j u_j)^2\right]^{(e)} dx$$

Since the integrand is a constant, the integral is easily evaluated as

$$U^{(e)} = \frac{1}{2} \left[\frac{EA}{L} (b_i u_i + b_j u_j)^2\right]^{(e)}$$

The term for node i is

$$\text{Node } i \quad \frac{\partial U^{(e)}}{\partial u_i} = \left[\frac{EA}{L} (b_i b_i u_i + b_i b_j u_j)\right]^{(e)}$$

while the term for node j is

$$\textit{Node } j \quad \frac{\partial U^{(e)}}{\partial u_j} = \left[\frac{EA}{L}(b_j b_i u_i + b_j b_j u_j)\right]^{(e)}$$

Writing this in element matrix form gives

$$[B]^{(e)} = \begin{bmatrix} \frac{EA}{L} b_i b_i & \frac{EA}{L} b_i b_j \\ \\ \frac{EA}{L} b_j b_i & \frac{EA}{L} b_j b_j \end{bmatrix}^{(e)} \tag{5.23}$$

Since EA/L is a common factor and $b_i = -1$, $b_j = 1$, the element matrix simplifies to

$$[B]^{(e)} = \left(\frac{EA}{L}\right)^{(e)} \begin{bmatrix} 1 & -1 \\ -1 & 1 \end{bmatrix}^{(e)} \tag{5.24}$$

which is the same as that obtained for the bar element in Section 4.4.

Also the external force terms are

$$W_C^{(e)} + W_D^{(e)} = -\left[\frac{1}{2}F_i u_i + \frac{1}{2}F_j u_j\right]^{(e)} - \int_{L^{(e)}} [p(N_i u_i + N_j u_j)]^{(e)} dx$$

Here one-half of the concentrated force F_i at node i is allocated to the element. Similarly one-half of the concentrated force F_j at node j is considered to act on the element. Integrating the second term gives

$$W_C^{(e)} + W_D^{(e)} = -\left[\frac{1}{2}F_i u_i + \frac{1}{2}F_j u_j\right]^{(e)} - \left[\frac{1}{2}pL\,(u_i + u_j)\right]^{(e)}$$

Again the term for node i is

$$\textit{Node } i \quad \frac{\partial W_C^{(e)}}{\partial u_i} + \frac{\partial W_D}{\partial u_i} = -\left[\frac{1}{2}F_i + \frac{1}{2}pL\right]^{(e)}$$

and for node j

$$\textit{Node } j \quad \frac{\partial W_C^{(e)}}{\partial u_j} + \frac{\partial W_D^{(e)}}{\partial u_j} = -\left[\frac{1}{2}F_j + \frac{1}{2}pL\right]^{(e)}$$

The term pL has the units of force per length times length, or force. The distributed force pL simply adds to the concentrated force F. The element column is

$$\{C\}^{(e)} = \begin{Bmatrix} \frac{1}{2}F_i + \frac{1}{2}pL \\ \frac{1}{2}F_j + \frac{1}{2}pL \end{Bmatrix}^{(e)} \tag{5.25}$$

Note that this form of the element column implies that the concentrated forces are already included in the element column. They should not be added to the global column as well. The full value F_i should be used for an element at the end of the rod.

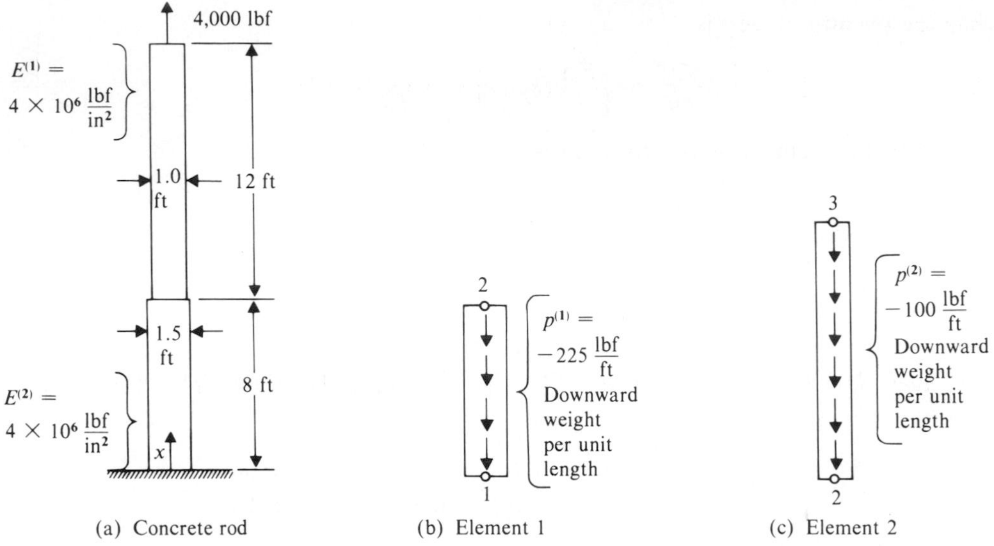

Figure 5.11 Problem specification (Example 5.4—Two-section concrete rod)

Example 5.4 *Two-Section Concrete Rod*

Given A two-section concrete rod has a square cross section. The thicker lower section has the properties

$$\text{Width of side} = 1.5 \text{ ft}$$
$$\text{Young's modulus, } E^{(1)} = 4 \times 10^6 \frac{\text{lbf}}{\text{in}^2}$$
$$\text{Length, } L^{(1)} = 8 \text{ ft}$$
$$\text{Specific weight, } \gamma^{(1)} = 100 \frac{\text{lbf}}{\text{ft}^3}$$

while the thinner upper section has the properties

$$\text{Width of side} = 1.0 \text{ ft}$$
$$\text{Young's modulus} = E^{(2)} = 4 \times 10^6 \frac{\text{lbf}}{\text{in}^2}$$
$$\text{Length, } L^{(2)} = 12 \text{ ft}$$
$$\text{Specific weight, } \gamma^{(2)} = 100 \frac{\text{lbf}}{\text{ft}^3}$$

Figure 5.11 illustrates the concrete rod. The rod is fixed at the bottom and subjected to an upward force of 4,000 lbf at the upper end.

Objective Use a two-element model to determine the elongation (or shortening) of the rod due to the weight of the rod and the upward force. Also determine the stresses in each section of the rod.

Solution The element stiffness for element 1 is

$$\left(\frac{EA}{L}\right)^{(1)} = 4 \times 10^6 \frac{\text{lbf}}{\text{in}^2} \times \frac{144 \text{ in}^2}{\text{ft}^2} \times (1.5 \text{ ft})^2 \times \frac{1}{8 \text{ ft}}$$

$$= 1.62 \times 10^8 \frac{\text{lbf}}{\text{ft}}$$

and the element matrix (5.24) is

$$[B]^{(1)} = (1.62 \times 10^8) \begin{matrix} & \begin{matrix} 1 & 2 \end{matrix} \\ \begin{matrix} 1 \\ 2 \end{matrix} & \begin{bmatrix} 1 & -1 \\ -1 & 1 \end{bmatrix}^{(1)} \end{matrix} \left(\frac{\text{lbf}}{\text{ft}}\right)$$

Now the weight per unit length p acting downward on the element is the specific weight times the cross sectional area:

$$p^{(1)} = \gamma^{(1)}A^{(1)} = -100 \frac{\text{lbf}}{\text{ft}^3} \times (1.5 \text{ ft})^2 = -225 \frac{\text{lbf}}{\text{ft}}$$

The element column is then given by (5.25) as

$$\frac{1}{2}pL = -\frac{1}{2} \times 225 \frac{\text{lbf}}{\text{ft}} \times 8 \text{ ft} = -900 \text{ lbf}$$

where one-half of the element weight (1,800 lbf) is applied at each node. The element column is

$$\{C\}^{(1)} = \begin{matrix} 1 \\ 2 \end{matrix} \begin{Bmatrix} -900 \\ -900 \end{Bmatrix}^{(1)}$$

The displacement at the fixed end (at node 1) is specified, so the force is unknown.

For element 2, the stiffness is

$$\left(\frac{EA}{L}\right)^{(2)} = 4 \times 10^6 \frac{\text{lbf}}{\text{in}^2} \times \frac{144 \text{ in}^2}{\text{ft}^2} \times (1.0 \text{ ft})^2 \times \frac{1}{12 \text{ ft}}$$

$$= 4.8 \times 10^7 \frac{\text{lbf}}{\text{ft}}$$

and the element matrix is

$$[B]^{(2)} = (4.8 \times 10^7) \begin{matrix} & \begin{matrix} 2 & 3 \end{matrix} \\ \begin{matrix} 2 \\ 3 \end{matrix} & \begin{bmatrix} 1 & -1 \\ -1 & 1 \end{bmatrix}^{(2)} \end{matrix}$$

The distributed load p is

$$p^{(2)} = \gamma^{(2)}A^{(2)} = -100 \frac{\text{lbf}}{\text{ft}^3} \times (1.0 \text{ ft})^2 = -100 \frac{\text{lbf}}{\text{ft}}$$

and

$$\frac{1}{2}p^{(2)}L^{(2)} = -\frac{1}{2} \times 100 \frac{\text{lbf}}{\text{ft}} \times 12 \text{ ft} = -600 \text{ lbf}$$

Again one-half of the element weight (1,200 lbf) is applied at each node. Thus the element column is

$$\{C\}^{(2)} = \begin{Bmatrix} -600 \\ -600 + 4{,}000 \end{Bmatrix} = \begin{matrix} 2 \\ 3 \end{matrix} \begin{Bmatrix} -600 \\ 3{,}400 \end{Bmatrix}^{(2)}$$

where the full upward force is applied to node 3. (It is not split between the two elements because there is no third element.)

Assembling the element matrices in the normal way gives the global equations

$$\begin{bmatrix} 1.62 \times 10^8 & -1.62 \times 10^8 & 0 \\ -1.62 \times 10^8 & 2.10 \times 10^8 & -0.48 \times 10^8 \\ 0 & -0.48 \times 10^8 & 0.48 \times 10^8 \end{bmatrix} \begin{Bmatrix} u_1 \\ u_2 \\ u_3 \end{Bmatrix} = \begin{Bmatrix} -900 \\ -1{,}500 \\ 3{,}400 \end{Bmatrix}$$

The known nodal displacement is $u_1 = 0.0$, and the global equations become

$$\begin{bmatrix} 1.62 \times 10^8 & 0 & 0 \\ 0 & 2.10 \times 10^8 & -0.48 \times 10^8 \\ 0 & -0.48 \times 10^8 & 0.48 \times 10^8 \end{bmatrix} \begin{Bmatrix} u_1 \\ u_2 \\ u_3 \end{Bmatrix} = \begin{Bmatrix} 0 \\ -1{,}500 \\ 3{,}400 \end{Bmatrix}$$

The solution is

$$\begin{Bmatrix} u_1 \\ u_2 \\ u_3 \end{Bmatrix} = \begin{Bmatrix} 0.0 \\ 1.173 \times 10^{-5} \\ 8.256 \times 10^{-5} \end{Bmatrix} \qquad \text{(ft)}$$

Because the 4,000 lbf upward force is larger than the weight, both nodal displacements are positive.

The element strains are

$$\epsilon^{(1)} = \frac{u_2 - u_1}{L^{(1)}} = 1.173 \times 10^{-5} \text{ ft} \times \frac{1}{8 \text{ ft}}$$

$$= 1.47 \times 10^{-6} \frac{\text{ft}}{\text{ft}}$$

$$\epsilon^{(2)} = \frac{u_3 - u_2}{L^{(2)}} = 7.083 \times 10^{-5} \text{ ft} \frac{1}{12 \text{ ft}}$$

$$= 5.90 \times 10^{-6} \frac{\text{ft}}{\text{ft}}$$

Also the stresses are

$$\sigma^{(1)} = E^{(1)}\epsilon^{(1)} = 5.88 \frac{\text{lbf}}{\text{in}^2} = 847 \frac{\text{lbf}}{\text{in}^2}$$

$$\sigma^{(2)} = E^{(2)}\epsilon^{(2)} = 23.6 \frac{\text{lbf}}{\text{in}^2} = 3{,}400 \frac{\text{lbf}}{\text{ft}^2}$$

These are constant over the element involved. The forces acting on the two elements are $R^{(1)} = 1{,}900$ lbf, $R^{(2)} = 3{,}400$ lbf.

5.4 Boundary Value Problems and Functionals

The previous two sections used the potential energy as a functional I for solid mechanics problems. Then the principle of minimum potential energy was employed to find the solution to the finite element analysis. While this method gives good physical insight into the variational method, a more general approach is required for other problems, such as fluid mechanics and heat transfer.

A. Differential Equations and Functionals

Engineering students are very familiar with differential equations and boundary conditions but not yet familiar with the variational method. The approach taken here is to relate a given functional, I, to its corresponding boundary value problem. One result of this is an equation called the Euler-Lagrange equation, which can then be related to the original boundary value problem. It is derived in Chapter 6.

After the functional I has been obtained, minimization equations are used to evaluate element matrices. A computer program, called FINONE (*FIN*ite element *ONE* dimensional solver), has been developed for solving this class of problems. The program is described more fully in Appendix A. It can be used to solve any second-order boundary value problem in one dimension of the type discussed in the next few pages.

One major disadvantage of the variational method is that it does not work for all problems. That is, there is no functional or variational method for some differential equations. Nearly all solid mechanics problems, as well as heat conduction problems, do have one. General fluid flow problems (Navier-Stokes equations), on the other hand, have no corresponding functional. When specific approximations are made and certain terms neglected, then variational methods can be used with fluid mechanics problems. One example is pipe flow (Pouiseille flow), when inertia forces can be set to zero and viscous forces dominate, as in Example 5.7. Then a variational method works. Another example is ideal flow where inertia forces are large and viscous forces are neglected. Again a functional can be found. The variational method is still widely used, so it should be understood by engineers interested in finite element methods.

B. Boundary Value Problems

Consider a differential equation of the general form

$$D\left(x, T, \frac{dT}{dx}\right) = 0 \tag{5.26}$$

where D may contain terms with up to first derivatives of both T and dT/dx. A fairly general linear second-order example might be

$$\frac{d}{dx}\left(K\frac{dT}{dx}\right) + PT + Q = 0 \tag{5.27}$$

where $K = K(x)$, $P = P(x)$, and $Q = Q(x)$ are known coefficient functions. This is called self-adjoint in mathematical terms. It is the most general differential equation for which

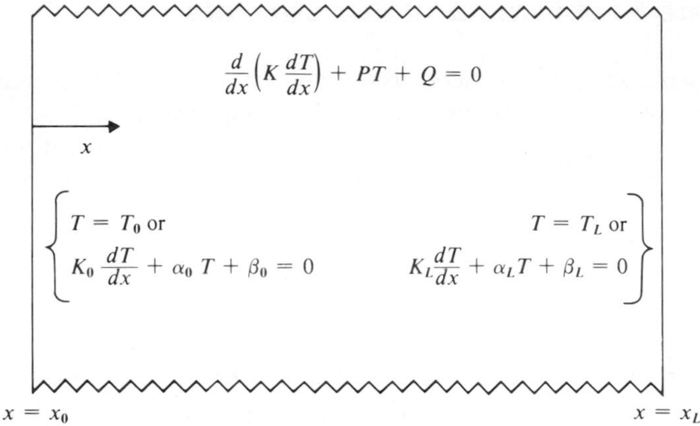

Figure 5.12 General boundary value problem

a functional exists. The most general equation contains an additional first derivative term of the form $M\ dT/dx$ (actually a transformation of variable can be used to convert one form to the other in one dimension, but not in two or three dimensions). This is analyzed using Galerkin's method in Chapter 8.

Figure 5.12 shows the function $T(x)$ over the solution domain $x_0 < x < x_L$. Two boundary conditions exist for a second-order equation. By the definition of a boundary value problem, one boundary condition is applied at one end, $x = x_0$, and the other is applied at the other end, $x = x_L$. Initial value problems—where, say, both conditions are specified at one end—are not considered here.

Boundary conditions are mathematically classified in two types: specified value (Dirichlet) and derivative (mixed). At $x = x_0$, either

$$
\begin{aligned}
&T = T_0, \quad \text{specified value (Dirichlet), or} \\
&K_0 \frac{dT}{dx} + \alpha_0 T + \beta_0 = 0, \quad \text{derivative (mixed)}
\end{aligned}
\tag{5.28}
$$

Here

$$
\begin{aligned}
T_0 &= \text{value of } T(x) \text{ if specified at } x = x_0 \\
K_0 &= \text{value of } K(x) \text{ at } x = x_0 \\
\alpha_0 &= \text{constant at } x = x_0 \\
\beta_0 &= \text{constant at } x = x_0
\end{aligned}
$$

Note that only *one* of these two boundary conditions can be used in any single boundary value problem.

At the other end, $x = x_L$, the boundary condition is either

$$
\begin{aligned}
&T = T_L, \quad \text{specified value (Dirichlet), or} \\
&K_L \frac{\partial T}{\partial x} + \alpha_L T + \beta_L = 0, \quad \text{derivative (mixed)}
\end{aligned}
\tag{5.29}
$$

where

$$T_L = \text{value of } T(x) \text{ if specified at } x = x_L$$
$$K_L = \text{value of } K(x) \text{ at } x = x_L$$
$$\alpha_L = \text{constant specified at } x = x_L$$
$$\beta_L = \text{constant specified at } x = x_L$$

Once again, only one of these two is actually used in any single boundary value problem.

C. Examples

Three examples are given here to illustrate general boundary value problems. Example 5.5 treats heat conduction in an infinite slab, with the resulting simple differential equation; two types of boundary conditions are considered. Example 5.6 treats heat conduction and convection in a thin fin; it yields a more complicated differential equation. Example 5.7 considers a fluid flow between two flat plates. All three of these problems are studied in typical undergraduate engineering programs.

Example 5.5 *Heat Conduction in Infinite Slab*

Given A solid slab has length L (mm) in the direction of heat flow. Let the geometric parameters in the problem be x = coordinate in heat flow direction (mm), y = coordinate perpendicular to heat flow direction (mm), and z = coordinate out of paper (mm). The slab is assumed very long in both the y and z directions so that no heat flow occurs in either direction. Figure 5.13 illustrates the geometry.

The Fourier heat conduction law is

$$q_x = -k\frac{dT}{dx}$$

where q_x = conduction heat flow per unit area (W/mm^2), k = conduction coefficient (W/mm-°C), and T = temperature (°C). This gives the conduction heat flow per unit area in the infinite slab. Usually k is a function of x, $k = k(x)$.

Considering the total conduction heat flow in the x direction gives

$$Q_x = -kA_x\frac{dT}{dx} \quad (Q_x = q_xA_x)$$

where

$$Q_x = \text{total conduction heat flow (W)}$$
$$A_x = \text{cross sectional area perpendicular to heat flow direction (mm}^2\text{)}$$

The units of heat flow, watts, are those of energy per unit time, 1 watt = 1 N·m/s, which are the units of power. Thus the heat flow is actually an expression of thermal power.

The boundary conditions are different at the two ends of the slab. At the left end, the temperature is specified; $T = T_0$, $x = 0$. At the right end, the heat flow is given with the value Q_L. Thus the boundary condition is $Q_x = Q_L$, $x = L$. This may be called a heat flow boundary condition.

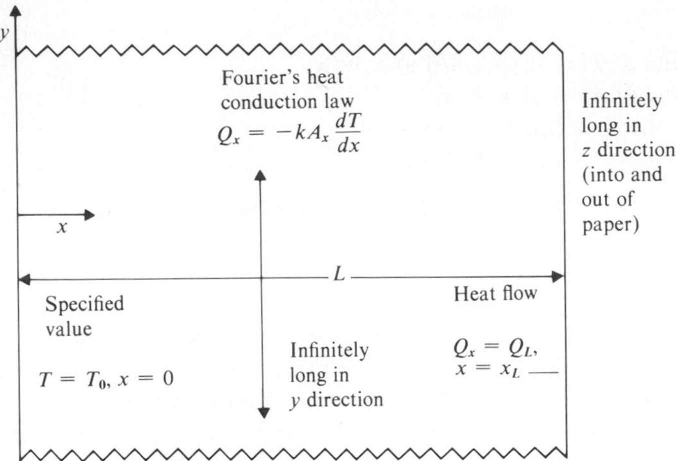

Figure 5.13 Problem geometry (Example 5.5—Heat conduction in infinite slab)

$Q_{x(\text{left face})}$

$= -\left(Q_x - \frac{dQ_x}{dx}\frac{dx}{2}\right)$ → dy • Q_x dx → $Q_{x(\text{right face})} = \left(Q_x + \frac{dQ_x}{dx}\frac{dx}{2}\right)$

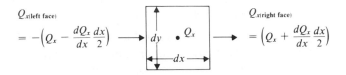

Figure 5.14 Differential cube (Example 5.5—Heat conduction in infinite slab)

Objective Obtain the governing differential equation and boundary conditions for the temperature in the slab as a function of x using differential analysis. Determine the coefficient functions $K(x)$, $P(x)$, $Q(x)$ and boundary conditions α_L, β_L in the general relations (5.27) and (5.29).

Solution A differential cube of the slab, shown in Figure 5.14, has dimensions dx, dy, and dz. No heat is added to or taken out of the interior of the slab—thus, the heat flow must be conserved. The total heat flow entering the left face must equal the total heat flow leaving the right face.

$$\begin{pmatrix}\text{Total heat conduction}\\ \text{through left face}\end{pmatrix} + \begin{pmatrix}\text{Total heat conduction}\\ \text{through right face}\end{pmatrix} = 0$$

Heat flow out of the cube is taken as positive. Thus

$$Q_{x(\text{left face})} + Q_{x(\text{right face})} = 0$$

The heat flow at the center of the cube is Q_x. Expand the heat flow in a Taylor's series about the center. At the left face of the cube, the heat flow is

$$-\left(Q_x - \frac{dQ_x}{dx}\frac{dx}{2}\right)$$

while the heat flow through the right face is

$$\left(Q_x + \frac{dQ_x}{dx}\frac{dx}{2}\right)$$

Summing the heat flows yields the total heat flow relation

$$-\left(Q_x - \frac{dQ_x}{dx}\frac{dx}{2}\right) + \left(Q_x + \frac{dQ_x}{dx}\frac{dx}{2}\right) = 0$$

Cancelling like terms and dividing by dx gives the equation

$$\frac{dQ_x}{dx} = 0$$

This leads to the final equation.

Fourier's conduction law is

$$Q_x = -kA_x \frac{dT}{dx}$$

where the cross sectional area is $A_x = dy\ dz$. Substituting this into the equation just derived yields

$$\frac{d}{dx}\left(kA_x \frac{dT}{dx}\right) = 0, \quad 0 < x < L$$

where the negative sign has been dropped. This is the second-order differential equation governing the temperature $T(x)$. It also governs the heat flow in an insulated thin axisymmetric rod of cross sectional area $A_x(x)$. Figure 5.15 shows the rod. Because the rod is insulated and no heat flows out of the sides of the rod, the temperature is a function of x alone.

The units of each quantity have already been indicated. The terms in the differential equation have units

$$\frac{d}{dx}\left(kA_x \frac{dT}{dx}\right) = \frac{1}{\text{mm}}\left(\frac{\text{W}}{\text{mm-°C}}\right)\text{mm}^2\left(\frac{\text{°C}}{\text{mm}}\right) = \frac{\text{W}}{\text{mm}}$$

The watt is a unit of thermal power, so the above terms are expressed as power per length dx. They represent the time rate of change of thermal energy ($\text{W} = \text{N}\cdot\text{m/s}$) per length dx.

This can be compared to the differential equation for the axial deformation of a bar obtained in Example 4.6.

$$\frac{d}{dx}\left(EA \frac{du}{dx}\right) = 0$$

Here the units (in the metric system) are

$$\frac{d}{dx}\left(EA \frac{du}{dx}\right) = \left(\frac{1}{\text{mm}}\right)\left(\frac{\text{kN}}{\text{mm}^2}\right)(\text{mm}^2)\left(\frac{\text{mm}}{\text{mm}}\right) = \frac{\text{kN}}{\text{mm}}$$

Thus the differential equation for the solid mechanics problem has the units of force per length rather than power per length as in the heat transfer problem.

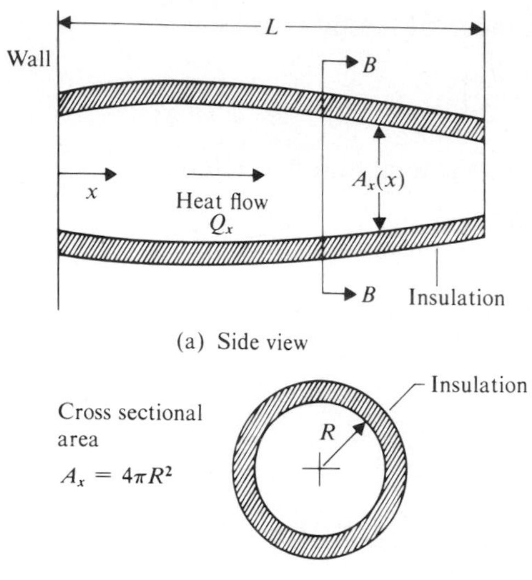

Figure 5.15 Rod geometry (Example 5.5—Heat conduction in infinite slab)

Comparing this to the general differential equation (5.27)

$$\frac{d}{dx}\left(K\frac{dT}{dx}\right) + PT + Q = 0$$

produces the known coefficient functions $K(x) = kA_x$, $P(x) = 0$, $Q(x) = 0$. Recall that $k = k(x)$ so the function $K(x)$ will vary with x in general. Only one coefficient function is used for this problem.

On the left-end boundary, the specified value of T is $T = T_0$, $x = 0$. At the other end, the heat flow is given as $Q_x = Q_L$, $x = L$. From Fourier's law again

$$-kA_x\frac{dT}{dx} = Q_L, \quad x = L$$

Transferring the term Q_L to the left side and dropping both minus signs (to make the term kA_x be consistent with the differential equation) yields

$$kA_x\frac{dT}{dx} + Q_L = 0, \quad x = L$$

Now it is easily seen that the heat flow boundary is a derivative boundary condition. The general derivative boundary condition (5.29) is

$$K_L\frac{dT}{dx} + \alpha_L T + \beta_L = 0, \quad x = L$$

which indicates that $K_L = kA_x$, $\alpha_L = 0$, $\beta_L = Q_L$. Note that these are known constants (evaluated at $x = L$) rather than known functions. Also note that the constant K_L must

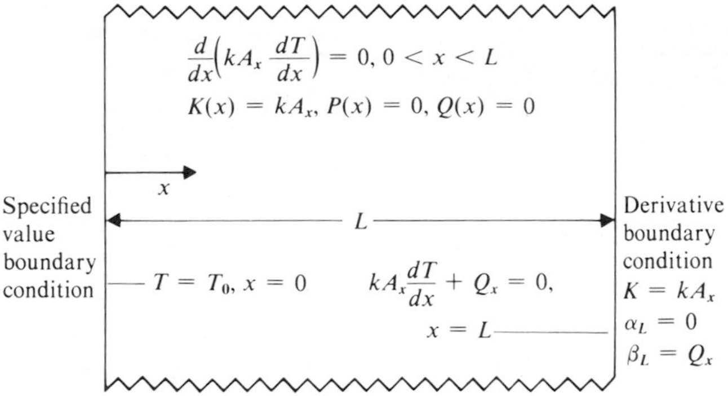

Figure 5.16 Boundary value problem (Example 5.5—Heat conduction in infinite slab)

have the same value $k\ A_x$ as the function $K(x)$ evaluated at $x = L$. (It does here.) The full boundary value problem is illustrated in Figure 5.16. A finite element numerical example is carried out in Example 5.8.

Example 5.6 Heat Conduction and Convection in Thin Fin

Given A thin fin has length L in the direction of heat flow. It has small thickness t in the z direction where $t/L << 1$ rather than being infinite in the z direction (as was the infinite slab in Example 5.5). Thus the temperature is approximately uniform over the fin thickness. Figure 5.17 illustrates the geometry. Figure 5.18 shows a perspective view. The fin has width W in the y direction where $t/W << 1$. Temperatures at both ends x_0 and x_L of the fin are independent of y, so the temperature in the fin is approximately a function of x only.

As in the previous example, the Fourier heat conduction law

$$q_x = -k\frac{dT}{dx}$$

governs heat flow per unit area in the x direction. The total conduction heat flow is

$$Q_x = -kA_x\frac{dT}{dx}$$

where the cross sectional area for the thin fin is $A_x = Wt$ rather than $dy\ dz$.

A fluid, at rest or moving, fills the space above and below the fin. The fluid is maintained at temperature T_∞. If T_∞ is lower than the temperature of the fin, heat convects out of the fin into the fluid at the rate of

$$q_z = h(T - T_\infty)_{\text{(out of top)}} + h(T - T_\infty)_{\text{(out of bottom)}}$$

where q_z = convection heat flow per unit area (W/mm^2), h = convection coefficient (W/mm^2-°C), and T_∞ = temperature of fluid at large distance from fin (°C). Heat flows by convection out of both the top and bottom of the fin. The convection coefficient, h, is

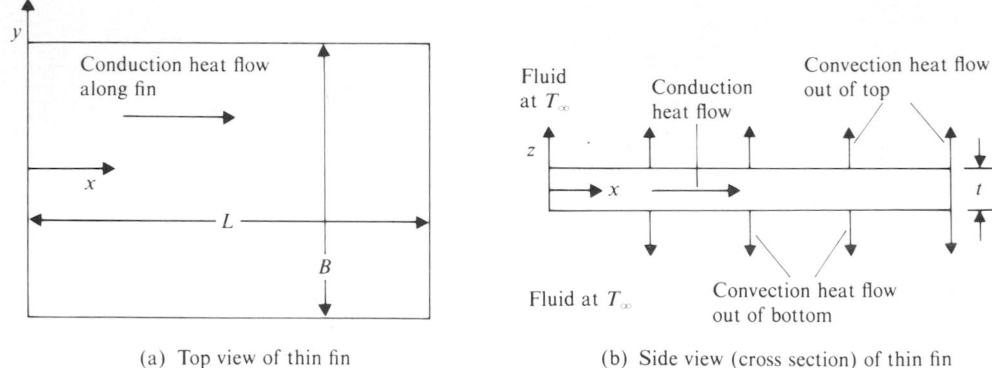

(a) Top view of thin fin (b) Side view (cross section) of thin fin

Figure 5.17 Problem geometry (Example 5.6—Heat conduction and convection in thin fin)

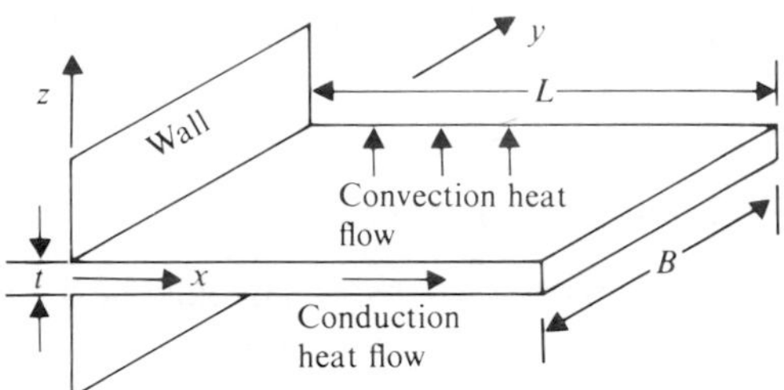

Figure 5.18 Perspective view (Example 5.6—Heat conduction and convection in thin fin)

measured; values are given in heat transfer books. It depends upon the fluid properties and velocity over the fin. The heat flow is proportional to the temperature difference $T - T_\infty$.

The total convection heat flow out of the top and bottom of the fin is $Q_z = 2h\ A_z \times (T - T_\infty)$ where Q_z = total convection heat flow (W) and A_z = cross sectional area perpendicular to heat flow direction (mm²). For the convection heat flow, $A_z = Wdx$.

At each end of the fin, the boundary conditions are specified: $T = T_0$, $x = 0$ and $T = T_L$, $x = L$. Other conditions, such as one specifying the heat flow, could also be used.

Objective Obtain the governing differential equation and compare it to the general boundary value problem.

Solution The differential section of the fin, shown in Figure 5.19, has length dx, width W, and thickness t. A heat balance on it yields the equation

$$\begin{pmatrix}\text{Total heat conduction}\\ \text{through left face}\end{pmatrix} + \begin{pmatrix}\text{Total heat conduction}\\ \text{through right face}\end{pmatrix}$$

$$+ \begin{pmatrix}\text{Total heat convection}\\ \text{through top face}\end{pmatrix} + \begin{pmatrix}\text{Total heat convection}\\ \text{through bottom face}\end{pmatrix} = 0$$

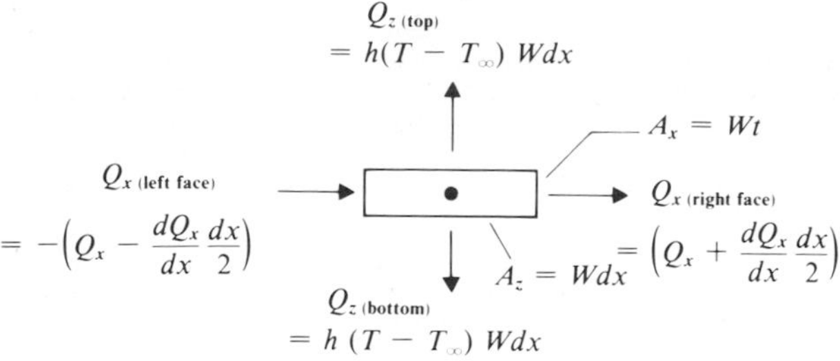

Figure 5.19 Heat balance on differential section (Example 5.6—Heat conduction and convection in thin fin)

The terms are

$$Q_{x(\text{left face})} + Q_{x(\text{right face})} + Q_{z(\text{top})} + Q_{z(\text{bottom})} = 0$$

Writing out all of the terms gives

$$-\left(Q_x - \frac{dQ_x}{dx}\frac{dx}{2}\right) + \left(Q_x + \frac{dQ_x}{dx}\frac{dx}{2}\right) + hA_z(T - T_\infty)$$
$$+ hA_z(T - T_\infty) = 0$$

Simplifying and dividing by dx yields

$$\frac{dQ_x}{dx} + 2hW(T - T_\infty) = 0$$

Using Fourier's law for Qx, this becomes

$$\frac{d}{dx}\left(kA_x \frac{dT}{dx}\right) - 2hW(T - T_\infty) = 0$$

where the entire expression has been multiplied by -1, making the second derivative term positive.

Writing out individual terms in this governing differential equation yields

$$\frac{d}{dx}\left(kWt \frac{dT}{dx}\right) - 2hWT + 2hWT_\infty = 0$$

Again this differential equation has the units of W/mm (power per length). The general differential equation (5.27) is

$$\frac{d}{dx}\left(K \frac{dT}{dx}\right) + PT + Q = 0$$

and the known coefficient functions are $K(x) = kA_x = kWt$, $P(x) = -2hW$, and $Q(x) = 2hWT_\infty$. All of the K, P, Q functions are used for this problem. Example 5.9 shows a full finite element numerical example.

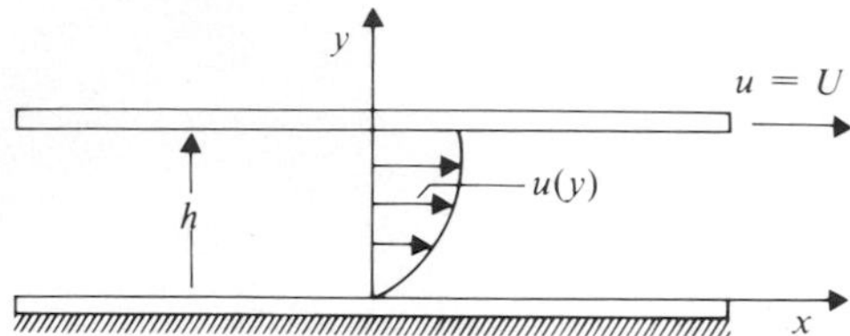

Figure 5.20 Problem geometry (Example 5.7—Fluid flow between flat plates)

Example 5.7 *Fluid Flow Between Two Flat Plates*

Given Consider laminar flow between two infinite parallel plates. Figure 5.20 shows the two plates separated by a constant distance h. Let x be the coordinate in the direction of flow, while y is taken across the flow.

x = horizontal position (ft)
y = vertical position (ft)
z = coordinate out of paper (ft)
h = distance between plates (ft)

Fluid thermodynamic properties are defined as

P = Pressure (lbf/ft^2)
μ = dynamic viscosity (lbf-sec/ft^2)
ρ = density (slug/ft^3)

The fluid velocity is u = horizontal velocity (ft/sec). It is the objective of fluid mechanics analysis to determine the fluid velocity, given the other problem parameters.

No variations occur into or out of the paper, thus the fluid properties are not functions of z. Also, the pressure P has a constant gradient in the x direction

$$\frac{\partial P}{\partial x} = \text{constant}$$

and has no variation with y. A decreasing pressure ($\partial P/\partial x < 0$) produces a positive fluid velocity. Continuity requires that the fluid velocity u does not vary in the x direction. Thus $u = u(y)$. No fluid velocity occurs in the y direction. Finally, the flow is assumed to be incompressible, fully developed, and steady.

The upper plate has velocity U in the positive x direction while the lower plate is fixed in space. No slip boundary conditions apply, so the boundary conditions are $u = 0$, $y = 0$ and $u = U$, $y = h$. These can be used directly.

Two special cases occur: Couette flow and Poiseuille flow. For Couette flow, the upper plate moves but the pressure gradient is zero:

$$\left.\begin{aligned} U &\neq 0 \\ \frac{\partial P}{\partial x} &= 0 \end{aligned}\right\} \qquad \text{Couette flow}$$

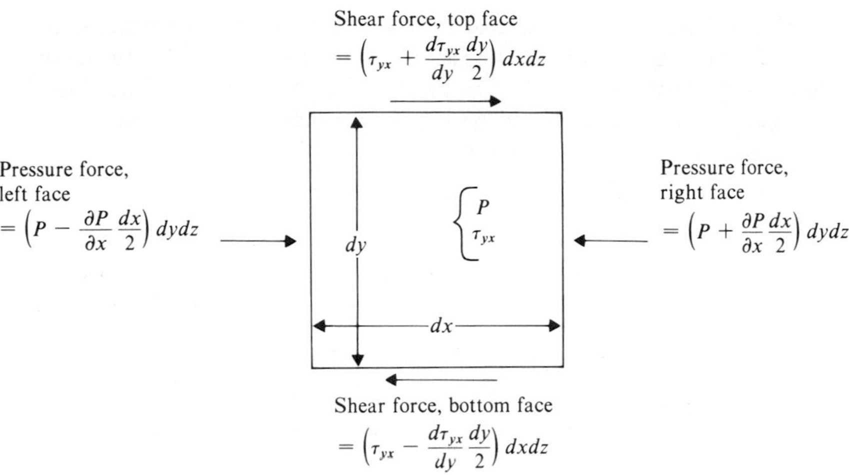

Figure 5.21 Forces acting on differential cube of fluid (Example 5.7—Fluid flow between flat plates)

The upper plate motion drives the flow. In Poiseuille flow, the upper plate is fixed but a pressure gradient is present:

$$\left.\begin{aligned} U &= 0 \\ \frac{\partial P}{\partial x} &\neq 0 \end{aligned}\right\} \qquad \text{Poiseuille flow}$$

Now the pressure gradient drives the flow.

The fluid is assumed to be Newtonian—that is, the fluid obeys Newton's law of viscosity.

$$\tau_{yx} = \mu \frac{du}{dy}$$

The shear stress is linearly related to the velocity gradient.

Objective Use differential analysis to obtain the differential equation for the velocity $u(y)$ in this problem. Compare the results to the general differential equation (5.27) and boundary conditions (5.28) and (5.29).

Solution A differential cube of fluid with dimensions dx, dy, dz is shown in Figure 5.21. For steady viscous flow, the sum of the forces in the x direction must equal zero: $\sum F_x = 0$. The four forces are

$$\sum F_x = F_{x\text{(pressure force on left face)}} + F_{x\text{(pressure force on right face)}} + F_{x\text{(shear force on bottom face)}} + F_{x\text{(shear force on top face)}} = 0$$

These forces can be summed to give the differential equation.

Let the pressure at the center of the cube be P. Taking a Taylor's series expansion of the pressure in the x direction gives pressure terms, as shown in Figure 5.21. Multiplying by the area of the faces $dy\,dz$ produces the forces on each face. Similarly the shear stress is expanded in a Taylor's series expansion in the y direction. Multiplying by the area of the faces $dx\,dz$ gives the forces. Summing the forces yields

$$\sum F_x = \left(P - \frac{\partial P}{\partial x}\frac{dx}{2}\right) dy\,dz - \left(P + \frac{\partial P}{\partial x}\frac{dx}{2}\right) dy\,dz$$
$$- \left(\tau_{yx} - \frac{d\tau_{yx}}{dy}\frac{dy}{2}\right) dx\,dz + \left(\tau_{yx} + \frac{d\tau_{yx}}{dy}\frac{dy}{2}\right) dx\,dz = 0$$

Cancelling terms and dividing by dy yields

$$-\frac{\partial P}{\partial x} A_y + \frac{d\tau_{yx}}{dy} A_y = 0$$

where the cross sectional area is taken as $A_y = dx\,dz$. Each term has units of force (in the x direction) per length (in the y direction). The units are the same as those for displacement of an axisymmetric rod—force/length.

Essentially the same information can be obtained after division by A_y. The result is

$$-\frac{\partial P}{\partial x} + \frac{d\tau_{yx}}{dy} = 0$$

This equation can be conveniently thought of as stress per length. The units are

$$\frac{d\tau_{yx}}{dy} = \left(\frac{\text{kN}}{\text{mm}^2}\right)\left(\frac{1}{\text{mm}}\right)$$

The differential equation is now modified to include the fluid velocity rather than the shear stress.

Newton's law of viscosity is

$$\tau_{yx} = \mu \frac{du}{dy}$$

Thus the differential equation becomes

$$\frac{d}{dy}\left(\mu \frac{du}{dy}\right) - \frac{\partial P}{\partial x} = 0, \quad 0 < y < h$$

where $\partial P/\partial x =$ constant. In the most general form, the viscosity can be a function of y. It may be solved for the velocity profile $u(y)$ when the pressure gradient is given.

This differential equation can be compared to the general form (5.27):

$$\frac{d}{dy}\left(K \frac{du}{dy}\right) + Pu + Q = 0$$

Here y and u are used instead of x and T. It is easily seen that the known coefficient functions are $K(y) = \mu$, $P(y) = 0$, and $Q(y) = -\partial P/\partial x$. Also, the general form of specified boundary condition (5.28) and (5.29) is $u = u_0 = 0$, $y = y_0 = 0$ and $u = u_L = U$, $y = y_L = h$. The example is complete. A full finite element numerical example is carried out in Example 5.10.

D. More on Boundary Value Problems

Linear boundary value problems with either specified values or derivative boundary conditions have unique solutions. This implies that once a solution has been obtained that satisfies the differential equation plus boundary conditions, no other can exist. A special case in which this is *not* true follows.

Actually another type of boundary conditions occurs when the derivative is specified on the boundary ($\alpha = 0$). Then,

$$K \frac{dT}{dx} + \beta = 0$$

This is called either the specified derivative or the Neumann boundary condition. If a boundary value problem has specified derivatives at both ends and also the differential equation does not contain the term PT, the solution will not be unique. Consider a solution $T(x)$. Because only derivative terms of T occur in both the differential equation and boundary conditions, $T(x) + C$ is also a solution (where C is any constant). Thus the solution is not unique. This case can occur in ideal flow problems and heat conduction problems.

At this point, it is useful to know that any second-order ordinary differential equation can be converted to the general form (5.27) by a suitable transformation of coefficients. For example, an equation of the form

$$A(x) \frac{d^2T}{dx^2} + B(x) \frac{dT}{dx} + C(x)T + D(x) = 0$$

may be converted to self-adjoint form by the relations

$$K(x) = e^{\int \frac{B(x)}{A(x)} dx}$$

$$P(x) = \frac{C(x)}{A(x)} e^{\int \frac{B(x)}{A(x)} dx}$$

Unfortunately, partial differential equations (in two or three dimensions) do not have this nice feature.

5.5 General Functional and Element Matrices

A finite element method can now be developed for the general boundary value problem just discussed. The approach taken in this section is simply to give the general functional (without derivation) and obtain the proper finite matrices. The derivation of the general functional is presented in Chapter 6. It is worthwhile to learn how to derive a functional for a given boundary value problem, but this knowledge is not essential to the use of finite element computer codes or, in many cases, even to the development of finite element methods. If a sufficiently general function has already been obtained, it may be employed for many different engineering problems. This is the approach taken in this chapter.

The general approach taken here leads directly to the coding of the computer program FINONE. It can be used to solve any of the problems formulated so far, as well as many others. Where possible, the variable names in the program match those used in the text. Appendix A gives more details of the program formulation and input/output formats.

A. General Functional

The general second-order self-adjoint boundary value problem (5.27), (5.28), and (5.29) is written again as

$$\frac{d}{dx}\left(K\frac{dT}{dx}\right) + PT + Q = 0$$

At $x = x_0$, either $T = T_0$ or

$$K_0\frac{dT}{dx} + \alpha_0 T + \beta_0 = 0$$

At $x = x_L$, either $T = T_L$ or

$$K_L\frac{dT}{dx} + \alpha_L T + \beta_L = 0$$

Several engineering examples were presented in the previous section.

The corresponding general functional is

$$I = \int_{x_0}^{x_L}\left[\frac{1}{2}K\left(\frac{dT}{dx}\right)^2 - \frac{1}{2}PT^2 - QT\right]dx - \underbrace{\left(\frac{1}{2}\alpha_0 T^2 + \beta_0 T\right)_{x_0}}_{\substack{\text{added to functional if derivative}\\ \text{boundary condition applied at } x = x_0}} + \underbrace{\left(\frac{1}{2}\alpha_L T^2 + \beta_L T\right)_{x_L}}_{\substack{\text{added to functional if derivative}\\ \text{boundary condition applied at } x = x_L}} \quad \textbf{(5.30)}$$

The first term (integrated from x_0 to x_L) corresponds to the differential equation, while the second and third terms are used only if the derivative boundary condition applies at x_0 or x_L, respectively. As noted in Section 5.1, this functional must be minimized with respect to each unknown nodal value of T.

Table 5.1
Unknown Nodal Values for Various Boundary Conditions

Case	Boundary Condition at Left End $x=x_0$	Boundary Condition at Right End $x=x_L$	Unknown Nodal Values	Range of m
1	Specified value $T_1 = T_0$	Specified value $T_N = T_L$	$T_2, T_3, \ldots, T_{N-1}$	$2,3, \ldots, N-1$
2	Derivative	Specified value $T_N = T_L$	$T_1, T_2, T_3, \ldots, T_{N-1}$	$1,2,3, \ldots, N-1$
3	Specified value $T_1 = T_0$	Derivative	$T_2, T_3, \ldots, T_{N-1}, T_N$	$2,3, \ldots, N-1, N$
4	Derivative	Derivative	$T_1, T_2, T_3, \ldots, T_{N-1}, T_N$	$1,2,3, \ldots, N-1, N$

A stationary value is obtained when the first variation is zero: $\delta I = 0$. A set of minimization equations can easily be derived as (5.4)

$$\delta I = \sum_{m=1 \text{ or } 2}^{N-1 \text{ or } N} \frac{\partial I}{\partial T_m} \delta T_m = 0 \tag{5.31}$$

where m is the range of the unknown nodal values T_m. Table 5.1 shows four possible cases, depending upon whether the boundary condition specifies the nodal value at that point or the derivative boundary condition is applied. For the derivative boundary condition, the nodal value is unknown and part of the problem solution.

B. Steps for Element Matrices

Element matrices similar to those for spring and bar elements are employed with the general variational method (functional). They are assembled in the standard manner for solution of the global set of nodal values. Once the general element matrix and column are obtained, any linear second-order differential equation of the form (5.30) can be solved.

The evaluation of the element matrices involves the following steps:

1. Divide the functional into element-size integrals.
2. Write the element integral in terms of element interpolation functions.
3. Apply minimization equations to element integrals.
4. Evaluate element integral.
5. Form element matrices.

These will now be carried out.

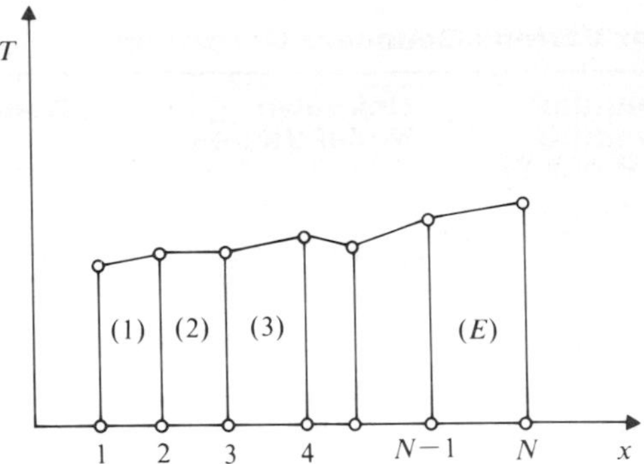

Figure 5.22 Division of variational principle into element-size integrals

C. Element Functional

Let the general functional I from equation (5.30) be divided into E elements and N nodes ($E = N - 1$), as shown in Figure 5.22.

$$I = I^{(1)} + I^{(2)} + I^{(3)} + \ldots + I^{(E)} = \sum_{e=1}^{E} I^{(e)}$$

The element integrals are

$$\begin{aligned} I = & \int_{x_1}^{x_2} \left[\frac{1}{2} K \left(\frac{dT}{dx} \right)^2 - \frac{1}{2} PT^2 - QT \right]^{(1)} dx \\ & + \int_{x_2}^{x_3} \left[\frac{1}{2} K \left(\frac{dT}{dx} \right)^2 - \frac{1}{2} PT^2 - QT \right]^{(2)} dx \\ & + \ldots \\ & + \int_{x_{N-1}}^{x_N} \left[\frac{1}{2} K \left(\frac{dT}{dx} \right)^2 - \frac{1}{2} PT^2 - QT \right]^{(E)} dx \\ & - (\alpha_0 T^2 + \beta_0 T)_{x_0}^{(1)} + (\alpha_L T^2 + \beta_L T)_{x_L}^{(E)} \end{aligned}$$

or for the eth element

$$\begin{aligned} I^{(e)} = & \int_{L^{(e)}} \left[\frac{1}{2} K \left(\frac{dT}{dx} \right)^2 - \frac{1}{2} PT^2 - QT \right]^{(e)} dx \\ & - \underbrace{\left(\frac{1}{2} \alpha_0 T^2 + \beta_0 T \right)_{x_0}^{(e)}}_{\substack{\text{included only if} \\ e=1 \text{ and} \\ \text{derivative} \\ \text{boundary condition} \\ \text{applied at } x=x_0}} + \underbrace{\left(\frac{1}{2} \alpha_L T^2 + \beta_L T \right)_{x_L}^{(e)}}_{\substack{\text{included only if} \\ e=E \text{ and derivative} \\ \text{boundary condition} \\ \text{applied at } x=x_L}} \end{aligned} \quad \textbf{(5.32)}$$

This is the contribution of element e to the total functional I. The two derivative boundary condition terms are included only if the conditions described below them apply. The role of the boundary terms is somewhat emphasized in the following derivation because students seem to have some difficulty with the concept of general derivative boundary conditions.

Within each element, the interpolation function for T is given by Equation (3.4) as

$$T^{(e)} = N_i^{(e)} T_i + N_j^{(e)} T_j$$

where, from Equation (3.6)

$$N_i^{(e)} = \frac{1}{L^{(e)}}\left(a_i^{(e)} + b_i^{(e)} x\right)$$

$$N_j^{(e)} = \frac{1}{L^{(e)}}\left(a_j^{(e)} + b_j^{(e)} x\right)$$

Also of use is the derivative, Equation (3.8)

$$\frac{dT^{(e)}}{dx} = \frac{1}{L^{(e)}}\left(b_i^{(e)} T_i + b_j^{(e)} T_j\right)$$

Let the functions $K(x)$, $P(x)$, $Q(x)$ be approximated by constant values within an element: $K = K^{(e)}$, $P = P^{(e)}$, $Q = Q^{(e)}$. Actually these represent the least accurate approximations for $K(x)$, $P(x)$, $Q(x)$ but keep the derivation simple. Linear approximations for these functions are considered in the problems.

The element integral becomes

$$\begin{aligned} I^{(e)} = & \int_{L^{(e)}} \left[\frac{1}{2} K^{(e)} \left(\frac{dT^{(e)}}{dx}\right)^2 - \frac{1}{2} P^{(e)} (T^{(e)})^2 - Q^{(e)} T^{(e)}\right] dx \\ & - \left[\frac{1}{2} \alpha_0 (T^{(e)})^2 + \beta_0 T^{(e)}\right]_{x_0} + \left[\frac{1}{2} \alpha_L (T^{(e)})^2 + \beta_L T^{(e)}\right]_{x_L} \end{aligned} \tag{5.33}$$

Substituting the appropriate expressions and taking the superscript (e) outside the brackets (it then applies to all terms inside the brackets),

$$\begin{aligned} I^{(e)} = & \int_{L^{(e)}} \left[\frac{1}{2} K \frac{1}{L^2} (b_i T_i + b_j T_j)^2 - \frac{1}{2} P (N_i T_i + N_j T_j)^2 \right. \\ & \left. - Q (N_i T_i + N_j T_j)\right]^{(e)} dx \\ & - \left(\frac{1}{2} \alpha_0 T_i^2 + \beta_0 T_i\right)_{x_0}^{(e)} + \left(\frac{1}{2} \alpha_L T_j^2 + \beta_L T_j\right)_{x_L}^{(e)} \end{aligned} \tag{5.34}$$

The boundary terms have been evaluated from the relations $N_i = 1$, $N_j = 0$ at $x = x_0 = X_i$ and $N_i = 0$, $N_j = 1$ at $x = x_L = X_j$. The integrals in this equation could be evaluated directly at this point, but a procedure more suited to computer implementation will be followed.

D. Minimization

As in Section 5.2, the element contribution to the minimization equations are

$$\frac{\partial I^{(e)}}{\partial T_i} = \frac{\partial}{\partial T_i}\left\{\int_{L^{(e)}}\left[\frac{1}{2}K\frac{1}{L^2}(b_iT_i + b_j T_j)^2 - \frac{1}{2}P(N_i T_i + N_j T_j)^2 - Q(N_i T_i + N_j T_j)\right]^{(e)} dx - \left(\frac{1}{2}\alpha_0 T_i^2 + \beta_0 T_i\right)_{x_0}^{(e)} + \left(\frac{1}{2}\alpha_L T_j^2 + \beta_L T_j\right)_{x_L}^{(e)}\right\} \tag{5.35a}$$

and

$$\frac{\partial I^{(e)}}{\partial T_j} = \frac{\partial}{\partial T_j}\left\{\int_{L^{(e)}}\left[\frac{1}{2}K\frac{1}{L^2}(b_i T_i + b_j T_j)^2 - \frac{1}{2}P(N_i T_i + N_j T_j)^2 - Q(N_i T_i + N_j T_j)\right]^{(e)} dx - \left(\frac{1}{2}\alpha_0 T_i^2 + \beta_0 T_i\right)_{x_0}^{(e)} + \left(\frac{1}{2}\alpha_L T_j^2 + \beta_L T_j\right)_{x_L}^{(e)}\right\} \tag{5.35b}$$

Differentiating with respect to the appropriate nodal T value yields

$$\frac{\partial I^{(e)}}{\partial T_i} = \int_{L^{(e)}}\left[K\frac{1}{L^2}b_i(b_i T_i + b_j T_j) - P N_i(N_i T_i + N_j T_j) - Q N_i\right]^{(e)} dx - (\alpha_0 T_i + \beta_0)_{x_0}^{(e)} \tag{5.36a}$$

and

$$\frac{\partial I^{(e)}}{\partial T_j} = \int_{L^{(e)}}\left[K\frac{1}{L^2}b_j(b_i T_i + b_j T_j) - P N_j(N_i T_i + N_j T_j) - Q N_j\right]^{(e)} dx + (\alpha_L T_j + \beta_L)_{x_L}^{(e)} \tag{5.36b}$$

The integrals are now easily evaluated with the aid of Equation (3.11).

The results, following integration, are

$$\frac{\partial I^{(e)}}{\partial T_i} = \left[\frac{K}{L}b_i(b_i T_i + b_j T_j) - PL\left(\frac{1}{3}T_i + \frac{1}{6}T_j\right) - \frac{1}{2}QL\right]^{(e)} - \underbrace{(\alpha_0 T_i + \beta_0)_{x_0}^{(e)}}_{\substack{\text{included only if} \\ e=1 \text{ and derivative} \\ \text{boundary condition} \\ \text{applied at } x=x_0}} \tag{5.37a}$$

and

$$\frac{\partial I^{(e)}}{\partial T_j} = \left[\frac{K}{L} b_j (b_i T_i + b_j T_j) - PL\left(\frac{1}{6} T_i + \frac{1}{3} T_j\right) - \frac{1}{2} QL\right]^{(e)} + \underbrace{(\alpha_L T_j + \beta_0)^{(e)}_{x_L}}_{\substack{\text{included only if} \\ e=N \text{ and derivative} \\ \text{boundary condition} \\ \text{applied at } x=x_L}} \tag{5.37b}$$

Both of these expressions could be rearranged into terms multiplied by T_i, those multiplied by T_j, and those remaining. Element matrices result.

E. Element Matrices

In element matrix form, these relations are written

$$\begin{Bmatrix} \dfrac{\partial I}{\partial T_i} \\ \dfrac{\partial I}{\partial T_j} \end{Bmatrix} = [B]^{(e)} \begin{Bmatrix} T_i \\ T_j \end{Bmatrix} - \{C\}^{(e)} \tag{5.38}$$

where

$$[B]^{(e)} = \underbrace{\begin{array}{c} \\ i \\ j \end{array} \overset{\begin{array}{cc} \qquad\qquad i \qquad\qquad & \qquad\qquad j \qquad\qquad \end{array}}{\begin{bmatrix} \dfrac{K}{L} b_i^2 - \dfrac{1}{3} PL - \alpha_0 & \dfrac{K}{L} b_i b_j - \dfrac{1}{6} PL \\ \dfrac{K}{L} b_i b_j - \dfrac{1}{6} PL & \dfrac{K}{L} b_j^2 - \dfrac{1}{3} PL + \alpha_L \end{bmatrix}^{(e)}}}_{\text{general element matrix } (2\times 2)}$$

and

$$\{C\} = \underbrace{\begin{array}{c} i \\ j \end{array} \begin{Bmatrix} \dfrac{1}{2} QL + \beta_0 \\ \dfrac{1}{2} QL - \beta_L \end{Bmatrix}^{(e)}}_{\text{general element column } (1\times 2)}$$

It is easily seen that the general element matrix $[B]$ is symmetric, as were the element matrices for the solid mechanics problems.

At the end points, the role of the constants $\alpha_0, \beta_0, \alpha_L, \beta_L$ has already been indicated in some detail. It is useful to do so again in summary form.

At $x = x_0$, α_0 and β_0 are included in the element matrix and column if and only if 1) it is the first element and 2) the derivative boundary condition is applied there.

At $x = x_L$, α_L and β_L are included in the element matrix and column if and only if 1) it is the last element and 2) the derivative boundary condition is applied there.

These principles are illustrated in the first example in the next section.

F. Derivation in Matrix Form

In order to give some guidance for future derivations that are not so easy, the derivation of the general element matrix is repeated in matrix form. The element functional (5.32) is

$$I^{(e)} = \int_{L^{(e)}} \left[\frac{1}{2} \frac{dT}{dx} K \frac{dT}{dx} - \frac{1}{2} TPT - QT \right]^{(e)} dx - \left[\frac{1}{2} T \, \alpha_0 \, T + \beta_0 \, T \right]_{x_0}^{(e)} + \left[\frac{1}{2} T \, \alpha_L \, T + \beta_L \, T \right]_{x_L}^{(e)}$$

The terms $T^{(e)}$ and $dT^{(e)}/dx$ are now to be written in matrix form.

Recall from Section 3.2 that the interpolation function is (3.9)

$$T^{(e)} = \{N\} \, \{T\}$$

where

$$\{N\} = \{N_i \quad N_j\}, \qquad \{T\} = \begin{Bmatrix} T_i \\ T_j \end{Bmatrix}$$

Also the derivative is (3.10).

$$\frac{dT^{(e)}}{dx} = \{D\} \, \{T\}$$

where $\{D\} = \dfrac{1}{L} \{b_i \quad b_j\}$. The element functional becomes (5.34)

$$\begin{aligned} I^{(e)} = \int_{L^{(e)}} \Bigg[& \frac{1}{2} \{T\}^T \{D\}^T K \, \{D\} \, \{T\} \\ & - \frac{1}{2} \{T\}^T \{N\}^T P \, \{N\} \, \{T\} - \{T\}^T \{N\}^T Q \Bigg] dx \\ & - \left[\frac{1}{2} \{T\}^T \{N\}^T \alpha_0 \, \{N\} \{T\} + \{T\}^T \{N\}^T \beta_0 \right]_{x_0}^{(e)} \\ & + \left[\frac{1}{2} \{T\}^T \{N\}^T \alpha_L \{N\} \{T\} + \{T\}^T \{N\}^T \beta_L \right]_{x_L}^{(e)} \end{aligned} \tag{5.39}$$

This is more complicated in appearance than (5.34), but the same form applies to higher order elements and other possibly useful applications.

Differentiating with respect to T_i and T_j (5.35) yields

$$\begin{Bmatrix} \dfrac{\partial I}{\partial T_i} \\ \dfrac{\partial I}{\partial T_j} \end{Bmatrix} = \int_{L^{(e)}} \left[\{D\}^T K \{D\} \{T\} - \{N\}^T P \{N\} \{T\} - \{N\}^T Q \right]^{(e)} dx$$

$$- \left[\{N\}^T \alpha_0 \{N\} \; \{T\} + \{N\}^T \beta_0 \right]_{x_0}^{(e)}$$

$$+ \left[\{N\}^T \alpha_L \{N\} \; \{T\} + \{N\}^T \beta_L \right]_{x_L}^{(e)}$$

These terms are grouped into those containing $\{T\}$, which give the element $[B]$ matrix, and those that are left, which form the element $\{C\}$ column.

Finally the element matrix is written in matrix form as

$$[B]^{(e)} = \int_{L^{(e)}} \left[\{D\}^T K \{D\} - \{N\}^T P \{N\} \right]^{(e)} dx - \left[\{N\}^T \alpha_0 \{N\} \right]_{x_0}^{(e)} + \left[\{N\}^T \beta_L \{N\} \right]_{x_L}^{(e)} \tag{5.40a}$$

The element column is

$$\{C\}^{(e)} = \int_{L^{(e)}} \left[\{N\}^T Q \right]^{(e)} dx - \left[\{N\}^T \beta_0 \right]_{x_0}^{(e)} + \left[\{N\}^T \beta_L \right]_{x_L}^{(e)} \tag{5.40b}$$

The integrals are easily evaluated with the resulting element matrices as given in (5.38).

5.6 Applications to Heat Transfer

One of the benefits of developing a finite element algorithm for the general boundary value problem just considered is that it is easily adapted to several applications. This section presents two heat transfer examples, one dealing with heat conduction in an infinite slab, the other with heat conduction and convection in a thin fin. These follow directly the boundary value Examples 5.5 and 5.6, where the governing differential equations and boundary conditions were derived. The second example, 5.9, indicates how input data are prepared for computer program FINONE.

Example 5.8 *Heat Conduction in Infinite Slab*

Given A solid slab of metal, described in some detail in Example 5.5, has length $L = 20$ mm and is infinitely long in the upward (y) direction as well as infinitely long in the direction into the paper (z). Figure 5.23 shows the geometry. The conduction coefficient is $k = 0.30$ W/mm-°C. At the left end, the temperature is maintained at $T_0 = 100$°C at $x = 0$, while at the right end there are three cases:

1. $T_L = 50$°C, $x = 20$ mm (specified temperature)
2. $Q_{xL} = 0$ W, $x = 20$ mm (insulated boundary condition)
3. $Q_{xL} = 0.60$ W, $x = 20$ mm (heat flow boundary condition)

Figure 5.24 illustrates the boundary conditions for each case.

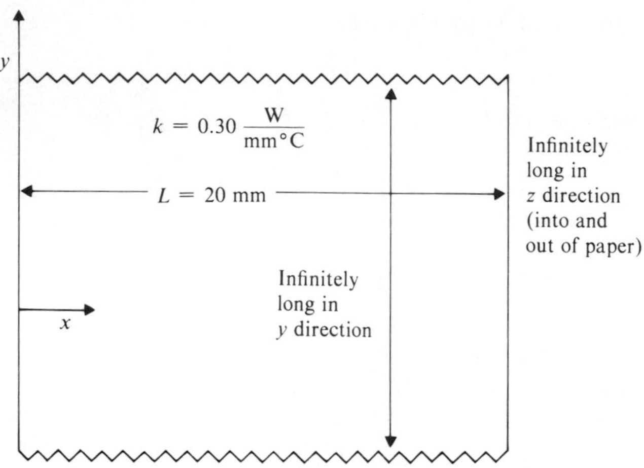

Figure 5.23 Geometry (Example 5.8—Heat conduction in infinite slab)

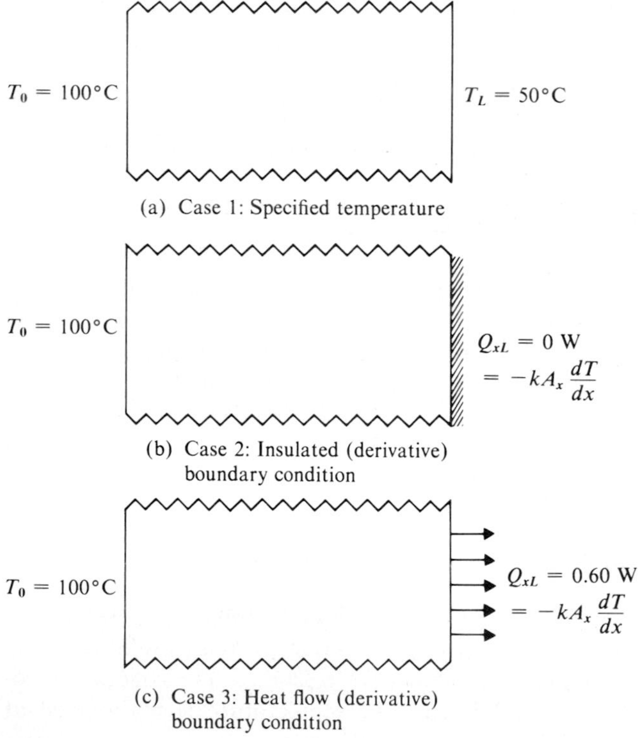

Figure 5.24 Three boundary-condition cases (Example 5.8—Heat conduction in infinite slab)

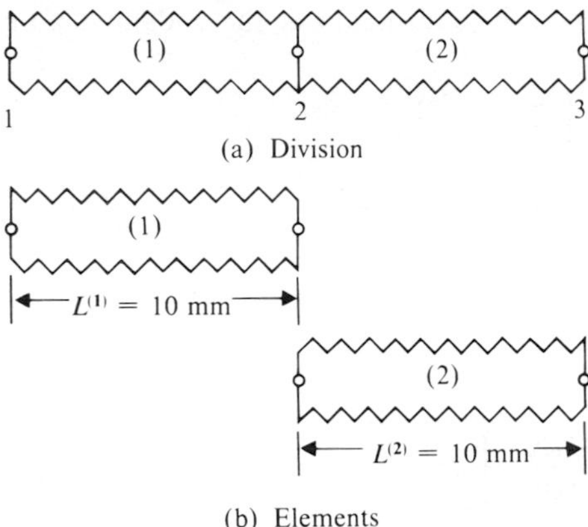

Figure 5.25 Division into elements (Example 5.8—Heat conduction in infinite slab)

Objective Determine the temperature distribution in the slab for all three boundary condition cases. Use a two-element model with simplex elements, as shown in Figure 5.25.

Solution The three cases are solved using the element matrix (5.38) and column (5.38) obtained in the previous section.

Case 1: Specified Temperature

The coefficient functions $K(x)$, $P(x)$, $Q(x)$ were determined in Example 5.5 to be $K(x) = kA_x = 0.30$ W-mm/°C, $P(x) = 0$, $Q(x) = 0$ where the unit cross sectional area A_x is taken as $A_x = dy\,dz = 1$ mm².

The corresponding functional is easily found to be (5.30)

$$I = \int_0^L \left[\frac{1}{2} kA_x \left(\frac{dT}{dx}\right)^2\right] dx$$

Here both end points are specified, so no extra terms are added to the functional. In dimensional units

$$I = \int_0^{20} \left[\frac{1}{2} 0.30 \left(\frac{dT}{dx}\right)^2\right] dx$$

Actually writing down the functional is not really necessary once the appropriate element matrices are derived.

For the heat transfer problem, the units are

$$I = \int_0^L \left[\frac{1}{2} kA_x \left(\frac{dT}{dx}\right)^2\right] dx = \left(\frac{\text{W}}{\text{mm-°C}}\right) (\text{mm}^2) \left(\frac{\text{°C}}{\text{mm}}\right)^2 \text{mm} = \text{W-°C}$$

The key parameters for heat conduction are thermal power (W) and temperature (°C). The functional I has the units of thermal power times temperature. This set of units does not seem to have any particularly obvious physical significance. It is similar to axial displacement of a rod in an interesting way.

Consider a rod with specified displacements at the ends and no applied loads. The functional (5.17) is

$$I = U = \int_0^L \left[\frac{1}{2} EA\, \epsilon^2\right] dx$$

The strain is

$$\epsilon = \frac{du}{dx}$$

so the functional becomes

$$I \int_0^L \left[\frac{1}{2} EA \left(\frac{du}{dx}\right)^2\right] dx$$

The units are

$$I = \left(\frac{\text{kN}}{\text{mm}^2}\right) (\text{mm}^2) \left(\frac{\text{mm}}{\text{mm}}\right)^2 (\text{mm}) = \text{kN-mm}$$

For the rod, the key parameters are force (kN) and displacement (mm). The functional has the units of force times displacement, which is similar to the heat conduction example. In the rod, however, the force times the displacement has the physical significance of potential energy. This analogy is developed further in Section 5.8.

The element lengths are equal, with value $L^{(1)} = 10$ mm, $L^{(2)} = 10$ mm. Now the element coefficients for element 1 are $K^{(1)} = 0.30$ W-mm/°C, $P^{(1)} = 0$, $Q^{(1)} = 0$. For element 2, the coefficients are $K^{(2)} = 0.30$ W-mm/°C, $P^{(2)} = 0$, $Q^{(2)} = 0$.

Let the element matrix (5.38) be rewritten here for clarity

$$B^{(e)} = \begin{bmatrix} \frac{K}{L} b_i^2 - \frac{1}{3} PL & \frac{K}{L} b_i\, b_j - \frac{1}{6} PL \\ \frac{K}{L} b_i\, b_j - \frac{1}{6} PL & \frac{K}{L} b_j^2 - \frac{1}{3} PL \end{bmatrix}^{(e)}$$

Both α_0 and α_L are omitted because derivative boundary conditions are not applied in Case 1. Recall that $b_i = -1$, $b_j = 1$. The element matrix has the units determined by K/L, which are

$$\frac{K}{L} = \left(\frac{\text{W-mm}}{°\text{C}}\right)\left(\frac{1}{\text{mm}}\right) = \frac{\text{W}}{°\text{C}}$$

This represents the rate of change of thermal power with respect to temperature. For a bar element, the units are those of stiffness or axial force per displacement. Thus the thermal element may be thought of as having the units of "thermal stiffness"—thermal power per temperature.

For element 1, the element matrix is

$$[B]^{(1)} = \begin{matrix} \\ 1 \\ 2 \end{matrix} \begin{bmatrix} \overset{1}{\frac{(0.30)}{10}(-1)(-1) - \frac{1}{3}(0)(10)} & \overset{2}{\frac{(0.30)}{10}(-1)(1) - \frac{1}{6}(0)(10)} \\ \frac{(0.30)}{10}(-1)(1) - \frac{1}{6}(0)(10) & \frac{(0.30)}{10}(1)(1) - \frac{1}{3}(0)(10) \end{bmatrix}^{(1)}$$

with the result

$$[B]^{(1)} = \begin{matrix} \\ 1 \\ 2 \end{matrix} \begin{matrix} \begin{matrix} 1 & \quad\quad 2 \end{matrix} \\ \begin{bmatrix} 0.030 & -0.030 \\ -0.030 & 0.030 \end{bmatrix}^{(1)} \end{matrix} \left(\frac{\text{W}}{°\text{C}}\right)$$

It is easily seen that all of the parameters for element 2 are exactly the same as those for element 1. Thus

$$[B]^{(2)} = \begin{matrix} \\ 2 \\ 3 \end{matrix} \begin{matrix} \begin{matrix} 2 & \quad\quad 3 \end{matrix} \\ \begin{bmatrix} 0.030 & -0.030 \\ -0.030 & 0.030 \end{bmatrix}^{(2)} \end{matrix} \left(\frac{\text{W}}{°\text{C}}\right)$$

Also the element columns (5.38) are

$$C^{(1)} = \begin{Bmatrix} \frac{1}{2} QL \\ \frac{1}{2} QL \end{Bmatrix}^{(1)} = \begin{Bmatrix} \frac{1}{2}(0)(10) \\ \frac{1}{2}(0)(10) \end{Bmatrix}^{(1)} = \begin{matrix} 1 \\ 2 \end{matrix} \begin{Bmatrix} 0 \\ 0 \end{Bmatrix}^{(1)}$$

and

$$C^{(2)} = \begin{Bmatrix} \frac{1}{2} QL \\ \frac{1}{2} QL \end{Bmatrix}^{(2)} = \begin{Bmatrix} \frac{1}{2}(0)(10) \\ \frac{1}{2}(0)(10) \end{Bmatrix}^{(2)} = \begin{matrix} 2 \\ 3 \end{matrix} \begin{Bmatrix} 0 \\ 0 \end{Bmatrix}^{(2)}$$

The element properties are complete for Case 1.

Assembling the element matrices in the normal fashion gives the global matrices

$$\begin{bmatrix} 0.030 & -0.030 & 0 \\ -0.030 & 0.060 & -0.030 \\ 0 & -0.030 & 0.030 \end{bmatrix} \begin{Bmatrix} T_1 \\ T_2 \\ T_3 \end{Bmatrix} = \begin{Bmatrix} 0 \\ 0 \\ 0 \end{Bmatrix}$$

Taking into account the two specified boundary conditions for Case 1 in the normal manner, $T = 100°\text{C}$, $x = 0$ mm and $T = 50°\text{C}$, $x = 20$ mm produces the final set of global equations for T_1, T_2, T_3 as

$$\begin{bmatrix} 0.030 & 0 & 0 \\ 0 & 0.060 & 0 \\ 0 & 0 & 0.030 \end{bmatrix} \begin{Bmatrix} T_1 \\ T_2 \\ T_3 \end{Bmatrix} = \begin{Bmatrix} 3.0 \\ 4.5 \\ 1.5 \end{Bmatrix}$$

The finite element solution at the node points is

$$\begin{Bmatrix} T_1 \\ T_2 \\ T_3 \end{Bmatrix} = \begin{Bmatrix} 100 \\ 75 \\ 50 \end{Bmatrix} \qquad \text{Case 1: Specified temperatures finite element solution}$$

This simple boundary value problem has the exact solution $T(x) = 100 - 2.5x$ with the evaluation at the node points

$$\begin{Bmatrix} T_1 \\ T_2 \\ T_3 \end{Bmatrix} = \begin{Bmatrix} 100 \\ 75 \\ 50 \end{Bmatrix} \qquad \text{Case 1: Exact solution}$$

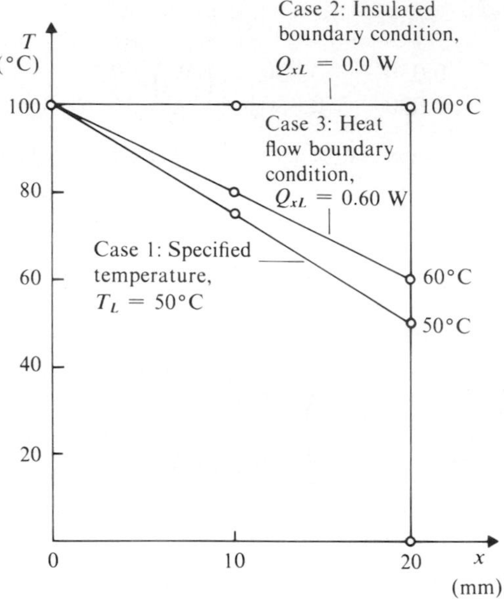

Figure 5.26 Temperature profiles (Example 5.8—Heat conduction in infinite slab)

They match exactly (as they should) when the analytical solution to the boundary value problem is linear. Figure 5.26 plots the result for Case 1.

Case 2: Insulated Boundary Condition

The insulated boundary condition at the right end is

$$Q_{xL} = -kA_x \frac{dT}{dx} = 0, \quad x = 20 \text{ mm}$$

Comparing this to the general derivative boundary condition

$$K_L \frac{dT}{dx} + \alpha_L T + \beta_L = 0$$

it is immediately apparent that $\alpha_L = 0, \beta_L = 0$ for the insulated boundary.

Element 1 has the same functional as in Case 2. In the special case of the insulated boundary, element 2 also has the same I value as in Case 1. Because both α_L and β_L have the value zero, the element functional is the same as in Case 1 (actually, a term of value zero is added to it at $x = x_L$). Similarly, the element matrices for element 2 have the same value as in Case 1 (actually, zero is added to the element $[B]$ matrix and subtracted from the element $\{C\}$ column). Thus the assembled global matrix is also the same as in Case 1.

The difference between Case 2 and Case 1 arises in the treatment of the specified nodal values. Only one is used: $T = 100°C$, $x = 0$ mm. Now the global equations are

$$\begin{bmatrix} 0.030 & 0 & 0 \\ 0 & 0.060 & -0.030 \\ 0 & -0.030 & 0.030 \end{bmatrix} \begin{Bmatrix} T_1 \\ T_2 \\ T_3 \end{Bmatrix} = \begin{Bmatrix} 3.0 \\ 3.0 \\ 0.0 \end{Bmatrix}$$

The finite element solution is

$$\begin{Bmatrix} T_1 \\ T_2 \\ T_3 \end{Bmatrix} = \begin{Bmatrix} 100 \\ 100 \\ 100 \end{Bmatrix}$$

Case 2: Insulated boundary condition finite element solution

Again, the exact solution is easy to obtain. It is $T(x) = 100$, which matches the finite element solution. The physical significance is simply that no sources (or sinks) of heat occur in the metal slab and no heat is allowed to pass through the right boundary. Thus the temperature must be constant and the heat flow Q_x must be zero. Figure 5.26 plots the solution to Case 2. It can be seen from this case that treating an insulated boundary is accomplished by doing nothing at x_L (adding $\alpha_L = 0$ and $\beta_L = 0$ to the element matrices does nothing).

Case 3: Heat Flow Boundary Condition

It is now desired to include a given value of heat flow at the right boundary

$$Q_{xL} = -kA_x \frac{dT}{dx} = 0.60 \text{ W}$$

and rearranging slightly yields

$$kA_x \frac{dT}{dx} + Q_{xL} = 0.030 \frac{dT}{dx} + 0.60 = 0$$

Comparing this to the general derivative boundary condition

$$K_L \frac{dT}{dx} + \alpha_L T + \beta_L = 0$$

it is apparent that $\alpha_L = 0$, $\beta_L = 0.60$ W. These are the derivative boundary condition constants.

The functional for element 1 is the same as in Cases 1 and 2 but it is different for element 2. The result is (5.32)

$$I^{(2)} = \int_{L^{(2)}} \left[\frac{1}{2} kA_x \left(\frac{dT}{dx} \right)^2 \right] dx + \left(\frac{1}{2} \alpha_L T^2 + \beta_L T \right)_{x_L}$$

or substituting numerical values

$$I^{(2)} = \int_{L^{(2)}} \left[\frac{1}{2} (0.030) \left(\frac{dT}{dx} \right)^2 \right] dx + (0.60T)_{x_L}$$

As noted earlier in this example, the function is not really required here, but the change in I due to the heat flow boundary is easily illustrated at this point.

In evaluating the element matrices, only one term (α_L) need be added to $[B]^{(2)}$

$$[B]^{(2)} = \begin{matrix} & 2 & 3 \\ 2 & 0.030 & -0.030 \\ 3 & -0.030 & 0.030 + \alpha_L \end{matrix}$$

and one term (β_L) be subtracted from $\{C\}$

$$\{C\}^{(2)} = \begin{matrix} 2 \\ 3 \end{matrix} \begin{Bmatrix} 0 \\ 0 - \beta_L \end{Bmatrix}^{(2)}$$

The numerical result is

$$[B]^{(2)} = \begin{bmatrix} 0.030 & -0.030 \\ -0.030 & 0.030 \end{bmatrix}^{(2)}$$

(since $\alpha_L = 0$, there is no change) and

$$\{C\}^{(2)} = \begin{Bmatrix} 0 \\ -0.60 \end{Bmatrix}^{(2)}$$

All element properties have been evaluated.

Assembling the global matrices for the third and last time yields

$$\begin{bmatrix} 0.030 & -0.030 & 0 \\ -0.030 & 0.060 & -0.030 \\ 0 & -0.030 & 0.030 \end{bmatrix} \begin{Bmatrix} T_1 \\ T_2 \\ T_3 \end{Bmatrix} = \begin{Bmatrix} 0 \\ 0 \\ -0.60 \end{Bmatrix}$$

Including the specified value $T_1 = 100$ in the normal way produces the global equations

$$\begin{bmatrix} 0.030 & 0 & 0 \\ 0 & 0.060 & -0.030 \\ 0 & -0.030 & 0.030 \end{bmatrix} \begin{Bmatrix} T_1 \\ T_2 \\ T_3 \end{Bmatrix} = \begin{Bmatrix} 3.0 \\ 3.0 \\ -0.60 \end{Bmatrix}$$

The finite element solution is

$$\begin{Bmatrix} T_1 \\ T_2 \\ T_3 \end{Bmatrix} = \begin{Bmatrix} 100 \\ 80 \\ 60 \end{Bmatrix}$$

Case 3: Heat flow boundary condition finite element solution

Again, the exact solution, $T(x) = 100 - 2x$, matches the finite element solution everywhere. Figure 5.26 plots the results for all three cases. Table 5.2 summarizes the results.

Several techniques are illustrated in this example. The use of general matrices is shown. The method of including derivative boundary conditions is developed. Finally, the overall solution procedure for finite elements, starting from the differential equation to the nodal temperatures, is worked out by hand. Usually this is done by computer, as illustrated in the next example.

Table 5.2
Temperatures for Three Cases
(Example 5.8: Heat Conduction in Infinite Slab)

Position *x* (mm)	Case 1: Specified Temperature *T* (°C)	Case 2: Insulated Boundary *T* (°C)	Case 3: Heat Flow Boundary *T* (°C)
0	100	100	100
10	75	100	80
20	50	100	60

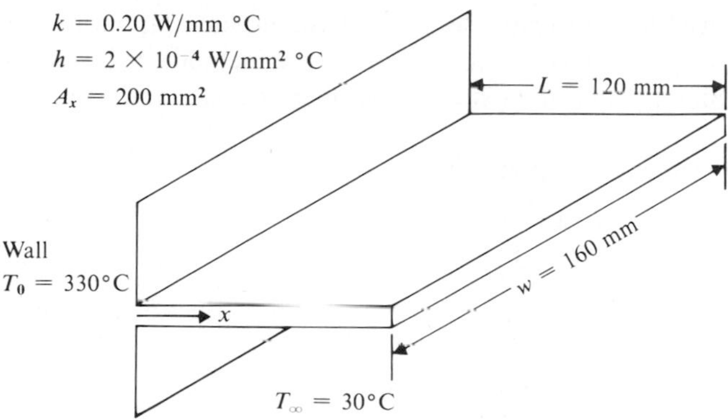

Figure 5.27 Schematic of fin (Example 5.9—Conduction and convection in thin fin)

Example 5.9 *Heat Conduction and Convection in Thin Fin*

Given The differential equation and boundary conditions for this problem were obtained in Example 5.6. Figure 5.27 illustrates a perspective view of the thin aluminum fin. It has thickness t and width W. Numerical parameters are $L = 120$ mm, $W = 160$ mm, $t = 1.25$ mm. This satisfies the assumptions of thinness

$$\frac{t}{L} = \frac{1.25}{120} << 1$$

$$\frac{t}{W} = \frac{1.25}{160} << 1$$

so the temperature is a function of x only. At any position x along the fin, the cross sectional area A_x is $A_x = Wt = (160\text{ mm})(1.25\text{ mm}) = 200\text{ mm}^2$. The conduction coefficient k for aluminum is $k = 0.20$ W/mm-°C, and the convection coefficient h for forced air convection over the fin is $h = 2 \times 10^{-4}$ W/mm²-°C. The temperature of the surrounding air is taken as $T_\infty = 30°\text{C}$. At the end points, the temperatures are $T_0 = 330°\text{C}$, $x = 0$, and $T_L = 30°\text{C}$, $x = L$.

The exact solution for this problem is

$$T = (T_0 - T_\infty)\,(\cosh mx - \frac{\cosh mL}{\sinh ml} \sinh mx) + T_\infty$$

where

$$m^2 = \frac{2h}{kt} = \frac{2hW}{kA_x}$$

This may be compared to the finite element solution.

Objective Use three equal-length finite elements to obtain the temperature profile in the fin. Compare this to the exact solution.

Solution The procedure followed for this example is to set the problem up either for solution by hand calculations (to better illustrate the details of the method) or by use of computer program FINONE. Simple input data such as number of nodes and elements is taken first.

The numbers of nodes and elements are given by

Number of nodes, NP = 4
Number of elements, NE = 3

Node numbers and nodal coordinates are given in Table 5.3(a), where the units are mm. Element/node connectivity data are presented in Table 5.3(b).

The differential equation is derived in Example 5.6 as

$$\underbrace{\frac{d}{dx}\left(kWt\,\frac{dT}{dx}\right)}_{\substack{\text{Heat conduction}\\ \text{along fin in}\\ x\text{ direction}}} - \underbrace{2hW\,(T - T_\infty)}_{\substack{\text{Heat convection out}\\ \text{of top and bottom of}\\ \text{fin in } z \text{ direction}}} = 0$$

Comparing this to the general differential equation (5.27)

$$\frac{d}{dx}\left[K(x)\,\frac{dT}{dx}\right] + P(x)T + Q(x) = 0$$

gives the general coefficient functions as $K(x) = kWt$, $P(x) = -2hW$, and $Q(x) = 2hWT_\infty$. Substituting numerical values for the coefficient functions yields

$$K(x) = kWt = \left(0.20\,\frac{\text{W}}{\text{mm-°C}}\right)(160\text{ mm})\,(1.25\text{ mm}) = 40\,\frac{\text{W-mm}}{\text{°C}}$$

$$P(x) = -2hW = -2\left(2 \times 10^{-4}\,\frac{\text{W}}{\text{mm}^2\text{-°C}}\right)(160\text{ mm})$$

$$= -0.064\,\frac{\text{W}}{\text{mm-°C}}$$

Table 5.3
Input Data for Program FINONE
(Example 5.9: Heat Conduction and Convection in Thin Fin)

(a) Nodal Coordinates

Global Node Number (INODE)	*x* Coordinate of Node (XNODE) (inches)
1	0.0
2	40.0
3	80.0
4	120.0

(b) Element/Node Connectivity Data

Element Number (NEL)	Global Node Number	
	Node i *(NODEI)*	*Node j* *(NODEJ)*
1	1	2
2	2	3
3	3	4

(c) Coefficient Functions (Only One Element Is Input Since All Have the Same Values)

Element Number (IEL)	Coefficient Functions		
	Value of K(x) in Element (KXE) (W-mm/°C)	*Value of P(x) in Element (PE) (W/mm-°C)*	*Value of Q(x) in Element (QE) (W/mm)*
1	40.0	−0.064	1.92

(d) Prescribed Nodal Values

Global Node Number (IBOUND)	Prescribed Nodal Value (TBOUND) (°C)
1	330.0
4	30.0

$$Q(x) = 2hWT_\infty = 2\left(2 \times 10^{-4} \frac{\text{W}}{\text{mm}^2\text{-}^\circ\text{C}}\right) (160 \text{ mm})\ (30^\circ\text{C})$$

$$= 1.92 \frac{\text{W}}{\text{mm}}$$

With these values, the differential equation becomes

$$40 \frac{d^2T}{dx^2} - 0.064\,T + 1.92 = 0$$

since all of the parameters are constant except T.

A coefficient parameter (ICO) in FINONE indicates that the coefficient functions have the same numerical values for each element: Coefficient parameter, ICO = 1. Table 5.3(c) then gives the input data. The number of prescribed nodal values, NBOUND, is 2. Node numbers and actual boundary values are presented in Table 5.3(d). There are no derivative boundary values for this example, so the total number of nodes for which derivative boundary values are given, NDERIV, is 0. Also the output parameter, NOUT, is taken as 0 for maximum output.

With these coefficient functions, the functional (5.30) can be immediately written as

$$I = \int_0^L \left[\frac{1}{2}(kWt)\left(\frac{dT}{dx}\right)^2 - \frac{1}{2}(-2hW)T^2 - (2hWT_\infty)T\right] dx$$

Note that no derivative boundary terms are included because the temperature T is specified at both ends. The general functional (5.30) is

$$I = \int_{x_0}^{x_L} \left[\frac{1}{2} K \left(\frac{dT}{dx}\right)^2 - \frac{1}{2} PT^2 - QT\right] dx$$

with the numerical result

$$I = \int_0^{120} \left[20 \left(\frac{dT}{dx}\right)^2 + 0.032T^2 - 1.92T\right] dx$$

The element matrices have the form

$$\begin{Bmatrix} \dfrac{\partial I}{\partial T_i} \\ \dfrac{\partial I}{\partial T_j} \end{Bmatrix}^{(e)} = [B]^{(e)} \begin{Bmatrix} T_i \\ T_j \end{Bmatrix}^{(e)} - \{C\}^{(e)}$$

where the element matrix is

$$[B]^{(e)} = \begin{bmatrix} \dfrac{K}{L} b_i^2 - \dfrac{1}{3} PL & \dfrac{K}{L} b_i b_j - \dfrac{1}{6} PL \\ \dfrac{K}{L} b_i b_j - \dfrac{1}{6} PL & \dfrac{K}{L} b_j^2 - \dfrac{1}{3} PL \end{bmatrix}^{(e)}$$

and the element column is

$$\{C\}^{(e)} = \begin{Bmatrix} \frac{1}{2} QL \\ \frac{1}{2} QL \end{Bmatrix}^{(e)}$$

Evaluating one term as an example gives

$$B_{ii} = \left(40 \frac{\text{W-mm}}{°\text{C}}\right)\left(\frac{1}{40 \text{ mm}}\right)(-1)(-1)$$

$$- \frac{1}{3}\left(-0.064 \frac{\text{W}}{\text{mm-}°\text{C}}\right)(40 \text{ mm})$$

$$= 1.000 \frac{\text{W}}{°\text{C}} + 0.8533 \frac{\text{W}}{°\text{C}} = 1.8533 \frac{\text{W}}{°\text{C}}$$

and

$$C_i = \frac{1}{2}\left(1.92 \frac{\text{W}}{\text{mm}}\right)(40 \text{ mm}) = 38.40 \text{ W}$$

All other terms are easily evaluated.

The element matrix is

$$[B]^{(e)} = \begin{matrix} & i & j \\ i & 1.8533 & -0.5733 \\ j & -0.5733 & 1.8533 \end{matrix}^{(e)} (\text{W}/°\text{C})$$

where all three element matrices are the same (for this problem—in general they will not be the same). Also the element column is

$$\{C\}^{(e)} = \begin{matrix} i \\ j \end{matrix} \begin{Bmatrix} 38.40 \\ 38.40 \end{Bmatrix}^{(e)} (\text{W})$$

The element columns are all the same for this example. There is not really any need to write out the element matrices for each element.

Assembling all of the element matrices into global form yields

$$\begin{bmatrix} 1.8533 & -0.5733 & 0 & 0 \\ -0.5733 & 3.7067 & -0.5733 & 0 \\ 0 & -0.5733 & 3.7067 & -0.5733 \\ 0 & 0 & -0.5733 & 3.7067 \end{bmatrix} \begin{Bmatrix} T_1 \\ T_2 \\ T_3 \\ T_4 \end{Bmatrix} = \begin{Bmatrix} 38.40 \\ 76.80 \\ 76.80 \\ 38.40 \end{Bmatrix}$$

After including the boundary conditions $T_1 = 330$, $T_4 = 30$, the global equations become

$$\begin{bmatrix} 1.8533 & 0 & 0 & 0 \\ 0 & 3.7067 & -0.5733 & 0 \\ 0 & -0.5733 & 3.7067 & 0 \\ 0 & 0 & 0 & 1.8533 \end{bmatrix} \begin{Bmatrix} T_1 \\ T_2 \\ T_3 \\ T_4 \end{Bmatrix} = \begin{Bmatrix} 611.60 \\ 266.00 \\ 94.00 \\ 55.60 \end{Bmatrix}$$

The solution is obtained by a standard solution technique.

Table 5.4
Comparison of Three-Element and Analytical Solutions (Example 5.9: Heat Conduction and Convection in Thin Fin)

Position *x* (mm)	Three-Element Solution °C	Analytical Solution °C	Error %
0	330	330	—
20	203.77	164.67	24
40	77.54	90.42	−14
60	57.45	57.30	0.4
80	37.35	42.22	−12
100	33.67	35.48	−5
120	30.00	30.00	—

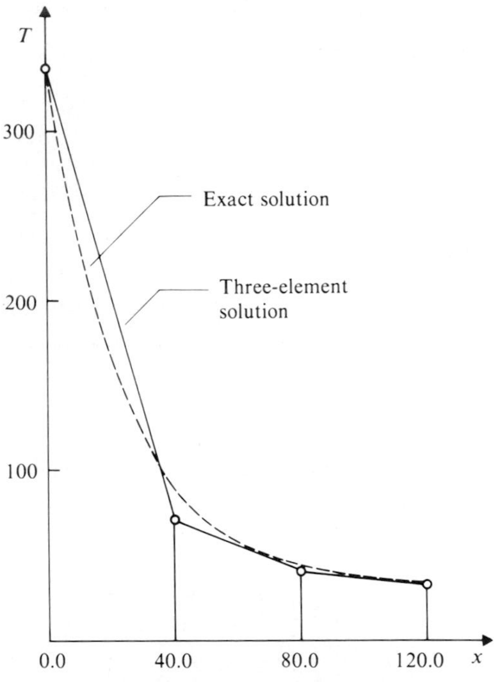

Figure 5.28 Comparison of three-element and analytical solution (Example 5.9—Conduction and convection in thin fin)

The final solution at all four node points is

$$\begin{Bmatrix} T_1 \\ T_2 \\ T_3 \\ T_4 \end{Bmatrix} = \begin{Bmatrix} 330.00 \\ 77.54 \\ 37.35 \\ 30.00 \end{Bmatrix} \quad \text{Three-element solution}$$

At the same node points, the exact solution is

$$\begin{Bmatrix} T_1 \\ T_2 \\ T_3 \\ T_4 \end{Bmatrix} = \begin{Bmatrix} 330.00 \\ 90.42 \\ 42.22 \\ 30.00 \end{Bmatrix} \quad \text{Analytical solution}$$

A more detailed comparison is shown in Table 5.4 and in Figure 5.28. The error is relatively large in the element where the slope changes rapidly. Of course, a linear approximation will not be too good in this region; additional elements should be used to produce a more accurate solution.

5.7 Applications to Fluid Mechanics and Solution Accuracy

This section presents an application to fluid flow between two flat plates with a pressure gradient (Poiseuille flow). The objective is to obtain the velocity profile between the plates. In the previous two examples, the problem parameters were constant. In this example the fluid viscosity is variable because the plates are heated to demonstrate the treatment of variable parameters in addition to an application to fluid mechanics. The example also illustrates the accuracy consideration discussed next.

A. Solution Accuracy

One of the most important considerations in finite element analysis is that of the accuracy of the solution. If the analytical (exact) solution to the problem is known, the accuracy is easy to determine. Nearly always the analytical solution is not known—this is really the reason for using finite elements. Two questions are typically asked: Have the right number of elements been used? Does the solution match the physical reality? The first question can be addressed here. The second can be answered only as follows: If the boundary value problem used accurately represents the physical process, then the finite element solution will match physical reality (assuming question 1 has been answered in the affirmative).

The question of how many elements is the right number depends upon several factors. These include the solution quantity desired, the element order, the type of differential equation, and the numerical accuracy of computer used. The technique for determining this is relatively straightforward. A plot of the desired quantity, say S_n, at some node point X_n versus the number of elements (nodes) is used. Figure 5.29 illustrates a typical plot

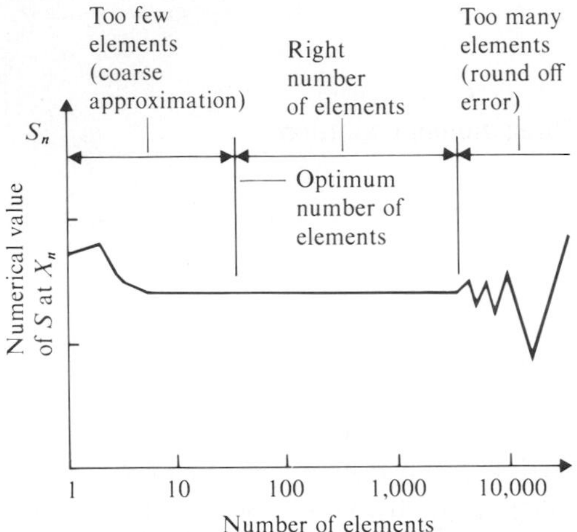

Figure 5.29 Numerical accuracy versus number of elements

of S_n as the number of elements increases. When the value of S_n changes less than the desired accuracy (say, by 0.1%) upon doubling the number of elements, then the correct number has been employed.

If fewer elements are employed, significant error occurs because of the coarseness of the approximation of the interpolation functions to the actual solution within each element. The plot may oscillate up and down somewhat or monotonically approach the correct solution. As the number of elements increases further, the nodal value chosen for scrutiny remains nearly constant. The optimum, shown in Figure 5.29, is the number of elements that gives the accuracy desired at the least computer cost. As the number of elements reaches some large value, the rounding off process in the computer begins to introduce errors; sometimes this problem can be eliminated by using double precision.

As indicated earlier, the size of the solution error depends upon the quantity desired. When the derivative $S = dT/dx$ is the objective, generally the error is larger. Thus, for a given number of elements, the heat flow through the thin fin in the previous example would be determined with lower accuracy than would the temperature. This is because small errors in T can produce large errors in slope. Alternatively, a quantity dependent upon the integral of T is calculated more accurately than T itself. Thus, the convection heat flow out through the upper and lower surfaces of the fin

$$S = Q_{\text{convection}} = \int_0^L 2hW(T - T_\infty)\, dx$$

is quite accurate, even for a small number of elements. This is illustrated in the next example.

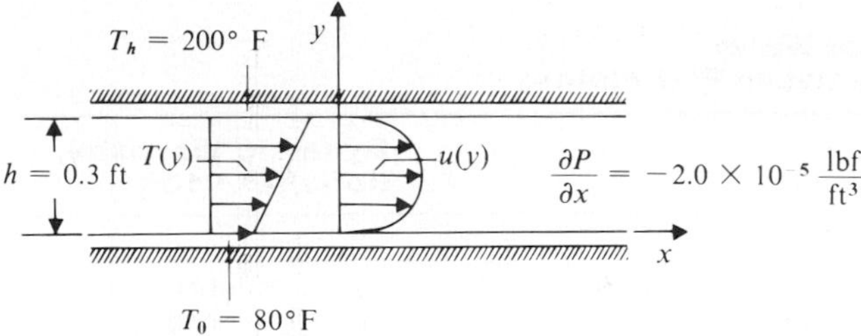

Figure 5.30 Problem geometry (Example 5.10—Fluid flow between heated flat plates)

B. Example

Example 5.10 *Fluid Flow Between Heated Flat Plates*

Given Laminar flow of a viscous fluid occurs between two infinitely long and wide flat plates, as shown in Figure 5.30. The governing differential equation was derived in Example 5.7 as

$$\frac{d}{dy}\left(\mu \frac{du}{dy}\right) - \frac{\partial P}{\partial x} = 0$$

For this example, both plates are taken as fixed in space. Thus the boundary conditions are $u = 0$ at $y = 0$ and $u = 0$ at $y = h$. The density ρ is constant. The numerical values (except for viscosity) are

$$h = 0.3 \text{ ft}$$

$$\frac{\partial P}{\partial x} = -2.0 \times 10^{-5} \text{ lbf/ft}^3$$

$$\rho = 1.90 \text{ lbf-sec}^2/\text{ft}^4$$

These are values for water at the appropriate temperature and for laminar flow (Reynolds number less than 2,000).

The viscosity varies because the temperature varies across the region between the plates: $\mu = \mu(T)$. In this example, the lower plate is maintained at $T_0 = 80°$F at $y = 0$ ft while the upper plate is maintained at $T_h = 200°$F at $y = 0.3$ ft. Table 5.5 gives a table of fluid viscosity vs. temperature over that temperature range, as obtained from an engineering handbook. The temperature distribution from $y = 0$ to $y = h$ is given as linear

$$T(y) = T_0 + (T_h - T_0)\frac{y}{h}$$

Thus the viscosity $\mu(y)$ can be obtained from the table at the start of the example.

Table 5.5
Viscosity vs. Temperature for Water
(Example 5.10: Fluid Flow Between Flat Plates)

Temperature, T (°F)	**Dynamic Viscosity, μ (lbf-s/ft^2 $\times 10^5$)**
80	1.60
120	1.40
160	1.10
200	0.70

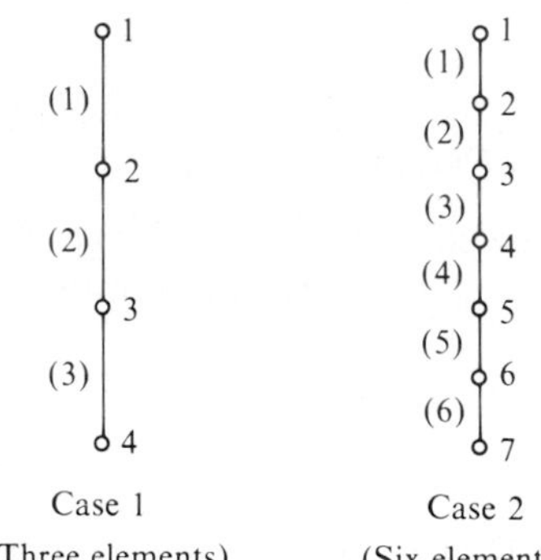

Figure 5.31 Division into elements (Example 5.10—Fluid flow between heated flat plates)

Objective Determine the functional for the problem. Solve for the velocity profile $u(y)$ using two divisions into elements. Also calculate the shear stress exerted by the fluid on both walls and the volume rate of flow between the plates.

Case 1: Three equal-length elements

Case 2: Six equal-length elements

Figure 5.31 shows the elements and nodes. Let the two accuracy quantities be

$$S = \tau_{yx} \text{ at } y = 0$$

$$Q = \int_0^h u\,dy$$

Solution The differential equation was derived in Example 5.7 as

$$\frac{d}{dy}\left(\mu \frac{du}{dy}\right) - \frac{\partial P}{\partial x} = 0$$

with the boundary conditions $u = 0, y = 0$ and $u = 0, y = h$. Comparing to the general equation

$$\frac{d}{dy}\left[K(y)\frac{dT}{dy}\right] + P(y)T + Q(y) = 0$$

yields the coefficient functions

$$K(y) = \mu$$

$$P(y) = 0.0$$

$$Q(y) = -\frac{\partial P}{\partial x}$$

As noted, the viscosity is a function of y.

It is worthwhile to consider the parameters involved in the differential equation. The term

$$\mu \frac{du}{dy} = \tau_{yx}$$

represents the shear stress acting on the face $dxdz$ of the fluid particle. The differential equation represents the change of this quantity over the vertical distance h.

$$\frac{d}{dy}\left(\mu \frac{du}{dy}\right) = \frac{d}{dy}(\tau_{yx}) = \frac{\text{Shear stress}}{\text{Length}} = \left(\frac{\text{lbf}}{\text{ft}^2}\right)\left(\frac{1}{\text{ft}}\right)$$

This is similar to the parameter for the axial displacement of a rod—force/length—where the shear stress is equivalent to the force.

The general functional (5.30) is

$$I = \int_{x_0}^{x_L} \left[\frac{1}{2} K \left(\frac{dT}{dx}\right)^2 - \frac{1}{2} PT^2 - QT\right] dx$$

and the specific functional for this example is

$$I = \int_0^h \left[\frac{1}{2} \mu \left(\frac{du}{dy}\right)^2 + \frac{\partial P}{\partial x}\right] dy$$

with the units

$$I = \left(\frac{\text{lbf-s}}{\text{ft}^2}\right)\left(\frac{\text{ft}}{\text{s}}\right)^2 (\text{ft}) = \left(\frac{\text{lbf}}{\text{ft}^2}\right)\left(\frac{\text{ft}}{\text{s}}\right)$$

These are the units of shear stress times velocity. Recall that the displacement of a rod has a functional that has the units of energy—force times displacement. This energy is stored inside the rod as potential energy, which is reclaimed when the displacement returns to zero while acting against some force. In viscous fluid flow, the energy is not stored but converted to heat through viscous action (dissipation). Thus the functional represents the shear stress (force per area) times velocity (displacement per time), or energy loss per area per time

$$I = \frac{\text{Force}}{\text{Area}} \times \frac{\text{Displacement}}{\text{Time}} = \frac{\text{Energy loss}}{\text{Area} \times \text{Time}}$$

If the parameters are multiplied by the area ($dxdz$), the functional would represent the time rate of energy loss.

As noted in Example 5.8, the element matrix has the units given by K/L. In this fluid mechanics problem, the parameters are

$$\frac{K}{L} = \frac{\mu}{h} = \left(\frac{\text{lbf-s}}{\text{ft}^2}\right)\left(\frac{1}{\text{ft}}\right)$$

These can be rearranged to give

$$\frac{K}{L} = \left(\frac{\text{lbf}}{\text{ft}^2}\right)\left(\frac{\text{s}}{\text{ft}}\right) = \frac{\text{Shear stress}}{\text{Velocity}}$$

Recall that the element matrix represents force per displacement or stiffness for solid mechanics problems. For viscous fluid mechanics, the element matrix represents the change of shear stress with respect to velocity. This may be thought of as a "viscous stiffness."

Two approaches are used here for the two cases. In Case 1, the work is carried out by hand to illustrate the technique. In Case 2, program FINONE is employed.

Case 1: Three Equal-Length Elements

The element values for the viscosity are (taking the average of the nodal values) $\mu^{(1)} = 1.5 \times 10^{-5}$, $\mu^{(2)} = 1.25 \times 10^{-5}$, $\mu^{(3)} = 0.9 \times 10^{-5}$, and the element values of K are $K^{(1)} = \mu^{(1)} = 1.5 \times 10^{-5}$, $K^{(2)} = \mu^{(2)} = 1.25 \times 10^{-5}$, $K^{(3)} = \mu^{(3)} = 0.9 \times 10^{-5}$. For element 1, with length 0.1, the coefficient matrix is obtained from Equation (5.38) as

$$[B]^{(1)} = \begin{matrix} & 1 & 2 \\ 1 & \\ 2 & \end{matrix} \begin{bmatrix} 1.5 & -1.5 \\ -1.5 & 1.5 \end{bmatrix}^{(1)} \times 10^{-4} \left(\frac{\text{lbf-s}}{\text{ft}^3}\right)$$

with the column matrix

$$\{C\}^{(1)} = \begin{matrix} 1 \\ 2 \end{matrix} \begin{bmatrix} 1.0 \\ 1.0 \end{bmatrix}^{(1)} \times 10^{-6} \left(\frac{\text{lbf}}{\text{ft}^2}\right)$$

For the other two elements

$$[B]^{(2)} = \begin{matrix} & 2 & 3 \\ 2 & \\ 3 & \end{matrix} \begin{bmatrix} 1.25 & -1.25 \\ -1.25 & 1.25 \end{bmatrix}^{(2)} \times 10^{-4}$$

$$\{C\}^{(2)} = \begin{matrix} 2 \\ 3 \end{matrix} \begin{bmatrix} 1.0 \\ 1.0 \end{bmatrix}^{(2)} \times 10^{-6}$$

$$[B]^{(3)} = \begin{matrix} & 3 & 4 \\ 3 & \\ 4 & \end{matrix} \begin{bmatrix} 0.90 & -0.90 \\ -0.90 & 0.90 \end{bmatrix}^{(3)} \times 10^{-4}$$

$$\{C\}^{(3)} = \begin{matrix} 3 \\ 4 \end{matrix} \begin{bmatrix} 1.0 \\ 1.0 \end{bmatrix}^{(3)} \times 10^{-6}$$

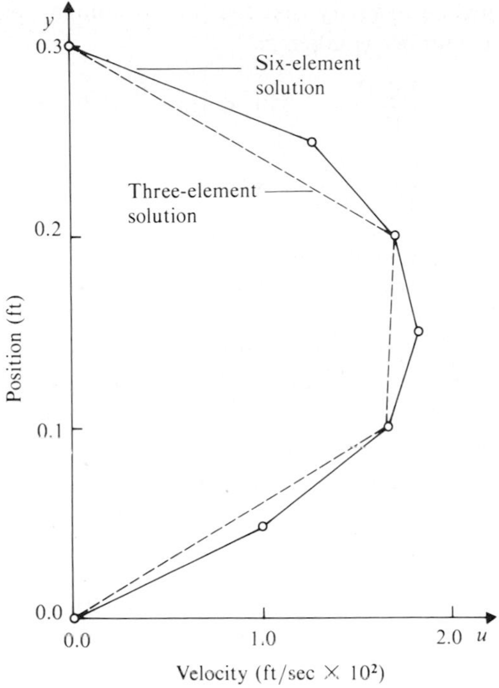

Figure 5.32 Velocity profiles (Example 5.10—Fluid flow between heated flat plates)

Assembling the global coefficient matrix by the usual procedure and including the prescribed nodal values yields

$$10^{-4} \times \begin{bmatrix} 1.0 & 0.0 & 0.0 & 0.0 \\ 0.0 & 2.75 & -1.25 & 0.0 \\ 0.0 & -1.25 & 2.15 & 0.0 \\ 0.0 & 0.0 & 0.0 & 1.0 \end{bmatrix} \begin{Bmatrix} u_1 \\ u_2 \\ u_3 \\ u_4 \end{Bmatrix} = \begin{Bmatrix} 0.0 \\ 2.0 \\ 2.0 \\ 0.0 \end{Bmatrix} \times 10^{-6}$$

The velocity profile, shown in Figure 5.32, is

$$\begin{Bmatrix} u_1 \\ u_2 \\ u_3 \\ u_4 \end{Bmatrix} = \begin{Bmatrix} 0.0 \\ 1.563 \\ 1.839 \\ 0.0 \end{Bmatrix} \times 10^{-2} \left(\frac{\text{ft}}{\text{sec}}\right) \qquad \begin{array}{l}\text{Case 1:} \\ \text{Three-element solution}\end{array}$$

As noted earlier, no exact solution is available for comparison. For a fluid with constant viscosity, μ, the velocity profile would be parabolic, so that u_2 would equal u_3. The lower viscosity near the upper plate leads to slightly higher velocities. The velocity profile is not very smooth because only three elements were used for the solution.

At this point, it is worthwhile to verify that the flow is laminar, as assumed in the beginning. The flow Reynolds number is taken as

$$\mathrm{Re} = \frac{\rho u h}{\mu} = 1.9\,\frac{\text{lbf-s}^2}{\text{ft}^4} \times 1.839 \times 10^{-2}\,\frac{\text{ft}}{\text{s}} \times 0.3\text{ ft}$$

$$\times\,\frac{1}{1.25 \times 10^{-5}}\,\frac{\text{ft}^2}{\text{lbf-s}} = 838$$

This is well below 2,000, so the flow is laminar.

At the lower plate, the accuracy parameter S = shear stress is

$$\tau_{xy}\bigg|_{y=0} = \mu\,\frac{du}{dy} = \mu^{(1)}\,\frac{du^{(1)}}{dy} = \mu^{(1)}\,\frac{u_2}{L^{(1)}}$$

Evaluating,

$$\tau_{xy}\bigg|_{y=0} = 1.5 \times 10^{-5}\,\frac{\text{lbf-s}}{\text{ft}^2} \times 0.01563\,\frac{\text{ft}}{\text{s}} \times \frac{1}{0.1}\,\frac{1}{\text{ft}} = 2.34 \times 10^{-6}\,\frac{\text{lbf}}{\text{ft}^2}$$

Similarly, at the upper plate

$$\tau_{xy}\bigg|_{y=h} = -0.9 \times 10^{-5}\,\frac{\text{lbf-s}}{\text{ft}^2} \times (-0.01839)\,\frac{\text{ft}}{\text{s}} \times \frac{1}{1.0}\,\frac{1}{\text{ft}}$$

$$= 1.66 \times 10^{-6}\,\frac{\text{lbf}}{\text{ft}^2}$$

The lower shear stress is larger because the viscosity is higher at the lower plate.

The other accuracy parameter, volume rate of flow per unit width of the plate Q, is given by

$$Q = \int_0^h u\,dy$$

From Equation (3.11), this is

$$Q = \frac{1}{2}\left[L^{(1)}(u_1 + u_2) + L^{(2)}(u_2 + u_3) + L^{(3)}(u_3 + u_4)\right]$$

and substituting,

$$Q = 0.00340\,\frac{\text{ft}^3}{\text{s-ft}}$$

The volume rate of flow is relatively small because $\frac{\partial P}{\partial x}$ is small.

Case 2: Six Equal-Length Elements

The input to FINONE for Case 2 starts with

Number of nodes, NP = 7
Number of elements, NE = 6

Table 5.6(a) gives the nodal coordinates, while Table 5.6(b) shows the element/node connectivity. The coefficient function $K(x) = \mu$ must be determined from Table 5.5. Assuming a linear variation for the viscosity between the temperatures given in the table, the values for K are

$$K^{(1)} = 1.55 \times 10^{-5} \qquad K^{(4)} = 1.175 \times 10^{-5}$$
$$K^{(2)} = 1.45 \times 10^{-5} \qquad K^{(5)} = 1.00 \times 10^{-5}$$
$$K^{(3)} = 1.325 \times 10^{-5} \qquad K^{(6)} = 0.8 \times 10^{-5}$$

Table 5.6(c) presents the element coefficient functions and Table 5.6(d) the prescribed nodal values. The solution, plotted in Figure 5.32, is

$$\begin{Bmatrix} u_1 \\ u_2 \\ u_3 \\ u_4 \\ u_5 \\ u_6 \\ u_7 \end{Bmatrix} = \begin{Bmatrix} 0.0 \\ 0.933 \\ 1.585 \\ 1.921 \\ 1.874 \\ 1.319 \\ 0.0 \end{Bmatrix} \times 10^{-2} \left(\frac{\text{ft}}{\text{s}}\right) \qquad \begin{array}{l}\text{Case 2:} \\ \text{Six-element solution}\end{array}$$

Note that at the points $x = 0.1$ and 0.2, the three-element solution is $u = 1.563 \times 10^{-2}$, $x = 0.1$ and $u = 1.839 \times 10^{-2}$, $x = 0.2$. The six-element solution gives $u = 1.585 \times 10^{-2}$, $x = 0.1$ and $u = 1.874 \times 10^{-2}$, $x = 0.2$. The results are very close at the node points but not in between. This will generally occur when the solution to the differential equation is nearly parabolic (as it would be if the viscosity were constant).

The accuracy parameter (shear stress on the lower plate) is

$$S = \tau_{xy}\Big|_{y=0} = 2.89 \times 10^{-6} \frac{\text{lbf}}{\text{ft}^2}$$

As seen from the three-element example, the change in S is 19%. On the upper plate

$$\tau_{xy}\Big|_{y=h} = 2.11 \times 10^{-6} \frac{\text{lbf}}{\text{ft}^2}$$

Usually a large number of elements must be used to obtain accurate derivatives. The volume rate of flow per unit length is

$$Q = \frac{1}{2}\left[L^{(1)}(u_1 + u_2) + \ldots + L^{(6)}(u_6 + u_7)\right] = 0.00381 \frac{\text{ft}^3}{\text{s-ft}}$$

where the change is only 11%. Thus integrals require fewer elements for a given level of accuracy than derivatives.

From the nearly parabolic shape of the six-element solution, it appears that this number of elements gives a relatively accurate solution. Calculation of shear stresses to less than 10% error may require more elements, but Q is already known to about this accuracy.

Table 5.6
Input Data for Program FINONE in Case 2—
Fluid Flow Between Heated Flat Plates

(a) Nodal Coordinates

Global Node Number (INODE)	Coordinate of Node (XNODE)
1	0.00
2	0.05
3	0.10
4	0.15
5	0.20
6	0.25
7	0.30

(b) Element/Node Connectivity Data

Element Number (NEL)	Global Node Number	
	Node i (NODEI)	*Node j (NODEJ)*
1	1	2
2	2	3
3	3	4
4	4	5
5	5	6
6	6	7

(c) Element Coefficient Functions

Element Number (IEL)	Coefficient Functions		
	Value of K(x) in Element (KXE) (lbf-s/ft^2 $\times$ 10^5)	*Value of P(x) in Element (PE)*	*Value of Q(x) in Element (QE) (lbf/ft^3 $\times$ 10^5)*
1	1.55	0.0	2.0
2	1.45	0.0	2.0
3	1.325	0.0	2.0
4	1.175	0.0	2.0
5	1.00	0.0	2.0
6	0.80	0.0	2.0

Table 5.6 *Continued*

(d) Prescribed Nodal Values

Global Node Number (IBOUND)	Prescribed Nodal Value (TBOUND) (°C)
1	0.0
7	0.0

Table 5.7
Accuracy Parameters vs. Number of Elements

Accuracy Parameter	Number of Elements in Solution			
	3	6	12	24
Shear stress at lower wall, τ_{xy} (lbf/ft^2 $\times$ 10^6)	2.34	2.89	3.15	3.28
Volume flow rate, Q(ft^3/s-ft $\times$ 10^3)	3.40	3.81	3.92	3.95

C. Accuracy Parameters

Table 5.7 shows the solution accuracy for Example 5.10 as the number of elements is extended up to 24 elements. The change gets smaller and smaller as the number of elements is extended from six to 12 and extended again to 24 elements. Figure 5.33 plots the accuracy parameters, shear stress τ_{xy} and volume rate of flow Q, vs. number of elements. The change between the last two values, expressed as a percentage, is

$$\text{Percent change in } \tau_{xy} = \frac{3.28 - 3.15}{3.28} \times 100 = 3.96\%$$

$$\text{Percent change in } Q = \frac{3.95 - 3.92}{3.95} \times 100 = 0.76\%$$

It is easily seen that, for large numbers of elements, the derivative property τ_{xy} converges to the correct solution much more slowly than the integral property Q (in this case, about five times slower).

5.8 Key Parameters

As indicated earlier in this chapter, one of the advantages of the general approach taken here is that solid mechanics, heat transfer, and fluid mechanics analyses can all be performed in the same manner. The boundary value problems just developed illustrate this. Certain key parameters can be identified for each type of problem.

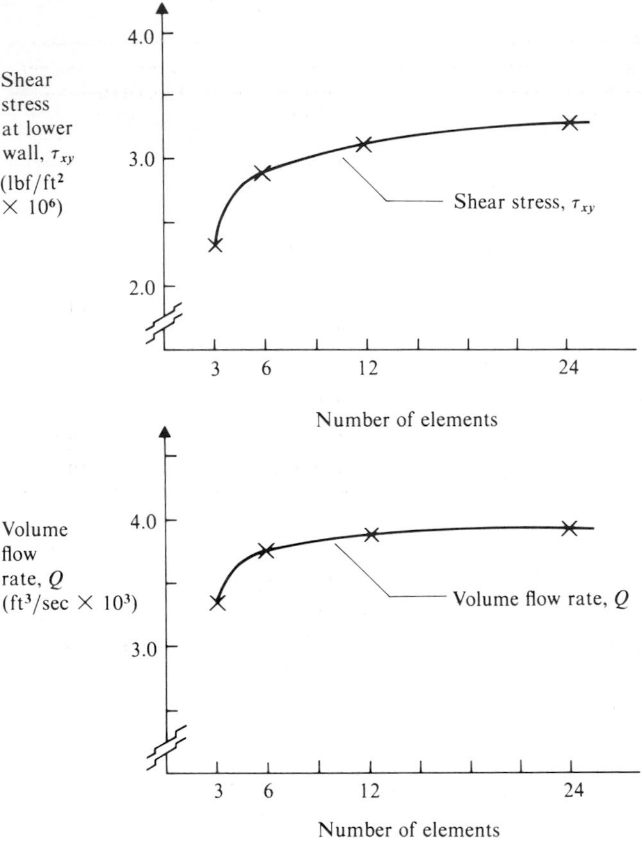

Figure 5.33 Shear stress and flow rate versus number of elements

A. Primary and Secondary Parameters

Two types of key parameters are conveniently used here. In a solid mechanics problem, the primary parameter would be the displacement u; the secondary parameter would be the force F. Table 5.8 indicates the primary and secondary parameters for various problems.

The primary parameter is the basic one for a given problem. For longitudinal deformation of a rod, it is the displacement; for heat transfer, it is temperature; for fluid flow, it is velocity. All three can be classified as "displacement-like" variables.

Secondary parameters are those that determine the governing factor in the problem. In solid mechanics, it is the summation of forces. Thus the force R is the key secondary parameter. For heat transfer, the heat flow Q_x (thermal power) is summed, as shown in Examples 5.5 and 5.6. The shear stress τ_{xy} acting on a fluid particle governs the viscous

Table 5.8
Key Primary and Secondary Parameters

Problem Type	Primary Parameter "Displacement"		Secondary Parameter "Force"	
	Symbol and Parameter	Units	Symbol and Parameter	Units
General	T	—	$S = K\frac{dT}{dx}$	—
Solid mechanics	u Displacement	mm	R Axial force	kN
Heat transfer	T Temperature	°C	Q_x Thermal power	W
Fluid mechanics	u Velocity	$\frac{\text{mm}}{\text{s}}$	τ_{yx} Shear stress	$\frac{\text{kN}}{\text{mm}^2}$

fluid flow problem developed in Example 5.7. All of the secondary parameters are classified here as "force-like" variables. Each is related to the coefficient function $K(x)$ times the derivative dT/dx. For convenience, define the general secondary parameter, used in this section only; as

$$S = K\frac{dT}{dx} \tag{5.41}$$

B. Differential Equations

For solid mechanics problems, the basic relation for the axial force is

$$R = EA\frac{du}{dx} \tag{5.42}$$

Similarly, the thermal power in heat transfer is given by

$$Q_x = -kA_x\frac{dT}{dx} \tag{5.43}$$

Finally, in the fluid mechanics example, the shear stress is

$$\tau_{yx} = \mu\frac{du}{dy} \tag{5.44}$$

These are all of the secondary parameters.

All of the differential equations were obtained by summing the secondary parameter over a differential element, as shown in Section 5.4. In each case the resulting differential

Table 5.9
Key Parameters in Differential Equations

Problem Type	Coefficient Function	Units for Coefficient Function	Differential Equation	Key Secondary Parameter Per Length	Units for Differential Equation
General	$K(x)$	—	$\frac{dS}{dx} = \frac{d}{dx}\left(K\frac{dT}{dx}\right) = 0$	$\frac{\text{S}}{\text{Length}}$	—
Solid mechanics	EA	kN	$\frac{dR}{dx} = \frac{d}{dx}\left(EA\frac{du}{dx}\right) = 0$	$\frac{\text{Force}}{\text{Length}}$	$\frac{\text{kN}}{\text{mm}}$
Heat transfer	kA_x	$\frac{\text{W-mm}}{°\text{C}}$	$\frac{dQ_x}{dx} = \frac{d}{dx}\left(kA_x\frac{dT}{dx}\right) = 0$	$\frac{\text{Thermal power}}{\text{Length}}$	$\frac{\text{W}}{\text{mm}}$
Fluid mechanics	μ	$\frac{\text{kN-s}}{\text{mm}^3}$	$\frac{d\tau_{yx}}{dy} = \frac{d}{dy}\left(\mu\frac{du}{dy}\right) = 0$	$\frac{\text{Stress}}{\text{Length}}$	$\left(\frac{\text{kN}}{\text{mm}^2}\right)\frac{1}{\text{mm}}$

equation is expressed in terms of the secondary parameter per length (in the coordinate direction). Table 5.9 indicates the parameters involved. Only the second-order term

$$\frac{dS}{dx} = \frac{d}{dx}\left(K\,\frac{dT}{dx}\right) = 0$$

is considered. Usually it is the most important one.

C. Functional

It is interesting to examine the physical interpretation and units of the functionals for these problems. Consider only the first term of the functional (5.30)

$$I = \int_{x_0}^{x_L}\left[\frac{1}{2}\,K\left(\frac{dT}{dx}\right)^2\right]dx$$

or using (5.41)

$$I = \int_{x_0}^{x_L}\left[\frac{1}{2}\,S\,\frac{dT}{dx}\right]dx$$

This can also be simplified by "cancelling" the dx terms

$$I = \int_{T_0}^{T_L}\left[\frac{1}{2}\,S\right]dT \tag{5.45}$$

In the solid mechanics problem this has the physical interpretation of stored potential energy. Table 5.10 shows the parameters for the general and three specific cases.

Table 5.10
Key Parameters in Functionals

Problem Type	Functional, I — Normal Form	Functional, I — Key Parameter	Key Parameter "Energy"	Units	Physical Interpretation
General	$\int_{x_0}^{x_L} \left[\frac{1}{2} K \left(\frac{dT}{dx}\right)^2\right] dx$	$\int_{T_0}^{T_L} \frac{1}{2} S\, dT$	ST	—	—
Solid mechanics	$\int_{x_0}^{x_L} \left[\frac{1}{2} EA \left(\frac{du}{dx}\right)^2\right] dx$	$\int_{u_0}^{u_L} \frac{1}{2} R\, du$	Fu Force × displacement	$kN\text{-}mm$	Stored potential energy
Heat transfer	$\int_{x_0}^{x_L} \left[\frac{1}{2} kA_x \left(\frac{dT}{dx}\right)^2\right] dx$	$\int_{T_0}^{T_L} \frac{1}{2} Q_x dT$	$Q_x T$ Thermal power × temperature	W-°C	?
Fluid mechanics	$\int_{y_0}^{y_L} \left[\frac{1}{2} \mu \left(\frac{du}{dx}\right)^2\right] dy$	$\int_{u_0}^{u_L} \frac{1}{2} \tau_{yx} du$	$\tau_{yx} u$ Shear stress × velocity	$\left(\frac{\text{kN}}{\text{mm}^2}\right) \frac{\text{mm}}{\text{s}}$	Power loss per area due to viscous effects

In each functional, the key parameters are multiplied together: I = Secondary parameter times primary parameter = ST. In the axial displacement of a rod, the functional is

$$I = \int_{x_0}^{x_L} \left[\frac{1}{2} EA \left(\frac{du}{dx}\right)^2\right] dx$$

Using the expression for the force (5.42) yields

$$I = \int_{x_0}^{x_L} \left[\frac{1}{2} R \frac{du}{dx}\right] dx$$

which can also be written as

$$I = \int_{u_0}^{u_L} \left[\frac{1}{2} R\right] du \tag{5.46}$$

Clearly this produces the force times displacement, or potential energy stored in the rod, which could be reclaimed.

In the heat transfer problem, the functional is

$$I = \int_{x_0}^{x_L} \left[\frac{1}{2} kA_x \left(\frac{dT}{dx}\right)^2\right] dx$$

This can also be written as (5.43)

$$I = \int_{x_0}^{x_L} \left[-\frac{1}{2} Q_x \frac{dT}{dx}\right] dx$$

or finally

$$I = \int_{T_0}^{T_L} \left[-\frac{1}{2} Q_x \right] dT \tag{5.47}$$

which yields the thermal power times the temperature. No physical interpretation is readily available.

The fluid mechanics problem has the functional

$$I = \int_{x_0}^{x_L} \left[\frac{1}{2} \mu \left(\frac{du}{dy} \right)^2 \right] dx$$

Again using the expression for shear stress (5.44), this becomes

$$I = \int_{u_0}^{u_L} \left[\frac{1}{2} \tau_{xy} \right] dx \tag{5.48}$$

The physical interpretation is the shear stress (force per area) times velocity (displacement per time). Thus it is the power (energy per time) per area. Because the process is a viscous one it dissipates energy. Thus the functional represents the viscous power loss.

D. Element Matrix

The element matrix $[B]$ is the basic quantity used in finite element analysis. Thus it is worth exploring the physical interpretation of the element matrix. The most important term has the form

$$[B]^{(e)} = \left[\frac{K}{L} b_i^2 \right]^{(e)}$$

It is derived by minimizing the functional with respect to nodal values such as T_n

$$\frac{\partial I}{\partial T_n} = 0$$

and then separating out the nodal values T_n in the form

$$\left\{ \frac{\partial I}{\partial T_n} \right\}^{(e)} = [B]^{(e)} \{T\}^{(e)} - \{C\}^{(e)}$$

This gives the result of the element matrix having the secondary parameter divided by the primary parameter.

$$[B]^{(e)} = \frac{\text{Secondary parameter}}{\text{Primary parameter}} = \frac{S}{T}$$

Now the element matrix can be seen as the rate of change of the secondary parameter with respect to the primary parameter.

Table 5.11 shows the element matrix terms and physical interpretation for the three problem types. Each represents a type of "stiffness" of the secondary parameter relative to the first. By analogy with solid mechanics, each can be thought of as a force per displacement or stiffness.

Table 5.11
Key Parameters in Element Matrices

Problem Type	Element Matrix	Physical Interpretation "Stiffness"	Units
General	$\frac{K}{L} b_i^2$	—	—
Solid mechanics	$\frac{EA}{L} b_i^2$	F/u Force/Displacement	$\frac{\text{kN}}{\text{mm}}$
Heat transfer	$\frac{kA_x}{L} b_i^2$	Q_x/T Thermal power/Temperature	$\frac{\text{W}}{°\text{C}}$
Fluid mechanics	$\frac{\mu}{L} b_i^2$	τ_{yx}/u Shear stress/Velocity	$\left(\frac{\text{kN}}{\text{mm}^2}\right)\left(\frac{\text{s}}{\text{mm}}\right)$

5.9 Problems

A. Functionals

5.1 A functional I is given by

$$I = -3\,u_1^2 + 2\,u_1$$

Determine the first and second variation with respect to u_1. Is the functional a minimum, maximum, or inflection point?

5.2 Consider the functional

$$I = 3.2\,T_1^2 + 4.7\,T_1 + 0.2\,T_2^2 + 3.4\,T_2 + 1.25$$

Determine whether this is a minimum, saddle point, or maximum by evaluating the first and second variations with respect to T_1 and T_2.

5.3 The functional I has the form

$$I = 4\,u_1 + 2\,u_1^2 + u_2 - 2\,u_2^2$$

Evaluate the first and second variation with respect to u_1 and u_2. Is the functional a minimum, maximum, or saddle point?

5.4 If a functional I has the general form

$$I = a\,T_1^2 + b\,T_2^2 + c\,T_1 + d\,T_2$$

where a, b, c, d are constants, determine the conditions under which the function is a minimum.

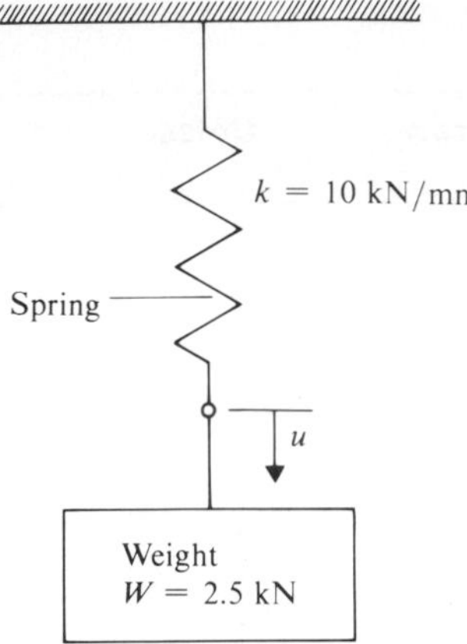

Figure 5.34 Diagram for Problem 5.5

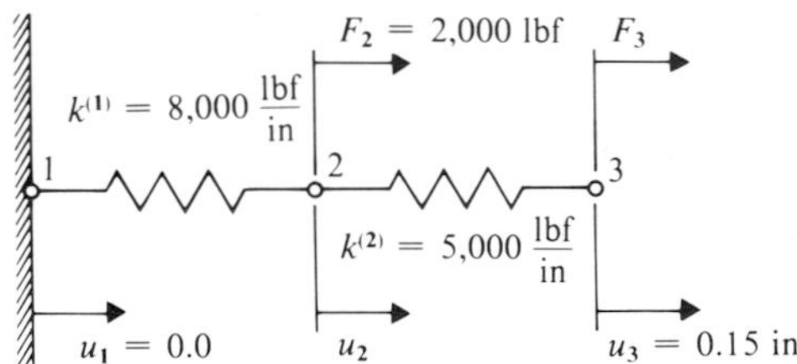

Figure 5.35 Diagram for Problem 5.6

B. Springs and Bars

5.5 A vertical spring is attached to a fixed support at the top and a mass at the bottom, as shown in Figure 5.34. The spring has stiffness $k = 10$ kN/mm and the mass has weight $W = 2.5$ kN. Determine the equilibrium displacement. Evaluate the total potential energy for the spring/mass system. Obtain the first variation of the total potential energy with respect to the displacement.

5.6 Two springs are shown in series in Figure 5.35. The stiffnesses are $k^{(1)} = 8{,}000$ lbf/in, $k^{(2)} = 5{,}000$ lbf/in. At the left end the springs are attached to the wall. The specified nodal displacement at node 3 is $u_3 = 0.15$ in and the applied force is $F_2 = 2{,}000$ lbf. Determine the functional for the problem, the nodal displacements, and the forces in each spring.

5.7 A three spring–three weight problem is shown in Figure 5.36. Write the total potential energy (functional) for the system. Determine the displacements of each of the weights (after the weights have been slowly applied). Calculate the reaction force in each spring.

5.8 Figure 5.37 illustrates a steel rod with lengths and diameters (circular cross section) as indicated. The force acting on the rod is 900 lbf to the left. Young's modulus is 3.0×10^7 lbf/in^2. Use a two-element model of the rod to determine the displacement, strain, and stresses along the rod. Solve by hand.

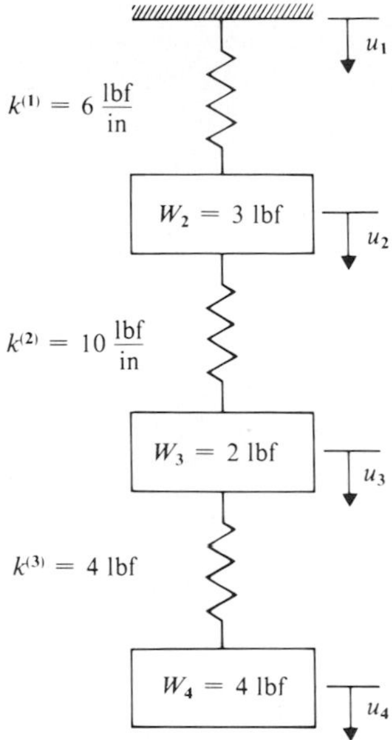

Figure 5.36 Diagram for Problem 5.7

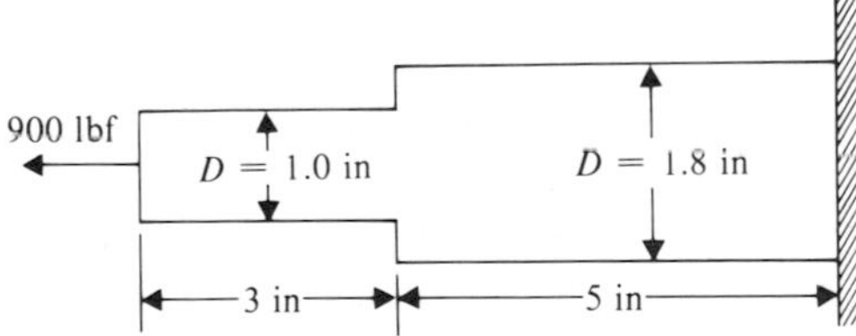

Figure 5.37 Diagram for Problem 5.8

5.9 A concrete column of square cross section, shown in Figure 5.38, has specific weight

$$\gamma = 110 \text{ lbf/ft}^3$$

and Young's modulus

$$E = 4.5 \times 10^6 \text{ lbf/in}^2$$

A 16,000-lbf force acts downward on the top of the column, as shown. Use a three-equal-length element model to determine the displacement, strain, and stress along the column. Carry out the analysis by hand except for solving the global set of equations.

5.10 An aluminum and steel composite rod is illustrated in Figure 5.39. Both sections are of square cross section. Young's moduli for the materials are given in Example 4.5. Determine the governing equation for this problem, using differential analysis similar to that used in Examples 5.5 and 5.6. Compare this to the general boundary value problem and evaluate the coefficient functions $K(x)$, $P(x)$, $Q(x)$ and associated boundary conditions. Also obtain the functional for this problem from Equation (5.30).

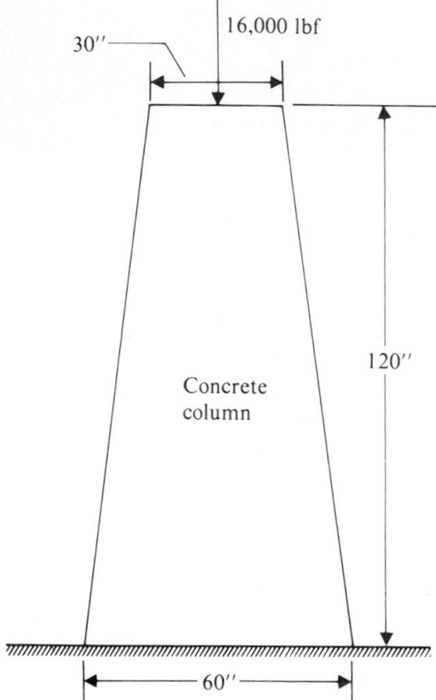

Figure 5.38 Diagram for Problem 5.9

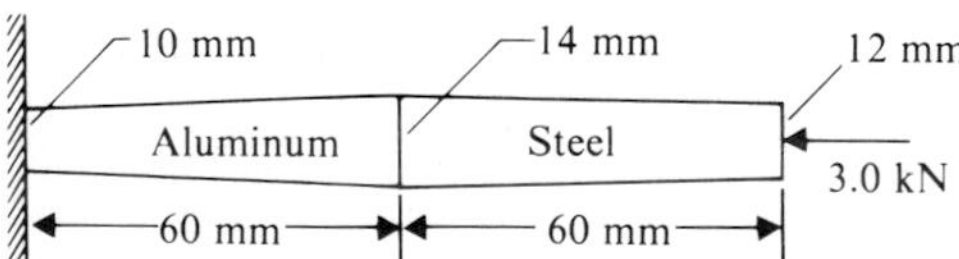

Figure 5.39 Diagram for Problem 5.10

5.11 Use computer program FINONE to solve a six-equal-length element model of the rod for the displacements, strains, stresses, and axial forces in each element for Problem 5.10. Plot the results vs. position x.

C. Heat Transfer

5.12 Redo Example 5.8 (Heat conduction in infinite slab) with a convection heat flow boundary condition at the right end. Figure 5.40 illustrates the problem geometry. The convection heat flow is

$$Q_{xL} = h\, A_x\, (T_L - T_\infty)$$

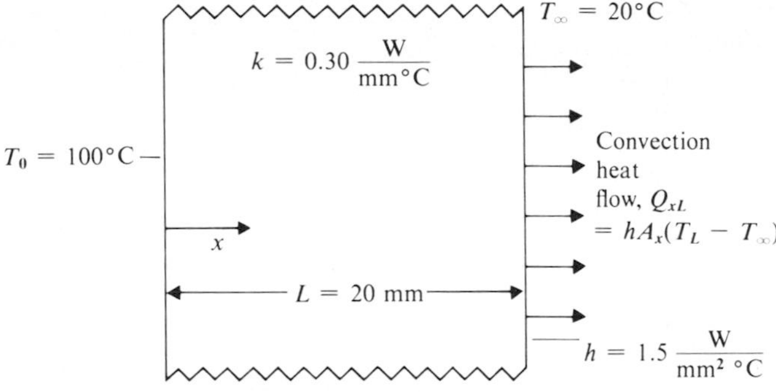

Figure 5.40 Diagram for Problem 5.12

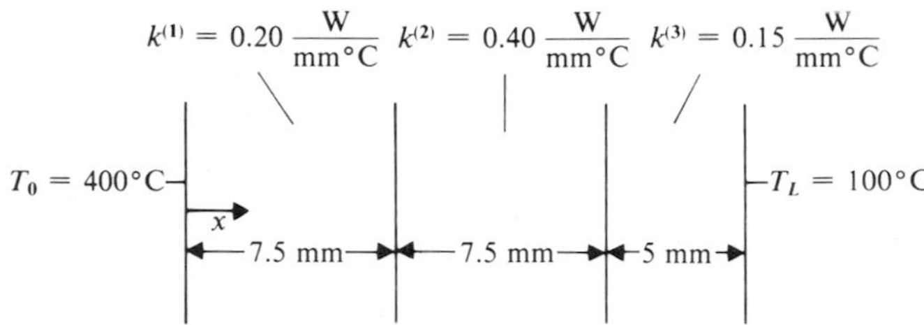

Figure 5.41 Diagram for Problem 5.13

where

$$
\begin{aligned}
h &= \text{Convection coefficient} = 1.5 \text{ W/mm}^2\text{-°C} \\
A_x &= \text{Cross sectional area} = dydz = 1 \text{ mm}^2 \\
T_\infty &= \text{Fluid temperature} = 20°\text{C}
\end{aligned}
$$

Use a two-element model to solve for the temperature distribution in the slab. Compare the heat flow at x_L to the values used in Example 5.8.

5.13 An infinite slab composed of three different materials is shown in Figure 5.41. Use a three-element model of the slab to determine the temperature profile and the heat flow per unit area through the slab. Solve with computer program FINONE.

5.14 An infinite slab similar to that in Example 5.5 has heat conduction in the x direction. It also has heat generation $\dot{q}$ per unit volume across the length of the slab where $\dot{q} = \dot{q}(x)$. Figure 5.42 shows a diagram. Determine the differential equation for the problem. Boundary conditions are a specified temperature at the left end and insulated at the right: $T = T_0$, $x = x_0$ and $Q_x = 0$, $x = x_L$. Compare this to the general boundary value problem in Section 5.4 to determine the coefficient functions $K(x)$, $P(x)$, $Q(x)$ and boundary parameters α_L, β_L. Solve the problem in Example 5.8, Case 2 with the addition of heat generation $\dot{q} = 0.2$ W/mm³. Solve by hand.

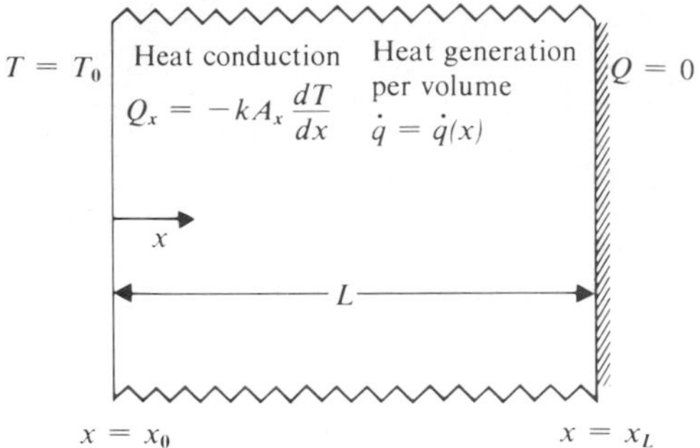

Figure 5.42 Diagram for Problem 5.14

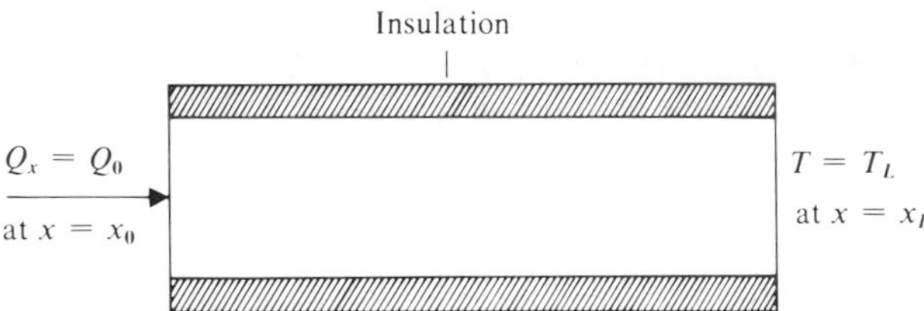

Figure 5.43 Diagram for Problem 5.15

5.15 An insulated round rod is illustrated in Figure 5.43. Heat conduction occurs in the rod with boundary conditions shown. Evaluate the coefficient functions $K(x)$, $P(x)$, $Q(x)$ and boundary conditions α_0, β_0 for the general boundary value problem discussed in Section 5.4.

5.16 The thin fin with heat conduction and convection in Examples 5.6 and 5.9 has the different numerical parameters $L = 400$ mm, $W = 200$ mm, $t = 3$ mm, $k = 0.15$, $h = 2.0 \times 10^{-5}$ W/mm²-°C, and $T_\infty = 40$°C. The wall at the left end is maintained at $T_0 = 250$°C, $x = 0$. It can be assumed that the fin is thin enough so that the heat lost through the end (at $x = L$) is negligible. Thus $Q_x = 0$, $x = L$. Use computer program FINONE to solve this problem for two cases.

Case 1: 8 equal-length elements
Case 2: 16 equal-length elements

After determining the temperature distribution in the fin, calculate the heat flow by convection through the rod at $x = 0$ and $x = L$. Evaluate the error using

$$\text{Error} = \frac{\text{Change of } T \text{ at } x = L/2}{\text{Value of } T \text{ at } x = L/2}$$

where the numerator is the change of T at $x = L/2$ from Case 1 to Case 2 and the denominator is the value of T at $x = L/2$ from Case 2.

5.17 Obtain the analytical solution to Problem 5.16. Compare the temperature distribution along the fin to both Case 1 and Case 2.

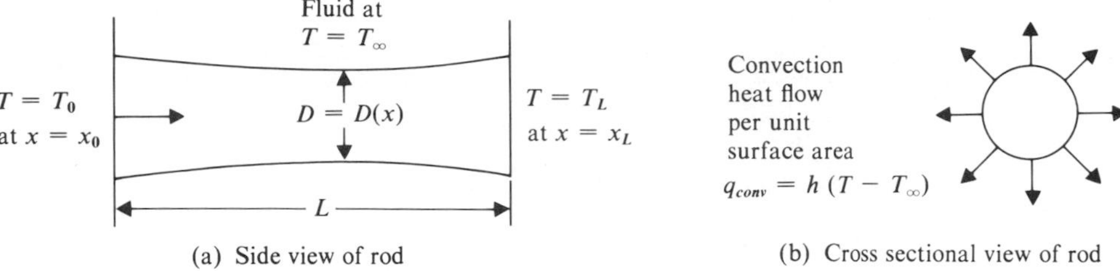

Figure 5.44 Diagram for Problems 5.18, 5.19, and 5.20

5.18 A thin cylindrical rod of diameter $D(x)$ has length L where $D/L << 1$. Thus it can be assumed that the temperature is uniform over the cross section and $T = T(x)$ alone. Heat conduction occurs along the axis of the rod. The material in the rod has conduction coefficient $k = k(x)$.

The rod is surrounded by a fluid at temperature T_∞. Figure 5.44 shows the diagram. Heat convection per unit surface area $q_{\text{conv}} = h(T - T_\infty)$ occurs from the rod into the fluid; the convection coefficient h is a constant. Let the temperatures be specified at each end. Derive the differential equation for this problem and compare it to the general boundary value problem. Also obtain a functional for this problem from Equation (5.30).

5.19 Problem 5.18 has the numerical parameters D = 16mm, L = 160 mm, k = 0.25 W/mm-°C, $h = 6.7 \times 10^{-5}$ W/mm²-°C, T_0 = 400°C, T_L = 100°C, and T_∞ = 100°C. Determine the problem differential equation and the coefficient functions $K(x)$, $P(x)$, $Q(x)$ for the general boundary value problem in Section 5.4.

5.20 Use a ten-equal-length element model of Problem 5.19 to solve for the temperature. Obtain the solution with program FINONE.

5.21 A thin fin with the same geometry as in Example 5.9 is to be considered. All other problem parameters are the same, except that the heat convection coefficient h varies from $h = 1.0 \times 10^{-4}$ W/mm²-°C at x = 0 mm where the air moves slowly (near the wall) to $h = 3.0 \times 10^{-4}$ W/mm²-°C at x = 120 mm where the air moves faster (near the tip of the fin). The variation is linear with position x. The governing equation and boundary conditions are the same as in Example 5.9. Solve, using three constant-property elements with program FINONE, and compare the results to Example 5.9.

5.22 Solve Problem 5.21 with 15 elements.

D. Fluid Mechanics

5.23 Resolve Example 5.10, Case 2 with the change that the upper plate moves with the velocity $U = -0.01$ ft/sec. The negative sign indicates that the plate moves from right to left.

a. Determine the velocity profile using six equal-length elements and program FINONE. Plot the profile.
b. Evaluate the shear stresses on the walls and the volume rate of flow.

5.24 Laminar viscous flow occurs between parallel flat plates a distance h apart. Figure 5.45 illustrates the geometry. The upper plate moves with velocity U, while the lower one is fixed. No pressure gradient exists in the x direction. Numerical values for the problem

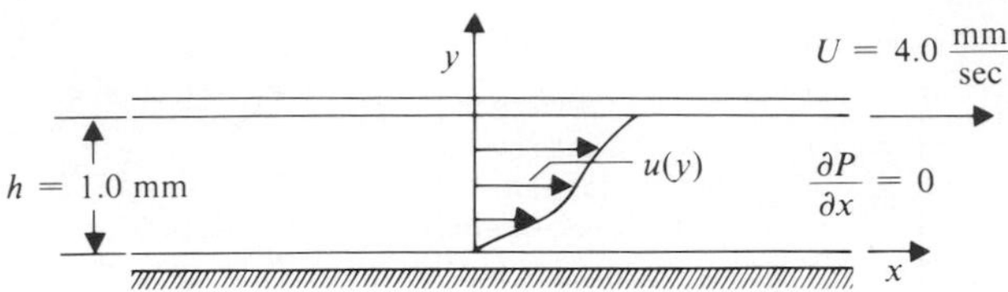

Figure 5.45 Diagram for Problem 5.24

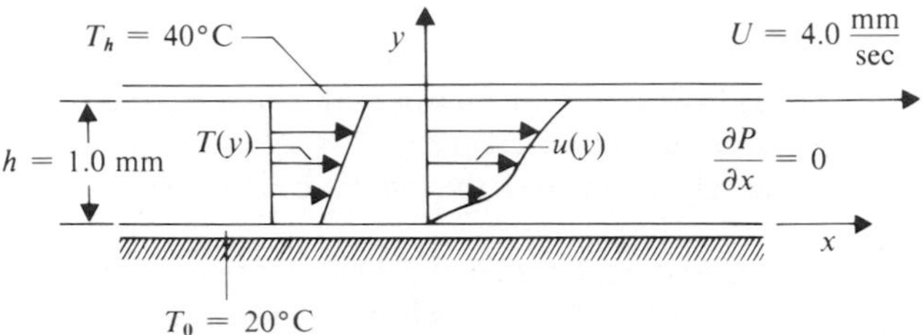

Figure 5.46 Sketch for Problem 5.25

parameters are $h = 1.0$ mm, $U = 4.0$ mm/sec, and $\mu = 1 \times 10^{-7}$ N-sec/mm². Use a four-element model to solve for the velocity profile between the plates. Employ program FINONE. Obtain the exact solution and compare to the finite element solution.

5.25 Solve Problem 5.24, shown in Figure 5.46, with the change that there is a linear temperature variation from the bottom plate at $T_1 = 20°C$ to the top plate at $T_2 = 40°C$. See Example 5.10 for a similar problem. Fluid viscosities at various temperatures are

T °C	μ N-sec/mm²
20	1.0×10^{-5}
30	8.0×10^{-6}
40	5.0×10^{-6}

a. Solve, using four equal-length elements and program FINONE.
b. Determine the shear stresses on the walls and the volume rate of flow between the plates.

E. Lubrication

5.26 A stepped bearing has the geometry shown in Figure 5.47. The inlet pressure P_i is 35 psia, while the exit pressure P_e is atmospheric at 14.7 psia. Because of the pressure gradient, the flow is from left to right. Consider the lubricant fluid to be incompressible with constant viscosity. The pressure in the bearing $P(x)$ is obtained by solving the differential equation

$$\frac{d}{dx}\left(h^3 \frac{dP}{dx}\right) = 0$$

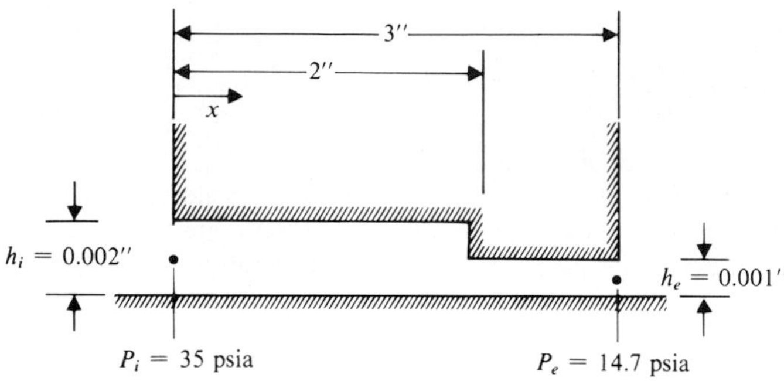

Figure 5.47 Stepped hydrostatic bearing for Problems 5.26 and 5.27

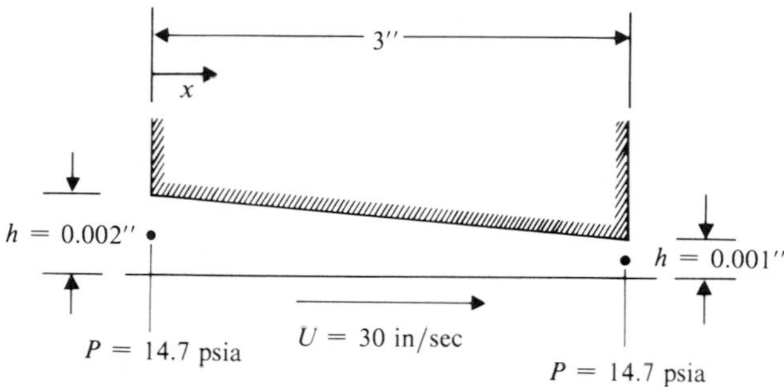

Figure 5.48 Plane slider bearing for Problem 5.28

with the boundary conditions indicated above. Here $h(x)$ is the film thickness, which can easily be obtained from the diagram. Because both the top and bottom of the bearing are fixed in space, it is called a hydrostatic bearing. The bearing is very long in the direction into and out of the paper.

Use a two-element model to obtain the pressure profile $P(x)$ in the hydrostatic bearing. Carry out the solution by hand. Determine the total force exerted on the upper part of the bearing.

5.27 Solve Problem 5.26 using 15 elements and program FINONE. Plot the pressure as a function of x.

5.28 A plane slider bearing is shown in Figure 5.48. The film thickness $h(x)$ is linear from the value of $h = 0.002$ inch to $h = 0.001$ inch. The upper part of the bearing is fixed in space, while the lower portion moves with velocity $U = 30$ in/sec. Both inlet and exit pressures are atmospheric at 14.7 psia. The lubricating fluid is incompressible with viscosity

$$\mu = 1.6 \times 10^{-6} \frac{\text{lbf-sec}}{\text{in}^2}$$

The governing differential equation for this problem is

$$\frac{d}{dx}\left(h^3 \frac{dP}{dx}\right) = 12\,\mu U \frac{dh}{dx}$$

Also, the bearing is very long into and out of the paper.

Determine the pressure profile $P(x)$ along the slider bearing, using 15 elements and program FINONE. Also obtain the total force (assume a one-inch length into the paper) exerted on the top part of the bearing. Plot the pressure vs. position. What position x gives the peak pressure?

F. General Problems

5.29 Derive element matrices similar to those in (5.38)

$$\begin{Bmatrix} \dfrac{\partial I}{\partial T_i} \\ \dfrac{\partial I}{\partial T_j} \end{Bmatrix}^{(e)} = [B]^{(e)} \begin{Bmatrix} T_i \\ T_j \end{Bmatrix}^{(e)} - \{C\}^{(e)}$$

but use linearly varying properties over the element of the form

$$\begin{aligned} K(x) &= N_i K_i + N_j K_j \\ P(x) &= N_i P_i + N_j K_j \\ Q(x) &= N_i Q_i + N_j Q_j \end{aligned}$$

Here K_i, K_j, P_i, P_j, Q_i, Q_j represent the values of the coefficient functions at the node points of an element (e).

5.30 Modify computer program FINONE to use the results for the linear property matrix and column just derived in Problem 5.29. Solve for the temperature distribution in the fin in Problem 5.21 using the modified version of FINONE. Compare the results to those in Problem 5.21.

5.31 Make up your own problem involving solid mechanics, heat transfer, or fluid mechanics. Use computer program FINONE to obtain a solution, using at least 10 elements.

5.32 Solve an ordinary differential equation and boundary conditions representing a physical problem of your choice. Use computer program FINONE. It should not be a solid mechanics, heat transfer, or fluid mechanics problem of the types covered in this chapter.

Chapter 6

Derivation of Functionals in One Dimension

6.1 Variation of a Functional

A. Method

Chapter 5 developed general finite element solutions for problems in solid mechanics, heat transfer, and fluid mechanics using the variational method. The concept of a general functional was introduced, but no method was given as to how to derive the functional. Chapter 6 shows how to do this for the general differential equation discussed in Section 5.4. Solid mechanics problems, where the functional may be obtained from the potential energy, do not require these methods, although they may still be used, if desired.

Three sections related to the derivation of functionals are presented in this chapter:

Section 6.1 Variation of a Functional

Section 6.2 Euler–Lagrange Equation

Section 6.3 General Functional

The first section, 6.1, examines a small variation of both a function $T(x)$ and functional $I(T)$. This provides an introduction to the actual derivation of the functional. In Section 6.2, a functional of a certain form is assumed at the beginning. An associated equation, which makes the first variation of the functional δI vanish, is obtained. It is called the Euler–Lagrange equation. In Section 6.3, this result is used to determine the general functional (5.30) for the general boundary value problem (5.27), (5.28), and (5.29). The steps may be summarized as

Step 1 Start with assumed functional form I

Step 2 Take first variation $\delta I = 0$

Step 3 Obtain Euler–Lagrange equation

Step 4 Determine general functional

The solution to the differential equation $T(x)$ produces a minimum (or maximum) value of I.

It should be noted that the objective here is not just the derivation of the functional for a second-order ordinary differential equation. The method developed works for higher order ordinary differential equations, such as those describing beam bending—see Chapter 7—and partial differential equations.

The derivation of functionals is not essential to the use of the variational finite element method, because functionals usually have already been derived by researchers in various fields. They may thus be simply used rather than derived. This chapter is for students (and researchers) who wish to know more about the derivation process.

B. General Functional

The variational method is concerned with variations of a functional, denoted I, defined over the interval x_0 to x_L. Let that functional have the general form

$$I(T) = \int_{x_0}^{x_L} F(x, T, T_x)dx - G_0(x, T)_{x_0} + G_L(x, T)_{x_L} \tag{6.1}$$

where, for convenience in writing,

$$T_x = \frac{dT}{dx}$$

Here the integrand F is a function of x, T, and the derivative of T with respect to x. The terms G_0 and G_L are associated with the derivative boundary conditions at the end points.

Compare this form of the functional to (5.30)

$$I = \int_{x_0}^{x_L} \left[\frac{1}{2} K (T_x)^2 - \frac{1}{2} PT^2 - QT\right] dx$$
$$- \left(\frac{1}{2} \alpha_0 T^2 + \beta_0 T\right)_{x_0} + \left(\frac{1}{2} \alpha_L T^2 + \beta_L T\right)_{x_L}$$

The integrand F and boundary functions G_0, G_L are then

$$\begin{aligned} F(x, T, T_x) &= \frac{1}{2} KT_x^2 - \frac{1}{2} PT^2 - QT \\ G_0(x, T)_{x_0} &= \left(\frac{1}{2} \alpha_0 T + \beta_0 T\right)_{x_0} \\ G_L(x, T)_{x_L} &= \left(\frac{1}{2} \alpha_L T + \beta_L T\right)_{x_L} \end{aligned} \tag{6.2}$$

Generally the boundary functions have derivatives of order one less than the integrand.

The functional $I(T)$ is *not* a function of x because I is a definite integral. It may be thought of in the following way. A function $T(x)$ is an input to an integration box with a functional (number) as an output. Figure 6.1 illustrates this. As any function $T(x)$ is substituted into the integrand, the integral evaluated, and the limits substituted, a different number I comes out for each separate $T(x)$.

C. First Variation

Let the sum of T and δT be denoted as the approximate function T^*:

$$T^*(x) = T(x) + \delta T(x) \tag{6.3}$$

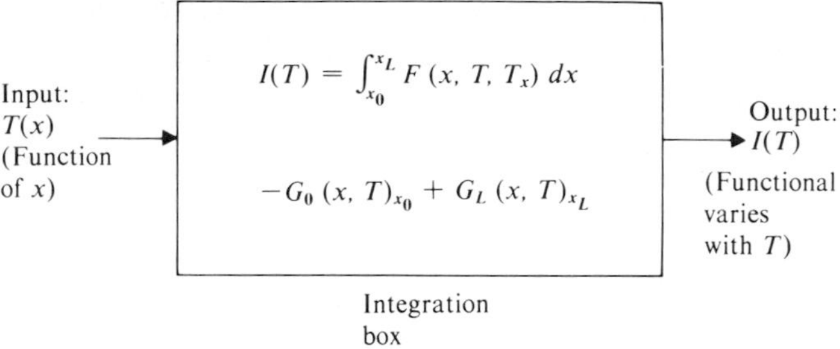

Figure 6.1 Concept of a functional as an integration box

Here it will be shown that

$$\begin{aligned} T^*(x) &= \text{approximate (finite element) solution} \\ T(x) &= \text{exact solution to differential equation} \\ \delta T(x) &= \text{first variation} \end{aligned}$$

if the first variation δI is set equal to zero for a given problem and a minimum results. Note that δI is zero—not the variation of the function δT. Thus the solution is still an approximation T^*, but the functional is minimized within the limits of the interpolation function chosen within the elements. Actually if the exact solution $T(x)$ is known, then $\delta T = 0$ when $\delta I = 0$. In general, for finite element solutions of engineering problems, the exact solution is not known, so δT will have some small, nonzero value over the solution region.

Suppose that the function $T(x)$ is substituted into the functional I. A small variation of T, denoted δT, away from that function produces a small variation of the functional I, denoted δI. The difference or variation δI is given by

$$\delta I = I(T^*) - I(T) = I(T + \delta T) - I(T) \tag{6.4}$$

Note that both the function $T(x)$ and the small variation of it $\delta T(x)$ are functions of x, but δI is not because the variation with x is integrated out.

At this point, it has not been determined what the specific form of the approximation function T^* should be. Figure 6.2 shows two possible choices:

$$\begin{aligned} T_1^*(x) &= T(x) + \delta T_1(x) \\ T_2^*(x) &= T(x) + \delta T_2(x) \end{aligned}$$

Here the end points are taken as fixed; that is, specified values (5.28) and (5.29) at the boundaries are considered here for simplicity. The variations must be zero at both ends for this case: $\delta T_1 = 0, \delta T_2 = 0$. The variation δT is arbitrary in the interior. The function T^* closest to the minimum of I could look like T_1^*, T_2^*, or another function.

Let the functional I be expanded in a Taylor's series for small values of δT

$$I(T + \delta T) = I(T) + \frac{\delta I}{\delta T}\delta T + \ldots$$

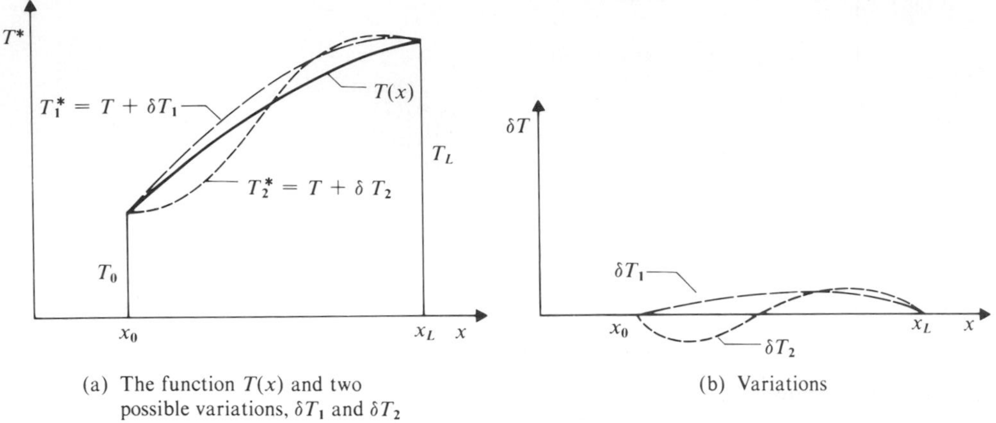

(a) The function $T(x)$ and two possible variations, δT_1 and δT_2

(b) Variations

Figure 6.2 Approximation functions and variations with specified boundary conditions

Terms of order $(\delta T)^2$ or higher are neglected. Then the variation is

$$\delta I = I(T + \delta T) - I(T) = \frac{\delta I}{\delta T}\,\delta T = 0$$

Because the variation δT is arbitrary (except possibly at the end points), δI vanishes only if

$$\frac{\delta I}{\delta T} = 0 \tag{6.5}$$

This indicates that $T(x)$ represents a stationary value of the functional $I(T)$, as discussed in Section 5.1. The variation δI is not zero for any variation δT away from $T(x)$.

The problem in engineering is that $T(x)$, the exact solution, is not known; otherwise, there would be no need for a finite element approximate solution. Thus the procedure is to find the approximation $T^*(x)$ that makes the variation δI as small as possible.

Example 6.1 *Differential Equation and Functional*

Given Consider the simple differential equation

$$\frac{d^2T}{dx^2} = 0$$

with the boundary conditions $T = 0, x = 0$ and $T = 1, x = 1$. Comparing this differential equation with (5.27) gives $K(x) = 1$, $P(x) = 0$, $Q(x) = 0$, and with the specified boundary conditions (5.28) and (5.29), $T_0 = 0$, $T_L = 1$. This is the same second-order differential equation as obtained for Example 5.5 (Heat conduction in infinite slab), where the conduction coefficient k is unity and boundary condition at x_L is different.

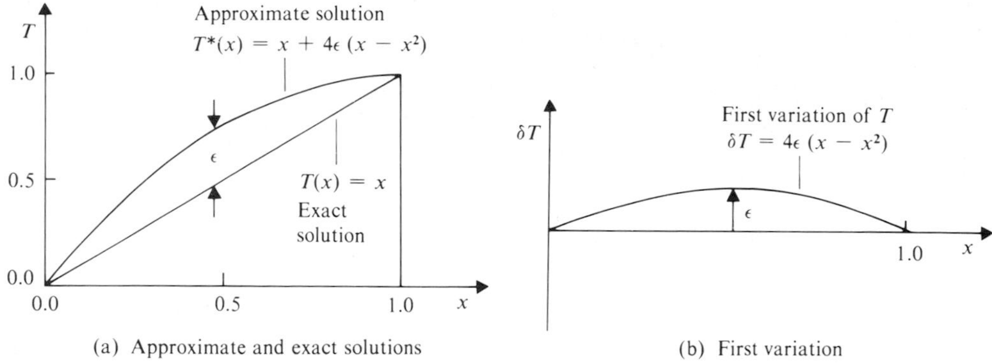

(a) Approximate and exact solutions

(b) First variation

Figure 6.3 Plots of T^*, T, δT (Example 6.1—Differential equation and functional)

The functional I for this problem is

$$I = \int_0^1 \frac{1}{2}\left(\frac{dT}{dx}\right)^2 dx$$

as will be shown in the next section.

Objective The exact solution to the differential equation is $T(x) = x$. Let the approximate solution have the form

$$T^*(x) = T(x) + \delta T(x)$$

where

$$\begin{aligned} T(x) &= x & &\text{(Exact solution)} \\ \delta T(x) &= 4\epsilon(x - x^2) & &\text{(First variation)} \end{aligned}$$

Here ϵ is a small constant ($\epsilon << 1$) whose meaning is examined later. Determine the functional I and its first variation δI. Set the first variation equal to zero and determine the value of ϵ that makes the functional a minimum.

Solution The approximate solution is illustrated in Figure 6.3. Note that δT vanishes at both end points, as required for the type of boundary conditions given.

At the center, $\delta T = \epsilon$, $x = 0.5$. Thus ϵ represents the "magnitude" of the variation of T. Here ϵ is a free parameter that acts in a manner similar to a nodal value of T in a finite element solution. It may be convenient to think of the variation δT (and ϵ) as acting in a direction perpendicular to the x axis. The magnitude ϵ of δT is independent of x.

For the exact solution, the functional $I(T)$ is given by

$$I(T) = \frac{1}{2}\int_0^1 \left(\frac{dT}{dx}\right)^2 dx$$

The derivative of T is

$$\frac{dT}{dx} = 1$$

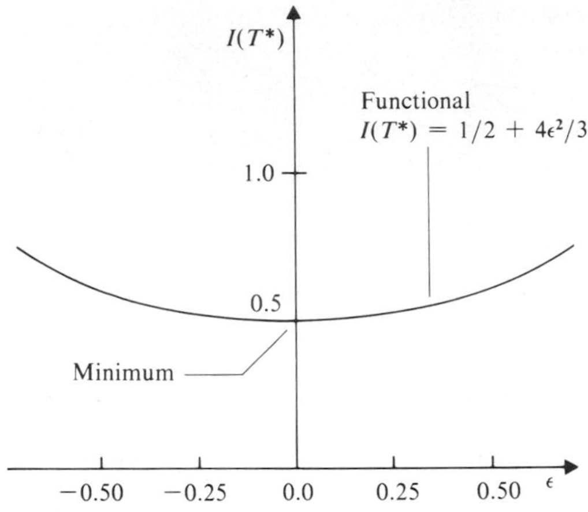

Figure 6.4 Plot of functional versus parameter ϵ (Example 6.1—Differential equation and functional)

and the functional is easily evaluated as

$$I(T) = \frac{1}{2}$$

Now consider the functional $I(T^*)$ for the approximate solution

$$I(T^*) = \frac{1}{2} \int_0^1 \left(\frac{dT^*}{dx}\right)^2 dx$$

Evaluating the derivative of T^* gives

$$\frac{dT^*}{dx} = \frac{d}{dx}[x + 4\epsilon(x - x^2)] = 1 + 4\epsilon(1 - 2x)$$

and the functional becomes

$$I(T^*) = \frac{1}{2} \int_0^1 [1 + 4\epsilon(1 - 2x)]^2 \, dx$$

The result is

$$I(T^*) = \frac{1}{2} + \frac{4}{3}\epsilon^2$$

which is a function of ϵ (but not of x).

Figure 6.4 plots the functional $I(T^*)$ versus ϵ. It is a quadratic in ϵ with a minimum value at $\epsilon = 0$. Clearly the exact solution to the differential equation, $T(x) = x$, produces the minimum value of the function. This conclusion has been arrived at by plotting the functional, but it is normally obtained using the first variations, as follows.

In the preceding discussion, the functional was expanded in a Taylor's series for small values of δT. For this example, $\delta T = 4\epsilon(x - x^2)$, so a small value of δT corresponds to a small value of the parameter ϵ. The Taylor's series is then

$$I(T + \delta T) = I(T) + \frac{\partial I}{\partial \epsilon} \delta T + \ldots$$

where higher order terms are neglected. The first variation of I is

$$\delta I = I(T + \delta T) - I(T) = \frac{\partial I}{\partial \epsilon} \delta T = 0$$

Once again, I or T varies with respect to the magnitude ϵ, not x. The above relation vanishes when

$$\frac{\partial I}{\partial \epsilon} = 0$$

or, for this example,

$$\frac{\partial I}{\partial \epsilon} = \frac{8}{3} \epsilon = 0$$

The solution is $\epsilon = 0$. Restating, the first variation of I is zero when $\epsilon = 0$. Therefore the exact solution to the differential equation $T(x) = x$ is the function that also makes the first variation of the functional zero. This conclusion can be obtained for any shape of δT that is continuous and satisfies the boundary conditions. Problems at the end of the chapter consider other simple forms for the variation δT.

6.2 Euler–Lagrange Equation

A functional (6.1) containing the terms T and T_x is assumed and a corresponding equation (called the Euler–Lagrange equation) obtained. Comparison of the differential equation to the Euler–Lagrange equation can then be used to determine the functional. The resulting procedure is not straightforward. It depends upon the proper initial guess of the form of the functional. Often some experience with the whole method is necessary to choose the proper initial form. The residual method, discussed in Chapter 8, starts with the boundary value problem, which is one advantage over the variational method.

A. First Variation of Functional

Start with the functional (6.1)

$$I(T) = \int_{x_0}^{x_L} F(x,T,T_x)\, dx - G_0(x,T)_{x_0} + G_L(x,T)_{x_L}$$

The first variation (6.4) is

$$\delta I = \int_{x_0}^{x_L} F(x,T^*,T_x^*)\,dx - G_0(x,T^*)_{x_0} + G_L(x,T^*)_{x_L}$$

$$- \int_{x_0}^{x_L} F(x,T,T_x)\,dx + G_0(x,T)_{x_0} - G_L(x,T)_{x_L}$$

The first integrand is expanded in a Taylor's series for small δT and δT_x.

$$F(x,T^*,T_x^*) = F(x,T + \delta T,\ T_x + \delta T_x)$$

$$= F(x,T,T_x) + \frac{\partial F}{\partial T}\delta T + \frac{\partial F}{\partial T_x}\delta T_x + \ldots$$

Higher order terms in δT and δT_x are neglected.

Similarly, the boundary terms can be expanded as

$$G(x,T^*) = G(x,T + \delta T) = G(x,T) + \frac{\partial G}{\partial T}\delta T + \ldots$$

where higher order terms in δT are neglected.

The term δT_x represents the variation of the derivative of T

$$\delta T_x = \delta\left(\frac{dT}{dx}\right)$$

This is the same as the derivative of the variation

$$\delta\left(\frac{dT}{dx}\right) = \frac{d}{dx}(\delta T)$$

The order of these two operations can be interchanged without changing the result. This is shown in Example 6.2.

Example 6.2 *Variation of a Derivative of a Function*

Given Consider the approximate solution to Example 6.2, $T^*(x) = x + 4\epsilon(x-x^2)$. Recall that the first part, x, is the exact solution, while the second part, $4\epsilon(x-x^2)$, is the first variation.

Objective Evaluate the terms

$$\delta\left(\frac{dT}{dx}\right) \quad \text{and} \quad \frac{d}{dx}(\delta T)$$

using the approximate solution.

Solution The first derivative is

$$\frac{dT^*}{dx} = \frac{d}{dx}[x + 4\epsilon(x - x^2)] = 1 + 4\epsilon\,(1 - 2x)$$

Here the first part is the derivative of the exact solution

$$\frac{dT}{dx} = 1$$

while the second part is the variation of the derivative

$$\delta\left(\frac{dT}{dx}\right) = 4\epsilon(1 - 2x) \tag{a}$$

Alternatively, the variation of T is $\delta T = 4\epsilon(x - x^2)$. Its derivative is

$$\frac{d}{dx}(\delta T) = 4\epsilon(1 - 2x) \tag{b}$$

Clearly, from (a) and (b),

$$\delta\left(\frac{dT}{dx}\right) = \frac{d}{dx}(\delta T)$$

so the two "operations" d/dx and δ are independent of one another.

Returning to the variation δI gives

$$\begin{aligned}\delta I = &\int_{x_0}^{x_L}\left[F(x,T,T_x) + \frac{\partial F}{\partial T}\delta T + \frac{\partial F}{\partial T_x}\frac{d}{dx}(\delta T)\right]dx\\ &-\left[G_0(x,T) + \frac{\partial G_0}{\partial T}\delta T\right]_{x_0} + \left[G_L(x,T) + \frac{\partial G_L}{\partial T}\delta T\right]_{x_L}\\ &-\int_{x_0}^{x_L} F(x,T,T_x)\,dx + G_0(x,T)_{x_0} - G_L(x,T)_{x_L}\end{aligned}$$

Subtracting like terms yields

$$\delta I = \int_{x_0}^{x_L}\left[\frac{\partial F}{\partial T}\delta T + \frac{\partial F}{\partial T_x}\frac{d}{dx}(\delta T)\right]dx + \left[\frac{\partial G}{\partial T}\delta T\right]_{x_0}^{x_L}$$

Here the boundary term is

$$\left[\frac{\partial G}{\partial T}\delta T\right]_{x_0}^{x_L} = -\left[\frac{\partial G_0}{\partial T}\delta T\right]_{x_0} + \left[\frac{\partial G_L}{\partial T}\delta T\right]_{x_L}$$

The second term in the integral is integrated by parts, as follows

$$\int_{x_0}^{x_L}\left[\frac{\partial F}{\partial T_x}\frac{d}{dx}(\delta T)\right]dx = \left[\frac{\partial F}{\partial T_x}\delta T\right]_{x_0}^{x_L} - \int_{x_0}^{x_L}\left[\frac{d}{dx}\left(\frac{\partial F}{\partial T_x}\right)\delta T\right]dx$$

The results can now be added together.

The first variation of I is then

$$\delta I = I(T + \delta T) - I(T) = \int_{x_0}^{x_L} \left[\frac{\partial F}{\partial T} - \frac{d}{dx}\left(\frac{\partial F}{\partial T_x}\right)\right] \delta T \, dx$$

$$+ \left[\frac{\partial F}{\partial T_x} \delta T \right]_{x_0}^{x_L} + \left[\frac{\partial G}{\partial T} \delta T \right]_{x_0}^{x_L}$$

Factoring out a negative sign from the integral and adding the boundary terms together yields

$$\delta I = - \int_{x_0}^{x_L} \left[\frac{d}{dx}\left(\frac{\partial F}{\partial T_x}\right) - \frac{\partial F}{\partial T} \right] \delta T \, dx + \left[\left(\frac{\partial F}{\partial T_x} + \frac{\partial G}{\partial T} \right) \delta T \right]_{x_0}^{x_L} \qquad \textbf{(6.6)}$$

The first variation is now in the desired form.

B. Euler–Lagrange Equation

The first variation δI must equal zero. Since δT is any arbitrary variation, the general integral is zero only if the integrand is zero.

$$\frac{d}{dx}\left(\frac{\partial F}{\partial T_x}\right) - \frac{\partial F}{\partial T} = 0 \qquad \text{(Euler–Lagrange equation)} \qquad \textbf{(6.7)}$$

The boundary term is set equal to zero as well

$$\left[\frac{\partial F}{\partial T_x} + \frac{\partial G}{\partial T} \right]_{x_0}^{x_L} = 0 \qquad \text{(Derivative boundary condition terms)} \qquad \textbf{(6.8)}$$

These are the Euler–Lagrange equation and associated boundary conditions.

Recall that the original differential equation (5.26) has the form $D(x, T, T_x) = 0$ with the associated boundary conditions. This original equation is the same as the Euler–Lagrange equation

$$D(x, T, T_x) = \frac{d}{dx}\left(\frac{\partial F}{\partial T_x}\right) - \frac{\partial F}{\partial T} = 0 \qquad \textbf{(6.9)}$$

Thus the integrand F of the functional I is related to the differential equation. Given any integrand F, the corresponding differential equation D can be found.

Example 6.3 *Simple Boundary Value Problem*

Given The simple differential equation

$$\frac{d^2T}{dx^2} = 0$$

with boundary conditions $T = 0$, $x = 0$ and $T = 1$, $x = 1$. This is the same boundary value problem as in Example 6.1.

Objective Use the Euler–Lagrange equation to determine the functional for the problem.

Solution The differential equation is expressed as

$$D(x,T,T_x) = \frac{d}{dx}(T_x) = 0$$

From (6.9)

$$\frac{d}{dx}(T_x) = \frac{d}{dx}\left(\frac{\partial F}{\partial T_x}\right) - \frac{\partial F}{\partial T}$$

Comparing the terms containing T_x yields

$$\frac{\partial F}{\partial T_x} = T_x$$

Integrating once with respect to T_x gives

$$F = \frac{1}{2} T_x^2$$

There is no term in the differential equation containing T, so this is the final result. The functional is

$$I = \int_0^1 F(x,T,T_x)\,dx = \int_0^1 \frac{1}{2} T_x^2\,dx$$

Usually this is written

$$I(T) = \frac{1}{2}\int_0^1 \left(\frac{dT}{dx}\right)^2 dx$$

This is the same functional as employed in Example 6.1. Thus it has been "derived."

A constant of integration could have been added when integrating with respect to T_x earlier. However, the constant of integration would not affect the result of setting the first variation to zero.

Example 6.4 *Derivative Boundary Condition*

Given The differential equation

$$\frac{d^2T}{dx^2} = 0$$

with the boundary conditions

$$T = 0, \quad x = 0$$

$$\frac{dT}{dx} - 2T + 0.5 = 0, \quad x = 1$$

This is the same as Example 6.2 except for the derivative boundary condition at $x = 1$.

Objective Determine the functional for the problem.

Solution As in Example 6.2, the integrand is

$$F = \frac{1}{2} T_x^2$$

The value of T at $x = 0$ is specified, so the function G_0 is ignored. At the other end, the boundary relation is

$$\frac{\partial F}{\partial T_x} + \frac{\partial G_L}{\partial T} = 0, \quad x = 1$$

From above

$$\frac{\partial F}{\partial T_x} = \frac{\partial}{\partial T_x}\left(\frac{1}{2} T_x^2\right) = T_x$$

with the result

$$T_x + \frac{\partial G_L}{\partial T} = 0, \quad x = 1$$

The actual derivative boundary condition is $T_x - 2T + 0.5 = 0$. Comparing these two last relations indicates that G_L is given by

$$\frac{\partial G_L}{\partial T} = -2T + 0.5$$

Integrating once yields $G_L = -T^2 + 0.5T$. The constant of integration is not important (it may be added to the functional or not).

The full functional is (6.6)

$$I = \int_{x_0}^{x_L} F(x,T,T_x)\, dx + G_L(x,T)\Big|_{x_L}$$

Substituting F and G_L gives

$$I = \int_0^1 \left[\frac{1}{2}\left(\frac{dT}{dx}\right)^2\right] dx + [-T^2 + 0.5T]\Big|_{x=1}$$

Adding the boundary term at $x = 1$ insures that the derivative boundary condition will be satisfied.

6.3 General Functional

A. General Boundary Value Problem

Consider the general boundary value problem (5.27)

$$\frac{d}{dx}\left(K\frac{dT}{dx}\right) + PT + Q = 0$$

and (5.28), (5.29) at $x = x_0$, either $T = T_0$ or

$$K_0 \frac{dT}{dx} + \alpha_0 T + \beta_0 = 0$$

and at $x = x_L$, either $T = T_L$ or

$$K_L \frac{dT}{dx} + \alpha_L T + \beta_L = 0$$

as discussed in Sections 5.4 and 5.5. The objective is to determine the functional for this problem, using the Euler–Lagrange equation.

Let the general differential equation be written as

$$D(x,T,T_x) = \frac{d}{dx}(KT_x) + PT + Q = 0$$

where

$$T_x = \frac{dT}{dx}$$

In similar fashion, the boundary conditions are at $x = x_0$, either $T = T_0$ or $K_0 T_x + \alpha_0 T + \beta_0 = 0$, and at $x = x_L$, either $T = T_L$ or $K_L T_x + \alpha_L T + \beta_L = 0$.

B. Comparison to Euler–Lagrange Equation

Comparing the first term of the Euler–Lagrange equation and the general differential equation

$$\frac{d}{dx}\left(\frac{\partial F}{\partial T_x}\right) = \frac{d}{dx}(KT_x)$$

indicates that both are functions of T_x. Integrating both sides with respect to x yields

$$\frac{\partial F}{\partial T_x} = KT_x$$

Now integrating with respect to T_x, holding T constant, produces

$$F = \frac{1}{2} K T_x^2 + C$$

where the constant of integration is a function of T, $C = C(T)$. The next step is to evaluate C.

The second term in the Euler–Lagrange equation is equated to the rest of the general equation

$$-\frac{\partial F}{\partial T} = PT + Q \qquad \textbf{(a)}$$

Substituting the expression for F just obtained

$$-\frac{\partial F}{\partial T} = -\frac{\partial}{\partial T}\left(\frac{1}{2} K T_x^2 + C\right) = -\frac{\partial C}{\partial T} \qquad \textbf{(b)}$$

Equating (a) and (b) gives

$$\frac{\partial C}{\partial T} = -PT - Q$$

and integrating with respect to T, while holding T_x constant, yields

$$C = -\frac{1}{2} PT^2 - QT$$

The final result for F is

$$F = \frac{1}{2} K\, T_x^2 - \frac{1}{2} PT^2 - QT$$

where the integrand is now completed.

The derivative boundary conditions (6.8) are

$$\left[\frac{\partial F}{\partial T_x} + \frac{\partial G_0}{\partial T}\right] = 0, \quad x = x_0$$

$$\left[\frac{\partial F}{\partial T_x} + \frac{\partial G_L}{\partial T}\right] = 0, \quad x = x_L$$

At $x = x_0$, the first term is

$$\frac{\partial F}{\partial T_x} = \frac{\partial}{\partial T_x}\left[\frac{1}{2} K\, T_x^2 + \frac{1}{2} PT^2 + QT\right]$$

$$= K_0 T_x, \quad x = x_0$$

Equating this to the general derivative boundary condition yields

$$K_0 T_x + \frac{\partial G_0}{\partial T} = K_0 T_x + \alpha_0 T + \beta_0 = 0$$

Here

$$\frac{\partial G_0}{\partial T} = \alpha_0 T + \beta_0$$

and integrating both sides with respect to T,

$$G_0 = \frac{1}{2} \alpha_0 T^2 + \beta_0 T, \quad x = x_0$$

Similarly at $x = x_L$,

$$G_L = \frac{1}{2} \alpha_L T^2 + \beta_L T, \quad x = x_L$$

This completes the derivative boundary terms added to the functional.

C. Result

Recall that the functional has the form (6.1)

$$I(T) = \int_{x_0}^{x_L} F(x,T,T_x)\, dx - G_0(x,T)_{x_0} + G_L(x,T)_{x_L}$$

The result for the general boundary value problem is then

$$I(T) = \int_{x_0}^{x_L} \left[\frac{1}{2} K\, T_x^2 - \frac{1}{2} PT^2 - QT \right] dx$$

$$- \left(\frac{1}{2} \alpha_0 T^2 + \beta_0 T \right)_{x_0} + \left(\frac{1}{2} \alpha_L T^2 + \beta_L T \right)_{x_L}$$

This is the desired result.

6.4 Problems

6.1 Redo Example 6.2, using an approximation of the form

$$T^*(x) - \begin{cases} x + 2\epsilon\, x, & 0 \le x \le \dfrac{1}{2} \\ x + 2\epsilon\,(1 - x), & \dfrac{1}{2} < x \le 1 \end{cases}$$

where ϵ may be either positive or negative. Plot the value of the functional I versus ϵ for this problem.

6.2 Redo Example 6.2, using an approximation of the form

$$T^*(x) = x + \epsilon \sin \pi x$$

where ϵ may be either positive or negative. Plot the value of the functional I versus ϵ for this problem.

6.3 Determine the functional for the boundary value problem

$$\frac{d}{dx}\left(x^2 \frac{dT}{dx} \right) + 3\, x\, T = 0$$

$$T = 4, \quad x = 0$$
$$T = 1, \quad x = 1$$

using the Euler–Lagrange equation.

6.4 Determine the functional for the boundary value problem

$$\frac{d^2T}{dx^2} + x^2 T + x^3 = 0$$

$$\frac{dT}{dx} + 3T = 0, \quad x = 0$$

$$\frac{dT}{dx} + T - 1 = 0, \quad x = 2$$

using the Euler–Lagrange equation.

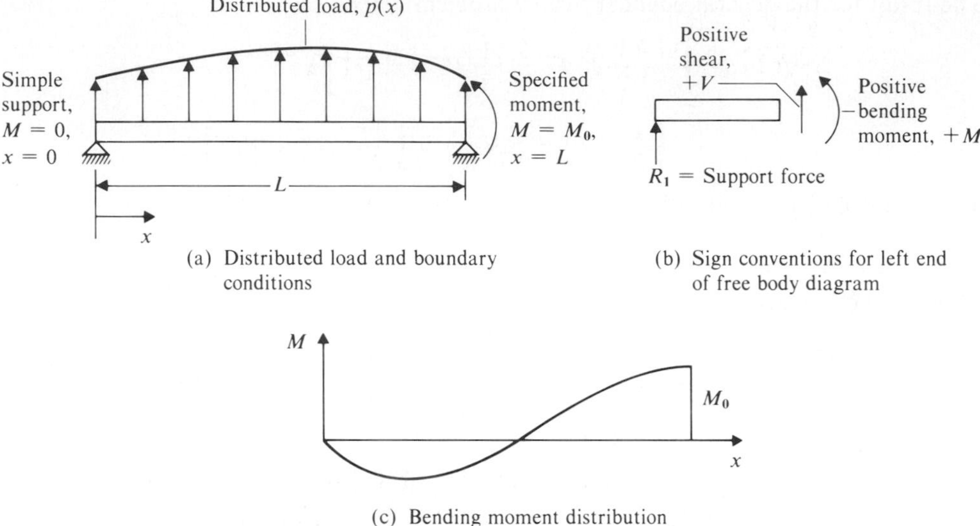

Figure 6.5 Diagram for Problem 6.5

6.5 A simple beam undergoes transverse loading. The shear force along the beam is given by

$$\frac{dV}{dx} + p = 0, \quad 0 \leq x \leq L$$

while the bending moment is determined from

$$\frac{dM}{dx} = -V, \quad 0 \leq x \leq L$$

Figure 6.5 shows a diagram including sign conventions. Here

$V(x)$ = shear force
$M(x)$ = bending moment
$p(x)$ = transverse distributed load (per unit length)

At the left end, the beam is simply supported, so the boundary condition is $M = 0$, $x = 0$. At the right end, the moment is specified as M_L, yielding the boundary condition $M = M_L$, $x = L$. Determine the functional for the moment distribution using the Euler–Lagrange equation.

6.6 The neutron flux ϕ (in neutrons/m^2-s) in a semi-infinite planar unshielded nuclear reactor is given by

$$\frac{d}{dx}\left(D\frac{d\phi}{dx}\right) - \frac{D}{L^2}\phi + S = 0$$

where

$D = D(x)$ = diffusion coefficient
$S = S(x)$ = neutron source strength
L = length in x direction

The boundary conditions are

$$\phi = 0, \quad x = -\frac{L}{2}$$

$$\phi = 0, \quad x = \frac{L}{2}$$

corresponding to a vacuum on both sides. Determine the functional for this problem, using the Euler–Lagrange equation.

Chapter 7

Deflection of Beams and Higher Order Equations

7.1 Introduction

A. Higher Order Elements

Chapters 5 and 6 considered second-order ordinary differential equations and their finite element solution. Bending of beams and other engineering problems lead to differential equations of fourth or even higher order. These can be treated by an extension of the methods just discussed. The second derivative of the transverse deflection occurs in the functional. The second derivative of a simplex linear interpolation function is zero, so simplex elements cannot be used, and new interpolation functions must be developed.

First, a functional is developed for the fourth-order differential equation describing Bernoulli–Euler bending of beams with appropriate boundary conditions. Second, the element interpolation functions are chosen to satisfy conditions of compatibility and completeness. For one-dimensional problems, this is not difficult. In the case of bending of two-dimensional plates and shells, choosing interpolation functions that are continuous across interelement boundaries is not easy. Finally, the element coefficient matrix and column can be obtained. In this case, it is the stiffness matrix for beam bending that will be derived.

The finite element method was originally developed for long, thin structural members in airframes where they were subject to axial displacements as well as transverse bending. Only transverse bending is considered in this chapter. Textbooks present many treatments of pure beam bending, using both the direct method of derivation (discussed in Sections 1.4 and 4.3) and the variational method. Only the variational method is given here.

B. Beam Theory

A Bernoulli–Euler beam (see Figure 7.1) is governed by the differential equation

$$\frac{d^2}{dx^2}\left(EI\,\frac{d^2w}{dx^2}\right) - p = 0, \quad 0 \le x \le L \tag{7.1}$$

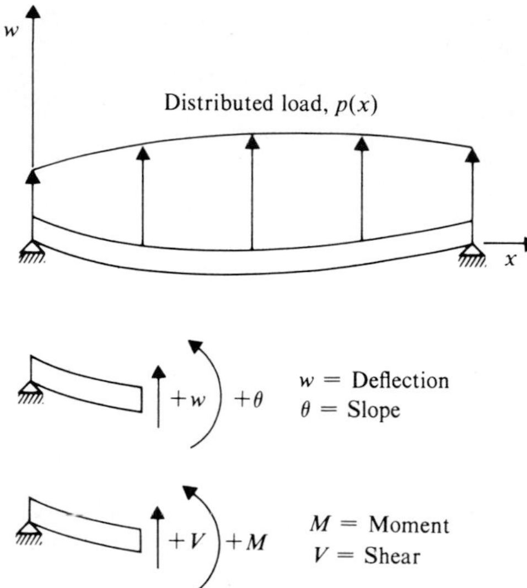

Figure 7.1 Deflection of beams—Sign conventions

where w = transverse deflection, E = Young's modulus, I = cross sectional area moment of inertia, and p = distributed load. The parameters E, I, p are all functions of x. Various quantities have physical significance, as follows:

$$\frac{dw}{dx} = \theta = \text{slope}$$

$$EI\frac{d^2w}{dx^2} = M = \text{bending moment}$$

$$-\frac{d}{dx}\left(EI\frac{d^2w}{dx^2}\right) = -\frac{dM}{dx} = V = \text{shear force}$$

If one end of the beam is simply supported, both the deflection and moment are zero:

$$\left.\begin{aligned} w &= 0 \\ M &= EI\frac{d^2w}{dx^2} = 0 \end{aligned}\right\} \text{Simply supported end} \tag{7.2}$$

When the beam end is fixed or cantilevered, the deflection and slope vanish:

$$\left.\begin{aligned} w &= 0 \\ \theta &= \frac{dw}{dx} = 0 \end{aligned}\right\} \text{Fixed end} \tag{7.3}$$

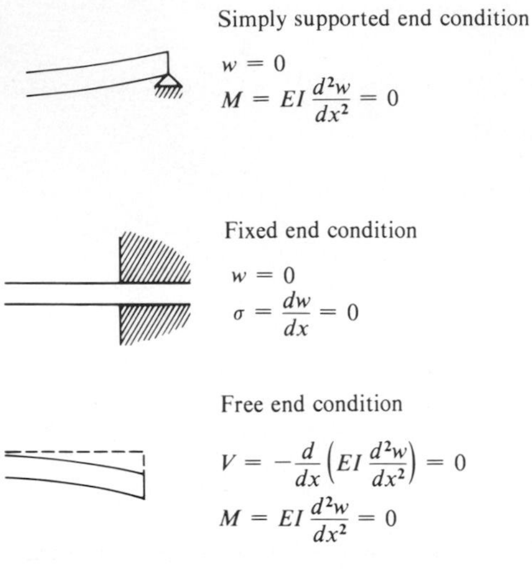

Simply supported end condition

$$w = 0$$
$$M = EI\frac{d^2w}{dx^2} = 0$$

Fixed end condition

$$w = 0$$
$$\sigma = \frac{dw}{dx} = 0$$

Free end condition

$$V = -\frac{d}{dx}\left(EI\frac{d^2w}{dx^2}\right) = 0$$
$$M = EI\frac{d^2w}{dx^2} = 0$$

Figure 7.2 Normal end conditions

A free beam end is specified by zero shear and moment:

$$\left.\begin{aligned} V &= -\frac{d}{dx}\left(EI\frac{d^2w}{dx^2}\right) = 0 \\ M &= EI\frac{d^2w}{dx^2} = 0 \end{aligned}\right\}\text{Free end} \tag{7.4}$$

Figure 7.2 shows all three cases. These are the normal boundary conditions for beams.

7.2 Derivation of Functional

A. Differential Equation

The fourth-order beam equation can be written in terms of second-order derivatives as

$$\frac{d^2}{dx^2}(EI\, w_{xx}) - p = 0 \tag{7.5}$$

Here, for convenience, w_{xx} is defined as

$$w_{xx} = \frac{d^2w}{dx^2}$$

The boundary conditions are discussed later.

B. Functional

The general form of the functional (6.1) used for second-order differential equations (with the variable w replacing T) is

$$I(w) = \int_0^L F(x,w,w_x)\, dx - G_0(x,w)_{x_0} + G_L(x,w)_{x_L}$$

where the integrand F is a function of x, w, w_x. With higher order equations, the integrand must include higher order terms such as w_{xx}: $F = F(x,w,w_x,w_{xx})$. The term w_x could be included throughout the derivation, but it corresponds to the second-order term in the differential equation, so it may be dropped here temporarily.

For the fourth-order beam equation, the form of the functional is taken as

$$I = \int_0^L F(x,w,w_{xx})\, dx \tag{7.6}$$

No boundary terms need to be added for the three boundary conditions just discussed, as shown in the following derivation.

C. First Variation

The first variation is $\delta I - I(w^*) - I(w)$. This is written out as

$$\delta I = \int_0^L F(x,w^*,w^*_{xx})\, dx - \int_0^L F(x,w,w_{xx})\, dx$$

Expanding the first integrand in a Taylor's series gives

$$\begin{aligned} F(x,w^*,w^*_{xx}) &= F(x,w + \delta w, w_{xx} + \delta w_{xx}) \\ &= F(x,w,w_{xx}) + \frac{\partial F}{\partial w}\delta w + \frac{\partial F}{\partial w_{xx}}\delta w_{xx} + \ . \ . \ . \end{aligned}$$

Thus the first variation is

$$\delta I = \int_0^L \Bigg[\underbrace{\frac{\partial F}{\partial x}\delta w}_{①} + \underbrace{\frac{\partial F}{\partial w_{ww}}\delta w_{xx}}_{②} \Bigg] dx \tag{7.7}$$

where like terms have cancelled out.

Integrating the second term by parts gives

$$\begin{aligned} ② &= \int_0^L \left[\frac{\partial F}{\partial w_{xx}} \frac{d}{dx}(\delta w_x) \right] dx \\ &= \left[\frac{\partial F}{\partial w_{xx}} \delta(w_x) \right]_0^L - \int_0^L \frac{d}{dx}\left(\frac{\partial F}{\partial w_{xx}} \right) \delta(w_x)\, dx \end{aligned}$$

where the relation

$$\delta(w_{xx}) = \frac{d}{dx}\delta(w_x)$$

has been used. Recall that differentiation and variation are independent of one another. Integrating by parts again,

$$② = \left[\frac{\partial F}{\partial w_{xx}}\,\delta(w_x)\right]_0^L - \left[\frac{d}{dx}\left(\frac{\partial F}{\partial w_{xx}}\right)\delta w\right]_0^L + \int_0^L \left[\frac{d^2}{dx^2}\left(\frac{\partial F}{\partial w_{xx}}\right)\right]\delta w\, dx = 0$$

This is now recombined with the first term.

The first variation (7.7) becomes

$$\delta I = \int_0^L \left[\frac{d^2}{dx^2}\left(\frac{\partial F}{\partial w_{xx}}\right) + \frac{\partial F}{\partial w}\right]\delta w\, dx + \left[\frac{\partial F}{\partial w_{xx}}\,\delta(w_x)\right]_0^L - \left[\frac{d}{dx}\left(\frac{\partial F}{\partial w_{xx}}\right)\delta w\right]_0^L = 0 \tag{7.8}$$

This is in the form suitable for the Euler–Lagrange equation. A negative sign has *not* been factored out of the integral term as one was in Chapter 6. This produces a stationary point which is a minimum for the beam.

D. Euler–Lagrange Equation

As in Chapter 6, the variation δw is arbitrary, so the integral term is zero only if the integrand vanishes.

$$\frac{d^2}{dx^2}\left(\frac{\partial F}{\partial w_{xx}}\right) + \frac{\partial F}{\partial w} = 0 \tag{7.9}$$

This is the Euler–Lagrange equation.

Comparing this to the general differential equation (7.5),

$$\frac{d^2}{dx^2}(EIw_{xx}) - p = 0$$

the first term yields

$$\frac{\partial F}{\partial w_{xx}} = EIw_{xx}$$

Integrating once while keeping w constant gives

$$F = \frac{1}{2}EI(w_{xx})^2 + C(w) \tag{a}$$

Comparing the remaining two terms,

$$\frac{\partial F}{\partial w} = -p$$

but from the expression for F just obtained,

$$\frac{dC}{dw} = -p$$

Thus

$$C = -pw \tag{b}$$

Finally, summing (a) and (b),

$$F = \frac{1}{2} EI(w_{xx})^2 - pw$$

This is the integrand in the functional.

The first variation (7.8) has several terms that now have a physical interpretation. In order, these are

$$w = \text{deflection}$$

$$w_x = \frac{dw}{dx} = \theta = \text{slope}$$

$$\frac{\partial F}{\partial w_{xx}} = EIw_{xx} = M = \text{moment}$$

$$-\frac{d}{dx}\left(\frac{\partial F}{\partial w_{xx}}\right) = -\frac{d}{dx}(EIw_{xx}) = V = \text{shear force}$$

Thus the first variation (7.8) can be written in these terms as

$$\begin{aligned} \delta I = \int_0^L \left[\frac{d^2}{dx^2}\left(EI\frac{d^2w}{dx^2}\right) - p\ w\right] \delta w\ dx \\ + M\delta\theta\Big|_0^L + V\delta w\Big|_0^L = 0 \end{aligned} \tag{7.10}$$

The terms evaluated at the end points 0 and L are obviously related to the boundary conditions. In fact, they indicate what type of boundary conditions are required for this differential equation and corresponding functional. If the deflection w is specified at an end, then the variation is zero: $\delta w = 0$. Also, if the slope θ is specified, the variation must vanish: $\delta\theta = 0$. The normal boundary conditions are expressed as $\delta w = 0$, $M = 0$ (simply supported end [7.2]); $\delta w = 0$, $\delta\theta = 0$ (fixed end [7.3]); $M = 0$, $V = 0$ (free end [7.4]). In all cases, the terms evaluated at the end points have the value zero.

The functional I is simply

$$I = \int_0^L \left[\frac{1}{2} EI\left(\frac{d^2w}{dx^2}\right)^2 - pw\right] dx \tag{7.11}$$

Note that no derivative boundary conditions are part of the functional I. Thus, if the deflection w is not specified as zero, the shear force is automatically set equal to zero. Also, if the slope is not specified as zero, the moment automatically vanishes. Actually what occurs is that the finite element solution that minimizes the functional will have the correct boundary condition. This circumstance has given the boundary conditions for the deflection and slope the name of *geometric* or rigid boundary conditions:

$$w = 0$$

$$\frac{dw}{dx} = \theta = 0$$

Also, the shear and moment conditions are called *natural* boundary conditions:

$$-\frac{d}{dx}\left(EI\,\frac{d^2w}{dx^2}\right) = V = 0$$

$$EI\,\frac{d^2w}{dx^2} = M = 0$$

A boundary condition specifying the shear could be applied by adding the proper terms in (7.10). Similarly, the moment could be set equal to some known value at one end. These boundary conditions are not discussed here.

E. Physical Interpretation

The functional I in (7.11) represents the potential energy stored in the beam. It is the sum of the strain energy U associated with the beam bending plus the potential W of the distributed load p to do work: $I = U + W$, or

$$I = \int_0^L \left[\frac{1}{2}\,EI\left(\frac{d^2w}{dx^2}\right)^2\right] dx - \int_0^L pw\, dx$$

Consider first the work in the beam due to the applied load $p(x)$. Assuming that p is held constant during the deflection of the beam, the work that can be reclaimed is $W = -\int_0^L pw\, dx$. Note that p has units of force per length and w has units of length. After integration over the length L, the units are force times length, or energy. The negative sign indicates that the work can be done by the beam on the surroundings (work done by an external force on an object is taken as positive when using Newton's law).

The strain energy term is written in terms of the bending moment M as

$$U = \int_0^L \frac{1}{2}\,M\left(\frac{d^2w}{dx^2}\right) dx, \qquad M = EI\,\frac{d^2w}{dx^2}$$

For small deformations of beams, the second derivative of w equals the curvature κ of the beam

$$\kappa = \frac{d^2w}{dx^2}$$

The curvature of the beam is the inverse of the radius of curvature R (Figure 7.3); thus the curvature has units of inverse length.

$$\frac{1}{R} = |\kappa| = \left|\frac{d^2w}{dx^2}\right|$$

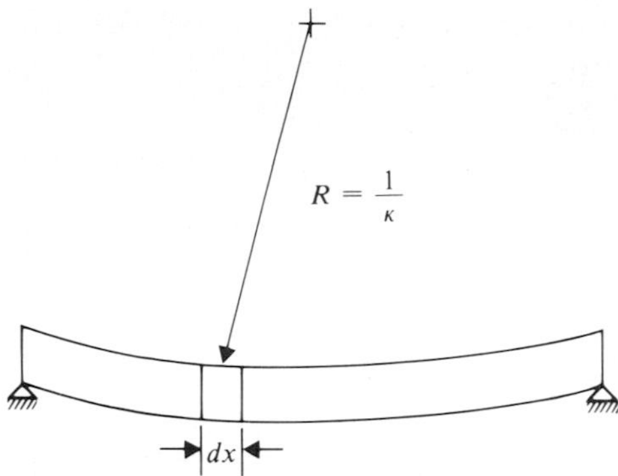

Figure 7.3 Radius of curvature

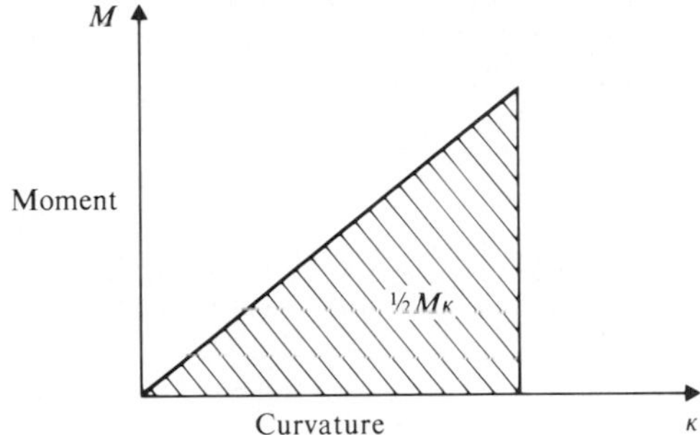

Figure 7.4 Moment versus curvature for deflection of beam

The moment is related to the curvature by the equation $M = EI\,\kappa$. Figure 7.4 plots M versus κ. The moment has units of force times length, while the curvature has units of inverse length. This represents the strain energy of beam bending per unit length

$$\frac{1}{2}\,M(x)\,\frac{d^2w(x)}{dx^2} = \frac{1}{2}\,M(x)\,\kappa(x)$$

Integration over the length gives units of force times length, or energy. The factor of 1/2 gives the area under the curve of moment versus curvature, as shown in Figure 7.4.

The potential energy in the beam represents a minimum principle. When the proper deformation function $w(x)$ is found that minimizes the potential energy, it is the actual deformation of the beam. Another way of putting it is that the first variation of the potential energy δI is zero for any arbitrary variation of the deformation δw: $\delta I = \delta U + \delta w = 0$. It is possible to start from this point and derive the Bernoulli–Euler beam equation, considering the physical significance of each step. That is left to the student.

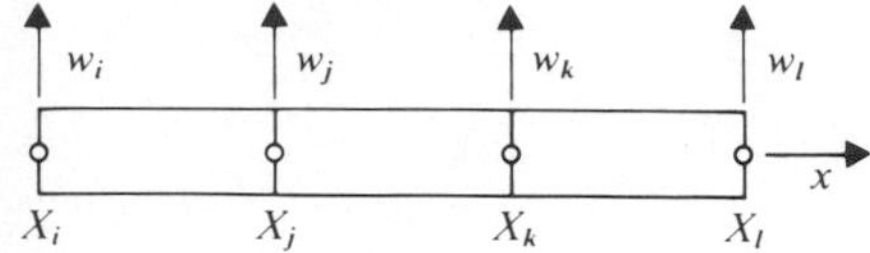

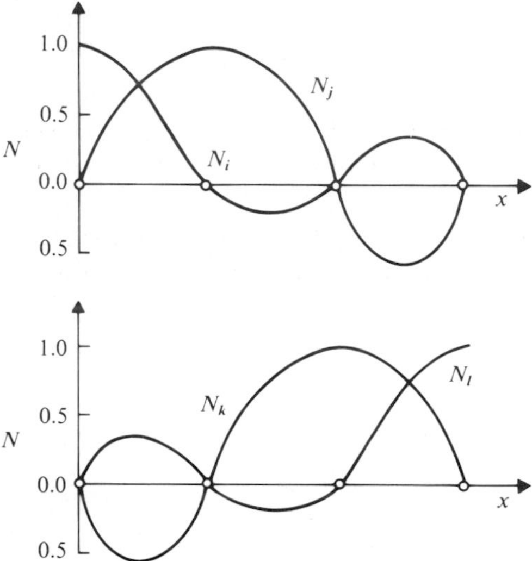

Figure 7.5 Beam element with Lagrange shape functions

7.3 Higher Order Elements

The linear polynomials used for interpolation functions in previous results clearly cannot be used for fourth-order differential equations and their associated variational methods. Second-order derivatives in the functional would all be zero. This chapter will consider two types of polynomials: Lagrange and Hermite. Lagrange interpolation polynomials represent a function in terms of the location of the zeros of the function. Alternatively, the Hermite polynomials are specified by zeros of both the function and its derivatives. These were briefly discussed in Section 3.6.

A. Lagrangian Polynomials

Four parameters are introduced during the solution of fourth-order differential equations. For the transverse deflection, assume an interpolation function of the form $w = \alpha_1 + \alpha_2 x + \alpha_3 x^2 + \alpha_4 x^3$. Four nodal points at X_i, X_j, X_k, and X_ℓ are located within the element; X_i and X_ℓ are the end points, as shown in Figure 7.5. As in Chapter 3, the values for α are determined by evaluating the polynomial at the four nodal points. The equations are

$$w_i = \alpha_1 + \alpha_2 X_i + \alpha_3 X_i^2 + \alpha_4 X_i^3$$

$$w_j = \alpha_1 + \alpha_2 X_j + \alpha_3 X_j^2 + \alpha_4 X_j^3$$

$$w_k = \alpha_1 + \alpha_2 X_k + \alpha_3 X_k^2 + \alpha_4 X_k^3$$

$$w_\ell = \alpha_1 + \alpha_2 X_\ell + \alpha_3 X_\ell^2 + \alpha_4 X_\ell^3$$

While these may be solved for the α values, there is a better approach.

The interpolation function for w is defined as

$$w = G_i w_i + G_j w_j + G_k w_k + G_\ell w_\ell \tag{7.12}$$

In this case the functions G are called shape functions rather than pyramid functions because they are cubic in form rather than linear. It is still required that each shape function G have the value unity at the corresponding node point and zero at the other three node points in the element.

General formulas are available for functions with the above-described properties. The polynomials themselves are called Lagrangian polynomials. In general, the nth order polynomial has the form

$$G_m = \frac{\displaystyle\prod_{\substack{p=1 \\ p\neq m}}^{n+1} (x - X_p)}{\displaystyle\prod_{\substack{p=1 \\ p\neq m}}^{n+1} (X_m - X_p)}$$

The specific Lagrange shape functions for the beam deflection are

$$G_i = \frac{(x - X_j)(x - X_k)(x - X_\ell)}{(X_i - X_j)(X_i - X_k)(X_i - X_\ell)}$$

$$G_j = \frac{(x - X_i)(x - X_k)(x - X_\ell)}{(X_j - X_i)(X_j - X_k)(X_j - X_\ell)}$$

$$G_k = \frac{(x - X_i)(x - X_j)(x - X_\ell)}{(X_k - X_i)(X_k - X_j)(X_k - X_\ell)}$$

$$G_\ell = \frac{(x - X_i)(x - X_j)(x - X_k)}{(X_\ell - X_i)(X_\ell - X_j)(X_\ell - X_k)}$$

Since the element stiffness matrix is independent of the actual position of the first node X_i, let this be zero. Also, take the distance between nodal points to be equal within the element, as shown in Figure 7.5.

$$X_i = 0, \qquad X_k = \frac{2}{3}L$$

$$X_j = \frac{1}{3}L, \quad X_\ell = L$$

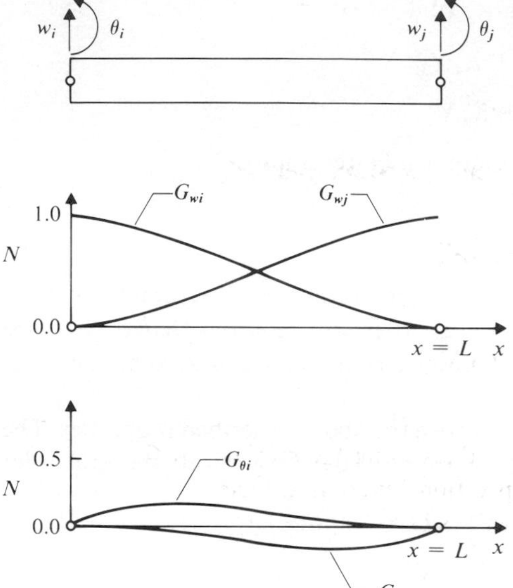

Figure 7.6 Beam element with Hermite shape functions

Then the shape functions are

$$\begin{aligned}
G_i &= -\frac{1}{6L^3}(3x - L)(3x - 2L)(3x - 3L) \\
G_j &= -\frac{1}{2L^3}(3x)(3x - 2L)(3x - 3L) \\
G_k &= -\frac{1}{6L^3}(3x)(3x - L)(3x - 3L) \\
G_\ell &= \frac{1}{2L^3}(3x)(3x - L)(3x - 2L)
\end{aligned} \tag{7.13}$$

Figure 7.5 shows graphs of the four shape functions.

B. Hermite Polynomials

In the transverse bending of beams considered in the last section, both the deflection and its first derivative (the slope) are very important. Hermite polynomials are very useful for this. They are defined so that both their value and first derivatives have the value unity or zero at the end points of the interval (X_i, X_j). The full interpolation function for the deflection w then has interelement continuity for both the displacement and the slope. Figure 7.6 shows an element.

Start with the cubic polynomial for w, $w = \alpha_1 + \alpha_2 x + \alpha_3 x^2 + \alpha_4 x^3$. The first derivative or slope is

$$\theta = \frac{dw}{dx} = \alpha_2 + 2\alpha_3 x + 3\alpha_4 x^2$$

Evaluating the deflection and slope at the element end points yields the equations for α_1, α_2, α_3, α_4 as

$$w_i = \alpha_1 + \alpha_2 X_i + \alpha_3 X_i^2 + \alpha_4 X_i^3$$

$$\theta_i = \alpha_2 + 2\alpha_3 X_i + 3\alpha_3 X_i^2$$

$$w_j = \alpha_1 + \alpha_2 X_j + \alpha_3 X_j^2 + \alpha_4 X_j^3$$

$$\theta_j = \alpha_2 + 2\alpha_3 X_j + 3\alpha_4 X_j^2$$

These four equations may be solved for the α values using the methods of Chapter 3.

It is most convenient to define the interpolation function in terms of the nodal deflections w_i, w_j, θ_i, θ_j rather than four nodal values of w (Figure 7.6).

$$w = G_{wi} w_i + G_{\theta i} \theta_i + G_{wj} w_j + G_{\theta j} \theta_j \tag{7.14}$$

Here the four functions G are shape functions associated with the four nodal values. In column form this is

$$w = \lfloor G_{wi} \quad G_{\theta i} \quad G_{wj} \quad G_{\theta j} \rfloor \begin{Bmatrix} w_i \\ \theta_i \\ w_j \\ \theta_j \end{Bmatrix} = \lfloor G \rfloor \{w\}$$

Each node has two degrees of freedom, w and θ.

The four Hermite polynomials are given by

$$\begin{aligned}
G_{wi} &= 1 - 3\left(\frac{x - X_i}{X_j - X_i}\right)^2 + 2\left(\frac{x - X_i}{X_j - X_i}\right)^3 \\
G_{\theta i} &= (x - X_i)\left[\left(\frac{x - X_i}{X_j - X_i}\right) - 1\right]^2 \\
G_{wj} &= \left(\frac{x - X_i}{X_j - X_i}\right)^2 \left[3 - 2\left(\frac{x - X_i}{X_j - X_i}\right)\right] \\
G_{\theta j} &= \frac{(x - X_i)^2}{X_j - X_i}\left[\left(\frac{x - X_i}{X_j - X_i}\right) - 1\right]
\end{aligned} \tag{7.15}$$

As noted for the Lagrangian polynomials, the resulting stiffness matrix is independent of X_i, so this can be taken as zero. Also, $X_j = L$. Then the shape functions simplify somewhat to

$$
\begin{aligned}
G_{wi} &= 1 - 3\left(\frac{x}{L}\right)^2 + 2\left(\frac{x}{L}\right)^3 \\
G_{\theta i} &= x\left[\left(\frac{x}{L}\right) - 1\right]^2 \\
G_{wj} &= \left(\frac{x}{L}\right)^2\left[3 - 2\left(\frac{x}{L}\right)\right] \\
G_{\theta j} &= \frac{x^2}{L}\left[\left(\frac{x}{L}\right) - 1\right]
\end{aligned}
\tag{7.16}
$$

These are easier to work with. Plots are given in Figure 7.6.

Because of the emphasis on the slope, it will be useful to compute the first derivative of each of these terms.

$$
\begin{aligned}
\frac{dG_{wi}}{dx} &= -\frac{6}{L}\left[\left(\frac{x}{L}\right) - \left(\frac{x}{L}\right)^2\right] \\
\frac{dG_{\theta i}}{dx} &= 1 - 4\left(\frac{x}{L}\right) + 3\left(\frac{x}{L}\right)^2 \\
\frac{dG_{wj}}{dx} &= \frac{6}{L}\left[\left(\frac{x}{L}\right) - \left(\frac{x}{L}\right)^2\right] \\
\frac{dG_{\theta j}}{dx} &= -2\left(\frac{x}{L}\right) + 3\left(\frac{x}{L}\right)^2
\end{aligned}
$$

Now the shape factors can be evaluated at the node points.

At $x = 0$, the shape function and its first derivatives are

$$
\begin{aligned}
G_{wi} &= 1, & G_{wj} &= 0 \\
G_{\theta i} &= 0, & G_{\theta j} &= 0 \\
\frac{dG_{wi}}{dx} &= 0, & \frac{dG_{wj}}{dx} &= 0 \\
\frac{dG_{\theta i}}{dx} &= 1, & \frac{dG_{\theta j}}{dx} &= 0
\end{aligned}
$$

Also, at $x = L$, similar terms are

$$
\begin{aligned}
G_{wi} &= 0, & G_{wj} &= 1 \\
G_{\theta i} &= 0, & G_{\theta j} &= 0 \\
\frac{dG_{wi}}{dx} &= 0, & \frac{dG_{wj}}{dx} &= 0
\end{aligned}
$$

$$\frac{dG_{\theta i}}{dx} = 0, \qquad \frac{dG_{\theta j}}{dx} = 1$$

Thus, both the shape functions and their first derivatives have the value unity at the appropriate node and zero at the other node.

In this case, the element derivative matrix of interest is not the first derivative. Let $[D]$ be defined by

$$\frac{d^2w}{dx^2} = [D]\{w\} \tag{7.17}$$

where

$$[D] = \begin{bmatrix} \frac{d^2G_{wi}}{dx^2} & \frac{d^2G_{\theta i}}{dx^2} & \frac{d^2G_{wj}}{dx^2} & \frac{d^2G_{\theta j}}{dx^2} \end{bmatrix}$$

Element derivative matrix (1×4)

For the Hermite beam element, this is

$$[D] = \frac{1}{L^2}\left[\left(-6 + 12\frac{x}{L}\right)\ \left(-4L + 6x\right)\ \left(6 - 12\frac{x}{L}\right)\ \left(-2L + 6x\right)\right] \tag{7.18}$$

The moment within the element is $M = EI\ [D]\{w\}$. Thus the moment is approximated in a linear fashion over each element. The moment at the node points can be written as

$$\begin{Bmatrix} M_i \\ M_j \end{Bmatrix} = \frac{EI}{L^2}\begin{bmatrix} -6 & -4L & 6 & -2L \\ 6 & 2L & -6 & 4L \end{bmatrix}\{w\}$$

7.4 Element Matrices

The element coefficient (stiffness) matrix and column are obtained by substituting the interpolation function for w into the variational principle and differentiating with respect to the nodal variable. In the case of the Lagrange polynomial, the nodal variables are the four nodal deflection values, while the two displacements and two slopes are used in the Hermite polynomial case.

The functional for a single element is

$$I^{(e)} = \int_{L^{(e)}} \left[\frac{1}{2} EI\left(\frac{d^2w}{dx^2}\right)^2 - pw\right]^{(e)} dx \tag{7.19}$$

Actually the limits of integration can be made 0 to L, where L is the element length. The functional is sometimes written in matrix form as

$$I^{(e)} = \int_{L^{(e)}} \left[\frac{1}{2} EI\ \{w\}^T[D]^T[D]\{w\} - p\{G\}\{w\}\right] dx \tag{7.20}$$

A. Lagrange Element

The element stiffness matrix and column are given by

$$\begin{Bmatrix} \dfrac{\partial I}{\partial w_i} \\ \dfrac{\partial I}{\partial w_j} \\ \dfrac{\partial I}{\partial w_k} \\ \dfrac{\partial I}{\partial w_\ell} \end{Bmatrix}^{(e)} = [B]^{(e)} \begin{Bmatrix} w_i \\ w_j \\ w_k \\ w_\ell \end{Bmatrix}^{(e)} - \{C\}^{(e)}$$

The Lagrange polynomials from Equations (7.13) can be substituted and the integrations carried out. This will not be done here because of the desirability of interelement continuity of the first derivative, which is not present in the Lagrangian interpolation functions. However, the method of developing a higher order element has been introduced. Now attention will be concentrated on the Hermite interpolation functions.

B. Hermite Element

Now the element stiffness matrix and force column take the form

$$\begin{Bmatrix} \dfrac{\partial I}{\partial w_i} \\ \dfrac{\partial I}{\partial \theta_i} \\ \dfrac{\partial I}{\partial w_j} \\ \dfrac{\partial I}{\partial \theta_j} \end{Bmatrix}^{(e)} = [B]^{(e)} \begin{Bmatrix} w_i \\ \theta_i \\ w_j \\ \theta_j \end{Bmatrix}^{(e)} - \{C\}^{(e)} \tag{7.21}$$

The first row in the element stiffness matrix and force column is given by

$$\begin{aligned} \frac{\partial I}{\partial w_i} &= \frac{\partial}{\partial w_i} \int_{L^{(e)}} \left[\frac{1}{2} EI \left(\frac{d^2 w}{dx^2} \right)^2 - pw \right] dx \\ &= \int_{L^{(e)}} \left\{ EI \frac{d^2 G_{wi}}{dx^2} \left[\frac{d^2 G_{wi}}{dx^2} w_i + \frac{d^2 G_{\theta i}}{dx^2} \theta_i + \frac{d^2 G_{wj}}{dx^2} w_j + \frac{d^2 G_{\theta j}}{dx^2} \theta_j \right] \right. \\ &\qquad \left. - pG_{wi} \right\} dx \end{aligned}$$

where $E = E^{(e)}$, $I = I^{(e)}$, and $p = p^{(e)}$ are constants in element (e). Substituting for the shape functions and their second derivatives yields

$$\frac{\partial I}{\partial w_i} = \int_0^L \Big\{ EI \Big[-\frac{6}{L^2}\Big(1 - \frac{2x}{L}\Big) \Big] \Big[-\frac{6}{L^2}\Big(1 - \frac{2x}{L}\Big) w_i - \frac{4}{L}\Big(1 - \frac{3}{2}x\Big)\theta_i + \frac{6}{L^2}\Big(1 - \frac{2x}{L}\Big) w_j - \frac{2}{L}\Big(1 - 3x\Big)\theta_j \Big] - p\Big(1 - \frac{3x^2}{L^2} + \frac{2x^3}{L^3}\Big) \Big\}^{(e)} dx$$

Evaluating the integrals yields

$$\frac{\partial I}{\partial w_i} = \Big[\frac{2EI}{L^3}(6w_i + 3L\theta_i - 6w_j + 6L\theta_j) - \frac{1}{2}PL \Big]^{(e)}$$

This gives the first row.

Evaluating the remaining three terms in similar fashion gives the element stiffness matrix

$$[B]^{(e)} - \begin{matrix} w_i \\ \theta_i \\ w_j \\ \theta_j \end{matrix} \; \frac{2EI}{L^3} \begin{matrix} \quad w_i & \theta_i & w_j & \theta_j \end{matrix} \begin{bmatrix} 6 & 3L & -6 & 3L \\ & 2L^2 & -3L & L^2 \\ & & 6 & -3L \\ \text{(Symmetric)} & & & 2L^2 \end{bmatrix}^{(e)} \tag{7.22}$$

and the element force column

$$\{C\}^{(e)} = \frac{1}{12}pL \begin{Bmatrix} 6 \\ L \\ 6 \\ -L \end{Bmatrix} \tag{7.23}$$

The name "element stiffness matrix" arises from the term $2EI/L^3$, which has the units of force per length. Also, pL has the units of force.

The element stiffness matrix can also be expressed in the integral form

$$[B]^{(e)} = \int_0^L EI\,[D]^T[D]\,dx \tag{7.24}$$

while the force column is

$$\{C\} = \int_0^L p\{G\}\,dx \tag{7.25}$$

These expressions permit the evaluation of other interpolation functions. An example might be a sixth-order interpolation function that has nodal values for deflection w, slope θ, and moment M.

Within an element the slope, moment, and shear can be expressed in terms of the nodal deflections and slopes from Equation (7.14) as

$$\begin{aligned}\theta = \frac{dw}{d\theta} &= -\frac{6}{L}\left[\left(\frac{x}{L}\right) - \left(\frac{x}{L}\right)^2\right] w_i \\ &+ \left[1 - 4\left(\frac{x}{L}\right) + 3\left(\frac{x}{L}\right)^2\right]\theta_i + \frac{6}{L}\left[\left(\frac{x}{L}\right) - \left(\frac{x}{L}\right)^2\right] w_j \\ &+ \frac{x}{L}\left[3\left(\frac{x}{L}\right) - 2\right]\theta_j\end{aligned} \tag{7.26}$$

$$\begin{aligned}M = EI\frac{d^2w}{dx^2} &= -\frac{6EI}{L^2}\left[1 - 2\left(\frac{x}{L}\right)\right] w_i \\ &- \frac{2EI}{L}\left[2 - 3\left(\frac{x}{L}\right)\right]\theta_i + \frac{6EI}{L^2}\left[1 - 2\left(\frac{x}{L}\right)\right] w_j \\ &+ \frac{2EI}{L}\left[3\left(\frac{x}{L}\right) - 1\right]\theta_j\end{aligned} \tag{7.27}$$

$$\begin{aligned}V = -\frac{d}{dx}\left(EI\frac{d^2w}{dx^2}\right) &= -\frac{12EI}{L^3} w_i \\ &- 6\frac{EI}{L^2}\theta_i + \frac{12EI}{L^3} w_j - 6\frac{EI}{L^2}\theta_j\end{aligned} \tag{7.28}$$

where EI is assumed constant within the element. It is easily seen that the shear is a constant over the element.

Concentrated loads may be applied to beams at various locations instead of or in addition to the distributed load $p(x)$. This problem is easily handled by placing a node at the location of the concentrated load. One half of the concentrated load is placed on each element at that node point. Note that the intersection of the w_i and w_j rows and columns of the stiffness matrix $[B]$ have units of force per length. These two terms multiplied together have units of force. Similarly, the w_i and w_j terms of the element force column $\{C\}$ have units of force. Thus the concentrated loads can be added directly onto the terms $pL/2$ in the element force column with the same sign convention as $p(x)$. For example, a force F at node j in an element would have an element force column of the form

$$\{C\} = \begin{matrix} w_i \\ \theta_i \\ w_j \\ \theta_j \end{matrix} \begin{Bmatrix} \frac{1}{2}pL \\ \frac{1}{12}pL^2 \\ \frac{1}{2}pL + \frac{1}{2}F \\ -\frac{1}{12}pL^2 \end{Bmatrix}$$

with the other half of F at node i of the following element.

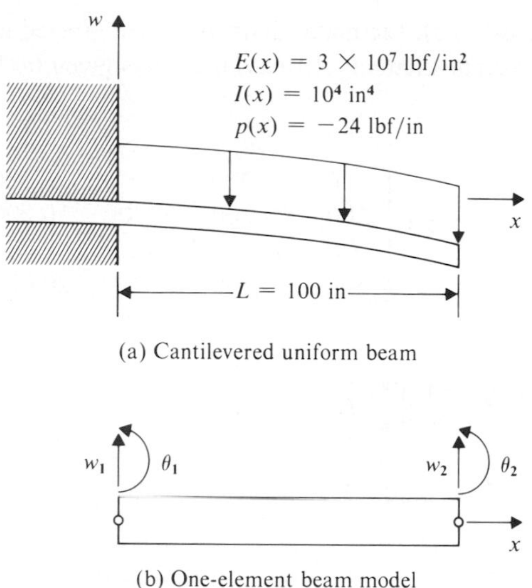

(a) Cantilevered uniform beam

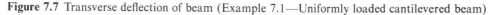

(b) One-element beam model

Figure 7.7 Transverse deflection of beam (Example 7.1—Uniformly loaded cantilevered beam)

7.5 Applications

Finite element solutions to uniform beam problems always produce the exact solution for the deflection at each nodal point. The exact solution is a polynomial of fourth order or lower that is matched at each node point by the Hermite interpolation function. This is similar to the second-order differential equations with second-order polynomial solutions. Beams with variable moment of inertia or other properties can be conveniently analyzed with finite elements.

Example 7.1 *Uniformly Loaded Cantilever Beam*

Given A cantilevered beam, shown in Figure 7.7(a), has the properties $E = 3.0 \times 10^7$ lbf/in², $I = 10^4$ in⁴, $p = -24$ lbf/in, and $L = 100$ in.

Objective Find the deflection in the beam, using one element of length L. Compare to the exact solution

$$w = \frac{p}{EI}\left(\frac{x^4}{24} - \frac{x^3 L}{6} + \frac{x^2 L^2}{4}\right)$$

Solution As shown in Figure 7.7(b), the nodal deflections are labeled w_1 and w_2, while the slopes are labeled θ_1 and θ_2. The element stiffness matrix is given by Equation (7.19) as

$$[B]^{(1)} = \begin{matrix} w_1 \\ \theta_1 \\ w_2 \\ \theta_2 \end{matrix} \quad 6 \times 10^5 \begin{matrix} w_1 & \theta_1 & w_2 & \theta_2 \end{matrix} \begin{bmatrix} 6 & 300 & -6 & 300 \\ 300 & 20{,}000 & -300 & 10{,}000 \\ -6 & -300 & 6 & -300 \\ 300 & 10{,}000 & -300 & 20{,}000 \end{bmatrix}^{(1)}$$

Also, the element load vector is

$$\{C\}^{(1)} = \begin{matrix} w_1 \\ \theta_1 \\ w_2 \\ \theta_2 \end{matrix} \begin{Bmatrix} -1{,}200 \\ -20{,}000 \\ -1{,}200 \\ 20{,}000 \end{Bmatrix}^{(1)}$$

This completes the element properties.

The assembly procedure is trivial in this case, resulting in the four equations

$$\begin{bmatrix} 6 & 300 & -6 & 300 \\ 300 & 20{,}000 & -300 & 10{,}000 \\ -6 & -300 & 6 & -300 \\ 300 & 10{,}000 & -300 & 20{,}000 \end{bmatrix} \begin{Bmatrix} w_1 \\ \theta_1 \\ w_2 \\ \theta_2 \end{Bmatrix} = \begin{Bmatrix} -2 \times 10^{-3} \\ -3.3333 \times 10^{-2} \\ -2 \times 10^{-3} \\ 3.3333 \times 10^{-2} \end{Bmatrix}$$

Here the equations have all been divided by 6×10^5 to keep the numbers in the stiffness matrix small.

Applying the fixed end boundary conditions at $x = 0$, $w_1 = 0$ and $\theta_1 = 0$, and these four equations reduce to $6w_2 - 300\theta_2 = -2 \times 10^{-3}$, $-300w_2 + 20{,}000\theta_2 = 3.3333 \times 10^{-2}$. Solving yields $w_2 = -1.0 \times 10^{-3}$ in, $\theta_2 = -1.3333 \times 10^{-5}$ rad. Substituting the beam parameters into the exact solution gives the same result at $x = 100$ in.

It is interesting to compute the slope, moment, shear, and load for this problem. The values are

$$\theta = \frac{dw}{dx} = \frac{p}{EI}\left(\frac{x^3}{6} - \frac{x^2 L}{2} + \frac{xL^2}{2}\right)$$

$$M = EI\frac{d^2w}{dx^2} = p\left(\frac{x^2}{2} - xL + \frac{L^2}{2}\right)$$

$$V = -\frac{d}{dx}\left(EI\frac{d^2w}{dx^2}\right) = -p\,(x - L)$$

$$p = \frac{d^2}{dx^2}\left(EI\frac{d^2w}{dx^2}\right) = p$$

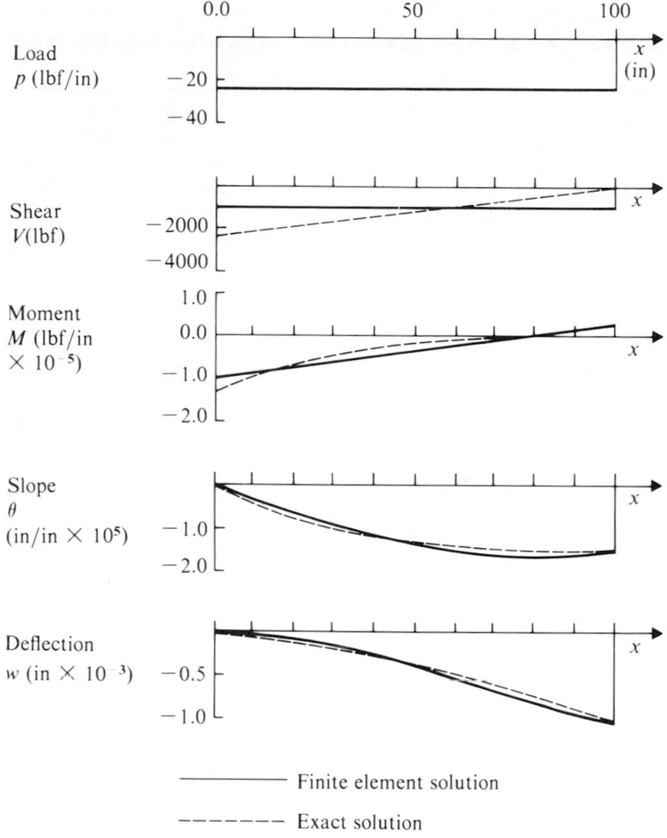

Figure 7.8 Load, shear, moment, slope, and deflection of beam (Example 7.1—Uniformly loaded cantilevered beam)

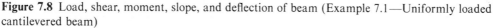

A comparison of the applied load, shear, moment, slope, and deflection is given in Figure 7.8 where the quantities are plotted versus x. Note that the exact solution is a fourth-order polynomial, while the finite element interpolation function is only third-order. Thus the exact solution is produced only at the node points for w and θ. Between the node points, the values are close but not exact. The moments do not match the exact solution at the node points as the deflections do. The shear is constant within the element, and the value $V = -1{,}200$ equals the average shear over the beam. Table 7.1 shows the numerical values at the ends and center of the beam.

Table 7.1
Beam Parameters at End Points and Center (Example 7.1: Uniformly Loaded Cantilevered Beam)

Beam Parameter	Left End $x = 0.0$ in		Center $x = 50$ in		Right End $x = 100$ in	
	Exact Solution	Finite Element	Exact Solution	Finite Element	Exact Solution	Finite Element
Load p (lbf/in)	−2,400	−2,400	−2,400	−2,400	−2,400	−2,400
Shear V (lbf)	−2,400	−1,200	−1,200	−1,200	0	−1,200
Moment M (lbf-in $\times 10^{-5}$)	−1.2	−1.0	−0.4	−0.3	0	2.0
Slope θ (in/in $\times 10^5$)	0	0	−1.167	−1.167	−1.33	−1.33
Displacement w (in $\times 10^3$)	0	0	−0.3542	−0.3333	−1.0	−1.0

It can be seen from this example that potential round-off error problems may occur with very long elements. In a Gauss elimination scheme to solve the algebraic equations, some terms are several orders of magnitude larger than others, such as $2L^2$ as compared to the term 6. Therefore, very long elements should be avoided.

Example 7.2 *Arched Beam with Distributed Load*

Given An arched beam with dimensions L = length, W = width, and H = height and $L = 200$ in, $W = 5$ in, and $H = 10 + (x^2/1{,}000)$ is shown in Figure 7.9(a). It is made of a homogeneous material (wood chips and resin) with a Young's modulus of $E = 2.0 \times 10^6$ lbf/in^2. The distributed load is $p = -10$ lbf/in applied over the middle portion of the beam. The beam specific weight is given by $\gamma = 0.078$ lbf/in^3. Consider the beam ends to be simply supported.

Objective Use a three-element model to determine the displacements in the beam. Take advantage of beam symmetry.

Solution The beam is simply supported at both ends but it is symmetric, so only half of it need be considered. If the right end is retained, the slope and shear stress vanish at the beam center.

$$\theta = \frac{dw}{dx} = 0, \quad x = 0$$

$$V = -\frac{d}{dx}\left(EI\,\frac{d^2w}{dx^2}\right) = 0, \quad x = 0$$

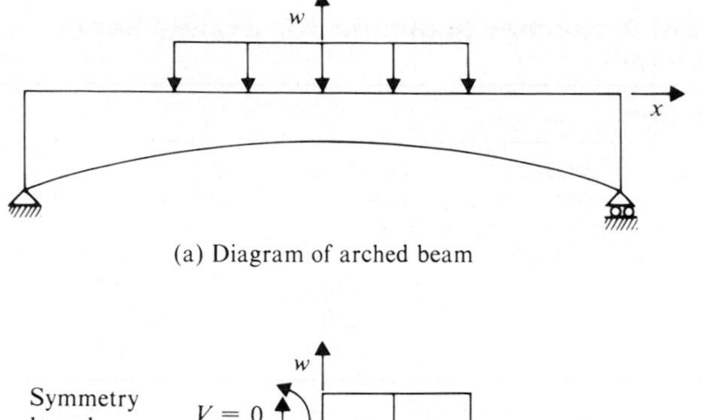

(a) Diagram of arched beam

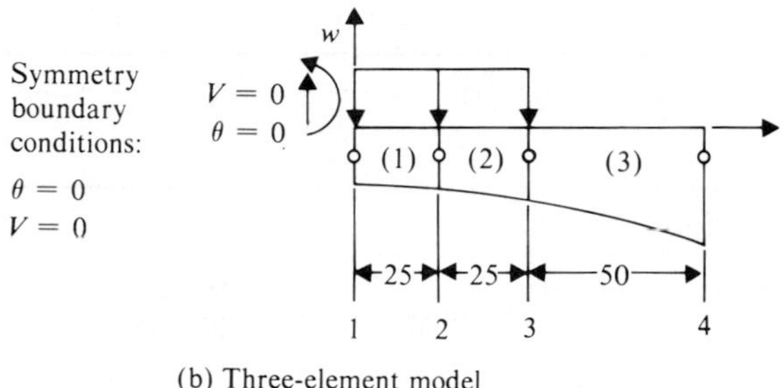

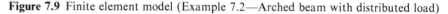

(b) Three-element model

Figure 7.9 Finite element model (Example 7.2—Arched beam with distributed load)

At $x = L/2$ (on the right end) the displacement and moment are zero.

$$w = 0, \quad x = L/2$$

$$M = EI\frac{d^2w}{dx^2} = 0, \quad x = L/2$$

These conditions will be incorporated into the nodal equations at the end of the problem.

Divide the half of the beam into three elements. As the displacements are likely to be the largest at the center, choose two elements of length 25 inches near the center and one of length 50 inches near the right end. Figure 7.9(b) shows the finite element model.

The beam property EI and the weight can be evaluated at the node points and then averaged over the element. The cross-sectional area moment of inertia I is given by $I = \frac{1}{12}WH^3$, while the weight per unit length is equal to γWH. Results are presented in Table 7.2.

Table 7.2
Nodal and Element Properties (Example 7.2: Arched Beam with Distributed Load)

(a) Nodal Properties

Node	Position *x* (in)	Width *W* (in)	Height *H* (in)	*EI* (lbf-in^2 × 10^{-8})	Weight/Length (lbf/in)
1	0	5	10.0	8.333	−3.90
2	25	5	10.36	10.010	−4.146
3	50	5	12.50	16.276	−4.68
4	100	5	20.0	66.667	−7.80

(b) Element Properties

Element	Length *L* (in)	*EI* (lbf-in^2 × 10^{-8})	Weight/Length (lbf/in)	Total Load *P* (lbf/in)
1	25	9.171	−4.023	−14.023
2	25	13.143	−4.413	−14.413
3	50	41.471	−6.24	−6.24

Using the notation introduced in Section 3.5 for vector problems, the nodal displacements and slopes are treated as one column matrix. The four displacements and four slopes are shown below

$$\begin{Bmatrix} w_1 \\ \theta_1 \\ w_2 \\ \theta_2 \\ w_3 \\ \theta_3 \\ w_4 \\ \theta_4 \end{Bmatrix} = \begin{Bmatrix} U_1 \\ U_2 \\ U_3 \\ U_4 \\ U_5 \\ U_6 \\ U_7 \\ U_8 \end{Bmatrix}$$

The element matrices and columns for element 1 are obtained from Equations (7.22) and (7.23) as

$$[B]^{(1)} = 1.1739 \times 10^5 \begin{matrix} \\ 1 \\ 2 \\ 3 \\ 4 \end{matrix} \begin{matrix} \begin{matrix} 1 & \quad 2 & \quad 3 & \quad 4 \end{matrix} \\ \begin{bmatrix} 6 & 75 & -6 & 75 \\ & 1{,}250 & -75 & 625 \\ & & 6 & -75 \\ (\text{Symmetric}) & & & 1{,}250 \end{bmatrix} \end{matrix}$$

$$
10^5 \begin{array}{c} \\ 1 \\ 2 \\ 3 \\ 4 \\ 5 \\ 6 \\ 7 \\ 8 \end{array}
\begin{bmatrix}
1 & 2 & 3 & 4 & 5 & 6 & 7 & 8 \\
7.0434 & 88.042 & -7.0434 & 88.042 & 0 & 0 & 0 & 0 \\
 & 1467.4 & -88.042 & 733.69 & & & 0 & 0 \\
 & & 17.137 & 38.130 & -10.094 & 126.17 & 0 & 0 \\
 & & & 3570.3 & -126.17 & 1051.4 & 0 & 0 \\
 & & & & 14.075 & -26.643 & -3.9813 & 99.529 \\
 & & & & & 5420.6 & -99.529 & 1658.8 \\
 & & & & & & 3.9812 & -99.529 \\
\text{(Symmetric)} & & & & & & & 3317.6
\end{bmatrix}
\begin{Bmatrix} U_1 \\ U_2 \\ U_3 \\ U_4 \\ U_5 \\ U_6 \\ U_7 \\ U_8 \end{Bmatrix}
=
\begin{Bmatrix} -175.29 \\ -730.36 \\ -355.45 \\ -20.31 \\ -336.16 \\ -549.33 \\ -156.0 \\ 1300.0 \end{Bmatrix}
$$

Figure 7.10 Global coefficient (stiffness) matrix and column (force vector) for Example 7.2

$$
\{C\}^{(1)} = \begin{array}{c} 1 \\ 2 \\ 3 \\ 4 \end{array}
\begin{Bmatrix} -175.29 \\ -730.36 \\ -175.29 \\ 730.36 \end{Bmatrix}
$$

Element 2 yields

$$
[B]^{(2)} = 1.6823 \times 10^5
\begin{array}{c} \\ 3 \\ 4 \\ 5 \\ 6 \end{array}
\begin{bmatrix}
3 & 4 & 5 & 6 \\
6 & 75 & -6 & 75 \\
 & 1{,}250 & -75 & 625 \\
 & & 6 & -75 \\
\text{(Symmetric)} & & & 1{,}250
\end{bmatrix}
$$

$$
\{C\}^{(2)} = \begin{array}{c} 3 \\ 4 \\ 5 \\ 6 \end{array}
\begin{Bmatrix} -180.16 \\ -750.67 \\ -180.16 \\ 750.67 \end{Bmatrix}
$$

while element 3 gives

$$
[B]^{(3)} = 0.66353 \times 10^5
\begin{array}{c} \\ 5 \\ 6 \\ 7 \\ 8 \end{array}
\begin{bmatrix}
5 & 6 & 7 & 8 \\
6 & 150 & -6 & 150 \\
 & 5{,}000 & -150 & 2{,}500 \\
 & & 6 & -150 \\
\text{(Symmetric)} & & & 5{,}000
\end{bmatrix}
$$

$$
\{C\}^{(3)} = \begin{array}{c} 5 \\ 6 \\ 7 \\ 8 \end{array}
\begin{Bmatrix} -156.0 \\ -1{,}300.0 \\ -156.0 \\ 1{,}300.0 \end{Bmatrix}
$$

These are now assembled, with the results shown in Figure 7.10, where the boundary conditions have not yet been included.

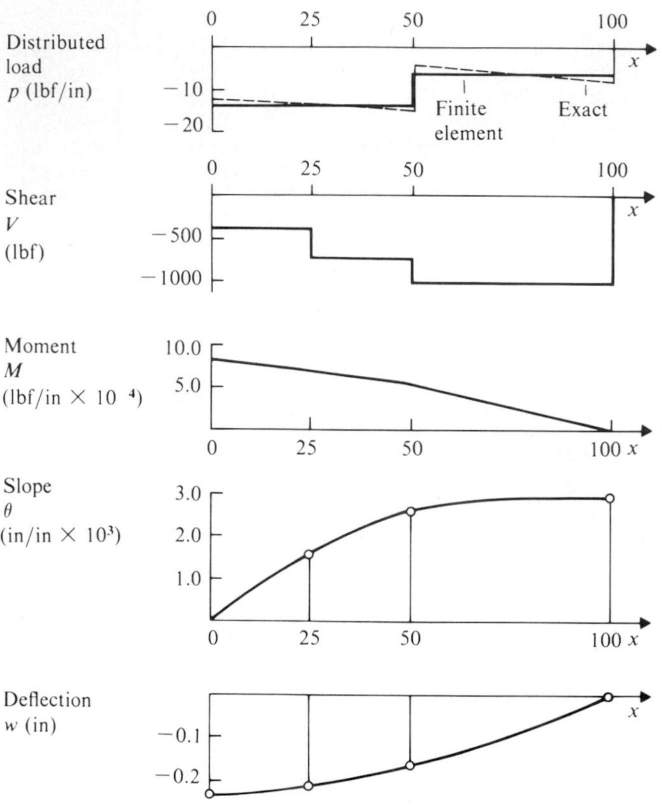

Figure 7.11 Nodal parameters (Example 7.2—Arched beam with distributed load)

Table 7.3
Nodal Parameters (Example 7.2: Arched Beam with Distributed Load)

	Node 1 $x = 0$ in	Node 2 $x = 25$ in	Node 3 $x = 50$ in	Node 4 $x = 100$ in
Distributed load (lbf/in)	−13.90	−14.146	−14.68	−7.80
Shear V (lbf)	0	−350	−710	−1,022
Moment M (lbf = in × 10^{-4})	7.76	6.885	5.11	0
Slope θ (in/in × 10^3)	0	1.639	2.614	2.894
Displacement w (in)	−0.2144	−0.1937	−0.1399	0

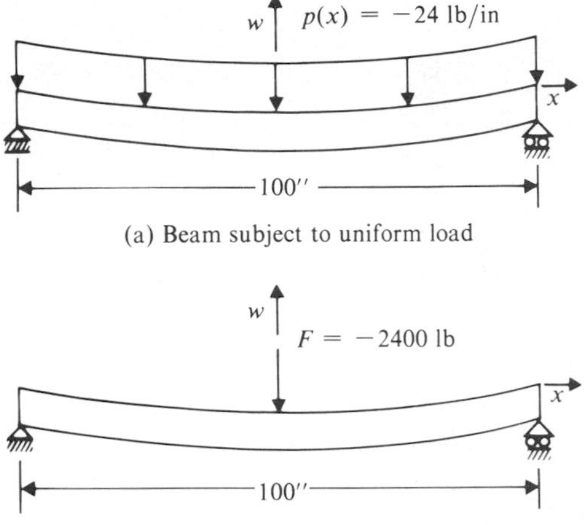

(a) Beam subject to uniform load

(b) Beam subject to concentrated load

Figure 7.12 Diagrams for Problems 7.1 and 7.2

The boundary conditions are given by $\theta_1 = U_2 = 0$, $w_4 = U_7 = 0$. These are incorporated into the global equations and a solution obtained by computer as

$$\begin{Bmatrix} w_1 \\ \theta_1 \\ w_2 \\ \theta_2 \\ w_3 \\ \theta_3 \\ w_4 \\ \theta_4 \end{Bmatrix} = \begin{Bmatrix} U_1 \\ U_2 \\ U_3 \\ U_4 \\ U_5 \\ U_6 \\ U_7 \\ U_8 \end{Bmatrix} = \begin{Bmatrix} -0.214405 \\ 0.0 \\ -0.193658 \\ 0.001639 \\ -0.139943 \\ 0.002614 \\ 0.0 \\ 0.002894 \end{Bmatrix}$$

Figure 7.11 plots and Table 7.3 gives the nodal values of the slope deflection and other nodal parameters. Because both of them are treated as nodal values, both curves are continuous, although the derivatives of the slope may not be continuous from one element to the next.

7.6 Problems

7.1 A uniformly loaded, simply supported beam, shown in Figure 7.12(a), is to be analyzed to determine the displacement at the center. Because of symmetry only one half of the beam needs to be analyzed. Use a single beam element. The problem parameters are $E = 3.0 \times 10^7$ lbf/in^2, $I = 10^4$ in^4, $p = -24$ lbf/in, and $L = 100$ in. Compare the deflection at the center to that given by the exact solution. Also compare the slope, moment, and shear for the finite element and exact solution.

7.2 The beam in the previous problem has a concentrated load of $-2{,}400$ lbf acting at the beam center instead of the uniform load of -24 lbf/in. Carry out a similar analysis. Figure 7.12(b) shows a diagram.

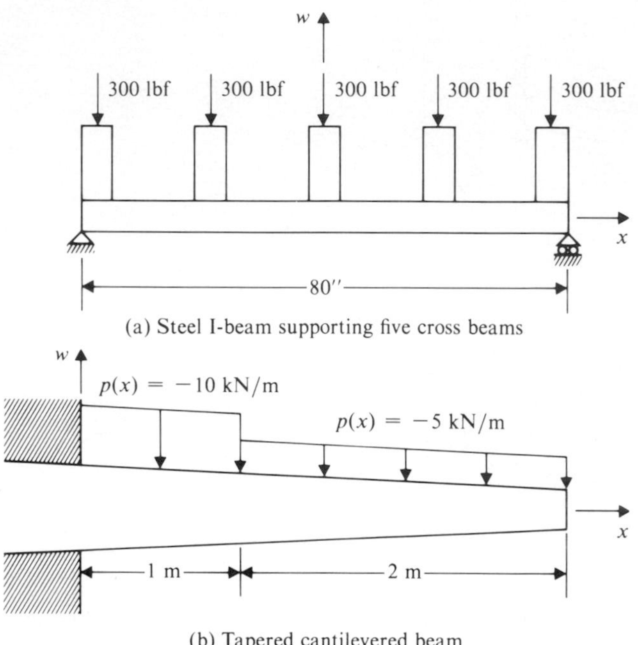

Figure 7.13 Diagrams for Problems 7.3 and 7.4

7.3 A steel I-beam, shown in Figure 7.13(a), supports five equally spaced wooden cross beams, each of which has a 300 lbf load on it. The I-beam has the following properties: $E = 3.0 \times 10^7$ lbf/in^2, $I = 2.0 \times 10^3$ in^4, and $L = 80$ in. Use a two-element model (taking advantage of problem symmetry) to determine the displacement and bending moment at the beam load points.

7.4 A beam of rectangular cross section has a width of 0.2 m; its height is 0.6 m at the wall but decreases linearly with x to 0.2 m at the cantilevered end. Figure 7.13(b) shows the beam length and loading. Use a two-element model to determine the deflection at the points $x = 1$ and $x = 3$. Young's modulus for the beam material is

$$E = 1.0 \times 10^7 \frac{\text{kN}}{\text{m}^2}$$

7.5 Derive an element stiffness matrix and column, using Lagrange polynomials for transverse bending of beams.

7.6 A Bernoulli–Euler beam, shown in Figure 7.14(a), is subject to transverse loading $p(x)$ as well as an axial load $F(x)$. The governing differential equation for transverse deflection is

$$\frac{d}{dx^2}\left(EI\frac{d^2w}{dx^2}\right) + \frac{d}{dx}\left(F\frac{dw}{dx}\right) - p = 0, \quad 0 \le x \le L$$

where w = transverse deflection, E = Young's modulus, I = cross sectional moment of inertia, p = distributed load, and F = axial load. Derive the element stiffness matrix for this problem. Note that only the axial loading term need be considered and added to the bending stiffness matrix.

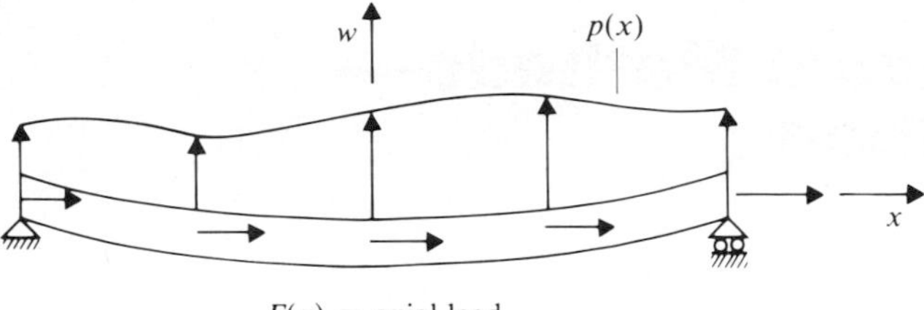

$F(x)$ = axial load

(a) Beam subject to axial load

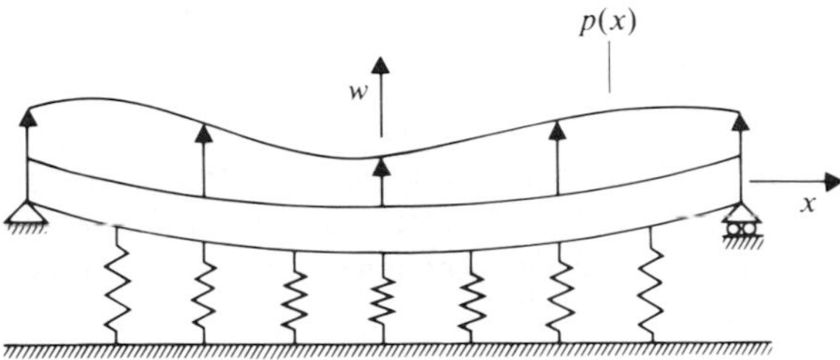

$k(x)$ = elastic foundation stiffness

(b) Beam on elastic foundation

Figure 7.14 Diagrams for Problems 7.6 and 7.7

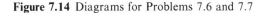

7.7 A beam is supported on an elastic foundation, with stiffness $k(x)$, as shown in Figure 7.14(b). The governing differential equation is

$$\frac{d}{dx^2}\left(EI\frac{d^2w}{dx^2}\right) + kw - p = 0, \quad 0 \leq x \leq L$$

Determine the element stiffness matrix for this problem. Only the elastic foundation effect need be considered and added to the bending stiffness matrix derived in Section 7.3.

7.8 A concentrated moment of magnitude M_c is applied to a node point in a finite element beam analysis. Determine the proper element force column for both elements adjacent to the node point.

Chapter 8

Weighted Residual Methods—Galerkin's Method

8.1 Introduction

The reason for using the method of weighted residuals has already been mentioned: many physical problems are described by differential equations for which no conventional variational method exists. A major area of importance is fluid mechanics, where the method of weighted residuals is most commonly used.

This chapter considers some simple engineering problems that illustrate the use of weighted residual finite elements. Very simple one-dimensional examples, governed by first-order ordinary differential equations, are solved in the first few sections; more complex problems are treated later on. In each case, the details of calculations are carried out by hand to demonstrate the approach, although a computer would be employed for practical problems.

A. First-Order Boundary Value Problem

Consider a first-order ordinary differential equation written in the general form

$$D[x,T] = 0 \tag{8.1}$$

where D may contain terms with up to first derivatives of T. This general form may be nonlinear. The first-order ordinary linear differential equation has the form

$$M\frac{dT}{dx} + PT + Q = 0 \tag{8.2}$$

where $M = M(x)$, $P = P(x)$, $Q = Q(x)$ are known coefficient functions. Let the boundary condition be specified, with the form

$$T = T_0 \text{ at } x = x_0 \tag{8.3}$$

This completes the boundary value problem.

Note that the first derivative term $M(dT/dx)$ is not present in the self-adjoint differential equation (5.27) discussed in Section 5.4. As noted in that section, a transformation of variables can be found to eliminate the first derivative term for any one-dimensional problem. Such a transformation cannot be found in two or three dimensions.

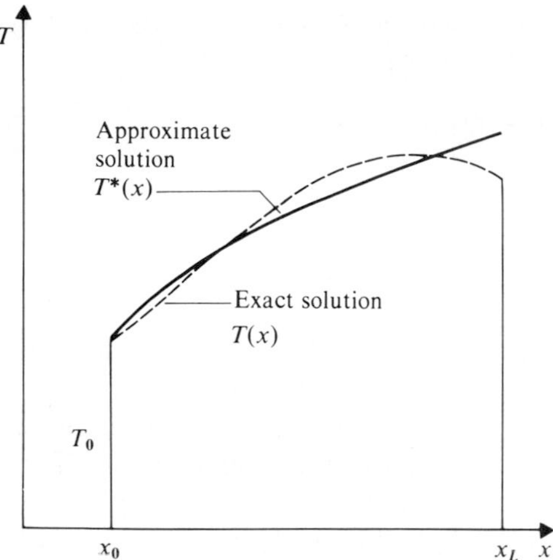

Figure 8.1 Exact and approximate solutions

The purpose of this chapter is to introduce the weighted residual method for more important two- and three-dimensional problems, so the first-order equation (8.1) is considered in some detail.

B. Errors and Residuals

An approximate solution is desired. Let $T^*(x)$ represent the approximate solution, which is close to the exact solution $T^*(x) \simeq T(x)$. Figure 8.1 illustrates the two functions. Substituting the approximate solution into the differential equation (8.1) gives

$$D[x, T^*] = r(x) \tag{8.4}$$

where $r(x)$ is the error. Note that the approximate solution does not satisfy the differential equation, so an error $r(x)$ occurs. Normally the function $T^*(x)$ is chosen so that it satisfies the boundary condition (8.3)

$$T^* = T_0 \text{ at } x = x_0$$

The error cannot be equal to zero at each point x. However, a residual R, which has in it the integral of $r(x)$ can be set to zero because it would represent some sort of average error over the interval x_0 to x_L. Let the residual have the general form

$$R = \int_{x_0}^{x_L} W(x)\, r(x)\, dx = 0 \tag{8.5}$$

where $W(x)$ is some sort of weighting function chosen by the user of the method, which is called the method of weighted residuals. This can also be expressed in terms of the differential equation as

$$R = \int_{x_0}^{x_L} W(x)\, D[x, T^*]\, dx = 0 \tag{8.6}$$

where the approximate solution T^* has been substituted into $D(x,T^*)$. Finally this may be written as

$$R = \int_{x_0}^{x_L} W\left[M\frac{dT^*}{dx} + PT^* + Q\right]dx = 0 \tag{8.7}$$

for the linear first-order equation.

C. Weighting Functions

The weighting function may take one of several forms—subdomain, Galerkin, least squares, collocation—as well as others not given here. Perhaps the simplest one is the subdomain method. It uses a weighting function of unity—usually defined over subregions of the full solution region. The Galerkin method employs weighting functions of the same type as the interpolation functions that approximate $T(x)$. The least squares method minimizes the square of the errors. The collocation method produces zero error at a number of specified points.

In this chapter the subdomain method is used as an introduction. Galerkin's method is one of the most popular techniques (as discussed in Chapter 2), and it is developed in some detail in Sections 8.3 and 8.4. The other techniques are left for more advanced texts.

8.2 Subdomain Finite Element Method

For the first-order linear differential equation $D[x,T] = 0$ with boundary condition $T = T_0$ at $x = x_0$, the subdomain finite element solution is relatively simple. The concepts are applied node by node or element by element. No element matrices or assembly procedures are required in this section.

A. Element Errors

The interval x_0 to x_L is broken into E elements, as in the variational method. Each element e has length $L^{(e)}$. The element interpolation function for each element has the form

$$T^{(e)} = \begin{cases} N_iT_i + N_jT_j, & x \text{ in element } e \\ 0, & x \text{ not in element } e \end{cases}$$

for a simplex element. Summing over all the elements gives the complete approximate function

$$T^* = T^{(1)} + T^{(2)} + \ldots T^{(E)} = \sum_{e=1}^{E} T^{(e)} \tag{8.8}$$

Figure 8.2 illustrates the element-by-element approximation of T^* to the exact solution $T(x)$ for three elements.

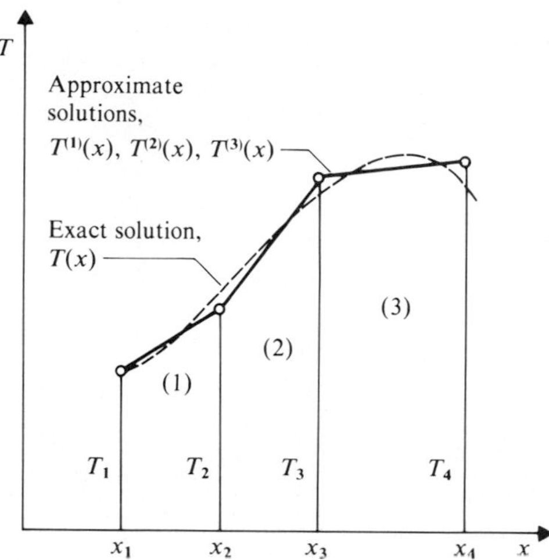

Figure 8.2 Exact and finite element solutions

Substituting the approximate solution T^* into the differential equation (8.1) gives the error: $D[x, T^*] = r(x)$. The approximate solution is actually the sum of $T^{(e)}$ over all of the elements, so the error can be expressed as the sum of each element error:

$$D[x, \sum_{e=1}^{E} T^{(e)}] = r^{(1)} + r^{(2)} + \ldots + r^{(E)} = \sum_{e=1}^{E} r^{(e)} \tag{8.9}$$

Each error $r^{(e)}$ has some finite value inside the element and the value zero outside the element.

B. Residuals and Weighting Functions

There are $N-1$ unknown nodal values: $T_2, T_3, \ldots, T_N$. Here $T_1 = T_0$ (from the boundary condition), so it is not an unknown. Define a set of weighting functions $W_2, W_3, \ldots, W_N$ which are functions of x. The forms of these functions are defined in a short while. The residuals for each unknown node have the form

$$\begin{aligned} R_2 &= \int_{x_0}^{x_L} W_2(x)\, r(x)\, dx = 0 \\ R_3 &= \int_{x_0}^{x_L} W_3(x)\, r(x)\, dx = 0 \\ &\ldots \\ R_N &= \int_{x_0}^{x_L} W_N(x)\, r(x)\, dx = 0 \end{aligned} \tag{8.10}$$

There are $N-1$ residuals and $N-1$ weighting functions. Each residual is set equal to zero, giving one equation for each unknown.

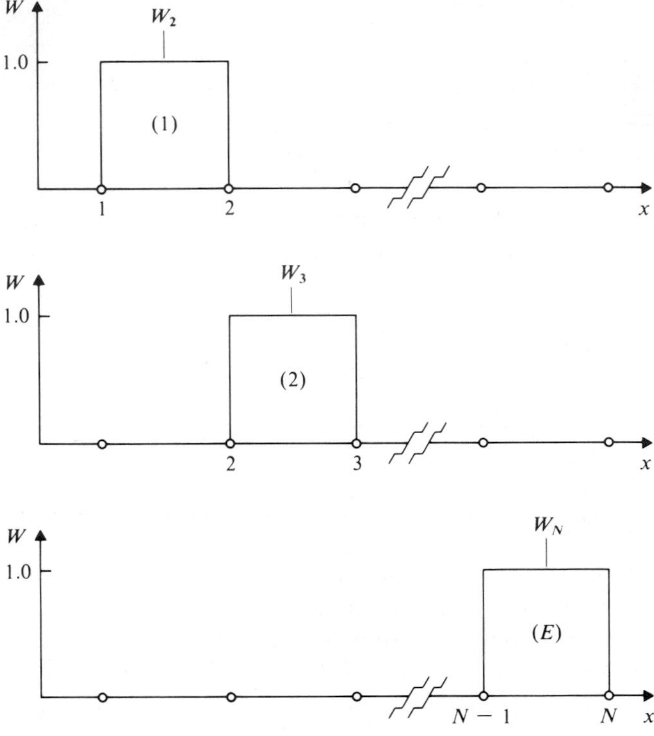

Figure 8.3 Subdomain weighting functions for finite elements

The weighting function is taken as unity over each element length for the subdomain method. For the finite element approach, the weighting function for each node n in element e is defined as

$$W_n = \begin{cases} 1.0, & x \text{ in element } e \\ 0.0, & x \text{ not in element } e \end{cases} \tag{8.11}$$

Here $n = e+1$ because the unknown node in each element is the right-hand one. These weighting functions are

$$\begin{aligned} W_2 &= \begin{cases} 1.0, & x \text{ in element } 1 \\ 0.0, & x \text{ not in element } 1 \end{cases} \\ W_3 &= \begin{cases} 1.0, & x \text{ in element } 2 \\ 0.0, & x \text{ not in element } 2 \end{cases} \\ &\cdots \\ W_N &= \begin{cases} 1.0, & x \text{ in element } E \\ 0.0, & x \text{ not in element } E \end{cases} \end{aligned} \tag{8.12}$$

for all of them. Figure 8.3 illustrates the weighting functions.

Each residual has the form

$$R_n = \int_{x_0}^{x_L} W_n(x)\, r(x)\, dx$$

$$= \int_{x_0}^{x_L} W_n(x)\, [r^{(1)} + r^{(2)} + \ldots + r^{(E)}]\, dx = 0$$

Now because the weighting function W_n has the value unity inside element e and zero outside, the limits of integration reduce to $L^{(e)}$. The residual becomes

$$R_n = \int_{L^{(e)}} r^{(e)}\, dx = 0 \tag{8.13}$$

For all of the residuals,

$$\begin{aligned} &\textit{Equation for } T_2 \qquad && R_2 = \int_{L^{(1)}} r^{(1)}\, dx = 0 \\ &\textit{Equation for } T_3 \qquad && R_3 = \int_{L^{(2)}} r^{(2)}\, dx = 0 \\ &\ldots \\ &\textit{Equation for } T_N \qquad && R_N = \int_{L^{(E)}} r^{(E)}\, dx = 0 \end{aligned} \tag{8.14}$$

Once again, R_1 is not considered because T_1 is a known nodal value. These equations form the subdomain finite element method for first-order ordinary differential equations.

C. Subdomain Examples

Two subdomain examples are now discussed. The first one uses the axial displacement of a rod to illustrate the method: obtaining the general formulation, solving a one-element problem, and solving a three-element problem. The second example formulates a similar heat transfer problem.

Example 8.1 *Subdomain Solution for Axial Displacement of Rod*

Given A rod of cross-sectional area $A(x)$ and Young's modulus $E(x)$, shown in Figure 8.4, is subject to an axial force F applied at the end. Figure 8.5 gives the rod geometry. If $u(x)$ represents the axial displacement, the differential equation is given by F = constant. Then, from simple solid mechanics analysis, $F = \sigma A$ where $\sigma(x)$ is the axial stress. Hooke's law is $\sigma = E\epsilon$ where $\epsilon(x)$ is the strain and $\epsilon = du/dx$. Substituting in this expression for F yields

$$EA\,\frac{du}{dx} - F = 0$$

Here F is taken as positive in the positive x direction when the force is acting on the left-hand portion of the bar. Take as the boundary condition $u = u_0$, $x = 0$. Figure 8.5 gives the problem parameters.

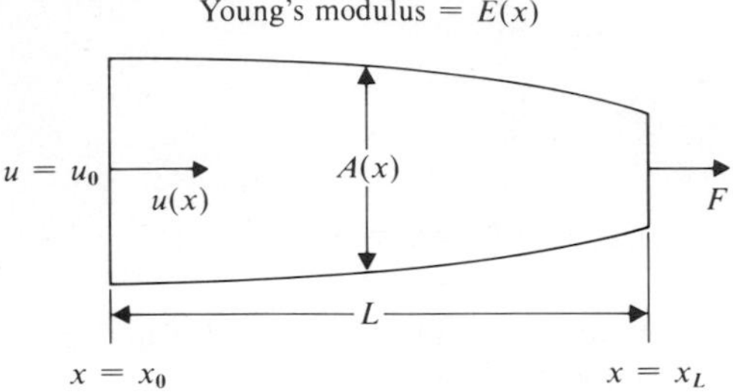

Figure 8.4 Example 8.1—Subdomain solution for axial rod

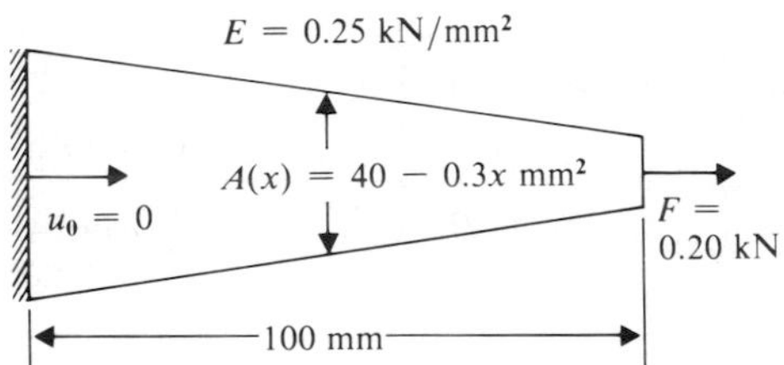

Figure 8.5 Rod geometry (Example 8.1—Subdomain solution for axial rod)

Objective Determine the element operator $L^{(e)}$, the element error $r^{(e)}(x)$, and the residual R_n. Consider two cases:

Case 1 One-element solution

Case 2 Three-element solution

where both use equal-length elements.

Solution The differential equation is

$$EA\frac{du}{dx} - F = 0$$

Then, the differential operator is

$$D[x,u] = EA\frac{du}{dx} - F$$

The differential element operator is

$$D[x,u^{(e)}] = E^{(e)}A^{(e)}\frac{du^{(e)}}{dx} - F^{(e)}$$

where $u^{(e)}$ = element axial displacement, $E^{(e)}$ = element Young's modulus, $A^{(e)}$ = element cross sectional area of rod, and $F^{(e)}$ = element axial force. Each of these element values is approximated using a simple polynomial. If $u^{(e)}$ is taken as linear (simplex element), the other coefficients $E^{(e)}$, $A^{(e)}$, $F^{(e)}$ are taken as constant within the element.

If u^* represents an approximate solution, the error $r(x)$ is

$$r(x) = EA \frac{du^*}{dx} - F$$

The element residual $r^{(e)}(x)$ is (8.9)

$$r^{(e)}(x) = \left[EA \frac{du^*}{dx} - F\right]^{(e)}$$

This will depend upon the approximations chosen for the coefficients (Young's modulus, area, and force).

The residual R for each node is

$$R_n = \int_0^L W_n \left[EA \frac{du^*}{dx} - F\right] dx = 0$$

where the weighting functions have the value 1.0 in the element preceding the node and zero elsewhere. Because of this, the integration limits reduce from the entire length of the rod to an element length. The residual (8.13) is

$$R_n = \int_{L^{(e)}} \left[EA \frac{du^*}{dx} - F\right]^{(e)} dx = 0$$

where each of the terms in square brackets is an *element value.*

For this application, the error $r(x)$ has the physical interpretation of a force. Given an approximate displacement, the error in the force over each element is

$$r^{(e)}(x) = \left[EA \frac{du^*}{dx} - F\right]^{(e)}$$

The units are

$$\left[EA \frac{du^*}{dx} - F\right]^{(e)} = \left(\frac{\text{Force}}{\text{Area}}\right) (\text{Area}) \left(\frac{\text{Length}}{\text{Length}}\right) - (\text{Force})$$

or $r^{(e)}(x)$ = Approximate element force − Actual element force. The residual R represents the average of this error over the length of the element.

$$R_n = \int_{L^{(e)}} W_n(x)\, r^{(e)}(x)\, dx = \text{Average element error in force} = 0$$

Case 1: One-Element Solution

A rod shown in Figure 8.5 has the properties $L = 100$ mm, $E = 0.25$ kN/mm^2, and $F = 0.2$ kN. The cross sectional area varies linearly with the form $A(x) = 40 - 0.3x$ mm^2. At the left end the displacement is specified as $u_0 = 0.0$ mm at $x = 0$. The objective is to find the axial displacement, using one simplex element with constant properties. Plot the approximate solution versus the exact solution. The differential equation is quite simple and may be solved by direct integration. The solution is

$$u = -2.667 \ln\left(\frac{400 - 3x}{400}\right)$$

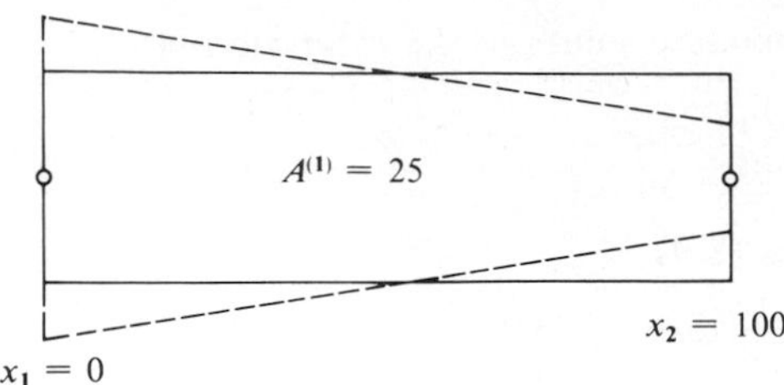

Figure 8.6 Finite element model (Example 8.1—Subdomain solution for axial rod; Case 1—One-element solution)

Consider a single simplex element with two nodes, as shown in Figure 8.6, with length equal to that of the rod. The nodal coordinates are $X_1 = 0$, $X_2 = 100$ and the element length is $L^{(1)} = 100$. Units are omitted for clarity.

The axial displacement in the element is given by (3.4)

$$u^{(1)} = N_1^{(1)} u_1 + N_2^{(1)} u_2$$

where u_1 and u_2 are the nodal values. Also the derivative is (3.8)

$$\frac{du^{(1)}}{dx} = \frac{1}{L^{(1)}} (b_1 u_1 + b_2 u_2)$$

where $b_1 = -1$, $b_2 = +1$. Then

$$\frac{du^{(1)}}{dx} = -\frac{1}{100} u_1 + \frac{1}{100} u_2$$

The coefficients are $E^{(1)} = 0.25$, $F^{(1)} = 0.2$, which are constant already; the cross sectional area is simply taken as the average of the two end points (nodal values of $A(x)$)

$$A^{(1)} = \frac{A_1 + A_2}{2} = \frac{40 + 10}{2} = 25$$

Now the element error is

$$r^{(1)}(x) = \left[EA \frac{du^*}{dx} - F \right]^{(1)}$$

$$= (0.25)(25) \left(-\frac{1}{100} u_1 + \frac{1}{100} u_2 \right) - 0.2$$

$$= -0.0625 \, u_1 + 0.0625 \, u_2 - 0.2$$

In this case, the residual is not a function of x, as it would normally be. For the residual

$$R_2 = \int_{L^{(1)}} r^{(1)}(x) \, dx = 0$$

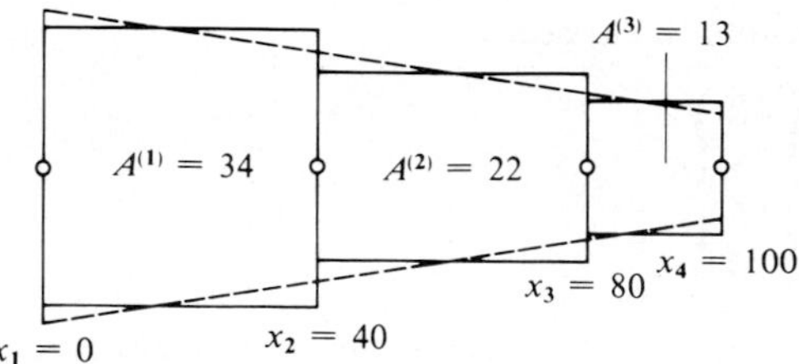

Figure 8.7 Finite element model (Example 8.1—Subdomain solution for axial rod; Case 2—Three-element solution)

which becomes

$$R_2 = \int_0^{100} (-0.0625\, u_1 + 0.0625\, u_2 - 0.2)\, dx = 0$$

After integration, $-6.25\, u_1 + 6.25\, u_2 - 20 = 0$, and this equation yields $u_2 = 3.2$.

Case 2: Three-Element Solution

A reasonable division scheme, which places a small element at the end of the rod, is given by the node points

$$\begin{Bmatrix} X_1 \\ X_2 \\ X_3 \\ X_4 \end{Bmatrix} = \begin{Bmatrix} 0 \\ 40 \\ 80 \\ 100 \end{Bmatrix}$$

Figure 8.7 shows the division. The element lengths are $L^{(1)} = 40$, $L^{(2)} = 40$, $L^{(3)} = 20$. Within each element, the displacement functions are

$$u^{(1)} = N_1^{(1)}\, u_1 + N_2^{(1)}\, u_2$$

$$u^{(2)} = N_2^{(2)}\, u_2 + N_3^{(2)}\, u_3$$

$$u^{(3)} = N_3^{(3)}\, u_3 + N_4^{(3)}\, u_4$$

with derivatives

$$\frac{du^{*(1)}}{dx} = \frac{1}{40}(-u_1 + u_2)$$

$$\frac{du^{*(2)}}{dx} = \frac{1}{40}(-u_2 + u_3)$$

$$\frac{du^{*(3)}}{dx} = \frac{1}{20}(-u_3 + u_4)$$

Here only the element length varies.

At each of these node points the area is

$$\begin{Bmatrix} A_1 \\ A_2 \\ A_3 \\ A_4 \end{Bmatrix} = \begin{Bmatrix} 40 \\ 28 \\ 16 \\ 10 \end{Bmatrix}$$

from the formula $A(x) = 40 - 0.3x$. The area for each element is taken as the average at the two appropriate node points

$$A^{(1)} = \frac{1}{2}(A_1 + A_2) = 34$$

$$A^{(2)} = \frac{1}{2}(A_2 + A_3) = 22$$

$$A^{(3)} = \frac{1}{2}(A_3 + A_4) = 13$$

Now the errors are

$$r^{(1)} = \left[EA\frac{du}{dx} - F\right]^{(1)} = [(0.25)(34)\frac{1}{40}(-u_1 + u_2) - 0.2]$$

$$= -0.2125\,u_1 + 0.2125\,u_2 - 0.2$$

$$r^{(2)} = \left[EA\frac{du}{dx} - F\right]^{(2)} = \left[(0.25)(22)\frac{1}{40}(-u_2 + u_3) - 0.2\right]$$

$$= -0.1375\,u_2 + 0.1375\,u_3 - 0.2$$

$$r^{(3)} = \left[EA\frac{du}{dx} - F\right]^{(3)} = \left[(0.25)(13)\frac{1}{20}(-u_3 + u_4) - 0.2\right]$$

$$= -0.1625\,u_3 + 0.1625\,u_4 - 0.2$$

Once again, the errors are not functions of x.

The residual for node 2 is

$$R_2 = \int_0^{40} (-0.2125\,u_1 + 0.2125\,u_2 - 0.2)\,dx$$

$$= -8.5\,u_1 + 8.5\,u_2 - 8 = 0$$

At $x = 0$, $u_1 = 0$, which gives the solution $u_2 = 0.9412$. For node 3

$$R_3 = \int_{40}^{80} (-0.1375\,u_2 + 0.1375\,u_3 - 0.2)\,dx$$

$$= -5.5\,u_2 + 5.5\,u_3 - 8 = 0$$

Table 8.1
Exact and Finite Element Solutions (Example 8.1: Subdomain Solution for Axial Displacement of Rod)

Position *x* (mm)	Exact Solution *u* (mm)	*Case 1* One-Element Solution *u* (mm)	*Case 2* Three-Element Solution *u* (mm)
0	0	0	0
20	0.4334	0.64	0.4706
40	0.9511	1.28	0.9412
60	1.5943	1.92	1.6685
80	2.4435	2.56	2.3957
100	3.6968	3.20	3.6265

Table 8.2
Solution Errors (Example 8.1: Subdomain Solution for Axial Displacement of Rod)

Position *x* (mm)	Error In One-Element Solution %	Error In Three-Element Solution %
0	0	0
20	47	8.6
40	35	−1.0
60	20	4.7
80	5	−2.0
100	−13	−1.9

Noting that u_2 has already been obtained, this yields $u_3 = 2.3957$. Finally, for node 4

$$R_4 = \int_{80}^{100} (-0.1625\, u_3 + 0.1625\, u_4 - 0.2)\, dx$$

$$= -3.25\, u_3 + 3.25\, u_4 - 4 = 0$$

Solving for u_4 gives $u_4 = 3.6265$. The nodal values are

$$\begin{Bmatrix} u_1 \\ u_2 \\ u_3 \\ u_4 \end{Bmatrix} = \begin{Bmatrix} 0 \\ 0.9412 \\ 2.3957 \\ 3.6265 \end{Bmatrix}$$

Table 8.1 presents the values of the exact and finite element solutions at six equally spaced points along the bar. The finite element data between node points are obtained from the interpolation function, which is linear. Clearly, the linear one-element solution can only approximate the curved exact solution fairly crudely. Figure 8.8 shows a graph of the results. The three-element solution approximates the exact solution pretty well. Table 8.2 gives the error in the two finite element solutions. The maximum error in the

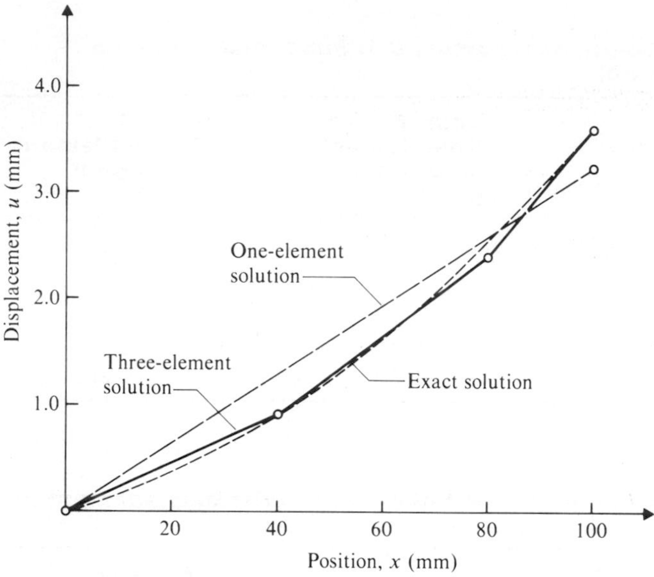

Figure 8.8 Displacement versus position (Example 8.1—Subdomain solution for axial rod)

one-element solution is about 47 percent while it is only about 9 percent for the three-element solution. Increasing the number of elements would further reduce the error at all points along the rod.

Example 8.2 Subdomain Formulation for Heat Conduction in Insulated Rod

Given Heat conduction occurs through an insulated axial rod with small cross sectional area $A(x)$. Because the cross sectional area is small, the temperature in the rod is uniform over the cross sectional area. It is a function of position x alone. Also, the rate of heat flow Q_x in the rod is kept constant by the insulation. The equation governing the flow of heat is given by

$$Q_x = -kA \frac{dT}{dx}$$

where Q_x = heat conduction (W), k = thermal conductivity (W/mm-°C), A = cross sectional area (mm²), and T = temperature (°C). At one end, the temperature is specified $T = T_Q$ at $x = x_Q$. Figure 8.9 shows the rod geometry.

Objective Find the differential operator L, the error $r(x)$, and the residual R.

Solution The differential operator is easily obtained from the differential equation as

$$D = kA \frac{dT}{dx} + Q_x$$

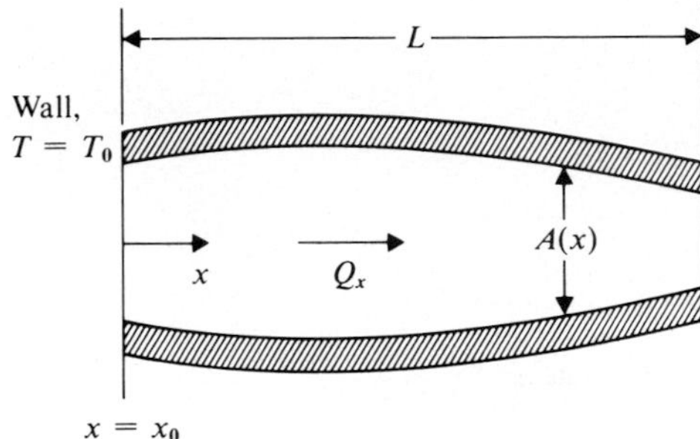

Figure 8.9 Rod geometry (Example 8.2—Heat conduction in insulated rod)

and the error is

$$r(x) = kA\frac{dT^*}{dx} + Q_x$$

The residual is

$$R = \int_Q^L W\left[kA\frac{dT^*}{dx} + Q_x\right] dx = 0$$

8.3 The Galerkin Method

The Galerkin, least squares, collocation, and other methods are used for solving engineering problems, but the Galerkin method (named after its Russian inventor) is more popular and less cumbersome to formulate than the others. Only it is discussed in this text. Advanced books on finite element techniques discuss least squares, etc., in some detail for those interested.

A. Nodal Weighting Functions

The Galerkin method involves taking the weighting functions as the same approximating functions as used for T^*. Let T^* have the normal finite element form

$$T \simeq T^* = \sum_{e=1}^{E} T^{(e)}$$

where

$$T^{(e)} = \begin{cases} N_i T_i + N_j T_j, & x \text{ in element } e \\ 0, & x \text{ not in element } e \end{cases}$$

For a given nodal value, denoted by T_n, the two pyramid functions are

$$N_j^{(e)} + N_i^{(e+1)}$$

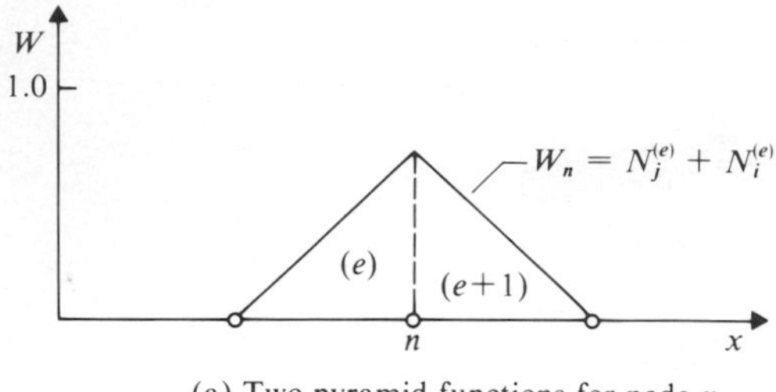

(a) Two pyramid functions for node n

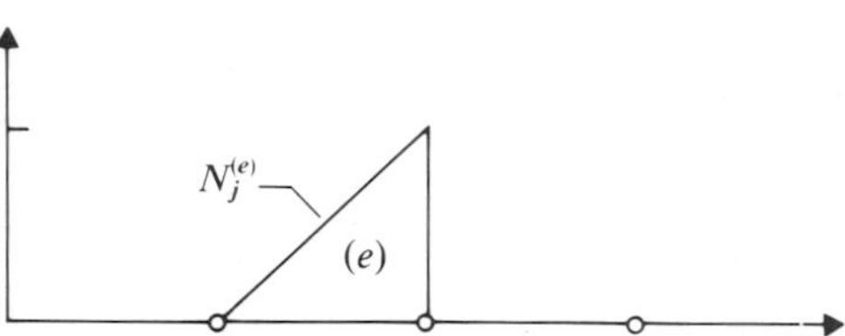

(b) Left pyramid function

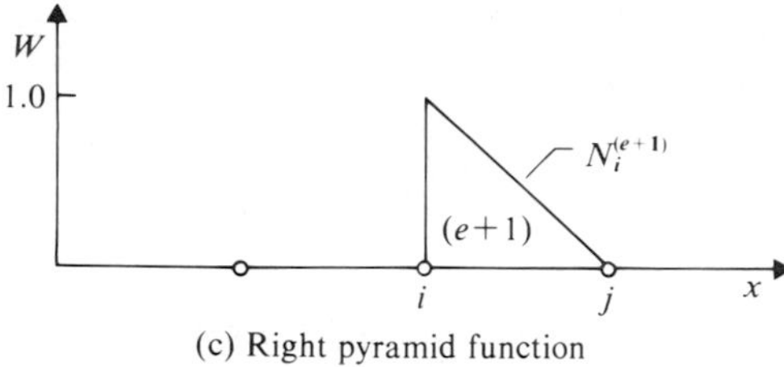

(c) Right pyramid function

Figure 8.10 Galerkin weighting function for node n

as shown in Figure 8.10. In this expression, the right pyramid function (in element e) has the local node number $j = n$. For the adjacent element $e+1$, the left pyramid function has the local node number $i = n$.

As noted above, the Galerkin method uses these approximating functions as weighting functions.

$$W_n = N_j^{(e)} + N_i^{(e+1)} \tag{8.15}$$

The value of W_n is zero outside of elements e and $e + 1$. Figure 8.10 indicates the weighting function W_n. Note that for node 1, the value of T_1 is known from the boundary condition $T_1 = T_0$, so this is not considered an unknown. Thus the weighting function W_1 is not defined. For the last node, only one element E is involved, so

$$W_N = N_N^{(e)} \tag{8.16}$$

This gives all of the weighting functions for a first-order ordinary differential equation.

B. Nodal Residuals

The finite element approximation is substituted into the differential equation

$$D\left[x, \sum_{e=1}^{E} T^{(e)}\right] = \sum_{e=1}^{E} r^{(e)}(x) = r^{(1)} + r^{(2)} + \ldots + r^{(E)}$$

A sum of element errors results, as in the subdomain method. The nodal residuals are

$$R_2 = -\int_{x_0}^{x_L} W_2 \, [r^{(1)} + r^{(2)} + \ldots + r^{(E)}] \, dx = 0$$

$$R_3 = -\int_{x_0}^{x_L} W_3 \, [r^{(1)} + r^{(2)} + \ldots + r^{(E)}] \, dx = 0 \tag{8.17}$$

$$\ldots$$

$$R_N = -\int_{x_0}^{x_L} W_N \, [r^{(1)} + r^{(2)} + \ldots + r^{(E)}] \, dx = 0$$

The negative sign in front of the integral simply gives the element matrices the same sign as obtained with the variational method. The weighting functions are shown in Figure 8.11 for these residuals.

For node n, the weighting function is nonzero only in the two adjacent elements e and $e + 1$, so the limits of integration reduce to $L^{(e)}$ and $L^{(e+1)}$ with the form

$$R_n = -\int_{L^{(e)}} N_n^{(e)} r^{(e)} \, dx - \int_{L^{(e+1)}} N_n^{(e+1)} r^{(e+1)} \, dx = 0 \tag{8.18}$$

Writing this out for all of the unknown nodes gives

$$R_2 = -\int_{L^{(1)}} N_2^{(1)} r^{(1)} \, dx - \int_{L^{(2)}} N_2^{(2)} r^{(2)} \, dx = 0$$

$$R_3 = -\int_{L^{(2)}} N_3^{(2)} r^{(2)} \, dx - \int_{L^{(3)}} N_3^{(3)} r^{(3)} \, dx = 0 \tag{8.19}$$

$$\ldots$$

$$R_N = -\int_{L^{(E)}} N_N^{(E)} r^{(E)} \, dx = 0$$

Each residual involves only two elements because the weighting function W_n is defined as zero outside of those elements. Also, the last residual involves only one element because the weighting function has only one pyramid function for that node (8.16).

For a general first-order differential equation of the form

$$M \frac{dT}{dx} + PT + Q = 0 \tag{8.20}$$

with the boundary condition

$$T = T_0 \text{ at } x = x_0 \tag{8.21}$$

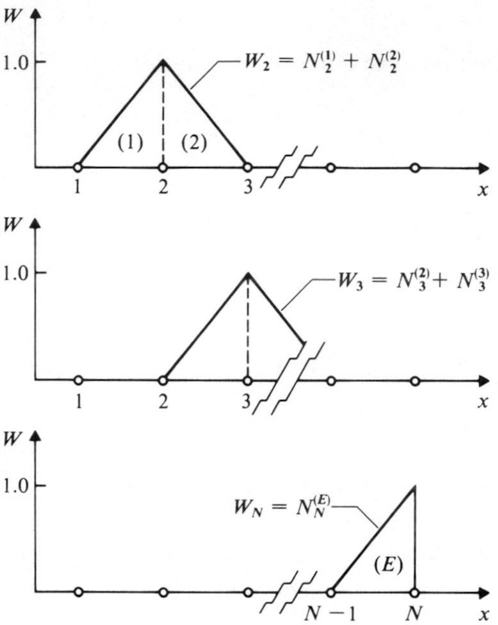

Figure 8.11 Galerkin weighting functions for residuals

the error for an element is

$$r^{(e)}(x) = \left[M\frac{dT}{dx} + PT + Q\right]^{(e)}$$

The residual for node n is

$$R_n = -\int_{L^{(e)}} \left\{N_n\left[M\frac{dT}{dx} + PT + Q\right]\right\}^{(e)} dx - \int_{L^{(e+1)}} \left\{N_n\left[M\frac{dT}{dx} + PT + Q\right]\right\}^{(e+1)} dx = 0$$

Evaluating the integrals gives

$$\begin{aligned} R_n = &\left[-\frac{M}{2}(b_{n-1}T_{n-1} + b_nT_n) - \frac{1}{6}PL(T_{n-1} + 2T_n) - \frac{1}{2}QL\right]^{(e)} \\ &+ \left[-\frac{M}{2}(b_nT_n + b_{n+1}T_{n+1}) - \frac{1}{6}PL(2T_n + T_{n+1}) - \frac{1}{2}QL\right]^{(e+1)} = 0 \end{aligned} \tag{8.22}$$

These are the nodal residuals that are used to give the solution (actually element matrices, discussed in the next section, are normally used). Note that the nodal value T_n is related to both the node before it, T_{n-1} and the node following it, T_{n+1}. Recall that in the subdomain method, T_n was related only to T_{n-1}. Thus the Galerkin method requires the solution of a set of simultaneous equations.

Example 8.3 *Galerkin Solution for Axially Deformed Rod*

Given The rod of Example 8.1 has the differential equation

$$EA\frac{du}{dx} - F = 0$$

The associated boundary condition is $u = u_0$ at $x = x_0$. As in several previous examples, $E = 0.25\ \text{kN/mm}^2$, $A = 40 - 0.3x\ \text{mm}^2$, $F = 0.2$ kN, $L = 100$ mm, and $u_0 = 0.0$ mm.

Objective Use three simplex elements, shown in Figure 8.7, to obtain a finite element solution by the Galerkin technique.

Solution As in the subdomain method, Example 8.1, the node points are located at

$$\begin{Bmatrix} X_1 \\ X_2 \\ X_3 \\ X_4 \end{Bmatrix} = \begin{Bmatrix} 0 \\ 40 \\ 80 \\ 100 \end{Bmatrix}$$

so that the element lengths are $L^{(1)} = 40$, $L^{(2)} = 40$, $L^{(3)} = 20$. The nodal areas are

$$\begin{Bmatrix} A_1 \\ A_2 \\ A_3 \\ A_4 \end{Bmatrix} = \begin{Bmatrix} 40 \\ 26 \\ 16 \\ 10 \end{Bmatrix}$$

and the element average areas are $A^{(1)} = 34$, $A^{(2)} = 22$, $A^{(3)} = 13$. Comparing this example with equation (8.20) gives $M = EA$, $P = 0$, $Q = -F$, and $T = u$. The residual becomes (8.22)

$$R_n = \left[\frac{EA}{2}(b_iu_i + b_ju_j) - \frac{FL}{2}\right]^{(e)} + \left[\frac{EA}{2}(b_iu_i + b_ju_j) - \frac{FL}{2}\right]^{(e+1)} = 0$$

This relation can now be used to determine the residual for each unknown node point.

For nodes 2,3,4 the equations are

$$R_2 = \left[\frac{(0.25)(34)}{2}(-u_1 + u_2) - \frac{(0.2)(40)}{2}\right]^{(1)} + \left[\frac{(0.25)(22)}{2}(-u_2 + u_3) - \frac{(0.2)(40)}{2}\right]^{(2)} = 0$$

$$R_3 = \left[\frac{(0.25)(22)}{2}(-u_2 + u_3) - \frac{(0.2)(40)}{2}\right]^{(2)}$$

$$+ \left[\frac{(0.25)(13)}{2}(-u_3 + u_4) - \frac{(0.2)(20)}{2}\right]^{(3)} = 0$$

$$R_4 = \left[\frac{(0.25)(13)}{2}(-u_3 + u_4) - \frac{(0.2)(20)}{2}\right]^{(3)} = 0$$

The three equations simplify to

Node 2 $\quad (-4.25\, u_1 + 4.25\, u_2 - 4.0)^{(1)} + (-2.75\, u_2 + 2.75\, u_3 - 4.0)^{(2)} = 0$

Node 3 $\quad (-2.75\, u_2 + 2.75\, u_3 - 4.0)^{(2)} + (-1.625\, u_3 + 1.625\, u_4 - 2.0)^{(3)} = 0$

Node 4 $\quad (-1.625\, u_3 + 1.625\, u_4 - 2.0)^{(3)} = 0$

Combining the terms from each element yields the three equations

$$-4.25\, u_1 + 1.5\, u_2 + 2.75\, u_3 = 8.0$$

$$-2.75\, u_2 + 1.125\, u_3 + 1.625\, u_4 = 6.0$$

$$-1.625\, u_3 + 1.625\, u_4 = 2.0$$

These three simultaneous equations are solved to obtain the nodal values

$$\begin{Bmatrix} u_1 \\ u_2 \\ u_3 \\ u_4 \end{Bmatrix} = \begin{Bmatrix} 0 \\ 0.9412 \\ 2.3957 \\ 3.6265 \end{Bmatrix}$$

The results are exactly the same as those for the subdomain method, Example 8.1. This agreement stems from the simplicity of the equations solved here.

C. Element Matrix and Column

The Galerkin finite element analysis shown so far has been carried out by hand to illustrate the details of the method. Each residual R_n (8.18) has in it integrals over two elements, e and $e + 1$. These residuals can be broken up into element-by-element calculations. Element integrals are evaluated and placed in a conventional element matrix and column.

For element e, the interpolation function has the form

$$T^{(e)} = N_i^{(e)}\, T_i + N_j^{(e)}\, T_j$$

where T_i and T_j are the element nodal values of T. Figure 8.12(a) illustrates the element. Let the element residuals be denoted $R_i^{(e)}$, $R_j^{(e)}$ where (8.18)

$$R_i^{(e)} = \int_{L^{(e)}} W_i^{(e)}\, r^{(e)}\, dx \tag{8.23}$$

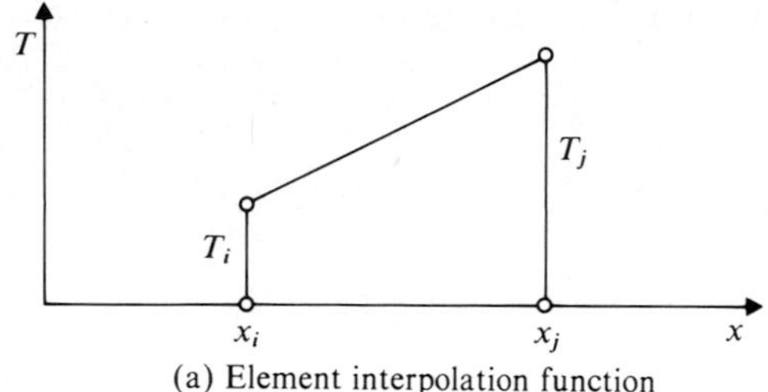

(a) Element interpolation function

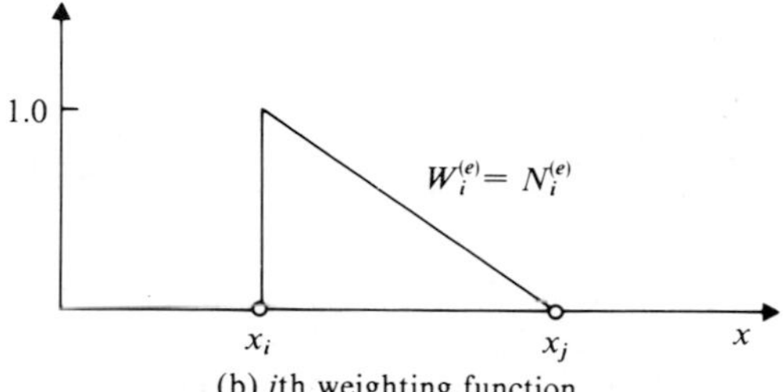

(b) ith weighting function

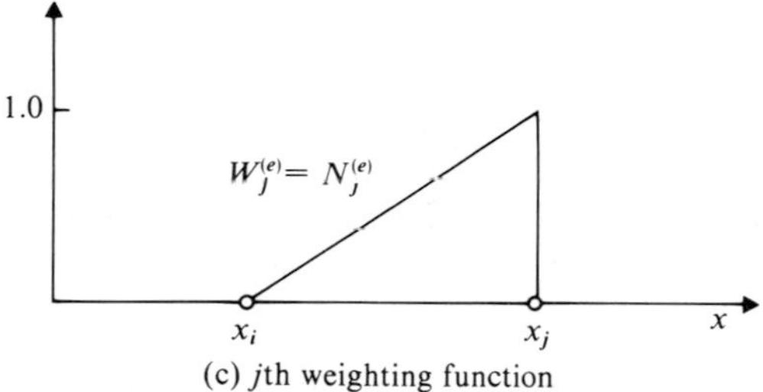

(c) jth weighting function

Figure 8.12 Element Galerkin weighting functions

$$R_j^{(e)} = \int_{L^{(e)}} W_j^{(e)} r^{(e)} \, dx$$

Here the element residuals are

$R_i^{(e)}$ = contribution of element e to residual in equation for node i

$R_j^{(e)}$ = contribution of element e to residual in equation for node j

It should be noted that $R_i^{(e)}$, $R_j^{(e)}$ are not set equal to zero but that the sum of all of the R_i, R_j for all elements is zero. The pyramid functions are shown in Figures 8.12(b) and (c) for the element. For the Galerkin finite element technique, the weighting functions are simply equal to the nodal pyramid functions: $W_i^{(e)} = N_i^{(e)}$, $W_j^{(e)} = N_j^{(e)}$.

In terms of element contributions to the residual, the full residual equations (8.19) are written out as

$$R_2 = R_2^{(1)} + R_2^{(2)} = 0$$
$$R_3 = R_3^{(2)} + R_3^{(3)} = 0$$
$$\cdots$$

$$\cdots$$

$$R_N = R_N^{(E)} = 0$$

The individual element contributions are

Element 1

$$R_1^{(1)} = \int_{L^{(1)}} N_1^{(1)} r^{(1)}\, dx \qquad \text{(not used because } T_1 = T_0\text{)}$$

$$R_2^{(1)} = \int_{L^{(1)}} N_2^{(1)} r^{(1)}\, dx$$

Element 2

$$R_2^{(2)} = \int_{L^{(2)}} N_2^{(2)} r^{(2)}\, dx \tag{8.24}$$

$$R_3^{(2)} = \int_{L^{(2)}} N_3^{(2)} r^{(2)}\, dx$$

$$\cdots$$

Element E

$$R_N^{(E)} = \int_{L^{(E)}} N_N^{(E)} r^{(E)}\, dx$$

Each element residual is not a function of x (because that is integrated out) but does contain the nodal values T_i and T_j.

The element residuals produce expressions of the form

$$\begin{Bmatrix} R_i \\ R_j \end{Bmatrix} = [B]^{(e)} \begin{Bmatrix} T_i \\ T_j \end{Bmatrix} - \{C\}^{(e)} \tag{8.25}$$

where

$$[B]^{(e)} = \begin{matrix} & i & j \\ i & B_{ii} & B_{ij} \\ j & B_{ji} & B_{jj} \end{matrix}^{(e)}$$

Element matrix

$$\{C\}^{(e)} = \begin{matrix} i \\ j \end{matrix}\begin{Bmatrix} C_i \\ C_j \end{Bmatrix}^{(e)}$$

Element column

For ordinary first- or second-order differential equations, the element matrix is (2×2), while the element column is (1×2). For first-order differential equations, the element matrix is not symmetric.

For the general equation (8.20)

$$M \frac{dT}{dx} + PT + Q = 0$$

the element matrix terms are given by

$$B_{ii} = \left[-\frac{1}{2} Mb_i - \frac{1}{3} PL \right]^{(e)}$$

$$B_{ij} = \left[-\frac{1}{2} Mb_j - \frac{1}{6} PL \right]^{(e)}$$

$$B_{ji} = \left[-\frac{1}{2} Mb_i - \frac{1}{6} PL \right]^{(e)}$$

$$B_{jj} = \left[-\frac{1}{2} Mb_j - \frac{1}{3} PL \right]^{(e)}$$

Also, the element column terms are

$$C_i = \left[\frac{1}{2} QL \right]^{(e)}$$

$$C_j = \left[\frac{1}{2} QL \right]^{(e)}$$

These are easily used.

Example 8.4 *Galerkin Method with Element Matrices*

Given The three-element model for the axially displaced rod solved in Example 8.3.

Objective Use the element matrix and assembly method to solve the problem.

Solution The element error is given by

$$r^{(e)} = \left[\frac{EA}{L} (b_i u_i + b_j u_j) - F \right]^{(e)}$$

For node i, the contribution to the residual R_i is

$$R_i^{(e)} = \int_{L^{(e)}} N_i^{(e)} \left[\frac{EA}{L} (b_i u_i + b_j u_j) - F \right]^{(e)} dx$$

and for node j, the contribution to the residual R_j is

$$R_j^{(e)} = \int_{L^{(e)}} N_j^{(e)} \left[\frac{EA}{L} (b_i u_i + b_j u_j) - F \right]^{(e)} dx$$

Separating the terms yields

$$R_i^{(e)} = \int_{L^{(e)}} \left[\frac{EA}{L} b_i N_i \, dx \right]^{(e)} u_i + \left[\int_{L^{(e)}} \frac{EA}{L} b_j N_i \, dx \right]^{(e)} u_j - \left[\int_{L^{(e)}} F \, N_i \, dx \right]^{(e)}$$

$$= \left[\frac{1}{2} EA \, b_i \right]^{(e)} u_i + \left[\frac{1}{2} EA \, b_j \right]^{(e)} u_j - \left[\frac{1}{2} FL \right]^{(e)}$$

and

$$R_j^{(e)} = \int_{L^{(e)}} \left[\frac{EA}{L} b_i N_j \, dx\right]^{(e)} u_i + \left[\int_{L^{(e)}} \frac{EA}{L} b_j N_j \, dx\right]^{(e)} u_j - \left[\int_{L^{(e)}} F \, N_j \, dx\right]^{(e)}$$

$$= \left[\frac{1}{2} EA \, b_i\right]^{(e)} u_i + \left[\frac{1}{2} EA \, b_j\right]^{(e)} u_j - \left[\frac{1}{2} FL\right]^{(e)}$$

Comparing this to the element matrix and column (8.25) in the form

$$R_i = B_{ii}u_i + B_{ij}u_j - C_i$$

$$R_j = B_{ji}u_i + B_{jj}u_j - C_j$$

gives an element matrix of the form

$$[B]^{(e)} = \begin{array}{c} \\ i \\ j \end{array} \begin{array}{c} \begin{array}{cc} \quad i & \quad j \end{array} \\ \begin{bmatrix} \frac{1}{2} EA \, b_i & \frac{1}{2} EA \, b_j \\ \frac{1}{2} EA \, b_i & \frac{1}{2} EA \, b_j \end{bmatrix}^{(e)} \end{array}$$

and an element column of the form

$$\{C\}^{(e)} = \begin{array}{c} i \\ j \end{array} \begin{bmatrix} \frac{1}{2} FL \\ \frac{1}{2} FL \end{bmatrix}^{(e)}$$

Noting some common factors, these simplify further to the element matrix

$$[B]^{(e)} = \left(\frac{1}{2} EA\right)^{(e)} \begin{array}{c} \\ i \\ j \end{array} \begin{array}{c} \begin{array}{cc} i & \quad j \end{array} \\ \begin{bmatrix} -1 & 1 \\ -1 & 1 \end{bmatrix}^{(e)} \end{array}$$

where $b_i = -1$ and $b_j = 1$, while the element column is

$$\{C\} = \left(\frac{1}{2} FL\right)^{(e)} \begin{array}{c} i \\ j \end{array} \begin{Bmatrix} 1 \\ 1 \end{Bmatrix}^{(e)}$$

These are the element matrices. It is easily seen that they are not symmetric (because the differential equation is of odd order).

For element 1, the numerical values are

$$[B]^{(1)} = \frac{1}{2} (0.25)(34) \begin{bmatrix} -1 & 1 \\ -1 & 1 \end{bmatrix}$$

$$= \begin{array}{c} \\ 1 \\ 2 \end{array} \begin{array}{c} \begin{array}{cc} 1 & \quad\quad 2 \end{array} \\ \begin{bmatrix} -4.25 & 4.25 \\ -4.25 & 4.25 \end{bmatrix}^{(1)} \end{array}$$

$$\{C\}^{(1)} = \frac{1}{2}(0.2)(40)\begin{Bmatrix}1\\1\end{Bmatrix} = \begin{matrix}1\\2\end{matrix}\begin{Bmatrix}4.0\\4.0\end{Bmatrix}^{(1)}$$

Element 2 gives

$$[B]^{(2)} = \frac{1}{2}(0.25)(22)\begin{bmatrix}-1 & 1\\-1 & 1\end{bmatrix} = \begin{matrix} \\ 2\\3\end{matrix}\begin{matrix}2 & 3\\ \begin{bmatrix}-2.75 & 2.75\\-2.75 & 2.75\end{bmatrix}^{(2)}\end{matrix}$$

$$\{C\}^{(2)} = \frac{1}{2}(0.2)(40)\begin{Bmatrix}1\\1\end{Bmatrix} = \begin{matrix}2\\3\end{matrix}\begin{Bmatrix}4.0\\4.0\end{Bmatrix}^{(2)}$$

Finally element 3 is

$$[B]^{(3)} = \frac{1}{2}(0.25)(13)\begin{bmatrix}-1 & 1\\-1 & 1\end{bmatrix} = \begin{matrix} \\ 3\\4\end{matrix}\begin{matrix}3 & 4\\ \begin{bmatrix}-1.625 & 1.625\\-1.625 & 1.625\end{bmatrix}^{(3)}\end{matrix}$$

$$\{C\}^{(3)} = \frac{1}{2}(0.2)(40)\begin{Bmatrix}1\\1\end{Bmatrix} = \begin{matrix}2\\3\end{matrix}\begin{Bmatrix}4.0\\4.0\end{Bmatrix}^{(2)}$$

Assembling these element values in the standard manner yields

$$\begin{bmatrix}-4.25 & 4.25 & 0.0 & 0.0\\-4.25 & 1.5 & 2.75 & 0.0\\0.0 & -2.75 & 1.125 & 1.625\\0.0 & 0.0 & -1.625 & 1.625\end{bmatrix}\begin{Bmatrix}u_1\\u_2\\u_3\\u_4\end{Bmatrix} = \begin{Bmatrix}4.0\\8.0\\6.0\\2.0\end{Bmatrix}$$

Adding the boundary condition to row 1 gives the final set of four equations in four unknowns:

$$\begin{bmatrix}-4.25 & 0.0 & 0.0 & 0.0\\0.0 & 1.5 & 2.75 & 0.0\\0.0 & -2.75 & 1.125 & 1.625\\0.0 & 0.0 & -1.625 & 1.625\end{bmatrix}\begin{Bmatrix}u_1\\u_2\\u_3\\u_4\end{Bmatrix} = \begin{Bmatrix}0.0\\8.0\\6.0\\2.0\end{Bmatrix}$$

The solution is the same as in Example 8.3 for this simple problem. Note that the global matrix is not symmetric.

The element and global matrices used in this chapter differ from those derived in Chapter 5 because they have been derived from a first-order differential equation,

$$EA\frac{du}{dx} - F = 0$$

rather than a second-order self-adjoint differential equation. The second-order differential equation can be obtained by differentiating the above equation

$$\frac{d}{dx}\left(EA\frac{du}{dx}\right) = 0$$

The force is constant, so $dF/dx = 0$. Applying Galerkin's method to this equation would produce the symmetric element matrix

$$[B] = \left(\frac{EA}{L}\right) \begin{bmatrix} 1 & -1 \\ -1 & 1 \end{bmatrix}^{(e)}$$

as developed in Chapter 4 for a bar. The displacements obtained for this problem are the same in either case.

8.4 Galerkin's Method for General Second-Order Equation

Consider a general second-order differential equation of the form

$$\frac{d}{dx}\left[K(x)\frac{dT}{dx}\right] + M(x)\frac{dT}{dx} + P(x)T + Q(x) = 0 \tag{8.26}$$

with boundary conditions

$$\left.\begin{array}{l} T = T_0, \text{ Specified value} \\ K_0 \dfrac{dT}{dx} + \alpha_0 T + \beta_0 = 0, \text{ Derivative} \end{array}\right\} \begin{array}{l} \text{At } x = x_0 \\ \text{(only one applies)} \end{array} \tag{8.27}$$

and

$$\left.\begin{array}{l} T = T_L, \text{ Specified value} \\ K_L \dfrac{dT}{dx} + \alpha_L T + \beta_L = 0, \text{ Derivative} \end{array}\right\} \begin{array}{l} \text{At } x = x_L \\ \text{(only one applies)} \end{array} \tag{8.28}$$

A coordinate transformation can be found for any second-order ordinary differential equation to make it self-adjoint (eliminate the first derivative term $M\,dT/dx$). In two or three dimensions such a coordinate transformation cannot be found. The treatment of the first derivative term is left in for its usefulness later on in this text.

A. Errors and Residuals

The approximation for $T(x)$ has the form

$$T^*(x) = \sum_{e=1}^{E} T^{(e)}(x)$$

When this approximation is substituted into the general differential equation, element errors occur of the form

$$r(x) = \sum_{e=1}^{E} r^{(e)}(x)$$

These have the form

$$r^{(e)} = \frac{d}{dx}\left[K^{(e)}\frac{dT^{(e)}}{dx}\right] + M^{(e)}\frac{dT^{(e)}}{dx} + P^{(e)}T^{(e)} + Q^{(e)} \tag{8.29}$$

where $K^{(e)}$, $M^{(e)}$, $P^{(e)}$, and $Q^{(e)}$ are constant coefficient functions within element e. Only constant property elements are discussed in this section, but variable property elements are developed with the same methods.

A series of weighting functions, $W_1(x)$, $W_2(x)$, $W_3(x)$, . . ., $W_N(x)$, is chosen, where N is the number of unknown nodal points. There will then be N residuals of the form

$$\begin{aligned}
R_1 &= -\int_{x_0}^{x_L} W_1 r\,dx = 0 \qquad \text{(not used if } T_1 = T_0\text{)}\\
R_2 &= -\int_{x_0}^{x_L} W_2 r\,dx = 0\\
R_3 &= -\int_{x_0}^{x_L} W_3 r\,dx = 0\\
&\;\;\cdots\\
R_N &= -\int_{x_0}^{x_L} W_N r\,dx = 0 \qquad \text{(not used if } T_N = T_L\text{)}
\end{aligned} \tag{8.30}$$

The negative sign produces element matrices that are the same as those for the variational method (when $M(x) = 0$).

Divide the weighting functions into element-size functions of the form

$$\begin{aligned}
W_1 &= \sum_{e=1}^{E} W_1^{(e)} = N_1^{(1)}\\
W_2 &= \sum_{e=1}^{E} W_2^{(e)} = N_2^{(1)} + N_2^{(2)}\\
W_3 &= \sum_{e=1}^{E} W_3^{(e)} = N_3^{(2)} + N_3^{(3)}\\
&\;\;\cdots\\
W_N &= \sum_{e=1}^{E} W_N^{(e)} = N_N^{(E)}
\end{aligned} \tag{8.31}$$

These are the same weighting functions as in (8.15) except for W_1, which is given by $W_1 = N_1^{(1)}$, and the last one W_N (8.16), which is $W_N = N_N^{(E)}$. These are illustrated in Figure 8.13.

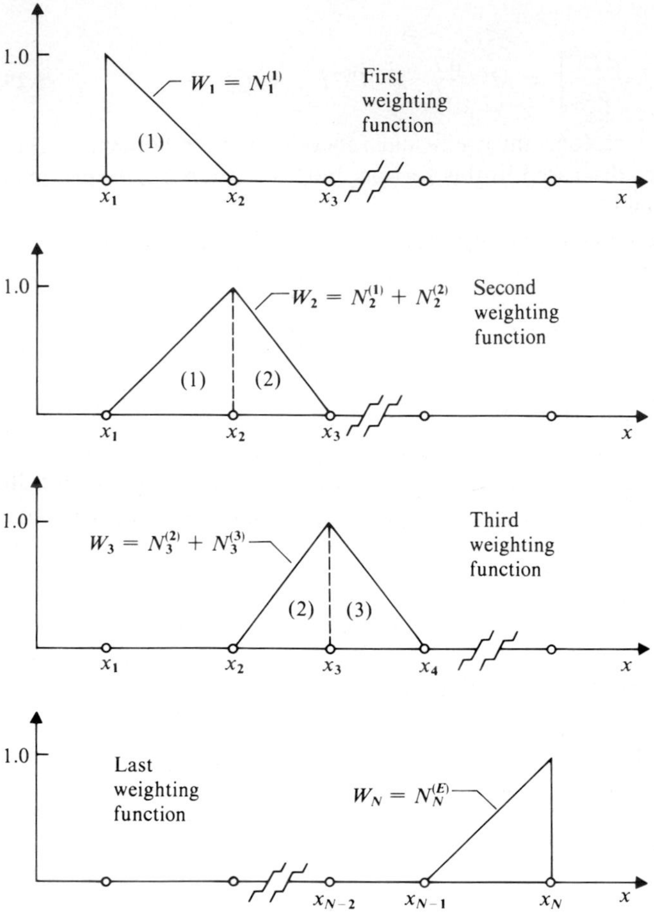

Figure 8.13 Galerkin weighting functions

The residuals become

$$R_1 = -\sum_{e=1}^{E} \int_{L^{(e)}} W_1^{(e)} r^{(e)} \, dx = 0$$

$$R_2 = -\sum_{e=1}^{E} \int_{L^{(e)}} W_2^{(e)} r^{(e)} \, dx = 0 \tag{8.32}$$

$$R_3 = -\sum_{e=1}^{E} \int_{L^{(e)}} W_3^{(e)} r^{(e)} \, dx = 0$$

. . .

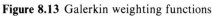

$$R_N = -\sum_{e=1}^{E} \int_{L^{(e)}} W_N^{(e)} r^{(e)} \, dx = 0$$

Here all terms have been written over an element.

B. Element Residuals

The element residuals R_i, R_j associated with the nodes i, j are

$$\begin{aligned} R_i^{(e)} &= -\int_{L^{(e)}} W_i^{(e)} r^{(e)} \, dx \\ R_j^{(e)} &= -\int_{L^{(e)}} W_j^{(e)} r^{(e)} \, dx \end{aligned} \qquad \textbf{(8.33)}$$

where $R_i^{(e)}$ = contribution of element e to residual in equation for node i, and $R_j^{(e)}$ = contribution of element e to residual in equation for node j. Substituting the pyramid functions N_i, N_j for the weighting functions W_i, W_j and the expression for the error yields

$$R_i = -\int_{L^{(e)}} \left\{ N_i \left[\frac{d}{dx}\left(K \frac{dT}{dx} \right) + M \frac{dT}{dx} + PT + Q \right] \right\}^{(e)} dx$$

$$R_j = -\int_{L^{(e)}} \left\{ N_j \left[\frac{d}{dx}\left(K \frac{dT}{dx} \right) + M \frac{dT}{dx} + PT + Q \right] \right\}^{(e)} dx$$

These are the element residuals used to evaluate the element matrices.

In a simplex element, the approximating function for $T(x)$ is linear. If this is substituted directly into the second-order derivative term

$$\frac{d}{dx}\left(K \frac{dT}{dx} \right)$$

and K is constant within the element, the result is zero. This problem is avoided by integrating the second-order term by parts.

The first residual is

$$R_i^{(e)} = -\int_{L^{(e)}} \left\{ N_i \left[\frac{d}{dx}\left(K \frac{dT}{dx} \right) + M \frac{dT}{dx} + PT + Q \right] \right\} dx$$

Note that the superscript (e) for the element is dropped for a while. The integration by parts formula employed here is

$$\int_{L^{(e)}} u dv = uv \Big|_{x_i}^{x_j} - \int_{L^{(e)}} v du$$

Considering only the second-order term in the previous residual, let the terms in the integration by parts formula be

$$u = N_i, \quad dv = \frac{d}{dx}\left(K \frac{dT}{dx} \right) dx$$

$$du = \frac{dN_i}{dx} dx, \quad v = K \frac{dT}{dx}$$

Evaluating the terms gives

$$\int_{L^{(e)}} \left\{ N_i \left[\frac{d}{dx} \left(K \frac{dT}{dx} \right) \right] \right\} dx = \left[N_i \, K \frac{dT}{dx} \right] \Bigg|_{x_i}^{x_j} - \int_{L^{(e)}} \left[\frac{dN_i}{dx} K \frac{dT}{dx} \right] dx$$

Now the terms on the right-hand side have only first derivatives in them. One derivative of T may be thought of as "transferred" over to the weighting function.

Incorporating all of this into the first residual gives

$$R_i^{(e)} = \int_{L^{(e)}} \left[\frac{dN_i}{dx} K \frac{dT}{dx} - N_i \left(M \frac{dT}{dx} + PT + Q \right) \right] dx - \left[N_i \, K \frac{dT}{dx} \right]_{x_i}^{x_j}$$

All of the terms evaluated at the end point (node) of one element cancel with the term evaluated at the end point (same node) of the adjacent element—except at the left boundary $x = x_0$. Only element 1 contributes to this term. If the value of T is specified at $x = x_0$, the weighting function W_1 is not defined, because T_1 is not an unknown. Thus the end point term vanishes. If the derivative boundary condition is given at $x = x_0$, it has the form (8.27)

$$K_0 \frac{dT}{dx} + \alpha_0 T + \beta_0 = 0 \text{ at } x = x_0$$

This can be solved for the first derivative term as

$$\left[K_0 \frac{dT}{dx} \right]_{x=x_0} = -\alpha_0 T - \beta_0$$

On the left boundary, the pyramid function has the value 1.0, so the end point term becomes

$$-\left[N_i \, K \frac{dT}{dx} \right]_{x_i}^{x_j} = \left[N_1 \, K_0 \frac{dT}{dx} \right]_{x=x_0} = \left[N_1 (-\alpha_0 T - \beta) \right]_{x=x_0}$$

$$= -\left[\alpha \; T + \beta \right]_{x=x_0}$$

The final expression for the residual is

$$R_i^{(e)} = \int_{L^{(e)}} \left[\frac{dN_i}{dx} K \frac{dT}{dx} - N_i \left(M \frac{dT}{dx} + PT + Q \right) \right] dx$$
$$- \left[\alpha_0 T + \beta_0 \right]_{x=x_0}$$

The term at $x = x_0$ is evaluated only if the derivative boundary condition applies at the left end.

C. Element Matrices

Carrying out the standard integration over the element after substituting a linear interpolation function for T yields

$$R_i^{(e)} = \left[\frac{K}{L} b_i b_i - \frac{1}{2} M b_i - \frac{1}{3} PL \right]^{(e)} T_i$$

$$
+ \left[\frac{K}{L} b_i b_j - \frac{1}{2} M b_j - \frac{1}{6} PL \right]^{(e)} T_j
- \left[\frac{1}{2} QL \right]^{(e)} - \left[\alpha_0 T_i + \beta_0 \right]^{(e)} \tag{8.34}
$$

This is the same expression as that obtained with the variational method in Section 5.7 except for the first derivative term (the term containing M). By a very similar derivation, R_j is found to be

$$
\begin{aligned}
R_j^{(e)} = {} & \left[\frac{K}{L} b_i b_j - \frac{1}{2} M b_i - \frac{1}{6} PL \right]^{(e)} T_i \\
& + \left[\frac{K}{L} b_j b_j - \frac{1}{2} M b_j - \frac{1}{3} PL \right]^{(e)} T_j \\
& - \left[\frac{1}{2} QL \right]^{(e)} + \left[\alpha_L T_j + \beta_L \right]^{(e)}
\end{aligned} \tag{8.35}
$$

Again, the result is the same as for the variational method except for the term containing M. Of course the residuals must be summed over all elements before being set equal to zero.

The previous relations are easily assembled as matrices. They are

$$
\begin{Bmatrix} R_i \\ R_j \end{Bmatrix} = [B]^{(e)} \begin{Bmatrix} T_i \\ T_j \end{Bmatrix} - \{C\}^{(e)} \tag{8.36}
$$

where the terms in the B matrix are

$$
\begin{aligned}
B_{ii} &= \left[\frac{K}{L} b_i b_i - \frac{1}{2} M b_i - \frac{1}{3} PL - \alpha_0 \right]^{(e)} \\
B_{ij} &= \left[\frac{K}{L} b_i b_j - \frac{1}{2} M b_j - \frac{1}{6} PL \right]^{(e)} \\
B_{ji} &= \left[\frac{K}{L} b_i b_j - \frac{1}{2} M b_i - \frac{1}{6} PL \right]^{(e)} \\
B_{jj} &= \left[\frac{K}{L} b_j b_j - \frac{1}{2} M b_j - \frac{1}{3} PL + \alpha_L \right]^{(e)}
\end{aligned}
$$

and the C column terms are

$$
\begin{aligned}
C_i &= \left[\frac{1}{2} QL + \beta_0 \right]^{(e)} \\
C_j &= \left[\frac{1}{2} QL - \beta_L \right]^{(e)}
\end{aligned}
$$

In each case the terms α and β are not evaluated unless the derivative boundary conditions are specified at the ith or jth node point. The element matrix is nonsymmetric because of the first derivative terms.

The nonsymmetry in the element matrix results in a nonsymmetry in the global matrix. It is still banded, but the terms above the diagonal and below the diagonal are stored. Thus the storage requirement is higher. If there are N node points, the storage for a second-order one-dimensional equation is increased from $2N$ to $3N$. Also, the number of Gauss elimination operations required to solve the set of nonsymmetric linear algebraic equations is larger than in the symmetric case. In spite of this increase, the actual computer time used in the solution of one-dimensional problems is quite small in either case.

D. Galerkin Example

A fluid flow example is solved with the Galerkin method. It is governed by an ordinary second-order differential equation with a first derivative term. It cannot be solved with a variational formulation unless some sort of transformation of variables is found to eliminate the first-order derivative term.

Example 8.5 *Fluid Flow Between Flat Permeable Plates*

Given The region between two infinitely long flat plates, shown in Figure 8.14, is filled with an incompressible isoviscous fluid (ρ = density, μ = viscosity). The upper plate at $y = h$ moves in the positive x direction with velocity U and the lower plate at $y = 0$ is fixed. A constant pressure gradient exists in the x direction. Both plates are permeable and a vertical fluid velocity $v = -U$ (equal in magnitude to the upper plate velocity) exists in the downward direction. No-slip boundary conditions apply at both upper and lower plates.

Objective Solve the differential equation for this problem with the Galerkin finite element method, using three elements in the vertical direction. Let the parameters have the values $h = 0.03$ ft, $\partial P/\partial x = -0.2$ lbf/ft^3, $\mu = 2.0 \times 10^{-3}$ lbf-sec/ft^2, $\rho = 1.0$ slug/ft^3, and $U = 0.1$ ft/sec. Compare the exact solution to the Galerkin finite element solution.

Solution The continuity and momentum equations are

$$\frac{\partial u}{\partial x} + \frac{\partial v}{\partial y} = 0$$

$$\rho u \frac{\partial u}{\partial x} + \rho v \frac{\partial u}{\partial y} = -\frac{\partial P}{\partial x} + \mu \left(\frac{\partial^2 u}{\partial x^2} + \frac{\partial^2 u}{\partial y^2} \right)$$

with boundary conditions $u = U, y = h$ and $u = 0, y = 0$. However, $v = -U$ (constant) so the continuity equation reduces to $\partial u/\partial x = 0$ and the momentum equation becomes

$$-\rho U \frac{du}{dy} = -\frac{\partial P}{\partial x} + \mu \frac{d^2 u}{dy^2}$$

Now u is a function only of y. Rearranging and using total derivatives gives

$$\frac{d}{dy}\left[\mu \frac{du}{dy}\right] + \rho U \frac{du}{dy} - \frac{\partial P}{\partial x} = 0$$

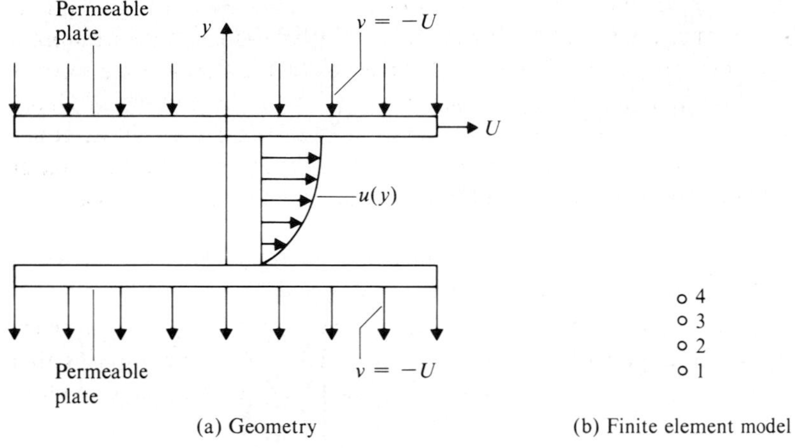

Figure 8.14 Problem geometry (Example 8.5—Fluid flow between flat permeable plates)

The solution is found as

$$u = U\left[1 - \left(\frac{h}{\rho U^2}\frac{\partial P}{\partial x}\right)\right] \frac{\left\{1 - \exp\left[-\left(\frac{\rho U h}{\mu}\right)\left(\frac{y}{h}\right)\right]\right\}}{\left\{1 - \exp\left[-\left(\frac{\rho U h}{\mu}\right)\right]\right\}}$$

$$+ U\left(\frac{h}{\rho U^2}\frac{\partial P}{\partial x}\right)\left(\frac{y}{h}\right)$$

Let the coefficients in Equation (8.26) be defined as $K(x) = \mu = 0.002$, $M(x) = \rho U = 0.1$, $P(x) = 0$, and $Q(x) = -\frac{\partial P}{\partial x} = 0.2$. The differential equation becomes

$$\frac{d}{dx}\left[0.002\frac{du}{dy}\right] + 0.1\frac{du}{dy} + 0.2 = 0$$

Again the exact solution is

$$u = 0.20595\left\{1 - \exp\left[-1.5\left(\frac{y}{h}\right)\right]\right\} - 0.06\left(\frac{y}{h}\right)$$

For element 1, the element coefficient matrices are given by Equation (8.36) with $L = 0.01$.

$$B_{11} = \left[\frac{0.002}{0.01}(-1)(-1) - \frac{1}{2}(0.1)(-1) - \frac{1}{2}(0)(0.01)\right]^{(1)} = 0.25$$

$$B_{12} = \left[\frac{0.002}{0.01}(-1)(1) - \frac{1}{2}(0.1)(1) - \frac{1}{6}(0)(0.01)\right]^{(1)} = 0.25$$

$$B_{21} = \left[\frac{0.002}{0.01}(-1)(1) - \frac{1}{2}(0.1)(-1) - \frac{1}{6}(0)(0.01)\right]^{(1)} = -0.15$$

$$B_{22} = \left[\frac{0.002}{0.01}(1)(1) - \frac{1}{2}(0.1)(1) - \frac{1}{3}(0)(0.01)\right]^{(1)} = 0.15$$

$$C_1 = \left[\frac{1}{2}(0.2)(0.01)\right]^{(1)} = 0.001$$

$$C_2 = \left[\frac{1}{2}(0.2)(0.01)\right]^{(1)} = 0.001$$

No derivative boundary conditions exist. All elements have the same element matrices.

Assembling the remaining equations in normal fashion yields

$$\begin{bmatrix} 0.25 & -0.25 & 0 & 0 \\ -0.15 & 0.40 & -0.25 & 0 \\ 0 & -0.15 & 0.40 & -0.25 \\ 0 & 0 & -0.15 & 0.15 \end{bmatrix} \begin{Bmatrix} u_1 \\ u_2 \\ u_3 \\ u_4 \end{Bmatrix} = \begin{Bmatrix} 0.001 \\ 0.002 \\ 0.002 \\ 0.001 \end{Bmatrix}$$

Using the conditions $u_1 = 0$ and $u_4 = 0.1$, the solution is

$$\begin{Bmatrix} u_1 \\ u_2 \\ u_3 \\ u_4 \end{Bmatrix} = \begin{Bmatrix} 0.0 \\ 0.06163 \\ 0.09061 \\ 0.1 \end{Bmatrix} \qquad \text{Galerkin finite element solution}$$

This may be compared with the exact solution evaluated at the same points.

$$\begin{Bmatrix} u_1 \\ u_2 \\ u_3 \\ u_4 \end{Bmatrix} = \begin{Bmatrix} 0.0 \\ 0.06104 \\ 0.09019 \\ 0.1 \end{Bmatrix} \qquad \text{Exact solution}$$

Figure 8.15 plots the two results, which are quite close. This is somewhat surprising, because the solution is not a simple second-order polynomial, which yielded such small errors for some earlier examples.

E. Numerical Oscillations

Sometimes the calculated finite element solution for differential equations with first- and second-order terms will have numerical errors. Figure 8.16 shows an example. The numerical errors produce oscillations, as shown in the plot. They are not present in the real solution.

These numerical problems occur when the first derivative term is large compared to the second derivative term. Let the first derivative term be approximated in the x direction by

$$M\frac{dT}{dx} \cong M\frac{\Delta T}{\Delta x}$$

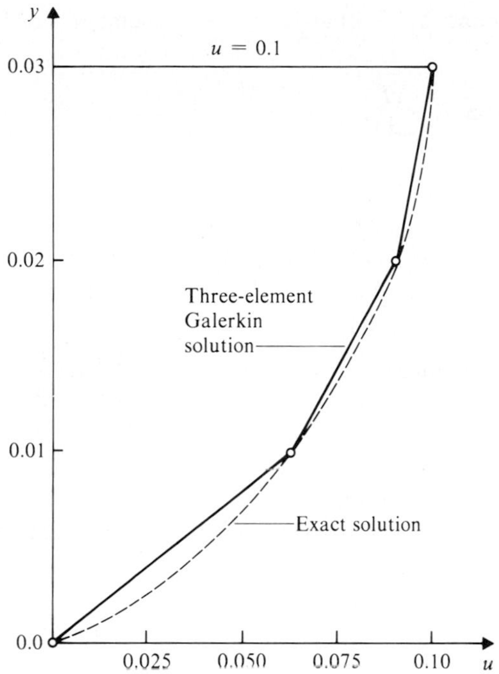

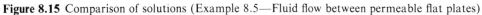

Figure 8.15 Comparison of solutions (Example 8.5—Fluid flow between permeable flat plates)

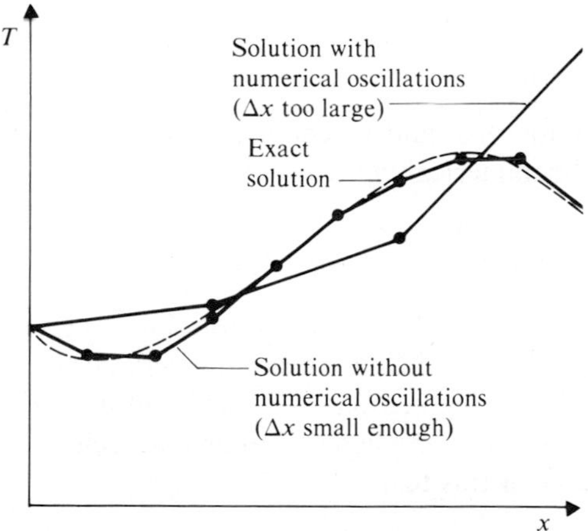

Figure 8.16 Numerical oscillations

where Δx = element length and ΔT = change in T over element. For the second-order term

$$\frac{d}{dx}\left(K\frac{dT}{dx}\right) \cong K\frac{\Delta T}{(\Delta x)^2}$$

The ratio of these two terms is

$$\frac{M\dfrac{dT}{dx}}{\dfrac{d}{dx}\left(K\dfrac{dT}{dx}\right)} = \frac{M\dfrac{\Delta T}{\Delta x}}{K\dfrac{\Delta T}{(\Delta x)^2}} = \frac{M(\Delta x)}{K}$$

This ratio is called the Péclet number (Pe)

$$\text{Pe} = \left[\frac{M(\Delta x)}{K}\right]^{(e)}$$

which is dependent on the element length.

Numerical oscillations do not occur when the Péclet number is less than or equal to 2. In terms of Δx,

$$\left[\frac{M(\Delta x)}{K}\right]^{(e)} \leq 2$$

If the element length is chosen small enough, the numerical oscillations do not occur. For the fluid flow example just done, the Péclet number is

$$\text{Pe} = \frac{\rho U\dfrac{du}{dy}}{\dfrac{d}{dy}\left(\mu\dfrac{du}{dy}\right)} \cong \frac{\rho U\dfrac{\Delta u}{\Delta y}}{\mu\dfrac{\Delta u}{(\Delta y)^2}} = \frac{\rho U(\Delta y)}{\mu}$$

Note that the Péclet number for this fluid mechanics problem is actually the element Reynolds number. This has the numerical value

$$\frac{\rho U\,(\Delta y)}{\mu} = \frac{0.1(0.01)}{0.002} = 0.5$$

The Péclet number is 0.5, which is less than 2, so the solution is quite smooth.

Sometimes the requirement for small Δx means that a very large number of elements and nodes must be used. The resulting computer storage and cost may be too large to make the solution practical. Another method of removing oscillations is called "upwinding." It is beyond the scope of this text.

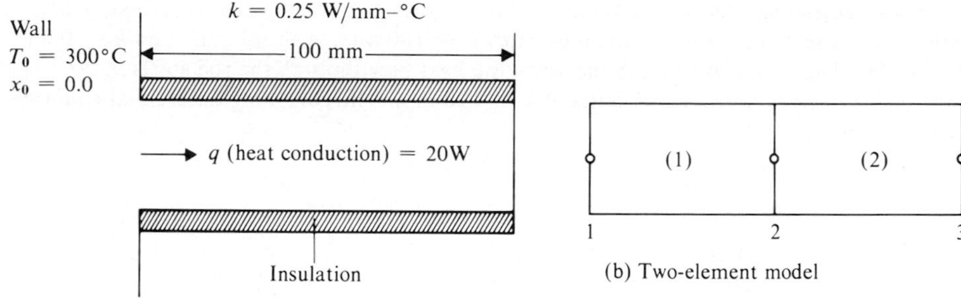

Figure 8.17 Geometry for Problem 8.4

8.5 Problems

A. Introductory Problems

8.1 A first-order differential equation is

$$\frac{dT}{dx} = 2x$$

with the boundary condition $T = 1$, $x = 0$. (*a*) Assume a single simplex element of length $L^{(1)} = 1$. Determine the error $r(x)$ in the element. (*b*) Evaluate the residual R, using the subdomain method. (*c*) Calculate both the error $r(x)$ and residual R, using the exact solution $T = x^2 + 1$.

8.2 Solve Problem 8.1, but employing the Galerkin method.

8.3 Resolve Example 8.5—Case 1, using a linear variation for the rod area. Compare the result to that for Example 8.5 with a constant-area element. Is it more accurate?

B. First-Order Equations

8.4 An insulated cylindrical rod, shown in Figure 8.17(a), has uniform heat conduction with the properties $L = 100$ mm, $A = 40$ mm², $k = 0.25$ W/mm-°C, and $q = 20$ W. The heat flow through the rod is 20 watts and the temperature at $x = 0$ is 300°C. The differential equation to be used for this problem is

$$kA\frac{dT}{dx} + q = 0$$

which is first-order. Note that it may be easily converted to a second-order differential equation (as discussed in Chapter 5) by differentiating it with respect to x. Use the subdomain method with two equal-length simplex elements to determine the temperature profile in the rod. Compare to the exact solution.

8.5 A cylindrical cooling rod, similar to that in Problem 8.4 but without the insulation, with uniform cross sectional area has the properties $L = 100$ mm, $A = 40$ mm^2, and $k = 0.25$ W/mm-°C. Due to losses through the sides, the heat flow through the rod starts at 20 watts and decreases linearly to 5 watts at $x = 100$ mm. The governing differential equation for the rod is

$$kA\frac{dT}{dx} + q = 0$$

where q is now a function of x. At the left end, $T = 300$°C. Use three equal-length, equal-property elements to obtain the temperature profile in the rod. Employ the subdomain method. Obtain the exact solution and compare to the finite element solution.

8.6 Solve Problem 8.5 using Galerkin's method.

8.7 The slope of a curve $y(x)$ is given by

$$\frac{dy}{dx} = 2x + 3y$$

If the curve passes through the point $x = 0$, $y = 0$, determine the equation of the curve $y(x)$ in the interval $x = 0$ to $x = 2$ using three equal-length elements. Employ the subdomain method. Compare the finite element solution to the exact solution

$$y = \frac{2}{9}\exp(3x) - \frac{2x}{3} - \frac{2}{9}$$

8.8 A tank is filled with 10 gallons of brine in which is dissolved 5 pounds of salt. Brine containing 3 pounds of salt per gallon enters the tank at 2 gallons per minute. The mixture is well stirred. Brine leaves the tank at 2 gallons per minute. Determine A, the number of pounds of salt in the tank, as a function of time for the first 10 minutes. Solve using three elements of length 2 min, 3 min, and 5 min, using the Galerkin method. Compare to the exact solution $A(t) = 30 - 25\exp(-t/5)$.

8.9 An insulated tapered rod, shown in Figure 8.18(a), has uniform heat conduction along its length. The cross sectional area of the rod is given by

$$A = \begin{cases} 40, & 0 \le x \le 50 \\ 70 - 0.6x, & 50 \le x \le 100 \end{cases}$$

where the area is in mm^2. Other problem parameters are $L = 100$ mm, $k = 0.25$ W/mm-°C, $q = 20$ W, and $T_0 = 300$°C at $x = 0$ mm. The governing differential equation is Fourier's law

$$kA\frac{dT}{dx} + q = 0$$

(no heat flow is lost through the insulated sides). Note that this is easily converted to a second-order differential equation (such as discussed in Chapter 5) by differentiating once with respect to x. Use three equal-area elements, as shown in Figure 8.18(b), and the

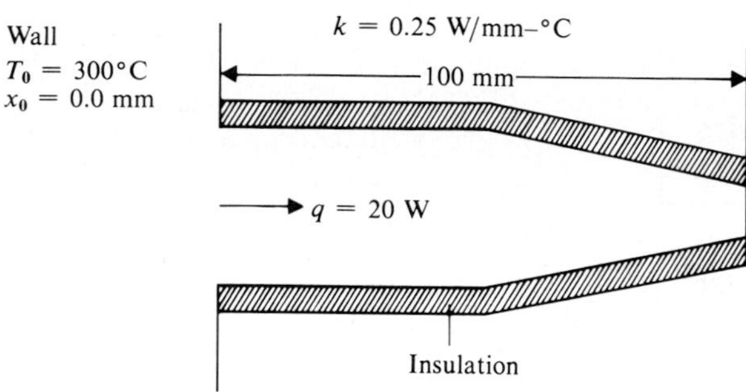

(a) Schematic of tapered cooling rod

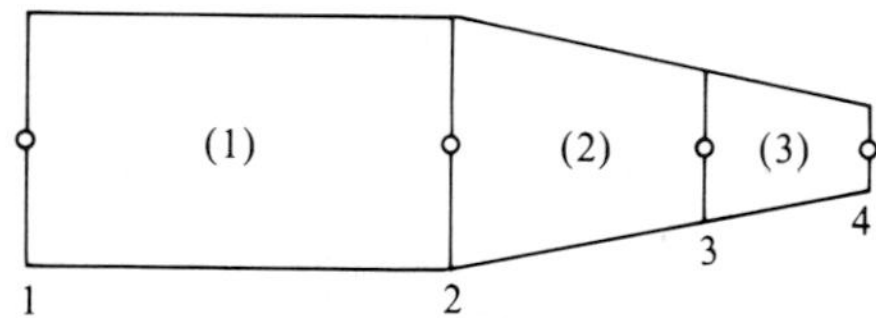

(b) Three-element model

Figure 8.18 Geometry for Problems 8.9 and 8.10

subdomain weighted residual method to obtain the temperature profile. Compare to the exact solution

$$T = \begin{cases} -2x + 300, & 0 \leq x \leq 50 \\ 133.33 \ln(70 - 0.6x) - 291.84, & 50 < x \leq 100 \end{cases}$$

at the node points.

8.10 Solve Problem 8.9, but employ the Galerkin weighted residual method.

C. Second-Order Equations

8.11 Re-solve Example 8.5 with a pressure gradient of

$$\frac{\partial P}{\partial x} = -2.0 \text{ lbf/ft}^3$$

Plot the result and compare to the exact solution, as given in Example 8.5.

8.12 A second-order differential equation is

$$\frac{d^2y}{dx^2} + \frac{dy}{dx} - 2y = 0$$

with the boundary conditions $y = 1$ at $x = 0$, $y = e$ at $x = 1$. Use a two-element model (with node at $x = 0.6$) to solve for the value of y at $x = 0.6$. Plot the finite element solution and the exact solution $y = e^x$ over the region $x = 0$ to $x = 1$.

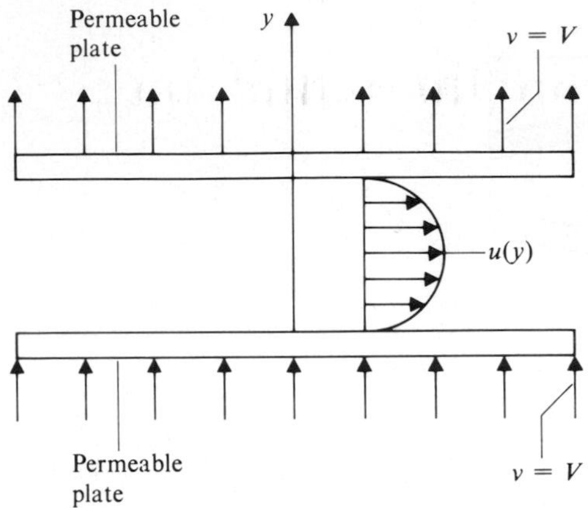

Figure 8.19 Geometry for Problem 8.13

8.13 The region between two infinitely long flat plates is filled with an incompressible isoviscous fluid. Both the upper and lower plates are fixed. A constant pressure gradient exists in the x direction. Both plates are permeable and a vertical velocity $u = V$ (upward velocity) exists through both plates. Figure 8.19 shows the geometry. Problem parameters are $h = 0.3$ ft, $\partial P/\partial x = -0.4$ lbf/ft^2, $\mu = 1.0 \times 10^{-3}$ lbf-sec/ft^2, $\rho = 1.0$ slug/ft^3, and $V = 0.8$ ft/sec. Solve for the velocity profile, using three equal-length elements and the Galerkin method.

Part 3
Partial Differential Equations

Chapter 9

Variational Methods—Two- and Three-Dimensional Field Problems

9.1 Introduction

Many of the principles used for one-dimensional problems are almost directly applicable to two- and three-dimensional field (scalar, not vector) problems. A general derivation is carried out, and then specific applications are considered.

Perhaps the simplest two-dimensional equation is that of steady heat conduction

$$\frac{\partial}{\partial x}\left(k\frac{\partial T}{\partial x}\right) + \frac{\partial}{\partial y}\left(k\frac{\partial T}{\partial y}\right) = 0$$

where there are no sources or sinks. For homogeneous materials (k = constant) this becomes simply Laplace's equation

$$\frac{\partial^2 T}{\partial x^2} + \frac{\partial^2 T}{\partial y^2} = 0, \text{ in } A \tag{9.1}$$

Let the temperature T be specified as T_0 on the boundary L, shown in Figure 9.1,

$$T = T_0 \tag{9.2}$$

where T_0 is a known function on the boundary. This is the differential equation to be solved with the boundary condition.

The equivalent functional I (given here without proof, for the moment) is

$$I = \iint_A \left[\frac{1}{2}\left(\frac{\partial T}{\partial x}\right)^2 + \frac{1}{2}\left(\frac{\partial T}{\partial y}\right)^2\right] dA \tag{9.3}$$

with the boundary condition

$$T = T_0 \text{ on } L \tag{9.4}$$

This must be minimized with respect to the unknown nodal temperatures, as in the one-dimensional case.

The basic principles involved here are the same as those discussed for the variational method in one dimension (Chapters 5 and 6). A reader not really interested in the details of the variational method should concentrate on Sections 9.2 and 9.5; possibly Section 9.6 on three-dimensional elements would be of use, as well.

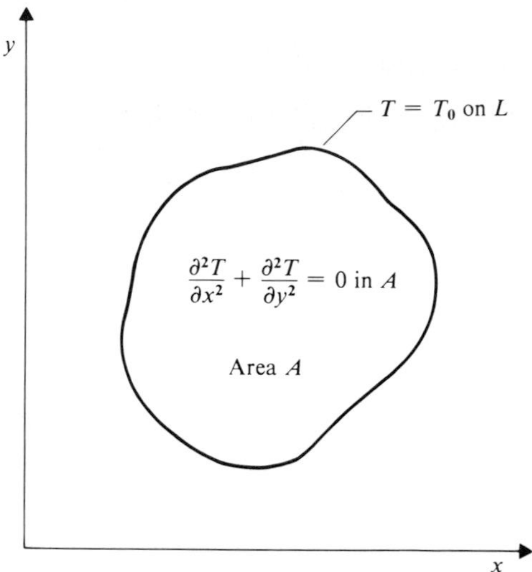

Figure 9.1 Heat conduction in two-dimensional body

9.2 General Boundary Value Problem

The general boundary value problem solved in preceding chapters for one dimension is here extended to two-dimensional problems. Now the general field variable T is a function of x and y, r and θ, or some other coordinates. Once the finite element formulation has been completed for the general field variable, many practical problems in heat transfer, fluid mechanics, solid mechanics, and other types of analysis may be found.

A. General Boundary Value Problem in Two Dimensions

Consider a general two-dimensional partial differential equation of the form

$$\frac{\partial}{\partial x}\left[K_x(x,y)\,\frac{\partial T}{\partial x}\right] + \frac{\partial}{\partial y}\left[K_y(x,y)\,\frac{\partial T}{\partial y}\right] + P(x,y)T + Q(x,y) = 0 \quad \textbf{(9.5)}$$

that is valid in an area A. This is an extension of equation (5.27). The functions K_x, K_y, P, Q are given functions of x and y. Typically, the functions K_x and K_y represent some material property, such as the heat conduction coefficient. If the material is homogeneous and isotropic, K_x and K_y are constants. P and Q represent different physical properties in different problems.

A boundary, labeled L, surrounds the area A, as shown in Figure 9.2. Boundary conditions of the two types discussed in Section 5.4 may exist. In this chapter, it is assumed that on some or all of the boundary, the temperature is specified and that this portion of the boundary is labeled L_1. Then on L_1

$$T = T_0(x,y) \quad \textbf{(9.6)}$$

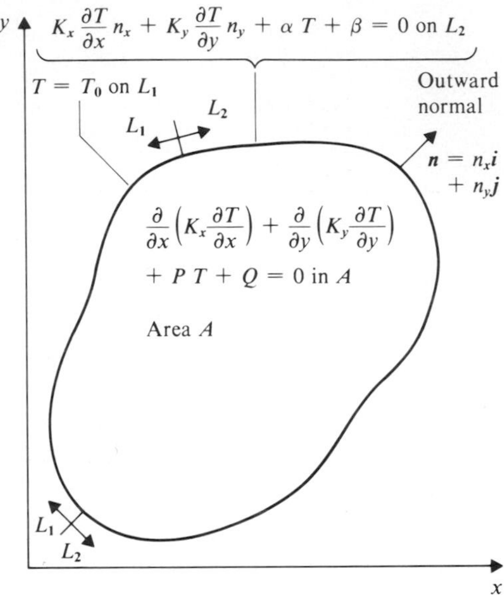

Figure 9.2 General two-dimensional boundary value problem

where T_0 is a known function of x and y. On the remainder of the boundary, labeled L_2, the general derivative boundary condition is specified in the form

$$K_x(x,y)\,\frac{\partial T}{\partial x}\,n_x + K_y(x,y)\,\frac{\partial T}{\partial y}\,n_y + \alpha(x,y)T + \beta(x,y) = 0 \qquad \textbf{(9.7)}$$

The functions α and β are specified along L_2. Also, the quantities n_x and n_y are the direction cosines of the outward normal to L_2. Of course, L_1 and L_2 must add together to form the complete boundary L.

The general elliptic boundary value problem is often classified according to its boundary condition, as shown in Table 9.1. If T is specified along the entire boundary (9.6), the boundary condition is called the first, or Dirichlet, kind. When the derivative boundary condition (9.7) is used, but α is zero, it is called the second, or Neumann, kind. Finally, the general derivative boundary condition is called either the third, or Robin, kind. In a mixed kind, some portions of the boundary have Dirichlet boundary conditions and others derivative boundary conditions.

Example 9.1 Boundary Value Problem for Inviscid Incompressible Flow

Given An incompressible inviscid flow, shown in Figure 9.3, occurs between two horizontal walls and around an elliptical body in the center. A velocity potential ϕ describes the problem.

Objective Find the differential equation and boundary conditions for this problem.

Table 9.1
Types of Boundary Value Problems

Differential Equation		
$\frac{\partial}{\partial x}\left(K_x \frac{\partial T}{\partial x}\right) + \frac{\partial}{\partial y}\left(K_y \frac{\partial T}{\partial y}\right) + PT + Q = 0$		
Boundary Conditions		
Equation	**Section of L**	**Type**
$T = T_0$	L_1	First or Dirichlet
$K_x \frac{\partial T}{\partial x} n_x + K_y \frac{\partial T}{\partial y} n_y + \beta = 0$	L_2	Second or Neumann
$K_x \frac{\partial T}{\partial x} n_x + K_y \frac{\partial T}{\partial y} n_y + \alpha T + \beta = 0$	L_2	Third or Robin

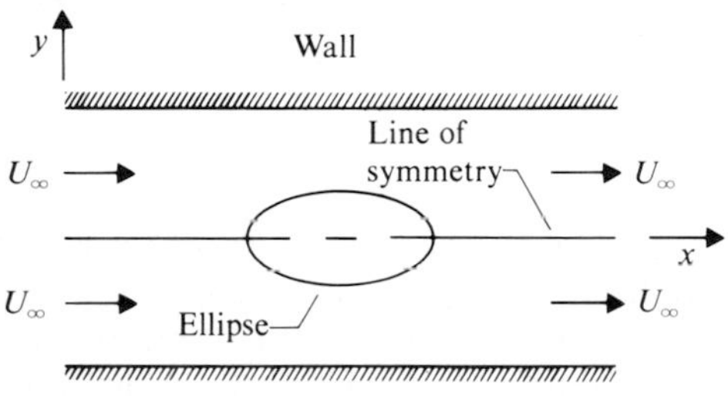

(*a*) Schematic

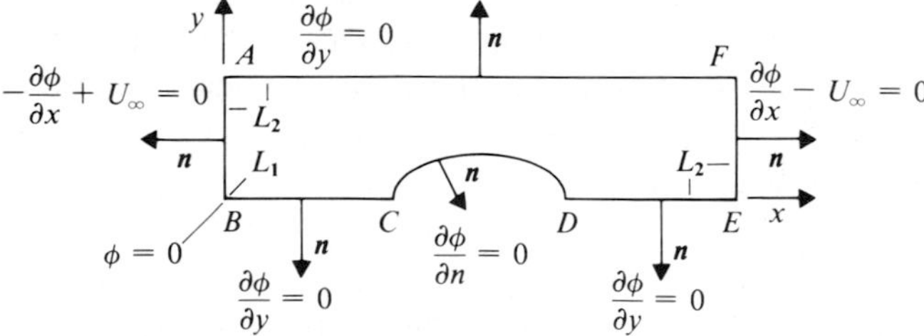

(*b*) Boundary conditions

Figure 9.3 Flow around ellipse (Example 9.1—Boundary value problem for inviscid incompressible flow)

Solution For potential flow, the velocity potential ϕ is related to the fluid velocity components by

$$u = \frac{\partial \phi}{\partial x}, \quad v = \frac{\partial \phi}{\partial y}$$

where u and v are the horizontal and vertical velocities, respectively. The potential must satisfy Laplace's equation

$$\frac{\partial^2 \phi}{\partial x^2} + \frac{\partial^2 \phi}{\partial y^2} = 0$$

This must be solved and then the velocity components calculated. In the general equation (9.5), $K_x(x,y) = 1.0$, $K_y(x,y) = 1.0$, $P(x,y) = 0$, and $Q(x,y) = 0$.

Along the line AB, the incoming velocity must be specified by

$$u = U_\infty = \frac{\partial \phi}{\partial x}$$

The outward normal is $\boldsymbol{n}_{AB} = -\boldsymbol{i}$, and the boundary condition on AB becomes

$$-\frac{\partial \phi}{\partial x} + U_\infty = 0$$

Then $\alpha(x,y) = 0$, $\beta(x,y) = U_\infty$.

In this type of problem, where only the derivative of the potential is specified, the result is not unique. If any constant is added to the velocity potential, the sum also satisfies the differential equation and boundary conditions. For convenience in this problem, the potential ϕ at point B is set equal to zero. Actually, any one point would do. At all other points along the boundary, the velocity potential is unknown. Since only derivatives of ϕ are used to calculate the velocity components, the value assigned to ϕ at B does not matter.

Along the line of symmetry BC, the outward normal is vertically downward and the boundary condition is $\partial\phi/\partial y = 0$. Fluid cannot penetrate the surface of the ellipse from C to D. If v_n is the velocity normal to the surface of the ellipse, the boundary condition on CD is

$$v_n = \frac{\partial \phi}{\partial n} = \frac{\partial \phi}{\partial x} n_x + \frac{\partial \phi}{\partial y} n_y = 0$$

Each point along CD has a different outward normal that is found from the equation for the ellipse. When the perimeter of the ellipse is approximated by the straight sides of triangular elements, the normal is determined for each segment. Along the rest of the line of symmetry DE, $\partial\phi/\partial y = 0$.

At the end of the flow region EF, the velocity must again be U_∞. The boundary condition is

$$\frac{\partial \phi}{\partial x} - U_\infty = 0, \quad \boldsymbol{n}_{EF} = \boldsymbol{i}$$

Along the wall, the vertical velocity must be zero

$$v = \frac{\partial \phi}{\partial y} = 0 \text{ on } FA$$

The boundary conditions are complete.

B. General Boundary Value Problem in Three Dimensions

In three dimensions, the general differential equation is

$$\begin{aligned}\frac{\partial}{\partial x}\left[K_x(x,y,z)\frac{\partial T}{\partial x}\right] + \frac{\partial}{\partial y}\left[K_y(x,y,z)\frac{\partial T}{\partial y}\right] \\ + \frac{\partial}{\partial z}\left[K_z(x,y,z)\frac{\partial T}{\partial z}\right] + P(x,y,z)T + Q(x,y,z) = 0\end{aligned} \tag{9.8}$$

valid over a volume V. A surface S bounds the volume. Boundary conditions along S specify either T:

$$T = T_0\,(x,y,z) \text{ on } S_1 \tag{9.9}$$

or the derivative boundary condition in the form

$$\begin{aligned}K_x(x,y,z)\,\frac{\partial T}{\partial x}\,n_x + K_y(x,y,z)\,\frac{\partial T}{\partial y}\,n_y + K_z(x,y,z)\,\frac{\partial T}{\partial z}\,n_z \\ + \alpha(x,y,z)T + \beta(x,y,z) = 0 \text{ on } S_2\end{aligned} \tag{9.10}$$

The functions α and β are given along S_2. Also, n_x, n_y, n_z are the components of the outward normal to S_2.

9.3 Derivation of Functional

This section derives the functional for a general three-dimensional field problem. The methods of Chapter 6 can be followed here. The steps are

1. Start with assumed functional form I
2. Take first variation $\delta I = 0$
3. Obtain Euler–Lagrange equation
4. Determine general functional

A two-dimensional functional is easily obtained by dropping the z terms from the three-dimensional relations.

A. Boundary Value Problem and Functional

The second-order differential equation is

$$\frac{\partial}{\partial x}\left(K\,\frac{\partial T}{\partial x}\right) + \frac{\partial}{\partial y}\left(K\,\frac{\partial T}{\partial y}\right) + \frac{\partial}{\partial z}\left(K\,\frac{\partial T}{\partial z}\right) + PT + Q = 0, \text{ in } V \tag{9.11}$$

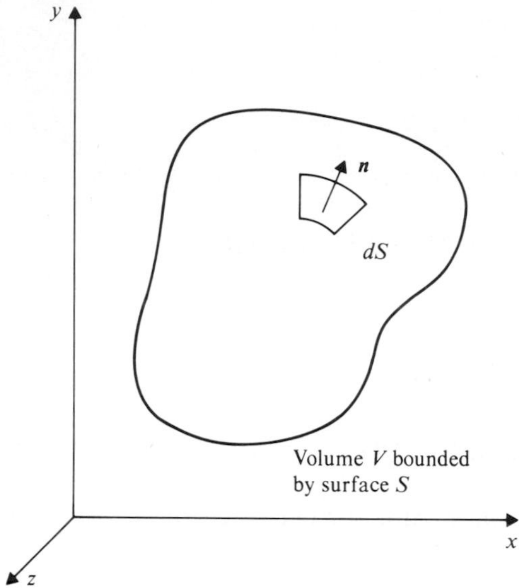

Figure 9.4 Three-dimensional problem

where K,P,Q are all functions of x,y,z defined in V. The associated boundary condition on S_1, where T is specified, is

$$T = T_0 \tag{9.12}$$

and on S_2, where a derivative boundary condition is given, by

$$K\frac{\partial T}{\partial x}n_x + K\frac{\partial T}{\partial y}n_y + K\frac{\partial T}{\partial z}n_z + \alpha T + \beta = 0 \tag{9.13}$$

Here α and β are functions of x,y,z defined on S_2, and n_x, n_y, n_z are components of outward normal to S_2. A schematic is shown in Figure 9.4.

Assume the general form of the functional as a simple extension of the functional for one-dimensional problems

$$I(T) = \iiint_V F(x,y,z,T,T_x,T_y,T_z)\,dV + \iint_{S_2} G(x,y,z,T)\,dS \tag{9.14}$$

Here, for convenience, the derivatives are written as

$$T_x = \frac{\partial T}{\partial x}, \quad T_y = \frac{\partial T}{\partial y}, \quad T_z = \frac{\partial T}{\partial z}$$

The first term corresponds to the differential equation, while the second yields the derivative boundary condition.

B. First Variation

Let the approximation to the exact solution be written as

$$T^*(x,y,z) = T(x,y,z) + \delta T(x,y,z) \quad \textbf{(9.15)}$$

where T^* = approximate solution, T = exact solution to boundary value problem, and δT = first variation. Then the first variation of the functional is $\delta I = I(T^*) - I(T) = I(T + \delta T) - I(T)$. Expanding this in a Taylor's series for small variations of δT gives

$$I(T + \delta T) = I(T) + \frac{\partial I}{\partial T}\delta T + \ldots$$

where terms above the first order are neglected. The first variation vanishes if

$$\frac{\partial I}{\partial T} = 0 \quad \textbf{(9.16)}$$

Thus $T(x)$ represents a stationary value of the functional $I(T)$.

The first variation of the general functional (9.14) is

$$\delta I = \iiint_V F(x,y,z,T^*,T_x^*,T_y^*,T_z^*)\, dV + \iint_{S_2} G(x,y,z,T^*)\, dS$$
$$-\iiint_V F(x,y,z,T,T_x,T_y,T_z)\, dV - \iint_{S_2} G(x,y,z,T)\, dS$$

A Taylor's series expansion of the first integrand for small variations δT, δT_x, δT_y, δT_z yields the relation

$$F(x,y,z,T^*,T_x^*,T_y^*,T_z^*)$$
$$= F(x,y,z,T + \delta T,T_x + \delta T_x,T_y + \delta T_y,T_z + \delta T_z)$$
$$= F(x,y,z,T,T_x,T_y,T_z)$$
$$+ \frac{\partial F}{\partial T}\delta T + \frac{\partial F}{\partial T_x}\delta T_x + \frac{\partial F}{\partial T_y}\delta T_y + \frac{\partial F}{\partial T_z}\delta T_z + \ldots$$

Similarly, the derivative boundary term is

$$G(x,y,z,T^*) = G(x,y,z,T + \delta T) = G(x,y,z,T) + \frac{\partial G}{\partial T}\delta T + \ldots$$

These give the variation δI.

Summing all of the terms in δI and subtracting like terms produces

$$\delta I = \iiint_V \left[\frac{\partial F}{\partial T}\delta T + \frac{\partial F}{\partial T_x}\delta T_x + \frac{\partial F}{\partial T_y}\delta T_y + \frac{\partial F}{\partial T_z}\delta T_z\right] dV$$
$$- \iint_{S_2} \frac{\partial G}{\partial T}\delta T\, dS$$

As shown in Chapter 6, the operations of differentiation and variation are independent of one another. Thus

$$\delta T_x = \delta\left(\frac{\partial T}{\partial x}\right) = \frac{\partial}{\partial x}(\delta T)$$

$$\delta T_y = \delta\left(\frac{\partial T}{\partial y}\right) = \frac{\partial}{\partial y}(\delta T)$$

$$\delta T_z = \delta\left(\frac{\partial T}{\partial z}\right) = \frac{\partial}{\partial z}(\delta T)$$

and the last three terms of the integral become Ⓐ, where

$$Ⓐ = \iiint_V \left[\frac{\partial F}{\partial T_x}\delta T_x + \frac{\partial F}{\partial T_y}\delta T_y + \frac{\partial F}{\partial T_z}\delta T_z\right] dV$$

$$= \iiint_V \left[\frac{\partial F}{\partial T_x}\frac{\partial}{\partial x}(\delta T) + \frac{\partial F}{\partial T_y}\frac{\partial}{\partial y}(\delta T) + \frac{\partial F}{\partial T_z}\frac{\partial}{\partial z}(\delta T)\right] dV$$

This must be transformed into the form of the Euler–Lagrange equations.

Use the identities

$$\frac{\partial}{\partial x}\left[\frac{\partial F}{\partial T_x}\delta T\right] = \frac{\partial}{\partial x}\left(\frac{\partial F}{\partial T_x}\right)\delta T + \frac{\partial F}{\partial T_x}\frac{\partial}{\partial x}(\delta T)$$

$$\frac{\partial}{\partial y}\left[\frac{\partial F}{\partial T_y}\delta T\right] = \frac{\partial}{\partial y}\left(\frac{\partial F}{\partial T_y}\right)\delta T + \frac{\partial F}{\partial T_y}\frac{\partial}{\partial y}(\delta T)$$

$$\underbrace{\frac{\partial}{\partial z}\left[\frac{\partial F}{\partial T_z}\delta T\right]}_{\text{Terms In Ⓑ}} = \underbrace{\frac{\partial}{\partial z}\left(\frac{\partial F}{\partial T_z}\right)}_{\text{Terms In Ⓒ}}\delta T + \underbrace{\frac{\partial F}{\partial T_z}\frac{\partial}{\partial z}}_{\text{Terms In Ⓐ}}(\delta T)$$

to obtain the term Ⓐ as Ⓐ = Ⓑ − Ⓒ where

$$Ⓐ = \underbrace{\iiint_V \left[\frac{\partial}{\partial x}\left(\frac{\partial F}{\partial T_x}\delta T\right) + \frac{\partial}{\partial y}\left(\frac{\partial F}{\partial T_y}\delta T\right) + \frac{\partial}{\partial z}\left(\frac{\partial F}{\partial T_z}\delta T\right)\right] dV}_{\text{Term Ⓑ}}$$

$$\underbrace{- \iiint_V \left[\frac{\partial}{\partial x}\left(\frac{\partial F}{\partial T_x}\right)\delta T + \frac{\partial}{\partial y}\left(\frac{\partial F}{\partial T_y}\right)\delta T + \frac{\partial}{\partial z}\left(\frac{\partial F}{\partial T_z}\right)\delta T\right] dV}_{\text{Term Ⓒ}}$$

The divergence theorem applies to term Ⓑ of the form

$$Ⓑ = \iint_S \left[\frac{\partial F}{\partial T_x}\delta T\, n_x + \frac{\partial F}{\partial T_y}\delta T\, n_y + \frac{\partial F}{\partial T_z}\delta T\, n_z\right] dS$$

where n_x, n_y, n_z are the components of the outward normal to the surface S. The divergence theorem converts an integral over the volume V into an integral over the bounding surface S.

Now δI becomes

$$\delta I = \iiint_V \left[\frac{\partial F}{\partial T} - \frac{\partial}{\partial x}\left(\frac{\partial F}{\partial T_x}\right) - \frac{\partial}{\partial y}\left(\frac{\partial F}{\partial T_y}\right) - \frac{\partial}{\partial z}\left(\frac{\partial F}{\partial T_z}\right)\right] \delta T\, dV$$
$$+ \iint_S \left[\frac{\partial F}{\partial T_x} n_x + \frac{\partial F}{\partial T_y} n_y + \frac{\partial F}{\partial T_z} n_z\right] \delta T\, dS$$
$$+ \iint_{S_2} \frac{\partial G}{\partial T} \delta T\, dS$$

Consider the second term over surface S. On the boundary S_1, the variation δT is zero, because T is known along that part of the boundary. Thus the integration reduces to the surface S_2, and the last two terms can be combined to give

$$\delta I = -\iiint_V \left[\frac{\partial}{\partial x}\left(\frac{\partial F}{\partial T_x}\right) + \frac{\partial}{\partial y}\left(\frac{\partial F}{\partial T_y}\right) + \frac{\partial}{\partial z}\left(\frac{\partial F}{\partial T_z}\right) - \frac{\partial F}{\partial T}\right] \delta T\, dV$$
$$+ \iint_{S_2} \left[\frac{\partial F}{\partial T_x} n_x + \frac{\partial F}{\partial T_y} n_y + \frac{\partial F}{\partial T_z} n_z + \frac{\partial G}{\partial T}\right] \delta T\, dS \tag{9.17}$$

Here a negative sign has been factored out of the first term.

C. Euler–Lagrange Equation

The first variation must equal zero. Here δT is arbitrary, so the general integral (9.17) vanishes only if the integrand is zero. Thus the Euler–Lagrange equation is

$$\frac{\partial}{\partial x}\left(\frac{\partial F}{\partial T_x}\right) + \frac{\partial}{\partial y}\left(\frac{\partial F}{\partial T_y}\right) + \frac{\partial}{\partial z}\left(\frac{\partial F}{\partial T_z}\right) - \frac{\partial F}{\partial T} = 0 \text{ in } V \tag{9.18}$$

and the derivative boundary term is

$$\frac{\partial F}{\partial T_x} n_x + \frac{\partial F}{\partial T_y} n_y + \frac{\partial F}{\partial T_z} n_z + \frac{\partial G}{\partial T} = 0 \text{ on } S_2 \tag{9.19}$$

These can now be employed to obtain the general functional.

D. General Functional

Equating the first three terms in the Euler–Lagrange equation (9.18) to the first three terms in the differential equation (9.11) gives

$$\frac{\partial}{\partial x}\left(\frac{\partial F}{\partial T_x}\right) + \frac{\partial}{\partial y}\left(\frac{\partial F}{\partial T_y}\right) + \frac{\partial}{\partial z}\left(\frac{\partial F}{\partial T_z}\right) = \frac{\partial}{\partial x}(KT_x) + \frac{\partial}{\partial y}(KT_y)$$
$$+ \frac{\partial}{\partial z}(KT_z)$$

Integrating each of the three terms with respect to T_x, T_y, T_z, respectively, yields

$$F = \frac{1}{2} KT_x^2 + \frac{1}{2} KT_y^2 + \frac{1}{2} KT_z^2 + C$$

where the constant of integration C is a function of T: $C = C(T)$. Here C is easily evaluated from the remainder of the Euler–Lagrange equation.

Set the last term in the Euler–Lagrange equation equal to the remaining terms in the differential equation

$$-\frac{\partial F}{\partial T} = PT + Q \tag{a}$$

Substituting the expression for F just obtained into $-\partial F/\partial T$ gives

$$-\frac{\partial F}{\partial T} = -\frac{\partial C}{\partial T} \tag{b}$$

Now equating (a) and (b)

$$-\frac{\partial C}{\partial T} = PT + Q$$

Integrating with respect to T yields

$$C = -\frac{1}{2} PT^2 - QT$$

The complete integrand is thus

$$F = \frac{1}{2} K\left(\frac{\partial T}{\partial x}\right)^2 + \frac{1}{2} K\left(\frac{\partial T}{\partial y}\right)^2 + \frac{1}{2} K\left(\frac{\partial T}{\partial z}\right)^2 - \frac{1}{2} PT^2 - QT$$

This is the general integrand for three-dimensional field problems.

The derivative boundary condition term (9.19) in the Euler–Lagrange equation is also equated to the derivative boundary condition (9.13):

$$\frac{\partial F}{\partial T_x} n_x + \frac{\partial F}{\partial T_y} n_y + \frac{\partial F}{\partial T_z} n_z + \frac{\partial G}{\partial T} = KT_x n_x + KT_y n_y + KT_z n_z + \alpha T + \beta$$

Most of the terms cancel out because the form for F has already been determined. The remaining terms are

$$\frac{\partial G}{\partial T} = \alpha T + \beta$$

Integrating with respect to T gives

$$G = \frac{1}{2} \alpha T^2 + \beta T$$

This completes the derivative boundary condition term.

Table 9.2
General Boundary Value Problem and Functionals for Two- and Three-Dimensional Field Problems

Dimensions	General Boundary Value Problem	General Functional	
Two	$\frac{\partial}{\partial x}\left(K_x \frac{\partial T}{\partial x}\right) + \frac{\partial}{\partial y}\left(K_y \frac{\partial T}{\partial y}\right) + PT + Q = 0$ in A $T = T_0$ on L_1 $K_x \frac{\partial T}{\partial x} n_x + K_y \frac{\partial T}{\partial y} n_y + \alpha T + \beta = 0$ on L_2	$\iint_A \left[\frac{1}{2} K_x \left(\frac{\partial T}{\partial x}\right)^2 + \frac{1}{2} K_y \left(\frac{\partial T}{\partial y}\right)^2 - \frac{1}{2} PT^2 - QT\right] dA + \int_{L_2} \left[\frac{1}{2}\alpha T^2 + \beta T\right] dL$	**(9.21)**
Three	$\frac{\partial}{\partial x}\left(K_x \frac{\partial T}{\partial x}\right) + \frac{\partial}{\partial y}\left(K_y \frac{\partial T}{\partial y}\right) + \frac{\partial}{\partial z}\left(K_z \frac{\partial T}{\partial z}\right) + PT + Q = 0$ in V $T = T_0$ on S_1 $K_x \frac{\partial T}{\partial x} n_x + K_y \frac{\partial T}{\partial y} n_y + K_z \frac{\partial T}{\partial z} n_z + \alpha T + \beta = 0$ on S_2	$\iiint_V \left[\frac{1}{2} K_x \left(\frac{\partial T}{\partial x}\right)^2 + \frac{1}{2} K_y \left(\frac{\partial T}{\partial y}\right)^2 + \frac{1}{2} K_z \left(\frac{\partial T}{\partial z}\right)^2 - \frac{1}{2} PT^2 - QT\right] dV + \iint_{S_2} \left[\frac{1}{2}\alpha T^2 + \beta T\right] dS$	**(9.22)**

Now the complete general functional is

$$I(T) = \iiint_V \left[\frac{1}{2} K \left(\frac{\partial T}{\partial x}\right)^2 + \frac{1}{2} K \left(\frac{\partial T}{\partial y}\right)^2 + \frac{1}{2} K \left(\frac{\partial T}{\partial z}\right)^2 - \frac{1}{2} PT^2 - QT\right] dV + \iint_{S_2} \left[\frac{1}{2}\alpha T^2 + \beta T\right] dS \qquad (9.20)$$

This corresponds to the general boundary value problem given at the beginning of this section.

The previous result can easily be made a bit more general to include three different coefficient functions, $K_x = K_x(x,y,z)$, $K_y = K_y(x,y,z)$, and $K_z = K_z(x,y,z)$. Table 9.2 gives the general boundary value problems and associated functionals for two-dimensional and three-dimensional field problems. These are used extensively in this chapter.

9.4 Element Functional for Simplex Two-Dimensional Element

This section derives the element functional for a two-dimensional simplex element. A more general derivation for two-dimensional elements is carried out later in the chapter. Finally, three-dimensional elements are discussed at the end of the chapter.

A. General Form

Steps similar to those in Section 5.5 are followed to determine the element contribution to the functional. The functional is broken up into element-size integrals $I = I^{(1)} + I^{(2)} + I^{(3)} + \ldots + I^{(E)}$ for E elements. For the eth element

$$I^{(e)} = \iint_{A^{(e)}} \left[\frac{1}{2} K_x \left(\frac{\partial T}{\partial x}\right)^2 + \frac{1}{2} K_y \left(\frac{\partial T}{\partial y}\right)^2 - \frac{1}{2} PT^2 - QT\right]^{(e)} dA^{(e)} + \int_{L_2^{(e)}} \left(\frac{1}{2} \alpha T^2 + \beta T\right)^{(e)} dL_2 \tag{9.23}$$

where the second integral is used only along the side or sides of the element that lie along L_2. This is most conveniently written out term by term as

$$\begin{aligned} I^{(e)} = {} & \underbrace{\iint_{A^{(e)}} \left[\frac{1}{2}\left(\frac{\partial T}{\partial x}\right) K_x \left(\frac{\partial T}{\partial x}\right)\right]^{(e)} dA}_{\textcircled{K_x}\ \text{term}} + \underbrace{\iint_{A^{(e)}} \left[\frac{1}{2}\left(\frac{\partial T}{\partial y}\right) K_y \left(\frac{\partial T}{\partial y}\right)\right]^{(e)} dA}_{\textcircled{K_y}\ \text{term}} \\ & - \underbrace{\iint_{A^{(e)}} \left[\frac{1}{2} TPT\right]^{(e)} dA}_{\textcircled{P}\ \text{term}} - \underbrace{\iint_{A^{(e)}} [TQ]^{(e)} dA}_{\textcircled{Q}\ \text{term}} \\ & + \underbrace{\int_{L_2^{(e)}} \left[\frac{1}{2} T \alpha T\right]^{(e)} dL_2}_{\textcircled{\alpha}\ \text{term}} + \underbrace{\int_{L_2^{(e)}} [T\beta]^{(e)} dL_2}_{\textcircled{\beta}\ \text{term}} \end{aligned} \tag{9.24}$$

In this section, the circled terms denote the contribution to the element functional. The integrands in each term have been reordered for matrix calculations. Each term is now evaluated for a simplex two-dimensional element.

B. Simplex Two-Dimensional Element

Within each element (shown in Figure 9.5), the interpolation function is given by (3.19): $T^{(e)} = \lfloor N \rfloor \{T\}$, where for a simplex element

$$\lfloor N \rfloor = \lfloor N_i \quad N_j \quad N_k \rfloor, \quad \{T\} = \begin{Bmatrix} T_i \\ T_j \\ T_k \end{Bmatrix}$$

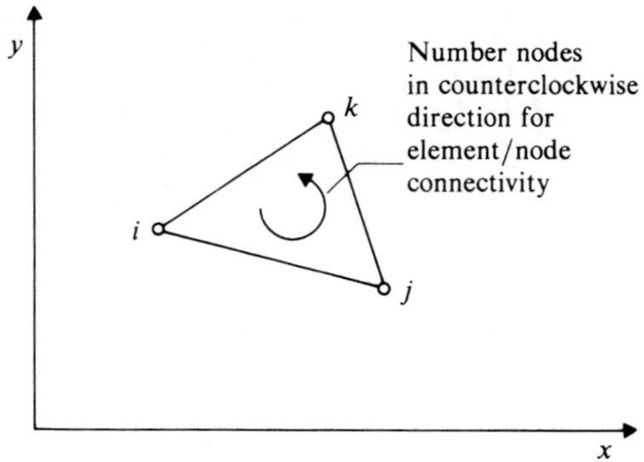

Figure 9.5 Two-dimensional simplex element

This is easily written out as

$$T^{(e)} = \{N\}\,\{T\} = \{N_i \quad N_j \quad N_k\} \begin{Bmatrix} T_i \\ T_j \\ T_k \end{Bmatrix}$$

$$= N_i T_i + N_j T_j + N_k T_k$$

Also the transpose form of this will be used in the following derivations: $T^{(e)} = \{T\}^T\,\{N\}^T$. It is left to the reader to verify that the result is the same.

The derivatives are given by

$$\frac{\partial T^{(e)}}{\partial x} = \{D_x\}\{T\}$$

$$\frac{\partial T^{(e)}}{\partial y} = \{D_y\}\{T\}$$

where the derivative rows are

$$\{D_x\} = \underbrace{\frac{1}{2A}\{b_i \quad b_j \quad b_k\}}_{x \text{ derivative column } (1\times 3)}$$

$$\{D_y\} = \underbrace{\frac{1}{2A}\{c_i \quad c_j \quad c_k\}}_{y \text{ derivative column } (1\times 3)}$$

Recall that the derivatives are simply

$$\frac{\partial T^{(e)}}{\partial x} = \{D_x\}\{T\} = \frac{1}{2A}\{b_i \quad b_j \quad b_k\} \begin{Bmatrix} T_i \\ T_j \\ T_k \end{Bmatrix}$$

$$= \frac{1}{2A}(b_i T_i + b_j T_j + b_k T_k)$$

$$\frac{\partial T^{(e)}}{\partial y} = \{D_y\}\{T\} = \frac{1}{2A}\{c_i \quad c_j \quad c_k\}\begin{Bmatrix} T_i \\ T_j \\ T_k \end{Bmatrix}$$

$$= \frac{1}{2A}(c_iT_i + c_jT_j + c_kT_k)$$

The transposes of these expressions are

$$\frac{\partial T^{(e)}}{\partial x} = \{T\}^T\{D_x\}^T, \qquad \frac{\partial T^{(e)}}{\partial y} = \{T\}^T\{D_y\}^T$$

Again the reader can verify these.

Let all of the coefficient functions K_x, K_y, P, Q, α, β be constant over the element. The element functional can now be obtained.

K_x Term

The K_x term from the element functional I is

$$\textcircled{K_x} = \iint_{A^{(e)}} \left[\frac{1}{2}\frac{\partial T}{\partial x} K_x \frac{\partial T}{\partial x}\right]^{(e)} dA$$

A circle denotes the contribution of the K_x term to the functional. Substituting the relation for $\partial T/\partial x$ gives

$$\textcircled{K_x} = \iint_{A^{(e)}} \left[\frac{1}{2}\{T\}^T\{D_x\}^T K_x \{D_x\}\{T\}\right]^{(e)} dA$$

The individual terms can be written out as

$$\textcircled{K_x} = \iint_{A^{(e)}} \left[\frac{1}{8A^2}(b_iT_i + b_jT_j + b_kT_k) K_x (b_iT_i + b_jT_j + b_kT_k)\right]^{(e)} dA$$

It is not evaluated further at this point.

K_y Term

This is evaluated in a manner very similar to that used for the K_x term.

$$\textcircled{K_y} = \iint_{A^{(e)}} \left[\frac{1}{2}\frac{\partial T}{\partial y} K_y \frac{\partial T}{\partial y}\right]^{(e)} dA$$

Substituting for $\partial T/\partial y$ yields

$$\textcircled{K_y} = \iint_{A^{(e)}} \left[\frac{1}{2}\{T\}^T\{D_y\}^T K_y \{D_y\}\{T\}\right]^{(e)} dA$$

which can be written out as

$$\textcircled{K_y} = \iint_{A^{(e)}} \left[\frac{1}{8A^2}(c_iT_i + c_jT_j + c_kT_k) K_y (c_iT_i + c_jT_j + c_kT_k)\right]^{(e)} dA$$

P Term

$$\textcircled{P} = -\iint_{A^{(e)}} \left[\frac{1}{2}\, T\, P\, T\right]^{(e)} dA$$

$$= -\iint_{A^{(e)}} \left[\frac{1}{2}\, \{T\}^T\, \{N\}^T\, P\, \{N\}\, \{T\}\right]^{(e)} dA$$

Individual terms can be written out as

$$\textcircled{P} = -\iint_{A^{(e)}} \left[\frac{1}{2}\, (N_i T_i + N_j T_j + N_k T_k) P \right.$$

$$\left. (N_i T_i + N_j T_j + N_k T_k)\right]^{(e)} dA$$

Q Term

$$\textcircled{Q} = -\iint_{A^{(e)}} \left[T\, Q\right]^{(e)} dA = -\iint_{A^{(e)}} \left[\{T\}^T\, \{N\}^T\, Q\right]^{(e)} dA$$

which is

$$\textcircled{Q} = -\iint_{A^{(e)}} \left[(N_i T_i + N_j T_j + N_k T_k) Q\right]^{(e)} dA$$

Now for the terms along the derivative boundary L_2.

α Term

$$\textcircled{\alpha} = \int_{L_2^{(e)}} \left[\frac{1}{2}\, T\, \alpha\, T\right] dL_2 = \int_{L_2^{(e)}} \left[\frac{1}{2}\, \{T\}^T\, \{N\}^T\, \alpha\, \{N\}\, \{T\}\right]^{(e)} dL_2$$

β Term

$$\textcircled{\beta} = \int_{L_2^{(e)}} \left[T\, \beta\right] dL_2 = \int_{L_2^{(e)}} \left[\{T\}^T\, \{N\}^T\, \beta\right]^{(e)} dL_2$$

The results from all of these terms are summed together in Figure 9.6.

9.5 Element Matrices for Simplex Two-Dimensional Element

This section obtains the element matrix $[B]$ and column $\{C\}$ for a simplex two-dimensional element. The element matrices are written out in two different formats for various uses.

The key to developing finite element analysis of a problem is the element coefficient matrix and column. It is very convenient to have completely general element matrices that can be reduced to specific problem requirements. In specific areas, the matrices have appropriate names, indicated in Table 9.3, descriptive of physical quantities.

$$I^{(e)} = \iint_{A^{(e)}} \Big[\underbrace{\frac{1}{2}\{T\}^T[D_x]^T K_x [D_x]\{T\}}_{\text{(}K_x\text{) term}} + \underbrace{\frac{1}{2}\{T\}^T[D_y]^T K_y [D_y]\{T\}}_{\text{(}K_y\text{) term}}$$

$$\underbrace{-\frac{1}{2}\{T\}^T\{N\}^T P [N]\{T\}}_{\text{(}P\text{) term}} - \underbrace{\{T\}^T\{N\}^T Q\Big]^{(e)} dA}_{\text{(}Q\text{) term}}$$

$$+ \int_{L_2^{(e)}} \Big[\underbrace{\frac{1}{2}\{T\}^T\{N\}^T\alpha\, [N]\{T\}}_{\text{(}\alpha\text{) term}} + \underbrace{\{T\}^T\{N\}^T\beta\Big]^{(e)} dL_2}_{\text{(}\beta\text{) term}}$$

Figure 9.6 Element functional for simplex element

Table 9.3
Element Matrices

Problem Area	Element Matrix [B]	Element Column {C}
Heat transfer	Element conduction matrix	Element source column
Fluid mechanics	Element fluidity matrix	Element source column
Solid mechanics	Element stiffness matrix	Element force column

A. Minimization of Functional

Differentiating the element functional with respect to the nodal values of T gives

$$\begin{Bmatrix} \dfrac{\partial I}{\partial T_i} \\ \dfrac{\partial I}{\partial T_j} \\ \dfrac{\partial I}{\partial T_k} \end{Bmatrix} = \iint_{A^{(e)}} \Big[[D_x]^T K_x [D_x]\{T\} + [D_y]^T K_y [D_y]\{T\} - \{N\}^T P[N]\{T\} - \{N\}^T Q \Big]^{(e)} dA + \int_{L_2^{(e)}} \Big[\{N\}^T \alpha [N]^T\{T\} + \{N\}^T\beta \Big]^{(e)} dL_2 \tag{9.25}$$

This leads to the minimization equations, as discussed in detail in Chapter 5.

The process of differentiation is illustrated, using the K_x term as an example.

$$\frac{\partial}{\partial T_i} \textcircled{K_x} = \frac{\partial}{\partial T_i} \iint_{A^{(e)}} \left[\frac{1}{2} \{T\}^T \{D_x\}^T K_x \{D_x\}\{T\} \right]^{(e)} dA$$

$$= \frac{\partial}{\partial T_i} \iint_{A^{(e)}} \left[\frac{1}{8A^2} (b_i T_i + b_j T_j + b_k T_k) K_x \right.$$

$$\left. (b_i T_i + b_j T_j + b_k T_k) \right]^{(e)} dA$$

$$= \iint_{A^{(e)}} \left[\frac{1}{8A^2} (b_i) K_x (b_i T_i + b_j T_j + b_k T_k) \right.$$

$$\left. + \frac{1}{8A^2} (b_i T_i + b_j T_j + b_k T_k) K_x (b_i) \right]^{(e)} dA$$

$$= \iint_{A^{(e)}} \left[\frac{1}{4A^2} (b_i) K_x (b_i T_i + b_j T_j + b_k T_k) \right]^{(e)} dA$$

Writing this in matrix form yields

$$\frac{\partial}{\partial T_i} \textcircled{K_x} = \iint_{A^{(e)}} \left[\frac{b_i}{4A^2} K_x \{D_x\}\{T\} \right]^{(e)} dA$$

Other terms can be modified by the reader.

B. Element Matrices

The element matrices are obtained from the relation

$$\begin{Bmatrix} \dfrac{\partial I}{\partial T_i} \\ \dfrac{\partial I}{\partial T_j} \\ \dfrac{\partial I}{\partial T_k} \end{Bmatrix}^{(e)} = [B]^{(e)} \{T\}^{(e)} - \{C\}^{(e)}$$

Comparison with (9.25) shows that the element matrix is

$$[B]^{(e)} = \underbrace{\iint_{A^{(e)}} \left[\{D_x\}^T K_x \{D_x\} \right]^{(e)} dA}_{\boxed{K_x} \text{ term}} + \underbrace{\iint_{A^{(e)}} \left[\{D_y\}^T K_y \{D_y\} \right]^{(e)} dA}_{\boxed{K_y} \text{ term}}$$

$$- \underbrace{\iint_{A^{(e)}} \left[\{N\}^T P \{N\} \right]^{(e)} dA}_{\boxed{P} \text{ term}} + \underbrace{\int_{L_2^{(e)}} \left[\{N\}^T \alpha \{N\} \right]^{(e)} dL_2}_{\boxed{\alpha} \text{ term}} \quad \textbf{(9.26)}$$

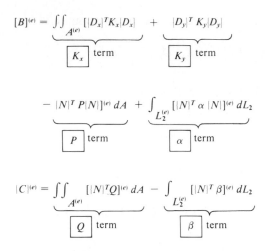

$$[B]^{(e)} = \underbrace{\iint_{A^{(e)}} [\{D_x\}^T K_x \{D_x\}}_{\boxed{K_x}\text{ term}} + \underbrace{\{D_y\}^T K_y \{D_y\}}_{\boxed{K_y}\text{ term}}$$

$$\underbrace{- \{N\}^T P \{N\}]^{(e)}\, dA}_{\boxed{P}\text{ term}} + \underbrace{\int_{L_2^{(e)}} [\{N\}^T \alpha \{N\}]^{(e)}\, dL_2}_{\boxed{\alpha}\text{ term}}$$

$$\{C\}^{(e)} = \underbrace{\iint_{A^{(e)}} [\{N\}^T Q]^{(e)}\, dA}_{\boxed{Q}\text{ term}} - \underbrace{\int_{L_2^{(e)}} [\{N\}^T \beta]^{(e)}\, dL_2}_{\boxed{\beta}\text{ term}}$$

Figure 9.7 Element matrices

where all terms containing $\{T\}$ in (9.25) have been included. In this section, a square denotes the contribution of that term to the element matrix. The element column is

$$\{C\}^{(e)} = \underbrace{\iint_{A^{(e)}} \left[\{N\}^T Q\right]^{(e)} dA}_{\boxed{Q}\text{ term}} + \underbrace{\int_{L_2^{(e)}} \left[\{N\}^T \beta\right]^{(e)} dL_2}_{\boxed{\beta}\text{ term}} \tag{9.27}$$

where terms not containing $\{T\}$ are the ones left in (9.25). Figure 9.7 rewrites the element matrices.

C. Evaluation of [*B*]

For a simplex two-dimensional element, the element matrix can be determined easily. All of the integrals can be evaluated by the methods from Sections 3.2 and 3.3. From (9.26)

K_x *Term*

$$\boxed{K_x} = \iint_{A^{(e)}} \left[\{D_x\}^T K_x \{D_x\}\right]^{(e)} dA$$

The square denotes the contribution to the element matrices

$$\boxed{K_x} = \iint_{A^{(e)}} \left[\frac{K_x}{4A^2} \begin{Bmatrix} b_i \\ b_j \\ b_k \end{Bmatrix} \{b_i \quad b_j \quad b_k\}\right]^{(e)} dA$$

Multiplying out the column and row gives

$$\boxed{K_x} = \left(\frac{K_x}{4A^2}\right)^{(e)} \begin{bmatrix} \iint b_ib_i\,dA & \iint b_ib_j\,dA & \iint b_ib_k\,dA \\ \iint b_jb_i\,dA & \iint b_jb_j\,dA & \iint b_jb_k\,dA \\ \iint b_kb_i\,dA & \iint b_kb_j\,dA & \iint b_kb_k\,dA \end{bmatrix}^{(e)}$$

All integrand terms are constant, so the result is

$$\boxed{K_x} = \left(\frac{K_x}{4A}\right)^{(e)} \begin{bmatrix} b_ib_i & b_ib_j & b_ib_k \\ b_jb_i & b_jb_j & b_jb_k \\ b_kb_i & b_kb_j & b_kb_k \end{bmatrix}^{(e)}$$

K_y Term

The K_y term is evaluated as

$$\boxed{K_y} = \left(\frac{K_y}{4A}\right)^{(e)} \begin{bmatrix} c_ic_i & c_ic_j & c_ic_k \\ c_jc_i & c_jc_j & c_jc_k \\ c_kc_i & c_kc_j & c_kc_k \end{bmatrix}^{(e)}$$

in a similar manner.

P Term

$$\boxed{P} = -\iint_{A^{(e)}} \left[\{N\}^T P \,\lfloor N\rfloor\right]^{(e)} dA$$

$$= -\iint_{A^{(e)}} \left[P \begin{Bmatrix} N_i \\ N_j \\ N_k \end{Bmatrix} \lfloor N_i \quad N_j \quad N_k\rfloor\right]^{(e)} dA$$

Again, multiplying out the column and row yields

$$\boxed{P} = -(P)^{(e)} \begin{bmatrix} \iint N_iN_i\,dA & \iint N_iN_j\,dA & \iint N_iN_k\,dA \\ \iint N_jN_i\,dA & \iint N_jN_j\,dA & \iint N_jN_k\,dA \\ \iint N_kN_i\,dA & \iint N_kN_j\,dA & \iint N_kN_k\,dA \end{bmatrix}^{(e)}$$

All of the integrals have the form

$$\iint_{A^{(e)}} N_i^2\,dA = \frac{1}{6}A^{(e)}$$

$$\iint_{A^{(e)}} N_iN_j\,dA = \frac{1}{12}A^{(e)}$$

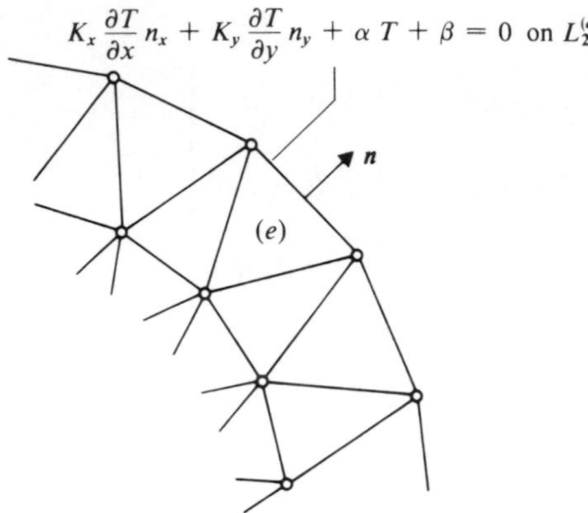

(a) Derivative boundary condition on one side of element

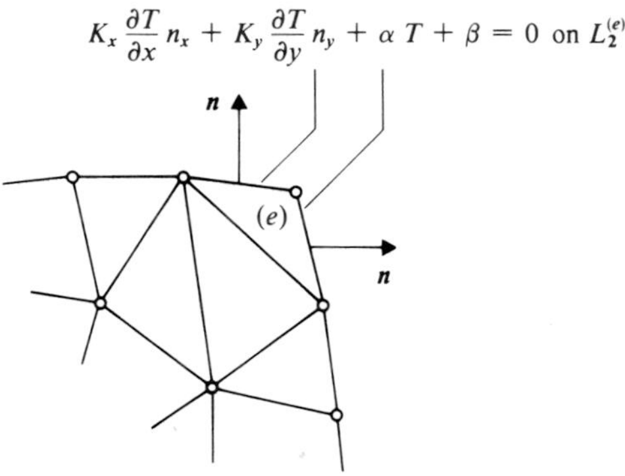

(b) Derivative boundary condition on two sides of element

Figure 9.8 Element derivative boundary conditions

and so on, evaluated by using (3.27). The result is

$$\boxed{P} = -\left(\frac{PA}{12}\right)^{(e)} \begin{bmatrix} 2 & 1 & 1 \\ 1 & 2 & 1 \\ 1 & 1 & 2 \end{bmatrix}^{(e)}$$

α Term

The α term is evaluated along the side of an element rather than in the interior. If a side of the element is located on the derivative boundary L_2, then the α term is evaluated. An element can have a derivative boundary condition on one side, as shown in Figure 9.8(a), or on two sides, as illustrated in Figure 9.8(b). In a simplex element, no more than two sides of any single element would be on L_2. If no derivative boundary conditions occur in the element, the α term is not added to the element matrix.

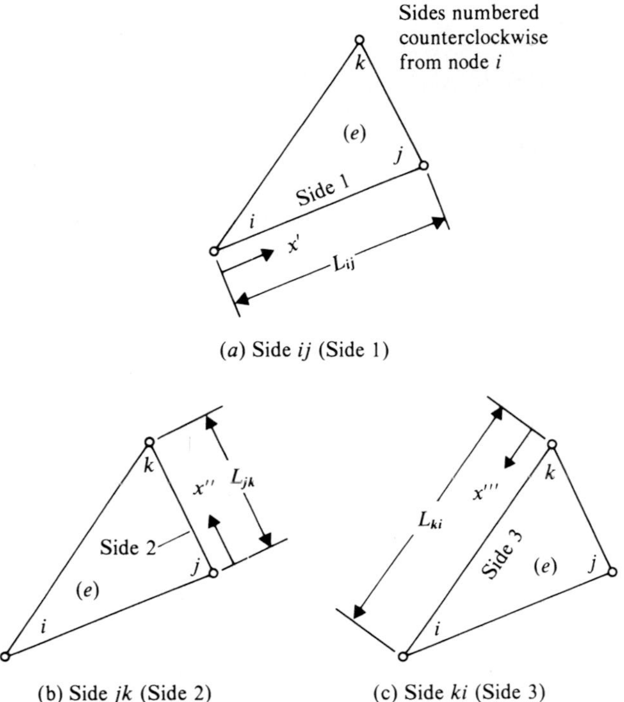

Figure 9.9 Element sides for derivative boundary conditions

Element side lengths are easily calculated from the nodal coordinates as

$$
\begin{aligned}
L_{ij} &= [(x_j - x_i)^2 + (y_j - y_i)^2]^{1/2} \qquad \text{(Side 1)}\\
L_{jk} &= [(x_k - x_j)^2 + (y_k - y_j)^2]^{1/2} \qquad \text{(Side 2)}\\
L_{ki} &= [(x_i - x_k)^2 + (y_i - y_k)^2]^{1/2} \qquad \text{(Side 3)}
\end{aligned}
\tag{9.28}
$$

Figure 9.9 illustrates the geometry. Again, these need to be calculated only if derivative boundary conditions occur on the edge of the element.

Writing out the α term gives

$$
\boxed{\alpha} = \int_{L_2^{(e)}} \left[\{N\}^T \alpha \, \{N\} \right]^{(e)} dL_2
$$

$$
= \int_{L_2^{(e)}} \left[\alpha \begin{Bmatrix} N_i \\ N_j \\ N_k \end{Bmatrix} \{N_i \quad N_j \quad N_k\} \right]^{(e)} dL_2
$$

All integral terms have the form

$$\int_{L_2^{(e)}} N_i^2 \, dL_2, \quad \int_{L_2^{(e)}} N_i \, N_j \, dL_2$$

Here the pyramid functions are functions of x, y: $N_i = N_i(x,y)$, $N_j = N_j(x,y)$, and $N_k = N_k(x,y)$.

Consider side ij (side 1) first. Let x' represent the coordinate along the line from i to j, as shown in Figure 9.9(a). The interpolation functions now reduce to the one-dimensional form.

$$T^{(e)}{}_{\text{on side } ij} = N_i(x') \, T_i + N_j(x') \, T_j$$

where

$$N_i(x') = \frac{1}{L_{ij}} (L_{ij} - x')$$

$$N_j(x') = \frac{1}{L_{ij}} (x')$$

Thus the α term on side ij becomes

$$\boxed{\alpha}_{\text{ on side } ij} = (\alpha)^{(e)} \begin{bmatrix} \int N_i^2 \, dL_2 & \int N_i N_j \, dL_2 & 0 \\ \int N_i N_i \, dL_2 & \int N_j^2 \, dL_2 & 0 \\ 0 & 0 & 0 \end{bmatrix}^{(e)}$$

The zeros occur on the row and column associated with node k, which is not involved in side ij (side 1). The integrals have the value (3.12)

$$\int_{L_{ij}^{(e)}} N_i^2 \, (x') \, dL_2 = \int_{L_{ij}^{(e)}} N_j^2 \, (x') \, dL_2 = \frac{1}{3} L_{ij}$$

$$\int_{L_{ij}^{(e)}} N_i(x') N_j(x') \, dL_2 = \frac{1}{6} L_{ij}$$

with the final result

$$\boxed{\alpha}_{\text{ on side } ij} = \left(\frac{L_{ij}}{6}\right)^{(e)} \begin{bmatrix} 2 & 1 & 0 \\ 1 & 2 & 0 \\ 0 & 0 & 0 \end{bmatrix}$$

The other two sides are obtained in a similar fashion.

Along side jk (side 2), shown in Figure 9.9(b), the interpolation function is

$$T^{(e)}{}_{\text{on side } jk} = N_j(x'') T_j + N_k(x'') \, T_k$$

where the x'' represents the coordinate along side 2.

$$N_j(x'') = \frac{1}{L_{jk}} (L_{jk} - x'')$$

$$[B]^{(e)} = \frac{K_x}{4A}\begin{bmatrix} b_ib_i & b_ib_j & b_ib_k \\ b_ib_j & b_jb_j & b_jb_k \\ b_ib_k & b_jb_k & b_kb_k \end{bmatrix}^{(e)} \qquad (K_x \text{ matrix})$$

$$+ \frac{K_y}{4A}\begin{bmatrix} c_ic_i & c_ic_j & c_ic_k \\ c_ic_j & c_jc_j & c_jc_k \\ c_ic_k & c_jc_k & c_kc_k \end{bmatrix}^{(e)} \qquad (K_y \text{ matrix})$$

$$- \frac{PA}{12}\begin{bmatrix} 2 & 1 & 1 \\ 1 & 2 & 1 \\ 1 & 1 & 2 \end{bmatrix}^{(e)} \qquad (P \text{ matrix})$$

$$+ \frac{\alpha L_{ij}}{6}\begin{bmatrix} 2 & 1 & 0 \\ 1 & 2 & 0 \\ 0 & 0 & 0 \end{bmatrix}^{(e)}_{\substack{\text{On side } ij \\ \text{(Side 1)}}} \qquad (\alpha \text{ matrix } ij)$$

$$+ \frac{\alpha L_{jk}}{6}\begin{bmatrix} 0 & 0 & 0 \\ 0 & 2 & 1 \\ 0 & 1 & 2 \end{bmatrix}^{(e)}_{\substack{\text{On side } jk \\ \text{(Side 2)}}} \qquad (\alpha \text{ matrix } jk)$$

$$+ \frac{\alpha L_{ki}}{6}\begin{bmatrix} 2 & 0 & 1 \\ 0 & 0 & 0 \\ 1 & 0 & 2 \end{bmatrix}^{(e)}_{\substack{\text{On side } ki \\ \text{(Side 3)}}} \qquad (\alpha \text{ matrix } ki)$$

Figure 9.10 Element matrix for simplex two-dimensional element, shown in matrix format

$$N_k(x'') = \frac{1}{L_{jk}}(x'')$$

The rest of the analysis follows the one for side ij. Side ki can be carried out by the reader if desired.

All of the contributions to the element matrix $[B]$ are added together. Figure 9.10 gives the result, illustrating the contribution of each term. This is called the matrix format, where the contribution of each term is written as a three by three matrix.

D. Evaluation of $\{C\}$

The terms in the element column are easily evaluated by using the same techniques just employed.

Q Term

From (9.27), the Q term is

$$\boxed{Q} = \iint_{A^{(e)}} \left[\{N\}^T Q \right]^{(e)} dA$$

Writing out the terms gives

$$\boxed{Q} = (Q)^{(e)} \begin{Bmatrix} \iint N_i \, dA \\ \iint N_j \, dA \\ \iint N_k \, dA \end{Bmatrix}^{(e)}$$

For the two-dimensional simplex element, this becomes

$$\boxed{Q} = \left(\frac{1}{3} QA \right)^{(e)} \begin{Bmatrix} 1 \\ 1 \\ 1 \end{Bmatrix}^{(e)}$$

β Term

$$\boxed{\beta} = -\int_{L_2^{(e)}} \left[\{N\}^T \beta \right]^{(e)} dL_2$$

The column integrals are

$$\boxed{\beta} = -(\beta)^{(e)} \begin{Bmatrix} \int N_i \, dL_2 \\ \int N_j \, dL_2 \\ \int N_k \, dL_2 \end{Bmatrix}^{(e)}$$

These integrals have the value

$$\int_{L_2^{(e)}} N(x') \, dL_2 = \frac{1}{2} L_2^{(e)}$$

along any particular side, as found by the same method as that employed for the α term. On side ij (side 1), the β term becomes

$$\boxed{\beta}_{\text{on side } ij} = -\frac{1}{2} \beta L_{ij} \begin{Bmatrix} 1 \\ 1 \\ 0 \end{Bmatrix}^{(e)}$$

The other two sides are easily evaluated.

Both Q and β terms are added to give the element $\{C\}$ column. Figure 9.11 shows the full element column.

$$\{C\}^{(e)} = \frac{QA}{3}\begin{Bmatrix}1\\1\\1\end{Bmatrix}^{(e)} \qquad (Q \text{ column})$$

$$-\frac{\beta L_{ij}}{2}\begin{Bmatrix}1\\1\\0\end{Bmatrix}^{(e)} \text{On side } ij \text{ (Side 1)} \qquad (\beta \text{ column } ij)$$

$$-\frac{\beta L_{jk}}{2}\begin{Bmatrix}0\\1\\1\end{Bmatrix}^{(e)} \text{On side } jk \text{ (Side 2)} \qquad (\beta \text{ column } jk)$$

$$-\frac{\beta L_{ki}}{2}\begin{Bmatrix}1\\0\\1\end{Bmatrix}^{(e)} \text{On side } ki \text{ (Side 3)} \qquad (\beta \text{ column } ki)$$

Figure 9.11 Element column for simplex two-dimensional element, shown in matrix format

These element matrices and columns are the basis for the solution of two-dimensional field problems. They are assembled into the global matrices and vectors by the same techniques developed earlier. By including the proper terms in the problem interior and along the boundary, a large variety of problems in heat transfer, fluid mechanics, and solid mechanics can be solved.

E. Term-by-Term Format

The element matrices have just been presented in matrix format. An alternative format, called here the term-by-term format, is more convenient for actual use in a computer program such as FINTWO. The element matrices are

$$[B]^{(e)} = \begin{matrix} & i & j & k \\ i \\ j \\ k \end{matrix} \underbrace{\begin{bmatrix} B_{ii} & B_{ij} & B_{ik} \\ B_{ji} & B_{jj} & B_{jk} \\ B_{ki} & B_{kj} & B_{kk} \end{bmatrix}^{(e)}}_{\text{Element matrix}}$$

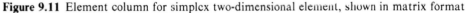

$$\{C\}^{(e)} = \begin{matrix} i \\ j \\ k \end{matrix} \underbrace{\begin{Bmatrix} C_i \\ C_j \\ C_k \end{Bmatrix}^{(e)}}_{\text{Element column}}$$

Each B component is obtained from Equation (9.29) as

$$B_{ii} = \left[\frac{K}{4A}(b_i^2 + c_i^2) - \frac{1}{6}PA + \left(\frac{1}{3}\alpha L_{ij}\right)_{\text{on side } ij} + \left(\frac{1}{3}\alpha L_{ki}\right)_{\text{on side } ki}\right]^{(e)}$$

$$B_{ij} = \left[\frac{K}{4A}(b_ib_j + c_ic_j) - \frac{1}{12}PA + \left(\frac{1}{6}\alpha L_{ij}\right)_{\text{on side } ij}\right]^{(e)}$$

$$B_{ik} = \left[\frac{K}{4A}(b_ib_k + c_ic_k) - \frac{1}{12}PA + \left(\frac{1}{6}\alpha L_{ki}\right)_{\text{on side } ki}\right]^{(e)}$$

$$B_{jj} = \left[\frac{K}{4A}(b_j^2 + c_j^2) - \frac{1}{6}PA + \left(\frac{1}{3}\alpha_{jk}\right)_{\text{on side } jk} + \left(\frac{1}{3}\alpha L_{ij}\right)_{\text{on side } ij}\right]^{(e)} \tag{9.31}$$

$$B_{jk} = \left[\frac{K}{4A}(b_jb_k + c_jc_k) - \frac{1}{12}PA + \left(\frac{1}{6}\alpha L_{jk}\right)_{\text{on side } jk}\right]^{(e)}$$

$$B_{kk} = \left[\frac{K}{4A}(b_k^2 + c_k^2) - \frac{1}{6}PA + \left(\frac{1}{3}\alpha L_{ki}\right)_{\text{on side } ki} + \left(\frac{1}{3}\alpha L_{jk}\right)_{\text{on side } jk}\right]^{(e)}$$

The symmetry of the element coefficient matrix gives the remaining B values. For the column (9.30) yields

$$C_i = \left[\frac{1}{3}QA - \left(\frac{1}{2}\beta L_{ij}\right)_{\text{on side } ij} - \left(\frac{1}{2}\beta L_{ki}\right)_{\text{on side } ki}\right]^{(e)}$$

$$C_j = \left[\frac{1}{3}QA - \left(\frac{1}{2}\beta L_{jk}\right)_{\text{on side } jk} - \left(\frac{1}{2}\beta L_{ij}\right)_{\text{on side } ij}\right]^{(e)} \tag{9.32}$$

$$C_k = \left[\frac{1}{3}QA - \left(\frac{1}{2}\beta L_{ki}\right)_{\text{on side } ki} - \left(\frac{1}{2}\beta L_{jk}\right)_{\text{on side } jk}\right]^{(e)}$$

These results are quite general in form. Note that the terms designated "on side . . ." are used only if the derivative boundary condition is specified along the corresponding side.

Example 9.2 *Derivative Boundary Conditions*

Given An element is taken in the upper left-hand corner of the region in which an ideal incompressible fluid flows around an ellipse (from Example 9.1). Figure 9.12 shows the element connectivity table and illustrates the boundary conditions. Nodal coordinates are $(X_3, Y_3) = (0, 2)$ m, $(X_2, Y_2) = (0, 1)$ m, and $(X_6, Y_6) = (2, 2)$ m, and the velocity is $U_\infty = 100$ m/sec.

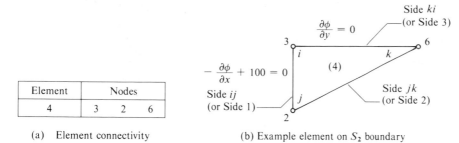

Element	Nodes		
4	3	2	6

(a) Element connectivity

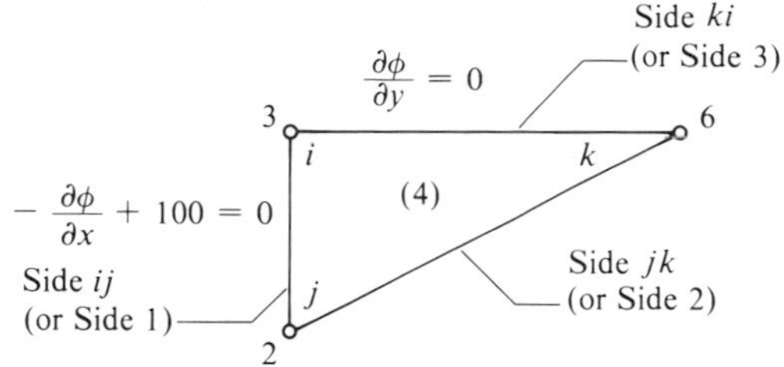

(b) Example element on S_2 boundary

Element	Side	α	β
4	1	0	100
4	3	0	0

(c) Element derivative boundary values

Figure 9.12 Element properties (Example 9.2—Derivative boundary condition)

Objective Find the element fluidity matrix and vector.

Solution From Equation (9.19), the element variational principle is

$$I^{(4)} = \int\int_{A^{(4)}} \left[\frac{1}{2}\left(\frac{\partial \phi}{\partial x}\right)^2 + \frac{1}{2}\left(\frac{\partial \phi}{\partial y}\right)^2 \right] dA^{(4)} + \int_{\text{side } ij} (U_\infty T)\, dy$$

where $K = 1$, $P = 0$, and $Q = 0$. Along side ij (side 1), $\alpha = 0$, $\beta = 100$, and along side ki (side 3), $\alpha = 0$, $\beta = 0$. In this case, no term is added for side ki because both α and β are zero. These values are given in the lower table in Figure 9.12.

The element properties yield $b_3 = -1$, $c_3 = 2$; $b_2 = 0$, $c_2 = -2$; $b_6 = 1$, $c_6 = 0$; and the area is 1.0. From Equation (9.31)

$$B_{33} = \left[\frac{K}{4A}(b_3^2 + c_3^2) - \frac{1}{6} PA \right]$$

9.6 Element Matrices for General Two-Dimensional Element

One of the major advantages of the finite element technique over other methods, such as the finite difference method, is the easy extension to higher order elements. This section provides the general expressions for the element matrices to do this.

A. Element Properties

Consider an element with nodes labeled $i, j, \ldots, p$. The element interpolation function is

$$T^{(e)} = \lfloor G \rfloor \{T\} \tag{9.33}$$

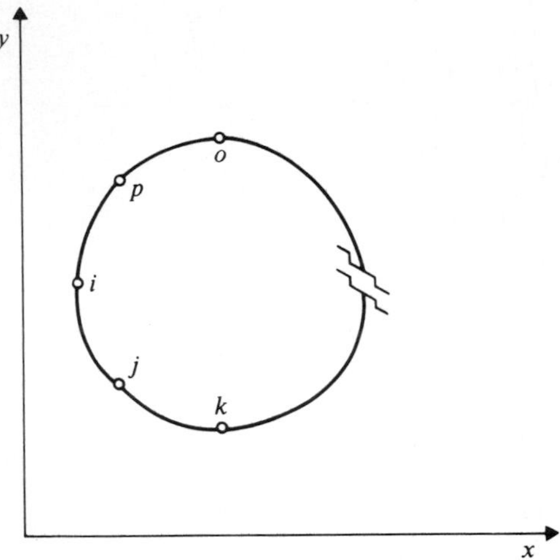

Figure 9.13 General two-dimensional element

where

$$\{G\} = \underbrace{\{G_i \quad G_j \quad \ldots \quad G_p\}}_{\text{Element shape functions}}$$

$$\{T\} = \underbrace{\begin{Bmatrix} T_i \\ T_j \\ \ldots \\ T_p \end{Bmatrix}}_{\text{Nodal values}}$$

Here G is used (rather than N) for the shape functions to distinguish between higher order and simplex elements. Figure 9.13 shows a schematic of a general element. The derivatives are

$$\begin{aligned} \frac{\partial T^{(e)}}{\partial x} &= \{D_x\} = \begin{Bmatrix} \dfrac{\partial G_i}{\partial x} & \dfrac{\partial G_j}{\partial x} & \ldots & \dfrac{\partial G_p}{\partial x} \end{Bmatrix} \\ \frac{\partial T^{(e)}}{\partial y} &= \{D_y\} = \begin{Bmatrix} \dfrac{\partial G_i}{\partial y} & \dfrac{\partial G_j}{\partial y} & \ldots & \dfrac{\partial G_p}{\partial y} \end{Bmatrix} \end{aligned} \quad \textbf{(9.34)}$$

which can be evaluated only when the exact form of G is specified.

B. Element Functional

With these relations, the element functional (9.24) becomes

$$I^{(e)} = \iint_{A^{(e)}} \left[\frac{1}{2} \{T\}^T \{D_x\}^T K_x \{D_x\} \{T\} + \frac{1}{2} \{T\}^T \{D_y\}^T K_y \{D_y\} \{T\} - \frac{1}{2} \{T\}^T \{G\}^T P \{G\} \{T\} - \{T\}^T \{G\}^T Q \right]^{(e)} dA + \int_{L_2^{(e)}} \left[\frac{1}{2} \{T\}^T \{G\}^T \alpha \{G\} \{T\} + \{T\}^T \{G\}^T Q \right]^{(e)} dL_2 \tag{9.35}$$

This is very similar to the simplex two-dimensional case.

C. Element Matrix

Differentiating with respect to the nodal values of T gives

$$\begin{Bmatrix} \dfrac{\partial I}{\partial T_i} \\ \dfrac{\partial I}{\partial T_j} \\ \cdots \\ \dfrac{\partial I}{\partial T_p} \end{Bmatrix}^{(e)} = \iint_{A^{(e)}} \left[\{D_x\}^T K_x \{D_x\} \{T\} + \{D_y\}^T K_y \{D_y\} \{T\} - \{N\}^T P \{N\} \{T\} - \{N\}^T Q \right]^{(e)} dA + \int_{L^{(e)}} \left[\frac{1}{2} \{G\}^T \alpha \{G\} \{T\} + \{G\}^T Q \right]^{(e)} dL_2 \tag{9.36}$$

The element matrix is obtained by comparison with the relation (9.36)

$$\begin{Bmatrix} \dfrac{\partial I}{\partial T_i} \\ \dfrac{\partial I}{\partial T_j} \\ \cdots \\ \dfrac{\partial I}{\partial T_p} \end{Bmatrix}^{(e)} = [B]^{(e)} \{T\}^{(e)} - \{C\}^{(e)}$$

Thus the element matrix is

$$[B]^{(e)} = \iint_{A^{(e)}} \left[\{D_x\}^T K_x \{D_x\} + \{D_y\}^T K_y \{D_y\} - \{G\}^T P \{G\} \right]^{(e)} dA + \int_{L_2^{(e)}} \left[\{G\}^T \alpha \{G\} \right]^{(e)} dL_2 \tag{9.37}$$

and the element column is

$$\{C\}^{(e)} = \iint_{A^{(e)}} \left[\{G\}^T Q \right]^{(e)} dA - \int_{L_2^{(e)}} \left[\{G\}^T \beta \right]^{(e)} dL_2 \tag{9.38}$$

These are the general expressions for the element matrices.

It is worthwhile to evaluate a term or two in detail. Consider the P term

$$\boxed{P} = - \begin{bmatrix} \iint PG_iG_i\, dA & \iint PG_iG_j\, dA & \dots & \iint PG_iG_p\, dA \\ \iint PG_jG_i\, dA & \iint PG_jG_j\, dA & \dots & \iint PG_jG_p\, dA \\ \dots & \dots & & \dots \\ \iint PG_pG_i\, dA & \iint PG_pG_i\, dA & \dots & \iint PG_pG_p\, dA \end{bmatrix}^{(e)}$$

and the Q term

$$\boxed{Q} = \begin{Bmatrix} \iint QG_i\, dA \\ \iint QG_j\, dA \\ \dots \\ \iint QG_p\, dA \end{Bmatrix}$$

Other terms are similar.

An example of a nonsimplex element is examined in Chapter 12. A rectangular element is developed. More advanced examples are shown in Chapter 13 on isoparametric elements.

9.7 Element Matrices in Three Dimensions

It is relatively easy to extend the development of element matrices and columns into three-dimensional problems. Both simplex and general elements are considered.

A. Simplex Three-Dimensional Element

The functional for element e is obtained from Equation (9.22) as

$$\begin{aligned} I^{(e)} = \iiint_{V^{(e)}} \Bigg[& \frac{1}{2} \{T\}^T \{D_x\}^T K_x \{D_x\} \{T\} \\ & + \frac{1}{2} \{T\}^T \{D_y\}^T K_y \{D_y\} \{T\} \\ & + \frac{1}{2} \{T\}^T \{D_z\}^T K_z \{D_z\} \{T\} \\ & - \frac{1}{2} \{T\}^T \{N\}^T P \{N\} \{T\} - \{T\}^T \{N\}^T Q \Bigg]^{(e)} dV \end{aligned} \tag{9.39}$$

Continued on next page

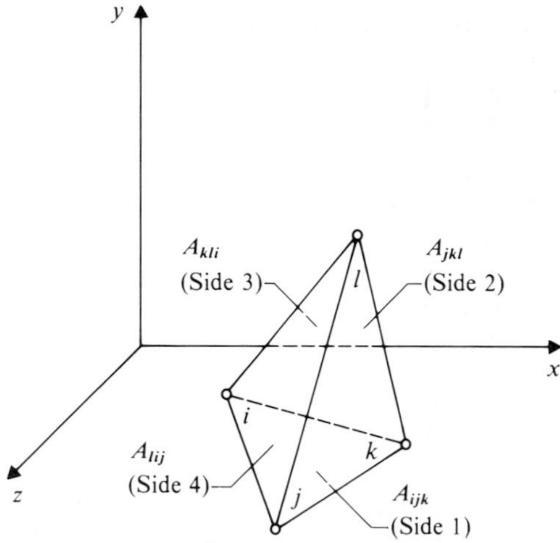

Figure 9.14 Three-dimensional simplex element. Areas of four sides A are used in derivative boundary conditions.

$$+ \iint_{S_2^{(e)}} \left[\frac{1}{2} \{T\}^T [N]^T \alpha [N] \{T\} + \{T\}^T [N]^T \beta \right]^{(e)} dS_2 \qquad \textbf{(9.39 cont.)}$$

Now the first integral is over the element volume V and the second is over the element surface when that surface is located along a derivative boundary S_2. Figure 9.14 shows a schematic.

The interpolation function for a three-dimensional simplex element is $T^{(e)} = [N]\{T\}$ where

$$[N] = [N_i \quad N_j \quad N_k \quad N_\ell], \qquad \{T\} = \begin{Bmatrix} T_i \\ T_j \\ T_k \\ T_\ell \end{Bmatrix}$$

Also, the derivatives are

$$\frac{\partial T^{(e)}}{\partial x} = [D_x] = \frac{1}{6V} [b_i \quad b_j \quad b_k \quad b_\ell]$$

$$\frac{\partial T^{(e)}}{\partial y} = [D_y] = \frac{1}{6V} [c_i \quad c_j \quad c_k \quad c_\ell]$$

$$\frac{\partial T^{(e)}}{\partial z} = [D_z] = \frac{1}{6V} [d_i \quad d_j \quad d_k \quad d_\ell]$$

Also the integration formula (3.38) is used.

Element minimization equations are obtained from

$$\begin{Bmatrix} \dfrac{\partial I}{\partial T_i} \\ \dfrac{\partial I}{\partial T_j} \\ \dfrac{\partial I}{\partial T_k} \\ \dfrac{\partial I}{\partial T_\ell} \end{Bmatrix} = [B]^{(e)} \begin{Bmatrix} T_i \\ T_j \\ T_k \\ T_\ell \end{Bmatrix} - \{C\}^{(e)} \tag{9.40}$$

The element coefficient matrix is a four by four square symmetric matrix and the element vector is a one by four matrix. Within an element, the functions K_x, K_y, P, Q, α, β are assumed constant.

Evaluation of the element contributions to the minimization equations is tedious and closely parallels the two-dimensional case. Only the results are given here in matrix form.

$$[B]^{(e)} = \frac{K_x}{36V} \begin{bmatrix} b_ib_i & b_ib_j & b_ib_k & b_ib_\ell \\ & b_jb_j & b_jb_k & b_jb_\ell \\ & & b_kb_k & b_kb_\ell \\ \text{(Symmetric)} & & & b_\ell b_\ell \end{bmatrix}^{(e)} \qquad (K_x \text{ matrix})$$

$$+ \frac{K_y}{36V} \begin{bmatrix} c_ic_i & c_ic_j & c_ic_k & c_ic_\ell \\ & c_jc_j & c_jc_k & c_jc_\ell \\ & & c_kc_k & c_kc_\ell \\ \text{(Symmetric)} & & & c_\ell c_\ell \end{bmatrix}^{(e)} \qquad (K_y \text{ matrix})$$

$$+ \frac{K_z}{36V} \begin{bmatrix} d_id_i & d_id_j & d_id_k & d_id_\ell \\ & d_jd_j & d_jd_k & d_jd_\ell \\ & & d_kd_k & d_kd_\ell \\ \text{(Symmetric)} & & & d_\ell d_\ell \end{bmatrix}^{(e)} \qquad (K_z \text{ matrix})$$

(9.41)

$$- \frac{PV}{20} \begin{bmatrix} 2 & 1 & 1 & 1 \\ & 2 & 1 & 1 \\ & & 2 & 1 \\ \text{(Symmetric)} & & & 2 \end{bmatrix}^{(e)} \qquad (P \text{ matrix})$$

$$+ \frac{\alpha A_{ijk}}{12} \begin{bmatrix} 2 & 1 & 1 & 0 \\ & 2 & 1 & 0 \\ & & 2 & 0 \\ \text{(Symmetric)} & & & 0 \end{bmatrix}^{(e)}_{\substack{\text{on side } ijk \\ \text{(Side 1)}}} \qquad (\alpha \text{ matrix } ijk)$$

$$+ \frac{\alpha A_{jk\ell}}{12} \begin{bmatrix} 0 & 0 & 0 & 0 \\ & 2 & 1 & 1 \\ & & 2 & 1 \\ \text{(Symmetric)} & & & 2 \end{bmatrix}^{(e)}_{\substack{\text{on side } jk\ell \\ \text{(Side 2)}}} \qquad (\alpha \text{ matrix } jk\ell)$$

$$+\frac{\alpha A_{k\ell i}}{12}\begin{bmatrix}2 & 0 & 1 & 1\\ & 0 & 0 & 0\\ & & 2 & 1\\ \text{(Symmetric)} & & & 2\end{bmatrix}^{(e)}_{\substack{\text{on side } k\ell i\\ \text{(Side 3)}}} \qquad (\alpha \text{ matrix } k\ell i)$$

(9.41 cont.)

$$+\frac{\alpha A_{\ell ij}}{12}\begin{bmatrix}2 & 1 & 0 & 1\\ & 2 & 0 & 1\\ & & 0 & 0\\ \text{(Symmetric)} & & & 2\end{bmatrix}^{(e)}_{\substack{\text{on side } \ell ij\\ \text{(Side 4)}}} \qquad (\alpha \text{ matrix } \ell ij)$$

Up to three sides may have derivative boundary conditions specified. Only when the derivative boundary condition is actually used on a particular side is the boundary matrix evaluated; otherwise, it is ignored.

The element column is similarly derived as

$$\{C\}^{(e)} = \frac{QV}{4}\begin{Bmatrix}1\\1\\1\\1\end{Bmatrix}^{(e)} \qquad \text{(element } Q \text{ column)}$$

$$-\frac{\beta A_{ijk}}{3}\begin{Bmatrix}1\\1\\1\\0\end{Bmatrix}^{(e)}_{\substack{\text{on side } ijk\\ \text{(Side 1)}}} \qquad (\beta \text{ column } ijk)$$

$$-\frac{\beta A_{jk\ell}}{3}\begin{Bmatrix}0\\1\\1\\1\end{Bmatrix}^{(e)}_{\substack{\text{on side } jk\ell\\ \text{(Side 2)}}} \qquad (\beta \text{ column } jk\ell) \qquad \textbf{(9.42)}$$

$$-\frac{\beta A_{k\ell i}}{3}\begin{Bmatrix}1\\0\\1\\1\end{Bmatrix}^{(e)}_{\substack{\text{on side } k\ell i\\ \text{(Side 3)}}} \qquad (\beta \text{ column } k\ell i)$$

$$-\frac{\beta A_{\ell ij}}{3}\begin{Bmatrix}1\\1\\0\\1\end{Bmatrix}^{(e)}_{\substack{\text{on side } \ell ij\\ \text{(Side 4)}}} \qquad (\beta \text{ column } \ell ij)$$

As with the element matrix, the β columns along the S_2 boundary are evaluated; the others are ignored.

Fortunately, the assembly procedure to construct the global matrices is nearly the same as for one- and two-dimensional problems. The equation bandwidth usually increases dramatically. Within this bandwidth many zeros occur, and techniques for efficient solution are normally required. These are largely beyond the scope of this text.

B. General Three-Dimensional Element

Consider a general element, such as that discussed in Section 9.6 but in three dimensions. By analogy, the element matrix is (9.37)

$$[B]^{(e)} = \iiint_{V^{(e)}} \Big[\{D_x\}^T K_x \{D_x\} + \{D_y\}^T K_y \{D_y\} + \{D_z\}^T K_z \{D_z\} - \{N\}^T P \{N\} \Big]^{(e)} dA + \iint_{S_2^{(e)}} \Big[\{G\}^T \alpha \{G\} \Big]^{(e)} dA \quad \textbf{(9.43)}$$

and the element column is (9.38)

$$\{C\}^{(e)} = \iiint_{V^{(e)}} \Big[\{G\}^T Q \Big]^{(e)} dA - \iint_{S_2^{(e)}} \Big[\{G\}^T \beta \Big]^{(e)} dL_2 \quad \textbf{(9.44)}$$

9.8 Problems

9.1 Derive the functional for the boundary value problem

$$\frac{\partial}{\partial x}\left(K_x \frac{\partial T}{\partial x}\right) + \frac{\partial}{\partial y}\left(K_y \frac{\partial T}{\partial y}\right) = 0 \text{ in } A$$

$$T = T_0 \text{ on } L$$

in two dimensions. The coefficient functions are not equal—$K_x \neq K_y$. Follow the method developed in Section 9.2.

9.2 Heat conduction occurs in the plane of a thin fin of irregular shape in the x,y plane. Heat convection takes place through the top and bottom surfaces of the fin to the atmosphere. The governing differential equation for the problem is

$$\frac{\partial}{\partial x}\left(kt \frac{\partial T}{\partial x}\right) + \frac{\partial}{\partial y}\left(kt \frac{\partial T}{\partial y}\right) - 2h(T - T_\infty) = 0$$

where k = thermal conductivity, h = convection coefficient, t = fin thickness, T = fin temperature, and T_∞ = fluid temperature. The temperature around the boundary is specified. Determine the functional for this boundary value problem, using the results presented in Table 9.2. Do not derive the functional. Figure 9.15 shows the geometry.

9.3 Starting with the expression for the element function (9.23), derive the $\textcircled{K_y}$ term for the simplex element. Write out all of the terms (do not use matrix notation). Then evaluate all of the components of the $\boxed{K_y}$ term in the element matrix, showing that the result in Equation (9.29) is correct.

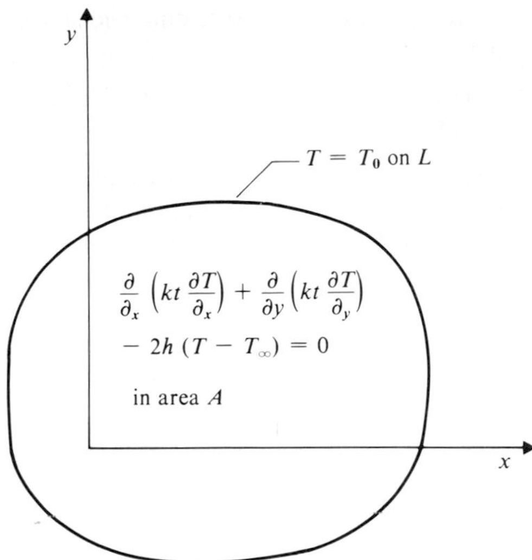

Figure 9.15 Diagram for Problem 9.2

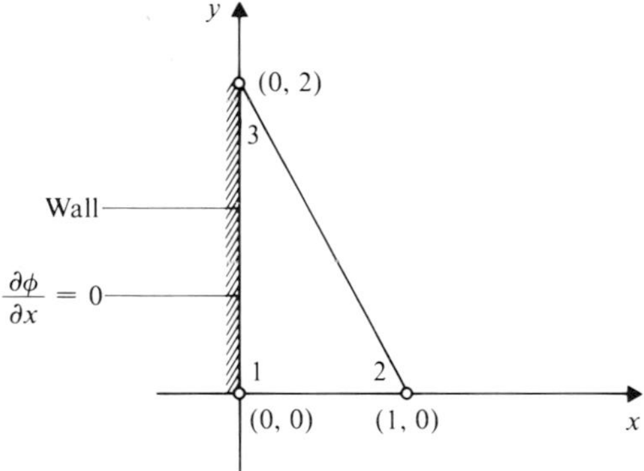

Figure 9.16 Element for Problem 9.5

9.4 Consider a general two-dimensional element with nodes labeled *i,j, . . . ,p*. Determine the form of the $\boxed{K_x}$ term in the element matrix, written out in a fashion similar to the $\boxed{P}$ shown immediately following Equation (9.38).

9.5 Ideal flow occurs near a wall. The differential equation is

$$\frac{\partial^2 \phi}{\partial x^2} + \frac{\partial^2 \phi}{\partial y^2} = 0$$

where ϕ = velocity potential. The boundary condition along the wall is (no flow through the wall) $\partial \phi / \partial x = 0$. Determine the element matrix for this problem, using a simplex element with the nodal coordinates shown in Figure 9.16.

9.6 Write a small computer program to evaluate the element matrix for three-dimensional heat conduction, governed by the differential equation

$$\frac{\partial}{\partial x}\left(k\frac{\partial T}{\partial x}\right)+\frac{\partial}{\partial y}\left(k\frac{\partial T}{\partial y}\right)+\frac{\partial}{\partial z}\left(k\frac{\partial T}{\partial z}\right)=0$$

where T = temperature, k = thermal conductivity = 0.30 W/mm-°C for a simplex element. Evaluate the element matrix for the element given in Example 3.5. Let the element dimensions have the units of mm. Assume that the element is an interior one, so that no boundary conditions apply.

Chapter 10

Applications—Two- and Three-Dimensional Field Problems

10.1 Heat Conduction and Convection in Thin Fin—Two-Element Example

This section carries out the finite element solution of heat convection and conduction in a trapezoidal two-dimensional fin. It illustrates the calculation of element conduction matrices involving derivative boundary conditions for both elements. The assembly of global equations is also shown in detail.

A more complicated heat conduction and convection problem is solved in the next section. Eight elements are employed. In that problem, the larger number of elements leads to a more accurate solution but makes the example somewhat more difficult to understand.

Example 10.1 *Conduction and Convection in Trapezoidal Fin*

Given A trapezoidal fin, similar to the rectangular fin in Example 5.6, has a small thickness in the vertical direction. The dimensions are given in Figure 10.1, where length = 120 mm, width at base = 160 mm, width at tip = 60 mm, and thickness, t = 1.25 mm. Other problem parameters are thermal conductivity = k = 0.20 W/mm-°C, convection coefficient = $h = 1 \times 10^{-5}$ W/mm^2-°C, and fluid temperature = T_∞ = 30°C.

At the base of the fin, the temperature decreases linearly from 330°C at one end to 250°C at the other end. Along the sides of the fin, it is assumed that the heat convection loss is very small (due to the small surface area compared to the fin top and bottom surface area). The side boundary is then considered insulated, and the boundary condition along it is $\dfrac{\partial T}{\partial n} = 0$, where $\partial T/\partial n$ is the derivative normal to the boundary.

Objective Determine the differential equation for the fin. Divide the fin into two simplex elements and determine the temperature at the node points. Also calculate the rate of heat loss by the fin.

Solution Because the fin is very thin, it may be considered two-dimensional with essentially uniform temperature in the vertical direction. Heat conduction within the fin is governed by the conduction law developed in the same manner as the thin fin in Example 5.6.

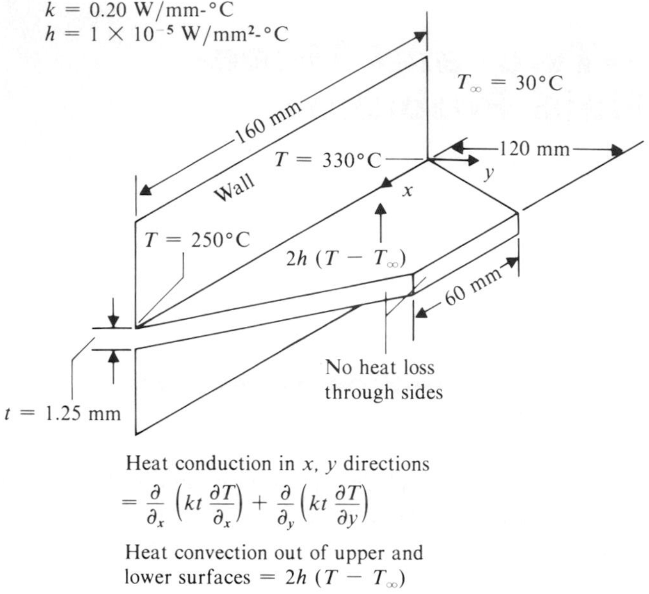

Figure 10.1 Fin geometry (Example 10.1—Heat conduction and convection in trapezoidal fin)

A. Differential Equation

A differential section of fin with dimensions dx by dy by t is shown in Figure 10.2. The total heat flow in the differential section is

$$\underbrace{-\left(Q_x - \frac{\partial Q_x}{\partial x}\frac{dx}{2}\right) + \left(Q_x + \frac{\partial Q_x}{\partial x}\frac{dx}{2}\right)}_{\text{Total heat conduction in } x \text{ direction}}$$

$$\underbrace{-\left(Q_y - \frac{\partial Q_y}{\partial y}\frac{dy}{2}\right) + \left(Q_y + \frac{\partial Q_y}{\partial y}\frac{dy}{2}\right)}_{\text{Total heat conduction in } y \text{ direction}}$$

$$+ \underbrace{h\,A_z(T - T_\infty)}_{\substack{\text{Total heat}\\ \text{convection}\\ \text{through top}\\ \text{face}}} + \underbrace{h\,A_z(T - T_\infty)}_{\substack{\text{Total heat}\\ \text{convection}\\ \text{through}\\ \text{bottom face}}} = 0$$

Fourier's heat conduction law in the two directions is

$$Q_x = -k\,A_x\frac{\partial T}{\partial x} = -kt\,dy\frac{\partial T}{\partial x}$$

$$Q_y = -k\,A_y\frac{\partial T}{\partial y} = -kt\,dx\frac{\partial T}{\partial y}$$

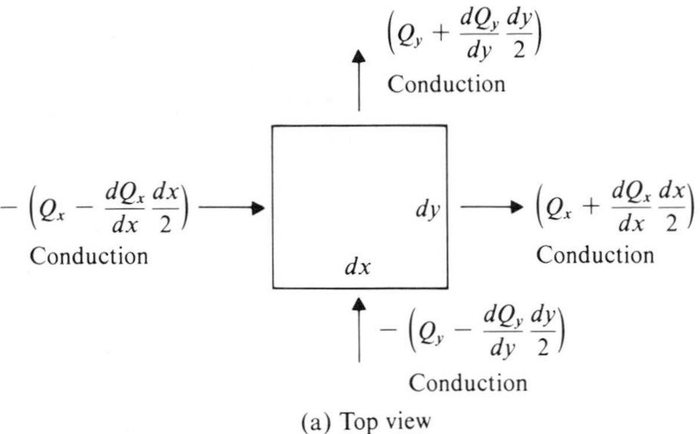

(a) Top view

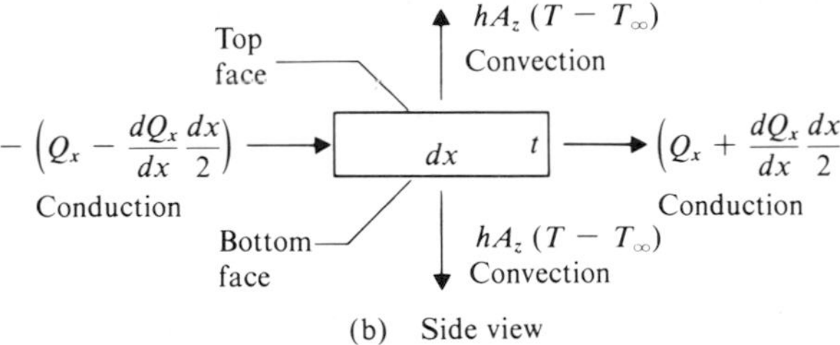

(b) Side view

Figure 10.2 Heat flow in differential section (Example 10.1—Conduction and convection in trapezoidal fin)

Also the differential area A_z on the top and bottom is given by $A_z = dxdy$. After cancelling like terms and dividing the differential equation by $dxdy$, the result is

$$\frac{\partial}{\partial x}\left[kt\,\frac{\partial T}{\partial x}\right] + \frac{\partial}{\partial y}\left[kt\,\frac{\partial T}{\partial y}\right] - 2h(T - T_\infty) = 0$$

This is the governing differential equation.

The general two-dimensional differential equation is (9.5)

$$\frac{\partial}{\partial x}\left[K_x\,(x,y)\,\frac{\partial T}{\partial x}\right] + \frac{\partial}{\partial y}\left[K_y(x,y)\,\frac{\partial T}{\partial y}\right] + P(x,y)T + Q(x,y) = 0$$

where

$$K_x(x,y) = K_y(x,y) = kt = 0.25 \text{ W/°C}$$
$$P(x,y) = -2h = -2 \times 10^{-5} \text{ W/mm}^2\text{-°C}$$
$$Q(x,y) = +2hT_\infty = 6 \times 10^{-4} \text{ W/mm}^2$$

Boundary conditions will be written on an element-by-element basis.

B. Element Properties

Let the trapezoidal fin be divided into two elements, as shown in Figure 10.3, where the diagonal is chosen from the highest boundary temperature to the opposite corner. This

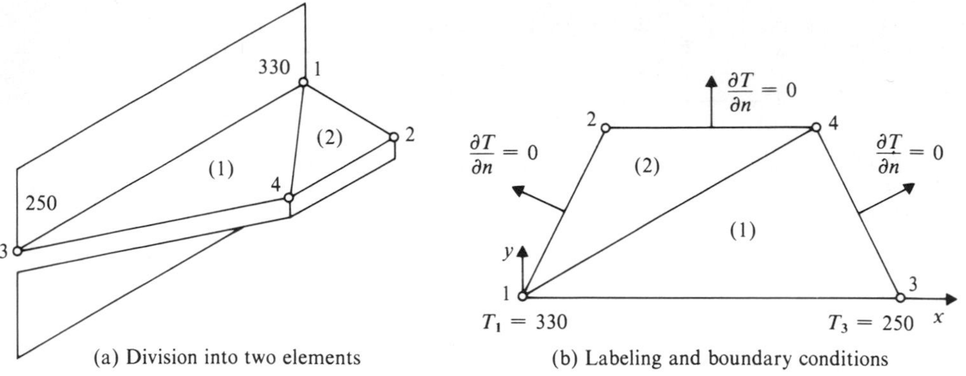

(a) Division into two elements (b) Labeling and boundary conditions

Figure 10.3 Division into elements (Example 10.1—Heat conduction and convection in trapezoidal fin)

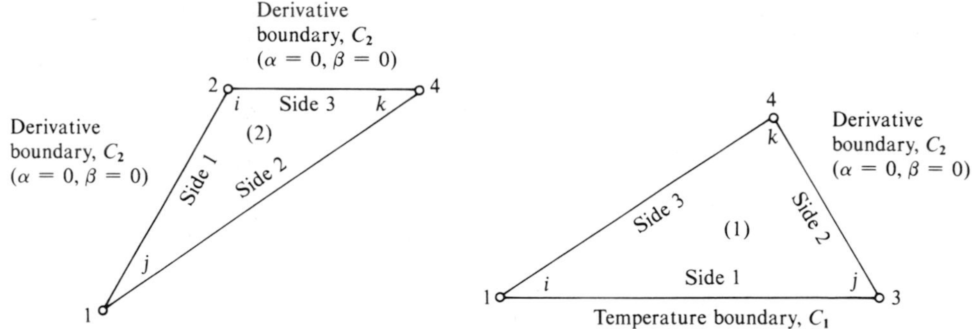

Figure 10.4 Derivative boundary conditions (Example 10.1—Heat conduction and convection in trapezoidal fin)

direction corresponds to the expected direction of largest temperature gradient; choosing this diagonal is usually good finite element practice. The node and element labeling system is shown.

It is often convenient to organize the input data into tables. This usually reduces errors in formulating the finite element model. For this example nodal coordinates are given in Table 10.1(a) and the element/node connectivity data are indicated in Table 10.1(b). Table 10.1(c) gives the values of the coefficient functions K_x, K_y, P, Q for each element in the model. Prescribed nodal temperatures are given in Table 10.1(d).

Element 1, shown in Figure 10.4, is labeled starting with node 1 and continuing counterclockwise to nodes 3 and 4. The corresponding nodal coordinates are (0,0), (160,0), (110,120). Along side 1, the temperatures are specified, while along side 2, derivative boundary conditions with $\alpha = 0$, $\beta = 0$ are given. The values of α and β are entered into Table 10.1(e) for computer input.

Table 10.1
Input Data (Example 10.1: Heat Conduction and Convection in Trapezoidal Fin)

(a) Nodal Coordinates

Node	X	Y
1	0	0
2	50	120
3	160	0
4	110	120

(b) Element/Node Connectivity Data

Element	Node		
	i	j	k
1	1	3	4
2	2	1	4

(c) Element Values of K_x, K_y, P, Q Coefficient Functions

Element	K_x	K_y	P	Q
1	0.25	0.25	-2×10^{-5}	6×10^{-4}
2	0.25	0.25	-2×10^{-5}	6×10^{-4}

(d) Prescribed Nodal Values

Node	Temperature
1	330
3	250

(e) Element Derivative Boundary Values

Element	Side	α	β
1	2	0	0
2	1	0	0
2	3	0	0

The element constants b and c are given by Equation (3.17) as

$$\begin{aligned}
b_1 &= Y_3 - Y_4 = 0 - 120 = -120 \text{ mm} \\
b_3 &= Y_4 - Y_1 = 120 - 0 = 120 \text{ mm} \\
b_4 &= Y_1 - Y_3 = 0 - 0 = 0 \text{ mm} \\
c_1 &= X_4 - X_3 = 110 - 160 = -50 \text{ mm} \\
c_3 &= X_1 - X_4 = 0 - 110 = -110 \text{ mm} \\
c_4 &= X_3 - X_1 = 160 - 0 = 160 \text{ mm}
\end{aligned}$$

while the element area A is 9,600 mm². Using the sum of simpler matrix form, as given by Equation (9.29),

$$[B]^{(1)} = \frac{K_x}{4A}\begin{bmatrix} b_1b_1 & b_1b_3 & b_1b_4 \\ b_1b_3 & b_3b_3 & b_3b_4 \\ b_1b_4 & b_3b_4 & b_4b_4 \end{bmatrix}^{(1)} \quad \text{(Element } x \text{ gradient matrix)}$$

$$+ \frac{K_y}{4A}\begin{bmatrix} c_1c_1 & c_1c_3 & c_1c_4 \\ c_1c_3 & c_3c_3 & c_3c_4 \\ c_1c_4 & c_3c_4 & c_4c_4 \end{bmatrix}^{(1)} \quad \text{(Element } y \text{ gradient matrix)}$$

$$- \frac{PA}{12}\begin{bmatrix} 2 & 1 & 1 \\ 1 & 2 & 1 \\ 1 & 1 & 2 \end{bmatrix}^{(1)} \quad \text{(Element } P \text{ matrix)}$$

where the boundary matrices are not included since $\alpha = 0$.

Substituting the proper terms

$$[B]^{(1)} = \frac{0.25}{4(9{,}600)}\begin{bmatrix} (-120)(-120) & (-120)(120) & (-120)(0) \\ (-120)(120) & (120)(120) & (120)(0) \\ (-120)(0) & (120)(0) & (0)(0) \end{bmatrix}^{(1)}$$

$$+ \frac{0.25}{4(9{,}600)}\begin{bmatrix} (-50)(-50) & (-50)(-110) & (-50)(160) \\ (-50)(-110) & (-110)(-110) & (-110)(160) \\ (-50)(160) & (-110)(160) & (160)(160) \end{bmatrix}^{(1)}$$

$$+ \frac{2\times 10^{-5}(9{,}600)}{12}\begin{bmatrix} 2 & 1 & 1 \\ 1 & 2 & 1 \\ 1 & 1 & 2 \end{bmatrix}^{(1)}$$

The result is

$$[B]^{(1)} = \begin{bmatrix} 0.09375 & -0.09375 & 0 \\ -0.09375 & 0.09375 & 0 \\ 0 & 0 & 0 \end{bmatrix}^{(1)}$$

$$+ \begin{bmatrix} 0.01628 & 0.03581 & -0.05208 \\ 0.03581 & 0.07878 & -0.11458 \\ -0.05208 & -0.11458 & 0.16667 \end{bmatrix}^{(1)}$$

$$+ \begin{bmatrix} 0.032 & 0.016 & 0.016 \\ 0.016 & 0.032 & 0.016 \\ 0.016 & 0.016 & 0.032 \end{bmatrix}^{(1)}$$

and adding the three matrices together,

$$[B]^{(1)} = \begin{matrix} \\ 1 \\ 3 \\ 4 \end{matrix} \begin{matrix} \begin{matrix} 1 & \quad 3 & \quad 4 \end{matrix} \\ \begin{bmatrix} 0.14203 & -0.04194 & -0.03608 \\ & 0.20453 & -0.09858 \\ \text{(Symmetric)} & & 0.19867 \end{bmatrix}^{(1)} \end{matrix}$$

This is the element conduction matrix.

The element column is obtained from Equation (9.30) as

$$\{C\}^{(1)} = \frac{QA}{3} \begin{Bmatrix} 1 \\ 1 \\ 1 \end{Bmatrix}^{(1)} \qquad \text{(Element column)}$$

and substituting

$$\{C\}^{(1)} = \frac{6\times10^{-4}(9{,}600)}{3} \begin{Bmatrix} 1 \\ 1 \\ 1 \end{Bmatrix}^{(1)}$$

The result is

$$\{C\}^{(1)} = \begin{matrix} 1 \\ 3 \\ 4 \end{matrix} \begin{Bmatrix} 1.92 \\ 1.92 \\ 1.92 \end{Bmatrix}^{(1)}$$

The boundary terms are all zero since $\beta = 0$.

Now consider element 2, with nodes taken in counterclockwise order 2, 1, 4, with nodal coordinates (50,120), (0,0), (110,120). The element constants b and c are

$$\begin{aligned} b_2 &= Y_1 - Y_4 = 0 - 120 = -120 \text{ mm} \\ b_1 &= Y_4 - Y_2 = 120 - 120 = 0 \text{ mm} \\ b_4 &= Y_2 - Y_1 = 120 - 0 = 120 \text{ mm} \\ c_2 &= X_4 - X_1 = 110 - 0 = 110 \text{ mm} \\ c_1 &= X_2 - X_4 = 50 - 110 = -60 \text{ mm} \\ c_4 &= X_1 - X_2 = 0 - 50 = -50 \text{ mm} \end{aligned}$$

with element area 3,600 mm^2. Both side 1 and side 3 have derivative boundary conditions specified but $\alpha = 0$ and $\beta = 0$ along the sides.

Equation (9.29) yields

$$[B]^{(2)} = \frac{0.25}{4(3{,}600)} \begin{bmatrix} (-120)(-120) & (-120)(0) & (-120)(120) \\ (-120)(0) & (0)(0) & (0)(120) \\ (-120)(120) & (0)(120) & (120)(120) \end{bmatrix}^{(2)}$$

$$+ \frac{0.25}{4(3{,}600)} \begin{bmatrix} (110)(110) & (110)(-60) & (110)(-50) \\ (110)(-60) & (-60)(-60) & (-60)(-50) \\ (110)(-50) & (-60)(-50) & (-50)(-50) \end{bmatrix}^{(2)}$$

$$+ \frac{2\times10^{-5}(3{,}600)}{12} \begin{bmatrix} 2 & 1 & 1 \\ 1 & 2 & 1 \\ 1 & 1 & 2 \end{bmatrix}^{(2)}$$

Adding the three matrices gives

$$[B]^{(2)} = \begin{array}{c} \\ 2 \\ 1 \\ 4 \end{array} \begin{array}{c} \begin{array}{ccc} 2 & 1 & 4 \end{array} \\ \begin{bmatrix} 0.47207 & -0.10858 & -0.33949 \\ & 0.07450 & 0.05808 \\ \text{(Symmetric)} & & 0.30540 \end{bmatrix}^{(2)} \end{array}$$

Also the element column is

$$\{C\}^{(2)} = \frac{6\times 10^{-4}(3{,}600)}{3} \begin{Bmatrix} 1 \\ 1 \\ 1 \end{Bmatrix}$$

$$\{C\}^{(2)} = \begin{array}{c} 2 \\ 1 \\ 4 \end{array} \begin{Bmatrix} 0.72 \\ 0.72 \\ 0.72 \end{Bmatrix}^{(2)}$$

C. Assembly and Solution

The two-element conduction matrices and element columns are assembled into the global equations, as shown in Figures 10.5 and 10.6. Each row and column from the element matrices is added to the corresponding row and column of the global matrix. In each square, the upper number comes from element 1 and the lower number comes from element 2. The final equations are

$$\begin{bmatrix} 0.21653 & -0.10858 & -0.04194 & 0.02200 \\ & 0.47207 & 0 & -0.33949 \\ & & 0.20453 & -0.09858 \\ \text{(Symmetric)} & & & 0.50407 \end{bmatrix} \begin{Bmatrix} T_1 \\ T_2 \\ T_3 \\ T_4 \end{Bmatrix} = \begin{Bmatrix} 2.64 \\ 0.72 \\ 1.92 \\ 2.64 \end{Bmatrix}$$

Only two temperatures, T_2 and T_4, are unknown. The two corresponding equations are $-0.10858T_1 + 0.47207T_2 + 0T_3 - 0.33949T_4 = 0.72$ and $0.02200T_1 - 0.33949T_2 - 0.09858T_3 + 0.50407T_4 = 2.64$. Substituting $T_1 = 330$, $T_3 = 250$, the solution is $T_2 = 205.56°\text{C}$, $T_4 = 178.17°\text{C}$. The temperatures at the node points are indicated in Figure 10.7.

D. Heat Flow

The x and y components of heat flow per unit area (by conduction) are

$$q_x = -kt\frac{\partial T}{\partial x}, \quad q_y = -kt\frac{\partial T}{\partial y}$$

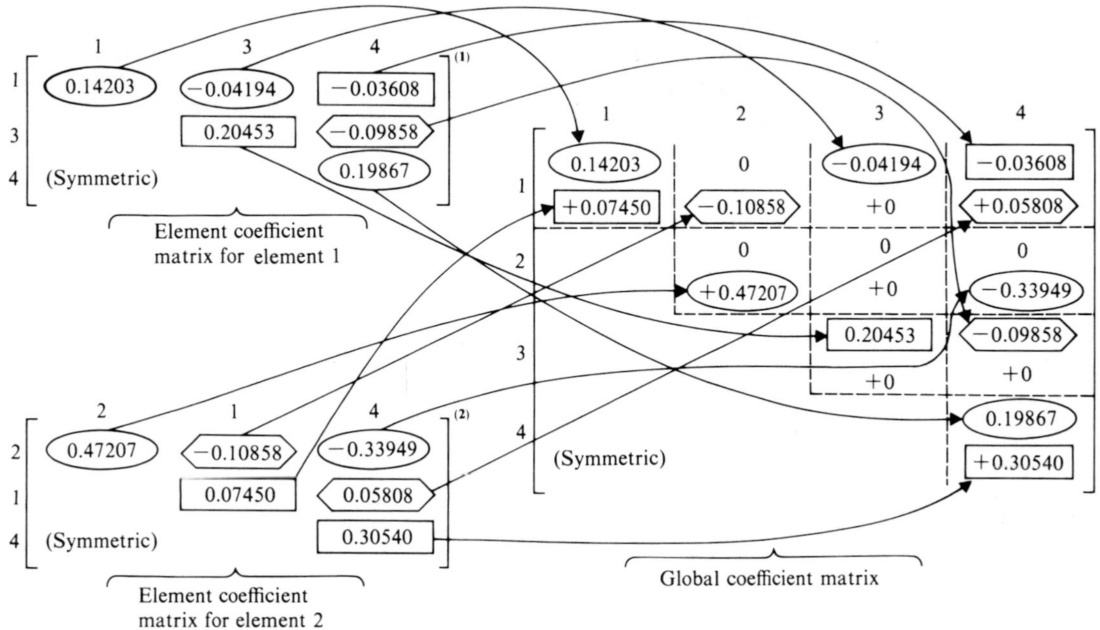

Figure 10.5 Assembly of global coefficient matrix (Example 10.1—Heat conduction and convection in trapezoidal fin)

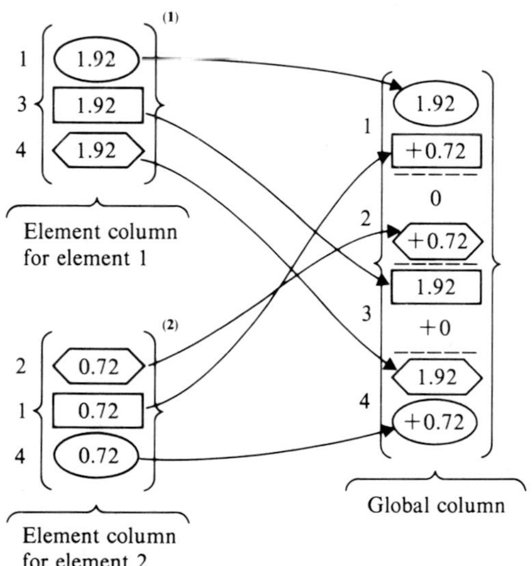

Figure 10.6 Assembly of global column (Example 10.1—Heat conduction and convection in trapezoidal fin)

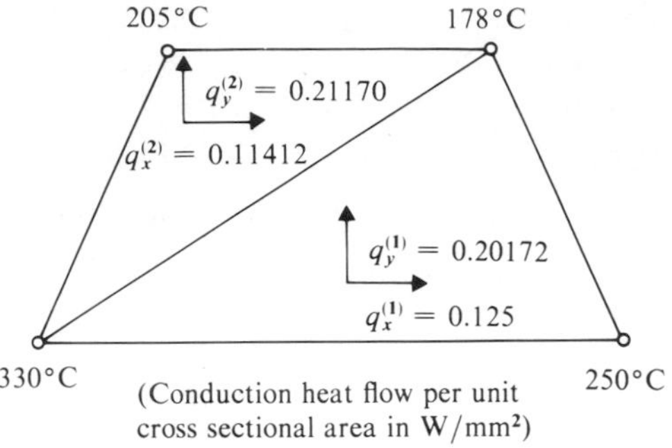

Figure 10.7 Nodal temperatures and heat flow (Example 10.1—Heat conduction and convection in trapezoidal fin)

For element 1,

$$q_x^{(1)} = \left[-\frac{kt}{2A}(b_1T_1 + b_3T_3 + b_4T_4)\right]^{(1)}$$

$$= -\frac{(0.20)(1.25)}{2\ (9{,}600)}[(-120)(330) + (120)(250) + (0)(178.17)]$$

$$= 0.125\ \text{W/mm}^2$$

$$q_y^{(1)} = \left[-\frac{kt}{2A}(c_1T_1 + c_3T_3 + c_4T_4)\right]^{(1)}$$

$$= -\frac{(0.20)(1.25)}{2\ (9{,}600)}[(-50)(330) + (-110)(250) + (160)(178.17)]$$

$$= 0.20172\ \text{W/mm}^2$$

The conduction heat flow is assumed constant within an element. Note that side 1 of element 1 forms the base of the trapezoidal fin. The heat flow through it is q_y, so that total heat loss through the fin is q_y times the base area $160t$, or the conduction heat loss through the base of the fin = 40.34 watts. Now the conduction heat flow per unit area in element 2 is

$$q_x^{(2)} = \left[-\frac{kt}{2A}(b_2T_2 + b_1T_1 + b_4T_4)\right]^{(2)}$$

$$= -\frac{(0.20)(1.25)}{2\ (3{,}600)}[(-120)(205.56) + (0)(330) + (120)(178.17)]$$

$$= 0.11412\ \text{W/mm}^2$$

$$q_y^{(2)} = \left[-\frac{kt}{2A}(c_2T_2 + c_1T_1 + c_4T_4\right]$$

$$= \frac{-(0.20)(1.25)}{2\ (3{,}600)}[(110)(205.56) + (-60)(330) + (-50)(178.17)]$$

$$= 0.21170\ \text{W/mm}^2$$

These conduction heat flows are indicated in Figure 10.7 by the arrows within the elements.

Another quantity that may be calculated is the convection heat loss through the top and bottom of each element:

$$\text{Total convective heat loss} = \iint_A [2h(T - T_\infty)]\, dA$$

For element 1, this becomes

$$(\text{Convective heat loss})^{(1)} = 2h \iint_{A^{(1)}} T^{(1)}\, dA - 2hT_\infty A^{(1)}$$

or

$$(\text{Convective heat loss})^{(1)} = \frac{2hA^{(1)}}{3}(T_1 + T_3 + T_4) - 2hT_\infty A^{(1)}$$

$$= 42.76 \text{ watts}$$

Also for element 2,

$$(\text{Convective heat loss})^{(2)} = 2h \iint_{A2} T^{(2)}\, dA^{(2)} - 2hT_\infty A^{(2)}$$

$$= \frac{2hA^{(2)}}{3}(T_2 + T_1 + T_4) - 2hT_\infty A^{(2)}$$

$$= 14.97 \text{ watts}$$

Thus the total convective heat loss is 57.13 watts.

Note that the conduction heat loss (40.34 watts) is not the same as the total convection heat loss (57.13 watts). This indicates that not enough elements were used to analyze the fin. If a large number of elements were used, the two quantities should match to several decimal places. A good question to ask is: Which of these two numbers is more likely to be close to the correct answer? Judging from Example 5.10, the calculation involving the integral is probably closer (57.13 watts).

10.2 Heat Conduction and Convection in Long Body—Eight-Element Example

This section discusses heat conduction in the interior of a very long body and various boundary conditions, including convection, around the body. The emphasis in this example is on the treatment of different types of boundary conditions, rather than convective loss from the internal region of a thin fin, as in Section 10.1.

Example 10.2 *Conduction and Convection in Two-Dimensional Body*

Given Heat conduction with constant conduction coefficient k occurs in the infinitely long body shown in Figure 10.8. No heat source or convection occurs inside the body. The boundary conditions on various sides are shown in the diagram. Let the thickness t be 1.0 mm for calculations. Parameters for the problem are $k = 0.25$ W/mm-°C, $h = 0.001$ W/mm^2-°C, $T_\infty = 50$°C, and $t = 1.0$ mm.

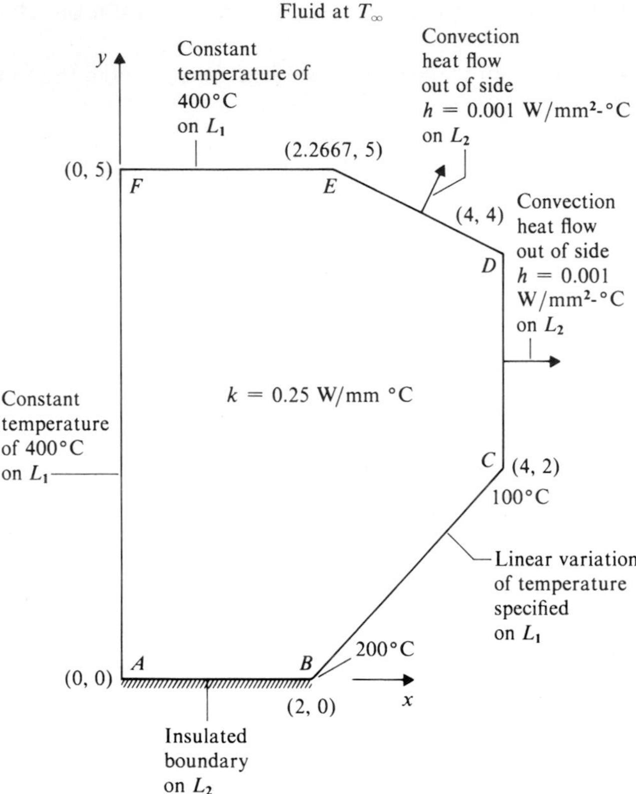

Figure 10.8 Problem geometry and boundary conditions (Example 10.2—Heat conduction and convection in two-dimensional body)

Objective Determine the differential equation and boundary conditions for the problem. Use eight elements to determine the temperature and the heat flow through the sides of the fin, which have the temperature specified. Divide the fin so as to produce a minimum bandwidth set of global equations.

Solution The body is infinitely long in the z direction (into and out of the paper). Thus the problem becomes a two-dimensional one in the x,y plane.

Heat conduction inside the body is governed by the differential equation

$$\frac{\partial}{\partial x}\left(kt\,\frac{\partial T}{\partial x}\right) + \frac{\partial}{\partial y}\left(kt\,\frac{\partial T}{\partial y}\right) = 0$$

as derived in Example 10.1, where the convection through the top and bottom of the thin fin is not present. Comparing this to the general differential equation (9.5) gives

$$\frac{\partial}{\partial x}\left[K_x(x, y)\,\frac{\partial T}{\partial x}\right] + \frac{\partial}{\partial y}\left[K_y(x, y)\,\frac{\partial T}{\partial y}\right] + P(x, y)\,T + Q(x, y) = 0$$

where $K_x(x,y) = K_y(x,y) = kt = 0.25$ W/°C, $P(x,y) = 0$, and $Q(x,y) = 0$. The differential equation is simpler than in the example discussed in the last section.

Along the boundary AB, insulation prevents heat conduction or convection, and the derivative boundary condition reduces to

$$kt \frac{\partial T}{\partial x} n_x + kt \frac{\partial T}{\partial y} n_y = 0$$

Comparing this to the general derivative boundary condition (9.7)

$$K_x \frac{\partial T}{\partial x} n_x + K_y \frac{\partial T}{\partial y} n_y + \alpha T + \beta = 0$$

the functions are $K_x(x,y) = kt = 0.25$ W/°C, $K_y(x,y) = kt = 0.25$ W/°C, $\alpha(x,y) = 0$, and $\beta(x,y) = 0$. The values for K_x and K_y were already known. Using the values of $\alpha = 0$ and $\beta = 0$ produces the insulated boundary along AB.

The boundary condition along BC is fairly obvious. The specified temperature is linear from 200°C to 100°C. A function $T_0(x,y)$ could be found, but this is not necessary. Specifying nodal temperatures along BC for simplex elements gives a linear temperature variation along BC.

Along CD, heat convection occurs to the outside fluid. The boundary condition is

$$kt \frac{\partial T}{\partial x} n_x + kt \frac{\partial T}{\partial y} n_y + ht(T - T_\infty) = 0$$

Again comparing to the general derivative boundary condition $K_x(x,y) = kt = 0.25$ W/°C, $K_y(x,y) = kt = 0.25$ W/°C, $\alpha(x,y) = ht = 0.001$ W/mm-°C, and $\beta(x,y) = -htT_\infty = -0.050$ W/mm. These values of α and β are used in the input to the computer program.

A similar result occurs for the heat convection through side DE. $K_x(x,y) = kt = 0.25$ W/°C, $K_y(x,y) = kt = 0.25$ W/°C, $\alpha(x,y) = ht = 0.001$ W/mm-°C, and $\beta(x,y) = -htT_\infty = -0.050$ W/mm. Again the α and β values are used directly.

Finally, the temperature along EF and FA is specified constant. Thus all nodes along those sides would set to 400°C.

The fin is divided up into eight elements, as shown in Figure 10.9. Table 10.2(a) and (b) gives the nodal coordinates and element/node connectivity data for this division scheme. The bandwidth is 4, which is a minimum. Table 10.2(c) gives the element values for K_x, K_y, P, Q coefficient functions.

Along the boundary L, the temperature and derivative boundaries are indicated along the sides of the elements by L_1 and L_2. Table 10.2(d) gives the prescribed nodal values. The derivative boundary sides and values of α, β are given in Table 10.2(e). These must be incorporated into the element matrices.

Consider first element 1 with the insulated boundary condition on side 1. The nodal coordinates are (0,0), (2,0), (0,3) and $b_1 = -3$, $c_1 = -2$, $b_4 = 3$, $c_4 = 0$, $b_2 = 0$, $c_2 = 2$. The element area is 3.0. From Equation (9.29), with $\alpha = 0$,

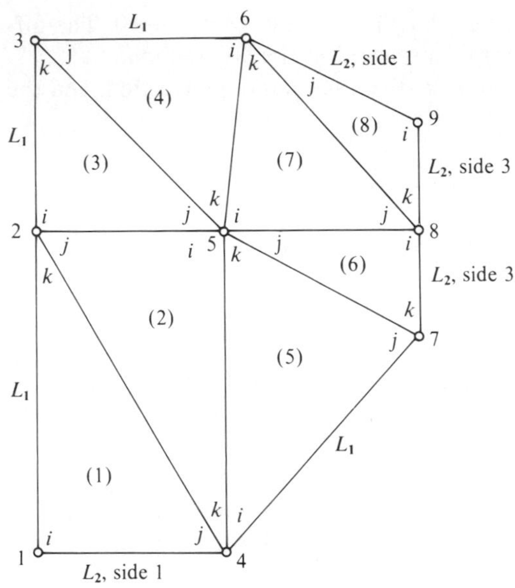

Figure 10.9 Division into elements (Example 10.2—Heat conduction and convection in two-dimensional body)

Table 10.2
Input Data (Example 10.2: Heat Conduction and Convection in Two-Dimensional Body)

(a) Nodal Coordinates

Node	*X*	*Y*
1	0.0	0.0
2	0.0	3.0
3	0.0	5.0
4	2.0	0.0
5	2.0	3.0
6	2.2667	5.0
7	4.0	2.0
8	4.0	3.0
9	4.0	4.0

Table 10.2 *Continued*

(b) Element/Node Connectivity Data

Element	Node		
	i	*j*	*k*
1	1	4	2
2	5	2	4
3	2	5	3
4	6	3	5
5	4	7	5
6	8	5	7
7	5	8	6
8	9	6	8

(c) Element Values for *K, P, Q* Coefficient Functions

Element	K_x	K_y	P	Q
(all)	0.25	0.25	0	0

(d) Prescribed Nodal Values

Node	Temperature °C
1	400.0
2	400.0
3	400.0
4	200.0
6	400.0
7	100.0

(e) Element Derivative Boundary Values

Element	Side	α	β
1	1	0	0
6	3	0.001	−0.050
8	1	0.001	−0.050
8	3	0.001	−0.050

$$[B]^{(1)} = \begin{array}{c} \\ 1 \\ 4 \\ 2 \end{array} \begin{array}{c} \begin{array}{ccc} 1 & 4 & 2 \end{array} \\ \begin{bmatrix} 0.27083 & -0.18750 & -0.08333 \\ -0.18750 & 0.18750 & 0 \\ -0.08333 & 0 & 0.08333 \end{bmatrix}^{(1)} \end{array}$$

and from Equation (9.27), with $\beta = 0$,

$$\{C\}^{(1)} = \begin{array}{c} 1 \\ 4 \\ 2 \end{array} \begin{Bmatrix} 0 \\ 0 \\ 0 \end{Bmatrix}^{(1)}$$

The elements 2, 3, 4, 5 have no derivative boundary conditions and are evaluated as in earlier examples.

Element 6 has nodal coordinates (4,3), (2,3), (4,2) and $b_8 = 1$, $c_8 = 2$, $b_5 = -1$, $c_5 = 0$, $b_7 = 0$, $c_7 = -2$, while the area is 1.0. With $\alpha = 0.001$ on side 3,

$$B_{88} = \left[\frac{1}{4A}(K\,b_8^2 + K\,c_8^2) - \frac{1}{6}PA\right]^{(6)} + \left(\frac{1}{3}\alpha\,L_1\right)_{\text{on side 1}}$$

$$+ \left(\frac{1}{3}\alpha\,L_3\right)_{\text{on side 3}}$$

$$= \frac{1}{4(1)}[(0.25)(1)^2 + (0.25)(2)^2] - \frac{1}{6}(0)(1)$$

$$+ \left[\frac{1}{3}(0.001)(1)\right]_{\text{on side 3}}$$

$$B_{88} = 0.31283$$

Note that the term on side 1 is not included because side 1 in element 6 is not on the boundary. With $\beta = -0.050$,

$$C_8 = \frac{1}{3}QA - \left(\frac{1}{2}\beta L_1\right)_{\text{on side 1}} - \left(\frac{1}{2}\beta L_3\right)_{\text{on side 3}}$$

$$= \frac{1}{3}(0)(1) - \left[\frac{1}{2}(-0.050)(1)\right]_{\text{on side 3}}$$

$$C_8 = 0.025$$

Again side 1 is excluded because it is not on the boundary. The complete results are

$$[B]^{(6)} = \begin{array}{c} \\ 8 \\ 5 \\ 7 \end{array} \begin{array}{c} \begin{array}{ccc} 8 & 5 & 7 \end{array} \\ \begin{bmatrix} 0.31283 & -0.06250 & -0.24983 \\ -0.06250 & 0.06250 & 0 \\ -0.24983 & 0 & 0.25033 \end{bmatrix}^{(6)} \end{array}$$

$$\{C\}^{(6)} = \begin{array}{c} 8 \\ 5 \\ 7 \end{array} \begin{Bmatrix} 0.0250 \\ 0 \\ 0.0250 \end{Bmatrix}^{(6)}$$

Element 7 does not have derivative boundary conditions specified.

Element 8 has nodal coordinates (4,4), (2.2667,5), (4,3) and $b_9 = 2$, $c_9 = 1.7333$, $b_6 = -1$, $c_6 = 0$, $b_8 = -1$, $c_8 = -1.7333$. The element area is 0.8667. For this element $\alpha = 0.001$ on sides 1 and 3:

$$B_{99} = \left\{\frac{1}{4(0.8667)}\left[(0.25)(2)^2 + (0.25)(1.7333)^2\right] - \frac{1}{6}(0)(0.8667)\right\}$$

$$+ \left[\frac{1}{3}(0.001)(2)\right]_{\text{on side 1}} + \left[\frac{1}{3}(0.001)(1)\right]_{\text{on side 3}}$$

$$B_{99} = 0.50613$$

The first component in the element column is

$$C_9 = \frac{1}{3}(0)(0.8667) - \left[\frac{1}{2}(-0.050)(2)\right]_{\text{on side 1}}$$

$$- \left[\frac{1}{2}(-0.050)(1)\right]_{\text{on side 3}}$$

$$C_9 = 0.07503$$

The complete element conduction matrix and column are

$$[B]^{(8)} = \begin{matrix} & 9 & 6 & 8 \\ 9 \\ 6 \\ 8 \end{matrix}\begin{bmatrix} 0.50613 & -0.14390 & -0.36073 \\ -0.14390 & -0.07278 & 0.07212 \\ -0.36073 & 0.07212 & 0.28911 \end{bmatrix}^{(8)}$$

$$\{C\}^{(8)} = \begin{matrix} 9 \\ 6 \\ 8 \end{matrix}\begin{Bmatrix} 0.07503 \\ 0.05003 \\ 0.02500 \end{Bmatrix}^{(8)}$$

The final global conduction matrix, shown in Table 10.3(a), is assembled by the methods illustrated earlier. The main diagonal terms are shown in the first column, the first diagonal above the main diagonal in the second column, and so on up to the full bandwidth of four. Table 10.3(b) gives the global column.

Actually only three nodal temperatures are unknown: T_5, T_8, T_9. These were obtained by computer solution and are given in Table 10.4(a). The element heat flows per unit area are calculated from

$$q_x = -k\frac{\partial T}{\partial x} = -\left[\frac{k}{2A}(b_iT_i + b_jT_j + b_kT_k)\right]^{(e)}$$

$$q_y = -k\frac{\partial T}{\partial y} = -\left[\frac{k}{2A}(c_iT_i + c_jT_j + c_kT_k)\right]^{(e)}$$

and given in Table 10.4(b). A schematic of the fin is shown in Figure 10.10 with nodal temperatures and element heat flows. In the horizontal direction the heat flows to the right, while in the vertical direction it flows downward (negative values indicate flow in the negative y direction).

Table 10.3
Global Conduction Matrix and Column
(Example 10.2: Heat Conduction and Convection in Two-Dimensional Body)

(a) Global Conduction Matrix (Written in Compact Column Form to Reduce Computer Storage)					(b) Global Column	
Row	Main Diagonal	First Diagonal Above Main	Second Diagonal Above Main	Third Diagonal Above Main	Row	Column
1	0.2708	−0.0833	0	−0.1875	1	0
2	0.5208	−0.1250	0	−0.3125	2	0
3	0.2373	0	−0.0167	−0.0956	3	0
4	0.3750	−0.1250	0	−0.0625	4	0
5	0.9856	−0.2333	−0.1250	−0.1731	5	0
6	0.4184	0	0.5545	−0.1434	6	0.05003
7	0.4378	−0.2498	0	0	7	0.02500
8	0.7292	−0.3607	0	0	8	0.05000
9	0.5061	0	0	0	9	0.07503

Table 10.4
Nodal Temperatures and Element Heat Flow
(Example 10.2: Heat Conduction and Convection in Two-Dimensional Body)

(a) Nodal Temperatures		(b) Heat Flow in Elements		
Node	Temperature (°C)	Element	Heat Flow q_x (W/mm^2)	Heat Flow q_y (W/mm^2)
1	400.00	1	25.00	0
2	400.00	2	12.23	− 8.51
3	400.00	3	12.23	0
4	200.00	4	0	−12.23
5	302.14	5	21.01	− 8.51
6	400.00	6	12.29	− 2.60
7	100.00	7	12.29	−13.87
8	203.82	8	12.34	−13.83
9	259.14			

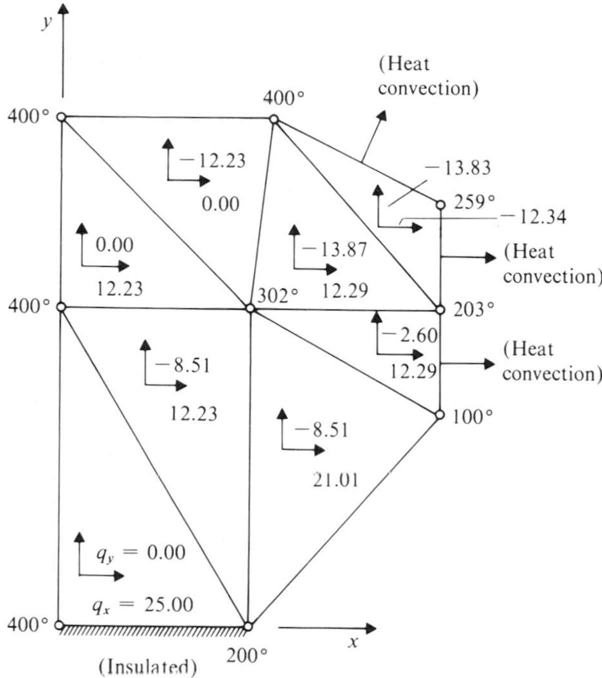

Figure 10.10 Nodal temperatures (°C) and element heat flows, shown by arrows within elements (Example 10.2—Heat conduction and convection in two-dimensional body)

10.3 Ideal Fluid Flow and Accuracy Considerations

Finite elements can also be employed to solve fluid flow problems. The analysis presented here is for steady, ideal, and incompressible flow in two dimensions. Figure 10.11 shows the geometry with flow around a circular cylinder. The term "steady" means that the flow pattern or streamlines do not change over time. The term "ideal" here indicates that the fluid has zero viscosity. This is also referred to as inviscid flow.

The three variables describing the fluid flow are u = horizontal velocity component, v = vertical velocity component, and P = pressure. Three differential equations describe the fluid motion. The first is the continuity equation for incompressible steady flow in two dimensions:

$$\frac{\partial u}{\partial x} + \frac{\partial v}{\partial y} = 0 \tag{10.1}$$

In the x direction, the momentum equation is

$$\rho u \frac{\partial u}{\partial x} + \rho v \frac{\partial u}{\partial y} = -\frac{\partial P}{\partial x} \tag{10.2a}$$

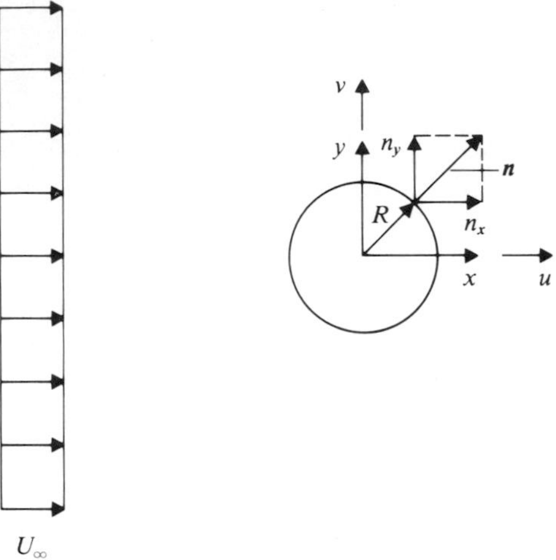

Figure 10.11 Ideal flow around a cylinder

In the y direction, the momentum equation is

$$\rho u \frac{\partial v}{\partial x} + \rho v \frac{\partial v}{\partial y} = -\frac{\partial P}{\partial y} \tag{10.2b}$$

These two equations are also called Euler's equation. Here the density ρ is simply a constant for the fluid.

At a solid boundary, the flow must not penetrate the wall. This can be written as

$$un_x + vn_y = 0 \tag{10.3}$$

where n_x = horizontal component of the outward normal to wall, and n_y = vertical component of the outward normal to the wall. In this section, it is assumed that the wall is stationary. Note that for viscous flow, the no-slip boundary condition holds. However, for zero viscosity conditions, the flow moves along the wall but not through it.

Let the vorticity, denoted by ω, be defined as

$$\omega = \frac{\partial v}{\partial x} - \frac{\partial u}{\partial y}$$

The vorticity is numerically equal to twice the angular speed of rotation of the fluid element about its own axis. Flows for which $\omega = 0$ are called irrotational. Figure 10.12 illustrates irrotational flow of a fluid particle around a cylinder.

A theorem, proved by Lord Kelvin, shows that the vorticity of each fluid particle in a fluid undergoing incompressible inviscid flow is preserved. This means that if a flow starts out as irrotational, it will remain irrotational. As an example, consider a fluid that is in uniform flow at a large distance away from a circular cylinder and then flows over it. A uniform flow is clearly irrotational (the fluid particles do not rotate about their own axes). Thus, as the fluid particles approach the cylinder and flow around it, they will remain irrotational.

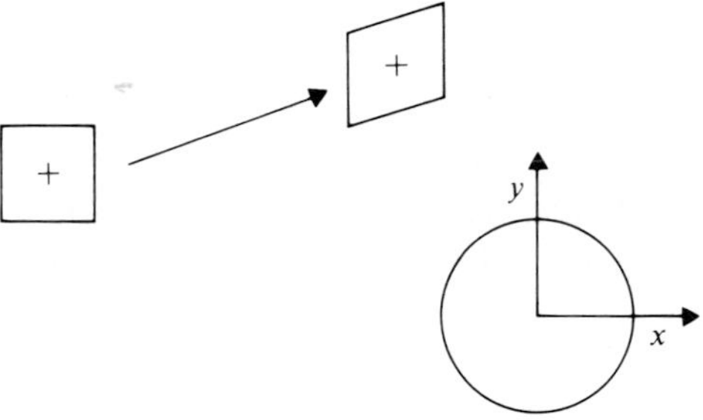

Figure 10.12 Irrotational motion of a fluid particle

The key point now is to define the velocity potential ϕ as

$$u = \frac{\partial \phi}{\partial x} \tag{10.4a}$$

$$v = \frac{\partial \phi}{\partial y} \tag{10.4b}$$

Then the vorticity ω is

$$\omega = \frac{\partial v}{\partial x} - \frac{\partial u}{\partial y} = \frac{\partial}{\partial x}\left(\frac{\partial \phi}{\partial y}\right) - \frac{\partial}{\partial y}\left(\frac{\partial \phi}{\partial x}\right) = 0$$

Thus the velocity potential identically satisfies the condition of irrotationality. There is no particular physical meaning associated with the velocity potential—it is more of a mathematical tool.

The flow must satisfy the continuity equation (10.1). Substituting the velocity potential into the continuity equation gives

$$\frac{\partial u}{\partial x} - \frac{\partial v}{\partial y} = \frac{\partial}{\partial x}\left(\frac{\partial \phi}{\partial x}\right) + \frac{\partial}{\partial y}\left(\frac{\partial \phi}{\partial y}\right) = 0$$

The result is

$$\frac{\partial^2 \phi}{\partial x^2} + \frac{\partial^2 \phi}{\partial y^2} = 0 \tag{10.5}$$

which is Laplace's equation. This is the differential equation that must be solved by using finite elements.

At a solid boundary, the flow cannot penetrate the wall. Equation (10.3) becomes

$$un_x + vn_y = \left(\frac{\partial \phi}{\partial x}\right) n_x + \left(\frac{\partial \phi}{\partial y}\right) n_y = 0$$

or

$$\frac{\partial \phi}{\partial x} n_x + \frac{\partial \phi}{\partial y} n_y = 0 \tag{10.6}$$

This can also be written as the derivative of ϕ in the normal direction:

$$\frac{\partial \phi}{\partial n} = 0$$

The two momentum equations (10.2a) and (10.2b) must now be considered. Again following fluid mechanics textbooks, the two momentum equations can be combined and integrated over the flow region. Bernoulli's equation

$$\frac{P}{\rho} + \frac{1}{2}(u^2 + v^2) = \text{constant}$$

results. Note that for irrotational flow, the same constant applies over the entire flow field. With the velocity potential, Bernoulli's equation becomes

$$\frac{P}{\rho} + \frac{1}{2}\left[\left(\frac{\partial \phi}{\partial x}\right)^2 + \left(\frac{\partial \phi}{\partial y}\right)^2\right] = \text{constant} \tag{10.7}$$

This is easily solved for the pressure once the velocity potential is found by solving Laplace's equation (10.5).

In addition to the illustration of the solution of ideal flow problems, the next few examples consider two questions:

Question 1 Do minor changes in the finite element division scheme produce large changes in the solution?

Question 2 How does the finite element solution compare to the exact (analytical) solution in various areas of the solution region?

Examples 10.3 and 10.4 compare two alternative division schemes that have the same number of nodes and nodal locations. Example 10.5 shows the effect of increasing the number of nodes and elements in the region above the cylinder.

Example 10.3 *Ideal Flow Around Cylinder—First Division Scheme*

Given Ideal flow takes place about a circular cylinder of radius 1 ft. The fluid extends to infinity away from the cylinder. The fluid velocity at infinity is 20 ft/sec from left to right. The velocity potential is presented in Section 9.2 with a discussion of ideal flow about an ellipse between two parallel walls. While the actual fluid region extends to infinity in all directions away from the cylinder, the finite element solution must be evaluated in a region of finite size. A solution region of 8 ft by 4 ft, shown in Figure 10.13, will be used as a reasonable approximation. Only the upper half-region need be analyzed because of symmetry about the x axis.

The exact solution to this problem is

$$\phi = U\left(r + \frac{R^2}{r}\right)\cos\theta + C$$

where C is an arbitrary constant. Here r, θ are polar coordinates of a point in space (θ is taken as counterclockwise from the positive x axis). The radial and tangential velocity components are

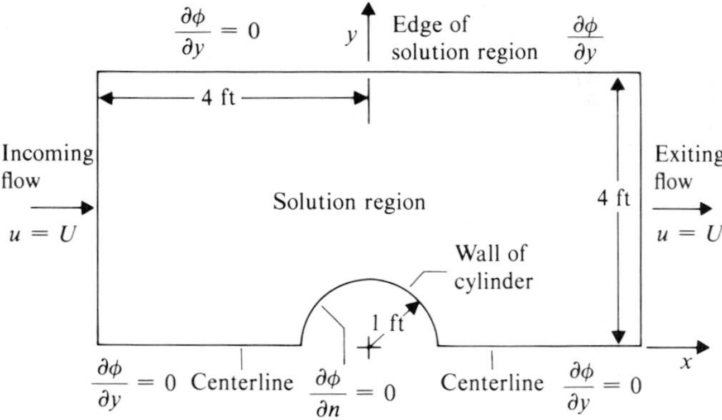

Figure 10.13 Geometry of solution region (Example 10.3—Ideal flow around cylinder, first division scheme)

$$v_r = U\left(1 - \frac{R^2}{r^2}\right)\cos\theta$$

$$v_\theta = -U\left(1 + \frac{R^2}{r^2}\right)\sin\theta$$

These velocity components can easily be resolved into Cartesian coordinates by $u = v_r\cos\theta - v_\theta\sin\theta$, $v = v_r\sin\theta + v_\theta\cos\theta$ for comparison to the finite element solution.

Objective Determine the velocity potential ϕ for the problem and the fluid velocity components in the elements, using computer program FINTWO. Compare the finite element solution to the exact solution near the front and on top of the cylinder.

Solution Figure 10.14 shows the division of the solution region into 24 elements. After division into 12 quadrilaterals, the diagonals (subdividing the quadrilaterals into triangles) were drawn towards the center of the cylinder. Note that the region is divided so as to be symmetric about the y axis, and the velocity component v is symmetric except for its sign. Thus only the first half of the solution region needs to be discussed at the end of this example.

As shown in Example 10.2, the coefficient functions are $K_x = 1.0$, $K_y = 1.0$, $P = 0.0$, and $Q = 0.0$. Along the left boundary, the incoming velocity is $u = 20$ or $\partial\phi/\partial x = 20$.

The derivative boundary condition is written (as in Example 9.1)

$$\frac{\partial\phi}{\partial x}n_x + \alpha\phi + \beta = 0, \quad \text{Left boundary (incoming flow)}$$

where $n_x = -1$. Thus the relation is

$$-\frac{\partial\phi}{\partial x} + 20 = 0$$

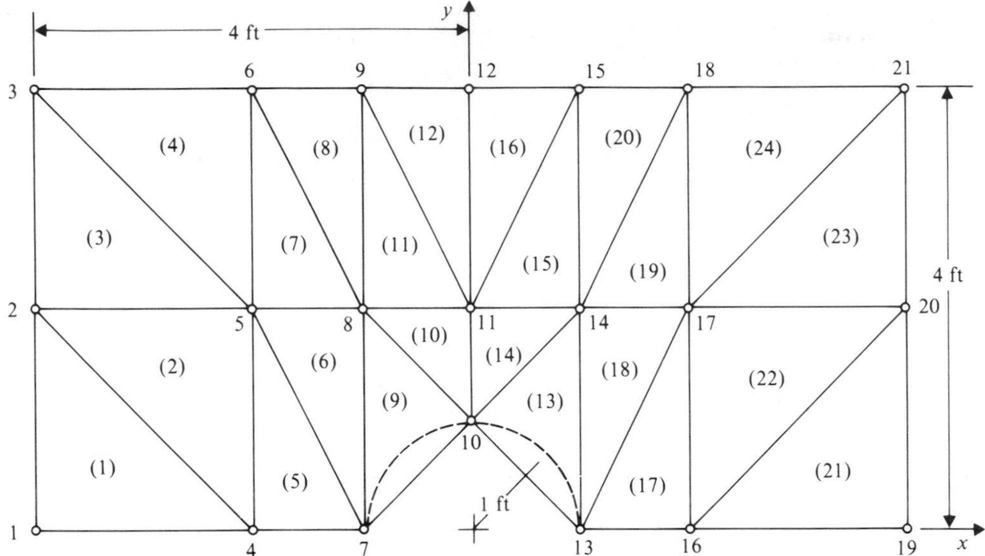

Figure 10.14 First division into nodes and elements (Example 10.3—Ideal flow around cylinder, first division scheme)

and the values of α and β are $\alpha = 0, \beta = 20$. Along the right boundary, the exiting velocity is also $u = 20$ or $\partial\phi/\partial x = 20$. Now the derivative boundary is

$$\frac{\partial \phi}{\partial x} n_x + \alpha\phi + \beta = 0, \quad \text{Right boundary (exiting flow)}$$

where $n_x = 1$. The relation becomes

$$\frac{\partial \phi}{\partial x} - 20 = 0$$

and α and β have the values $\alpha = 0, \beta = -20$. On the remaining sides no flow occurs through the boundaries, so $\alpha = 0$ and $\beta = 0$.

The region has 21 nodes with nodal coordinates given in Table 10.5(a). Element/node connectivity data are presented in Table 10.5(b). The nodes within an element are numbered counterclockwise starting with the node opposite the diagonal. The maximum difference R is 4, yielding a bandwidth of 5. Only one nodal value for ϕ is specified (see Table 10.5(d)) as $\phi_1 = 0$. Element derivative boundary values are given in Table 10.5(e) including both zero gradient and nonzero values for β along the lines $x = -4$ and $x = 4$. The sides are labeled 1,2,3, in accordance with the order of the nodes shown in the element/node connectivity table. For example, in element 1, side 1 occurs between nodes 1 and 4, while side 3 connects node 2 with node 1.

Table 10.5
Input Data (Example 10.3: Ideal Flow Around Cylinder, First Division Scheme)

(a) Nodal Coordinates

Node	x	y
1	−4	0
2	−4	2
3	−4	4
4	−2	0
5	−2	2
6	−2	4
7	−1	0
8	−1	2
9	−1	4
10	0	1
11	0	2
12	0	4
13	1	0
14	1	2
15	1	4
16	2	0
17	2	2
18	2	4
19	4	0
20	4	2
21	4	4

The element matrices are obtained by the same procedure as in Example 10.2, so the details are omitted here. These matrices and other element properties are conveniently produced by computer program FINTWO. Table 10.6 gives the velocity potentials at the node points and the velocity components $u,v,$ where

$$u = \frac{\partial \phi}{\partial x}, \quad v = \frac{\partial \phi}{\partial y}$$

The velocity components are taken within an element since the derivatives are constant over a single element. It can be seen that the horizontal velocity is symmetric about the line $x = 0$, while the vertical velocity is antisymmetric about the line $x = 0$ (positive on the left and negative on the right).

A sketch of the first half of the flow region, Figure 10.15, indicates the nodal potentials and the velocity components. The inlet velocity is close to 20 fps in both elements 1 and 3 with small vertical components. As noted earlier, only the first half need be shown.

Table 10.5 *Continued*

(b) Element/Node Connectivity Data

Element	Node		
	i	*j*	*k*
1	1	4	2
2	5	2	4
3	2	5	3
4	6	3	5
5	4	7	5
6	8	5	7
7	5	8	6
8	9	6	8
9	10	8	7
10	11	8	10
11	8	11	9
12	12	9	11
13	10	13	14
14	11	10	14
15	14	15	11
16	12	11	15
17	16	17	13
18	14	13	17
19	17	18	14
20	15	14	18
21	19	20	16
22	17	16	20
23	20	21	17
24	18	17	21

(c) Element Values for K_x, K_y, P, Q Coefficient Functions

Element	K_x	K_y	P	Q
1	1	1	0	0

(d) Prescribed Nodal Values

Node	Value
1	0.0

Table 10.5 *Continued*

(e) Element Derivative Boundary Values

Element	Side	α	β
1	1	0	0
1	3	0	20
3	3	0	20
4	1	0	0
5	1	0	0
8	1	0	0
9	3	0	0
12	1	0	0
13	1	0	0
16	3	0	0
17	3	0	0
20	3	0	0
21	1	0	−20
21	3	0	0
23	1	0	−20
24	3	0	0

Comparison to the velocity potential can be carried out after the value of C is obtained for this problem. To set $\phi = 0$ at node 1, let $\phi = 0$ at $r = 4, \theta = 180°$. From the exact solution

$$\phi = U\left(r + \frac{R^2}{r}\right)\cos\theta + C$$

$$0 = 20\left(4 + \frac{(1)^2}{4}\right)\cos(180) + C$$

Solving for C, $C = 85$, and for this problem

$$\phi = 20\left(r + \frac{1}{r}\right)\cos\theta + 85$$

Note that the velocities are independent of the value of C.

It is desired to look at the two solutions near the front and above the cylinder. Compare the velocity potential at the nodes 4, 7, 10 as given in Table 10.7(a). At node 4, approaching the front of the cylinder, the error is only 10 percent. At the stagnation point (node 7) the velocity components are zero ($u = 0$ and $v = 0$), and the error is rather large at 23 percent. This results from the large velocity change, from nearly 20 to zero, which cannot be accurately modeled with this small number of elements. The velocity potential at the top of the cylinder yields no error.

Element 5 is the closest to the cylinder. The element horizontal velocity, shown in Table 10.7(b), is 16.8, which is fairly close to the velocity of 15.0 at node 4. As noted earlier, the element velocities are constant, and it is somewhat difficult to decide how to

Table 10.6
Velocity Potential and Element Velocities
(Example 10.3: Ideal Flow Around Cylinder, First Division Scheme)

(a) Nodal Velocity Potentials		**(b) Element Velocity Components**		
Node	**Velocity Potential**	**Element**	**Horizontal Velocity** $u = \frac{\partial \phi}{\partial x}$	**Vertical Velocity** $v = \frac{\partial \phi}{\partial y}$
1	0.0	1	19.3	0.7
2	1.5	2	20.2	1.7
3	2.1	3	20.2	0.3
4	38.5	4	20.3	0.4
5	41.9	5	16.8	1.7
6	42.7	6	21.1	3.9
7	55.3	7	21.1	0.4
8	63.0	8	21.0	0.3
9	63.7	9	25.8	3.9
10	85.0	10	21.9	0.0
11	85.0	11	21.9	0.3
12	85.0	12	21.3	0.0
13	114.6	13	25.8	−3.9
14	106.9	14	21.9	0.0
15	106.2	15	21.9	−0.3
16	131.4	16	21.3	0.0
17	128.0	17	16.8	−1.7
18	127.2	18	21.1	−3.9
19	169.9	19	21.1	−0.4
20	168.5	20	21.0	−0.3
21	167.8	21	19.3	−0.7
		22	20.2	−1.7
		23	20.2	−0.3
		24	20.3	−0.4

compare these velocities to the exact solution. On the top of the cylinder, element 9 has a horizontal velocity of 25.8, which is well below the exact horizontal velocity of 40.0 at $r = 1.0$ and $\theta = 90°$. While these two are not exactly at the same spatial location, this element does predict the highest velocity with a 35 percent error. Element 10 is somewhat above the actual cylinder top.

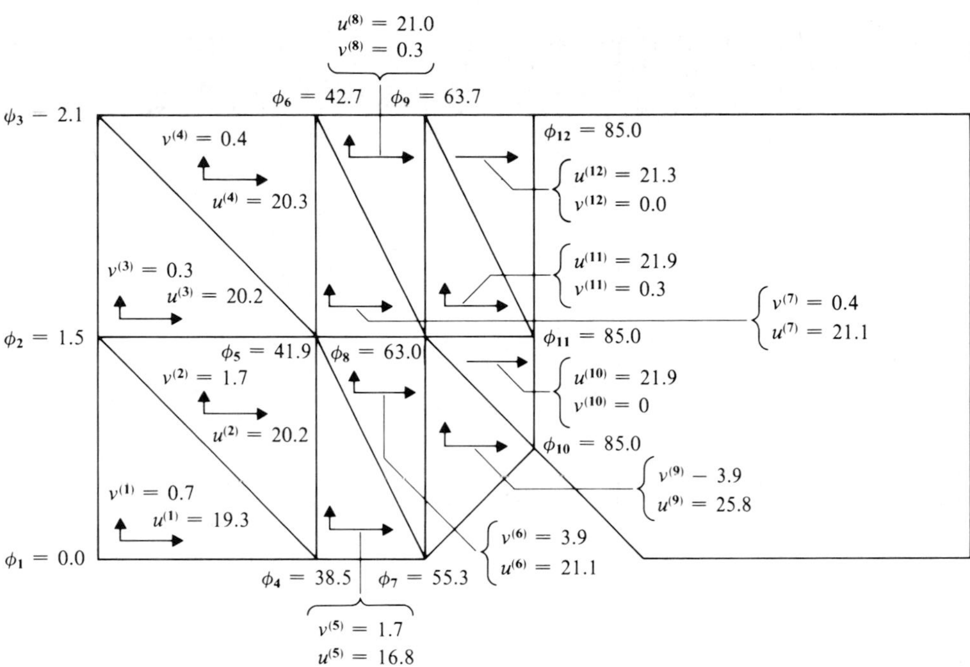

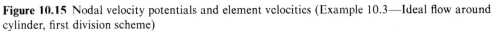

Figure 10.15 Nodal velocity potentials and element velocities (Example 10.3—Ideal flow around cylinder, first division scheme)

Table 10.7
Comparison with Exact Solution (Example 10.3: Ideal Flow Around Cylinder, First Division Scheme)

(a) Velocity Potentials

Node	Approximate ϕ	Exact ϕ	Error %	Location
4	38.5	35.0	10	Approaching cylinder
7	55.3	45.0	23	Stagnation point
10	85.0	85.0	0	Top of cylinder

(b) Velocities

Element	Approximate u	Approximate v	Node	Exact u	Exact v
5	16.8	1.7	4	15.0	0
9	25.8	3.9	7	0	0
10	21.9	0	10	40.0	0

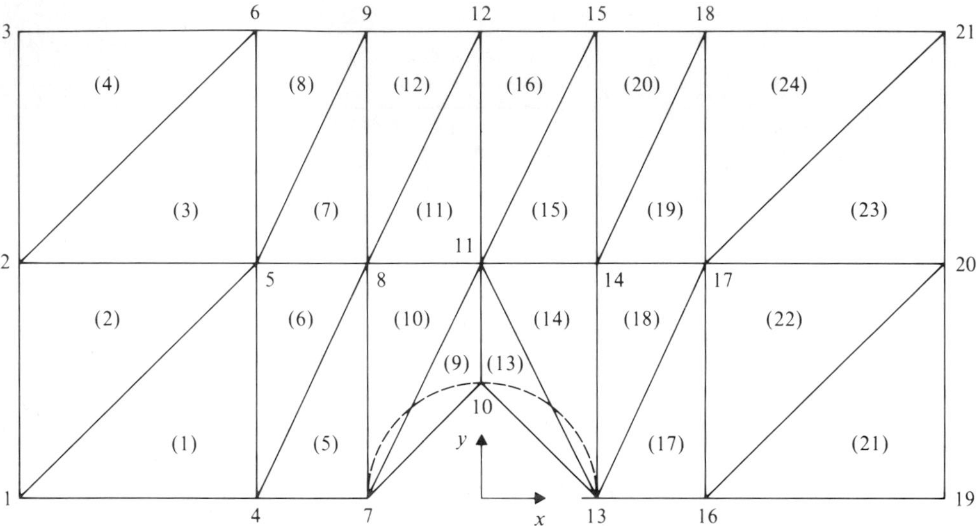

Figure 10.16 Second division into nodes and elements (Example 10.4—Ideal flow around cylinder, second division scheme)

Example 10.4 *Ideal Flow Around Cylinder—Second Division Scheme*

Given The same problem as Example 10.3 is considered with a division scheme, shown in Figure 10.16, chosen in no particular pattern but with the same number of node points and nodal locations. The connections between these node points differ from those in the previous example.

Objective Find the potential and velocity components. Compare the results with the previous example. Also answer Question 1.

Solution The nodal coordinates and the prescribed nodal potential are the same as in the previous example. Table 10.8 gives the new element/node connectivity data and derivative boundary conditions. Results are shown in Table 10.9. It is easily seen that the horizontal velocity components are no longer symmetric about the line $x = 0$, since the elements are not.

Again the finite element velocity potential at nodes 4, 7, and 10 is compared with the exact solution evaluated at nodes 4, 7, and 10. The results are shown in Table 10.10(a). In spite of the division scheme changes, there is very little change in the error. Similarly, the velocity components in Table 10.10(b) for the second division scheme produce about the same errors as the first division scheme (Example 10.3); for example, in the element with the highest velocity, element 9, the error is 29 percent, as compared to 35 percent.

Table 10.8
Input Data
(Example 10.4: Ideal Flow Around Cylinder, Second Division Scheme)

(a) Element/Node Connectivity Data

Element	Node i	Node j	Node k
1	4	5	1
2	2	1	5
3	5	6	2
4	3	2	6
5	7	8	4
6	5	4	8
7	8	9	5
8	6	5	9
9	10	11	7
10	8	7	11
11	11	12	8
12	9	8	12
13	10	13	11
14	14	11	13
15	14	15	11
16	12	11	15
17	16	17	13
18	14	13	17
19	17	18	14
20	15	14	18
21	19	20	16
22	17	16	20
23	20	21	17
24	18	17	21

(b) Element Derivative Boundary Conditions

Element	*Side*	α	β
1	3	0	0
2	1	0	20
4	1	0	20
4	3	0	0
5	3	0	0
8	3	0	0
9	3	0	0
12	3	0	0
13	1	0	0
16	3	0	0
17	3	0	0
20	3	0	0
21	1	0	−20
21	3	0	0
23	1	0	−20
24	3	0	0

This result answers the first question:

Question 1 Do minor changes in the finite element division scheme produce large changes in the solution?

Answer 1 If the division scheme simply rearranges the element subdivision without substantially increasing the number of nodes and elements in a region, the errors remain about the same.

It can also be seen that simply reconnecting the nodal points without making a significant change in the number of nodes does *not* indicate whether enough nodes have been employed or not. This is an important conclusion.

Table 10.9
Nodal Potentials and Element Velocities
(Example 10.4: Ideal Flow Around Cylinder, Second Division Scheme)

(a) Nodal Velocity Potentials		(b) Element Velocity Components		
Node	Velocity Potential, ϕ	Element	Horizontal Velocity $u = \partial\phi/\partial x$	Vertical Velocity $v = \partial\phi/\partial y$
1	0.0	1	19.4	1.4
2	1.2	2	20.2	0.6
3	1.8	3	20.2	0.4
4	38.8	4	20.3	0.3
5	41.5	5	17.4	3.1
6	42.4	6	20.9	1.4
7	56.1	7	20.9	0.4
8	62.4	8	20.9	0.4
9	63.3	9	28.5	0.0
10	84.6	10	22.2	3.1
11	84.6	11	22.2	0.0
12	84.6	12	21.3	0.4
13	113.1	13	28.5	0.0
14	106.8	14	22.2	−3.1
15	105.9	15	22.2	−0.4
16	130.4	16	21.3	0.0
17	127.7	17	17.4	−1.4
18	126.9	18	20.9	−3.1
19	169.2	19	20.9	−0.4
20	168.0	20	20.9	−0.4
21	167.4	21	19.4	−0.6
		22	20.2	−1.4
		23	20.2	−0.3
		24	20.3	−0.4

A critical question with any numerical method concerns the solution accuracy. The number of elements used and their location has a strong effect. Usually the number of elements cannot simply be increased without limit. Either computer costs or machine size limitations require the programmer to make various choices.

Normally a small parametric study of the solution accuracy versus the number of elements must be made for any problem. The temperature, velocity, or other field variable must change less than the desired percentage when the number of elements is changed substantially. Only then is the required number of elements known.

This section considers the effect of increasing the number of elements for ideal flow around a cylinder. The number is increased in the region just above the cylinder. A large change occurs, indicating that the 24 elements in Examples 10.3 and 10.4 were not enough for an accurate solution.

Table 10.10
Comparison with Exact Solution (Example 10.4: Ideal Flow Around Cylinder, Second Division Scheme)

(a) Velocity Potentials

Node	Approximate ϕ	Exact ϕ	Error %	Location
4	38.8	35.0	11	Approaching cylinder
7	56.1	45.0	25	Stagnation point
10	84.6	85.0	0.5	Top of cylinder

(b) Velocities

Element	Approximate u	Approximate v	Node	Exact u	Exact v
6	20.9	1.4	4	15.0	0
5	17.4	3.1	7	0	0
9	28.5	0.0	10	40.0	0

Example 10.5 *Ideal Flow Around Cylinder—32 Elements*

Given The same problem is considered again.

Objective It is desired to increase the accuracy of the analysis between the cylinder and the wall to obtain a better estimate of the peak velocity. Place extra nodes along the lines $x = \pm 0.5$ to increase the number of elements around the cylinder. The new division scheme is shown in Figure 10.17. Determine the velocity potential and velocity components. Compare the more detailed solution near the front and on top of the cylinder to that in Example 10.3. Also answer Question 2.

Solution The results are given in Table 10.11. Figure 10.17 shows the nodal potentials and velocity components for the first half of the flow region. In the entrance region the velocity in element 1 is very close to that in Figure 10.17: $u_1 = 19.3$, $v_1 = 0.7$ with 24 elements; $u_1 = 19.0$, $v_1 = 0.9$ with 32 elements. Not much change occurs here.

Table 10.12(a) gives the error at several node points of interest. At nodes 4 and 7, the errors with 24 elements (Example 10.3) are 10 percent and 23 percent, while the 32-element solution has errors of 9 percent and 20 percent. Again, the two solutions produce about the same errors for the region in which the fluid is approaching the cylinder.

Table 10.11
Nodal Potentials and Element Velocities
(Example 10.5: Ideal Flow Around Cylinder, 32 Elements)

(a) Nodal Velocity Potentials		**(b) Element Velocity Components**		
Node	Velocity Potential, ϕ	Element	Horizontal Velocity $u = \partial\phi/\partial x$	Vertical Velocity $v = \partial\phi/\partial y$
1	0.0	1	19.0	0.9
2	2.0	2	20.2	2.2
3	2.9	3	20.2	0.5
4	38.1	4	20.5	0.7
5	42.4	5	15.7	2.2
6	43.9	6	21.3	5.0
7	53.8	7	21.3	0.7
8	63.8	8	21.6	0.8
9	65.4	9	21.6	0.8
10	68.8	10	23.4	5.9
11	75.5	11	23.4	0.8
12	76.5	12	22.2	0.5
13	87.8	13	36.5	5.9
14	87.8	14	24.6	0.0
15	87.8	15	24.6	0.5
16	106.8	16	22.5	0.0
17	100.1	17	36.5	−5.9
18	99.0	18	24.6	0.0
19	121.8	19	24.6	−0.5
20	111.8	20	22.5	0.0
21	110.1	21	21.2	−5.0
22	137.5	22	23.4	−5.9
23	133.1	23	23.4	−0.8
24	131.7	24	22.2	−0.5
25	175.6	25	15.7	−2.2
26	173.6	26	21.3	−5.0
27	172.6	27	21.3	−0.7
		28	21.6	−0.8
		29	19.0	−1.0
		30	20.2	−2.2
		31	20.2	−0.5
		32	20.5	−0.7

On top of the cylinder, the element of interest is 9 (in Example 10.3) and 13 here. The velocity components are $u_9 = 25.8$, $v_9 = 3.9$ (24-element solution); $u_{13} = 36.5$, $v_{13} = 5.9$ (32-element solution).

The magnitude of velocity and the error, relative to the exact solution of 40.0 at the top, is $|u_9| = 26.1$, error $= 35\%$ (24-element solution); $|u_{13}| = 37.0$, error $= 7\%$ (32-element solution). In this region, the 32-element solution is a great deal more accurate. Table 10.12 shows the velocity in some other elements.

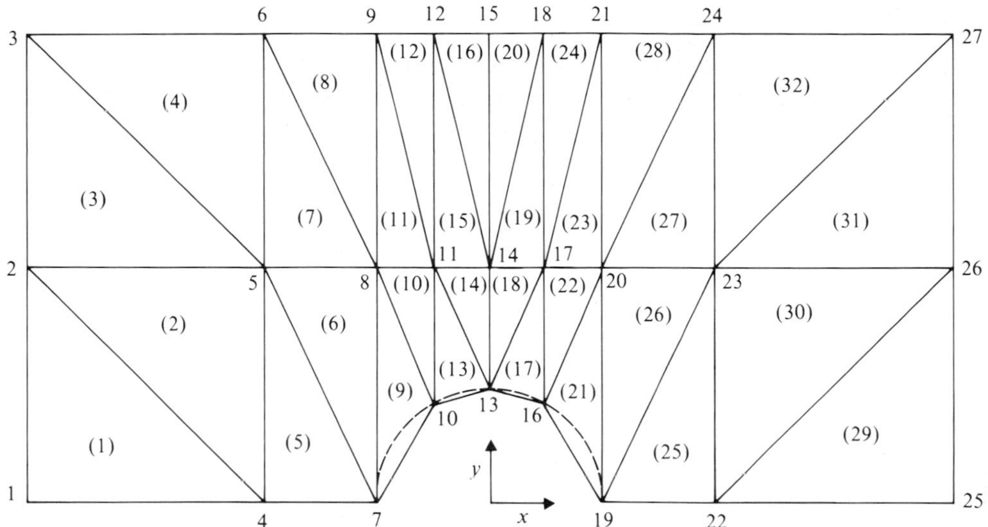

Figure 10.17 Division into nodes and elements (Example 10.5—Ideal flow around cylinder, 32 elements)

Table 10.12
Comparison with Exact Solution (Example 10.5: Ideal Flow Around Cylinder, 32 Elements)

(a) Velocity Potentials

Node	Approximate ϕ	Exact ϕ	Error %	Location
4	38.1	35.0	9	Approaching cylinder
7	53.8	45.0	20	Stagnation point
13	87.8	85.0	3	Top of cylinder

(b) Velocities

Element	Approximate u	Approximate v	Node	Exact u	Exact v
5	20.5	0.7	4	15.0	0.0
9	21.2	5.0	7	0.0	0.0
13	36.5	5.9	13	40.0	0.0

Recall the second question

Question 2 How does the finite element solution compare to the exact (analytical) solution in various areas of the solution region?

Answer 2 Substantially increasing the number of nodes (or elements) in a solution region improves the accuracy.

The further question of how much accuracy is required cannot be answered here. That must be an engineering decision balancing the cost of the analysis versus the accuracy desired.

10.4 Torsion in Bars

This section presents an engineering problem that can be solved using a second-order partial differential equation of the type discussed in Chapter 9. First, the governing equations are derived. Second, an example involving a triangular bar is carried out, and the finite element solution is compared to the exact solution. The objective is to determine the shear stress distribution over the bar cross section so that the peak stress magnitude and direction can be obtained.

In this torsion boundary value problem, a long bar of uniform cross section is twisted by couples at the ends. Figure 10.18 illustrates the geometry. A three-dimensional coordinate system x,y,z is set up at one end of the bar with z as the coordinate along the axis (centroid) of the shaft. As the shaft is twisted about the x axis by the couples, the displacements in the x,y plane are

$$\begin{aligned} u &= \theta zy = \text{horizontal displacement} \\ v &= \theta zx = \text{vertical displacement} \end{aligned} \tag{10.8}$$

where θ is the angle of rotation of the shaft. Axial displacements w occur as well: $w = w(x,y)$ = axial displacement. These are not known at this point. Cross sections of the bar do not remain planar but warp.

Actually, the axial displacements are not of interest here—the stresses are. The following analysis uses a stress function $\phi(x,y)$ to eliminate the axial displacement w from the equations. At the end of the solution for ϕ, the axial displacements $w(x,y)$ can be obtained, if desired.

The shear strains in each x,y plane along the bar (see Chapter 14 for a detailed derivation) are

$$\begin{aligned} \gamma_{xz} &= \frac{\partial u}{\partial z} + \frac{\partial w}{\partial x} \\ \gamma_{yz} &= \frac{\partial v}{\partial z} + \frac{\partial w}{\partial y} \end{aligned} \tag{10.9}$$

With the displacements (10.8), these become

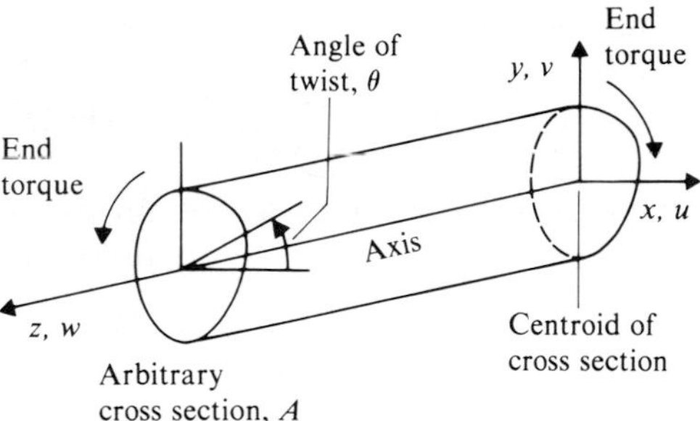

Figure 10.18 Torsion in bar

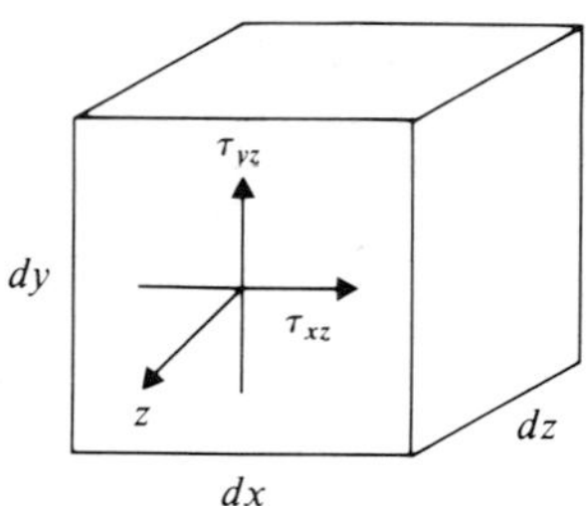

Figure 10.19 Shear stresses acting on cross sectional face of differential cube

$$
\begin{aligned}
\gamma_{xz} &= -\theta y + \frac{\partial w}{\partial x} \\
\gamma_{yz} &= \theta x + \frac{\partial w}{\partial y}
\end{aligned}
\tag{10.10}
$$

The corresponding shear stresses in each x,y plane are related to the shearing strains by

$$
\begin{aligned}
\tau_{xz} &= G\,\gamma_{xz} \\
\tau_{yz} &= G\,\gamma_{yz}
\end{aligned}
\tag{10.11}
$$

where G is the shear modulus. Figure 10.19 illustrates these stress components on a differential cube $dxdydz$ acting on a cross sectional face. Then

$$
\begin{aligned}
\tau_{xz} &= -G\theta y + G\,\frac{\partial w}{\partial x} \\
\tau_{yz} &= G\theta x + G\,\frac{\partial w}{\partial y}
\end{aligned}
\tag{10.12}
$$

The shear stresses are related to the axial displacement.

The governing differential equation for torsion in a bar is

$$\frac{\partial \tau_{xz}}{\partial x} + \frac{\partial \tau_{yz}}{\partial y} = 0 \tag{10.13}$$

It is obtained by setting the torque on a differential rectangle of dimensions dx by dy to zero (again see Chapter 14 if more details are desired). Thus it is called the equilibrium equation. Define a stress function $\phi(x,y)$ of the form

$$\begin{aligned} \tau_{xz} &= \frac{\partial \phi}{\partial y} \\ \tau_{yz} &= -\frac{\partial \phi}{\partial x} \end{aligned} \tag{10.14}$$

Substituting these expressions into the above equilibrium equation yields

$$\frac{\partial^2 \phi}{\partial x \partial y} - \frac{\partial^2 \phi}{\partial x \partial y} = 0$$

which is satisfied automatically.

Returning to the shear stress relations (10.12),

$$\tau_{xz} = \frac{\partial \phi}{\partial y} = -G\theta y + G\frac{\partial w}{\partial x}$$

$$\tau_{yz} = -\frac{\partial \phi}{\partial x} = G\theta x + G\frac{\partial w}{\partial y}$$

The objective is to eliminate w from these two equations. Differentiate the first equation with respect to y and the second equation with respect to x. Subtracting the second result from the first gives

$$\frac{\partial^2 \phi}{\partial x^2} + \frac{\partial^2 \phi}{\partial y^2} + 2G\theta = 0 \tag{10.15}$$

This is the field equation for the stress function ϕ.

Along the boundary of the bar, the shear stress must vanish. This may be written as

$$\tau_{xz}n_x + \tau_{yz}n_y = 0 \tag{10.16}$$

Using the stress function, the boundary condition is

$$\frac{\partial \phi}{\partial y}n_x - \frac{\partial \phi}{\partial x}n_y = 0$$

Let s represent the coordinate along the boundary. Then

$$n_x = -\frac{dy}{ds}$$

$$n_y = \frac{dx}{ds}$$

and the boundary condition becomes

$$\frac{\partial \phi}{\partial y}\left(-\frac{dy}{ds}\right) - \frac{\partial \phi}{\partial x}\left(\frac{dx}{ds}\right) = \frac{d\phi}{ds} = 0$$

Thus the stress function must be constant along the boundary of the cross section. The value of the constant is arbitrary, so it will be taken here as zero:

$$\phi = 0 \text{ on } L_1 \tag{10.17}$$

This is the boundary condition for the torsion problem.

Once the stress function $\phi(x,y)$ is obtained from the finite element solution, the stresses in each element are given by (10.14)

$$\tau_{xz} = \frac{\partial \phi}{\partial y}, \quad \tau_{yz} = -\frac{\partial \phi}{\partial x}$$

There are no normal stresses, so the magnitude of the shear stress is simply

$$\tau_m = \sqrt{\tau_{xz}^2 + \tau_{yz}^2} \tag{10.18}$$

If desired, the axial displacements can be determined from the differential relations (10.12). Finally the torque T required to produce the axial twist θ is evaluated with

$$T = 2 \int\int_A \phi(x,y)\, dA \tag{10.19}$$

The derivation for this is given in textbooks on the theory of elasticity.

Example 10.6 *Torsion of Triangular Bar*

Given Consider a triangular bar with the dimensions shown in Figure 10.20(a). The cross section is that of an equilateral triangle with the origin of the coordinate system at the centroid. Let the problem parameters be $L = 10$ in, $G = 11 \times 10^6$ lbf/in^2, and $\theta = 0.05$ deg/in $= 8.727 \times 10^{-4}$ rad/in. Thus the total twist angle is 0.5 degree over the full length of the bar.

The exact solution is

$$\phi = -G\theta\left[\frac{1}{2}(x^2 + y^2) + \frac{1}{2a}(x^3 - 3xy^2) - \frac{2a^2}{27}\right]$$

where, for this problem $a = 2$ inches. The stresses are

$$\tau_{xz} = \frac{\partial \phi}{\partial y} = -G\theta\left[y - \frac{3xy}{a}\right]$$

$$\tau_{yz} = -\frac{\partial \phi}{\partial x} = G\theta\left[x + \frac{3}{2a}(x^2 - y^2)\right]$$

and the largest stress occurs at the middle of one side

$$\tau_{max} = \frac{G\theta a}{2} = 9{,}600 \text{ lbf/in}^2$$

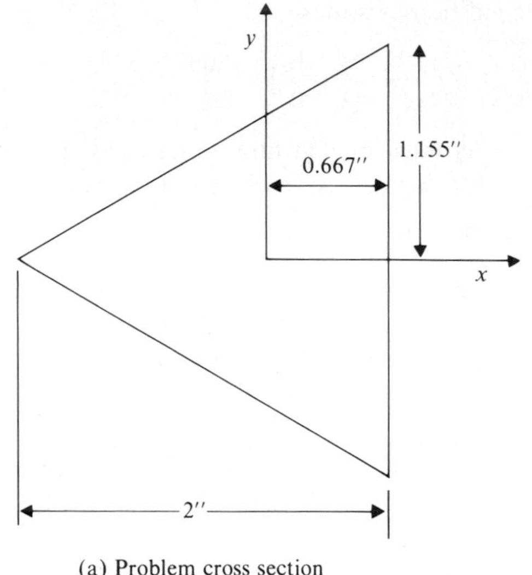

(a) Problem cross section

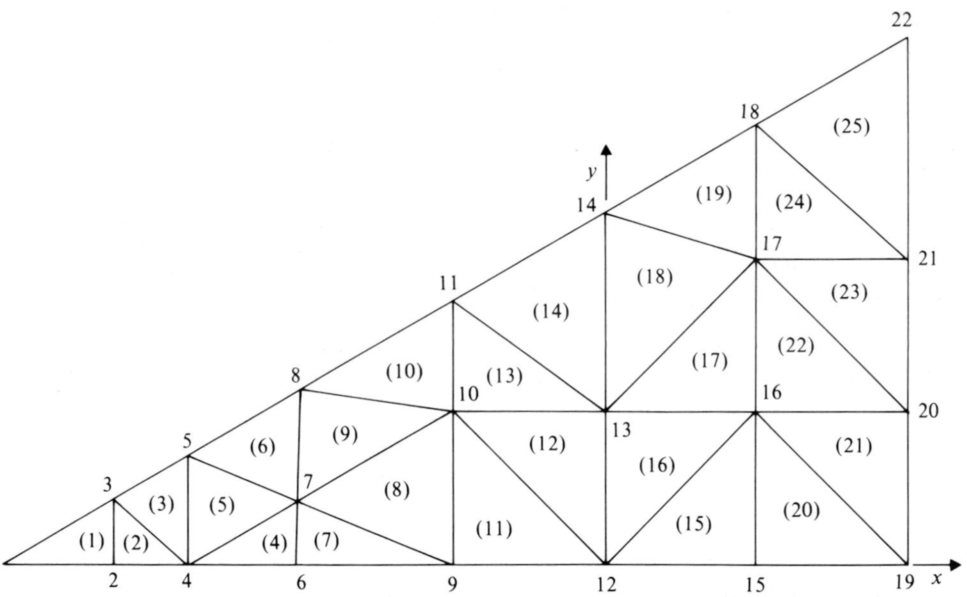

(b) Division of one half of bar into nodes and elements (shown enlarged)

Figure 10.20 Geometry for triangular bar (Example 10.6—Torsion of triangular bar)

Table 10.13
Nodal Coordinates (Example 10.6: Torsion of Triangular Bar)

Node	*x* Coordinate	*y* Coordinate
1	−1.33	0.0
2	−1.05	0.0
3	−1.05	0.14
4	−0.90	0.0
5	−0.90	0.24
6	−0.65	0.0
7	−0.65	0.15
8	−0.65	0.37
9	−0.30	0.0
10	−0.30	0.33
11	−0.30	0.59
12	0.00	0.0
13	0.00	0.33
14	0.00	0.67
15	0.33	0.0
16	0.33	0.34
17	0.33	0.67
18	0.33	0.97
19	0.67	0.0
20	0.67	0.34
21	0.67	0.67
22	0.67	1.15

Evaluating the torque on the cylinder required to produce this angle of twist is

$$T = 2\iint_A \phi(x,y)\, dA = \frac{G\theta a^4}{15\sqrt{3}} = 5{,}911.9 \text{ lbf-in}$$

Objective Use approximately 25 elements to solve for the stress function in the triangular bar. Also determine the element stresses and torque on the bar. Compare to the exact solution. Note that the x axis is a line of symmetry.

Solution Figure 10.20(b) shows the division into nodes and elements. The nodal coordinates are given in Table 10.13. Element/node connectivity data are easily obtained from Figure 10.20(b). The coefficient functions are

$$K_x(x,y) = 1.0$$

$$K_y(x,y) = 1.0$$

$$P(x,y) = 0.0$$

$$Q(x,y) = 2G\theta = 2\left(11 \times 10^6 \frac{\text{lbf}}{\text{in}^2}\right)\left(8.727 \times 10^{-4} \frac{\text{rad}}{\text{in}}\right)$$

$$= 1.920 \times 10^4 \text{ lbf/in}^3$$

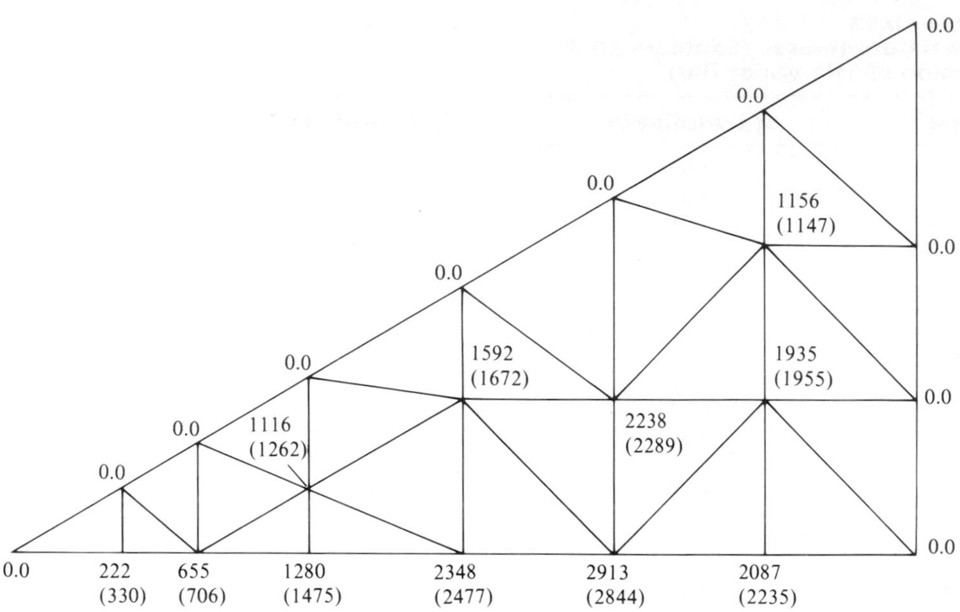

Figure 10.21 Stress function values, ϕ, for finite element solution (upper numbers) and exact solution (lower numbers) (Example 10.6—Torsion of triangular bar)

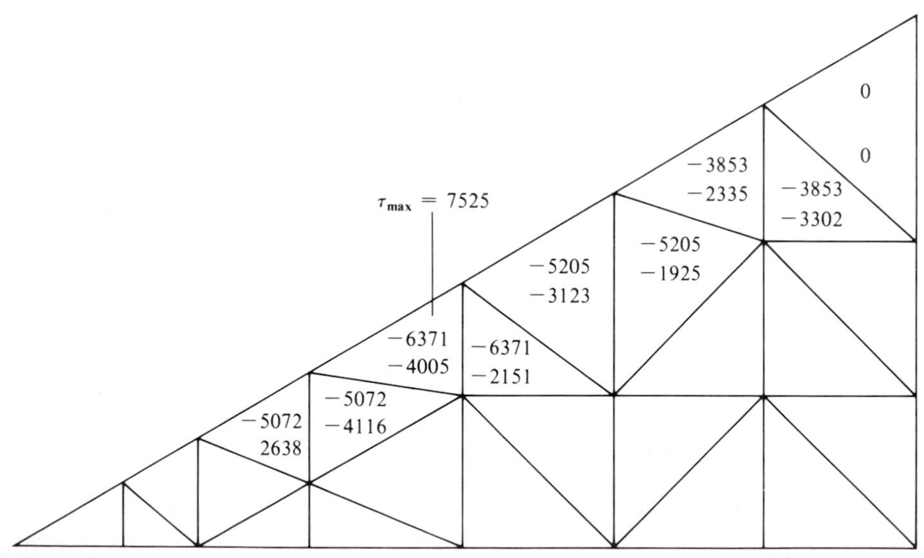

Figure 10.22 Element shear stresses (upper value is τ_{xz}, lower is τ_{yz} and peak stress (τ_{max}); units of stress are lbf/in^2 (Example 10.6—Torsion of triangular bar)

All elements have the same value. Similarly, all boundary nodes except those along the x axis have the boundary condition $\phi = 0$, while along the x axis, $\alpha = 0$ and $\beta = 0$ due to symmetry.

The nodal values of the stress function ϕ, as obtained from the finite element solution, are shown in Figure 10.21. Below each of these values, the exact solution is shown in parentheses. Also, some of the larger element stresses are indicated in Figure 10.22. The magnitude of the largest stress

$$\tau_{max} = \sqrt{(-6{,}371)^2 + (-4{,}005)^2} = 7{,}525 \text{ lbf/in}^2$$

occurs in element 10. This is 22 percent lower than the exact value of 9,600 lbf/in². The torque on the bar as calculated from the finite element solution is $T = 5{,}043$ lbf-in, which is 15 percent below the exact solution.

10.5 Problems

A. Heat Transfer in Thin Fins

10.1 Find the temperatures at nodes 2 and 4 in Example 10.1 (Heat conduction and convection in trapezoidal fin) if the nodes 1 and 3 are maintained at 280°C and 165°C. All other parameters are the same. Do not use program FINTWO. Also find the rate of heat flow by conduction through the wall and the heat loss by convection through the top and bottom of the fin.

10.2 Do Problem 10.1, using program FINTWO.

10.3 Heat conduction and convection occur in a trapezoidal fin, shown in Figure 10.23, with a length of 90 mm and width (at the tip) of 130 mm. It is desired to use this fin to replace the fin in Example 10.1. The new fin is shorter but wider. Using a two-element model similar to that in Example 10.1 (with new nodal coordinates for nodes 2 and 4), calculate the temperatures in the fin and the heat convection through the top and bottom of the fin. Perform the calculations by hand (except for a calculator), showing the steps. Do not use computer program FINTWO. Compare the results to those obtained for the fin in Example 10.1.

10.4 Solve for the temperature distribution in the fin in Example 10.1 (Heat conduction in trapezoidal fin), but employ the 16-element division shown in Figure 10.24. Use program FINTWO, presented in Appendix B. Compare the tip temperature distribution to the two-element analysis in Example 10.1. Calculate the rate of heat convection through the top and bottom of the fin, and compare the results to those of Example 10.1.

10.5 Heat conduction and convection occurs from the thin fin shown in Figure 10.25. The fin has the unusual shape shown because it is attached to a hot wall at 200°F along sides *FA*, *AB*, *CD* and to a circular pipe, also at 200°F, along side *BC*. The fin has thickness (in the z direction) $t = 0.1$ in. Along sides *DE* and *EF*, the fin is thin, so no heat flows through these sides. The fin is made of aluminum with $k = 9.8$ Btu/hr-in-°F, and the convection coefficient for the top and bottom of the fin is $h = 0.015$ Btu/hr-in²-°F. Use at least 12 elements, and program FINTWO to solve for the temperature distribution. Sketch the location of the node points and write in the nodal temperatures. Also calculate the heat flow from the wall into the fin.

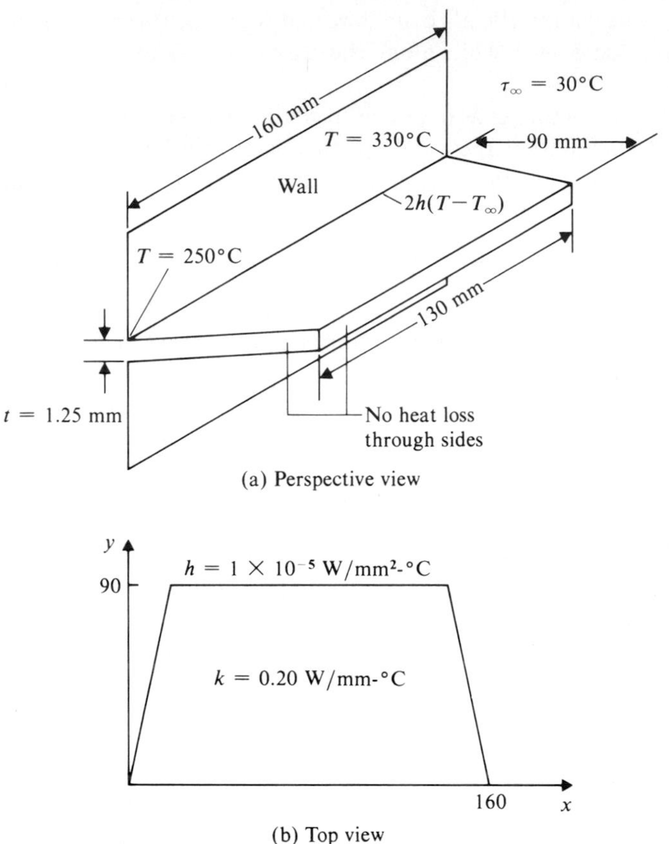

Figure 10.23 Heat conduction and convection in trapezoidal fin for Problem 10.3

10.6 Solve Problem 10.5 with at least 24 elements, keeping as many node points at the same location as possible. Compare the results for this problem with the nodal temperatures for Problem 10.5.

10.7 Make up a thin fin problem of your own choosing, including fin geometry and heat transfer properties. Use at least 10 elements to solve the problem. In your write-up of the problem, describe it in full and indicate the nodal temperatures in a sketch.

B. Heat Transfer in Very Long Bodies

10.8 A very long (in the z direction) wall that forms the side of a building is made of concrete with thermal conductivity $k = 1.0$ W/m-°C. The dimensions are shown in Figure 10.26. At the bottom of the wall, the ground temperature is taken as 22°C at the left edge, −9°C at the right edge; it varies linearly between the two so the wall temperature matches that. The top of the wall is insulated. On the left side, the air inside the building is maintained at 23°C and the convection coefficient is $h = 200$ W/m²-°C. On the outside wall, the air temperature is −10°C and the same convection coefficient applies.

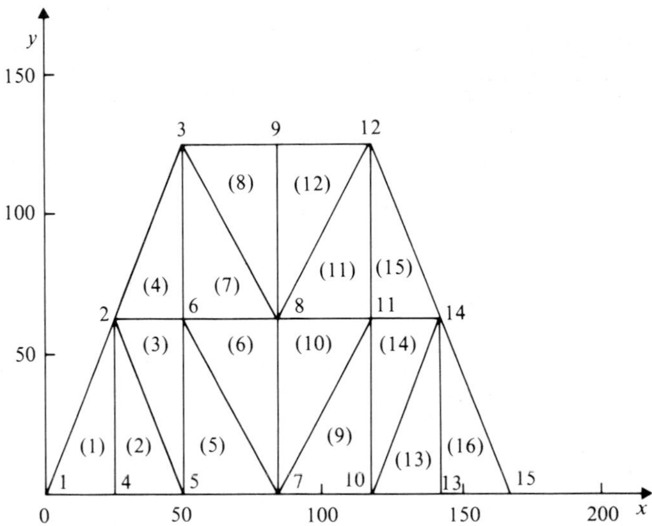

Figure 10.24 Division scheme for Problem 10.4 (same fin geometry as Example 10.1)

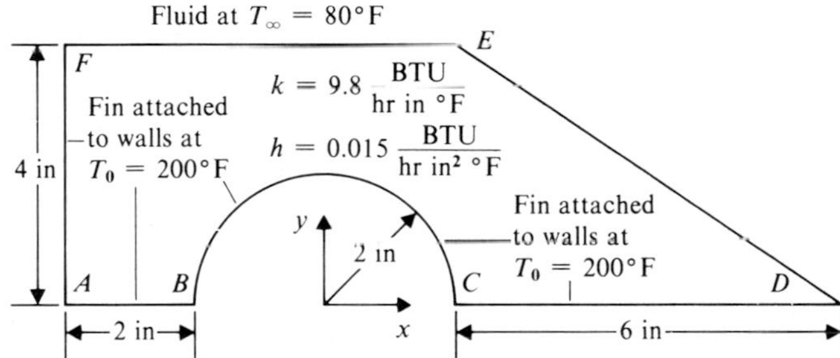

Figure 10.25 Fin geometry for Problems 10.5 and 10.6

Use computer program FINTWO with at least 10 elements to determine the temperature distribution in the wall. Sketch the wall and indicate nodal temperatures. Also determine the total heat flow from the building to the wall.

10.9 Redo Problem 10.8 but with earth insulation piled next to the wall, as shown in Figure 10.27. The bottom of the earth insulation extends 1.2 m away from the wall and its upper surface forms a 45° angle with the wall. The thermal conductivity of the earth is $k = 0.06$ W/m-°C, while the thermal conductivity of the concrete doesn't change. Assume that the bottom of the wall is at 22°C and that the temperature of the bottom of the earth insulation decreases linearly from 22°C at the left edge to −9°C at the right edge. At the outside surface of the earth, heat flows by convection to the outside air as shown. Assume perfect conduction through the concrete–earth interface.

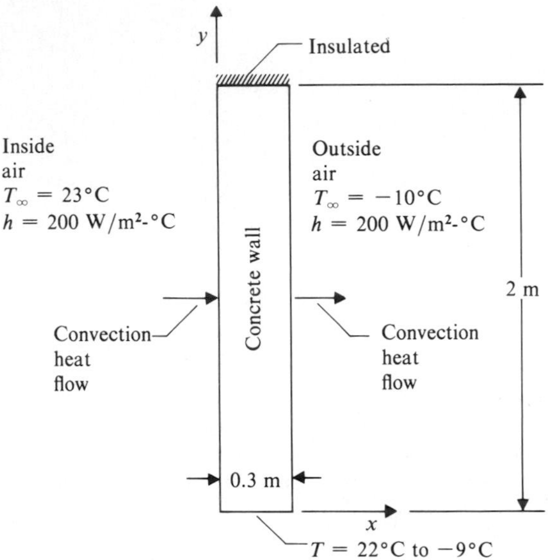

Figure 10.26 Concrete wall for Problem 10.8

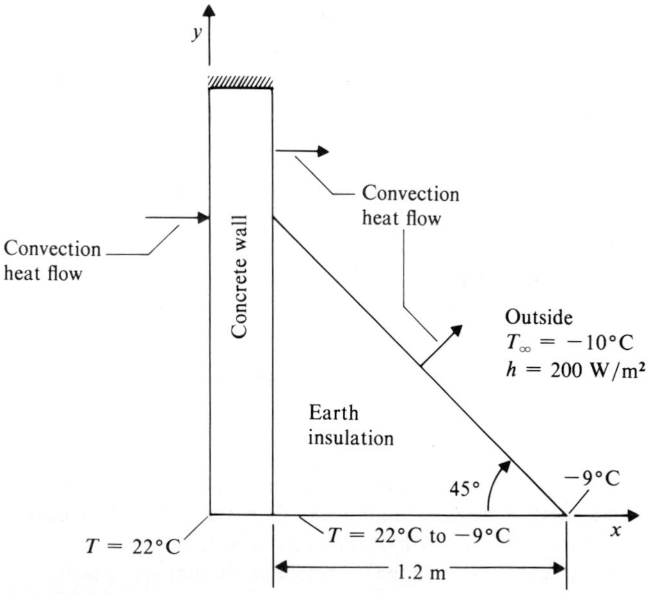

Figure 10.27 Concrete wall and earth insulation for Problem 10.9

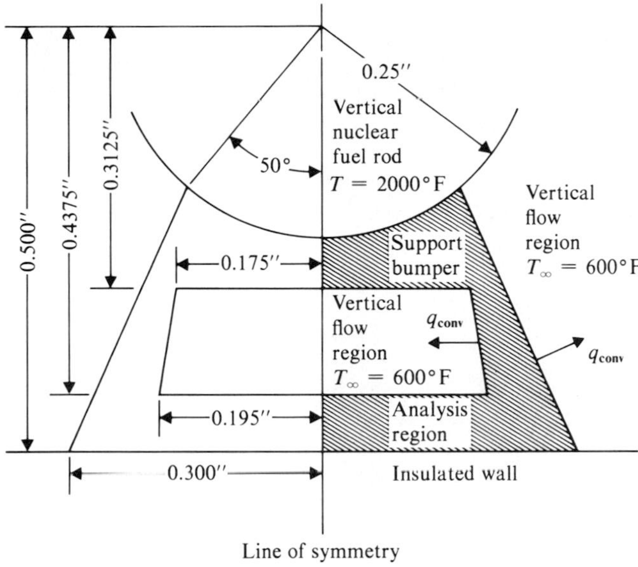

Figure 10.28 Problem geometry for Problem 10.10

Use at least five more elements to model the earth insulation area, and combine these with the elements used in Problem 10.8. Sketch the wall and earth, including nodal temperatures. Calculate the total heat flow from the building to the wall, and compare to the results from Problem 10.8.

10.10 A nuclear fuel rod whose axis is vertical is supported by a bumper, as shown in Figure 10.28. A liquid coolant at 600°F flows out of the plane of the paper through the center of the bumper support. The coolant also flows out of the plane of the paper along the side of the support bumper. Convection heat transfer cools the support. The support is assumed to be rigidly attached to the fuel rod and to the wall. Dimensions are shown in Figure 10.28.

The thermal conductivity of the bumper support is taken as 3.6 BTU/hr-in-°F. The bumper support may be considered infinitely long in the dimensions out of the plane of the paper.

The fluid flowing out of the plane of the paper is at a temperature of $T_\infty = 600$°F. The convection coefficient is taken as 0.13 BTU/hr-in^2-°F. It is assumed that the surface of the fuel rod is at 2,000°F and that the wall is insulated. The differential equation for the problem is

$$\frac{\partial}{\partial x}\left(kt\,\frac{\partial T}{\partial x}\right) + \frac{\partial}{\partial y}\left(kt\,\frac{\partial T}{\partial y}\right) = 0$$

It is desired to determine the maximum temperature along the bumper support/wall interface. Find both the location and its value.

By symmetry considerations, only one-half of the bumper support need be analyzed. Use approximately 25 nodes and 25 elements to create a relatively coarse division scheme. Try to make the elements approximately equal in size around the bumper.

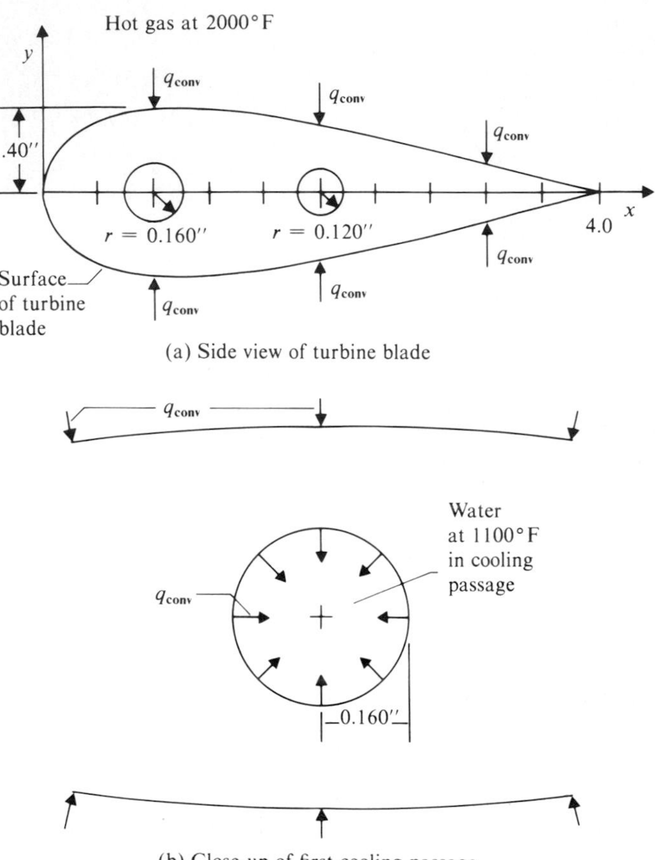

Figure 10.29 Turbine blade with cooling passage for Problem 10.11

Analyze the case in which the temperature of the fuel rod has increased to 2,200° F. Determine the location and increase in maximum temperature along the bumper support/wall interface under these conditions. Assume that all other conditions remain the same.

10.11 A turbine blade from a jet engine is often subject to very high temperatures from the hot gases flowing past it. Figure 10.29(a) illustrates the geometry. The gas temperature is 2,000° F. Cooling water flows through two passages in the blade, as shown (actually several passages are used, but two are considered for this problem). The cooling water is at 1,100° F. Thus heat enters through the surface of the blade by convection, flows in the interior by pure conduction, and then leaves the blade via two cooling passages. The blade axis (out of the paper) is long compared to the blade cross sectional length of 4.0 inches.

The aluminum blade has the thermal conductivity of $k = 9.8$ Btu/hr-in-° F. On the blade surface, the convection coefficient for the gas is $h_{gas} = 0.050$ Btu/hr-in-° F, while the convection coefficient for the cooling water is $h_{water} = 0.080$ Btu/hr-in-° F. There are no specified temperatures anywhere in the blade for this problem.

Table 10.14
Blade Thickness for Problem 10.11

Horizontal Position, x (Inches)	Blade Thickness, y (Inches)
0.0	0.0
0.20	0.277
0.40	0.360
0.80	0.400
1.20	0.384
1.60	0.344
2.00	0.286
2.40	0.220
2.80	0.154
3.20	0.096
3.60	0.056
4.00	0.0

The 4-inch-long symmetric turbine blade has thickness given in Table 10.14. It is a NACA profile of standard type, with 0.80 inch total thickness at 20 percent of the chord length. The first cooling passage has a radius of 0.160 inch, with its center at $x = 0.80$ inch. The second cooling passage has radius 0.120 inch, with center at $x = 2.00$ inches.

Analyze the interior and surface of the blade, using finite elements. Use approximately 30 to 40 elements and try to employ three elements with sides along the cooling passages. Consider only the upper half of the blade. Determine the peak value of the temperature and its location in the blade or on the surface. Sketch the blade with the division into elements; and indicate the temperatures over the surface of the blade and at nodes on the boundary of the cooling passages.

10.12 Redo the previous problem, but change the cooling water temperature to 900°F.

C. Ideal Flow

10.13 An incompressible, ideal flow occurs between two horizontal walls 4 m apart and around an ellipse with major horizontal axis of 2 m and minor vertical axis of 1 m. The incoming and outgoing flow has velocity $U_\infty = 100$ m/sec. The general boundary value problem was set up in Example 9.1. Let the flow region in the upper half-plane, Figure 10.30, be divided into elements as shown in Figure 10.31, using 16 elements. After division into eight quadrilaterals, the diagonals were drawn in the approximate direction toward the center of the ellipse. For convenience, the solution region is taken as 8 m long. Actually the velocity is 100 at $x = \pm\infty$, but a finite-length solution region must be used on the computer. Determine the velocity potential ϕ for the problem and the fluid velocity components in the elements.

10.14 The same problem as Problem 10.13 is considered but with a division scheme, shown in Figure 10.32, chosen in no particular pattern but with the same node points, connected differently. Find the potential and velocity components. Compare the results with those of the previous example.

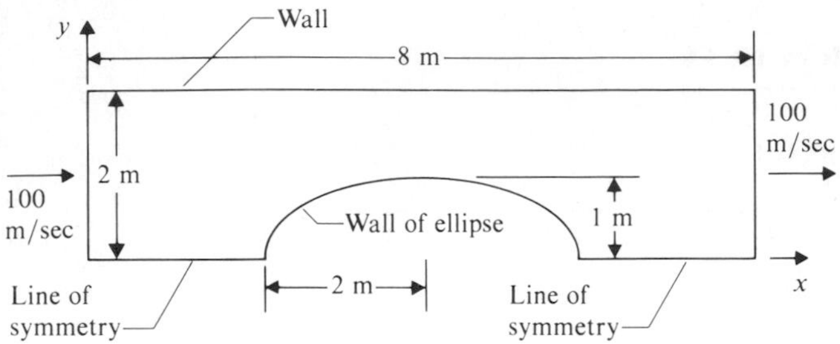

Figure 10.30 Solution region (Problem 10.13—Ideal flow past ellipse, one division scheme)

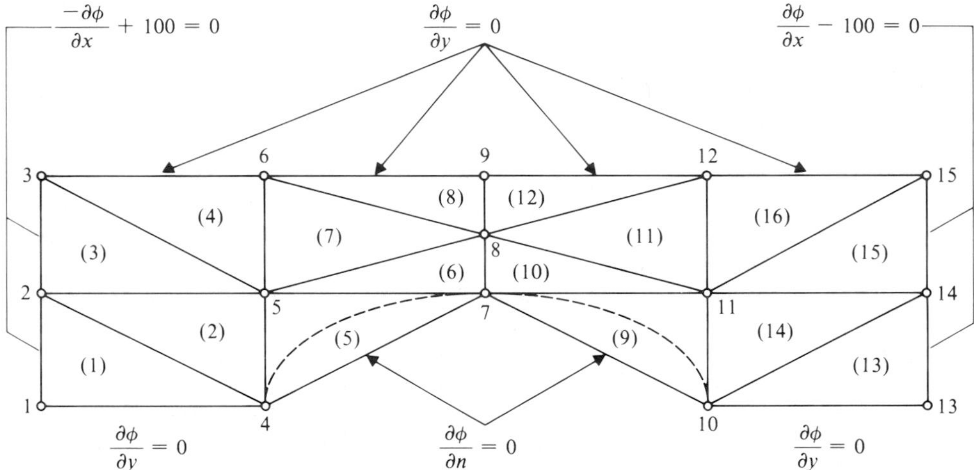

Figure 10.31 Division into elements (Problem 10.13—Ideal flow around ellipse, one division scheme)

10.15 Flows through and around rows of cylinders are often of interest in engineering. Such rows are present in heat exchangers, boilers, oil drilling rigs, and piles supporting buildings. For such applications, the flow velocities between the cylinders must be determined. Two staggered rows, with many cylinders along the rows in each direction, are shown in Figure 10.33. Assume that ideal flow occurs with incoming and exiting velocity of 20 ft/sec. By symmetry, only one passage between two cylinders need be considered. Also, the solution region must be finite in the flow direction. The nodes and elements are shown in Figure 10.34. Find the fluid velocities in each element and indicate the element with largest magnitude.

10.16 Determine the velocities for the previous problem but with the cylinders increased in diameter to 3 ft. The spacing between the centerlines of the cylinders remains the same. Find the element with the largest magnitude of velocity.

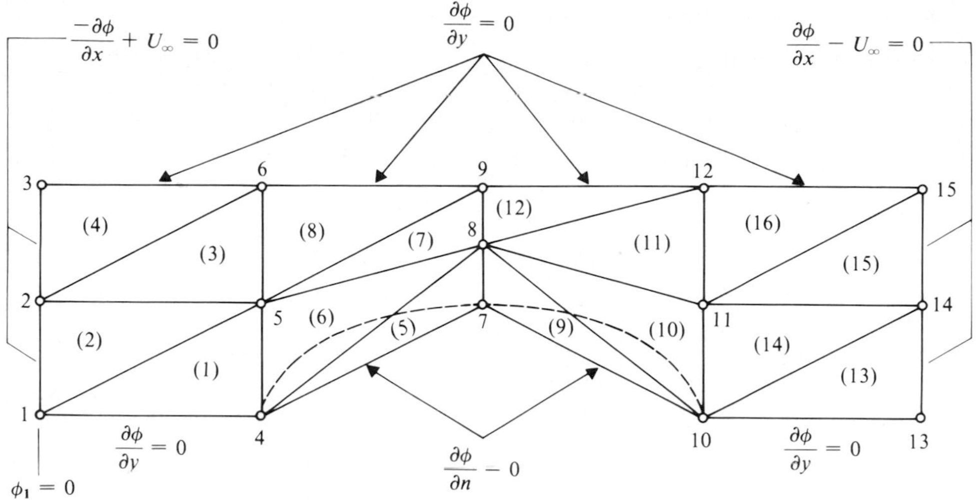

Figure 10.32 Division into elements (Problem 10.14—Ideal flow around ellipse, alternative division scheme)

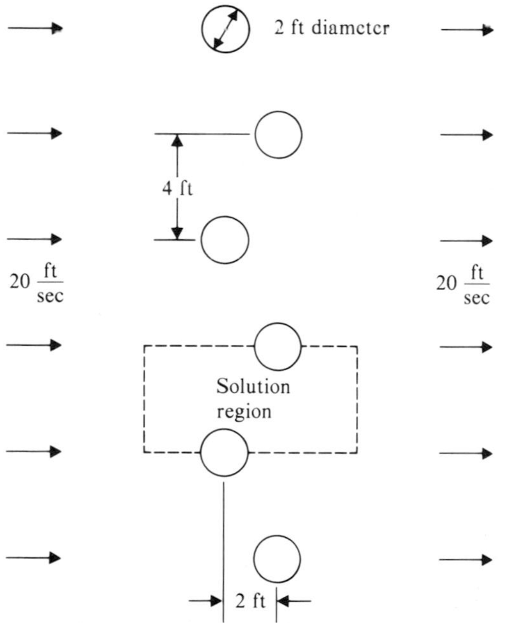

Figure 10.33 Geometry for ideal flow around staggered rows of cylinders for Problem 10.15

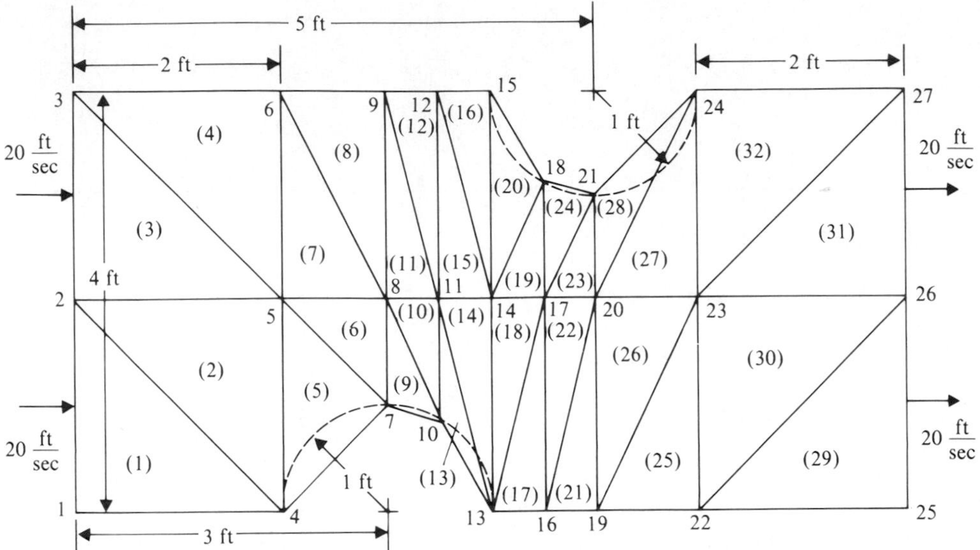

Figure 10.34 Nodes and elements for ideal flow around staggered rows of cylinders in Problem 10.15

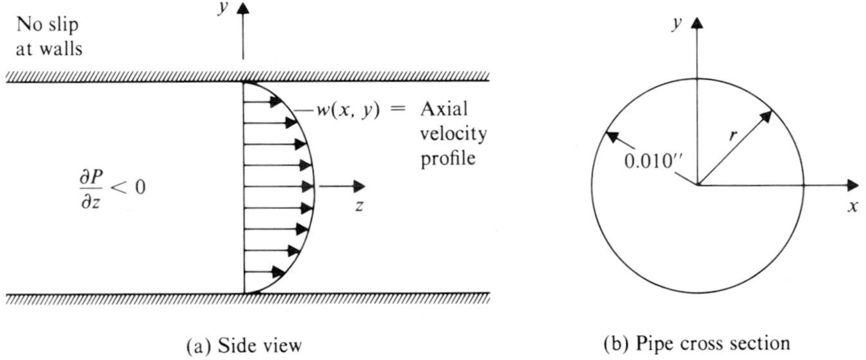

Figure 10.35 Geometry for viscous flow, Problem 10.17

D. Viscous Flow

10.17 A viscous fluid flows through the capillary circular pipe parallel to the pipe axis shown in Figure 10.35. A negative pressure gradient $\partial P/\partial z$ (pressure decreases in the positive z direction) drives the flow in the positive z direction with axial velocity $w(x,y)$. Problem parameters are

$$\frac{\partial P}{\partial z} = -5.0 \text{ lbf/in}^3$$
$$\mu = 1.0 \times 10^{-7} \text{ lbf-sec/in}^2 \text{ (water)}$$
$$\rho = 1.0 \times 10^{-3} \text{ slug/in}^3$$

The flow is fully developed and laminar, with a velocity profile as shown in Figure 10.35(a).

The actual governing differential equation for the velocity profile can be written in terms of radius r as

$$\frac{1}{r}\frac{\partial}{\partial r}\left(\mu r \frac{\partial w}{\partial r}\right) - \frac{\partial P}{\partial z} = 0$$

because the flow has symmetry about the z axis. The no-slip boundary condition along the pipe wall is $w = 0$, $r = R = 0.010$ in. Solving for the exact solution yields

$$w = -\frac{R^2}{4\mu}\left(\frac{\partial P}{\partial z}\right)\left[1 - \left(\frac{r}{R}\right)^2\right]$$

and the volume flow rate through the pipe is

$$Q = \int\int_A w dA = -\frac{\pi R^4}{8\mu}\left(\frac{\partial P}{\partial z}\right)$$

The average velocity over the cross section is

$$\overline{w} = \frac{Q}{A}$$

Then the flow Reynolds number is

$$\text{Re} = \frac{\rho \overline{w}(2R)}{\mu}$$

which can be employed to determine whether the flow is indeed laminar (Re $<$ 2,000).

Solve this problem for the axial velocity profile $w(x,y)$ in the x,y plane, using at least 20 elements. The governing equation is

$$\frac{\partial}{\partial x}\left(\mu \frac{\partial w}{\partial x}\right) + \frac{\partial}{\partial y}\left(\mu \frac{\partial w}{\partial y}\right) - \frac{\partial P}{\partial z} = 0$$

with the no-slip boundary condition $w = 0$, $r = R$. Consider a 45° solution region (taking into account lines of symmetry). Determine the nodal velocities and compare to the exact solution. Evaluate the finite element volume flow rate Q and compare that to the exact solution as well. Verify that the flow is laminar by calculating the Reynolds number Re.

10.18 Axial viscous flow occurs in a square pipe shown in Figure 10.36. The length of each side of the square pipe, $L = 0.017724$ in, is chosen so that the cross sectional area is the same as the circular pipe in Problem 10.17: $A = 3.14159 \times 10^{-4}$ in^2. All other problem parameters are the same as Problem 10.17.

Use approximately the same number of elements as in Problem 10.17 to analyze a 45 degree slice of the square duct. Compare the peak velocity (at the origin) and average velocity $\overline{w}$ for the square duct to the analytical or finite element solution (if you have done Problem 10.17) for the circular pipe in Problem 10.17. Differences occur because of the larger surface for viscous drag.

10.19 Axial viscous flow occurs between a square pipe of side length 0.019817 inch and an inner circular pipe of radius $R = 0.005$ inch. The cross sectional area A is the same as in the previous two problems. Perform an analysis similar to that performed for the square pipe (Problem 10.18) for a 45-degree slice of the pipe flow area. No-slip boundary conditions

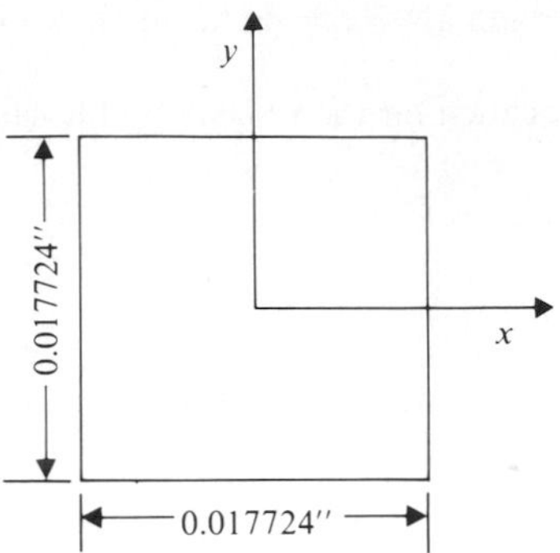

Figure 10.36 Square pipe cross section for Problem 10.18

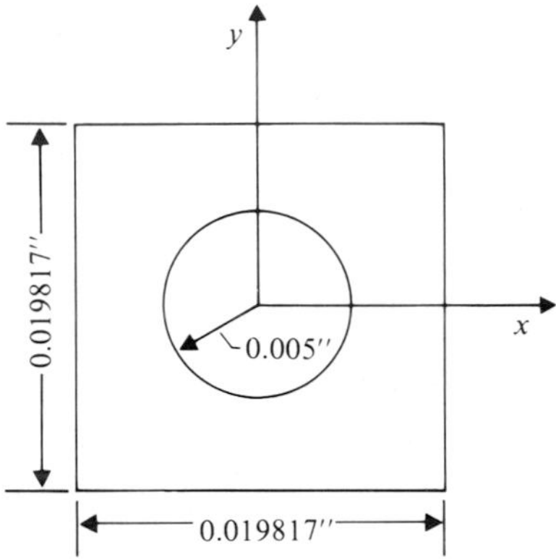

Figure 10.37 Square pipe cross section with inner pipe for Problem 10.19

apply along the inner side of the square pipe and the outer side of the circular pipe. Again the differences in the various cases result from larger surfaces for viscous drag on the flow. Compare the average flow velocity $\overline{w}$ for this case to that given by the exact solution for the circular pipe and other finite element solutions (if you have been asked to carry them out). Figure 10.37 shows the geometry.

10.20 A plane slider bearing, typical of the geometries found in industrial journal bearings, is indicated in Figure 10.38. The purpose of the design is to have the rotating shaft drag the viscous oil into the bearing. This creates high pressures inside the bearing and results in a force supporting the shaft in the bearing.

The dimensions of the slider are 2.0 by 3.0 inches, which represent a circular bearing that has been "unrolled" for analysis. Where the oil flows in, the clearance h_1 is 0.004 in, and the clearance h_2 where the oil flows out is 0.002 in. The slider (actually the unrolled shaft) moves at velocity U of 1,100 in/sec relative to the fixed bearing. The fluid viscosity is

$$\mu = 2.0 \times 10^{-6} \text{ lbf-sec/in}^2$$

which is typical of oils used for bearings.

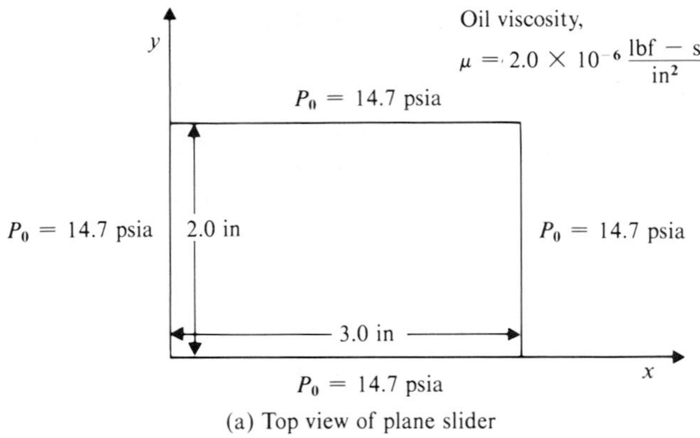

(a) Top view of plane slider

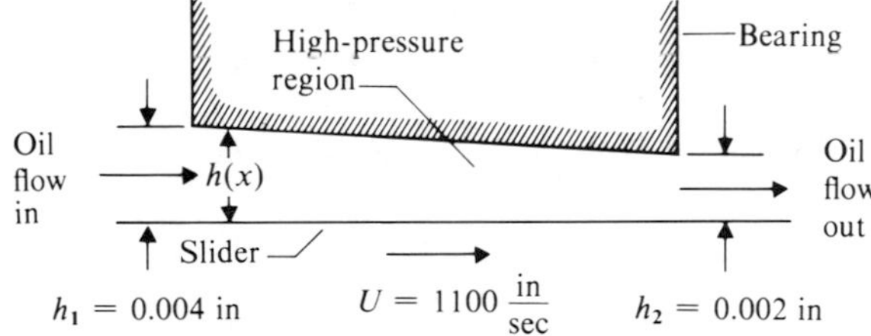

(b) Side view of plane slider

Figure 10.38 Slider geometry for Problem 10.20

The differential equation for the pressure $P(x,y)$ in this problem is derived in books on lubrication. It is

$$\frac{\partial}{\partial x}\left(h^3 \frac{\partial P}{\partial x}\right) + \frac{\partial}{\partial y}\left(h^3 \frac{\partial P}{\partial y}\right) = 6U\mu \frac{\partial h}{\partial x}$$

where

$$h = h(x) = \text{oil film thickness}$$

$$\frac{\partial h}{\partial x} = \text{oil film thickness gradient}$$

The film thickness is independent of y. Note that the plane slider has a constant film thickness gradient $\partial h/\partial x$. The film thickness $h(x)$ varies over the slider. It has been shown by many lubrication analyses that the pressure is constant across the film thickness (from slider to bearing in the z direction).

Along the boundaries of the bearing, the pressure is atmospheric. Thus $P_0 = 14.7$ psia on all four sides.

Use at least 10 elements to determine the pressure in the plane slider. Evaluate the film thickness cubed term h^3 at the centroid of each element. Sketch the location of the

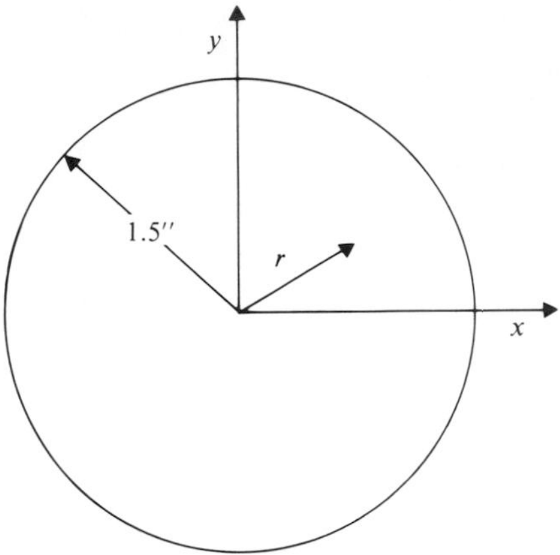

Figure 10.39 Circular shaft for Problem 10.21

node points and indicate the pressures at each node. The force generated by the fluid film is given by

$$F = \int\int_A P\, dA$$

Employ the integral of the pressure over each element to determine the value of this force.

E. Torsion Problems

10.21 An aluminum circular shaft of radius $R = 1.5$ inches is shown in Figure 10.39. It has the parameters $L = 10$ in, $G = 3.8 \times 10^6$ lbf/in^2, and $\theta = 0.02$ deg/in. Use at least 20 elements to analyze a 45° pie-shaped sector (sufficient because of symmetry) of the circular cross section. Also evaluate the element shear stresses τ_{xz}, τ_{yz} and torque T on the shaft.

The exact solution ϕ to this problem is a function of the radial coordinate r:

$$r = \sqrt{x^2 + y^2}$$

The equation for the stress function becomes

$$\frac{1}{r}\frac{\partial}{\partial r}(r\,\phi) = 0$$

Obtain the analytical solution $\phi\ (r)$ to this as well as the shear stresses and torque. Compare these results to the finite element solution.

10.22 Carry out the previous problem with at least 40 elements.

10.23 Circular shafts in rotating machines often have keyways cut in them so they can be connected to other machines. Figure 10.40 shows the geometry of a typical shaft. High stresses often occur in the keyway region. Parameters include $L = 10$ in, $G = 3.8 \times 10^6$ lbf/in^2 (aluminum), and $\theta = 0.02$ deg/in. Note that the centroid is no longer at the center

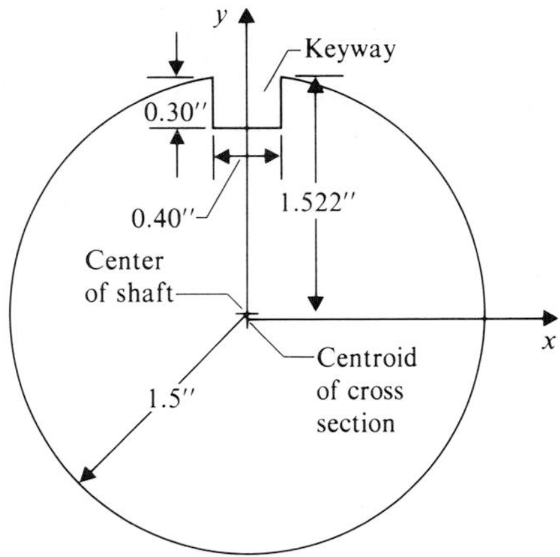

Figure 10.40 Geometry for Problem 10.22

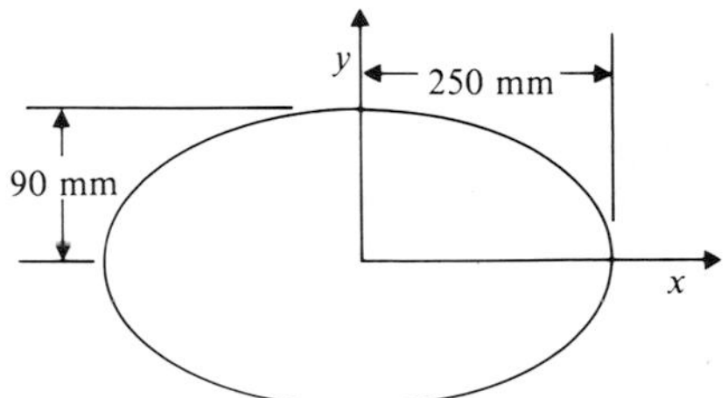

Figure 10.41 Elliptical bar for Problem 10.24

of the circle. Use at least 30 elements, with many concentrated around the keyway, to solve for the shear stresses in the right half of the shaft. Determine the magnitude and location of the peak stress.

Use the analytical solution described in Problem 10.21 for a circular shaft to determine the maximum stress in a shaft without a keyway. Compare this value to the peak finite element value.

10.24 A long brass bar with elliptical cross section is shown in Figure 10.41. The equation for the ellipse is

$$\frac{x^2}{a^2} + \frac{y^2}{b^2} - 1 = 0$$

where $a = 250$ mm and $b = 90$ mm. Other parameters are $L = 1{,}000$ mm, $G = 40\ \text{kN/mm}^2$, and $\theta = 0.001$ deg/mm. Solve, using at least 20 elements in one-fourth of the ellipse (taking lines of symmetry into account). Also obtain the shear stresses and torque.

The exact solution is

$$\phi = -2G\theta \frac{a^2b^2}{2(a^2+b^2)}\left(\frac{x^2}{a^2} + \frac{y^2}{b^2} - 1\right)$$

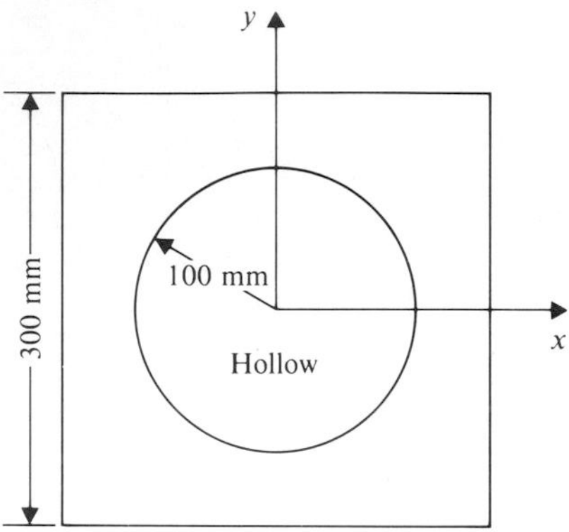

Figure 10.42 Hollow square shaft for Problem 10.25

Compare this to the finite element solution at the node points. The torque is given by

$$T = G\theta \frac{\pi a^3 b^3}{(a^2 + b^2)}$$

Compare to the finite element solution. Determine the magnitude and location of the maximum stress from the finite element solution and compare to the exact solution of

$$\tau_{max} = \frac{2T}{\pi a b^2} \text{ at } x = 0, y = b$$

10.25 A square steel shaft with dimensions 300 mm by 300 mm is shown in Figure 10.42. It has a circular hole of radius 100 mm bored throughout its length, creating a hollow shaft. Other parameters are $L = 1{,}000$ mm, $G = 76$ kN/mm^2, and $\theta = 0.002$ deg/mm. Use at least 20 elements to analyze a 45-degree segment of the shaft (taking into account lines of symmetry). Determine the element stresses and torque on the shaft.

10.26 A steel circular shaft, of radius 2.0 inches, used in a two-pole electric motor, has two flat sections cut in opposite sides of the shaft. Figure 10.43 shows the geometry. The shaft must be analyzed to determine the location of maximum shear stresses caused by torque acting on the shaft. As lines of symmetry exist along the x and y axes, only one-fourth of the shaft cross section needs to be analyzed.

Consider the problem parameters $L = 100$ in, $G = 11 \times 10^6$ lbf/in^2, and $\theta = 0.02$ degree/in. The finite element division scheme shown in Figure 10.43 has 23 simplex elements. The y coordinates for nodes 16, 18, 19 can be calculated from the circle of radius 2.0 inches. Find the stress function, shear stresses, and torque on the shaft.

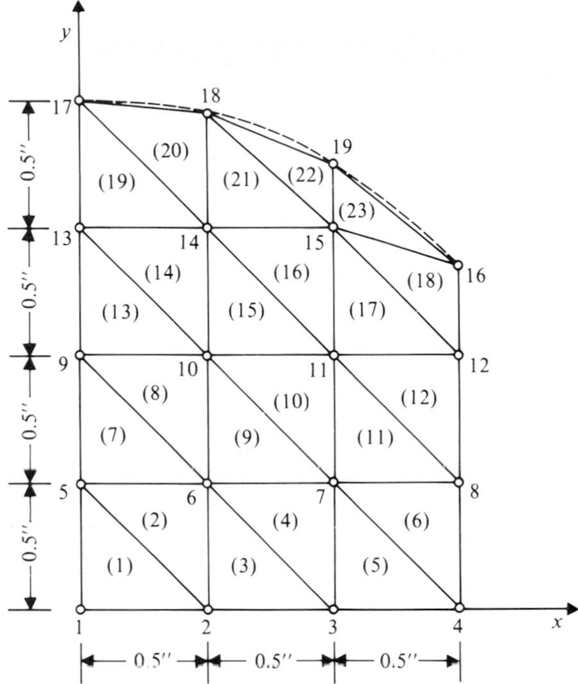

Figure 10.43 Finite element model for Problem 10.26

F. General Problems

10.27 Choose a problem involving the solution of a second-order partial differential equation suitable for a finite element solution similar to those carried out in Chapters 9 and 10. Describe the physical significance of the problem to be solved, and give the governing differential equation with boundary conditions. Present the solution with a plot of the results and its significance. Grading of this problem will be based in part on the use of the finite element solution to illustrate a certain point—such as the physical significance of some problem parameters as compared to others—or the accuracy of finite element results compared to the exact solution of the differential equation.

Chapter 11

Programming and Solution Methods

11.1 Introduction

The finite element solution of engineering problems is made possible by the computer. It should be apparent by this time that the large numbers of repetitive algebraic calculations that are required can only be done in a reasonable time on a modern computer. Fortunately the algebraic equations encountered in most of the heat transfer, fluid mechanics, and solid mechanics problems treated in this text are linear with symmetric, positive definite, and banded global matrices. All of these properties should be used to increase solution efficiency. Both computer time and storage space should be minimized as an integral part of the finite element solution procedure.

Many techniques that increase either the ease of implementation or the efficiency of solution are available. As finite element analysis became more and more widespread, certain methods came into common use. Some of the most important are discussed in this chapter to permit the reader to use them. They also are utilized in the example computer programs given in the Appendixes. This chapter is thus necessary to a complete understanding of the example programs.

A general finite element program called FINTWO has been developed for use with this text. It will solve any two-dimensional differential equation such as those discussed in Chapters 9 and 10. As the finite element techniques must be implemented on the computer, the following discussion will be directly related to the internal structure of FINTWO (the program is given in Appendix B). The program is divided into the following sections:

1. Input and output of problem data
2. Calculation of element properties and assembly of global matrices
3. Inclusion of prescribed nodal values into global matrix and column
4. Calculation of nodal T values, element derivatives, and integrals

11.2 Input and Output

One of the major areas of interest in finite elements is the input and output of data because there is usually a large volume of it. Finite element programs have the advantage of being quite flexible with regard to geometry, etc., so the user must be able to input a large number of problem parameters. Similarly, the output must include the input data (to verify that the input operation was correct) and the results of the calculations, which may include element derivatives as well as the nodal solution values. Clearly some care must be exercised in doing all of this. An overall description of the input data is given in the comment cards in the beginning of computer program FINTWO.

In FINTWO the input is dealt with in Section 1 of the program, which consists of the following subsections:

1.1. Title and global parameters
1.2. Nodal coordinates
1.3. Element/node connectivity
1.4. Calculate bandwidth
1.5. Coefficient functions
1.6. Prescribed nodal values
1.7. Element derivative boundary values

These will be discussed in some detail with regard to storage techniques and nomenclature.

First, two title lines, the number of node points, NP, and number of elements, NE, are read in Section 1.1. The node number INODE, nodal x coordinate XNODE, and nodal y coordinate YNODE are entered next for each node. These are stored in three column matrices of order NP as they are read in Section 1.2. Up to 150 node points are allowed, but this can easily be increased in the program. In the present form of FINTWO, these must be input by the user (usually calculated and typed into a terminal by hand). A much more efficient technique would be to write a small subroutine to generate these for a given problem or class of problems. One must be careful to number the nodes so as to keep the bandwidth small. Some of the nomenclature is given in Table 11.1. Because many examples have been given earlier, none are given here.

Section 1.3 reads the element/node connectivity data with the element number NEL, the first node NODEI, the second node NODEJ, and the third node NODEK stored in four column matrices of order NE. Recall from Chapter 3 that the nodes should be numbered in the counterclockwise direction to obtain the correct results. As with the number of nodes, the columns are limited to 100 elements, but this can be extended by changing the appropriate dimension statements. Section 1.4 then calculates the global bandwidth by determining the largest difference between node numbers for an element.

The element coefficients are read in Section 1.5. In many problems, the coefficients are the same for all of the elements. If the parameter ICO equals 1, only one data card needs to be read and all element coefficients are set equal to this. When ICO is not equal to 1, a separate card must input IEL, KXE, KYE, PE, and QE for each element. Again, it would be much more efficient for the user to write a short subroutine to calculate the coefficients for a given class of problems than to input these by hand when they vary from element to element.

Section 1.6 reads the number of prescribed nodal values NBOUND, the global node number IBOUND, and the prescribed nodal value TBOUND for node IBOUND. Note that the values of IBOUND and TBOUND are stored in columns of order NBOUND (not NP). For example, Example 10.2 has NP = 9 but only six prescribed nodal values,

Table 11.1
Input Nomenclature

1.2 Nodal Coordinates	NP = Number of nodes INODE = Column of node numbers (of order NP) XNODE = Column of nodal x coordinates (of order NP) YNODE = Column of nodal y coordinates (of order NP)
1.3 Element/Node Connectivity	NE = Number of elements NEL = Column of element numbers (of order NE) NODEI = Column of first node numbers in element NEL (of order NE) NODEJ = Column of second node numbers in element NEL (of order NE) NODEK = Column of third node numbers in element NEL (of order NE)
1.5 Coefficient Functions	IEL = Element number KXE = Column of element values of K_x (of order NE) KYE = Column of element values of K_y (of order NE) PE = Column of element values of P (of order NE) QE = Column of element values of Q (of order NE)
1.6 Prescribed Nodal Values	NBOUND = Number of prescribed nodal values IBOUND = Column of node numbers of prescribed nodes (of order NBOUND) TBOUND = Column of prescribed nodal values (of order NBOUND)
1.7 Derivative Boundary Values	NDERIV = Number of element sides on which derivative boundary conditions are specified IELDIR = Column of element numbers of derivative boundary conditions (of order NDERIV) NSIDE = Column of local numbers of sides on boundary (of order NDERIV) ALPHAE = Column of element values of α on boundary side (of order NDERIV) BETAE = Column of element values of β on boundary side (of order NDERIV)

so NBOUND = 6. Thus only six positions in IBOUND and NBOUND are occupied. Section 1.7 reads the value of NDERIV, number of elements on which the derivative boundary value is given IELDIR, local number of side on boundary NSIDE, value of alpha on that side ALPHAE, and value of beta on that side BETAE. Again, only NDERIV storage locations are used in the columns IELDIR, NSIDE, ALPHAE, and BETAE.

Consider Example 10.2 as an illustration. The input data are shown in Table 11.2. The overall problem parameters are NP = 9, NE = 8, NBOUND = 6, and NDERIV = 4. Note that NDERIV is the number of sides (4), *not* the number of elements on which boundary conditions are specified (3).

Each of the input data sections is also output immediately. This enables the user to check his input data for errors. As errors are often made, particularly in the element/node connectivity data and derivative boundary conditions, this is very important.

Table 11.2
Input Data (Example 10.2: Heat Conduction and Convection in a Two-Dimensional Body)

(a) Nodal Coordinates (NP = 9)

Node	x	y
INODE	XNODE	YNODE
1	0.0	0.0
2	0.0	3.0
3	0.0	5.0
4	2.0	0.0
5	2.0	3.0
6	2.2667	5.0
7	4.0	2.0
8	4.0	3.0
9	4.0	4.0

(b) Element/Node Connectivity Data (NE = 8)

Element		Nodes	
NEL	NODEI	NODEJ	NODEK
1	1	4	2
2	5	2	4
3	2	5	3
4	6	3	5
5	4	7	5
6	8	5	7
7	5	8	6
8	9	6	8

(c) Element Values for K_x, K_y, P, Q Coefficient Functions (NE = 8)

Element	K_x	K_y	P	Q
IEL	KXE	KYE	PE	QE
1 (same for all eight elements)	0.25	0.25	0	0

Table 11.2 *Continued*

(d) Prescribed Nodal Values (NBOUND = 6)

Node	Temperature
IBOUND	TBOUND
1	400.0
2	400.0
3	400.0
4	200.0
6	400.0
7	100.0

(e) Element Derivative Boundary Values (NDERIV = 4)

Element	Side	α	β
IELDIR	NSIDE	ALPHAE	BETAE
1	1	0	0
6	3	0.001	−0.050
8	1	0.001	−0.050
8	3	0.001	−0.050

11.3 Efficient Assembly and Matrix Storage Techniques

The actual assembly of the element coefficient matrices into the global coefficient matrices has already been illustrated in several places. Section 4.3 discussed the assembly of algebraic nodal equations for one-dimensional problems. Extensive examples were given in Chapter 4. All of these problems result in a global matrix with a bandwidth of 2. Two-dimensional assembly techniques were presented in examples given in Chapters 9 and 10. The basic ideas are largely presented there. The bandwidth varies widely from problem to problem.

In general the assembly algorithm uses the element/node connectivity data to enter the element matrices into the global matrices. Each element is considered in turn and the element nodal numbers matched to the proper location in the global matrices. Consider a simplex element in two dimensions where the element coefficient matrix [EBM] is 3 $\times$ 3 and the element column {EC} is of order 3. Let the global set of equations be denoted as $[A]\{T\} = \{F\}$ where the global coefficient matrix $[A]$ is of order NP $\times$ NP, the column of unknowns $\{T\}$ is of order NP, and the global column $\{F\}$ is of order NP. Here NP is the number of node points. A summary of all of these definitions is given in Table 11.3.

Table 11.3
Element and Global Labeling Definitions

Element Matrices	[EBM] = Element coefficient matrix [B] (of order 3 $\times$ 3) {EC} = Element column {C} (of order 3)
Global Matrices	NP = Number of node points NBW = Bandwidth [A] = Global coefficient matrix (of order NP $\times$ NP) [ACOND] = Condensed global coefficient matrix (of order NBW $\times$ NP) {T} = Global column of unknowns (of order NP) {F} = Global column (of order NP)

Before the solution techniques are discussed, one must consider the storage requirements. As seen in earlier chapters, finite element analysis generates a matrix [A], which is banded. It is desired to store [A] so that the zeros outside the bandwidth are not used and also to take advantage of the symmetry of [A]. Not all global coefficient matrices are symmetric (see Chapters 8 and 15), but many are. Perhaps the most common and easiest to understand is storage in rectangular form. Let the condensed global coefficient matrix be [ACOND], which is of order NBW $\times$ NP.

In condensed form, the main diagonal of [A] is placed in the first column of [ACOND]. Table 11.4 gives an example taken from the matrix generated in Example 10.2. The next diagonal above the main in [A] is stored in column two of [ACOND]. Since this diagonal contains one less element than the main diagonal, a zero is entered into the last position in column two of [ACOND]. This procedure continues with each above-main diagonal in [A] until the bandwidth is reached. All of the rest of the terms in [A] are zero, so they are ignored. All terms in [A] below the main diagonal are symmetric and are not stored in [ACOND]. Clearly [ACOND] must be of order NBW $\times$ NP. Table 11.5 gives the condensed matrix.

Some simple calculations indicate the savings in storage. Table 11.6 shows a comparison for small, medium, large, and very large problems, as judged by the number of node points. Typical bandwidths are also used. For the small problem, the storage requirements are of the same order, and the condensed form is not really required. As the medium problem is encountered, the savings become quite dramatic—10,000 locations versus 1,500. The 10,000 figure is still well within the central memory range of modern computers, but the cost is usually much larger if a large fraction of the core is used. In the case of large problems, the memory required would severely tax the computer capacity (100,000 to 1,000,000 words) of most mainframe computers. There may not be any room left in the machine for actually running the program.

In the case of very large programs, even the condensed matrix is usually too large to go on the machine. Iterative techniques for solving the equations must nearly always be used, where only the part of the equations actually being used is in the central memory

Table 11.4
Global Matrices
(Example 10.2: Heat Conduction and Convection in a Two-Dimensional Body)

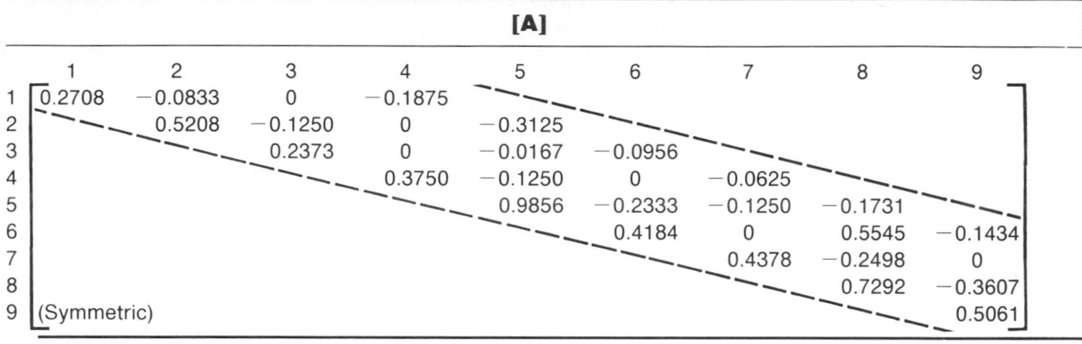

[A]

	1	2	3	4	5	6	7	8	9
1	0.2708	−0.0833	0	−0.1875					
2		0.5208	−0.1250	0	−0.3125				
3			0.2373	0	−0.0167	−0.0956			
4				0.3750	−0.1250	0	−0.0625		
5					0.9856	−0.2333	−0.1250	−0.1731	
6						0.4184	0	0.5545	−0.1434
7							0.4378	−0.2498	0
8								0.7292	−0.3607
9	(Symmetric)								0.5061

Table 11.5
Condensed Global Matrix and Column

(a) Global [ACOND] Matrix (Written in Compact Column Form to Reduce Computer Storage)					**(b) Global {*F*} Column**	
Row	**Main Diagonal**	**First Diagonal Above Main**	**Second Diagonal Above Main**	**Third Diagonal Above Main**	**Row**	**Column**
1	0.2708	−0.0833	0	−0.1875	1	0
2	0.5208	−0.1250	0	−0.3125	2	0
3	0.2373	0	−0.0167	−0.0956	3	0
4	0.3750	−0.1250	0	−0.0625	4	0
5	0.9856	−0.2333	−0.1250	−0.1731	5	0
6	0.4184	0	0.5545	−0.1434	6	0.05003
7	0.4378	−0.2498	0	0	7	0.02500
8	0.7292	−0.3607	0	0	8	0.05000
9	0.5061	0	0	0	9	0.07503

at any one time. Overlay techniques are used to store the remaining parts of the program on disk or tape until needed. Large-scale solid mechanics programs such as NASTRAN, SPAR, etc., are designed to do this routinely.

Before the element matrices can be added to the *A* matrix and *F* column, they must be set to zero. Section 2.1 of computer program FINTWO does this. A simple FORTRAN algorithm to accomplish this task might be

```
          DO 120 I=1,NP
             DO 130 J=1,NP
                A(I,J)=0.0
130          CONTINUE
             F(I)=0.0
120       CONTINUE
```

Table 11.4 *Continued*

$\{T\}$		$\{F\}$

$$\times \begin{Bmatrix} T_1 \\ T_2 \\ T_3 \\ T_4 \\ T_5 \\ T_6 \\ T_7 \\ T_8 \\ T_9 \end{Bmatrix} = \begin{Bmatrix} 0 \\ 0 \\ 0 \\ 0 \\ 0 \\ 0.05003 \\ 0.02500 \\ 0.05000 \\ 0.07503 \end{Bmatrix}$$

Table 11.6
Matrix Storage Requirements for Various Size Linear Problems

Problem Size	Node Points, NP	Band-width, NBW	Storage for [*A*]	Condensed Storage for [ACOND]	Comments
Small	10	4	100	40	Solve by Gauss elimination—No condensation required
Medium	100	15	10,000	1,500	Solve by banded Gauss elimination—condensation desirable
Large	1,000	40	10^6	40,000	Solve by banded Gauss elimination—condensation required
Very large	10^4	100	10^8	10^6	Requires matrix iteration solution—Too large for central memory of most computers

Table 11.7
Local and Global Element/Node Connectivity Data Definitions

Element Matrices	IEL = Local element number = NEL(I) {NN} = Local nodal coordinate column (of order 3)
Global Matrices	NE = Total number of elements {NEL} = Column of element numbers (of order NE) {NODEI} = Column of first node number in element NEL (of order NE) {NODEJ} = Column of second node number in element NEL (of order NE) {NODEK} = Column of third node number in element NEL (of order NE)

where NP is the number of nodal points. Actually the above routine must be modified slightly because of the condensed storage to read

```
        DO 120 I=1,NP
           DO 130 J=1,NBW
                 ACOND(I,J)=0.0
130        CONTINUE
           F(I)=0.0
120     CONTINUE
```

where NBW is the bandwidth. Now the global matrices have been zeroed.

Return now to the assembly algorithm, keeping in mind that the global matrix must be condensed for all but the smallest finite element solutions (very large problems are beyond the scope of this text). The element/node connectivity data have been entered into FINTWO as input or calculated by a mesh generation subroutine of some kind. For each element the element number is stored in the column {NEL}, while the three node numbers are stored in the three columns {NODEI}, {NODEJ}, and {NODEK}. Table 11.7 shows the nomenclature.

Each element must be placed into the global matrix, so a large DO loop over all elements is set up in Section 2.2 of FINTWO. For convenience within a given element, a small column {NN} of order 3 stores the local nodal coordinates, where IEL = NEL(I), NN(1) = NODEI (IEL), NN(2) = NODEJ (IEL), and NN(3) = NODEK (IEL). This occurs in Section 2.3 of FINTWO.

The next step is to calculate the element matrices. Section 2.3 also determines the nodal coordinates x and y, both columns of order 3. Local element coefficient functions KX, KY, P, Q are assigned in Section 2.4. Derivative boundary conditions are treated in Section 2.5. First the local element values of ISIDE, ALPHA, and BETA (all columns of order 3) are set to zero. Then all of the stored derivative boundary conditions are checked to determine if the element number IEL matches the derivative element number IELDIR. If there is a match, the values of ALPHAE and BETAE are placed in the local variables ALPHA and BETA for the appropriate side of the element.

Section 2.6 calls the subroutine ELEMNT, which calculates [EBM] and {EC} for element IEL. ELEMNT is used to evaluate the element matrices in a fairly obvious fashion. It will not be discussed in detail here. Note that the element coefficient matrices are not stored, but are assembled directly into the global matrices. This saves a large amount of storage space.

The actual assembly routine is relatively simple. If the $[A]$ global matrix were assembled, a simple FORTRAN routine to do so is

```
        DO 170 J=1,3
           DO 180 K=1,3
              IROW=NN(J)
              ICOL=NN(K)
              A(IROW,ICOL)=A(IROW,ICOL)+EBM(J,K)
180        CONTINUE
           F(IROW)=F(IROW)+EC(J)
170     CONTINUE
```

where IROW = Row of $[A]$ matrix, ICOL = Column of $[A]$ matrix.

Again, the condensed storage must be taken into account. Only the column location of the components of $[A]$ are moved. Let JCOL represent the new location, where

JCOL=ICOL−IROW+1
= Column in [ACOND] matrix

which gives the proper shifting to get the diagonal component of $[A]$ into the first column of [ACOND]. The above FORTRAN routine is simply modified as

```
        DO 170 J=1,3
           DO 180 K-1,3
              IROW=NN(J)
              ICOL=NN(K)
              ICOL=ICOL-IROW+1
              IF(JCOL .LT. 1) GO TO 181
              ACOND(IROW,JCOL)=ACOND(IROW,JCOL)+EBM(J,K)
181           CONTINUE
180        CONTINUE
           F(IROW)=F(IROW)+EC(J)
170     CONTINUE
```

Note that if the term JCOL is negative, this corresponds to a term below the diagonal in $[A]$. Because of symmetry, this is not stored in [ACOND]. The above routine is used in Section 2.7 of FINTWO. Finally, Section 2.8 writes out the global matrix and column just after assembly but before the prescribed nodal values are included.

11.4 Treatment of Boundary Conditions

In the preceding chapters two types of boundary conditions have been considered: prescribed nodal values and derivative boundary conditions. One method of including the prescribed nodal values into the global equations (either in $[A]$ or condensed [ACOND] form) is discussed here. The derivative boundary conditions are actually included in the element matrices [EBM] and {EC} (see Chapters 7 and 9). These are taken into account in the assembly just discussed. Thus only the prescribed nodal values need to be treated.

The method described here involves modifying the $[A]$, or equivalently [ACOND], matrix and $\{F\}$ column to include the given nodal values {TBOUND} at node points {IBOUND}. This is done without changing the size of $[A]$ or $\{F\}$. An alternative method involves renumbering and partitioning the matrices into two parts: unknown and known. Both methods work efficiently, but the second is not discussed.

Consider a general banded matrix $[A]\{T\} = \{F\}$ just assembled. Let the known nodal value be $T_i = \theta_i$. An outline of the steps to be followed is

1. Set all coefficients a_{ij} in row i to zero and replace f_i by $a_{ii}\theta_i$.
2. Add the value $a_{ji}\theta_i$ to the existing value of f_j in row j and then set a_{ji} to zero.

The resulting matrix remains symmetric and banded. Section 4.3 has discussed this previously, but the method must be modified somewhat to include condensed storage.

Example 11.1 *Prescribed Nodal Value Technique*

Given Consider the five linear equations with bandwidth 3 shown in Figure 11.1(a). Let the known nodal value be $T_2 = \theta_2$.

Objective Modify the set of equations using the technique just described.

Solution First consider equation 2, which is $a_{21}T_1 + a_{22}T_2 + a_{23}T_3 + a_{24}T_5 = f_2$. This satisfies the boundary condition if $a'_{21} = 0$, $a'_{23} = 0$, $a'_{24} = 0$, and $f'_2 = a_{22}\theta_2$ where the primed numbers indicate new values. Equation 2 then becomes $a_{22}T_2 = a_{22}\theta_2$, which clearly gives back $T_2 = \theta_2$. The new set of equations is shown in Figure 11.1(b). Dotted lines outline the areas of change. Note that it is not desirable to divide by a_{22}: if a_{22} is very large or very small, an ill-conditioned $[A]$ matrix may result.

At this point, only row 2 has been changed. However, the other equations also contain T_2, which is known. Equation 1 is $a_{11}T_1 + a_{12}T_2 + a_{13}T_3 = f_1$. The T_2 term can be transferred to the right side and then a_{12} set to zero. New parameters are defined as $f'_1 = f_1 - a_{12}\theta_2$, $a'_{12} = 0$. Both equations 3 and 4 contain T_2 as well. They may be changed by writing $f'_3 = f_3 - a_{32}\theta_2$, $a'_{32} = 0$ and $f'_4 = f_4 - a_{42}\theta_2$, $a'_{42} = 0$. The final results are shown in Figure 11.1(c). Dotted lines indicate both the row and column changes.

$$\begin{bmatrix} a_{11} & a_{12} & a_{13} & 0 & 0 \\ a_{21} & a_{22} & a_{23} & a_{24} & 0 \\ a_{31} & a_{32} & a_{33} & a_{34} & a_{35} \\ 0 & a_{42} & a_{43} & a_{44} & a_{45} \\ 0 & 0 & a_{53} & a_{54} & a_{55} \end{bmatrix} \begin{Bmatrix} T_1 \\ T_2 \\ T_3 \\ T_4 \\ T_5 \end{Bmatrix} = \begin{Bmatrix} f_1 \\ f_2 \\ f_3 \\ f_4 \\ f_5 \end{Bmatrix}$$

(a) Original equations

$$\begin{bmatrix} a_{11} & a_{12} & a_{13} & 0 & 0 \\ 0 & a_{22} & 0 & 0 & 0 \\ a_{31} & a_{32} & a_{33} & a_{34} & a_{35} \\ 0 & a_{42} & a_{43} & a_{44} & a_{45} \\ 0 & 0 & a_{53} & a_{54} & a_{55} \end{bmatrix} \begin{Bmatrix} T_1 \\ T_2 \\ T_3 \\ T_4 \\ T_5 \end{Bmatrix} = \begin{Bmatrix} f_1 \\ a_{22}\,\theta_2 \\ f_3 \\ f_4 \\ f_5 \end{Bmatrix}$$

(b) Equations after step 1

$$\begin{bmatrix} a_{11} & 0 & a_{13} & 0 & 0 \\ 0 & a_{22} & 0 & 0 & 0 \\ a_{31} & 0 & a_{33} & a_{34} & a_{35} \\ 0 & 0 & a_{43} & a_{44} & a_{45} \\ 0 & 0 & a_{53} & a_{54} & a_{55} \end{bmatrix} \begin{Bmatrix} T_1 \\ T_2 \\ T_3 \\ T_4 \\ T_5 \end{Bmatrix} = \begin{Bmatrix} f_1 - a_{12}\,\theta_2 \\ a_{22}\,\theta_2 \\ f_3 - a_{32}\,\theta_2 \\ f_4 - a_{42}\,\theta_2 \\ f_5 \end{Bmatrix}$$

(c) Equations after step 2

$$(\text{ACOND}) = \begin{bmatrix} a_{11} & 0 & a_{13} \\ a_{22} & 0 & 0 \\ a_{33} & a_{34} & a_{35} \\ a_{44} & a_{45} & 0 \\ a_{55} & 0 & 0 \end{bmatrix}$$

(d) (ACOND) matrix

Figure 11.1 Modification of global equations for prescribed nodal value

Several features can be observed. The resulting [A] matrix is symmetric and banded, just as it was originally. This means that the matrix can be stored in condensed form [ACOND], which is very desirable. Also, the changes only occur a distance of NBW -1 terms away from the a_{22} term. This completes the inclusion of $T_2 = \theta_2$ into the global equations.

$$\begin{bmatrix} 0.21653 & -0.10858 & -0.04194 & 0.02200 \\ -0.10858 & 0.47207 & 0 & -0.33949 \\ -0.04194 & 0 & 0.20453 & -0.09858 \\ 0.02200 & -0.33949 & -0.09858 & 0.50407 \end{bmatrix} \begin{Bmatrix} T_1 \\ T_2 \\ T_3 \\ T_4 \end{Bmatrix} = \begin{Bmatrix} 2.64 \\ 0.72 \\ 1.92 \\ 2.64 \end{Bmatrix}$$

(a) Equations before including prescribed nodal values

$$\begin{bmatrix} 0.21653 & 0 & 0 & 0 \\ 0 & 0.47207 & 0 & -0.33949 \\ 0 & 0 & 0.20453 & -0.09858 \\ 0 & -0.33949 & -0.09858 & 0.50407 \end{bmatrix} \begin{Bmatrix} T_1 \\ T_2 \\ T_3 \\ T_4 \end{Bmatrix} = \begin{Bmatrix} 71.455 \\ 36.551 \\ 15.760 \\ -4.62 \end{Bmatrix}$$

(b) Equations after adding $T_1 = 330$

$$\begin{bmatrix} 0.21653 & 0 & 0 & 0 \\ 0 & 0.47207 & 0 & -0.33949 \\ 0 & 0 & 0.20453 & 0 \\ 0 & -0.33949 & 0 & 0.50407 \end{bmatrix} \begin{Bmatrix} T_1 \\ T_2 \\ T_3 \\ T_4 \end{Bmatrix} = \begin{Bmatrix} 64.959 \\ 33.294 \\ 51.132 \\ 20.025 \end{Bmatrix}$$

(c) Equations after adding $T_3 = 250$

Figure 11.2 Modification of Example 10.1

Note that a_{22} in Example 11.1 is not set equal to 1.0. If the other terms in the matrix are very large or very small (say, $a_{11} = 10^{10}$, $a_{22} = 10^{11}$), then round-off error problems may occur during the solution of the equations.

Example 11.2 Numerical Example

Given Start with the assembled global matrix and column from Example 10.1—Conduction and convection in trapezoidal fin. The equations are shown in Figure 11.2(a).

Objective Include the prescribed nodal values $T_1 = 330$, $T_3 = 250$ in the equations for this problem, using the method just described.

Solution Consider row 1 first. All coefficients in that row are set to zero except 0.21653, and (0.21653)(330)= 71.455 replaces 2.64. For row 2, the right-hand side term 0.72 has added to it $-(-0.10858)(330)$ to yield 36.551. Then the term (-0.10858) is set to zero. Row 3 has 1.92, which gets $-(-0.04194)(330)$ added to it to give 15.760. Again the term (-0.04194) is set to zero. Row 4 is operated on in a similar manner.

The results of including $T_1 = 330$ are shown in Figure 11.2(b). A symmetric matrix of the same order as originally used occurs.

Now for $T_3 = 250$, consider row 3 first. All coefficients in that row are set to zero and 15.760 replaced by (0.20453)(250) = 51.132. Only row 4 has a nonzero term in column 3, so only it is changed. The term -4.62 has $-(-0.09858)(250)$ added to it to give 20.025. Then (-0.09858) is set to zero.

The final equations are again symmetric and banded. Actually, the computer program FINITE does these operations with the condensed matrix. The actual coding is given in Section 3 of the program.

11.5 Banded Gauss Elimination Solution of Linear Equations

The Gauss elimination method of solving linear simultaneous equations is widely used because it is relatively easy to understand but still efficient. It is one of several direct methods of solution including Gauss-Jordan, Cholesky, and others. Iterative methods such as Gauss-Seidel may also be used, but these will not be investigated here. These are usually employed when extremely large numbers of unknowns are encountered and are beyond this text. This section develops the method for a general banded symmetric system, which is then used in computer program FINTWO.

When Gauss elimination is used to solve a full (nonbanded) set of linear equations, it is not strictly the most efficient, direct technique. One of the factorization methods, such as the Cholesky or triple factoring, requires fewer computer operations. However, for positive definite banded symmetric matrices encountered in finite element solutions, the methods are relatively close in required solution times. A good discussion of the number of operations for full matrices is given in other texts.

A set of N banded, symmetric equations in N unknowns is reduced by Gauss elimination to an equivalent upper triangular banded set of equations. Here an equivalent set of equations is one that has the same solution as the original set. Once the upper matrix has been obtained, a back substitution technique is used to determine the unknowns. As an example, consider five equations with a bandwidth of three, of the form

$$
\begin{aligned}
a_{11}T_1 + a_{12}T_2 + a_{13}T_3 \phantom{+ a_{34}T_4 + a_{35}T_5} &= f_1 \\
a_{21}T_1 + a_{22}T_2 + a_{23}T_3 + a_{24}T_4 \phantom{+ a_{35}T_5} &= f_2 \\
a_{31}T_1 + a_{32}T_2 + a_{33}T_3 + a_{34}T_4 + a_{35}T_5 &= f_3 \\
a_{42}T_2 + a_{43}T_3 + a_{44}T_4 + a_{45}T_5 &= f_4 \\
a_{53}T_3 + a_{54}T_4 + a_{55}T_5 &= f_5
\end{aligned}
\tag{11.1}
$$

where the coefficient matrix is symmetric ($a_{21} = a_{12}$ and so on). This is reduced to the equivalent upper triangular banded equations

$$\begin{aligned} a_{11}T_1 + a_{12}T_2 + a_{13}T_3 \qquad\qquad\qquad\qquad &= f_1 \\ a'_{22}T_2 + a'_{23}T_3 + a'_{24}T_4 \qquad\qquad &= f'_2 \\ a''_{33}T_3 + a''_{34}T_4 + a''_{35}T_5 &= f''_3 \\ a'''_{44}T_4 + a'''_{45}T_5 &= f'''_4 \\ a''''_{55}T_5 &= f''''_5 \end{aligned} \tag{11.2}$$

which is clearly no longer symmetric. The prime superscripts in equations 2 through 5 indicate new values of a and f, which are obtained in the reduction process. Back substitution consists of starting with the last equation (which is easily solved for T_5) and proceeding from last equation to first to obtain the unknown nodal values.

The first step in the Gauss elimination procedure is to eliminate T_1 terms in all equations except the first. For equation 2, this is accomplished by multiplying equation 1 by $-a_{21}/a_{11}$ and adding it to equation 2 (while leaving equation 1 in its original state). The two equations are (if $a_{11} \neq 0$)

$$a_{11}T_1 + a_{12}T_2 + a_{13}T_3 = f_1$$

$$\left(a_{21} - \frac{a_{21}}{a_{11}}a_{11}\right)T_1 + \left(a_{22} - \frac{a_{21}}{a_{11}}a_{12}\right)T_2$$

$$+\left(a_{23} - \frac{a_{21}}{a_{11}}a_{13}\right)T_3 + a_{24}T_4 = \left(f_2 - \frac{a_{21}}{a_{11}}f_1\right)$$

Clearly the second equation reduces to

$$a'_{22}T_2 + a'_{23}T_3 + a'_{24}T_4 = f'_2 \tag{11.3}$$

where

$$a'_{22} = a_{22} - \frac{a_{21}}{a_{11}}a_{12}$$

$$a'_{23} = a_{23} - \frac{a_{21}}{a_{11}}a_{13}$$

$$a'_{24} = a_{24}$$

$$f'_2 = f_2 - \frac{a_{21}}{a_{11}}f_1$$

Note that the term farthest to the right is unchanged. Here equation 1 is sometimes called the pivot equation.

Continuing on with equation (row) 3, multiply equation 1 by $-a_{31}/a_{11}$ and add it to equation 3. Now the third equation becomes

$$\left(a_{31} - \frac{a_{31}}{a_{11}} a_{11}\right) T_1 + \left(a_{32} - \frac{a_{31}}{a_{11}} a_{12}\right) T_2$$

$$+ \left(a_{33} - \frac{a_{31}}{a_{11}} a_{13}\right) T_3 + a_{34}T_4 + a_{35}T_5 = \left(f_3 - \frac{a_{31}}{a_{11}} f_1\right)$$

or

$$a'_{32}T_2 + a'_{33}T_3 + a'_{34}T_4 + a'_{35}T_5 = f'_3 \tag{11.4}$$

where, from above,

$$a'_{32} = a_{32} - \frac{a_{31}}{a_{11}} a_{12}$$

$$a'_{33} = a_{33} - \frac{a_{31}}{a_{11}} a_{13}$$

$$a'_{34} = a_{34}$$

$$a'_{35} = a_{35}$$

$$f'_3 = f_3 - \frac{a_{31}}{a_{11}} f_1$$

There is no need to modify the rest of the equations (rows) because the T_1 terms are zero. Again note that both a'_{34} and a'_{35} are simply equal to the original values.

The completed first step of removing all T_1 terms except in equation 1 results in the set of equations

|←—— NBW ——→|

$$\begin{aligned}
a_{11}T_1 + a_{12}T_2 + a_{13}T_3 \qquad\qquad\qquad\qquad &= f_1 \\
\boxed{a'_{22}T_2 + a'_{23}T_3} + a'_{24}T_4 \qquad\qquad &= \boxed{f'_2} \\
\boxed{a'_{32}T_2 + a'_{33}T_3} + a'_{34}T_4 + a'_{35}T_5 &= \boxed{f'_3} \\
a'_{42}T_2 + a'_{43}T_3 + a'_{44}T_4 + a'_{45}T_5 &= f'_4 \\
a'_{53}T_3 + a'_{54}T_4 + a'_{55}T_5 &= f'_5
\end{aligned} \tag{11.5}$$

Only the terms a and f within the dotted areas are changed by the process just described. Clearly, if a_{11} is zero, a problem arises. Fortunately, finite element solutions produce zero diagonal terms only under very unusual circumstances, which are not considered here.

An important matter is the symmetry of the resulting system. The first row and column are no longer symmetric, but they are in the upper triangular banded form desired. In the remainder of the terms, $a'_{32} = a'_{23}$, which is seen from Equations (11.3) and (11.4), where

$$a'_{23} = a_{23} - \frac{a_{21}}{a_{11}} a_{13}$$

$$a'_{32} = a_{32} - \frac{a_{31}}{a_{11}} a_{12}$$

The original coefficient matrix was symmetric, so the primed matrix must be also. All primed terms outside the dotted square are the same as the original matrix. This means that only the upper half of the matrix must be stored during computations.

The second step in the Gauss elimination procedure consists of removing all T_2 terms below equation (row) 2, now the pivot row. For equation (row) 3, multiply equation 2 by $-a'_{32}/a'_{22}$ and add it to equation 3. Also, multiply equation 2 by $-a'_{42}/a'_{22}$ and add it to equation (row) 4. The resulting equations have the form

$$\begin{aligned}
a_{11}T_1 + a_{12}T_2 + a_{13}T_3 \qquad\qquad\qquad\qquad &= f_1 \\
a'_{22}T_2 + a'_{23}T_3 + a'_{24}T_4 \qquad\qquad\quad &= f'_2 \\
\boxed{a''_{33}T_3 + a''_{34}T_4} + a''_{35}T_5 &= \boxed{f''_3} \\
\boxed{a''_{43}T_3 + a''_{44}T_4} + a''_{45}T_5 &= \boxed{f''_4} \\
a''_{53}T_3 + a''_{54}T_4 + a''_{55}T_5 &= f''_5
\end{aligned} \tag{11.6}$$

where the new a and f terms are shown in the dotted areas. The values are

$$a''_{33} = a'_{33} - \frac{a'_{32}}{a'_{22}}\, a'_{23}$$

$$a''_{34} = a'_{34} - \frac{a'_{32}}{a'_{22}}\, a'_{24}$$

$$f''_3 = f'_3 - \frac{a'_{32}}{a'_{22}}\, f'_2$$

$$a''_{43} = a'_{43} - \frac{a'_{32}}{a'_{22}}\, a'_{23}$$

$$a''_{44} = a'_{44} - \frac{a'_{32}}{a'_{22}}\, a'_{24}$$

$$f''_4 = f'_4 - \frac{a'_{32}}{a'_{22}}\, 128'_2$$

Again a'_{22} must not be zero. This general procedure is continued until the final Equation (11.2) in upper triangular form is arrived at.

A general form for the reduction process can be fairly easily found. Let i represent the pivot row to be considered, and the NBW$-$1 (bandwidth minus 1) coefficients below and to the right of the a_{ii} term must be modified. The new coefficients are given by

$$a_{jk}^{(\text{new})} = a_{jk} - \frac{a_{ji}}{a_{ii}}\, a_{ik} \qquad \begin{cases} i = 1, \quad N-1 \\ j = i+1, \quad i + \text{NBW (also } j \le N) \\ k = i+1, \quad i + \text{NBW (also } k \le N) \end{cases}$$

It is convenient to deal only with the upper terms in what follows (for minimum storage requirements). Since the matrices are symmetric, $a_{ji} = a_{ij}$, so the above expression can be written as

$$a_{jk}^{(\text{new})} = a_{jk} - \frac{a_{ij}}{a_{ii}} a_{ik} \quad \begin{cases} i = 1, \quad N-1 \\ j = i+1, \quad i + \text{NBW (also } j \leq N) \\ k = j, \quad i + \text{NBW (also } k \leq N) \end{cases} \tag{11.7}$$

Similarly the f terms are given by

$$f_j^{(\text{new})} = f_j - \frac{a_{ij}}{a_{ii}} f_i \quad \begin{cases} i = 1, \quad N-1 \\ j = i+1, \quad i + \text{NBW (also } j \leq N) \end{cases} \tag{11.8}$$

These two formulas can easily be used to write a forward reduction computer program.

The final upper triangular banded equations have the form

$$\begin{aligned}
a_{11}T_1 + a_{12}T_2 + a_{13}T_3 \qquad\qquad\qquad\qquad &= f_1 \\
a_{22}T_2 + a_{23}T_3 + a_{24}T_4 \qquad\qquad &= f_2 \\
a_{33}T_3 + a_{34}T_4 + a_{35}T_5 &= f_3 \\
a_{44}T_4 + a_{45}T_5 &= f_4 \\
a_{55}T_5 &= f_5
\end{aligned}$$

The back-substitution now starts. Clearly the last equation is immediately solved for T_5 as $T_5 = f_5/a_{55}$. The next equation (moving up) gives T_4 as $T_4 = (f_4 - a_{45}T_5)/a_{44}$, while the next equation, for T_3, becomes $T_3 = (f_3 - a_{34}T_4 - a_{35}T_5)/a_{33}$. This is continued up through equation 1.

If i represents a row, the general formula for back-substitution is

$$T_i = \left(f_i - \sum_{j=i+1}^{i+\text{NBW}-1} a_{ij}T_j\right)\Big/a_{ii}, \quad \begin{cases} i = N-1, 1 \\ j \leq N \end{cases} \tag{11.9}$$

Again, a computer program is easily formulated.

Example 11.3 Banded Gauss Elimination of Five Unknowns

Given A set of five linear symmetric algebraic equations with bandwidth 3 is shown below.

|← NBW=3 →|

$$\begin{bmatrix} 2 & 3 & 1 & 0 & 0 \\ 3 & 4 & 5 & -2 & 0 \\ 1 & 5 & 1 & 0 & 3 \\ 0 & -2 & 0 & 2 & -4 \\ 0 & 0 & 3 & -4 & 1 \end{bmatrix} \begin{Bmatrix} T_1 \\ T_2 \\ T_3 \\ T_4 \\ T_5 \end{Bmatrix} = \begin{Bmatrix} 11 \\ 18 \\ 29 \\ -16 \\ -2 \end{Bmatrix}$$

Objective Find the solution, using banded Gauss elimination.

Solution The first step consists of eliminating all T_1 terms below the main diagonal. Multiply equation (row) 1 by $-3/2$ and add it to equation (row) 2. Row 2 becomes, from equation (11.3)

$$\left[3 - \frac{3}{2}(2)\right] T_1 + \left[4 - \frac{3}{2}(3)\right] T_2 + \left[5 - \frac{3}{2}(1)\right] T_3$$

$$+ \left[-2 - \frac{3}{2}(0)\right] T_4 = \left[18 - \frac{3}{2}(11)\right]$$

or simplifying, $-0.5\ T_2 + 3.5\ T_3 - 2\ T_4 = 1.5$. Now multiply equation (row) 1 by $-1/2$ and add it to equation (row) 3. Then row 3, from Equation (11.4), is

$$\left[1 - \frac{1}{2}(2)\right] T_1 + \left[5 - \frac{1}{2}(3)\right] T_2 + \left[1 - \frac{1}{2}(1)\right] T_3$$

$$+ \left[0 - \frac{1}{2}(0)\right] T_4 + \left[3 - \frac{1}{2}(0)\right] T_5 = \left[29 - \frac{1}{2}(11)\right]$$

and simplifying, $3.5\ T_2 + 0.5\ T_3 + 0\ T_4 + 3\ T_5 = 23.5$. This completes the first step.

Writing out the matrix equation after one step gives

$$\begin{bmatrix} 2 & 3 & 1 & 0 & 0 \\ 0 & -0.5 & 3.5 & -2 & 0 \\ 0 & 3.5 & 0.5 & 0 & 3 \\ 0 & -2 & 0 & 2 & -4 \\ 0 & 0 & 3 & -4 & 1 \end{bmatrix} \begin{Bmatrix} T_1 \\ T_2 \\ T_3 \\ T_4 \\ T_5 \end{Bmatrix} = \begin{Bmatrix} 11 \\ 1.5 \\ 23.5 \\ -16 \\ -2 \end{Bmatrix}$$

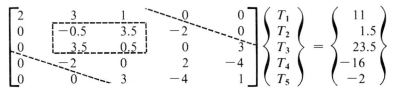

The areas with changes are indicated by dotted rectangles. Note that the matrix (except for row 1 and column 1) is still symmetric.

Step 2 consists of eliminating T_2 terms below the main diagonal from equations (rows) 3 and 4. Multiply row 2 by $-3.5/(-0.5)$ and add it to row 3, see Equation (11.8),

$$\left[3.5 - \frac{3.5}{-0.5}(-0.5)\right] T_2 + \left[0.5 - \frac{3.5}{-0.5}(3.5)\right] T_3$$

$$+ \left[0 - \frac{3.5}{-0.5}(-2)\right] T_4 + \left[3 - \frac{3.5}{-0.5}(0)\right] T_5$$

$$= \left[23.5 - \frac{3.5}{-0.5}(1.5)\right]$$

with the result $25\ T_3 - 14\ T_4 + 3\ T_5 = 34$. Multiply row 2 by $-(-2)/(-0.5)$ and add it to row 3, giving

$$\left[-2 - \frac{-2}{-0.5}(-0.5)\right] T_2 + \left[0 - \frac{-2}{-0.5}(3.5)\right] T_3$$

$$+ \left[2 - \frac{-2}{-0.5}(-2)\right] T_4 + \left[-4 - \frac{-2}{-0.5}(0)\right] T_5$$

$$= \left[-16 \frac{-2}{-0.5}(1.5)\right]$$

and reducing, $-14\ T_3 + 10\ T_4 - 4\ T_5 = -22$.

After two steps, the matrix equation is

$$\begin{bmatrix} 2 & 3 & 1 & 0 & 0 \\ 0 & -0.5 & 3.5 & -2 & 0 \\ 0 & 0 & 25 & -14 & 3 \\ 0 & 0 & -14 & 10 & -4 \\ 0 & 0 & 3 & -4 & 1 \end{bmatrix} \begin{Bmatrix} T_1 \\ T_2 \\ T_3 \\ T_4 \\ T_5 \end{Bmatrix} = \begin{Bmatrix} 11 \\ 1.5 \\ 34 \\ -22 \\ -2 \end{Bmatrix}$$

Again the changed terms are indicated by the dotted rectangles.

Completing the full process yields the matrix equation

|←—— NBW ——→|

$$\begin{bmatrix} 2 & 3 & 1 & 0 & 0 \\ 0 & -0.5 & 3.5 & -2 & 0 \\ 0 & 0 & 25 & -14 & 3 \\ 0 & 0 & 0 & 2.16 & -2.32 \\ 0 & 0 & 0 & 0 & -1.8519 \end{bmatrix} \begin{Bmatrix} T_1 \\ T_2 \\ T_3 \\ T_4 \\ T_5 \end{Bmatrix} = \begin{Bmatrix} 11 \\ 1.5 \\ 34 \\ -2.96 \\ -9.2593 \end{Bmatrix}$$

This is the banded upper triangular form shown in Equation (11.2) as desired.

Solving the last equation for T_5, $T_5 = (-9.2593)/(-1.8519) = 4.9998$. The next equation is solved by back-substitution as

$$T_4 = \left[-2.96 - (-2.32)(4.9998)\right]/2.16 = 3.9998,$$

with the result for T_3,

$$T_3 = \left[34 - (-14)(3.998) - (3)(4.998)\right]/25$$

$$T_3 = 3.0002$$

Solving for the other two values, $T_2 = 2.0020$, $T_1 = 0.9969$. The final values are

$$\begin{Bmatrix} T_1 \\ T_2 \\ T_3 \\ T_4 \\ T_5 \end{Bmatrix} = \begin{Bmatrix} 0.9969 \\ 2.0020 \\ 3.0002 \\ 3.9998 \\ 4.9998 \end{Bmatrix} \qquad \text{Gauss elimination solution}$$

The exact solution is $T_1 = 1$, $T_2 = 2$, $T_3 = 3$, $T_4 = 4$, and $T_5 = 5$. In this example only five significant figures were retained, leading to the errors shown above. Note that the error accumulates as the back substitution procedure goes up the column.

The details of the banded Gauss elimination technique can be determined from computer program FINTWO. Subroutine GAUSS actually does the computations. Note that the process must be somewhat modified to use the condensed form of storage for the $[A]$ matrix. Subroutine WRITEM writes out the results just before and just after the Gauss elimination technique.

In large finite element codes such as ANSYS, the storage of even a condensed matrix becomes impossible. A method called a wave front solver is used. In this method, most of the matrix is stored out of the core of the computer on tape or other external storage. Only a small part of the elements are being assembled in core at any one time. Details of this method are left for advanced texts.

11.6 Problems

11.1. Consider the set of algebraic equations

$$\begin{bmatrix} 3 & -1 & 4 & 0 \\ -1 & 5 & -2 & -3 \\ 4 & -2 & 6 & 1 \\ 0 & -3 & 1 & 2 \end{bmatrix} \begin{Bmatrix} T_1 \\ T_2 \\ T_3 \\ T_4 \end{Bmatrix} = \begin{Bmatrix} 6 \\ -2 \\ 4 \\ 3 \end{Bmatrix}$$

Modify these equations to take into account the prescribed nodal values $T_1 = 2.5$, $T_3 = -3$. Then write the result in condensed storage form.

11.2. A set of algebraic equations with a bandwidth of 2 (called a tridiagonal set of equations) is

$$\begin{bmatrix} 2.0 & -1.2 & 0 & 0 & 0 \\ -1.2 & 2.4 & -0.8 & 0 & 0 \\ 0 & -0.8 & 1.6 & -1.1 & 0 \\ 0 & 0 & -1.1 & 2.2 & 0.5 \\ 0 & 0 & 0 & 0.5 & 1.0 \end{bmatrix} \begin{Bmatrix} T_1 \\ T_2 \\ T_3 \\ T_4 \\ T_5 \end{Bmatrix} = \begin{Bmatrix} 3.0 \\ 2.6 \\ -1.7 \\ 3.2 \\ 1.8 \end{Bmatrix}$$

Solve by hand, using Gauss elimination.

11.3. A general tridiagonal set of algebraic equations has the form

$$\begin{bmatrix} a_{11} & a_{12} & 0 & 0 & \cdot & 0 \\ a_{21} & a_{22} & a_{23} & 0 & \cdot & 0 \\ 0 & a_{32} & a_{33} & a_{34} & \cdot & 0 \\ 0 & 0 & a_{43} & a_{44} & \cdot & 0 \\ \cdot & \cdot & \cdot & \cdot & \cdot & \cdot \\ 0 & 0 & 0 & 0 & \cdot & a_{NN} \end{bmatrix} \begin{Bmatrix} T_1 \\ T_2 \\ T_3 \\ T_4 \\ \cdot \\ T_N \end{Bmatrix} = \begin{Bmatrix} f_1 \\ f_2 \\ f_3 \\ f_4 \\ \cdot \\ f_N \end{Bmatrix}$$

This is symmetric: $a_{ij} = a_{ji}$. Develop a general method for solving this set of equations, using Gauss elimination but taking full advantage of the bandwidth of two (do not multiply or add the zero terms outside the bandwidth). Write a small computer program to solve for equations of this form. Solve the equations in Problem 11.2 with the program.

11.4. Solve the following set of algebraic equations by hand using Gauss elimination

$$\begin{bmatrix} 4 & 2 & -4 \\ 2 & 3 & -1 \\ -4 & -1 & 2 \end{bmatrix} \begin{Bmatrix} T_1 \\ T_2 \\ T_3 \end{Bmatrix} = \begin{Bmatrix} 2 \\ 1 \\ -7 \end{Bmatrix}$$

11.5 A banded set of algebraic equations is

$$\begin{bmatrix} 2 & -2 & 0.5 & 0 & 0 \\ -2 & 3.5 & 1 & -1 & 0 \\ 0.5 & 1 & 1 & 2 & -2 \\ 0 & -1 & 2 & 3 & 1.5 \\ 0 & 0 & -2 & 1.5 & 2 \end{bmatrix} \begin{Bmatrix} T_1 \\ T_2 \\ T_3 \\ T_4 \\ T_5 \end{Bmatrix} = \begin{Bmatrix} 2.5 \\ -4 \\ -4.25 \\ 1 \\ 6 \end{Bmatrix}$$

Solve by hand, using Gauss elimination.

Chapter 12

More Variational Methods—Two and Three Dimensions

12.1 Rectangular Elements

An element very often used is the rectangular element, as discussed in Section 3.7. This is so for two reasons:

1. Many analysis regions are at least partly rectangular in form.
2. Dividing regions up into rectangular elements may require less of an engineer's time than division of the region into other elements (triangles, say).

It is relatively easy to understand the properties of the rectangular element as compared to some of the other higher order elements, so it is considered here.

A. Element Interpolation Function

As shown in Figure 12.1, the element is aligned so that the sides are parallel to the x and y coordinate axes. The sides are of length L and W in the x and y directions, respectively. Let node points be located at each of the corners and labeled T_i, T_j, T_k, T_ℓ (counterclockwise).

The element interpolation function has the form

$$T^{(e)} = \alpha_1 + \alpha_2 x + \alpha_3 y + \alpha_4 xy \tag{12.1}$$

This form has all the linear terms of the two-dimensional simplex (triangular) element, plus one quadratic term necessary to account for the fourth node. It is called a bilinear function of x and y, so the element may be called a bilinear rectangular element. The other possible quadratic terms x^2 and y^2 have been omitted in the interpolation function, so it is an incomplete function as well. The bilinear term xy is chosen because it produces a linear variation of $T^{(e)}$ along any side parallel to a coordinate axis. Thus the interpolation functions for two adjacent bilinear rectangular elements will be continuous across the interelement boundary. Also adjacent triangular elements will have the same property. For this reason, these elements are sometimes called serendipity elements.

Consider the local coordinate system x,y with the origin at the center of the element. If the center of the element has global coordinates x_0, y_0, the global coordinates are related to the local coordinates by $x' = x + x_0, y' = y + y_0$. Generally element coefficient matrices in Cartesian coordinates are independent of the location of the origin.

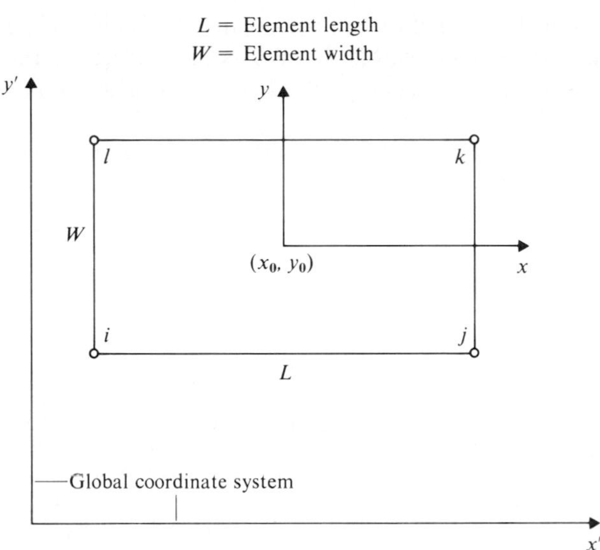

Figure 12.1 Rectangular element

It is convenient to start immediately with an interpolation function of the form

$$T^{(e)} = G_iT_i + G_jT_j + G_kT_k + G_\ell T_\ell \tag{12.2}$$

where the shape functions are

$$G_i = \frac{1}{LW}\left(\frac{L}{2} - x\right)\left(\frac{W}{2} - y\right)$$

$$G_j = \frac{1}{LW}\left(\frac{L}{2} + x\right)\left(\frac{W}{2} - y\right)$$

$$G_k = \frac{1}{LW}\left(\frac{L}{2} + x\right)\left(\frac{W}{2} + y\right)$$

$$G_\ell = \frac{1}{LW}\left(\frac{L}{2} - x\right)\left(\frac{W}{2} + y\right)$$

These have the normal shape function properties of value unity at the associated node and zero at the other three nodes. Of course, the area A is LW.

Consider the interpolation function along the x axis where $y = -\frac{W}{2}$. The pyramid functions reduce to

$$G_i = \frac{1}{L}\left(\frac{L}{2} - x\right), \quad G_j = \frac{1}{L}\left(\frac{L}{2} + x\right), \quad G_k = 0, \quad G_\ell = 0$$

and the interpolation function is

$$T^{(e)} = \frac{1}{L}\left(\frac{L}{2} - x\right)T_i + \frac{1}{L}\left(\frac{L}{2} + x\right)T_j$$

This gives a linear relation for T from node i to node j. Similar results are found for the other three sides. It is easily seen that interelement continuity of T is satisfied across the boundary between two rectangular elements or a rectangular element and a triangular element.

In higher order elements the gradients are not constant, as they are in simplex elements. The derivatives are

$$\frac{\partial T^{(e)}}{\partial x} = \{D_x\}\{T\}$$

$$\frac{\partial T^{(e)}}{\partial y} = \{D_y\}\{T\}$$

where

$$\{D_x\} = \frac{1}{LW}\left\{-\left(\frac{W}{2} - y\right) \quad \left(\frac{W}{2} - y\right) \quad \left(\frac{W}{2} + y\right) \quad -\left(\frac{W}{2} + y\right)\right\}$$

$$\{D_y\} = \frac{1}{LW}\left\{-\left(\frac{L}{2} - x\right) \quad -\left(\frac{L}{2} + x\right) \quad \left(\frac{L}{2} + x\right) \quad \left(\frac{L}{2} - x\right)\right\}$$

The x derivative is a linear function of y, and the y derivative is a linear function of x.

B. Element Matrices

The element matrices are given by (9.37) as

$$\left\{\begin{matrix} \dfrac{\partial I}{\partial T_i} \\ \dfrac{\partial I}{\partial T_j} \\ \dfrac{\partial I}{\partial T_k} \\ \dfrac{\partial I}{\partial T_\ell} \end{matrix}\right\}^{(e)} = [B]^{(e)}\{T\}^{(e)} - \{C\}^{(e)}$$

where the element matrix is

$$[B]^{(e)} = \iint_{A^{(e)}} \left[\{D_x\}^T K_x \{D_x\} + \{D_y\}^T K_y \{D_y\} - \{G\}^T P \{G\}\right]^{(e)} dA + \int_{L_2^{(e)}} \left[\{G\}^T \alpha \{G\}\right]^{(e)} dL_2$$

and the element column is

$$\{C\}^{(e)} = \iint_{A^{(e)}} \left[\{G\}^T Q\right]^{(e)} dA - \int_{L_2^{(e)}} \left[\{G\}^T \beta\right]^{(e)} dL_2$$

Each of the terms must be evaluated. Consider the P terms, which have the form

$$\begin{bmatrix} \iint PG_iG_i\,dA & \iint PG_iG_j\,dA & \iint PG_iG_k\,dA & \iint PG_iG_\ell\,dA \\ \iint PG_jG_i\,dA & \iint PG_jG_j\,dA & \iint PG_jG_k\,dA & \iint PG_jG_\ell\,dA \\ \iint PG_kG_i\,dA & \iint PG_kG_j\,dA & \iint PG_kG_k\,dA & \iint PG_kG_\ell\,dA \\ \iint PG_\ell G_i\,dA & \iint PG_\ell G_j\,dA & \iint PG_\ell G_k\,dA & \iint PG_\ell G_\ell\,dA \end{bmatrix}$$

These terms are relatively easily evaluated.

Consider constant property elements, where K_x, K_y, P, Q, α, β are all constant within the element. The integrals can be evaluated as

$$\iint_{A^{(e)}} G_\beta^2\,dA = \frac{1}{9}LW, \quad \beta = i,j,k,\ell$$

$$\iint_{A^{(e)}} G_iG_j\,dA = \frac{1}{18}LW, \quad \iint_{A^{(e)}} G_iG_k dA = \frac{1}{36}LW$$

$$\iint_{A^{(e)}} G_iG_\ell\,dA = \frac{1}{18}LW$$

When the shape functions of adjacent nodes are in the integrand, the value is $LW/18$, while nodes at opposite corners yield $LW/36$. Evaluating all of the terms in the element matrices give the general form similar to (9.29) as

$$\begin{aligned}
[B]^{(e)} &= \frac{K_xW}{6L}\begin{bmatrix} 2 & -2 & -1 & 1 \\ & 2 & 1 & -1 \\ & & 2 & -2 \\ \text{(Symmetric)} & & & 2 \end{bmatrix}^{(e)} && (K_x \text{ matrix}) \\
&+ \frac{K_yL}{6W}\begin{bmatrix} 2 & 1 & -1 & -2 \\ & 2 & -2 & -1 \\ & & 2 & 1 \\ \text{(Symmetric)} & & & 2 \end{bmatrix}^{(e)} && (K_y \text{ matrix}) \\
&- \frac{PLW}{36}\begin{bmatrix} 4 & 2 & 1 & 2 \\ & 4 & 2 & 1 \\ & & 4 & 2 \\ \text{(Symmetric)} & & & 4 \end{bmatrix}^{(e)} && (P \text{ matrix}) \\
&+ \frac{\alpha_{ij}L}{6}\begin{bmatrix} 2 & 1 & 0 & 0 \\ & 2 & 0 & 0 \\ & & 0 & 0 \\ \text{(Symmetric)} & & & 0 \end{bmatrix}_{\substack{\text{on side } ij \\ \text{(side 1)}}} && (\alpha \text{ matrix } ij)
\end{aligned} \qquad \textbf{(12.3)}$$

Continued on next page

$$+\frac{\alpha_{jk}L}{6}\begin{bmatrix} 0 & 0 & 0 & 0 \\ & 2 & 1 & 0 \\ & & 2 & 0 \\ \text{(Symmetric)} & & & 0 \end{bmatrix}_{\substack{\text{on side } jk \\ \text{(side 2)}}} \qquad (\alpha \text{ matrix } jk)$$

$$+\frac{\alpha_{k\ell}L}{6}\begin{bmatrix} 0 & 0 & 0 & 0 \\ & 0 & 0 & 0 \\ & & 2 & 1 \\ \text{(Symmetric)} & & & 2 \end{bmatrix}_{\substack{\text{on side } k\ell \\ \text{(side 3)}}} \qquad (\alpha \text{ matrix } k\ell) \qquad \textbf{(12.3 cont.)}$$

$$+\frac{\alpha_{\ell i}L}{6}\begin{bmatrix} 2 & 0 & 0 & 1 \\ & 0 & 0 & 0 \\ & & 0 & 0 \\ \text{(Symmetric)} & & & 2 \end{bmatrix}_{\substack{\text{on side } \ell i \\ \text{(side 4)}}} \qquad (\alpha \text{ matrix } \ell i)$$

Also the element column, expressed in a form similar to Equation (9.30), is

$$\{C\}^{(e)} = \frac{QLW}{4}\begin{Bmatrix} 1 \\ 1 \\ 1 \\ 1 \end{Bmatrix}^{(e)} \qquad (Q \text{ column})$$

$$-\frac{1}{2}\beta_{ij}L\begin{Bmatrix} 1 \\ 1 \\ 0 \\ 0 \end{Bmatrix}_{\substack{\text{on side } ij \\ \text{(side 1)}}} \qquad (\beta \text{ column } ij)$$

$$-\frac{1}{2}\beta_{jk}W\begin{Bmatrix} 0 \\ 1 \\ 1 \\ 0 \end{Bmatrix}_{\substack{\text{on side } jk \\ \text{(side 2)}}} \qquad (\beta \text{ column } jk) \qquad \textbf{(12.4)}$$

$$-\frac{1}{2}\beta_{k\ell}L\begin{Bmatrix} 0 \\ 0 \\ 1 \\ 1 \end{Bmatrix}_{\substack{\text{on side } k\ell \\ \text{(side 3)}}} \qquad (\beta \text{ column } k\ell)$$

$$-\frac{1}{2}\beta_{\ell i}W\begin{Bmatrix} 1 \\ 0 \\ 0 \\ 1 \end{Bmatrix}_{\substack{\text{on side } \ell i \\ \text{(side 4)}}} \qquad (\beta \text{ column } \ell i)$$

The element matrices are assembled into global equations in normal fashion.

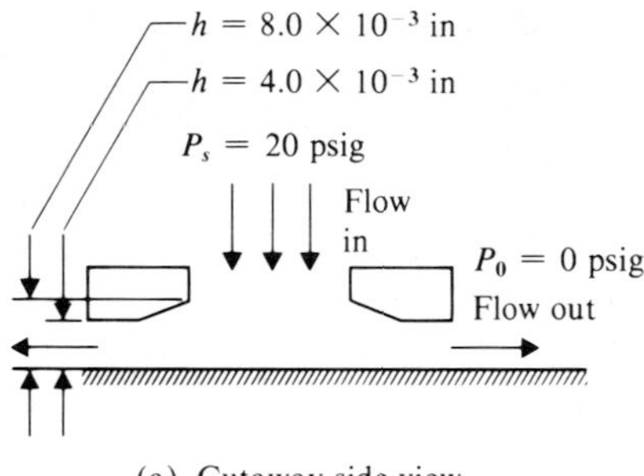

(a) Cutaway side view

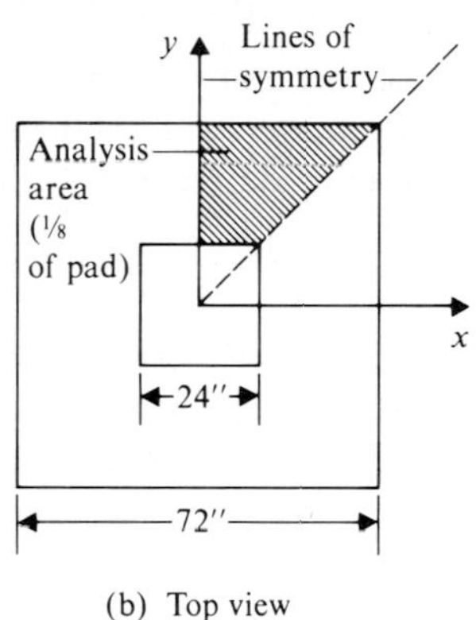

(b) Top view

Figure 12.2 Pad geometry (Example 12.1—Hydrostatic pad analysis with two element types)

C. Example

Modern finite element computer programs normally use several element types within the same program. Triangular elements may be used in regions where the geometry is complex so as to accurately model the real problem. Rectangular elements can then be used in rectangular regions or in regions away from boundaries. This makes the division scheme easier for the engineer. This section considers an example using the two types of two-dimensional elements developed so far.

Example 12.1 *Hydrostatic Pad Analysis with Two Element Types*

Given A hydrostatic pad, shown in Figure 12.2, is to be used to lift the seating sections in a stadium. With the resulting low coefficient of friction under the seats once the bearing is pressurized, the seats can be moved to create a stadium ideal for watching football or baseball. It will be assumed that the pressures are small enough so that the air under the pad can be considered incompressible and isoviscous. The supply pressure is $p_s = 20$ psig and the ambient pressure is $p_a = 0$ psig. Absolute viscosity for air is $\mu = 2.5 \times 10^{-9}$ lbf-sec/in^2. The film thickness $h(x,y)$, shown in Figure 12.2(a), is linearly tapered for the first half of the pad and then constant.

The pressure under the pad is governed by the Reynolds equation

$$\frac{\partial}{\partial x}\left(h^3 \frac{\partial p}{\partial x}\right) + \frac{\partial}{\partial y}\left(h^3 \frac{\partial p}{\partial y}\right) = 0$$

Also the flow rates under the pad are given by

$$u = -\frac{h^2}{12\mu}\frac{\partial p}{\partial x}$$

$$v = -\frac{h^2}{12\mu}\frac{\partial p}{\partial y}$$

These can be used to estimate the total flow rate under the pad and to size a supply pump.

Objective Use both triangular and rectangular elements to find the pressure under the hydrostatic pad. Take full advantage of lines of symmetry in the pad. Integrate the pressure to obtain the load-lifting capacity of the hydrostatic pad. Calculate the flow rate under the pad.

Solution Lines of symmetry for this problem occur along the x and y axes as well as along the lines $x = y$ and $x = -y$. The smallest region that can be treated is one-eighth of the pad. Figure 12.2(b) shows the part of pad to be considered. The pressure in the 24″ supply region is 20 psig, which will be added to the load capacity calculation at the end of the problem.

Comparing the governing differential equation for the hydrostatic pad to the general differential equation (9.5)

$$\frac{\partial}{\partial x}\left[K_x(x,y)\frac{\partial T}{\partial x}\right] + \frac{\partial}{\partial y}\left[K_y(x,y)\frac{\partial T}{\partial y}\right] + P(x,y)T + Q(x,y) = 0$$

yields the coefficient functions $K_x(x,y) = h^3$, $K_y(x,y) = h^3$, $P(x,y) = 0.0$, and $Q(x,y) = 0.0$. The boundary conditions (9.6) of general form are $T = T_0$ on L_1 and (9.7)

$$K_x(x,y)\frac{\partial T}{\partial x}n_x + K_y(x,y)\frac{\partial T}{\partial y}n_y + \alpha(x,y)T + \beta(x,y) = 0 \text{ on } L_2$$

For the hydrostatic pad, the nodal values are specified along the inner and outer edges $p = p_s = 20.0$ on $y = 12.0$, $p = p_0 = 0.0$ on $y = 36.0$. Along lines of symmetry, the pressure gradient must vanish, giving the boundary parameters $\alpha(x,y) = 0.0$, $\beta(x,y) = 0.0$ on $x = 0.0$, and $\alpha(x,y) = 0.0$, $\beta(x,y) = 0.0$ on the line $x = y$.

The division into elements is shown in Figure 12.3, where three rectangular and two triangular elements are conveniently used. Nodal coordinates and element/node connectivity data are given in Table 12.1. As noted earlier the element coefficient function K (equal to h^3) is computed and then averaged over the four node points in a rectangular element or three node points in a triangular element. The results are given in Table 12.1. Boundary values are also given for α and β.

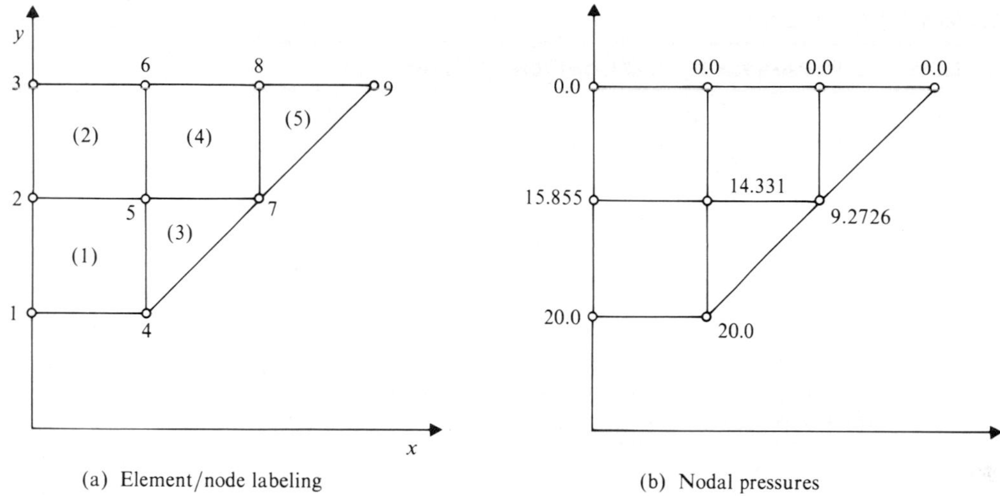

(a) Element/node labeling

(b) Nodal pressures

Figure 12.3 Division into elements and solution (Example 12.1—Hydrostatic pad analysis with two element types)

Table 12.1
Input Data (Example 12.1: Hydrostatic Pad Analysis with Two Element Types)

(a) Nodal Coordinates

Node	*x*	*y*
1	0.0	12.0
2	0.0	24.0
3	0.0	36.0
4	12.0	12.0
5	12.0	24.0
6	12.0	36.0
7	24.0	24.0
8	24.0	36.0
9	36.0	36.0

(b) Element/Node Connectivity Data and Element Type

Element	Nodes	Types
1	1 4 5 2	Rectangular
2	2 5 6 3	Rectangular
3	4 7 5	Triangular
4	5 7 8 6	Rectangular
5	7 9 8	Triangular

Table 12.1 *Continued*

(c) Element Values for *K, P, Q* Coefficient Functions

Element	$K \times 10^8$	P	Q
1	28.8	0.0	0.0
2	6.4	0.0	0.0
3	21.333	0.0	0.0
4	6.4	0.0	0.0
5	6.4	0.0	0.0

(d) Prescribed Nodal Values

Node	Pressure
1	20.0
3	0.0
4	20.0
6	0.0
8	0.0
9	0.0

(e) Element Derivative Boundary Conditions

Element	Side	α	β
1	2–1	0.0	0.0
2	3–2	0.0	0.0
3	4–7	0.0	0.0
5	7–9	0.0	0.0

For element 1, the element coefficient matrix (12.3) has the element x and y gradient matrices

$$[B]^{(1)} = \frac{28.8 \times 10^{-8}(12)}{6(12)} \begin{bmatrix} 2 & -2 & -1 & 1 \\ & 2 & 1 & -1 \\ & & 2 & -2 \\ \text{(Symmetric)} & & & 2 \end{bmatrix}^{(1)}$$

$$+ \frac{28.8 \times 10^{-8}(12)}{6(12)} \begin{bmatrix} 2 & 1 & -1 & -2 \\ & 2 & -2 & -1 \\ & & 2 & 1 \\ \text{(Symmetric)} & & & 2 \end{bmatrix}^{(1)}$$

where the remaining matrices are zero. Adding these two together yields

$$[B]^{(1)} = 4.80 \times 10^{-8} \begin{array}{c} \\ 1 \\ 4 \\ 5 \\ 2 \end{array} \begin{array}{c} \begin{array}{cccc} 1 & 4 & 5 & 2 \end{array} \\ \begin{bmatrix} 4 & -1 & -2 & -1 \\ & 4 & -1 & -2 \\ & & 4 & -1 \\ \text{(Symmetric)} & & & 4 \end{bmatrix}^{(1)} \end{array}$$

The element coefficient matrices for elements 2 and 4 (both rectangular with the same length and width) have the same square matrix except for the value of K used in front of the matrix. The two values are

$$[B]^{(2)} = 1.0667 \times 10^{-8} \begin{array}{c} \\ 2 \\ 5 \\ 6 \\ 3 \end{array} \begin{array}{c} \begin{array}{cccc} 2 & 5 & 6 & 3 \end{array} \\ \begin{bmatrix} 4 & -1 & -2 & -1 \\ & 4 & -1 & -2 \\ & & 4 & -1 \\ \text{(Symmetric)} & & & 4 \end{bmatrix}^{(2)} \end{array}$$

and

$$[B]^{(3)} = 1.0667 \times 10^{-8} \begin{array}{c} \\ 5 \\ 7 \\ 8 \\ 6 \end{array} \begin{array}{c} \begin{array}{cccc} 5 & 7 & 8 & 6 \end{array} \\ \begin{bmatrix} 4 & -1 & -2 & -1 \\ & 4 & -1 & -2 \\ & & 4 & -1 \\ \text{(Symmetric)} & & & 4 \end{bmatrix}^{(3)} \end{array}$$

All of the element columns have zero values.

For the two triangular elements, the element coefficient matrices are

$$[B]^{(3)} = 10.6667 \times 10^{-8} \begin{array}{c} \\ 4 \\ 7 \\ 5 \end{array} \begin{array}{c} \begin{array}{ccc} 4 & 7 & 5 \end{array} \\ \begin{bmatrix} 1 & 0 & -1 \\ & 1 & -1 \\ \text{(Symmetric)} & & 2 \end{bmatrix}^{(3)} \end{array}$$

$$[B]^{(5)} = 3.2 \times 10^{-8} \begin{array}{c} \\ 7 \\ 9 \\ 8 \end{array} \begin{array}{c} \begin{array}{ccc} 7 & 9 & 8 \end{array} \\ \begin{bmatrix} 1 & 0 & -1 \\ & 1 & -1 \\ \text{(Symmetric)} & & 2 \end{bmatrix}^{(5)} \end{array}$$

Both element columns are zero.

The element coefficient matrices are assembled as usual, with the results shown in Figure 12.4. A bandwidth of 5 can be seen in the global matrix. Actually only nodes 2, 5, and 7 are unknown. Solving for these gives $p_2 = 15.855$, $p_5 = 14.331$, $p_7 = 9.2726$. The complete set of nodal pressures is shown in Figure 12.3(b), including the known pressures along the boundary.

$$
10^{-8}\begin{array}{c}
 \\ 1 \\ 2 \\ 3 \\ 4 \\ 5 \\ 6 \\ 7 \\ 8 \\ 9
\end{array}
\begin{array}{ccccccccc}
1 & 2 & 3 & 4 & 5 & 6 & 7 & 8 & 9 \\
19.20 & -4.80 & 0 & -4.80 & -9.60 & 0 & 0 & 0 & 0 \\
 & 23.4667 & -1.0667 & -9.60 & -5.8667 & -2.1333 & 0 & 0 & 0 \\
 & & 4.2667 & 0 & -2.1333 & -1.0667 & 0 & 0 & 0 \\
 & & & 29.8667 & -15.4667 & 0 & 0 & 0 & 0 \\
 & & & & 49.0667 & -2.1333 & -11.7333 & -2.1333 & 0 \\
 & & & & & 8.5333 & -2.1333 & -1.0667 & 0 \\
 & & & & & & 18.133 & -4.2667 & 0 \\
 & & & & & & & 10.6667 & -3.2 \\
\text{(Symmetric)} & & & & & & & & 3.2
\end{array}
\begin{Bmatrix} P_1 \\ P_2 \\ P_3 \\ P_4 \\ P_5 \\ P_6 \\ P_7 \\ P_8 \\ P_9 \end{Bmatrix}
=
\begin{Bmatrix} 0 \\ 0 \\ 0 \\ 0 \\ 0 \\ 0 \\ 0 \\ 0 \\ 0 \end{Bmatrix}
$$

Figure 12.4 Global equations (Example 12.1—Hydrostatic pad analysis with two element types)

The load carried by the hydrostatic pad is obtained by integrating the pressure over the total pad area. Consider each of the elements in order

$$\text{Load}^{(1)} = \iint_{A^{(1)}} p^{(1)}\,dA = \frac{A^{(1)}}{4}\,[p_1 + p_4 + p_5 + p_2] = 2{,}526 \text{ lbf}$$

$$\text{Load}^{(2)} = \frac{A^{(2)}}{4}\,[p_2 + p_5 + p_6 + p_3] = 108 \text{ lbf}$$

$$\text{Load}^{(3)} = \frac{A^{(3)}}{3}\,[p_4 + p_7 + p_5] = 1{,}046 \text{ lbf}$$

$$\text{Load}^{(4)} = \frac{A^{(4)}}{4}\,[p_5 + p_7 + p_8 + p_6] = 850 \text{ lbf}$$

$$\text{Load}^{(5)} = \frac{A^{(5)}}{3}\,[p_7 + p_8 + p_9] = 223 \text{ lbf}$$

Adding these and multiplying by eight (for eight identical analysis areas) gives 45,856 lbf. Further adding the load area at the center (20 psig times 576 sq in) yields a total load of 57,400 lbf for each hydrostatic pad.

Evaluating the rate of flow is easiest done by considering the flow through side 1–4 of element 1. The flow rate Q is given by

$$Q = \int_{L_{14}} [hu]_{\text{on side 14}}\,dx = -\int_{L_{14}} \left[\frac{h^3}{12\mu}\frac{\partial p}{\partial y}\right]_{\text{on side 14}} dx$$

The derivative is

$$\begin{aligned}
\frac{\partial p^{(1)}}{\partial y} &= \{D_y\}\,\{p\} \\
&= \frac{1}{LW}\left\{-\left(\frac{L}{2} - x\right) \quad -\left(\frac{L}{2} + x\right) \quad \left(\frac{L}{2} + x\right) \quad \left(\frac{L}{2} - x\right)\right\} \begin{Bmatrix} p_1 \\ p_4 \\ p_5 \\ p_2 \end{Bmatrix}
\end{aligned}$$

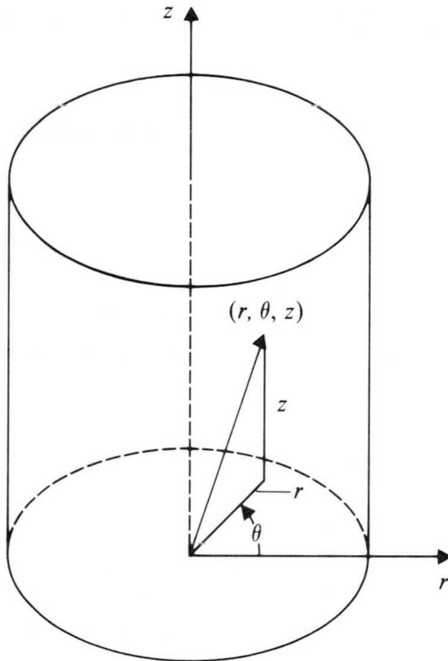

Figure 12.5 Geometry and coordinate system for axisymmetric problems

and integrating to obtain the flow rate yields

$$Q = -8.533\ [-p_1 - p_4 + p_5 + p_2]$$

Evaluating, $Q = 83.74$ in^3/sec. For all eight analysis areas, $Q = 670$ in^3/sec for the full pad. This figure indicates the size of the supply compressor.

12.2 Axisymmetric Field Problems

Problems discussed so far have primarily been two-dimensional in Cartesian (x,y) coordinates. Another important two-dimensional class of problems includes those that are symmetric about an axis. In cylindrical coordinates, the coordinates are (r, θ, z), but for axisymmetric problems none of the problem parameters are functions of the angular position θ. Figure 12.5 illustrates the geometry.

A. General Boundary Value Problem and Functional

The general boundary value problem is given in vector form by $\nabla \cdot (K \nabla T) + PT + Q = 0$ in V (convenient for non-Cartesian coordinates). The boundary conditions are specified $T = T_0$ on S_1 or derivative $(K \nabla T) \cdot n + \alpha T + \beta = 0$ on S_2, where n = outward normal to S_2. These specify the problem completely.

Expressing the general equation in cylindrical coordinates (r,z) gives

$$\frac{1}{r} \frac{\partial}{\partial r} \left(rK \frac{\partial T}{\partial r} \right) + \frac{\partial}{\partial z} \left(K \frac{\partial T}{\partial z} \right) + PT + Q = 0 \text{ in } A \tag{12.5}$$

where A is now the two-dimensional area. The specified boundary condition is

$$T = T_0 \text{ on } L_1 \tag{12.6}$$

where L_1 is part or all of the curve bounding A and the derivative boundary condition becomes

$$K \frac{\partial T}{\partial r} n_r + K \frac{\partial T}{\partial z} n_z + \alpha T + \beta = 0 \text{ on } L_2 \tag{12.7}$$

This is the axisymmetric field boundary value problem to be solved with finite elements.

Again using the expressions similar to those in Chapter 9, the functional is in vector form

$$I = \iiint_V \left\{ \frac{1}{2} K \nabla T \cdot \nabla T - PT^2 - QT \right\} dV + \iint_{S_2} \left(\frac{1}{2} \alpha T^2 + \beta T \right) dS$$

In this axisymmetric problem, the differential volume is $dV = 2\pi r\, dA$, while the differential surface area is $dS_2 = 2\pi r\, dL_2$. The functional is then

$$I = \iint_A \left\{ \frac{1}{2} K \left[\left(\frac{\partial T}{\partial r} \right)^2 + \left(\frac{\partial T}{\partial z} \right)^2 \right] - \frac{1}{2} PT^2 - QT \right\} 2\pi r\, dA + \int_{L_2} \left(\frac{1}{2} \alpha T^2 + \beta T \right) 2\pi r\, dL_2 \tag{12.8}$$

Here the constant term 2π can be taken outside the integral if desired.

A more general differential equation, used here, has the form

$$\frac{1}{r} \frac{\partial}{\partial r} \left(r K_{rr} \frac{\partial T}{\partial r} \right) + \frac{\partial}{\partial z} \left(K_{zz} \frac{\partial T}{\partial z} \right) + PT + Q = 0 \text{ in } A \tag{12.9}$$

with the boundary conditions

$$T = T_0 \text{ on } L_1 \tag{12.10}$$

or

$$K_{rr} \frac{\partial T}{\partial r} n_r + K_{\theta\theta} \frac{\partial T}{\partial z} n_z + \alpha T + \beta = 0 \text{ on } L_2 \tag{12.11}$$

The corresponding functional is

$$I = \iint_A \left\{ \frac{1}{2} K_{rr} \left(\frac{\partial T}{\partial r} \right)^2 + \frac{1}{2} K_{zz} \left(\frac{\partial T}{\partial z} \right)^2 - \frac{1}{2} PT^2 - QT \right\} 2\pi r\, dA + \int_{L_2} \left(\frac{1}{2} \alpha T^2 + \beta T \right) 2\pi r\, dL_2$$

In matrix form, it is

$$I = \iint_A (2\pi r)\left[\frac{1}{2}\left\{\frac{\partial T}{\partial r} \quad \frac{\partial T}{\partial z}\right\}[K]\begin{Bmatrix}\dfrac{\partial T}{\partial r}\\ \dfrac{\partial T}{\partial z}\end{Bmatrix} - \frac{1}{2}PT^2 - QT\right]dA + \int_{L_2}(2\pi r)\left(\frac{1}{2}\alpha T^2 + \beta T\right)dL_2 \tag{12.12}$$

where

$$[K] = \begin{bmatrix} K_{rr} & 0 \\ 0 & K_{\theta\theta}\end{bmatrix}$$

This is suitable for element matrix development.

B. Simplex Ring Elements

Simplex elements for this problem are axisymmetric rings whose properties are independent of the angle θ. In a way, this is similar to a two-dimensional element in Cartesian coordinates (x,y) that is infinitely long in the z direction. Figure 12.6(a) shows the cross section of the simplex ring element in the r,z plane. The element numbering must be counterclockwise for the following formulas to apply. A perspective of the ring element is shown in Figure 12.6(b).

Element interpolation functions are taken as linear, of the form

$$T^{(e)} = N_iT_i + N_jT_j + N_kT_k \tag{12.13}$$

where the pyramid functions are

$$N_i = \frac{1}{2A}(a_i + b_ir + c_iz), \quad \begin{cases} a_i = R_jZ_k - R_kZ_j \\ b_i = Z_j - Z_k \\ c_i = R_k - R_j \end{cases}$$

$$N_j = \frac{1}{2A}(a_j + b_jr + c_jz), \quad \begin{cases} a_j = R_kZ_i - R_iZ_k \\ b_j = Z_k - Z_i \\ c_j = R_i - R_k \end{cases}$$

$$N_k = \frac{1}{2A}(a_k + b_kr + c_kz), \quad \begin{cases} a_k = R_iZ_j - R_jZ_i \\ b_k = Z_i - Z_j \\ c_k = R_j - R_i \end{cases}$$

and $2A = b_ic_j - b_jc_i$. These expressions are exactly the same as those for Cartesian coordinates with r replacing x and z replacing y.

The element interpolation function T is

$$T^{(e)} = \lfloor N \rfloor \{T\} \tag{12.14}$$

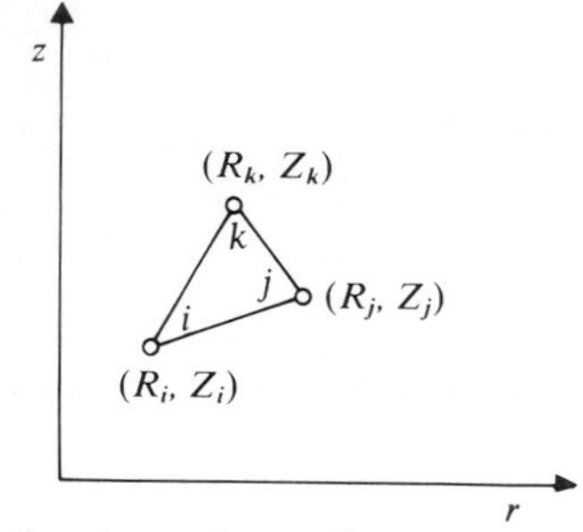

(a) Simplex element in r, z plane

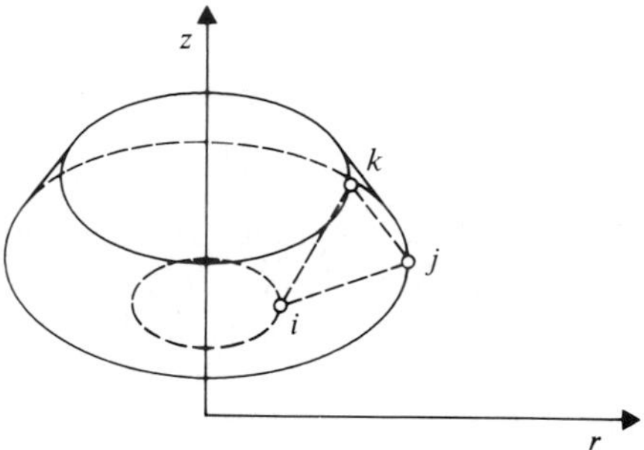

(b) Perspective view of simplex ring element

Figure 12.6 Geometry of axisymmetric simplex ring element

in terms of the shape functions $\lfloor N \rfloor = \lfloor N_i, N_j, N_k \rfloor$ and nodal values

$$\{T\} = \begin{Bmatrix} T_i \\ T_j \\ T_k \end{Bmatrix}$$

The element derivatives are

$$\begin{Bmatrix} \dfrac{\partial T}{\partial r} \\ \dfrac{\partial T}{\partial z} \end{Bmatrix} = [D] \{T\} \qquad \textbf{(12.15)}$$

where the element derivative matrix is

$$\underset{(2,3)}{[D]} = \frac{1}{2A} \begin{bmatrix} b_i & b_j & b_k \\ c_i & c_j & c_k \end{bmatrix}$$

This has the same form as in Cartesian coordinates (3.25).

For the simplex ring element, the functional for each element is

$$I^{(e)} = \iint_{A^{(e)}} (2\pi r) \left[\frac{1}{2} \{T\}^T [D]^T [K][D] \{T\} - \frac{1}{2} P \{T\}^T \{N\}^T \{N\} \{T\} - Q\{N\}\{T\} \right] dA + \int_{L_2^{(e)}} (2\pi r) \left[\frac{1}{2} \alpha \{T\}^T \{N\}^T \{N\} \{T\} + \beta \{N\} \{T\} \right] dL_2$$

The element minimization equations are

$$\left\{ \begin{array}{c} \dfrac{\partial I}{\partial T_i} \\ \dfrac{\partial I}{\partial T_j} \\ \dfrac{\partial I}{\partial T_k} \end{array} \right\}^{(e)} = [B]^{(e)} \{T\}^{(e)} - \{C\}^{(e)} \tag{12.16}$$

where the element matrix is

$$[B]^{(e)} = \iint_{A^{(e)}} (2\pi r) \left[[D]^T [K][D] - P \{N\}^T \{N\} \right] dA + \int_{L^{(e)}} (2\pi r) \, \alpha \, \{N\}^T \{N\} \, dL_2 \tag{12.17}$$

and the element column is

$$\{C\}^{(e)} = \iint_{A^{(e)}} (2\pi r) \, Q \, \{N\}^T \, dA + \int_{L_2^{(e)}} (2\pi r) \beta \, \{N\}^T \, dL_2 \tag{12.18}$$

These relations can now be evaluated for a simplex ring element.

C. Radial Approximation

It is easily seen that the integrals have the radial term $2\pi r$ in the integrand. This term produces the primary difference between plane two-dimensional problems and axisymmetric problems. Two approaches are commonly taken to approximate this radial term:

1. Centroidal radial approximation

$$r = \bar{r} = \frac{1}{3} (R_i + R_j + R_k) \tag{12.19}$$

2. Linear radial approximation

$$r = \{N\} \{R\} = \{N_i, N_j, N_k\} \left\{ \begin{array}{c} R_i \\ R_j \\ R_k \end{array} \right\} \tag{12.20}$$

Figure 12.7 shows the centroidal approximation. Note that these two approximations apply to the integral over the element area (first integral) and not to the integral over the element side (second integral).

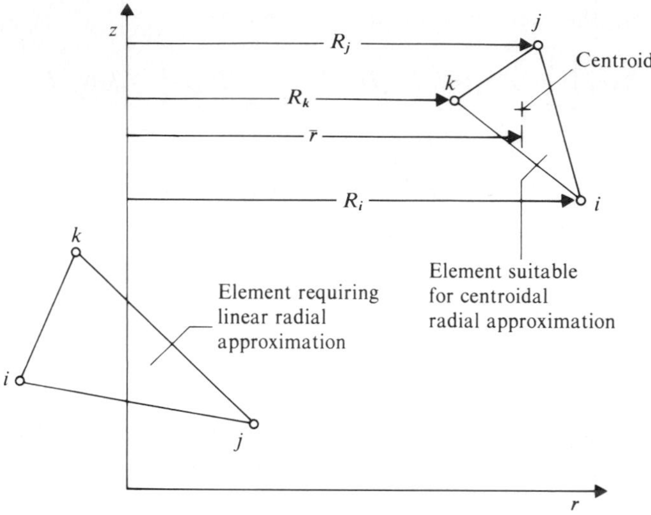

Figure 12.7 Axisymmetric simplex elements for centroidal and linear radial approximations

Either case gives an approximation to the actual integral in the element matrices. However, all finite elements are an approximation. This is just one more that is fairly consistent with the level of approximation used for the interpolation function.

The centroidal approximation to the radial term is just the average of the three radius values. It is less accurate than the linear approximation but easier to evaluate. In the case of axisymmetric problems where all elements are relatively far from the z axis, the centroidal approximation is a good one. This may be expressed as the ratio of the centroid $\bar{r}$ to the maximum radial difference for the element.

$$\left| \frac{\bar{r}}{\text{max diff } (R_i, R_j, R_k)} \right|^{(e)} >> 1 \qquad \textbf{(12.21)}$$

If an element is near the z axis, the linear or some other (better) approximation must be used. Only the centroidal approximation is discussed in this section, for simplicity. As such an element gets smaller and smaller, the maximum radial difference decreases and the approximation gets better.

Example 12.2 *Centroidal Radial Approximation for Axisymmetric Elements*

Given Consider the two elements illustrated in Figure 12.7. The first element has nodal coordinates $(R_i, Z_i) = (4.1, 2.7)$, $(R_j, Z_j) = (4.0, 4.0)$, $(R_k, Z_k) = (3.6, 3.5)$. The second element has nodal coordinates $(R_i, Z_i) = (-1.0, 1.0)$, $(R_j, Z_j) = (1.0, 0.5)$, $(R_k, Z_k) = (0.3, 4.0)$.

Objective Determine whether the centroidal radial approximation is appropriate for either of these two elements.

Solution The two element centroids are

$$\bar{r}^{(1)} = \frac{1}{3}(4.1 + 4.0 + 3.6) = 3.9$$

$$\bar{r}^{(2)} = \frac{1}{3}(-1.0 + 1.0 - 0.6) = -0.2$$

while the maximum differences are max diff$^{(1)}$ = 0.5, max diff$^{(2)}$ = 2.0. The ratios are

$$\left|\frac{\bar{r}}{\text{max diff}}\right|^{(1)} = \frac{3.9}{0.5} = 7.8$$

$$\left|\frac{\bar{r}}{\text{max diff}}\right|^{(2)} = \frac{0.2}{2.0} = 0.1$$

The first element is suitable for a centroidal radial approximation, while the second is not. As noted earlier, the centroidal approximation is improved as the element size becomes smaller in the radial direction.

D. Element Matrices

In the case of the simplex element with a centroidal radial approximation, the radial term $2\pi\bar{r}$ simply comes outside of the element integrals. The result is

$$[B]^{(e)} = 2\pi\bar{r}\iint_{A^{(e)}}\left[[D]^T[K][D] - P\,\lfloor N\rfloor^T\lfloor N\rfloor\right]dA + 2\pi\bar{r}_s\int_{L^{(e)}}\alpha\lfloor N\rfloor^T\lfloor N\rfloor\,dL_2$$

and

$$\{C\}^{(e)} = 2\pi\bar{r}\iint_{A^{(e)}} Q\,\lfloor N\rfloor^T dA + 2\pi\bar{r}_s\int_{L_2^{(e)}}\beta\,\lfloor N\rfloor^T dL_2$$

Here r_s denotes the centroid of the side. The integrals that remain are the same as those in Cartesian coordinates.

For constant property elements, the element matrix becomes

$$[B]^{(e)} = \frac{2\pi\bar{r}K_{rr}}{4A}\begin{bmatrix} b_ib_i & b_ib_j & b_ib_k \\ b_ib_j & b_jb_j & b_jb_k \\ b_ib_k & b_jb_k & b_kb_k \end{bmatrix}^{(e)} \qquad (K_{rr}\text{ matrix})$$

$$+ \frac{2\pi\bar{r}\,K_{zz}}{4A}\begin{bmatrix} c_ic_i & c_ic_j & c_ic_k \\ c_ic_j & c_jc_j & c_jc_k \\ c_ic_k & c_jc_k & c_kc_k \end{bmatrix}^{(e)} \qquad (K_{zz}\text{ matrix}) \qquad \textbf{(12.22)}$$

$$- \frac{2\pi\bar{r}\,PA}{12}\begin{bmatrix} 2 & 1 & 1 \\ 1 & 2 & 1 \\ 1 & 1 & 2 \end{bmatrix}^{(e)} \qquad (P\text{ matrix})$$

$$+ \frac{2\pi}{6}(\alpha\bar{r}L)_{ij}\begin{bmatrix} 2 & 1 & 0 \\ 1 & 2 & 0 \\ 0 & 0 & 0 \end{bmatrix}^{(e)}_{\substack{\text{on side } ij \\ \text{(side 1)}}} \qquad (\alpha \text{ matrix } ij)$$

$$+ \frac{2\pi}{6}(\alpha\bar{r}L)_{jk}\begin{bmatrix} 0 & 0 & 0 \\ 0 & 2 & 1 \\ 0 & 1 & 2 \end{bmatrix}^{(e)}_{\substack{\text{on side } jk \\ \text{(side 2)}}} \qquad (\alpha \text{ matrix } jk) \qquad \textbf{(12.22 cont.)}$$

$$+ \frac{2\pi}{6}(\alpha\bar{r}L)_{ki}\begin{bmatrix} 2 & 0 & 1 \\ 0 & 0 & 0 \\ 1 & 0 & 2 \end{bmatrix}^{(e)}_{\substack{\text{on side } ki \\ \text{(side 3)}}} \qquad (\alpha \text{ matrix } ki)$$

and the element column is

$$\{C\}^{(e)} = \frac{2\pi\bar{r}QA}{3}\begin{Bmatrix} 1 \\ 1 \\ 1 \end{Bmatrix}^{(e)} \qquad (Q \text{ column})$$

$$- \frac{2\pi(\beta\bar{r}L)_{ij}}{2}\begin{Bmatrix} 1 \\ 1 \\ 0 \end{Bmatrix}^{(e)}_{\substack{\text{on side } ij \\ \text{(side 1)}}} \qquad (\beta \text{ column } ij) \qquad \textbf{(12.23)}$$

$$- \frac{2\pi(\beta\bar{r}L)_{jk}}{2}\begin{Bmatrix} 0 \\ 1 \\ 1 \end{Bmatrix}^{(e)}_{\substack{\text{on side } jk \\ \text{(side 2)}}} \qquad (\beta \text{ column } jk)$$

$$- \frac{2\pi(\beta\bar{r}L)_{ki}}{2}\begin{Bmatrix} 1 \\ 0 \\ 1 \end{Bmatrix}^{(e)}_{\substack{\text{on side } ki \\ \text{(side 3)}}} \qquad (\beta \text{ column } ki)$$

On each side, the term $\bar{r}$ denotes the centroid of that side. As in the normal two-dimensional problem, the element matrix $[B]$ is a three by three matrix, and the element column $\{C\}$ is a three component column. Again note that the element numbering (as given in the element/nodal connectivity data) must be counterclockwise.

For the terms evaluated along the sides of elements, β is taken to be constant within the element. The other quantities that must be found are the side lengths. They are given by

$$\begin{aligned} L_{ij} &= [(r_i - r_j)^2 + (z_i - z_j)^2]^{1/2} \\ L_{jk} &= [(r_j - r_k)^2 + (z_j - z_k)^2]^{1/2} \\ L_{ki} &= [(r_k - r_i)^2 + (z_k - z_i)^2]^{1/2} \end{aligned} \qquad \textbf{(12.24)}$$

Only if derivative boundary conditions are to be imposed on a certain side are the derivative boundary matrix and column included in the appropriate element matrix and column. In computer programs, the sides are normally labeled 1, 2, 3 in the following manner: *ij* is 1, *jk* is 2, and *ki* is 3, again proceeding in counterclockwise fashion from the first node, chosen as node *i*. Figure 12.8 illustrates the labeling.

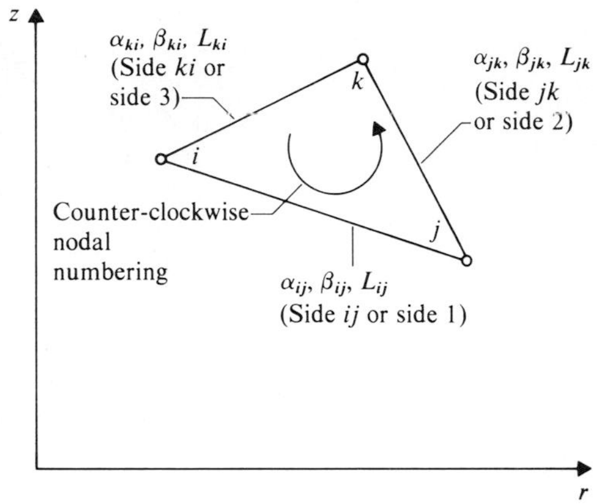

Figure 12.8 Nodal labeling, side labeling, and derivative boundary condition constants

12.3 Point Sources and Sinks

Sources or sinks concentrated at a point in space are encountered in many physical problems. Heat can be generated by a pipe (running in a direction perpendicular to the plane of a two-dimensional problem), sources or sinks in a fluid mechanics problem such as a well or pump, and forces in a solid mechanics problem. Actually these quantities are not localized at a point in space, but the radius of the pipe or area of application of a force is small compared with the overall problem dimensions. Thus they may be considered mathematically as point sources and sinks. Section 7.3 introduced the basic concepts for one-dimensional problems, and Section 8.4 extended the results to one-dimensional beam problems.

A. Source in an Element

A source of strength q^* is located at a point (in Cartesian coordinates) (x_0, y_0), shown in Figure 12.9(a). The unit impulse functions or dirac delta functions gives the source function $q = q^* \delta (x - X_0, y - Y_0)$. As in Section 7.3, the dirac delta function has the value zero except at $x = X_0$ and $y = Y_0$, where it has value infinity but in such a way that the area under the δ curve has value unity. Over an element of area A

$$\int_{A^{(e)}} \delta(x - X_0, y - Y_0)\, dA = \begin{cases} 1, & (X_0, Y_0) \text{ in } e \\ 0, & (X_0, Y_0) \text{ not in } e \end{cases}$$

The Q term in the element variational principle (9.23) is

$$\int_{A^{(e)}} [-QT]^{(e)}\, dA = \int_{A^{(e)}} [-q^*\delta(x - X_0, y - Y_0)T]^{(e)}\, dA$$

where X_0, Y_0 is inside the element (if not, the above result is zero).

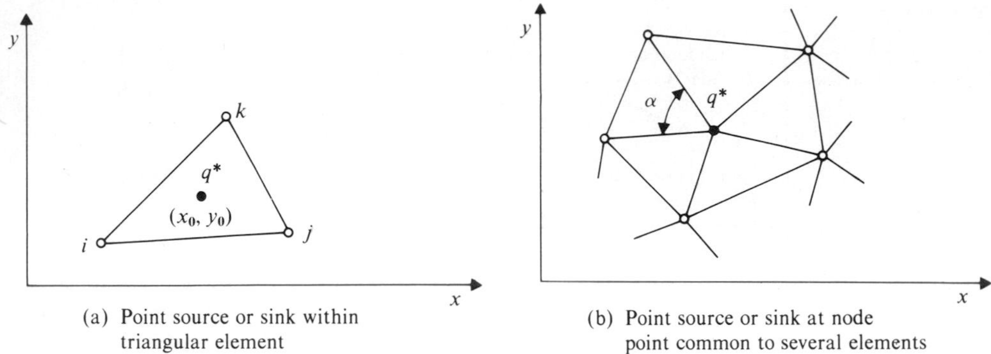

(a) Point source or sink within triangular element

(b) Point source or sink at node point common to several elements

Figure 12.9 Point sources and sinks

B. Simplex Element

Consider a triangular simplex element and differentiate with respect to T_i:

$$\frac{\partial}{\partial T_i}\int_{A^{(e)}} [-QT]^{(e)}\, dA$$

$$= \frac{\partial}{\partial T_i}\int_{A^{(e)}} [-q^*\delta(x - X_0, y - Y_0)(N_iT_i + N_jT_j + N_kT_k)]^{(e)}\, dA$$

$$= -q^* \int_{A^{(e)}} [\delta(x - X_0, y - Y_0)N_i]^{(e)}\, dA$$

The value of N_i at (X_0, Y_0) is picked out by the above integral

$$\int_{A^{(e)}} [\delta(x - X_0, y - Y_0)N_i]^{(e)}\, dA = N_i\Big|_{\substack{x = X_0 \\ y = Y_0}}$$

Thus the contribution of the source to the equation for T_i is

$$-q^*N_i\Big|_{\substack{x = X_0 \\ y = Y_0}}$$

Evaluating the terms for T_j and T_k yields the element column

$$\{C\}^{(e)} = q^* \begin{Bmatrix} N_i \\ N_j \\ N_k \end{Bmatrix}_{\substack{x = X_0 \\ y = Y_0}} \qquad \textbf{(12.25)}$$

If other Q terms arise from the differential equation, the above values are simply added. The sum of the pyramid functions is unity, $N_i + N_j + N_k = 1$, so the total value q^* is distributed among the three node points.

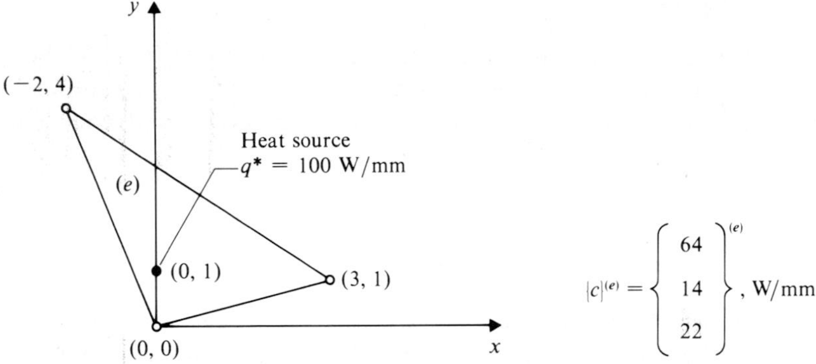

Figure 12.10 Geometry and result (Example 12.3—Heat source in triangular element)

C. Rectangular Element

For a rectangular element, the element column becomes

$$\{C\}^{(e)} = q^* \begin{Bmatrix} G_i \\ G_j \\ G_k \\ G_\ell \end{Bmatrix}_{\substack{x = X_0 \\ y = Y_0}} \tag{12.26}$$

These shape functions are the rectangular element functions (12.2). Again the sum of the shape functions equals unity at any point x,y:

$$G_i + G_j + G_k + G_\ell = 1$$

so exactly the value of q^* is added to the four element nodes.

Example 12.3 *Heat Source in Triangular Element*

Given A triangular simplex element has nodal coordinates (0,0), (3,1), and (−2,4). A heat source of strength $q^* = 100$ W/mm is located at the point (0,1). All length units are in mm. Figure 12.10 shows the geometry.

Objective Find the element column for this element.

Solution From Example 3.3, the pyramid functions are

$$\begin{aligned} N_i &= (14 - 3x - 5y)/14 \\ N_j &= (4x + 2y)/14 \\ N_k &= (-x + 3y)/14 \end{aligned}$$

At the source, $X_0 = 0$ mm, $Y_0 = 1$ mm, and evaluating at that point, $N_i = 0.64$, $N_j = 0.14$, and $N_k = 0.22$. Then the element column is (12.25)

$$\{C\}^{(e)} = \begin{Bmatrix} 64 \\ 14 \\ 22 \end{Bmatrix}^{(e)} \quad \text{W/mm}$$

D. Source at Node Point

A special case occurs when the source is located at a node point, shown in Figure 12.9(b). Then the source strength must be divided among the elements in the ratio that the angle α within the element (at the node) bears to 360°. Thus, if the source is at node k in element e, the element column becomes

$$\{C\}^{(e)} = \frac{\alpha}{360} q^* \begin{Bmatrix} 0 \\ 0 \\ 1 \end{Bmatrix}$$

It is easily seen that the sum of all of the elements containing the source add up to the total value of q^* in the global equations after assembly. Adding the value of q^* directly to the global column (but ignoring the source up to that point) avoids calculation of the angles α for each element. A similar technique works for rectangular elements.

12.4 Linear Property Triangular Elements

The previous element matrices have assumed constant values for the coefficient functions K_x, K_y, P, Q, where

$$\left.\begin{aligned} K_x^{(e)} &= \frac{1}{3}(K_{xi} + K_{xj} + K_{xk}) \\ K_y^{(e)} &= \frac{1}{3}(K_{yi} + K_{yj} + K_{yk}) \\ P^{(e)} &= \frac{1}{3}(P_i + P_j + P_k) \\ Q^{(e)} &= \frac{1}{3}(Q_i + Q_j + Q_k) \end{aligned}\right\} \text{constant property element} \qquad \mathbf{(12.27)}$$

Here the terms on the right are the values of K_x, K_y, P, Q evaluated at the element node points.

A. Linear Coefficient Functions

Linear interpolation approximations for the coefficient functions, shown in Figure 12.11, are

$$\left.\begin{aligned} K_x^{(e)} &= N_iK_{xi} + N_jK_{xj} + N_kK_{xk} \\ K_y^{(e)} &= N_iK_{yi} + N_jK_{yj} + N_kK_{yk} \\ P^{(e)} &= N_iP_i + N_jP_j + N_kP_k \\ Q^{(e)} &= N_iQ_i + N_jQ_j + N_kQ_k \end{aligned}\right\} \text{linear property elements} \qquad \mathbf{(12.28)}$$

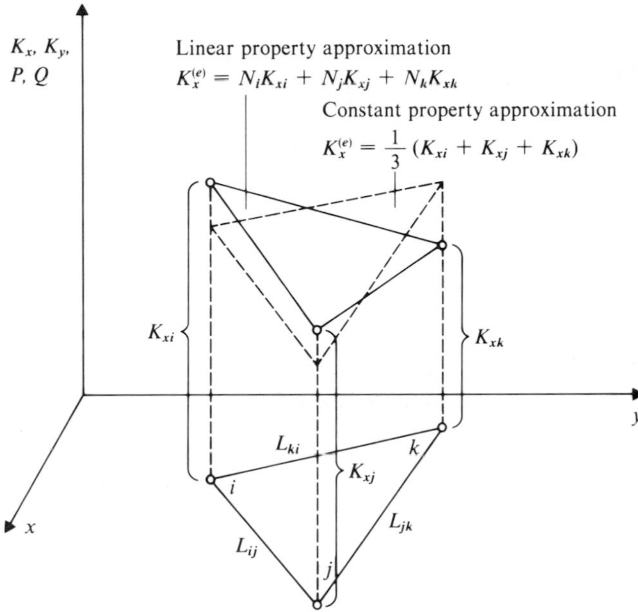

Figure 12.11 Constant and linear property approximations within a two-dimensional triangular element

These equations are a simple extension of the one-dimensional case considered in Section 7.4. Also along the derivative boundary, if one is present,

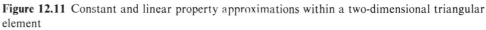

$$\begin{aligned}
[\alpha]^{(e)}_{\text{on side } ij} &= [N_i\alpha_i + N_j\alpha_j]^{(e)}_{\text{on side } ij} \\
[\beta]^{(e)}_{\text{on side } ij} &= [N_i\beta_i + N_j\beta_j]^{(e)}_{\text{on side } ij} \\
[\alpha]^{(e)}_{\text{on side } ik} &= [N_i\alpha_i + N_k\alpha_k]^{(e)}_{\text{on side } ik} \\
[\beta]^{(e)}_{\text{on side } ik} &= [N_i\beta_i + N_k\beta_k]^{(e)}_{\text{on side } ik} \\
[\alpha]^{(e)}_{\text{on side } jk} &= [N_j\alpha_j + N_k\alpha_k]^{(e)}_{\text{on side } jk} \\
[\beta]^{(e)}_{\text{on side } jk} &= [N_j\beta_j + N_k\beta_k]^{(e)}_{\text{on side } jk}
\end{aligned} \tag{12.29}$$

The individual values of α and β must be found for each of the three sides. Thus α_i for side ij is not necessarily the same value as α_i for side ik.

The element integral is

$$I^{(e)} = \iint_{A^{(e)}} \left\{ \frac{1}{8A^2} \left[(N_iK_{xi} + N_jK_{xj} + N_kK_{xk})\,(b_iT_i + b_jT_j + b_kT_k)^2 \right.\right.$$
$$\left. + (N_iK_{yi} + N_jK_{yj} + N_kK_{yk})\,(c_iT_i + c_jT_j + c_kT_k)^2 \right]$$
$$- \frac{1}{2}(N_iP_i + N_jP_j + N_kP_k)\,(N_iT_i + N_jT_j + N_kT_k)^2$$
$$\left. -(N_iQ_i + N_jQ_j + N_kQ_k)\,(N_iT_i + N_jT_j + N_kT_k) \right\}^{(e)} dA^{(e)}$$
$$+ \left\{ \int_{L_2^{(e)}} \left[\frac{1}{2}(N_i\alpha_i + N_j\alpha_j)\,(N_iT_i + N_jT_j)^2 \right.\right.$$
$$\left.\left. + (N_i\beta_i + N_j\beta_j)\,(N_iT_i + N_jT_j) \right] dL_2^{(e)} \right\} \text{ on side } ij \qquad \textbf{(12.30)}$$
$$+ \left\{ \int_{L^{(e)}} \left[\frac{1}{2}(N_i\alpha_i + N_k\alpha_k)\,(N_iT_i + N_kT_k)^2 \right.\right.$$
$$\left.\left. + (N_i\beta_i + N_k\beta_k)\,(N_iT_i + N_kT_k) \right] dL_2^{(e)} \right\} \text{ on side } ik$$
$$+ \left\{ \int_{L_2^{(e)}} \left[\frac{1}{2}(N_j\alpha_j + N_k\alpha_k)\,(N_jT_j + N_kT_k)^2 \right.\right.$$
$$\left.\left. + (N_j\beta_j + N_k\beta_k)\,(N_jT_j + N_kT_k) \right] dL_2^{(e)} \right\} \text{ on side } jk$$

This can be differentiated with respect to T_i, T_j, T_k and the integrals evaluated as in Section 9.4.

B. Element Matrices

Organizing the results into an element matrix yields

$$[B]^{(e)} = \frac{(K_{xi} + K_{xj} + K_{xk})}{12A} \begin{bmatrix} b_ib_i & b_ib_j & b_ib_k \\ b_ib_j & b_jb_j & b_jb_k \\ b_ib_k & b_jb_k & b_kb_k \end{bmatrix}^{(e)}$$
$$+ \frac{(K_{yi} + K_{yj} + K_{yk})}{12A} \begin{bmatrix} c_ic_i & c_ic_j & c_ic_k \\ c_ic_j & c_jc_j & c_jc_k \\ c_ic_k & c_jc_k & c_kc_k \end{bmatrix}^{(e)} \qquad \textbf{(12.31)}$$
$$- \frac{A}{60} \begin{bmatrix} (6P_i + 2P_j + 2P_k) & (2P_i + 2P_j + P_k) & (2P_i + P_j + 2P_k) \\ (2P_i + 2P_j + P_k) & (2P_i + 6P_j + 2P_k) & (P_i + 2P_j + 2P_k) \\ (2P_i + P_j + 2P_k) & (P_i + 2P_j + 2P_k) & (2P_i + 2P_j + 6P_k) \end{bmatrix}^{(e)}$$
$$+ \frac{L_{ij}}{12} \begin{bmatrix} (3\alpha_i + \alpha_j) & (\alpha_i + \alpha_j) & 0 \\ (\alpha_i + \alpha_j) & (\alpha_i + 3\alpha_j) & 0 \\ 0 & 0 & 0 \end{bmatrix}^{(e)}_{\text{on side } ij}$$
$$+ \frac{L_{ik}}{12} \begin{bmatrix} (3\alpha_i + \alpha_k) & 0 & (\alpha_i + \alpha_k) \\ 0 & 0 & 0 \\ (\alpha_i + \alpha_k) & 0 & (\alpha_i + 3\alpha_k) \end{bmatrix}^{(e)}_{\text{on side } ik}$$

Continued on next page

$$+\frac{L_{jk}}{12}\begin{bmatrix} 0 & 0 & 0 \\ 0 & (3\alpha_j + \alpha_k) & (\alpha_j + \alpha_k) \\ 0 & (\alpha_j + \alpha_k) & (\alpha_j + 3\alpha_k) \end{bmatrix}^{(e)}_{\text{on side } jk} \qquad \textbf{(12.31 cont.)}$$

Here the individual matrices are written as in Equation (9.29) with meanings discussed in some detail there. The element column is similarly obtained as

$$\{C\}^{(e)} = \frac{A}{12}\begin{Bmatrix} 2Q_i + Q_j + Q_k \\ Q_i + 2Q_j + Q_k \\ Q_i + Q_j + 2Q_k \end{Bmatrix}^{(e)} - \frac{L_{ij}}{6}\begin{Bmatrix} 2\beta_i + \beta_j \\ \beta_i + 2\beta_j \\ 0 \end{Bmatrix}^{(e)}_{\text{on side } ij}$$
$$- \frac{L_{ik}}{6}\begin{Bmatrix} 2\beta_i + \beta_k \\ 0 \\ \beta_i + 2\beta_k \end{Bmatrix}^{(e)}_{\text{on side } ik} - \frac{L_{jk}}{6}\begin{Bmatrix} 0 \\ 2\beta_j + \beta_k \\ \beta_j + 2\beta_k \end{Bmatrix}^{(e)}_{\text{on side } jk} \qquad \textbf{(12.32)}$$

This is similar to Equation (9.30).

It is easily seen that the x and y gradient terms merely average the values K_{xi}, K_{xj}, K_{xk} and K_{yi}, K_{yj}, K_{yk} over the element. Thus, the linear property element and constant property elements would give the same result for those terms. The other P, Q, α, β terms would yield different values.

12.5 Problems

12.1 A rectangular element has nodes (0,0), (4,0), (4,1), (0,1). Heat conduction occurs through the element and heat convection occurs through the sides. The governing differential equation is

$$\frac{\partial}{\partial x}\left(kt\frac{\partial T}{\partial x}\right) + \frac{\partial}{\partial y}\left(kt\frac{\partial T}{\partial y}\right) - 2h(T - T_\infty) = 0$$

where $k = 0.50$ W/mm-°C, $h = 1.5 \times 10^{-5}$ W/mm-°C, $t = 3.0$ mm, and $T_\infty = 20$°C. Find the element conduction matrix. It is an interior element, so no boundary conditions need be considered.

12.2 Modify computer program FINTWO to solve the differential equation

$$\frac{\partial}{\partial x}\left(K_x\frac{\partial T}{\partial x}\right) + \frac{\partial}{\partial y}\left(K_y\frac{\partial T}{\partial y}\right) + PT + Q = 0$$
$$T = T_0 \text{ on } L_1$$
$$\frac{\partial T}{\partial n} = 0 \text{ on } L_2$$

using rectangular elements.

12.3 (a) A thin fin has the dimensions $L = 120$ mm, $W = 160$ mm, and $t = 1.25$ mm, and other parameters $k = 0.20$ W/mm-°C, $h = 2 \times 10^{-4}$ W/mm²-°C, $T_{\text{wall}} = 300$°C, and $T_\infty = 50$°C. Use at least 16 rectangular elements to solve for the temperature distribution in the fin (use the program developed in Problem 12.2).

(b) Solve the same problem using simplex elements and the normal FINTWO with node points at exactly the same points as with the rectangular elements. Compare the results.

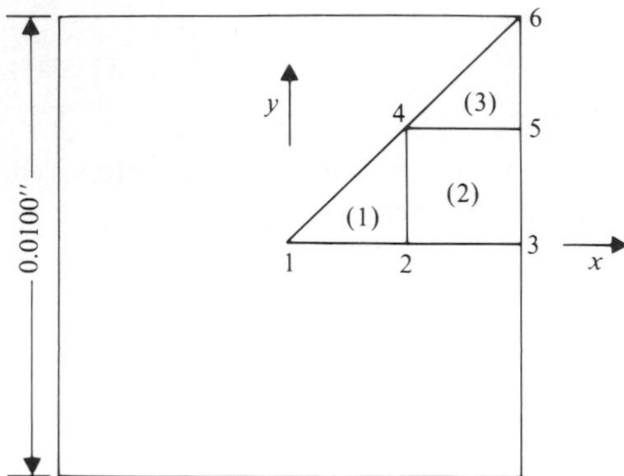

Figure 12.12 Geometry for Problem 12.4

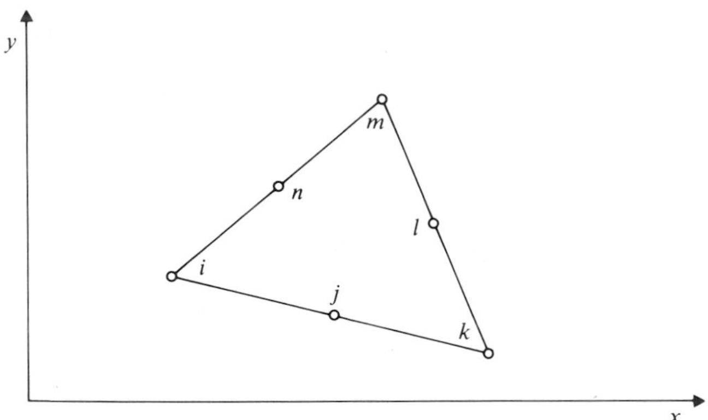

Figure 12.13 Six-node triangular element for Problem 12.6

12.4 Axial viscous flow occurs in a square capillary pipe of side length 0.0100 inch. The fluid density and viscosity are $\rho = 1.0 \times 10^{-3}$ slug/in^3, $\mu = 1.0 \times 10^{-7}$ lbf-sec/in^2, while the axial pressure gradient is $\partial P/\partial z = -2.0$ lbf/in^3. Use the three-element model shown in Figure 12.12 (two triangular and one rectangular) of one-eighth of the pipe to solve for the axial velocity at the unknown node points. (See Problems 10.15 and 10.16 for the appropriate differential equation and boundary conditions.) Solve by hand.

12.5 Modify computer program FINTWO to use rectangular elements rather than triangular elements. Solve a two-dimensional boundary value problem with the resulting program, using only rectangular elements.

12.6 A six-node triangular element, shown in Figure 12.13, has the corner coordinates (X_i, Y_i), (X_k, Y_k), (X_m, Y_m). The shape functions for this element are (see Section 3.7)

$$G_i = N_i(2N_i - 1) \qquad G_\ell = 4N_kN_m$$
$$G_j = 4N_iN_k \qquad G_m = N_m(2N_m - 1)$$
$$G_k = N_k(2N_k - 1) \qquad G_n = 4N_mN_i$$

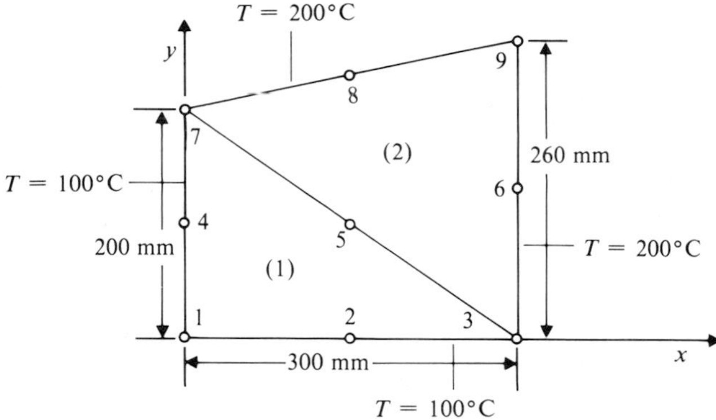

Figure 12.14 Heat conduction in trapezoidal long body for Problem 12.7

where N_i, N_k, N_m are the normal linear pyramid functions in terms of the corner coordinates.

Determine the element matrix for this element, corresponding to the general differential equation in two dimensions in area A

$$\frac{\partial}{\partial x}\left(K_x \frac{\partial T}{\partial x}\right) + \frac{\partial}{\partial y}\left(K_y \frac{\partial T}{\partial y}\right) = 0 \text{ in } A$$

and boundary condition $T = T_0$ on L where L is the boundary of A. Both K_x and K_y are general functions of x,y to be taken as constant in the element. (Hint: Evaluate the derivatives $\partial T/\partial x$, $\partial T/\partial y$ in terms of N_i, N_k, N_m and use the integral relation (3.27) to obtain the terms in the element matrix.)

12.7 This problem uses the element matrices developed in Problem 12.6 for a six-node triangular element. Heat conduction is governed by the differential equation

$$\frac{\partial^2 T}{\partial x^2} + \frac{\partial^2 T}{\partial y^2} = 0$$

and boundary conditions are $T = 100°\text{C}$ along bottom and left sides, $T = 200°\text{C}$ along top and right sides. Use the two-element model shown in Figure 12.14 to obtain the temperature at the node in the center.

12.8 An axisymmetric ring is shown in Figure 12.15(a). It has a trapezoidal cross section in the r,z plane as indicated in Figure 12.15(b). The ring is made of aluminum with thermal conductivity $k = 9.8$ BTU/hr-in-°F. Along the top of the ring, the temperature is 150°F. The inner wall is insulated, while the outer wall and bottom have convection heat flow with $T_\infty = 60°\text{F}$, $h = 0.020$ BTU/hr-in^2-°F. Use the two-element model with the centroidal approximation shown in Figure 12.15 to obtain the nodal temperatures at nodes 1 and 2. Solve by hand to become familiar with the technique.

12.9 Modify computer program FINONE to solve axisymmetric problems using simplex elements with the centroidal approximation. Employ the resulting program to solve for the temperature distribution in the hollow axisymmetric body shown in Figure 12.16. Use at least 20 elements.

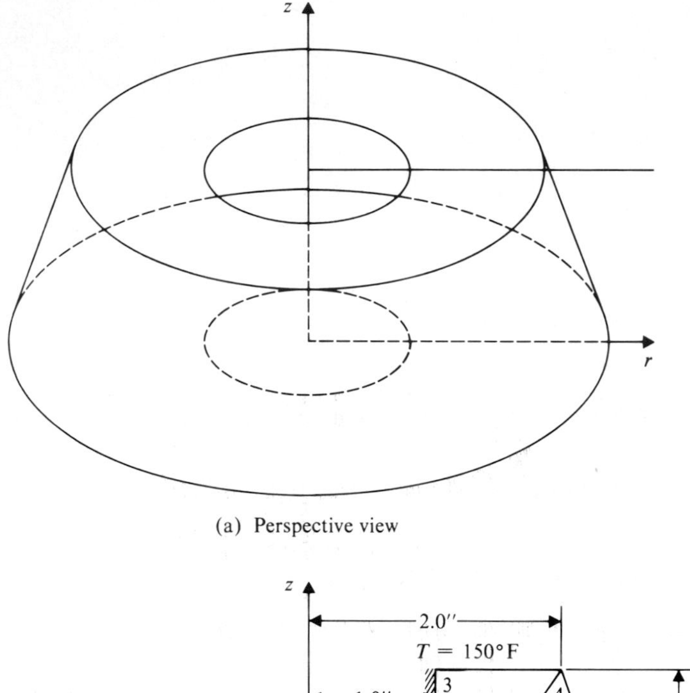

(a) Perspective view

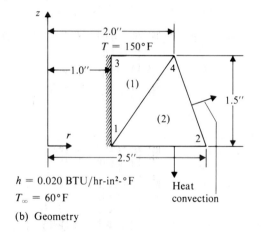

(b) Geometry

Figure 12.15 Axisymmetric ring for Problem 12.8

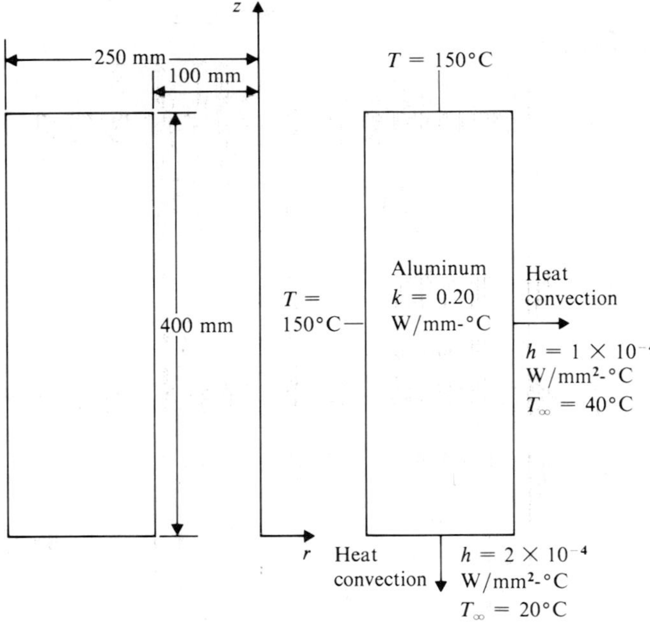

Figure 12.16 Axisymmetric solid for Problem 12.9

12.10 Obtain the element matrix for an axisymmetric field problem with the governing differential equation

$$\frac{1}{r}\frac{\partial}{\partial r}\left(r\,K_r\frac{\partial T}{\partial r}\right) + \frac{\partial}{\partial z}\left(K_z\frac{\partial T}{\partial z}\right) = 0$$

with boundary condition $T = T_0$ on L using the linear approximation $r = N_iR_i + N_jR_j + N_kR_k$ as discussed in Section 12.2. (Hint: Employ the integral relation 3.27.) Assume K_r, K_z are constant in the element.

12.11 A rectangular element with node points (1,0), (3,0), (3,1), (1,1) has a source of strength 100 W/mm located at the point (2.5, 0.5). Find the element column.

12.12 A concrete wall (similar to that in Problem 10.8) has the dimensions shown in Figure 12.17. The thermal conductivity of the concrete is $k = 1.0$ W/m-°C. It is proposed to heat the wall by electrical wires running in the z direction through the wall. There are five wires spaced 1/3 m apart. These provide a heat source of strength (per wire) $q^* = 5$ W/m where the heat is in watts per meter of length in the z direction.

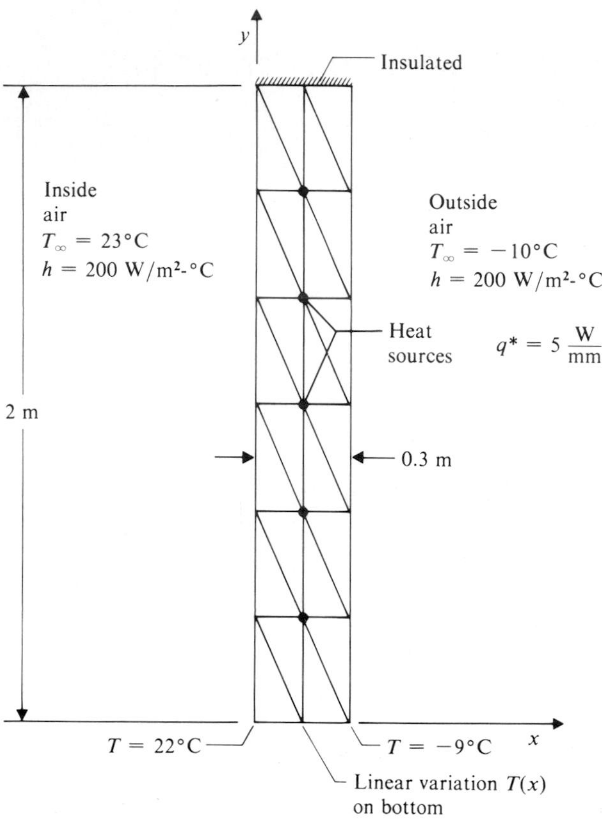

Figure 12.17 Heat sources in concrete wall for Problem 12.12

The division into triangular simplex elements is shown in Figure 12.17. The sources are placed at node points. Use computer program FINTWO to solve the problem. The program itself can easily be modified to include the source terms in the global column $\{F\} = q^*$. These terms are added just after prescribed nodal values are taken into account, but before the Gauss elimination solution of the algebraic equations. Sketch the wall, with node points, and indicate the calculated temperatures at the node points. Calculate the total heat flow from the inside air to the wall.

Chapter 13

Isoparametric and Other Elements

13.1 Introduction

Up to this point, only simplex and rectangular elements have been considered. This chapter considers useful higher order elements of various two-dimensional shapes. One method of generating higher order elements is to increase the order of the interpolating polynomial while keeping the element shape the same. Another method consists of altering the shape of the elements as well.

A relatively new element of the second type is called an isoparametric element. The name "isoparametric" comes from the concept that the same interpolation functions used for the problem unknown (T in the field problems or u,v in vector problems) are used to define the element shape. Thus the element shape may be changed to better correspond to actual problem geometry. This results in more accurate solutions for a given number of elements. Figure 13.1 shows triangular element examples.

Another important feature desired of elements is interelement continuity, as discussed in Sections 3.6 and 3.7. Higher order elements that have interelement continuity are called "serendipity" elements. The rectangular element and quadratic curve-sided triangular elements are serendipity elements. Also isoparametric elements are serendipity elements.

Higher order elements normally employ two coordinate systems: a global system (x,y) and a local system (r,s). The problem geometry is specified in the global system, but the local system is used to determine element properties. For the triangular elements shown in Figure 13.1(a) (a triangular element with a linear interpolation), no change in shape occurs from global to local coordinates, although the three node points change position. In the quadratic elements shown in Figure 13.1(b), the sides are formed by a second-order polynomial in the global system. This transforms into the regular triangle in local coordinates. An interpolation function of the form $\alpha_1 + \alpha_2 x + \alpha_3 y + \alpha_4 x^2 + \alpha_5 xy + \alpha_6 y^2$ is used with six node points.

Quadratic elements are shown in Figure 13.2. Linear quadratic elements, in Figure 13.2(a), are the simplest elements that change shape when transformed from global to local coordinate systems. Figure 13.2(b) yields a set of curved sides and has an interpolation function of the form $\alpha_1 + \alpha_2 x + \alpha_3 y + \alpha_4 x^2 + \alpha_5 xy + \alpha_6 y^2 + \alpha_7 x^2 y + \alpha_8 xy^2$. Eight nodes are required.

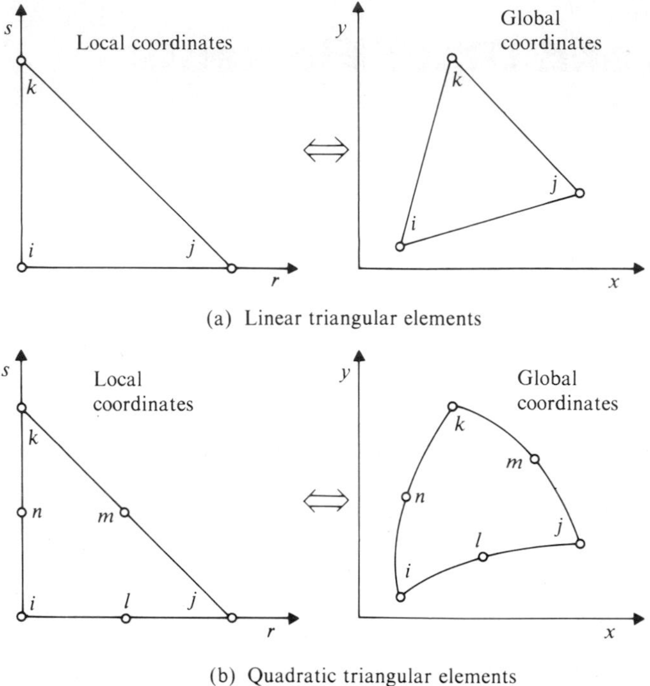

Figure 13.1 Linear and quadratic triangular isoparametric elements

The shape functions used to relate the global x,y coordinate system to the r,s coordinate system may be of lower, equal, or higher order than the interpolation function used for T. If a lower order is used, the element is called subparametric. When equal orders are used, it is isoparametric. Finally, higher order results in a superparametric element, which is not used (for reasons discussed in more advanced texts). Table 13.1 gives some examples.

Isoparametric (serendipity) elements are often used for automatic grid generating routines in advanced finite element codes. It is easy to subdivide the square shown in Figure 13.2(b) into any number of smaller elements. The techniques developed in this chapter are then used to map grid points in the local coordinate system over to the global coordinate system. Grid generation schemes are discussed in other texts.

One of the objectives of using higher order elements is to increase solution accuracy. It can be shown that the error is related to the highest complete polynomial in an interpolation or shape function. A complete linear polynomial is $\alpha_1 + \alpha_2 x + \alpha_3 y$ as employed in simplex triangular elements. The bilinear function $\alpha_1 + \alpha_2 x + \alpha_3 y + \alpha_4 xy$, as discussed for the rectangular element, is complete only to the first order because the second-order terms x^2 and y^2 are missing. The complete second-order polynomial is $\alpha_1 + \alpha_2 x + \alpha_3 y + \alpha_4 x^2 + \alpha_5 xy + \alpha_6 y^2$, which would be used for the quadratic triangular element in Figure 13.1(b).

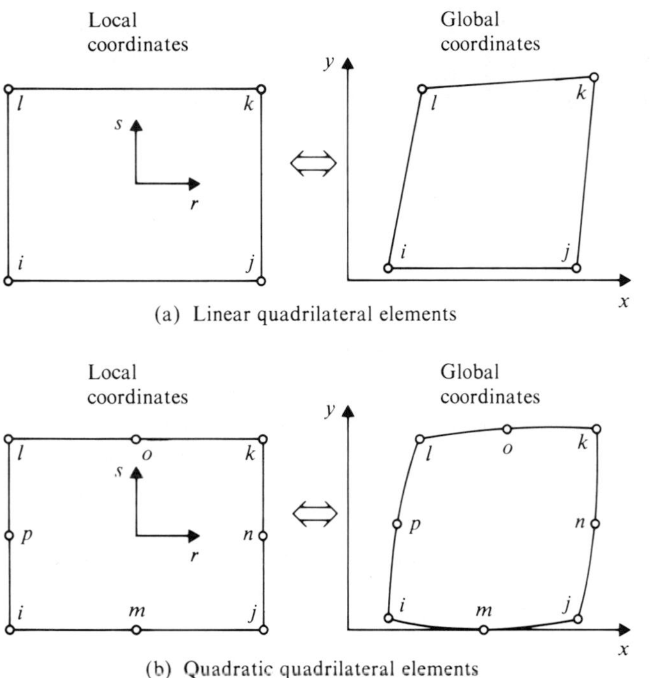

Figure 13.2 Linear and quadratic quadrilateral isoparametric elements

Table 13.1
Classification of Elements by Interpolation Function

Order of Interpolation Function for Coordinate System, $(x,y) \rightarrow (r,s)$	Order of Interpolation Function for Unknown Function $T(x,y)$	Element Classification (Superparametric not recommended for use)
Linear	Linear	Isoparametric
Quadratic	Linear	Superparametric
Linear	Quadratic	Subparametric
Quadratic	Quadratic	Isoparametric
Cubic	Quadratic	Superparametric
Linear	Cubic	Subparametric
Quadratic	Cubic	Subparametric
Cubic	Cubic	Isoparametric
Quartic	Cubic	Superparametric

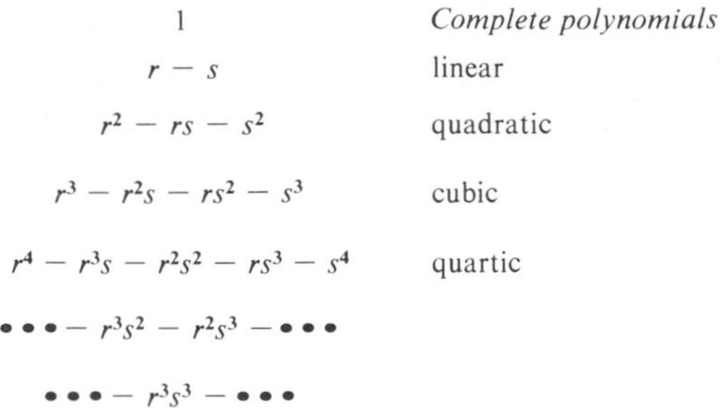

1
$r - s$
$r^2 - rs - s^2$
$r^3 - r^2s - rs^2 - s^3$
$r^4 - r^3s - r^2s^2 - rs^3 - s^4$
$\bullet\bullet\bullet - r^3s^2 - r^2s^3 - \bullet\bullet\bullet$
$\bullet\bullet\bullet - r^3s^3 - \bullet\bullet\bullet$
Complete polynomials
linear
quadratic
cubic
quartic

Figure 13.3 Pascal's triangle for determining complete polynomials

A fairly simple device for determining complete polynomials is Pascal's triangle. Figure 13.3 shows it with complete polynomials up to fourth (quartic) order. A finite element is classified according to the highest complete polynomial used in the interpolation function—not the highest order term. More discussion on this point occurs in later sections.

Isoparametric elements are useful elements that utilize higher order interpolation functions. Generally the higher order elements produce more accurate solutions to differential equations (given the same number of elements as used, say, with simplex elements). Unfortunately, this is not the only result. The evaluation of element properties now becomes much more complex and costly in terms of computer time. For example, a numerical integration scheme must be used to evaluate the element integrals. The general formulas given in Chapter 3 no longer apply.

Early in the development of higher order elements, including isoparametric ones, it was hoped that they would be much more efficient than the simplest possible element for a given problem. This has not proved the case, often because the gains in accuracy are somewhat offset by increased computer time. In many cases, however, a major cost factor is the time required of an engineer to set up the finite element model. In this case, the accuracy of higher order elements permits the use of small numbers of nodes and elements. Thus, higher order elements are well justified.

13.2 Simplest Isoparametric Element—Linear Quadrilateral

A. Shape Functions

The simplest isoparametric element in two dimensions is a quadrilateral with a bilinear interpolation function. Figure 13.4 gives a diagram. In the global coordinate system, an interpolation function of the bilinear form

$$T^{(e)}(x,y) = \alpha_1 + \alpha_2 x + \alpha_3 y + \alpha_4 xy \qquad \textbf{(13.1)}$$

is used, as in the rectangular element. This gives interelement continuity (or a serendipity element). At each of the four node points, (X_i,Y_i), (X_j,Y_j), (X_k,Y_k), (X_ℓ,Y_ℓ), this relation must be satisfied. The four resulting equations can be written as

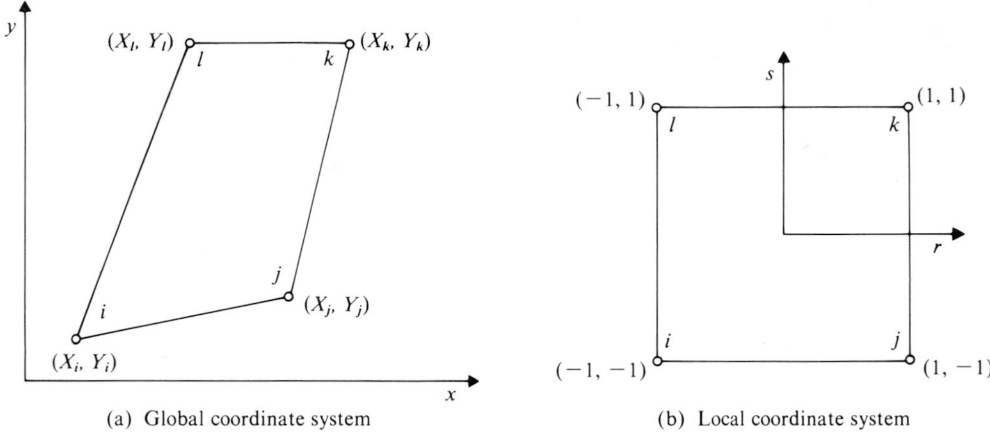

Figure 13.4 Linear quadrilateral isoparametric element

$$T_\beta = \alpha_1 + \alpha_2 X_\beta + \alpha_3 Y_\beta + \alpha_4 X_\beta Y_\beta, \quad \beta = i,j,k,\ell \tag{13.2}$$

where the T_β are the four nodal T values. With isoparametric elements the interpolation functions are not usually written explicitly in global coordinates.

In the local coordinate system, the element interpolation function is also taken as bilinear with the form

$$T^{(e)}(r,s) = \alpha_1 + \alpha_2 r + \alpha_3 s + \alpha_4 rs \tag{13.3}$$

Here the α values are determined from the nodal coordinates $(-1,-1)$, $(1,-1)$, $(1,1)$, $(-1,1)$. Clearly the geometry is much simpler in the local coordinate system than in the global one.

Evaluating the α constants yields

$$T^{(e)}(r,s) = G_i T_i + G_j T_j + G_k T_k + G_\ell T_\ell \tag{13.4}$$

where the shape functions are given by

$$G_i = \frac{1}{4}(1-r)(1-s)$$

$$G_j = \frac{1}{4}(1+r)(1-s)$$

$$G_k = \frac{1}{4}(1+r)(1+s)$$

$$G_\ell = \frac{1}{4}(1-r)(1+s)$$

Here the nomenclature G has been used rather than N to indicate that the coordinate system (r,s) is different. The shape functions $G(r,s)$ have the value one at the appropriate node and zero at the other three. Also the sum of the shape functions is unity at any point (R, S) inside the element.

It is also convenient to express the shape functions as

$$G_\beta = \frac{1}{4}(1 + rR_\beta)(1 + sS_\beta), \quad \beta = i,j,k,\ell \tag{13.5}$$

Here the values of R_β and S_β are the nodal coordinates as shown in Figure 13.4(a). Taken in order i,j,k,ℓ, they are $R_\beta = -1,1,1,-1$, and $S_\beta = -1,-1,1,1$. This form is convenient for computer calculations.

B. Global to Local Coordinates

Substituting Equation (13.2) into Equation (13.4) yields

$$\begin{aligned} T^{(e)}(r,s) = {} & G_i(\alpha_1 + \alpha_2 X_i + \alpha_3 Y_i + \alpha_4 X_i Y_i) \\ & + G_j(\alpha_1 + \alpha_2 X_j + \alpha_3 Y_j + \alpha_4 X_j Y_j) \\ & + G_k(\alpha_1 + \alpha_2 X_k + \alpha_3 Y_k + \alpha_4 X_k Y_k) \\ & + G_\ell(\alpha_1 + \alpha_2 X_\ell + \alpha_3 Y_\ell + \alpha_4 X_\ell Y_\ell) \end{aligned}$$

Grouping like terms in α_1,α_2, etc., gives

$$\begin{aligned} T^{(e)}(r,s) = {} & \alpha_1(G_i + G_j + G_k + G_\ell) \\ & + \alpha_2(G_i X_i + G_j X_j + G_k X_k + G_\ell X_\ell) \\ & + \alpha_3(G_i Y_i + G_j Y_j + G_k Y_k + G_\ell Y_\ell) \\ & + \alpha_4(G_i X_i Y_i + G_j X_j Y_j + G_k X_k Y_k + G_\ell X_\ell Y_\ell) \end{aligned}$$

Comparing this with Equation (13.1) for global coordinates, to make $T(r,s) = T(x,y)$, produces the equation for α_1, as $1 = G_i + G_j + G_k + G_\ell$. This is already known. For α_2 and α_3, the equations become

$$\begin{aligned} x &= G_i X_i + G_j X_j + G_k X_k + G_\ell X_\ell \\ y &= G_i Y_i + G_j Y_j + G_k Y_k + G_\ell Y_\ell \end{aligned} \tag{13.6}$$

These relations provide the coordinate transformation from global to local coordinates. The final equation for α_4 is not important here.

Because of the large number of terms involved in isoparametric elements, matrix notation is very useful for the calculations. Let the coordinate transformation, Equation (13.6), be written as

$$x = \underset{(1\times 4)}{\lfloor G \rfloor}\ \underset{(4\times 1)}{\{X\}}, \qquad y = \underset{(1\times 4)}{\lfloor G \rfloor}\ \underset{(4\times 1)}{\{Y\}} \tag{13.7}$$

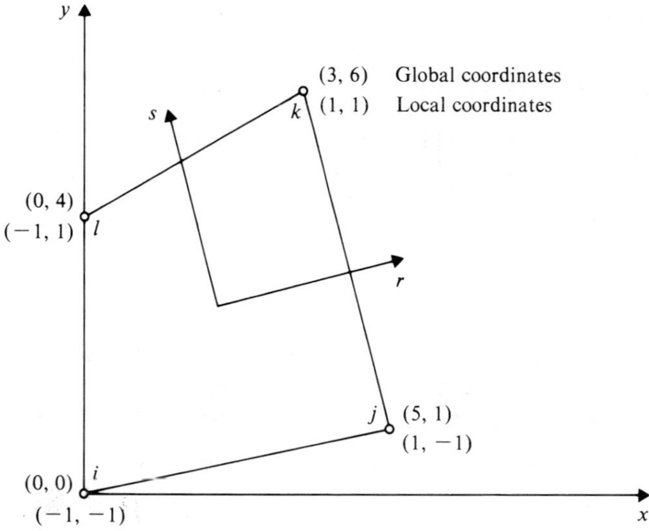

Figure 13.5 Coordinate systems for Example 13.1—Quadrilateral element

(the order of the matrix is indicated by the numbers in parentheses) where

$$\underset{(1\times 4)}{\{G\} = \{G_i \quad G_j \quad G_k \quad G_\ell\}},$$

$$\{X\} = \underset{(4\times 1)}{\begin{Bmatrix} X_i \\ X_j \\ X_k \\ X_\ell \end{Bmatrix}}, \quad \{Y\} = \underset{(4\times 1)}{\begin{Bmatrix} Y_i \\ Y_j \\ Y_k \\ Y_\ell \end{Bmatrix}}$$

Example 13.1 *Quadrilateral Element*

Given A bilinear isoparametric, serendipity quadrilateral element has the four nodal coordinates: (0,0), (5,1), (3,6), (0,4). Figure 13.5 shows a schematic.

Objective Find the coordinate transformation from global to local coordinates.

Solution Equation (13.7) gives

$$\{G\} = \frac{1}{4}\{\,(1-r)(1-s) \quad (1+r)(1-s) \quad (1+r)(1+s) \quad (1-r)(1+s)\}$$

$$\{X\} = \begin{Bmatrix} 0 \\ 5 \\ 3 \\ 0 \end{Bmatrix}, \quad \{Y\} = \begin{Bmatrix} 0 \\ 1 \\ 6 \\ 4 \end{Bmatrix}$$

and multiplying out the terms gives

$$x = \frac{5}{4}(1+r)(1-s) + \frac{3}{4}(1+r)(1+s)$$

$$y = \frac{1}{4}(1+r)(1-s) + \frac{3}{2}(1+r)(1+s) + (1-r)(1+s)$$

13.3 Derivatives

A. Jacobian

It is easily seen that derivatives in the local coordinate system r,s must be related to derivatives in the global coordinate system. The derivatives of some function (not specified yet) in r,s are obtained from the chain rule as

$$\frac{\partial}{\partial r} = \frac{\partial}{\partial x}\frac{\partial x}{\partial r} + \frac{\partial}{\partial y}\frac{\partial y}{\partial r}$$

$$\frac{\partial}{\partial s} = \frac{\partial}{\partial x}\frac{\partial x}{\partial s} + \frac{\partial}{\partial y}\frac{\partial y}{\partial s}$$

This can be expressed in matrix form as

$$\begin{Bmatrix} \dfrac{\partial}{\partial r} \\ \dfrac{\partial}{\partial s} \end{Bmatrix} = [J] \begin{Bmatrix} \dfrac{\partial}{\partial x} \\ \dfrac{\partial}{\partial y} \end{Bmatrix} \tag{13.8}$$

where the Jacobian $[J]$ is

$$\underset{(2\times 2)}{[J]} = \begin{bmatrix} \dfrac{\partial x}{\partial r} & \dfrac{\partial y}{\partial r} \\ \dfrac{\partial x}{\partial s} & \dfrac{\partial y}{\partial s} \end{bmatrix} \tag{13.9}$$

For calculations let the terms in $[J]$ be denoted

$$[J] = \begin{bmatrix} J_{11} & J_{12} \\ J_{21} & J_{22} \end{bmatrix} \tag{13.10}$$

Since the Jacobian is a matrix of order 2×2 for all elements, not just linear quadrilaterals, it is easy to work with.

Recall that the global x,y coordinates are related to the local r,s coordinates by Equation (13.6). Substituting these expressions into the Jacobian $[J]$ yields

$$[J] = \underset{(2\times 4)}{[F]} \; \underset{(4\times 2)}{[\{X\},\{Y\}]} \tag{13.11}$$

where $[F]$ is

$$[F] = \begin{bmatrix} \{F_1\} \\ \{F_2\} \end{bmatrix} = \begin{bmatrix} \dfrac{\partial}{\partial r} \{G\} \\ \dfrac{\partial}{\partial s} \{G\} \end{bmatrix}$$

$$(2\times 4) \qquad\qquad (2\times 4)$$

Writing $[F]$ in terms of the derivatives of $\{G\}$ from Equation (13.4) gives

$$\begin{aligned} \{F_1\} &= \frac{1}{4}\{-(1-s) \quad (1-s) \quad (1+s) \quad -(1+s)\} \\ \{F_2\} &= \frac{1}{4}\{-(1-r) \quad -(1+r) \quad (1+r) \quad (1-r)\} \end{aligned} \tag{13.12}$$

Now the terms in the Jacobian are

$$\begin{aligned} J_{11} &= \{F_1\}\{X\}, \quad J_{12} = \{F_1\}\{Y\} \\ J_{21} &= \{F_2\}\{X\}, \quad J_{22} = \{F_2\}\{Y\} \end{aligned} \tag{13.13}$$

The general derivative expression is

$$\begin{Bmatrix} \dfrac{\partial}{\partial r} \\ \dfrac{\partial}{\partial s} \end{Bmatrix} = [J] \begin{Bmatrix} \dfrac{\partial}{\partial x} \\ \dfrac{\partial}{\partial y} \end{Bmatrix}$$

where the derivatives may apply to any function. This represents a set of two simultaneous equations that may be inverted to yield

$$\begin{Bmatrix} \dfrac{\partial}{\partial x} \\ \dfrac{\partial}{\partial y} \end{Bmatrix} = [J]^{-1} \begin{Bmatrix} \dfrac{\partial}{\partial r} \\ \dfrac{\partial}{\partial s} \end{Bmatrix} \tag{13.14}$$

where $[J]^{-1}$ is the matrix inverse of the Jacobian. Since $[J]$ is of order 2, the inverse is easily evaluated as

$$[J]^{-1} = \frac{1}{\det J} \begin{bmatrix} \dfrac{\partial y}{\partial s} & -\dfrac{\partial y}{\partial r} \\ -\dfrac{\partial x}{\partial s} & \dfrac{\partial x}{\partial r} \end{bmatrix} \tag{13.15}$$

Here the determinant of J is

$$\det J = \frac{\partial x}{\partial r}\frac{\partial y}{\partial s} - \frac{\partial x}{\partial s}\frac{\partial y}{\partial r} \tag{13.16}$$

Usually both the determinant and inverse must be found numerically. Since $[J]$ is a matrix of order 2×2, it is most easily evaluated by a simple routine in a computer program.

For this linear quadrilateral element, the inverse of the Jacobian is found from Equation (13.4) as

$$[J]^{-1} = \frac{1}{\det J}\begin{bmatrix} J_{22} & -J_{12} \\ -J_{21} & J_{11} \end{bmatrix} \tag{13.17}$$

It can also be shown that the determinant is given by

$$\det J = J_{11}J_{22} - J_{12}J_{21} \tag{13.18}$$

These are the expressions required for calculating the element properties.

B. Derivative Matrix

Evaluation of the minimization equations involves integration over the element area. This is most conveniently done in the local coordinate system. The derivatives shown in the variational principle I are related to the r,s coordinate system by the chain rule

$$\frac{\partial T}{\partial x} = \frac{\partial T}{\partial r}\frac{\partial r}{\partial x} + \frac{\partial T}{\partial s}\frac{\partial s}{\partial x}$$

$$\frac{\partial T}{\partial y} = \frac{\partial T}{\partial r}\frac{\partial r}{\partial y} + \frac{\partial T}{\partial s}\frac{\partial s}{\partial y}$$

or in matrix form

$$\begin{Bmatrix} \dfrac{\partial T}{\partial x} \\ \dfrac{\partial T}{\partial y} \end{Bmatrix} = [J]^{-1} \begin{Bmatrix} \dfrac{\partial T}{\partial r} \\ \dfrac{\partial T}{\partial s} \end{Bmatrix} \tag{13.19}$$

However, T is expressed in terms of shape functions as $T = G_iT_i + G_jT_j + G_kT_k + G_\ell T_\ell$. Then from the definition of $[F]$, Equation (13.12),

$$\begin{Bmatrix} \dfrac{\partial T}{\partial r} \\ \dfrac{\partial T}{\partial s} \end{Bmatrix} = \underset{(2\times4)}{[F]}\ \underset{(4\times1)}{\{T\}} \tag{13.20}$$

where

$$\{T\} = \begin{Bmatrix} T_i \\ T_j \\ T_k \\ T_\ell \end{Bmatrix}$$

Combining all of these expressions yields

$$\begin{Bmatrix} \dfrac{\partial T}{\partial x} \\ \dfrac{\partial T}{\partial y} \end{Bmatrix} = \underset{(2\times2)}{[J]^{-1}}\ \underset{(2\times4)}{[F]}\ \underset{(4\times1)}{\{T\}} \tag{13.21}$$

Define the derivative matrix $[D]$ as

$$\underset{(2\times4)}{[D]} = \underset{(2\times2)}{[J]^{-1}}\ \underset{(2\times4)}{[F]}$$

and the derivatives are given by

$$\begin{Bmatrix} \dfrac{\partial T}{\partial x} \\ \dfrac{\partial T}{\partial y} \end{Bmatrix} = [D]\ \{T\} \tag{13.22}$$

The terms in $[D]$ can be obtained from Equations (13.13) and (13.12). It is of order (2×4). For elements with, say, eight nodes $[D]$ would have order (2×8).

Example 13.2 *Element Derivative Properties*

Given The linear isoparametric serendipity element as in Example 13.1 has nodal coordinates (0,0), (5,1), (3,6), (0,4).

Objective Determine the Jacobian of the transformation at the point $(1/\sqrt{3}, 1/\sqrt{3})$ in local coordinates. Also find the partial derivatives $\partial T/\partial x$ and $\partial T/\partial y$ at that point from equation (13.22).

Solution At the point $r = 0.577$, $s = 0.577$, the corresponding shape function and x,y coordinate values are

$$\begin{aligned} \{G\} &= \{0.0447;0.1668;0.6217;0.1668\} \\ x &= \{G\}\ \{X\} = 2.6992 \\ y &= \{G\}\ \{Y\} = 4.5642 \end{aligned}$$

The $[F]$ matrix is given by Equation (13.12) as

$$[F] = \begin{bmatrix} -0.1057 & 0.1057 & 0.3942 & -0.3942 \\ -0.1057 & -0.3942 & 0.3942 & 0.1057 \end{bmatrix}$$

and the Jacobian from Equation (13.11) as

$$[J] = [F]\ [\{X\}, \{Y\}]$$

$$[J] = \begin{bmatrix} -0.1057 & 0.1057 & 0.3942 & -0.3942 \\ -0.1057 & -0.3942 & 0.3942 & 0.1057 \end{bmatrix} \begin{bmatrix} 0 & 0 \\ 5 & 1 \\ 3 & 6 \\ 0 & 4 \end{bmatrix}$$

Evaluating

$$[J] = \begin{bmatrix} J_{11} & J_{12} \\ J_{21} & J_{22} \end{bmatrix} = \begin{bmatrix} 1.7111 & 0.8941 \\ -0.7884 & 2.3938 \end{bmatrix}$$

The determinant is $\det J = J_{11}J_{22} - J_{12}J_{21} = 4.8009$ and the inverse is

$$[J]^{-1} = \frac{1}{\det J} \begin{bmatrix} 2.3938 & -0.8941 \\ 0.7884 & 1.7111 \end{bmatrix}$$

$$[J]^{-1} = \begin{bmatrix} 0.4986 & -0.1862 \\ 0.1642 & 0.3564 \end{bmatrix}$$

Now the derivatives are given by

$$\begin{Bmatrix} \dfrac{\partial T}{\partial x} \\ \dfrac{\partial T}{\partial y} \end{Bmatrix} = [D] \begin{Bmatrix} T_i \\ T_j \\ T_k \\ T_\ell \end{Bmatrix}$$

with $[D] = [J]^{-1}[F]$

$$[D] = \begin{bmatrix} 0.4986 & -0.1862 \\ 0.1642 & 0.3564 \end{bmatrix}$$

$$\times \begin{bmatrix} -0.1057 & 0.1057 & 0.3942 & -0.3942 \\ -0.1057 & -0.3942 & 0.3942 & 0.1057 \end{bmatrix}$$

Evaluating

$$[D] = \begin{bmatrix} -0.03302 & 0.12610 & 0.12315 & -0.21623 \\ -0.05503 & -0.12313 & 0.2052 & -0.02706 \end{bmatrix}$$

These numbers illustrate matrices that are useful in the next section.

13.4 Element Matrices—Linear Quadrilateral

A. Laplace's Equation

The next matter to be considered is the element matrices. If these can be obtained, the isoparametric quadrilateral element can easily be incorporated into computer programs such as FINTWO along with simplex and other elements. For simplicity in the derivation, the following boundary value problem (Laplace's equation) is used.

$$\frac{\partial^2 T}{\partial x^2} + \frac{\partial^2 T}{\partial y^2} = 0 \text{ in } A$$

$$T = T_0 \text{ on } L$$

This will illustrate the development of isoparametric element matrices. The variational principle for this problem is

$$I = \iint_A \frac{1}{2}\left[\left(\frac{\partial T}{\partial x}\right)^2 + \left(\frac{\partial T}{\partial y}\right)^2\right] dA$$

Writing the element variational principle in matrix form as

$$I^{(e)} = \frac{1}{2}\iint_{A^{(e)}} \left\{\frac{\partial T}{\partial x}, \frac{\partial T}{\partial y}\right\} \begin{Bmatrix} \dfrac{\partial T}{\partial x} \\ \dfrac{\partial T}{\partial y} \end{Bmatrix} dA$$

which can be written from Equation (13.21) as

$$I^{(e)} = \frac{1}{2} \iint_{A^{(e)}} \{T\}^T [D]^T [I] [D] \{T\} dA$$

Here $[I]$ is

$$[I] = \begin{bmatrix} 1 & 0 \\ 0 & 1 \end{bmatrix}$$

The minimization equations are given by

$$\begin{Bmatrix} \dfrac{\partial I}{\partial T_i} \\ \dfrac{\partial I}{\partial T_j} \\ \dfrac{\partial I}{\partial T_k} \\ \dfrac{\partial I}{\partial T_\ell} \end{Bmatrix} = [B] \begin{Bmatrix} T_i \\ T_j \\ T_k \\ T_\ell \end{Bmatrix} - \{C\}$$

Evaluation of the element matrix yields

$$[B] = \iint_{A^{(e)}} [D]^T [D] \, dA \tag{13.23}$$

with $\{C\}$ having all zero entries (for this boundary value problem).

B. Integration in Local Coordinates

Integrals over the element area are necessary to determine element matrices. The integrals are very difficult to perform in the global x,y coordinate system but can be done reasonably easily numerically in the local r,s coordinate system. Let the unit vectors in the x,y coordinate system be $\boldsymbol{i}$, $\boldsymbol{j}$ while the unit vectors in the r,s coordinate system are $\boldsymbol{e}_r$, $\boldsymbol{e}_s$. The differential area is the vector product of the differential lengths $dx\,\boldsymbol{i}$ and $dy\,\boldsymbol{j}$: $(dx\,\boldsymbol{i}) \times (dy\,\boldsymbol{j}) = dA\,\boldsymbol{k}$. Figure 13.6 illustrates the unit vectors. Each of the differential lengths can be expressed in terms of the local system as

$$dx\,\boldsymbol{i} = \frac{\partial x}{\partial r} dr\,\boldsymbol{e}_r + \frac{\partial x}{\partial s} ds\,\boldsymbol{e}_s$$

$$dy\,\boldsymbol{j} = \frac{\partial y}{\partial r} dr\,\boldsymbol{e}_r + \frac{\partial y}{\partial s} ds\,\boldsymbol{e}_s$$

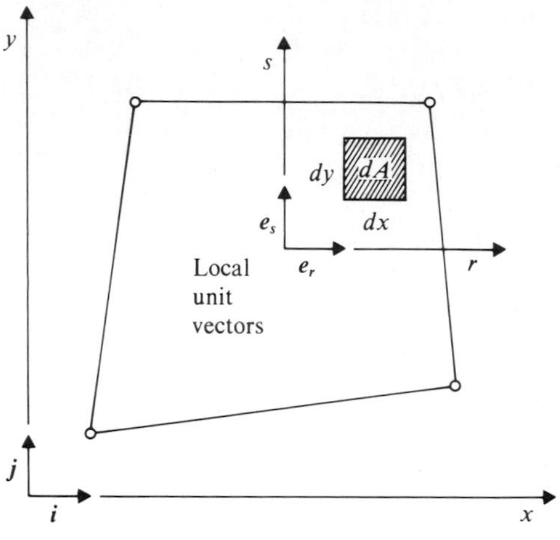

Figure 13.6 Vector relations for differential area

The vector product is

$$
\begin{aligned}
&(dx\, \boldsymbol{i}) \times (dy\, \boldsymbol{j}) \\
&= \left(\frac{\partial x}{\partial r}\, dr\, \boldsymbol{e}_r + \frac{\partial x}{\partial s}\, ds\, \boldsymbol{e}_s\right) \times \left(\frac{\partial y}{\partial r}\, dr\, \boldsymbol{e}_r + \frac{\partial y}{\partial s}\, ds\, \boldsymbol{e}_s\right) \\
&= \left(\frac{\partial x}{\partial r}\frac{\partial y}{\partial s}\, \boldsymbol{e}_r \times \boldsymbol{e}_s + \frac{\partial x}{\partial s}\frac{\partial y}{\partial r}\, \boldsymbol{e}_s \times \boldsymbol{e}_r\right) drds \\
&= \left(\frac{\partial x}{\partial r}\frac{\partial y}{\partial s} - \frac{\partial x}{\partial s}\frac{\partial y}{\partial r}\right) drds\, \boldsymbol{k} = dA\boldsymbol{k}
\end{aligned}
$$

Thus the differential area dA is given by

$$
dA = \left(\frac{\partial x}{\partial r}\frac{\partial y}{\partial s} - \frac{\partial x}{\partial s}\frac{\partial y}{\partial r}\right) drds
$$

The quantity in brackets represents the magnitude of the vector product of dx and dy, whose numerical value equals the area of a parallelogram (in local coordinates) with sides dx and dy. This term is also equal to the determinant of the Jacobian, so that $dA = \det J\, drds$ where

$$
\det J = \frac{\partial x}{\partial r}\frac{\partial y}{\partial s} - \frac{\partial x}{\partial s}\frac{\partial y}{\partial r}
$$

An integral in the global coordinate system can be related to one in the local coordinate system using

$$
\iint_{A^{(e)}} \Big[\qquad \Big] dA = \int_{-1}^{1} \int_{-1}^{1} \Big[\qquad \Big] \det J\, drds \tag{13.24}
$$

where the function contained in the brackets is expressed in the appropriate global or local coordinate system.

The element matrix, Equation (13.8), becomes

$$[B] = \int_{-1}^{1} \int_{-1}^{1} [D]^T [D] \det J \, drds \tag{13.25}$$

The integrand is a quite complicated function of r and s that must be evaluated numerically for nearly all elements except simplex ones. Numerical integration procedures for the linear quadrilateral and other elements are given in the next section.

C. General Boundary Value Problem

It is also useful to consider the more general case, where the differential equation is

$$\frac{\partial}{\partial x}\left(K_x \frac{\partial T}{\partial x}\right) + \frac{\partial}{\partial y}\left(K_y \frac{\partial T}{\partial y}\right) + PT + Q = 0$$

with the boundary condition $T = T_0$ on L_1 and

$$K_x \frac{\partial T}{\partial x} n_x + K_y \frac{\partial T}{\partial y} n_y + \alpha T + \beta = 0 \text{ on } L_2$$

The element variational principle is written in matrix form as

$$I^{(e)} = \int\int_{A^{(e)}} \left[\frac{1}{2} \left\{ \frac{\partial T}{\partial x}, \frac{\partial T}{\partial y} \right\} [K] \begin{Bmatrix} \frac{\partial T}{\partial x} \\ \frac{\partial T}{\partial y} \end{Bmatrix} - PT^2 - QT \right] dA + \int_{L_2^{(e)}} \left(\frac{1}{2} \alpha T^2 + \beta T \right)^{(e)} dL_2$$

where

$$[K] = \begin{bmatrix} K_x & 0 \\ 0 & K_y \end{bmatrix}$$

Substituting from Equation (13.11) gives

$$I^{(e)} = \int\int_{A^{(e)}} \left[\frac{1}{2} \{T\}^T[D]^T[K][D]\{T\} - P\,\{T\}^T\{G\}^T\{G\}\{T\} - Q\{G\}\{T\} \right]^{(e)} dA + \int_{L^{(e)}} \left[\frac{1}{2} \alpha\{T\}^T\{G\}^T\{G\}\{T\} + \beta\{G\}\{T\} \right]^{(e)} dL_2$$

The minimization equations yield

$$[B]^{(e)} = \iint_{A^{(e)}} \Big[[D]^T[K]\,[D] - P\,\{G\}^T\{G\} \Big]\, dA$$

$$+ \int_{L^{(e)}} \Big[\alpha\{G\}^T\{G\} \Big]\, dL_2$$

$$\{C\}^{(e)} = \iint_{A^{(e)}} \Big[Q\{G\}^T \Big]\, dA - \int_{L^{(e)}} \Big[\beta\{G\}^T \Big]\, dL_2$$

The terms evaluated over the element area are easily found in terms of local coordinates as before. Along the derivative boundary L_2, some additional material must be discussed. Consider side ij (side 1), where s is a constant, as the example. In local coordinates, the differential dL_2 is transformed as

$$dL_2 = \left[\left(\frac{dx}{dr}\right)^2 + \left(\frac{dy}{dr}\right)^2 \right]^{1/2} dr \quad \text{on side } ij$$

or, from Equation (13.9),

$$dL_2 = [J_{11}^2 + J_{12}^2]^{1/2}\, dr$$

Similarly along side jk (side 2), where r is a constant,

$$dL_2 = \left[\left(\frac{dx}{ds}\right)^2 + \left(\frac{dy}{ds}\right)^2 \right]^{1/2} ds \quad \text{on side } jk$$

or

$$dL_2 = [J_{21}^2 + J_{22}^2]^{1/2}\, ds$$

The full results, including all four side boundaries, are

$$[B]^{(e)} = \int_{-1}^{1}\int_{-1}^{1} \Big[[D]^T[K][D] - P\,\{G\}^T\{G\} \Big] \det J\, drds$$

$$+ \int_{-1}^{1} \{\alpha\}^T\{G\}^T\{G\}\{dL_2\}$$

$$\{C\}^{(e)} = \int_{-1}^{1}\int_{-1}^{1} \Big[Q\{G\}^T \Big] \det J\, drds + \int_{-1}^{1} \{\beta\}\{G\}^T\{dL_2\}$$

where

$$\{\alpha\} = \begin{Bmatrix} \alpha_{ij} \\ \alpha_{jk} \\ \alpha_{k\ell} \\ \alpha_{\ell i} \end{Bmatrix}, \qquad \beta = \begin{Bmatrix} \beta_{ij} \\ \beta_{jk} \\ \beta_{k\ell} \\ \beta_{\ell i} \end{Bmatrix}$$

$$\{dL_2\} = \begin{Bmatrix} [J_{11}^2 + J_{12}^2]^{1/2}\, dr \\ [J_{21}^2 + J_{22}^2]^{1/2}\, ds \\ [J_{11}^2 + J_{12}^2]^{1/2}\, dr \\ [J_{21}^2 + J_{22}^2]^{1/2}\, ds \end{Bmatrix}$$

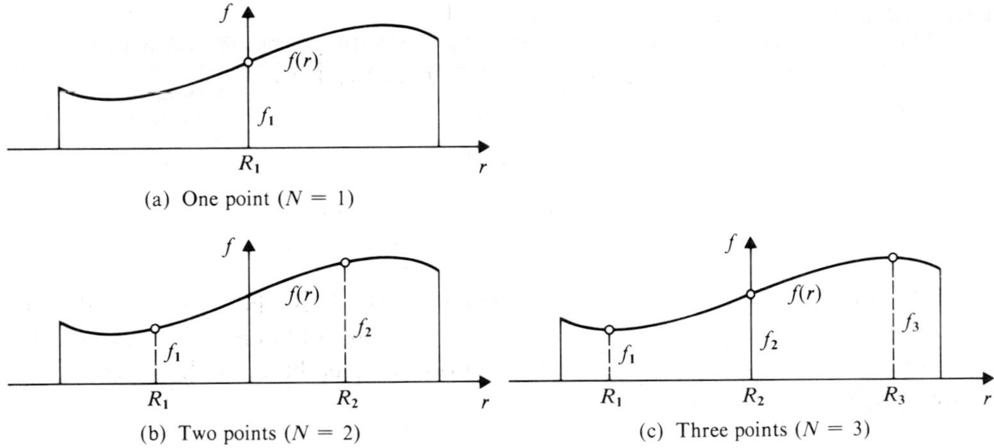

Figure 13.7 Gauss quadrature

13.5 Numerical Integration

There are several schemes used for numerical integration, but the Gauss quadrature approach is the most widely found. It will be the only one considered here. Generally quadrature formulas apply simple polynomial interpolation functions to approximate the integrand between the limits of integration. All integration is done in local coordinates (r,s).

A. One Dimension

In one dimension, the integral of interest is

$$\text{Int} = \int_{-1}^{1} f(r)\, dr \tag{13.26}$$

The simplest approximation is one point $f_1 = f(r)$, and the integral has the value Int $= 2f_1$, as shown in Figure 13.7(a). Generalizing this to two or more sampling points gives

$$\text{Int} = \int_{-1}^{1} f(r)\, dr = \sum_{i=1}^{N} W_i f_i \tag{13.27}$$

where f_i = value of f at sampling point, W_i = weighting factor, and N = number of sample points. The location of the sampling points R_i, illustrated in Figure 13.7, is based on Hermite polynomials, which give much less error than equally spaced sampling points. Numerical values of N, R_i, and W_i are shown in Table 13.2.

Table 13.2
Location of Sampling Points and Weight Factors for Gauss Quadrature

Number of Points N	Location of Sampling Points R_i	Associate Weight Factors W_i
1	0.0	2.0
2	± 0.5773502692	1.0
3	0.0	8/9 = 0.8888888889
	± 0.7745966692	5/9 = 0.5555555556

Sampling points are located symmetrically with respect to the center of the interval, but the end points are not included. Thus the Gauss method is called an open-interval method. When N is the number of points, the integral is exact if the integrand is a polynomial of degree $2N-1$. For example, three data points yield the exact integral of a fifth-order polynomial. Achieving the same accuracy using equally spaced sampling points, including the end points (known as the Newton-Cotes method), requires five points.

B. Two Dimensions

In two dimensions, both quadrilaterals and triangles must be considered. Start with quadrilaterals (actually squares) with an integral of the form

$$\text{Int} = \int_{-1}^{1}\int_{-1}^{1} f(r,s)\, dr ds \tag{13.28}$$

This can be integrated first in one direction

$$\text{Int} = \int_{-1}^{1}\left[\sum_{i=1}^{N} W_i f(R_i,s)\right] ds$$

and again in the other direction

$$\text{Int} = \sum_{j=1}^{N} W_j \left[\sum_{i=1}^{N} W_i f(R_i,S_j)\right]$$

with the final result

$$\text{Int} = \sum_{i=1}^{N}\sum_{j=1}^{N} W_i W_j f(R_i,S_j) \tag{13.29}$$

Fortunately, the weighting functions and location of sampling points that apply in two dimensions are the same as in one dimension. An example, using a four-point integration, is

$$\begin{aligned}\text{Int} = {} & (1.0)(1.0)f(R_1,S_1) + (1.0)(1.0)f(R_2,S_2) \\ & + (1.0)(1.0)f(R_3,S_3) + (1.0)(1.0)f(R_4,S_4)\end{aligned}$$

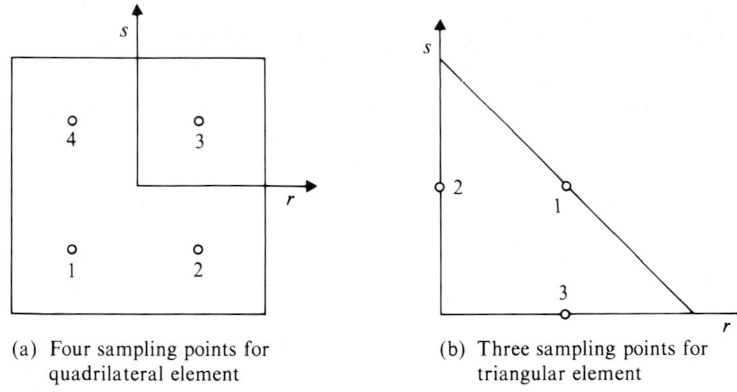

(a) Four sampling points for quadrilateral element

(b) Three sampling points for triangular element

Figure 13.8 Gauss quadrature in two-dimensional elements

Table 13.3
Location of Sampling Points and Weight Factors for Triangular Elements

Total Number of Points N	Location of Sampling Points R_i, S_i		Weighting Factor (W_i/Area)	Error
1	0.33333	0.33333	1.0	$0(h^2)$
3	0.5	0.5	0.33333	$0(h^3)$
	0.0	0.5	0.33333	
	0.5	0.0	0.33333	
4	0.33333	0.33333	0.56250	$0(h^4)$
	0.73333	0.13333	0.52083	
	0.13333	0.73333	0.52083	
	0.13333	0.13333	0.52083	

where the sampling points are

$$(R_i, S_j) = \pm 0.57735 \ldots = \pm 1/\sqrt{3}$$

This is illustrated in Figure 13.8(a)

For a triangle in local coordinates, the same formula, Equation (13.29), can be used, but with different sampling points and weighting factors. Table 13.3 gives the numerical values. This procedure is called Hammer's formula. Figure 13.8(b) illustrates the sample points when $N = 3$. The error term indicates that the order $0(h^i)$ gives an exact answer if the integrand is of order $i-1$.

Element	*Recommend sample points*	*Exact sample points*
Rectangular 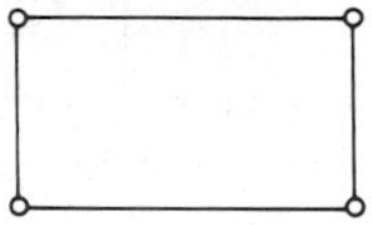	2 × 2	2 × 2
Linear quadrilateral isoparametric 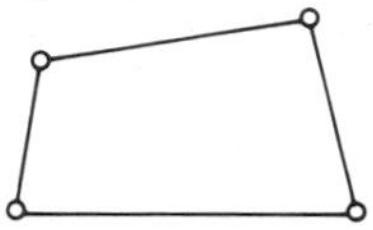	2 × 2	3 × 3
Eight-node rectangular 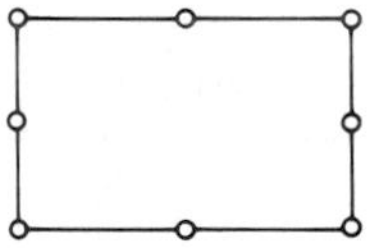	2 × 2	3 × 3
Quadratic quadrilateral isoparametric 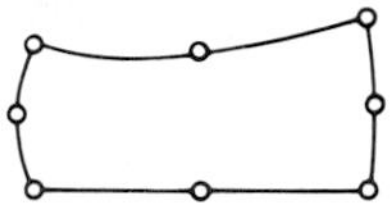	3 × 3	4 × 4

Figure 13.9 Sample points for two-dimensional elements

Numerical integration can also be carried out for three-dimensional elements—both cubic and tetrahedral in shape. Cubic elements yield the integration formula

$$\text{Int} = \int_{-1}^{1}\int_{-1}^{1}\int_{-1}^{1} f(r,s,p)\, drdsdp$$

$$= \sum_{i=1}^{N}\sum_{j=1}^{N}\sum_{k=1}^{N} W_iW_jW_kf(R_i,S_j,P_k)$$

with the same sampling point locations and weighting factors as in the two-dimensional case. Formulas for tetrahedral elements can be found in other references. The amount of computer time required to generate nodal properties for higher order three-dimensional elements can be quite large. That is beyond the scope of this text.

It has been shown from much experience that the number of sample points used for an element varies. Figure 13.9 gives the recommended number for two-dimensional elements. As an element becomes distorted relative to a rectangular shape (isoparametric elements), a larger number of sample points is required. The exact integration indicated in the second column will not necessarily yield better results because of the errors inherent in approximating functions in a piecewise fashion, as we do with finite elements.

Example 13.3 *Numerical Integration for Quadrilateral Element*

Given The quadrilateral element from Examples 13.1 and 13.2 is considered again.

Objective Determine the element matrix for Laplace's equation using four sample points.

Solution Consider four sampling points taken in order 1 through 4, as shown in Figure 13.8(a) for the quadrilateral element. Let the $[D]$ matrix for sample point 1 be denoted by $[D]_1$. The matrix product $[D]_1^T\,[D]_1$ is a 4×4 matrix (the same size as the element $[B]$ matrix). There will also be a value of det J_1 for sample point 1.

Each of the other three sample points has similar quantities. The element $[B]$ is given by Equation (13.25) as

$$
\begin{aligned}
[B] &= \int_{-1}^{1}\int_{-1}^{1} [D]^T[D] \det J \, dr ds \\
&= W(R_1)W(S_1)[D]_1^T[D]_1 \det J_1 \\
&\quad + W(R_2)W(S_2)[D]_2^T[D]_2 \det J_2 \\
&\quad + W(R_3)W(S_3)[D]_3^T[D]_3 \det J_3 \\
&\quad + W(R_4)W(S_4)[D]_4^T[D]_4 \det J_4
\end{aligned}
$$

13.6 Quadratic Quadrilateral Element

The linear quadrilateral element developed in earlier sections has straight sides. One of the benefits of isoparametric elements is accurate representation of curved boundaries. This can be done by using a quadratic polynomial for the interpolation function. Again consider a quadrilateral element, but add four midside nodes to the four corner nodes. The sides are then curved. Figure 13.10 shows a schematic.

A. Shape Functions

The interpolation function has the form

$$
\begin{aligned}
T^{(e)}(x,y) &= \alpha_1 + \alpha_2 x + \alpha_3 y + \alpha_4 x^2 + \alpha_5 xy + \alpha_6 y^2 \\
&\quad + \alpha_7 x^2 y + \alpha_8 xy^2
\end{aligned}
\qquad \textbf{(13.30)}
$$

It is complete only to the second-order terms. This relation must hold at all eight node points i through p, which yields the relation

$$
\begin{aligned}
T_\beta &= \alpha_1 + \alpha_2 X_\beta + \alpha_3 Y_\beta + \alpha_4 X_\beta^2 + \alpha_5 X_\beta Y_\beta + \alpha_6 Y_\beta^2 \\
&\quad + \alpha_7 X_\beta^2 Y_\beta + \alpha_8 X_\beta Y_\beta^2,
\end{aligned}
\qquad \textbf{(13.31)}
$$

$$\beta = i, \ldots, p$$

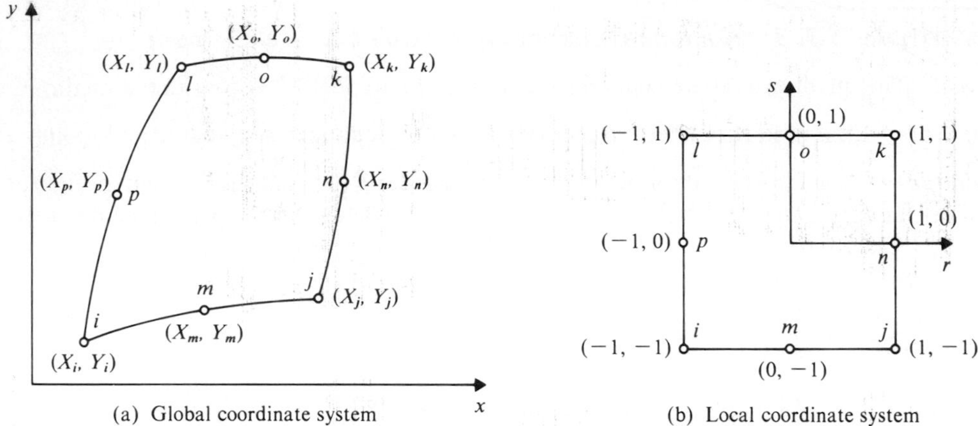

Figure 13.10 Quadratic isoparametric quadrilateral element

Similarly the interpolation function in local (r,s) coordinates is

$$\begin{aligned} T^{(e)}(r,s) &= \alpha_1 + \alpha_2 r + \alpha_3 s + \alpha_4 r^2 + \alpha_5 rs + \alpha_6 s^2 \\ &\quad + \alpha_7 r^2 s + \alpha_8 rs^2 \end{aligned} \tag{13.32}$$

Here the eight nodal coordinates are obvious.

In the local coordinate system with

$$\begin{aligned} T^{(e)}(r,s) &= G_i T_i + G_j T_j + G_k T_k + G_\ell T_\ell + G_m T_m \\ &\quad + G_n T_n + G_o T_o + G_p T_p \end{aligned} \tag{13.33}$$

the shape functions are

$$\begin{aligned} G_i &= \frac{1}{4}(1-r)(1-s)(-1-r-s) \\ G_j &= \frac{1}{4}(1+r)(1-s)(-1+r-s) \\ G_k &= \frac{1}{4}(1+r)(1+s)(-1+r+s) \\ G_\ell &= \frac{1}{4}(1-r)(1+s)(-1-r+s) \\ G_m &= \frac{1}{2}(1-r^2)(1-s) \\ G_n &= \frac{1}{2}(1-s^2)(1+r) \end{aligned} \tag{13.34}$$

Continued on next page

$$G_o = \frac{1}{2}(1-r^2)(1+s)$$

$$G_p = \frac{1}{2}(1-s^2)(1-r) \qquad \textbf{(13.34 cont.)}$$

As usual, the shape functions have value unity at the appropriate node and zero at all the other node points. This can be conveniently written in compact form as

$$G_\beta = \frac{1}{4}(1+rR_\beta)(1+sS_\beta)(-1+rR_\beta+sS_\beta), \qquad \beta = i,j,k,\ell$$

$$G_\beta = \frac{1}{2}(1-r^2)(1+sS_\beta), \qquad \beta = m,o \qquad \textbf{(13.35)}$$

$$G_\beta = \frac{1}{2}(1+rR_\beta)(1-s^2), \qquad \beta = n,p$$

where the eight nodal values are taken in order i,j,k,ℓ,m,n,o,p as $R_\beta = -1,1,1,-1,0,1,0,-1$ and $S_\beta = -1,-1,1,1,-1,0,1,0$. These are the nodal coordinates taken in order.

B. Transformation to Local Coordinates

Following a derivation similar to that for linear quadrilateral elements produces the coordinate transformation equations (from α_2 and α_3)

$$\begin{aligned} x &= G_iX_i + G_jX_j + G_kX_k + G_\ell X_\ell + G_mX_m + G_nX_n \\ &\quad + G_oX_o + G_pX_p \\ y &= G_iY_i + G_jY_j + G_kY_k + G_\ell Y_\ell + G_mY_m + G_nY_n \\ &\quad + G_oY_o + G_pY_p \end{aligned} \qquad \textbf{(13.36)}$$

In matrix form this is

$$\begin{aligned} x &= \lfloor G \rfloor \{X\} \\ y &= \lfloor G \rfloor \{Y\} \end{aligned} \qquad \textbf{(13.37)}$$

where

$$\underset{(1\times 8)}{\lfloor G \rfloor = \lfloor G_i \; G_j \; G_k \; . \; . \; . \; G_p \rfloor},$$

$$\{X\} = \underset{(8\times 1)}{\begin{Bmatrix} X_i \\ X_j \\ X_k \\ \cdot \\ \cdot \\ \cdot \\ X_p \end{Bmatrix}}, \qquad \{Y\} = \underset{(8\times 1)}{\begin{Bmatrix} Y_i \\ Y_j \\ Y_k \\ \cdot \\ \cdot \\ \cdot \\ Y_p \end{Bmatrix}}$$

Now the Jacobian is

$$\underset{(2\times2)}{[J]} = \begin{bmatrix} \dfrac{\partial x}{\partial r} & \dfrac{\partial y}{\partial r} \\ \dfrac{\partial x}{\partial s} & \dfrac{\partial y}{\partial s} \end{bmatrix} = \begin{bmatrix} J_{11} & J_{12} \\ J_{21} & J_{22} \end{bmatrix}$$

where

$$\begin{aligned} J_{11} &= \underset{(1\times8)}{\{F_1\}}\ \underset{(8\times1)}{\{X\}}, \quad & J_{12} &= \underset{(1\times8)}{\{F_1\}}\ \underset{(8\times1)}{\{Y\}} \\ J_{21} &= \underset{(1\times8)}{\{F_2\}}\ \underset{(8\times1)}{\{X\}}, \quad & J_{22} &= \underset{(1\times8)}{\{F_2\}}\ \underset{(8\times1)}{\{Y\}} \end{aligned} \qquad \textbf{(13.38)}$$

The $[F]$ matrix is given by

$$\underset{(2\times8)}{[F]} = \begin{bmatrix} \{F_1\} \\ \{F_2\} \end{bmatrix}$$

where

$$\begin{aligned} \underset{(1\times8)}{\{F_1\}} &= \frac{\partial}{\partial r}\{G\} \\ &= \left\{ \frac{\partial G_i}{\partial r} \quad \frac{\partial G_j}{\partial r} \quad \frac{\partial G_k}{\partial r} \quad \frac{\partial G_\ell}{\partial r} \quad \frac{\partial G_m}{\partial r} \quad \frac{\partial G_n}{\partial r} \quad \frac{\partial G_o}{\partial r} \quad \frac{\partial G_p}{\partial r} \right\} \\ \underset{(1\times8)}{\{F_2\}} &= \frac{\partial}{\partial s}\{G\} \\ &= \left\{ \frac{\partial G_i}{\partial s} \quad \frac{\partial G_j}{\partial s} \quad \frac{\partial G_k}{\partial s} \quad \frac{\partial G_\ell}{\partial s} \quad \frac{\partial G_m}{\partial s} \quad \frac{\partial G_n}{\partial s} \quad \frac{\partial G_o}{\partial s} \quad \frac{\partial G_p}{\partial s} \right\} \end{aligned} \qquad \textbf{(13.39)}$$

Equation (13.35) can be used to develop general formulas for these derivatives. The result is

$$\left.\begin{aligned} \frac{\partial G_\beta}{\partial r} &= \frac{1}{4} R_\beta (1 + sS_\beta)(2rR_\beta + sS_\beta) \\ \frac{\partial G_\beta}{\partial s} &= \frac{1}{4} S_\beta (1 + rR_\beta)(rR_\beta + 2sS_\beta) \end{aligned}\right\} \quad \beta = i,j,k,\ell$$

$$\left.\begin{aligned} \frac{\partial G_\beta}{\partial r} &= -r(1+sS_\beta) \\ \frac{\partial G_\beta}{\partial s} &= \frac{1}{2}(1-r^2)S_\beta \end{aligned}\right\} \quad \beta = m,o$$

$$\left.\begin{aligned} \frac{\partial G_\beta}{\partial r} &= \frac{1}{2} R_\beta(1-s^2) \\ \frac{\partial G_\beta}{\partial s} &= -(1+rR_\beta)s \end{aligned}\right\} \quad \beta = n,\ p$$

These give the terms in $[F]$.

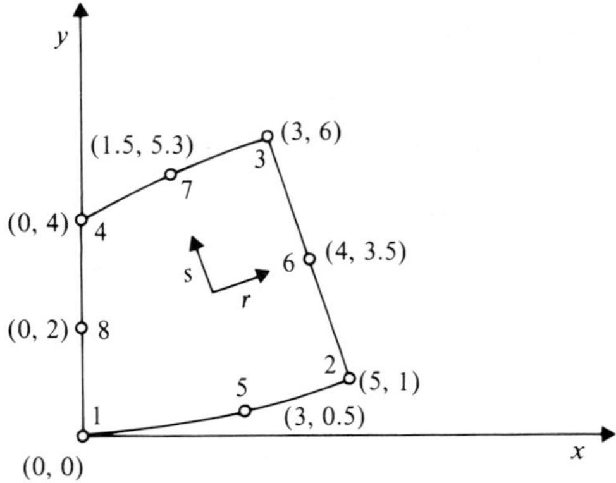

Figure 13.11 Element for Example 13.4—Quadratic quadrilateral element

The inverse of the Jacobian is

$$[J]^{-1} = \frac{1}{\det J}\begin{bmatrix} \dfrac{\partial y}{\partial s} & -\dfrac{\partial y}{\partial r} \\ -\dfrac{\partial x}{\partial s} & \dfrac{\partial x}{\partial r} \end{bmatrix}$$

where the determinant is

$$\det J = \frac{\partial x}{\partial r}\frac{\partial y}{\partial s} - \frac{\partial x}{\partial s}\frac{\partial y}{\partial r}$$

These expressions can be written, from Equation (13.17), as

$$[J]^{-1} = \frac{1}{\det J}\begin{bmatrix} J_{22} & -J_{12} \\ -J_{21} & J_{11} \end{bmatrix}$$

and $\det J = J_{11}J_{22} - J_{12}J_{21}$.

The element matrices are calculated using the same formulas derived in Section 13.4. Each matrix is now of order 8 for the quadratic element rather than of order 4 in the linear element case.

Example 13.4 *Quadratic Quadrilateral Isoparametric Element*

Given A quadratic quadrilateral has the same corners as the element considered in Examples 13.1, 13.2, and 13.3—(0,0), (5,1), (3,6), (0,4). The midpoints are given by (3,0.5), (4,3.5), (1.5,5.3), (0,2). The element is shown in Figure 13.11.

Objective Determine the element matrix for Laplace's equation.

$$[B] = \begin{bmatrix} 1.11 & 0.5159 & 0.5914 & 0.6428 & -0.6914 & -0.4103 & -0.7587 & -0.9997 \\ & 1.0952 & 0.4406 & 0.6753 & -1.0472 & -0.4274 & -0.6311 & -0.6213 \\ & & 1.1996 & 0.6679 & -0.5805 & -0.3897 & -1.2011 & -0.7282 \\ & & & 1.6192 & -0.8065 & -0.6157 & -1.1319 & -1.0511 \\ & & & & 2.5528 & -0.2784 & 0.7330 & 0.1183 \\ & & & & & 1.9161 & -0.1262 & 0.3317 \\ & & & & & & 2.4875 & 0.6286 \\ \text{(Symmetric)} & & & & & & & 12.3217 \end{bmatrix}$$

Figure 13.12 Element matrix for Example 13.4

Solution Using four sample points

$$\begin{aligned}[B] = {} & W(R_1)W(S_1)[D]_1^T[D]_1\det J_1 + W(R_2)W(S_2)[D]_2^T[D]_2\det J_2 \\ & + W(R_3)W(S_3)[D]_3^T[D]_3\det J_3 \\ & + W(R_4)W(S_4)[D]_4^T[D]_4\det J_4\end{aligned}$$

where $[D]^T[D]$ is an 8×8 matrix now. The element matrix is given in Figure 13.12.

13.7 Problems

13.1 A linear quadrilateral isoparametric element has corner nodes with global coordinates $(-1,0)$, $(0,-1)$, $(1,1)$, $(0,2)$. Find the coordinate transformation between the global and local coordinates. Use it to obtain the global coordinates x,y of the point $r = 0.5$, $s = 0.5$ of the element in local coordinates.

13.2 Evaluate the element derivative matrix $[D]$ for the linear quadrilateral element from Problem 13.1. Determine the numerical value for the point $r = 0.5$, $s = 0.5$.

13.3 The concept of an isoparametric element can be used in one dimension as well as in two or three dimensions. Consider a quadratic element for a one-dimensional problem with node points at X_i, X_j, X_k. Obtain the shape functions $G_i(r)$, $G_j(r)$, $G_k(r)$ in the local coordinate system r where the element extends from $r = -1$ to $r = 1$. If the middle node point is located in the center of the element, determine the element matrix to solve the differential equation

$$\frac{d^2T}{dx^2} + 2T = 0$$

using a two-point Gauss integration scheme.

13.4 A linear quadrilateral isoparametric element has corner nodes with global coordinates $(0,0)$, $(2,0)$, $(2,1)$, $(-1,2)$. Obtain the coordinate transformation from local to global coordinates.

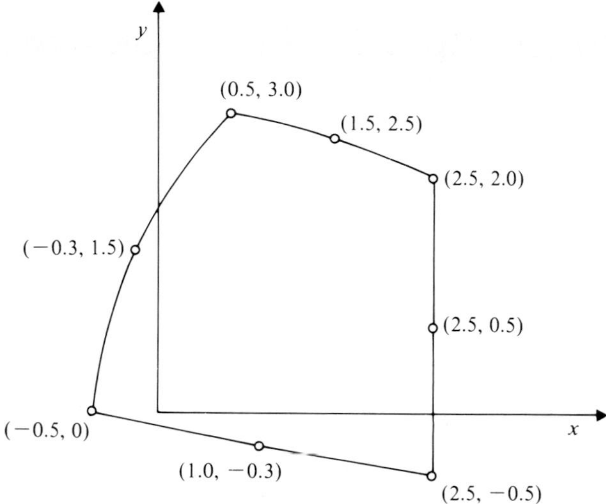

Figure 13.13 Quadratic quadrilateral for Problem 13.5

13.5 Determine the element matrix for the element described above. The differential equation is

$$\frac{\partial^2 T}{\partial x^2} + \frac{\partial^2 T}{\partial y^2} = 0$$

with the boundary condition $T = T_0$ on L. Use a four-point Gauss quadrature numerical integration scheme.

13.6 A quadratic quadrilateral isoparametric element is shown in Figure 13.13. Find the coordinate transformation between the local and global coordinates. Determine the global coordinates of the point $r = 0, s = 0$.

13.7 Modify computer program FINTWO so that it employs linear quadrilateral isoparametric elements rather than triangular ones. Use the resulting program to solve Example 10.2 with four quadrilateral elements. Compare the nodal temperatures and heat flows to those in Example 10.2. (Note: This problem requires quite a bit of programming but will provide the programmer with a potentially useful finite element analysis package.)

13.8 Do the previous problem, but rewrite computer program FINTWO to accept either triangular elements or quadratic ones. (Note: Both of the above comments are more strongly true.)

Chapter 14

Deformation–Stress Analysis in Solids

14.1 Introduction

A major difference exists between the scalar thermal and ideal flow problems considered so far and deformation–stress analysis to be presented in this chapter. The vector nature of the field variables—displacements, strains, and stresses—enters into the equations. Thus the independent coordinates increase (say from x to x,y,z in three dimensions) but the dependent variables increase as well (say displacements from u to u,v,w in three dimensions). The displacements in one direction are related by Poisson's law to those in another direction.

A. Two-Dimensional Analysis

Any real body has a three-dimensional nature to it, as do the displacements, strains, and stresses. However, in many bodies, the geometry of the body and the applied displacements and forces are such that the analysis is largely two-dimensional. A simplified analysis is the result. These simple cases are used in this chapter to illustrate the methods of finite element stress analysis without all of the terms involved in full three-dimensional problems.

Four two-dimensional special cases can typically be considered:

1. Plane stress
2. Plane strain
3. Axial symmetry
4. Radial strain

The details of two of these special case analyses will be developed in the remainder of this chapter, but the basic characteristics are given in Table 14.1. The radial strain case is often analyzed as a plane strain problem in cartesian x,y coordinates rather than polar r,θ coordinates.

Table 14.1
Special Cases for Two-Dimensional Deformation–Stress Analysis

Type	Deformation–Stress Assumption	Coordinate System	Geometry of Body	Applied Displacements and Forces
Plane stress	Stress varies with x,y only	x,y	Thin plate	On edges of plate
Plane strain	Displacement and strain vary with x,y only	x,y	Very long body	Independent of z
Axial symmetry	Displacement and strain vary with r,z only	r,z	Solid of revolution	Independent of θ
Radial strain	Displacement and strain vary with r,θ only	r,θ	Very long body	Independent of z

B. Formulation Methods

Several approaches to formulating finite element stress analysis are currently used: displacement, force (stress), and mixed. In the displacement method, the form of the displacement is assumed within an element, and all equations are developed from this. Virtual displacements may be used to derive the principle of minimum potential energy. The displacement method assumes displacement functions such that internal continuity and compatibility between elements (across interelement boundaries) are assured. Thus it is sometimes called the compatibility method, and the elements called compatible elements. In the force method, the complementary energy is used with an assumed stress field rather than assumed displacements. These stress functions are chosen to satisfy force equilibrium equations instead of compatibility. Finally, Reissner's principle can be employed, with elements assuming a combination of displacements and stress functions for a mixed approach.

Generally, the displacement or compatibility method is the easiest to apply and is by far the most widely used. Some problems are much better suited to the other approaches, but these are discussed in books specializing in elasticity applications. Only the displacement method is given here.

14.2 Displacements and Strains

A. Displacements in Two Dimensions

Any general three-dimensional body subject to externally applied displacements and forces will deform. The displacements of each point in the body away from the original position are labeled u,v,w. Only very small displacements are considered in this text such that linear stress-strain relations (see Section 14.3) hold. As the general three-dimensional displacement field is somewhat complicated, the two-dimensional case is considered first. The full three-dimensional displacement case is discussed at the end of this section. Stresses and stress–strain laws are presented in the next section.

Consider a two-dimensional solid (actually a solid that is infinitely long in the z direction) with displacements u,v in the two coordinate directions x,y. This corresponds to the case of plane strain where the displacements and strains are all in the x,y plane. The displacements are normally written in vector form as

$$u = \begin{Bmatrix} u \\ v \end{Bmatrix} \tag{14.1}$$

Displacement vector for plane strain (2×1)

B. Strains in Two Dimensions

The strains ϵ in plane strain are

$$\left.\begin{aligned} \epsilon_{xx} &= \frac{\partial u}{\partial x} \\ \epsilon_{yy} &= \frac{\partial v}{\partial y} \end{aligned}\right\} \quad \text{normal strains} \tag{14.2}$$

$$\epsilon_{xy} = \epsilon_{yx} = \frac{1}{2}\left(\frac{\partial u}{\partial y} + \frac{\partial v}{\partial x}\right) \quad \Big\} \quad \text{shearing strains}$$

Although there are four strain components, $\epsilon_{xy} = \epsilon_{yx}$, so only three are independent. Rigid body motions, both translations and rotations, are neglected.

The unit (or "true") shear strain γ is normally used in engineering and will be employed in this text. The unit shear is defined as

$$\gamma_{xy} = \gamma_{yx} = \frac{\partial u}{\partial y} + \frac{\partial v}{\partial x} \quad \Big\} \quad \text{unit shear strain} \tag{14.3}$$

These are just twice the shear strains ϵ_{xy}, ϵ_{yx}. A unit shear strain represents the actual angular deformation (change in angle between sides of the differential rectangle), but the shear strain represents one-half of this. Figure 14.1 shows the normal and shearing strains in the xy plane.

A more detailed analysis yields a derivation for the strain relations just presented. Figure 14.2 shows an undeformed differential element $ABCD$, shown by the dashed lines, and its deformed shape $A'B'C'D'$, shown by the solid lines. The normal strain ϵ_{xx} is defined as the change in length of side AB divided by its original length

$$\epsilon_{xx} = \frac{A'B' - AB}{AB}$$

The original length is $AB = dx$. The displacement at A' is u, while at B' it is expanded in a Taylor's series as

$$u_{B'} = u + \frac{\partial u}{\partial x}\,dx$$

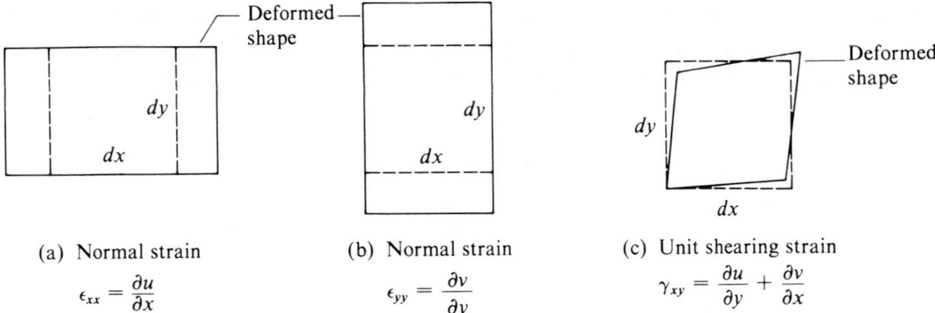

Figure 14.1 Strains in xy plane

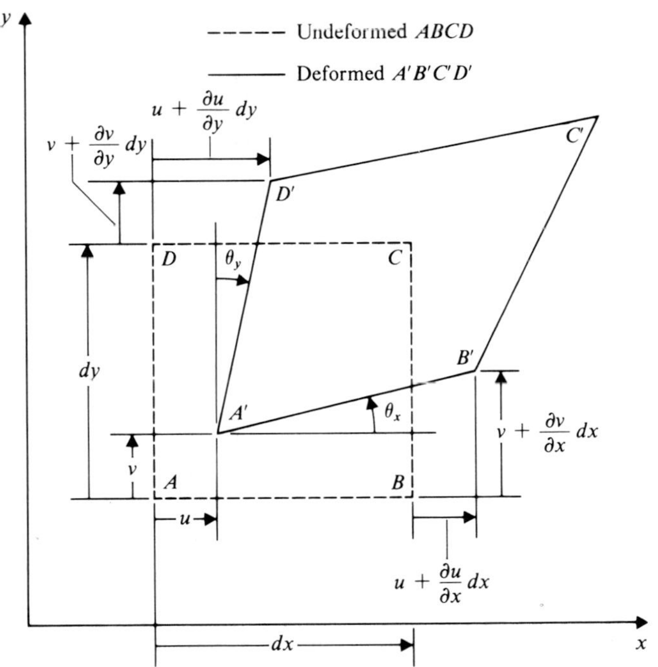

Figure 14.2 Deformation of differential rectangle in two dimensions

Evaluating the normal strain yields

$$\epsilon_{xx} = \frac{\left[dx + \left(u + \dfrac{\partial u}{\partial x}\,dx\right) - u\right] - dx}{dx} = \frac{\partial u}{\partial x}$$

The other normal strain is

$$\epsilon_{yy} = \frac{A'D' - AD}{AD}$$

$$= \frac{\left[dy + \left(v + \dfrac{\partial v}{\partial y}\,dy\right) - v\right] - dy}{dy} = \frac{\partial v}{\partial y}$$

by a similar analysis. Determining the angular deformation $\theta_x + \theta_y$ gives the unit shear strain $\gamma_{xy} = \theta_x + \theta_y$. For small angles,

$$\gamma_{xy} = \frac{B'B - A'A}{A'B'} + \frac{D'D - A'A}{A'D'}$$

$$= \frac{\left(v + \dfrac{\partial v}{\partial x}\,dx\right) - v}{\left[dx + \left(u + \dfrac{\partial u}{\partial x}\,dx\right) - u\right]} + \frac{\left(u + \dfrac{\partial u}{\partial y}\,dy\right) - u}{\left[dy + \left(v + \dfrac{\partial v}{\partial y}\,dy\right) - v\right]}$$

Finally, for small dx and dy

$$\gamma_{xy} = \frac{\partial v}{\partial x} + \frac{\partial u}{\partial y}$$

which completes the derivations.

In finite element analysis, the strains are written in column form as

$$\{\epsilon\} = \begin{Bmatrix} \epsilon_{xx} \\ \epsilon_{yy} \\ \gamma_{xy} \end{Bmatrix} \tag{14.4}$$

where γ_{yx} is simply equal to γ_{xy}, as noted earlier. Often elasticity textbooks write strains in tensor form, but that is not as convenient for computer implementation as the form above. In terms of derivatives of displacements, this is

$$\{\epsilon\} = \begin{Bmatrix} \dfrac{\partial u}{\partial x} \\ \dfrac{\partial v}{\partial y} \\ \dfrac{\partial u}{\partial y} + \dfrac{\partial v}{\partial x} \end{Bmatrix} \tag{14.5}$$

If the displacement field u,v is specified, the strains are easily found. This may also be written as

$$\underset{(3\times1)}{\{\epsilon\}} = \underset{(3\times2)}{\begin{bmatrix} \dfrac{\partial}{\partial x} & 0 \\ 0 & \dfrac{\partial}{\partial y} \\ \dfrac{\partial}{\partial y} & \dfrac{\partial}{\partial x} \end{bmatrix}} \underset{(2\times1)}{\begin{Bmatrix} u \\ v \end{Bmatrix}} \tag{14.6}$$

which applies for any two-dimensional element in solid mechanics.

C. Simplex Element in Two Dimensions

In a simplex element, the displacement column is related to the shape functions and nodal displacements by the relation (3.44)

$$\{u\} = [N]\,\{U\} \tag{14.7}$$

where

$$[N] = \underset{\text{Shape function matrix } (2\times6)}{\begin{bmatrix} N_i & 0 & N_j & 0 & N_k & 0 \\ 0 & N_i & 0 & N_j & 0 & N_k \end{bmatrix}}$$

$$\{U\} = \begin{Bmatrix} u_i \\ v_i \\ u_j \\ v_j \\ u_k \\ v_k \end{Bmatrix} = \underset{\text{Element displacement column } (6\times1)}{\begin{Bmatrix} U_{2i-1} \\ U_{2i} \\ U_{2j-1} \\ U_{2j} \\ U_{2k-1} \\ U_{2k} \end{Bmatrix}}$$

The nodal displacements illustrated in Figure 14.3 are placed in column form.

The element strain is (14.6)

$$\{\epsilon\} = \begin{bmatrix} \dfrac{\partial}{\partial x} & 0 \\ 0 & \dfrac{\partial}{\partial y} \\ \dfrac{\partial}{\partial y} & \dfrac{\partial}{\partial x} \end{bmatrix} \{u\} \tag{14.8}$$

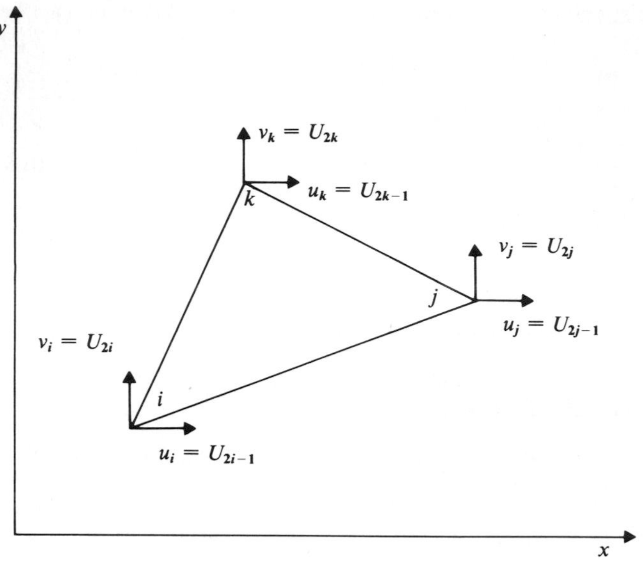

Figure 14.3 Simplex element for plane stress or plane strain. Label nodes counterclockwise.

However, $\{u\} = [N]\{U\}$, so this becomes

$$\{\epsilon\} = \begin{bmatrix} \dfrac{\partial}{\partial x} & 0 \\ 0 & \dfrac{\partial}{\partial y} \\ \dfrac{\partial}{\partial y} & \dfrac{\partial}{\partial x} \end{bmatrix} \begin{bmatrix} N_i & 0 & N_j & 0 & N_k & 0 \\ 0 & N_i & 0 & N_j & 0 & N_k \end{bmatrix} \{U\}$$

Let the element derivative matrix $[D]$ be defined by the expression

$$\{\epsilon\} = [D]\{U\} \tag{14.9}$$

where

$$[D] = \begin{bmatrix} \dfrac{\partial}{\partial x} & 0 \\ 0 & \dfrac{\partial}{\partial y} \\ \dfrac{\partial}{\partial y} & \dfrac{\partial}{\partial x} \end{bmatrix} \begin{bmatrix} N_i & 0 & N_j & 0 & N_k & 0 \\ 0 & N_i & 0 & N_j & 0 & N_k \end{bmatrix}$$

Element derivative matrix (3×6)

This definition of the element derivative matrix is consistent with usage in Section 3.3 on two-dimensional element properties and Section 4.4 on bar elements.

Evaluating all of the terms in the derivative matrix for a simplex element yields

$$[D] = \frac{1}{2A}\begin{bmatrix} b_i & 0 & b_j & 0 & b_k & 0 \\ 0 & c_i & 0 & c_j & 0 & c_k \\ c_i & b_i & c_j & b_j & c_k & b_k \end{bmatrix} \quad \textbf{(14.10)}$$

Element derivative matrix (3×6)

It is easily seen that the derivative matrix is independent of x,y. Thus the simplex element is a constant strain element.

Example 14.1 Two-Dimensional Simplex Element

Given A simplex element, shown in Figure 14.4, has nodal coordinates

$$\{X\} = \begin{Bmatrix} 0.0 \\ 2.0 \\ 0.0 \end{Bmatrix}, \qquad \{Y\} = \begin{Bmatrix} 0.0 \\ 0.0 \\ 1.0 \end{Bmatrix}$$

where the units are inches. The displacements are given as

$$\begin{Bmatrix} u_4 \\ v_4 \\ u_5 \\ v_5 \\ u_2 \\ v_2 \end{Bmatrix} = \begin{Bmatrix} 0.0 \\ 0.0 \\ -2.0 \\ -2.0 \\ 0.0 \\ 3.0 \end{Bmatrix} \times 10^{-4} \text{ (inches)}$$

Objective Determine the element strains.

Solution The nodal displacement column is

$$\{U\} = \begin{Bmatrix} U_7 \\ U_8 \\ U_9 \\ U_{10} \\ U_3 \\ U_4 \end{Bmatrix} = \begin{Bmatrix} 0.0 \\ 0.0 \\ -2.0 \\ -2.0 \\ 0.0 \\ 3.0 \end{Bmatrix} \times 10^{-4} \text{ (inches)}$$

Evaluating the element parameters b, c and area A in the standard manner, as in Section 3.4, gives the element derivative matrix as (14.10).

$$D = \frac{1}{2(1)}\begin{bmatrix} -1 & 0 & 1 & 0 & 0 & 0 \\ 0 & -2 & 0 & 0 & 0 & 2 \\ -2 & -1 & 0 & 1 & 2 & 0 \end{bmatrix}$$

The element strain is then (14.9)

$$\{\epsilon\} = [D]\{U\}$$

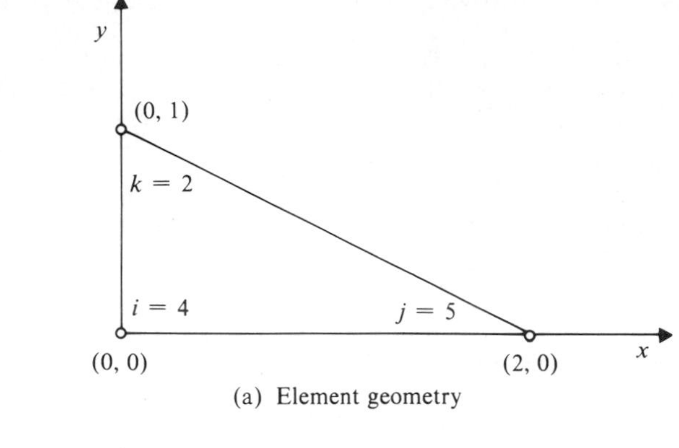

(a) Element geometry

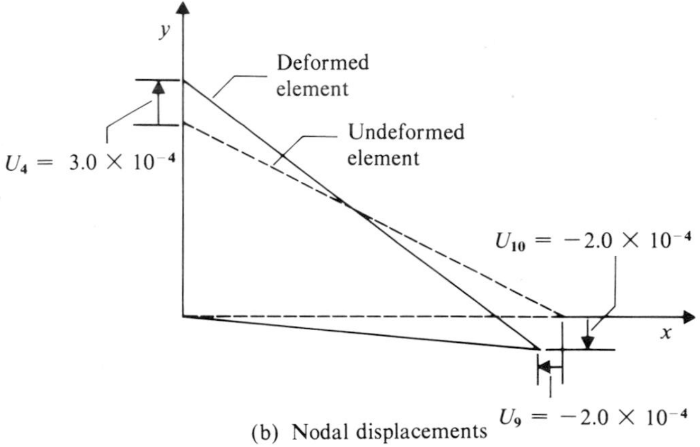

(b) Nodal displacements

Figure 14.4 Element geometry and nodal displacements (Example 14.1—Two-dimensional simplex element)

which yields

$$\begin{Bmatrix} \epsilon_{xx} \\ \epsilon_{yy} \\ \gamma_{xy} \end{Bmatrix} = \begin{Bmatrix} -1.0 \\ 3.0 \\ -1.0 \end{Bmatrix} \times 10^{-4} \text{ (inch/inch)}$$

Again note that the strains are constant within the element.

Figure 14.5 shows the deformed element and an infinitesimally small differential square in that element of dimensions dx by dy. It is easy to see that compression of the element in the x direction produces the negative strain $\epsilon_{xx} = -1.0 \times 10^{-4}$. In the y direction, the element is elongated, giving a positive strain $\epsilon_{yy} = 3.0 \times 10^{-4}$. The shear strain $\epsilon_{xy} = -1.0 \times 10^{-4}$ tends to displace element side 1 (between nodes 4 and 5) so that it forms a negative angle with respect to the x axis. This is negative shear strain.

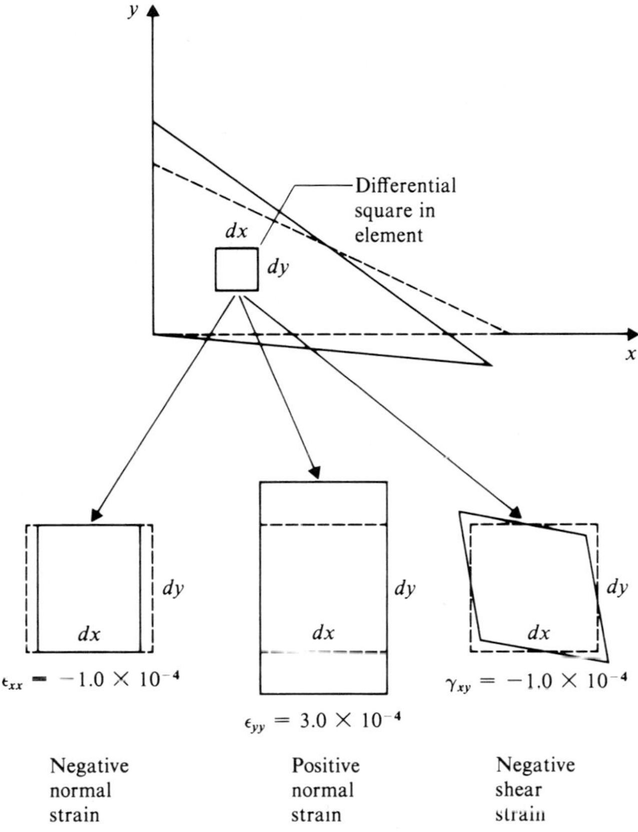

Figure 14.5 Strains in element (Example 14.1—Two-dimensional simplex element)

D. Strains in Three Dimensions

In three dimensions, the displacements are written in vector form as

$$\{u\} = \begin{Bmatrix} u \\ v \\ w \end{Bmatrix} \tag{14.11}$$

Figure 14.6 shows the coordinates and displacements for a differential cube.

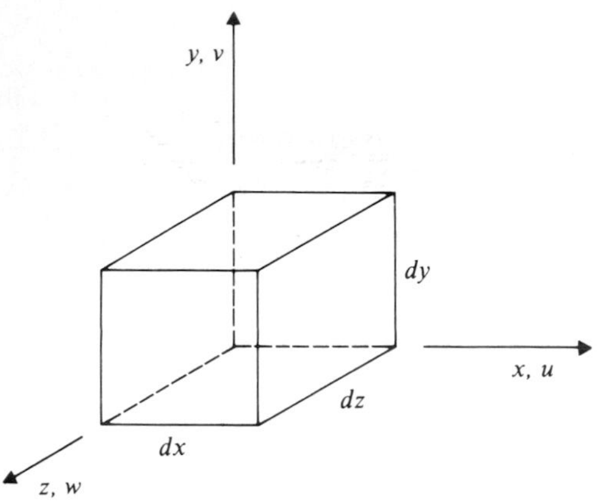

Figure 14.6 Differential cube in three dimensions

The strains are given by the relations

$$\left.\begin{aligned}\epsilon_{xx} &= \frac{\partial u}{\partial x}\\ \epsilon_{yy} &= \frac{\partial v}{\partial y}\\ \epsilon_{zz} &= \frac{\partial w}{\partial z}\end{aligned}\right\}\quad\text{normal strains}$$

$$\left.\begin{aligned}\gamma_{xy} &= \gamma_{yx} = \frac{\partial u}{\partial y} + \frac{\partial v}{\partial x}\\ \gamma_{yz} &= \gamma_{zy} = \frac{\partial v}{\partial z} + \frac{\partial w}{\partial y}\\ \gamma_{zx} &= \gamma_{xz} = \frac{\partial w}{\partial x} + \frac{\partial u}{\partial z}\end{aligned}\right\}\quad\text{unit shear strains} \tag{14.12}$$

As in the two-dimensional case, the unit shear strains are used, rather than the shear strains. Only six of the nine strain components are independent, as shown by this equation—the shear strains are related to one another.

In finite elements, these strains are written in column form as

$$\{\epsilon\} = \begin{Bmatrix}\epsilon_{xx}\\ \epsilon_{yy}\\ \epsilon_{zz}\\ \gamma_{xy}\\ \gamma_{yz}\\ \gamma_{zx}\end{Bmatrix} \tag{14.13}$$

Strain column in three dimensions (6×1)

In terms of derivatives of displacements, these are

$$\{\epsilon\} = \begin{Bmatrix} \dfrac{\partial u}{\partial x} \\ \dfrac{\partial v}{\partial y} \\ \dfrac{\partial w}{\partial z} \\ \dfrac{\partial v}{\partial x} + \dfrac{\partial u}{\partial y} \\ \dfrac{\partial w}{\partial y} + \dfrac{\partial v}{\partial z} \\ \dfrac{\partial u}{\partial z} + \dfrac{\partial w}{\partial x} \end{Bmatrix} \qquad \textbf{(14.14)}$$

or in matrix form as

$$\{\epsilon\} = \underset{(6\times 3)}{\begin{bmatrix} \dfrac{\partial}{\partial x} & 0 & 0 \\ 0 & \dfrac{\partial}{\partial y} & 0 \\ 0 & 0 & \dfrac{\partial}{\partial z} \\ \dfrac{\partial}{\partial y} & \dfrac{\partial}{\partial x} & 0 \\ 0 & \dfrac{\partial}{\partial z} & \dfrac{\partial}{\partial y} \\ \dfrac{\partial}{\partial z} & 0 & \dfrac{\partial}{\partial x} \end{bmatrix}} \underset{(3\times 1)}{\{u\}} \qquad \textbf{(14.15)}$$

This general form can be used for any three-dimensional element, once the shape functions are specified for the displacements.

E. Simplex Element in Three Dimensions

The element displacements for a simplex three-dimensional element are given by $\{u\} = [N]\{U\}$ where the terms are written out in Section 3.5. Following the methods just developed for two dimensions, the element strains are

$$\{\epsilon\} = [D]\{U\} \qquad \textbf{(14.16)}$$

where the element derivative matrix can be evaluated from the expression in Figure 14.7.

$$[D] = \begin{bmatrix} \frac{\partial}{\partial x} & 0 & 0 \\ 0 & \frac{\partial}{\partial y} & 0 \\ 0 & 0 & \frac{\partial}{\partial x} \\ \frac{\partial}{\partial y} & \frac{\partial}{\partial x} & 0 \\ 0 & \frac{\partial}{\partial z} & \frac{\partial}{\partial y} \\ \frac{\partial}{\partial z} & 0 & \frac{\partial}{\partial x} \end{bmatrix} \begin{bmatrix} N_i & 0 & 0 & N_j & 0 & 0 & N_k & 0 & 0 \\ 0 & N_i & 0 & 0 & N_j & 0 & 0 & N_k & 0 \\ 0 & 0 & N_i & 0 & 0 & N_j & 0 & 0 & N_k \end{bmatrix}$$

Figure 14.7 Element derivative matrix in three-dimensional simplex element

14.3 Stresses and Stress–Strain Law

A. Stresses in Two Dimensions

When a body is deformed, stresses occur as well as displacements and strains. The internal stresses in a three-dimensional body have nine components similar to the nine strain components. As with the strains, it is easier to consider a two-dimensional stress field before discussing the three-dimensional analysis. This is then related to the strains by Hooke's law.

For two-dimensional states of stress, the stress components are

$$\left.\begin{matrix} \sigma_{xx} \\ \sigma_{yy} \end{matrix}\right\} \quad \text{normal stresses}$$

$$\tau_{xy} = \tau_{yx}\} \quad \text{shear stresses}$$

Figure 14.8(a) shows a differential square, with the stresses shown on each side. The first subscript indicates the direction normal to the face, while the second subscript indicates the direction of the force. Only three of the four stresses are independent.

The stress column is

$$\{\sigma\} = \begin{Bmatrix} \sigma_{xx} \\ \sigma_{yy} \\ \tau_{xy} \end{Bmatrix} \tag{14.17}$$

Stress column in two dimensions (3×1)

This is similar in format to the strain column. Two special cases—plane stress and plane strain—are now considered.

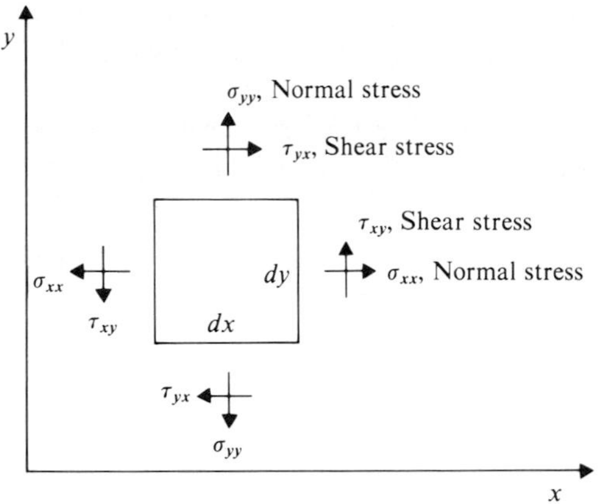

Figure 14.8 Stresses on differential square in two dimensions

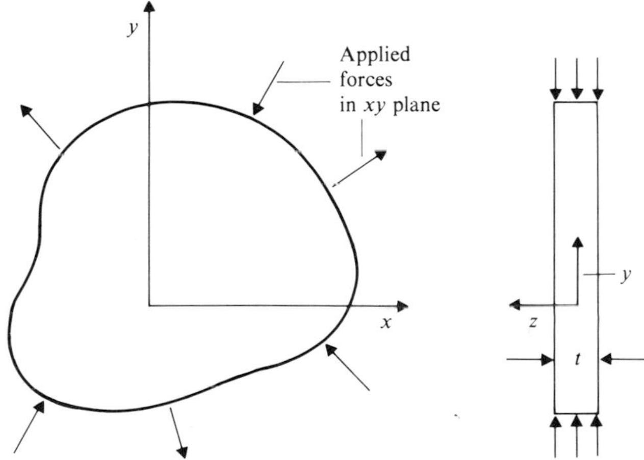

Figure 14.9 Thin plate in plane stress. No stresses can occur in z direction because plate is free to expand or contract.

B. Stress–Strain Law for Plane Stress

As initially indicated in Section 14.1, plane stress occurs in a very thin plate as shown in Figure 14.9. Let t denote the thickness of the plate in the z direction, where t is very small compared to the dimensions of the plate in either the x or y direction. It is assumed that the thin plate is free to expand or contract in the z direction as much as it wishes to. Thus no stresses can occur in the z direction.

$$\left.\begin{aligned} \sigma_{zz} &= 0 \\ \tau_{xz} &= \tau_{zx} = 0 \\ \tau_{yz} &= \tau_{zy} = 0 \end{aligned}\right\} \quad \text{plane stress in thin plate}$$

This is the key assumption for plane stress.

Materials that are linearly elastic and isotropic obey Hooke's law for small deformations. Hooke's law states that the stresses are linearly related to the strains. A derivation of this law is summarized here.

Let a differential square of a thin plate be subjected to a uniform normal stress σ_{xx} on the two faces shown in Figure 14.10(b). Thus the state of stress in the xy plane is

$$\left.\begin{aligned}\sigma_{xx} &= \sigma_{xx}\\ \sigma_{yy} &= 0\\ \tau_{xy} &= 0\end{aligned}\right\} \quad \text{applied normal stress } \sigma_{xx} \tag{14.18}$$

Also the normal stress in the z direction σ_{zz} is zero (the assumption of plane stress). The resulting normal strain is related to the normal stress by

$$\epsilon_{xx} = \frac{1}{E}\sigma_{xx}$$

where E = Young's modulus. Also, the rectangle contracts in the y direction with the normal strain

$$\epsilon_{yy} = -\frac{\nu}{E}\sigma_{xx}$$

where ν = Poisson's ratio. With this applied stress distribution, the shear strain is zero: $\gamma_{xy} = 0$. Thus, the state of strain in the xy plane is

$$\begin{aligned}\epsilon_{xx} &= \frac{1}{E}\sigma_{xx}\\ \epsilon_{yy} &= -\frac{\nu}{E}\sigma_{xx}\\ \gamma_{xy} &= 0\end{aligned} \tag{14.19}$$

Finally, the thin plate is free to deform in the z direction. It contracts to produce the strain

$$\epsilon_{zz} = -\frac{\nu}{E}\sigma_{xx} \tag{14.20}$$

This completes the state of deformation–stress for the applied stress σ_{xx} alone.

Now applying a uniform normal stress σ_{yy} in the y direction, shown in Figure 14.10(c), gives the states of stress as

$$\left.\begin{aligned}\sigma_{xx} &= 0\\ \sigma_{yy} &= \sigma_{yy}\\ \tau_{xy} &= 0\end{aligned}\right\} \quad \text{applied normal stress } \sigma_{yy} \tag{14.21}$$

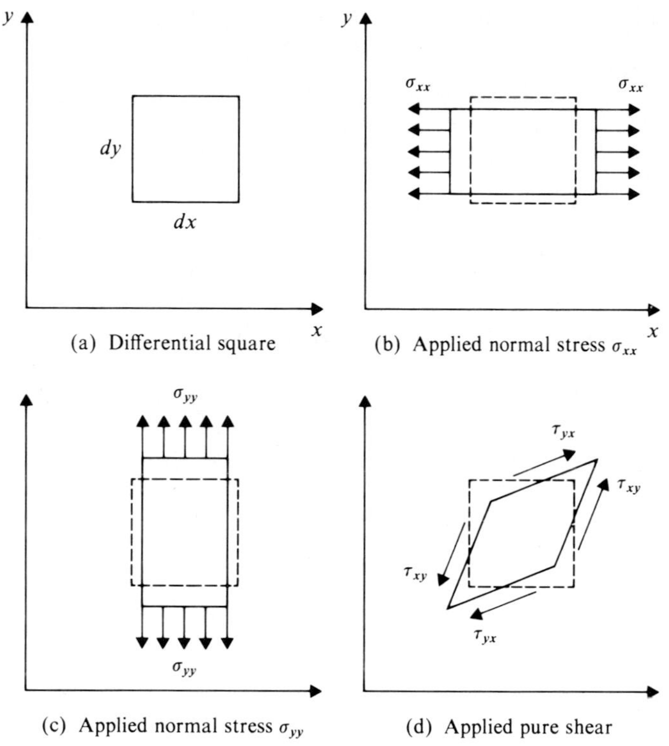

Figure 14.10 Differential square for plane stress

The resulting strains in the xy plane are

$$\epsilon_{xx} = -\frac{\nu}{E}\sigma_{yy}$$

$$\epsilon_{yy} = \frac{1}{E}\sigma_{yy} \tag{14.22}$$

$$\gamma_{xy} = 0$$

by an analysis similar to that just completed. Also, the thin plate deforms in the z direction to give the strain

$$\epsilon_{zz} = -\frac{\nu}{E}\sigma_{yy} \tag{14.23}$$

Applying a state of pure shear stress

$$\left.\begin{aligned} \sigma_{xx} &= 0 \\ \sigma_{yy} &= 0 \\ \tau_{xy} &= \tau_{xy} \end{aligned}\right\} \quad \text{applied pure shear } \tau_{xy} \tag{14.24}$$

as shown in Figure 14.10(d). The resulting state of strain in the xy plane is

$$\begin{aligned} \epsilon_{xx} &= 0 \\ \epsilon_{yy} &= 0 \\ \gamma_{xy} &= \frac{2(1+\nu)}{E}\tau_{xy} \end{aligned} \tag{14.25}$$

No contraction is produced in the z direction by pure shear.

Hooke's law can now be written for plane stress in a thin plate. If all three states of stress (14.18), (14.21), (14.24)—σ_{xx}, σ_{yy}, τ_{xy}—are applied to the thin plate at the same time, the resulting state of strain in the xy plane is obtained by linear superposition of (14.19), (14.22), (14.25)

$$\left.\begin{aligned} \epsilon_{xx} &= \frac{1}{E}[\sigma_{xx} - \nu\,\sigma_{yy}] \\ \epsilon_{yy} &= \frac{1}{E}[-\nu\,\sigma_{xx} + \sigma_{yy}] \\ \gamma_{xy} &= \frac{2(1+\nu)}{E}\tau_{xy} \end{aligned}\right\} \text{ plane stress} \tag{14.26}$$

Usually these relations are solved for the stresses as

$$\left.\begin{aligned} \sigma_{xx} &= \frac{E}{1-\nu^2}[\epsilon_{xx} + \nu\,\epsilon_{yy}] \\ \sigma_{yy} &= \frac{E}{1-\nu^2}[\nu\,\epsilon_{xx} + \epsilon_{yy}] \\ \tau_{xy} &= \frac{E}{2(1+\nu)}\gamma_{xy} \end{aligned}\right\} \text{ plane stress} \tag{14.27}$$

This is usually written in matrix form as

$$\{\sigma\} = [E]_{\substack{\text{plane}\\ \text{stress}}}\{\epsilon\} \tag{14.28}$$

where

$$[E]_{\substack{\text{plane}\\ \text{stress}}} = \underbrace{\frac{E}{1-\nu^2}\begin{bmatrix} 1 & \nu & 0 \\ \nu & 1 & 0 \\ 0 & 0 & \frac{1-\nu}{2} \end{bmatrix}}_{\substack{\text{Property matrix for plane}\\ \text{stress } (3\times 3)}}$$

This is the matrix that relates the stress to the strain for a plane stress problem. It represents Hooke's law for linear isotropic elastic materials.

A secondary equation is the relation for the strain in the z direction, obtained from superposing (14.20) and (14.23) as

$$\left.\epsilon_{zz} = -\frac{\nu}{E}(\sigma_{xx} + \sigma_{yy})\right\} \quad \text{plane stress} \tag{14.29}$$

This relation may be used after the displacements, strains, and stresses are determined in the xy plane.

In a plane stress element, the stress is given by (14.9)

$$\{\sigma\} = [E]_{\substack{\text{plane}\\\text{stress}}}\{\epsilon\} = [E]_{\substack{\text{plane}\\\text{stress}}}[D]\{U\} \tag{14.30}$$

This expression applies to any element with the appropriate property matrix $[E]$ for that element. A simplex element is a constant strain element, so it must also be a constant stress element. Example 14.2 (presented just after the plane strain development) illustrates the use of this equation for both plane stress and plane strain.

C. Stress–Strain Law for Plane Strain

Now an alternative geometry—plane strain—is considered. In this case, let the thickness of the body in the z direction be very long (infinite). Figure 14.11 shows some example problems, hydrostatic pressure on a long dam and applied force on a long solid body. Once again a thickness t is used in the analysis, but here it is a unit thickness—used just so that the plane strain problem has the same units as the plane stress problem. The applied displacements and forces must be constant in the z direction, so that the analysis is basically two-dimensional in nature. Then no displacement occurs in the z direction and no derivatives of displacement exist in the z direction.

$$\left.\begin{aligned} w &= 0 \\ \frac{\partial}{\partial z} &= 0 \end{aligned}\right\} \quad \text{plane strain} \tag{14.31}$$

These are the key assumptions for plane strain.

The strains in a general three-dimensional body are given by (14.12). Because the displacement w is zero, the normal strain is

$$\left.\epsilon_{zz} = \frac{\partial \overset{0}{\cancel{w}}}{\cancel{\partial} z} = 0\right\} \quad \text{plane strain in very long body}$$

Also, because the derivatives of displacement $\partial/\partial z$ are zero, two shearing strains involving the z direction are

$$\left.\begin{aligned} \gamma_{yz} = \gamma_{zy} &= \frac{\partial \overset{0}{\cancel{v}}}{\cancel{\partial} z} + \frac{\partial \overset{0}{\cancel{w}}}{\cancel{\partial} y} = 0 \\ \gamma_{zx} = \gamma_{xz} &= \frac{\partial \overset{0}{\cancel{w}}}{\cancel{\partial} x} + \frac{\partial \overset{0}{\cancel{u}}}{\cancel{\partial} z} = 0 \end{aligned}\right\} \quad \begin{array}{l}\text{plane strain}\\ \text{in very}\\ \text{long body}\end{array}$$

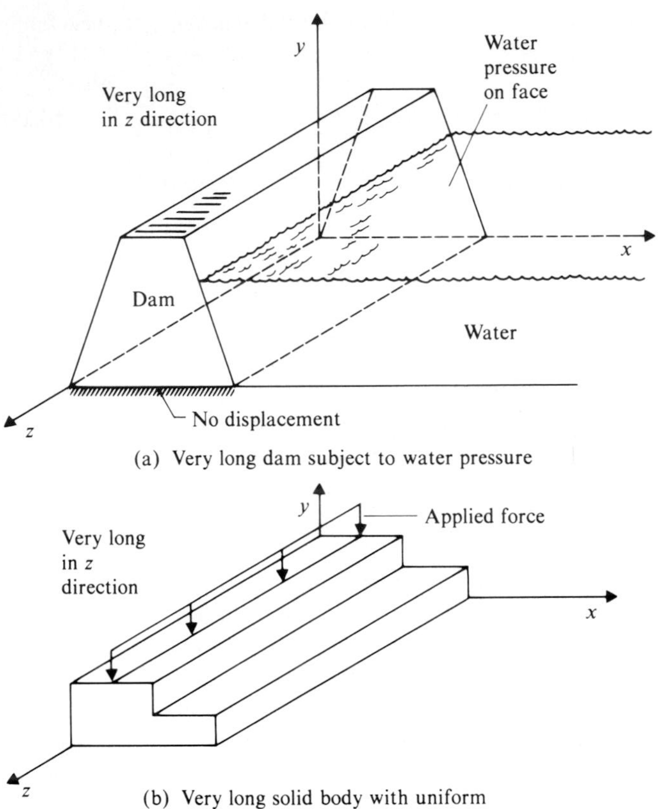

Figure 14.11 Examples of plane strain. Strains occur only in *xy* plane because body is very long.

The remaining nonzero strains are

$$\{\epsilon\} = \begin{Bmatrix} \epsilon_{xx} \\ \epsilon_{yy} \\ \gamma_{xy} \end{Bmatrix}$$

Strain column
for plane strain
(3×1)

It is easily seen that all of the strains occur in the xy plane, thus leading to the name *plane strain.* Note that unlike the plane stress problem, where the stress σ_{zz} is zero, the strain ϵ_{zz} vanishes for plane strain. However, the stress σ_{zz} is not zero. This may be thought of as the stress applied at the ends of a long body to keep it from having any displacement or strain in the z direction. If walls exist at each end of the body which prevent any z displacement, the stress σ_{zz} would be taken by the walls.

A linearly elastic, isotropic material undergoing small deformations is considered again. The stress–strain law is formulated in the same manner as just employed for plane stress, with the superposition of two normal stresses and a shear stress. The resulting stress–strain relations are

$$\left.\begin{aligned} \epsilon_{xx} &= \frac{1}{E}\,[\sigma_{xx} - \nu\,\sigma_{yy} - \nu\,\sigma_{zz}] \\ \epsilon_{yy} &= \frac{1}{E}\,[-\nu\,\sigma_{xx} + \sigma_{yy} - \nu\,\sigma_{zz}] \\ \gamma_{xy} &= \frac{1+\nu}{E}\,\tau_{xy} \end{aligned}\right\} \text{plane strain} \tag{14.32}$$

These are solved for the stresses to give

$$\left.\begin{aligned} \sigma_{xx} &= \frac{E}{(1+\nu)(1-2\nu)}\,[(1-\nu)\,\epsilon_{xx} + \nu\,\epsilon_{yy}] \\ \sigma_{yy} &= \frac{E}{(1+\nu)(1-2\nu)}\,[\nu\,\epsilon_{xx} + (1-\nu)\,\epsilon_{yy}] \\ \tau_{xy} &= \frac{E}{2(1+\nu)}\,\gamma_{xy} \end{aligned}\right\} \begin{array}{l}\text{plane}\\ \text{strain}\end{array} \tag{14.33}$$

The stress–strain law is written as

$$\{\sigma\} = [E]_{\substack{\textbf{plane}\\\textbf{strain}}}\{\epsilon\} \tag{14.34}$$

where

$$[E]_{\substack{\textbf{plane}\\\textbf{strain}}} = \underbrace{\frac{E}{(1+\nu)(1-2\nu)}\begin{bmatrix} (1-\nu) & \nu & 0 \\ \nu & (1-\nu) & 0 \\ 0 & 0 & \dfrac{1-2\nu}{2} \end{bmatrix}}_{\text{Property matrix for plane strain } (3\times 3)}$$

This gives a plane strain property matrix which is different from that for plane stress.

After the plane strain problem is solved in the xy plane, the end stress is obtained from

$$\sigma_{zz} = \frac{E}{(1+\nu)(1-2\nu)}\,[\nu\,\epsilon_{xx} + \nu\,\epsilon_{yy}] \tag{14.35}$$

It is not part of the normal solution procedure.

As in (14.30), the element stress for plane strain is given by

$$\{\sigma\}^{(e)} = [E]^{(e)}_{\substack{\textbf{plane}\\\textbf{strain}}}\{\epsilon\}^{(e)} = [E]^{(e)}_{\substack{\textbf{plane}\\\textbf{strain}}}[D]^{(e)}\{U\}^{(e)} \tag{14.36}$$

A simplex plane strain element has constant strain, so it must have constant stress as well. The numerical values for the stresses in plane stress and plane strain elements are different.

Example 14.2 Stress in Simplex Two-Dimensional Element

Given The simplex element described in Example 14.1. Assume the material properties $E = 3.0 \times 10^7$ lbf/in² and $\nu = 0.29$.

Objective Determine the stresses for the element in plane stress and plane strain, as well as the z-direction strains and stresses.

Solution The strains in the xy plane from Example 14.1 are

$$\begin{Bmatrix} \epsilon_{xx} \\ \epsilon_{yy} \\ \gamma_{xy} \end{Bmatrix}^{(e)} = \begin{Bmatrix} -1.0 \\ 3.0 \\ -1.0 \end{Bmatrix} \times 10^{-4} \text{ (inch/inch)}$$

A. Plane Stress

The property matrix is

$$[E]^{(e)}_{\substack{\text{plane}\\ \text{stress}}} = 3.275 \times 10^7 \begin{bmatrix} 1 & 0.29 & 0 \\ 0.29 & 1 & 0 \\ 0 & 0 & 0.3550 \end{bmatrix} \quad (\text{lbf/in}^2)$$

and the stresses are given by (14.28)

$$\{\sigma\}^{(e)} = [E]^{(e)}_{\substack{\text{plane}\\ \text{stress}}} \{\epsilon\}^{(e)}$$

Using the strains from Example 14.1, the stresses are

$$\begin{Bmatrix} \sigma_{xx} \\ \sigma_{yy} \\ \tau_{xy} \end{Bmatrix}^{(e)}_{\substack{\text{plane}\\ \text{stress}}} = \begin{Bmatrix} -426 \\ 8{,}875 \\ -1{,}163 \end{Bmatrix} \quad (\text{lbf/in}^2)$$

Figure 14.12 illustrates the stresses acting on an infinitesimally small differential square dx by dy in the elements. For plane stress, the compressive strain ϵ_{xx} produces a compressive (negative) stress σ_{xx}. Also the elongation strain ϵ_{yy} gives a tensile (positive) stress σ_{yy}. A negative shear strain yields a negative shear stress. The z-direction strain in the thin plate is (14.29)

$$\epsilon_{zz}^{(e)} = -\frac{\nu}{E}(\sigma_{xx} + \sigma_{yy})$$

$$\epsilon_{zz}^{(e)} = -8.167 \times 10^{-5} \text{ (inch/inch)}$$

This represents the strain required so that the stress σ_{zz} is zero. The compressive stresses in the xy plane produce an expansion (positive strain) in the z direction. Figure 14.12(c) illustrates this.

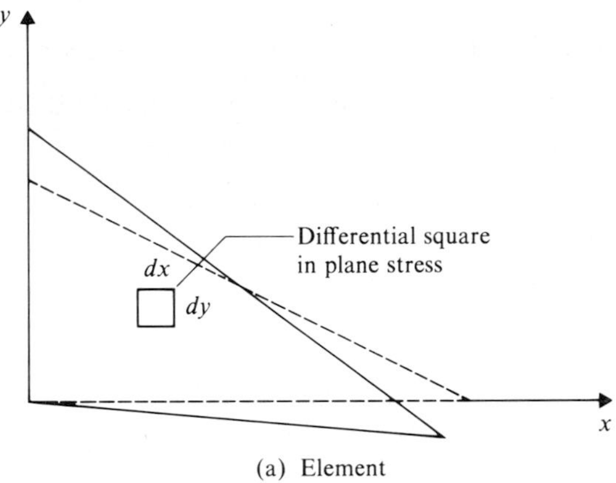

(a) Element

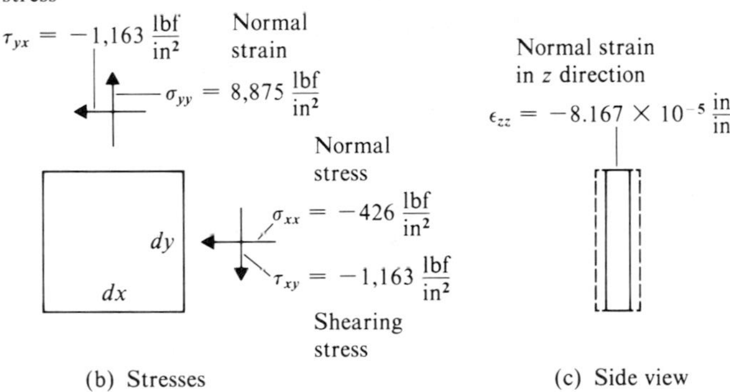

(b) Stresses (c) Side view

Figure 14.12 Element stresses in plane stress (Example 14.2—Stress in simplex two-dimensional element)

B. Plane Strain

The property matrix for plane strain (very long body) is

$$[E]^{(e)}_{\substack{\textbf{plane}\\\textbf{strain}}} = 5.537 \times 10^{7} \begin{bmatrix} 0.71 & 0.29 & 0 \\ 0.29 & 0.71 & 0 \\ 0 & 0 & 0.210 \end{bmatrix} \quad (\text{lbf/in}^2)$$

and the resulting stresses from (14.36) are

$$\begin{Bmatrix} \sigma_{xx} \\ \sigma_{yy} \\ \tau_{xy} \end{Bmatrix}^{(e)}_{\substack{\textbf{plane}\\\textbf{strain}}} = \begin{Bmatrix} 886 \\ 10{,}188 \\ -1{,}163 \end{Bmatrix} \quad (\text{lbf/in}^2)$$

Figure 14.13 illustrates these values.

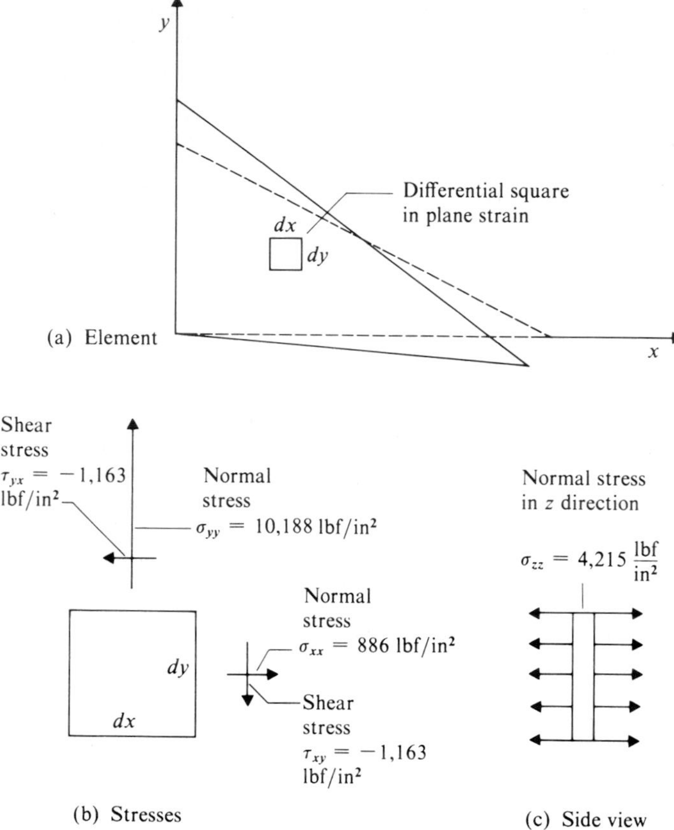

Figure 14.13 Element stresses in plane strain (Example 14.2—Stress in simplex two-dimensional element)

For plane strain, the z-direction stress required to maintain the state of two-dimensional strain is (14.35)

$$\sigma_{zz}^{(e)} = \frac{E}{(1 + \nu)(1 - 2\nu)} [\nu \, \epsilon_{xx} + \nu \, \epsilon_{yy}]$$

$$\sigma_{zz}^{(e)} = 4{,}215 \text{ lbf/in}^2$$

The positive sign indicates a tensile stress is required to prevent the long body from contracting in the z direction. This stress would tend to produce a contraction of the element in the xy plane, but the strains (and displacements) are already specified. Thus the required xy-plane stresses σ_{xx} and σ_{yy} must be larger, in the sense of more positive, to correspond to these additional requirements.

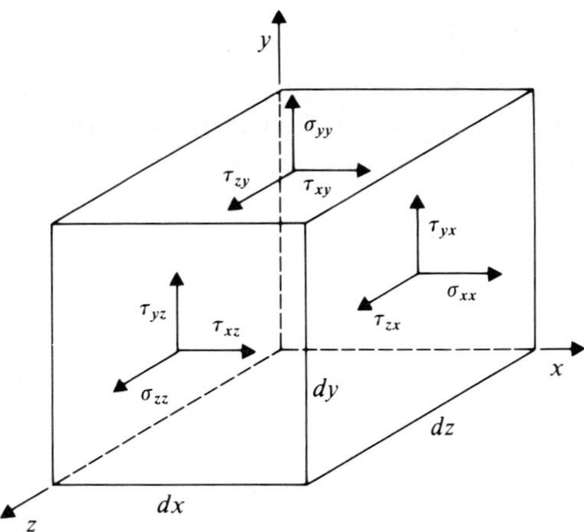

Figure 14.14 Stress components on differential cube

D. Stresses in Three Dimensions

In three dimensions, there are nine stress components.

$$\left.\begin{matrix}\sigma_{xx}\\ \sigma_{yy}\\ \sigma_{zz}\end{matrix}\right\} \quad \text{normal stress}$$

$$\left.\begin{matrix}\tau_{xy} = \tau_{yx}\\ \tau_{yz} = \tau_{zy}\\ \tau_{zx} = \tau_{xz}\end{matrix}\right\} \quad \text{shear stresses}$$

Figure 14.14 shows a differential cube with stresses on the faces. Only six of the nine components are independent.

In finite elements, the six independent stresses are written in column form as

$$\{\sigma\} = \begin{Bmatrix} \sigma_{xx} \\ \sigma_{yy} \\ \sigma_{zz} \\ \tau_{xy} \\ \tau_{yz} \\ \tau_{zx} \end{Bmatrix} \tag{14.37}$$

The stress column is similar to the strain column in format.

E. Stress–Strain Law for Three Dimensions

The stress–strain law for three-dimensional solids is obtained in a manner similar to that for plane stress and plane strain, by superposition of strains applied to the differential cube. If all three sides are simultaneously subject to normal stresses, $\sigma_{xx}, \sigma_{yy}, \sigma_{zz}$ uniformly distributed over the sides, the strain components are superposed to give

$$\begin{aligned}
\epsilon_{xx} &= \frac{1}{E}\,[\sigma_{xx} - \nu(\sigma_{yy} + \sigma_{zz})] \\
\epsilon_{yy} &= \frac{1}{E}\,[\sigma_{yy} - \nu(\sigma_{xx} + \sigma_{zz})] \\
\epsilon_{zz} &= \frac{1}{E}\,[\sigma_{zz} - \nu(\sigma_{xx} + \sigma_{yy})]
\end{aligned} \tag{14.38a}$$

Applying shearing strains yields the relations

$$\begin{aligned}
\gamma_{xy} &= \frac{1+\nu}{E}\,\tau_{xy} \\
\gamma_{yz} &= \frac{1+\nu}{E}\,\tau_{yz} \\
\gamma_{zx} &= \frac{1+\nu}{E}\,\tau_{zx}
\end{aligned} \tag{14.38b}$$

These are the strain–stress relations for an isotropic elastic solid. Solving these equations for the stresses yields

$$\begin{aligned}
\sigma_{xx} &= \frac{E}{(1+\nu)(1-2\nu)}\,[(1-\nu)\epsilon_{xx} + \nu(\epsilon_{yy} + \epsilon_{zz})] \\
\sigma_{yy} &= \frac{E}{(1+\nu)(1-2\nu)}\,[(1-\nu)\epsilon_{yy} + \nu(\epsilon_{zz} + \epsilon_{xx})] \\
\sigma_{zz} &= \frac{E}{(1+\nu)(1-2\nu)}\,[(1-\nu)\epsilon_{zz} + \nu(\epsilon_{xx} + \epsilon_{yy})] \\
\tau_{xy} &= \frac{E}{2(1+\nu)}\,\gamma_{xy} \\
\tau_{xz} &= \frac{E}{2(1+\nu)}\,\gamma_{xz} \\
\tau_{yz} &= \frac{E}{2(1+\nu)}\,\gamma_{yz}
\end{aligned} \tag{14.39}$$

These are the stress–strain relations in three dimensions.

In matrix form, the stress–strain relation can be written as

$$\{\sigma\} = [E]_{3\text{-D}}\,\{\epsilon\} \tag{14.40}$$

where $\{\sigma\}$ = stress column, $[E]_{3\text{-D}}$ = property matrix, and $\{\epsilon\}$ = strain column. The property matrix $[E]$ is easily found from (14.39) as

$$[E]_{3\text{-D}} = \underbrace{\frac{E}{(1+\nu)(1-2\nu)}\begin{bmatrix} 1-\nu & \nu & \nu & 0 & 0 & 0 \\ \nu & 1-\nu & \nu & 0 & 0 & 0 \\ \nu & \nu & 1-\nu & 0 & 0 & 0 \\ 0 & 0 & 0 & \dfrac{1-2\nu}{2} & 0 & 0 \\ 0 & 0 & 0 & 0 & \dfrac{1-2\nu}{2} & 0 \\ 0 & 0 & 0 & 0 & 0 & \dfrac{1-2\nu}{2} \end{bmatrix}}_{\text{Property matrix for three dimensions } (6\times 6)} \tag{14.41}$$

The stress in a three-dimensional element is given by

$$\{\sigma\}^{(e)} = [E]^{(e)}_{3\text{-D}}\,\{\epsilon\} \tag{14.42}$$

Or, in terms of the nodal displacements,

$$\{\sigma\}^{(e)} = [E]^{(e)}_{3\text{-D}}\,[D]\{U\}^{(e)} \tag{14.43}$$

Here the strain–displacement relation (14.16) has been used.

14.4 Total Potential Energy

A. Internal Energy and Applied Forces

The deformation of a three-dimensional elastic body follows the same principle as that for a one-dimensional elastic body—the minimization of potential energy. The concept was introduced in Chapter 5 (Section 5.2 for spring elements and Section 5.3 for bar elements). This section derives the potential energy, while the next section evaluates the element matrix. Again the total potential energy I is the sum of two parts: the internal strain energy U due to elastic deformation plus the potential W of the external loads to do work. Then

$$I = U + W \tag{14.44}$$

This must now be generalized to the three-dimensional problem considered in this chapter.

The internal strain energy is distributed over the solid body. This quantity is determined in a manner similar to that for bar elements in Chapter 5. States of initial strain and stress are treated later in the chapter.

Three different types of forces (loadings) are commonly applied to deformable bodies. They are:

1. Body forces (per unit volume), $F_B(x,y,z)$
2. Distributed surface forces (per unit area), $F_S(x,y,z)$
3. Concentrated forces, $F_1, F_2, \ldots, F_N$

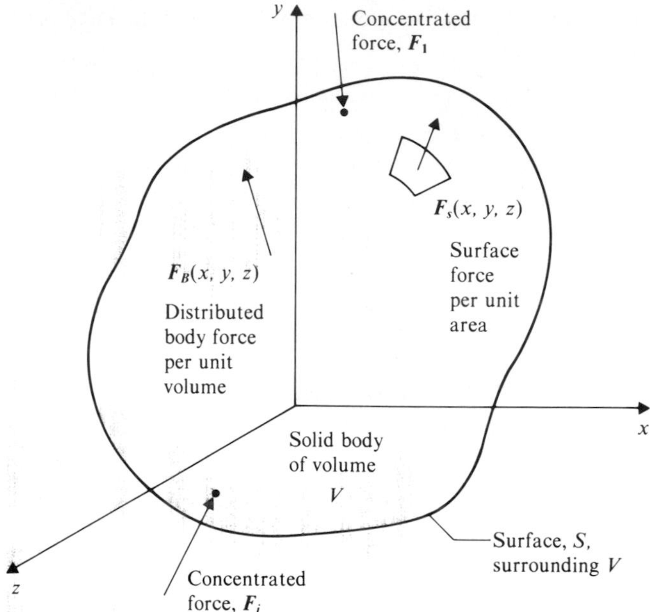

Figure 14.15 Solid body in three dimensions with applied forces

Figure 14.15 illustrates the three types of forces. The first two have a spatial *xyz* distribution and may vary from element to element. Concentrated forces are applied at a point, usually a node point. The total potential to do work is $W = W_B + W_S + W_C$ where W_B = potential of body force to do work, W_S = potential of surface force to do work, and W_C = potential of concentrated force to do work. These are now examined in turn.

B. Internal Strain Energy in Two Dimensions

The internal strain energy U for a linearly elastic body is the work that could be recovered if the deformed body were allowed to return to its undeformed position. Recall that the internal strain energy per unit volume (5.15) for a one-dimensional bar is

$$\text{Internal strain energy per volume in 1-D (one-dimensional bar)} = \int_0^{\epsilon} \sigma \, d\epsilon$$

The recoverable work at a position x is the integral of the internal stress σ over the strain ϵ. In two dimensions this is generalized to

$$\text{Internal strain energy per volume in 2-D} = U/\text{Volume} \tag{14.45}$$

$$= \int_0^{\epsilon_{xx}} \sigma_{xx} \, d\epsilon_{xx} + \int_0^{\epsilon_{yy}} \sigma_{yy} \, d\epsilon_{yy} + \int_0^{\gamma_{xy}} \tau_{xy} \, d\gamma_{xy}$$

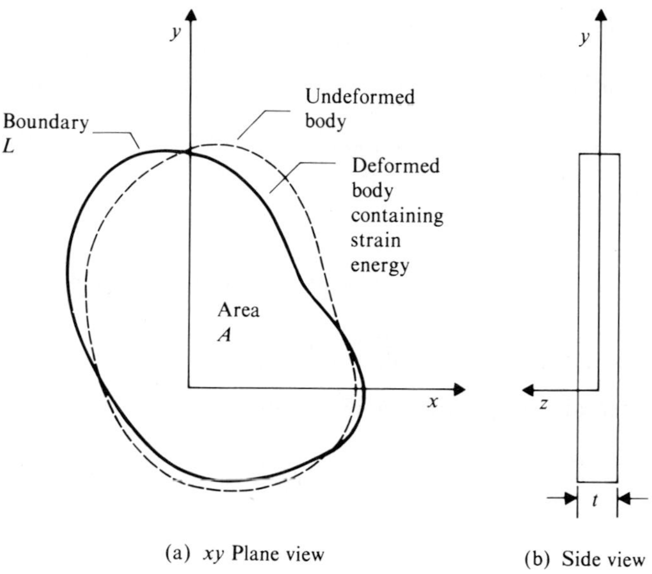

Figure 14.16 Internal strain energy in two-dimensional solid

for the deformed body shown in Figure 14.16. Only those stresses and strains in the same direction (say σ_{xx} and ϵ_{xx}) produce work or potential energy. Stresses and strains that are perpendicular to one another (such as σ_{yy} and ϵ_{xx}) produce no work.

Potential energy due to strains and stresses in the z direction is zero. Note that in plane stress, all stresses in the z direction are zero, so no internal strain energy terms involving these occur, even if z-direction strains exist. Similarly, in plane strain, no z-direction terms occur because all z-direction strains are zero. In matrix form, this is expressed as

$$U/\text{Volume} = \int_0^{\{\epsilon\}} \{\sigma\}^T \, d\{\epsilon\} \tag{14.46}$$

where

$$d\{\epsilon\} = \begin{Bmatrix} d\epsilon_{xx} \\ d\epsilon_{yy} \\ d\gamma_{xy} \end{Bmatrix}$$

Here both the stress column $\{\sigma\}$ and strain column $\{\epsilon\}$ have the normal two-dimensional form, as given in the two previous sections.

In linearly elastic materials undergoing small deformations, the stress–strain law is expressed as $\{\sigma\} = [E]\,\{\epsilon\}$ where the property matrix $[E]$ has already been obtained for both plane stress (14.28) and plane strain (14.36). Rewriting this in terms of the transpose of the stress column gives $\{\sigma\}^T = \{\epsilon\}^T\,[E]$. Then the internal strain energy per unit volume is

$$U/\text{Volume} = \int_0^{\{\epsilon\}} \{\epsilon\}^T[E] \, d\{\epsilon\} \tag{14.47}$$

Integrating this expression with respect to each ϵ term yields

$$U/\text{Volume} = \frac{1}{2} \{\epsilon\}^T[E]\{\epsilon\} \tag{14.48}$$

which is analogous to the result for the one-dimensional bar

$$U/\text{Volume} = \int_0^{\epsilon} \epsilon E \, d\epsilon = \frac{1}{2} \epsilon E \epsilon$$

For the one-dimensional bar, the total internal strain energy in the bar is (5.16)

$$U = \begin{matrix}\text{Total internal strain energy} \\ \text{(one-dimensional bar)}\end{matrix} = \int_V \left[\int_0^{\epsilon} \sigma \, d\epsilon \right] dV$$

In two dimensions, the volume integral is used with (14.45) to obtain

$$U = \begin{matrix}\text{Total internal strain} \\ \text{energy in 2-D}\end{matrix} = \int_V \left[\int_0^{\epsilon_{xx}} \sigma_{xx} \, d\epsilon_{xx} \right] dV$$

$$+ \int_V \left[\int_0^{\epsilon_{yy}} \sigma_{yy} \, d\epsilon_{yy} \right] dV + \int_V \left[\int_0^{\gamma_{xy}} \tau_{xy} \, d\gamma_{xy} \right] dV$$

Utilizing the matrix expression for U/Volume (14.48) just derived yields

$$U = \int_V \frac{1}{2} \{\epsilon\}^T[E]\{\epsilon\} \, dV$$

For a two-dimensional body of length t in the z direction, $dV = t \, dA$, and this simplifies to

$$U = \frac{1}{2} \int_A \{\epsilon\}^T[E]\{\epsilon\} t \, dA \tag{14.49}$$

where t is small for the thin body in plane stress and large (or of unit length) for plane strain.

C. Body Forces in Two Dimensions

Body forces are distributed over a body. Examples are gravitational forces and electromagnetic forces. In two dimensions, the body forces at any point x,y in the body have two components

$$\{F_B\} = \begin{Bmatrix} F_{Bx} \\ F_{By} \end{Bmatrix} \tag{14.50}$$

Distributed body forces
per unit volume (2×1)

where the units are force per unit volume. Figure 14.17(a) illustrates the body forces. The components F_{Bx}, F_{By} are taken as positive in the positive coordinate directions x,y.

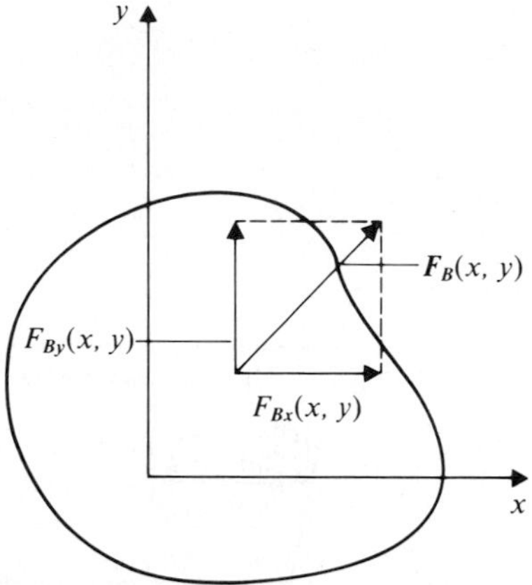

(a) Body force per unit volume applied over interior of solid body

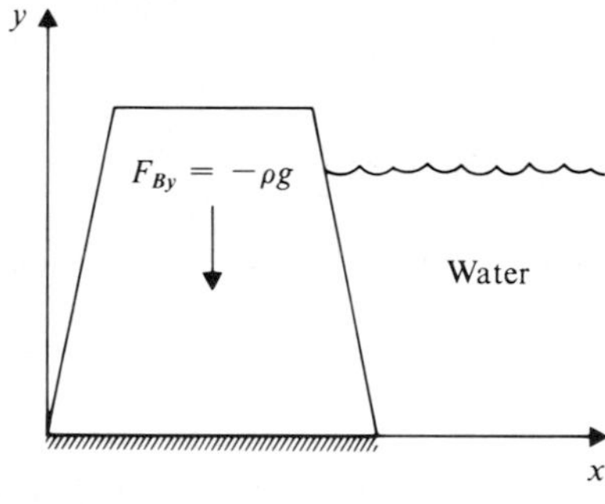

(b) Gravitational force in concrete dam

Figure 14.17 Distributed body forces in two dimensions

If a gravitational field is acting in the negative y direction, the body force column is

$$\{F_B\} = \begin{Bmatrix} 0 \\ -\rho g \end{Bmatrix}$$

such as in the concrete dam, Figure 14.17(b). This must be placed in the form of the potential of the external body force to do work.

The work per unit volume stored in the deformed body, with displacements $\{u\}$, due to the body force, is

$$\begin{matrix}\text{Potential of body} \\ \text{force to do work} \\ \text{per unit volume}\end{matrix} = -\{u,v\}\begin{Bmatrix} F_{Bx} \\ F_{By} \end{Bmatrix} = -\{u\}^T\{F_B\}$$

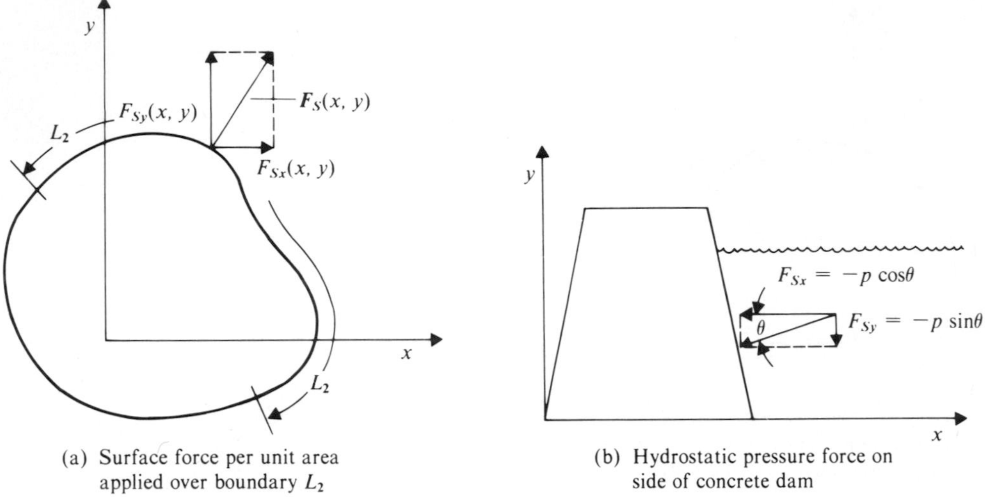

(a) Surface force per unit area applied over boundary L_2

(b) Hydrostatic pressure force on side of concrete dam

Figure 14.18 Distributed surface force in two dimensions

Integrating over the volume of the body gives the total potential as

$$W_B = \begin{matrix}\text{Potential of body} \\ \text{force to do work}\end{matrix} = -\int_A \{u\}^T\{F_B\}t\,dA \tag{14.51}$$

This is the body force term.

D. Surface Forces in Two Dimensions

Distributed surface forces act over the surface of the body. Examples include hydrostatic pressures, shear stresses, and other surface forces. At any point on the surface L_2, the surface force has two components

$$\{F_S\} = \begin{Bmatrix} F_{Sx} \\ F_{Sy} \end{Bmatrix} \tag{14.52}$$

Distributed surface forces per unit area (surface tractions) (2×1)

They are called surface tractions and have units of force per unit area. Figure 14.18(a) illustrates the surface force components. If the surface of the body is the side of the dam shown in Figure 14.18(b), consider the example of fluid pressure p acting in the direction normal to that surface. The surface force column is

$$\{F_S\} = \begin{Bmatrix} -p\cos\theta \\ -p\sin\theta \end{Bmatrix}$$

and the units are clearly force per unit area.

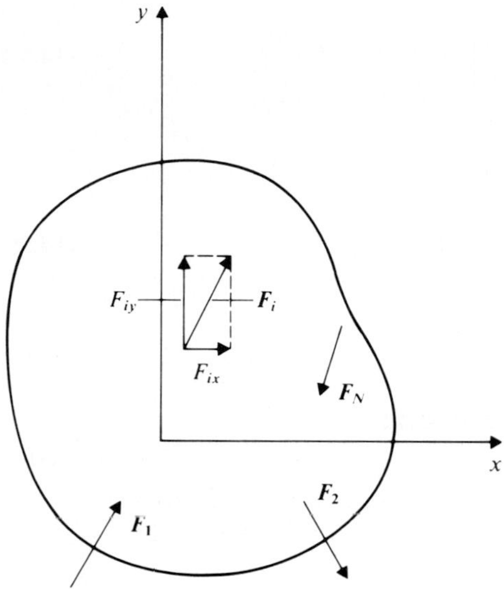

Figure 14.19 Concentrated force in two dimensions

The potential of the distributed surface force to do work, per unit area, in the deformed body with displacements $\{u\}$ is

$$\begin{matrix}\text{Potential of distributed}\\ \text{surface force to do}\\ \text{work per unit area}\end{matrix} = -\{u,v\}\begin{Bmatrix}F_{Sx}\\ F_{Sy}\end{Bmatrix} = -\{u\}^T\{F_S\}$$

Here the displacements are on the boundary L_2. Over the whole surface, the total potential is

$$W_S = \begin{matrix}\text{Potential of distributed}\\ \text{surface force to do work}\end{matrix} = -\int_{L_2} \{u\}^T\{F_S\}t\; dL_2 \qquad \textbf{(14.53)}$$

This is the surface force term.

E. Concentrated Forces in Two Dimensions

Concentrated external forces per unit length in the z direction act at various points on the body, either interior or exterior. Each force has two components.

$$F_i = \begin{Bmatrix}F_{xi}\\ F_{yi}\end{Bmatrix}, \qquad i = 1, 2, \ldots, N$$

Concentrated
force in two dimensions
(2×1)

as shown in Figure 14.19. There are N external forces. The potential of force $\{F_i\}$ to do work that has displacement $\{u_i\}$ is

$$\begin{array}{l}\text{Potential of}\\ \text{concentrated}\\ \text{force } i \text{ to}\\ \text{do work}\end{array} = -\{u_i, v_i\} \begin{Bmatrix} F_{ix} \\ F_{iy} \end{Bmatrix} = -\{u_i\}^T\{F_i\} \tag{14.54}$$

The total potential of all N forces is

$$W_C = \begin{array}{l}\text{Potential of } N\\ \text{concentrated forces}\\ \text{to do work}\end{array} = -\sum_{i=1}^{N} \{u_i\}^T\{F_i\}t \tag{14.55}$$

This is the concentrated force term.

F. Total Potential Energy in Two Dimensions

The total potential energy is the sum of four terms (14.44)

$$I = U + W_B + W_S + W_C$$

Writing out all of the terms derived gives

$$I = \underbrace{\frac{1}{2}\int_A \{\epsilon\}^T[E]\{\epsilon\}t\, dA}_{\substack{\text{Internal strain}\\ \text{energy term}}} - \underbrace{\int_A \{u\}^T\{F_B\}t\, dA}_{\substack{\text{Distributed body}\\ \text{force term}}}$$
$$- \underbrace{\int_{L_2} \{u\}^T\{F_S\}t\, dL_2}_{\substack{\text{Distributed surface}\\ \text{force term}}} - \underbrace{\sum_{i=1}^{N} \{u_i\}^T\{F_i\}t}_{\substack{\text{Concentrated force}\\ \text{term}}} \tag{14.56}$$

This expression is suitable for the derivation of an element matrix.

14.5 Element Matrix

A. Element Contribution to Potential Energy in Two Dimensions

The general expression for the total potential energy in a two-dimensional solid (14.56) is expressed for a single element as

$$I^{(e)} = \frac{1}{2}\int_{A^{(e)}} \{\epsilon\}^T\, [E]\{\epsilon\}t\, dA - \int_{A^{(e)}} \{u\}^T\{F_B\}t\, dA$$
$$- \int_{L_2^{(e)}} \{u\}^T\{F_S\}\; t\, dL_2 \tag{14.57}$$

Here the concentrated forces $\{F_i\}$ are not included in the element matrix and are added into the global equations by an assembly procedure. Recall that the displacements are given by (14.7)

$$\{u\} = [N]\{U\}$$

and that the strain $\{\epsilon\}$ can be expressed in terms of the nodal displacements as (14.9)

$$\{\epsilon\} = [D]\{U\}$$

where $\{U\}$ = nodal displacements in element (e), $[N]$ = shape function matrix, and $[D]$ = derivative matrix. Also the transpose of these expressions is $\{u\}^T = \{U\}^T[N]^T$, $\{\epsilon\}^T = \{U\}^T[D]^T$. Both of these are used in the potential energy expression.

The final result for the potential energy in the element is

$$I^{(e)} = \frac{1}{2}\int_{A^{(e)}} \{U\}^T[D]^T[E][D]\{U\}t\,dA - \int_{A^{(e)}} \{U\}^T[N]^T\{F_B\}t\,dA - \int_{L_2^{(e)}} \{U\}^T[N]^T\{F_S\}t\,dA \qquad \textbf{(14.58)}$$

All terms now involve the unknown nodal displacements $\{U\}$, while all other quantities—such as the material property matrix $[E]$, derivative matrix $[D]$, body force $\{F_B\}$, and surface force $\{F_S\}$—are known.

B. Minimization of Potential Energy in Two Dimensions

As with all previous element matrices, the element contribution to the minimization equations is given by differentiating I with respect to each nodal displacement. If there are nodes $i, j, \ldots, p$, this is expressed as

$$\begin{Bmatrix} \dfrac{\partial I}{\partial U_{2i-1}} \\ \dfrac{\partial I}{\partial U_{2i}} \\ \cdots \\ \dfrac{\partial I}{\partial U_{2p}} \\ (2p\times 1) \end{Bmatrix}^{(e)} = [B]^{(e)} \begin{Bmatrix} U_{2i-1} \\ U_{2i} \\ \cdots \\ U_{2p} \\ (2p\times 1) \end{Bmatrix}^{(e)} - \{C\}^{(e)} \qquad \textbf{(14.59)}$$

where the element matrices are of size

$[B]^{(e)}$	$\{C\}^{(e)}$
Element matrix	Element column
$(2p\times 2p)$	$(2p\times 1)$

In matrix form this can be written as

$$\frac{\partial I^{(e)}}{\partial\{U\}^{(e)}} = [B]^{(e)}\{U\}^{(e)} - \{C\}^{(e)} \qquad \textbf{(14.60)}$$

$$[B]^{(e)} = \frac{Et}{4A(1-\nu^2)} \begin{bmatrix} b_i & 0 & c_i \\ 0 & c_i & b_i \\ b_j & 0 & c_j \\ 0 & c_j & b_j \\ b_k & 0 & c_k \\ 0 & c_k & b_k \end{bmatrix} \begin{bmatrix} 1 & \nu & 0 \\ \nu & 1 & 0 \\ 0 & 0 & \frac{1-\nu}{2} \end{bmatrix} \begin{bmatrix} b_i & 0 & b_j & 0 & b_k & 0 \\ 0 & c_i & 0 & c_j & 0 & c_k \\ c_i & b_i & c_j & b_j & c_k & b_k \end{bmatrix}$$

Figure 14.20 Element matrix for simplex two-dimensional element in plane stress

Evaluating the minimization terms gives

$$\begin{aligned} \frac{\partial I^{(e)}}{\partial \{U\}^{(e)}} = \int_{A^{(e)}} [D]^T[E][D]\{U\}t\,dA \\ - \int_{A^{(e)}} [N]^T\{F_B\}t\,dA - \int_{L_2^{(e)}} [N]^T\{F_S\}t\,dL_2 \end{aligned} \tag{14.61}$$

Thus the element matrices are obtained by comparison with (14.59) as

$$[B]^{(e)} = \int_{A^{(e)}} [D]^T[E][D]t\,dA \tag{14.62}$$

$$\{C\}^{(e)} = \int_{A^{(e)}} [N]^T\{F_B\}t\,dA + \int_{L_2^{(e)}} [N]^T\{F_S\}t\,dL_2 \tag{14.63}$$

The element matrices are evaluated by multiplying several matrices together and then integrating over the area or boundary (as appropriate). In general, the derivative matrix $[D]$ and shape function matrix $[N]$ are functions of x,y.

C. Simplex Element Matrices in Two Dimensions

In a simplex element, the displacements are taken as linear. The derivative matrix $[D]$ is constant over the element (14.10), and the material property matrix is, as well. Thus the integrand in the element matrix is a constant. The element matrix then becomes $[B]^{(e)} = [D]^T[E][D]tA$. Figure 14.20 shows the element matrix for a simplex two-dimensional element in plane stress. Similarly, the plane strain equivalent is given in Figure 14.21.

The element column is composed of two parts: distributed body force term and distributed surface force term. Let these be written as

$$[B]^{(e)} = \frac{Et}{4A(1+\nu)(1-2\nu)} \begin{bmatrix} b_i & 0 & c_i \\ 0 & c_i & b_i \\ b_j & 0 & c_j \\ 0 & c_j & b_j \\ b_k & 0 & c_k \\ 0 & c_k & b_k \end{bmatrix} \begin{bmatrix} (1-\nu) & \nu & 0 \\ \nu & (1-\nu) & 0 \\ 0 & 0 & \frac{1-2\nu}{2} \end{bmatrix} \begin{bmatrix} b_i & 0 & b_j & 0 & b_k & 0 \\ 0 & c_i & 0 & c_j & 0 & c_k \\ c_i & b_i & c_j & b_j & c_k & b_k \end{bmatrix}$$

Figure 14.21 Element matrix for simplex two-dimensional element in plane strain

$\{C\}$	$=$	$\{C_B\}$	$+$	$\{C_S\}$
Element force column (6×1)		Distributed body force column (6×1)		Distributed surface force column (6×1)

These can now be evaluated for two-dimensional simplex elements.

Next consider the distributed body force term over the element

$$\{C_B\}^{(e)} = \int_{A^{(e)}} [N]^T\{F_B\}t\, dA$$

Let the body forces be taken as constant over the element. Recall the expression for the shape functions $[N]$ from (14.7). Multiplying out the $[N]^T$ and $\{F_B\}$ matrices gives the expression

$$\{C_B\}^{(e)} = \int_{A^{(e)}} \begin{Bmatrix} N_iF_{Bx} \\ N_iF_{By} \\ N_jF_{Bx} \\ N_jF_{By} \\ N_kF_{Bx} \\ N_kF_{By} \end{Bmatrix} t\, dA$$

Each of these integrals is easily evaluated, giving the result presented in Figure 14.22. Note that the body force terms have the units of force per volume. Multiplying by the element volume At then produces the units of force.

It is apparent that the distributed body force column for a simplex two-dimensional element adds one-third of the body forces F_{Bx} and F_{By} to each node. Actually, for this element, the same element body force column is produced by treating the body force as concentrated forces (of value $F_{Bx}At/3$ and $F_{By}At/3$) applied at each element node.

$$\{C_B\}^{(e)} = \frac{1}{3}At \begin{Bmatrix} F_{Bx} \\ F_{By} \\ F_{Bx} \\ F_{By} \\ F_{Bx} \\ F_{By} \end{Bmatrix}$$

Figure 14.22 Distributed body force column for simplex two-dimensional element (6 × 1). F_{Bx}, F_{By} are constant in element and have units of force per unit volume.

$$\{C_s\} = \frac{1}{2}L_{ij}t \begin{Bmatrix} F_{Sx} \\ F_{Sy} \\ F_{Sx} \\ F_{Sy} \\ 0 \\ 0 \end{Bmatrix} + \frac{1}{2}L_{jk}t \begin{Bmatrix} 0 \\ 0 \\ F_{Sx} \\ F_{Sy} \\ F_{Sx} \\ F_{Sy} \end{Bmatrix} + \frac{1}{2}L_{ki}t \begin{Bmatrix} F_{Sx} \\ F_{Sy} \\ 0 \\ 0 \\ F_{Sx} \\ F_{Sy} \end{Bmatrix}$$

On side ij (Side 1) — On side jk (Side 2) — On side ki (Side 3)

Figure 14.23 Distributed surface force column for simplex two-dimensional element (6 × 1). F_{Sx}, F_{Sy} are constant in element and have units of force per unit area.

The surface force term along part of the boundary L_2 (14.53) is

$$\{C_S\}^{(e)} = \int_{L_2^{(e)}} \{N\}^T\{F_S\}t\, dL$$

Again take constant surface forces over the element, and the term becomes

$$\{C_S\}^{(e)} = \{N\}^T\{F_S\}\, tL$$

These integrals along the boundary are evaluated in exactly the same manner as the boundary integrals in Chapter 9; the result is written out in Figure 14.23. Surface forces F_{Sx} and F_{Sy} have units of force per area. Multiplying by the surface area of the side tL yields units of force.

Once again the effect of the distributed surface force column for this two-dimensional simplex element is to add one-half of the surface force to each node on that side. As in the case of the body force column, the same result could be obtained by treating the distributed surface forces on an element side as concentrated forces (of numerical value $F_{Sx}Lt/2$ and $F_{Sy}Lt/2$) to each node on that side.

Concentrated forces are assumed to occur at a node point in the body. They are taken into account in the global equations after assembly. The nodal forces F_{ix}, F_{iy} simply replace the two entries in the global force column associated with the node i (actually displacements u_i, v_i). Thus, they do not enter into the element matrices.

14.6 Applications

This section considers two examples—one in plane stress and one in plane strain. The first is a simple two-element example with nodal applied forces only. The other example concerns the deformation–stress analysis of a long concrete dam subjected to the hydrostatic force of water along one wall. Both are carried out with program STRESS, but more detail is given in the first example.

Example 14.3 *Thin Plate in Plane Stress*

Given A thin plate, shown in Figure 14.24(a), is made of steel with the properties $E = 3.0 \times 10^7$ lbf/in^2 and $\nu = 0.29$. The plate has the geometry shown in the xy plane and thickness $t = 0.1$ in. The plate is rigidly held along the left edge. A concentrated force of 5,000 lb/in (per unit length into the paper) is applied at the outer corner, and another concentrated force of 3,000 lb/in is applied vertically downward at the top of the plate.

Objective Determine the nodal displacements, element strains, and element stresses. Use a two-element model and computer program STRESS.

Solution This problem is one of plane stress because the plate is very thin. Figure 14.25 shows the two-element model of the plate. The initial input parameters for STRESS are: number of nodes, NP = 4; number of elements, NE = 2; and type of analysis, NAN = 1. Table 14.2(a) gives the nodal coordinates and Table 14.2(b) the element/node connectivity.

Element properties are input next. All elements have the same properties, so the coefficient parameters, ICO = 1, and the properties are given in Table 14.2(c). The number of prescribed nodal forces is NPF = 3, with actual values shown in Table 14.2(d). At node 2, the applied force components are

$$(\text{TFORCE})_x = \frac{3}{5}\left(-5{,}000\ \frac{\text{lbf}}{\text{in}}\right)(0.1\ \text{in}) = -300.0\ \text{lbf}$$

$$(\text{TFORCE})_y = \frac{4}{5}\left(-5{,}000\ \frac{\text{lbf}}{\text{in}}\right)(0.1\ \text{in}) = -400.0\ \text{lbf}$$

while at node 4, the vertical component is

$$(\text{TFORCE})_y = \left(-3{,}000\ \frac{\text{lbf}}{\text{in}}\right)(0.1\ \text{in}) = -300\ \text{lbf}$$

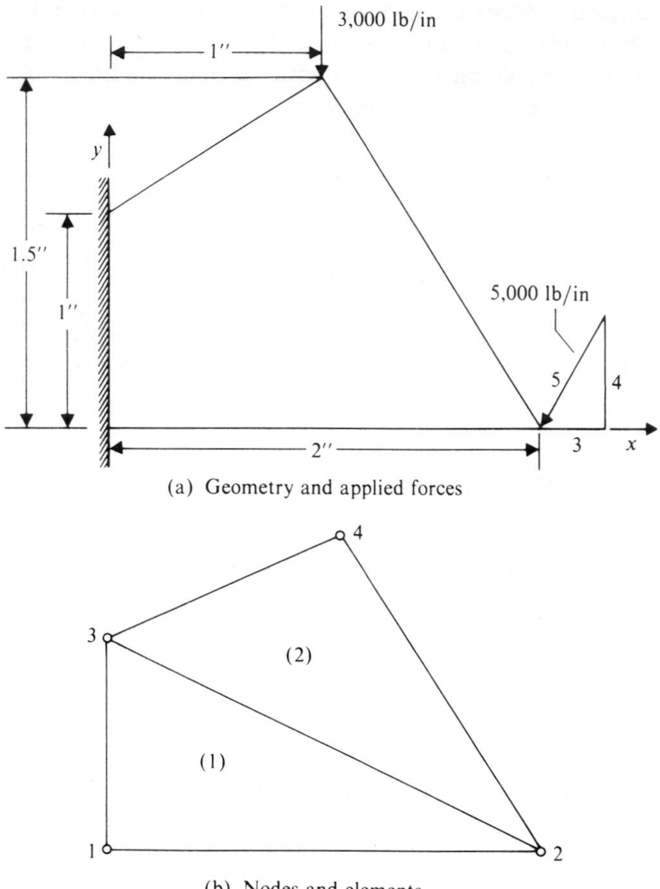

Figure 14.24 Problem specification (Example 14.3—Thin plate in plane stress)

The number of prescribed nodal displacements is NPD = 4, and the zero displacements are indicated in Table 14.2(e). Finally, the output parameter is taken as NOUT = 0. This completes the input.

Nodal displacements are illustrated in Figure 14.25(a). Table 14.3(a) gives the nodal values. Element strains are presented in Table 14.3(b). The element stresses are shown in Figure 14.25(b).

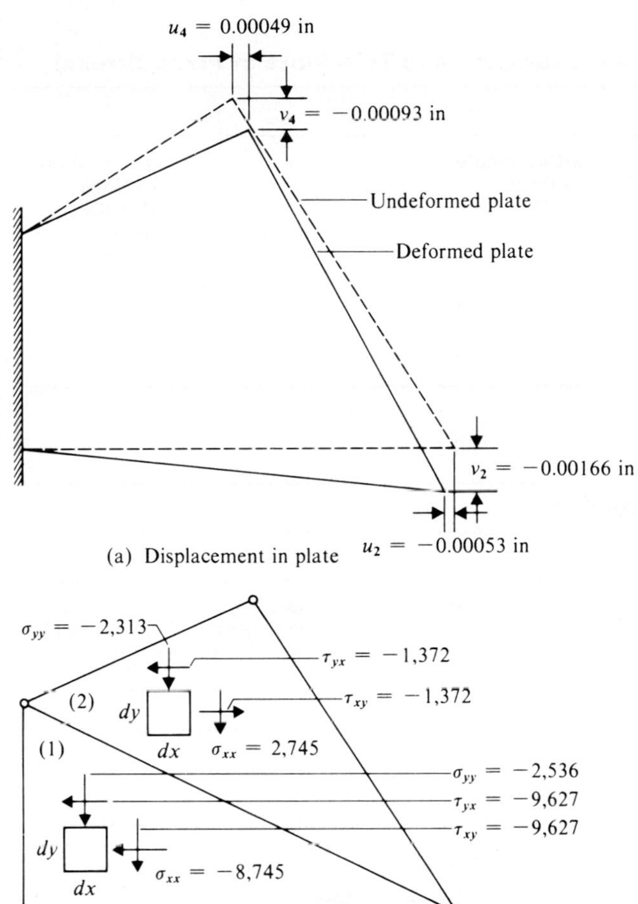

Figure 14.25 Results (Example 14.3—Thin plate in plane stress)

Table 14.2
Input Data for Program STRESS (Example 14.3: Thin Plate in Plane Stress)

(a) Nodal Coordinates

Global Node Number (INODE)	*x*-Coordinate of Node (XNODE) (inches)	*y*-Coordinate of Node (YNODE) (inches)
1	0.0	0.0
2	2.0	0.0
3	0.0	1.0
4	1.0	1.5

(b) Element/Node Connectivity Data

Element Number (NEL)	Element Type (NETYP)	Global Node Number			
		Node *i* (NODEI)	Node *j* (NODEJ)	Node *k* (NODEK)	Node ℓ (NODEL)
1	31	1	2	3	—
2	31	3	2	4	—

(c) Element Properties

Element Number (IEL)	Element Properties		
	Young's Modulus (YME) (lbf/in^2)	Poisson's Ratio (MUE)	Thickness (TE) (inches)
1	3.0×10^7	0.29	0.1

(d) Prescribed Nodal Forces

Global Node Number (IFORCE)	Global Direction of Force ($1 = x$, $2 = y$) (NFORCE)	Component of Prescribed Force (TFORCE) (lbf)
2	1	−300.0
2	2	−400.0
4	2	−300.0

Table 14.2 *Continued*

(e) Prescribed Nodal Displacements

Global Node Number (IDISP)	Direction of Displacement ($1=x$, $2=y$) (NDISP)	Component of Prescribed Nodal Displacement (TDISP) (inches)
1	1	0.0
1	2	0.0
3	1	0.0
3	2	0.0

Table 14.3
Results (Example 14.3: Thin Plate in Plane Stress)

(a) Nodal Displacements

Node	x Nodal Displacement (inches)	y Nodal Displacement (inches)
1	0.0	0.0
2	−0.0005340	−0.0016559
3	0.0	0.0
4	0.0004947	−0.0009316

(b) Element Strains

Element	Element Strains			
	ϵ_{xx} (in/in $\times 10^4$)	ϵ_{yy} (in/in $\times 10^4$)	ϵ_{zz} (in/in $\times 10^4$)	γ_{xy} (in/in $\times 10^4$)
1	−2.670	0.0	1.091	−8.280
2	1.139	−1.037	−0.042	−1.180

(c) Element Stresses

Element	Element Stresses		
	σ_{xx} (lbf/in^2)	σ_{yy} (lbf/in^2)	τ_{xy} (lbf/in^2)
1	−8,745	−2,536	−9,627
2	2,745	−2,313	−1,372

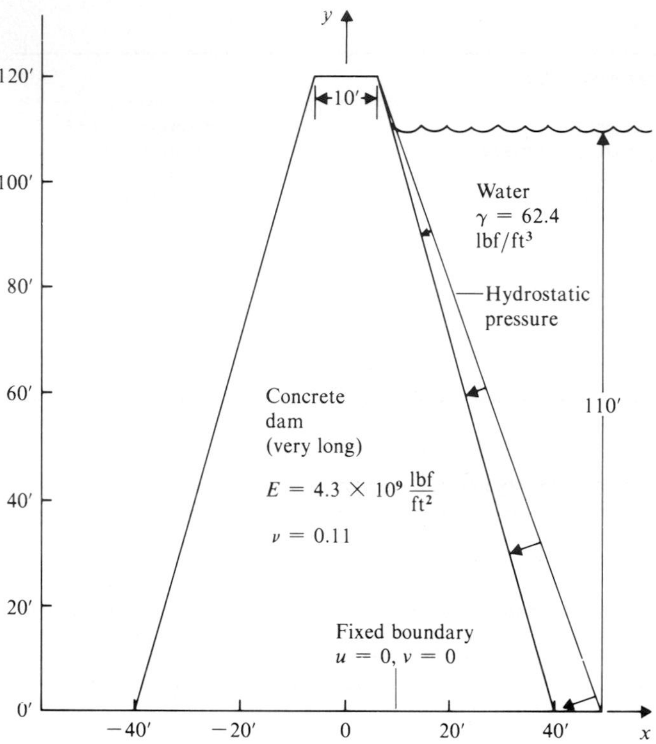

Figure 14.26 Geometry of dam in plane strain (Example 14.4—Concrete dam with hydrostatic pressure)

Example 14.4 *Concrete Dam with Hydrostatic Pressure*

Given A very long (into the plane of the paper) solid concrete dam is shown in Figure 14.26. The concrete has the properties $E = 4.3 \times 10^9$ lbf/ft² and $\nu = 0.11$, and the weight of the concrete is assumed to be neglected in this example. The water has a depth of 110 ft and a specific weight of $\gamma_{\text{water}} = 62.4$ lbf/ft³. All node points along the bottom of the dam can be taken as rigid: $u = v = 0$.

Objective Use program STRESS to determine the nodal displacements and element stresses in the dam, subject only to the hydrostatic water pressure. Employ at least 20 elements.

Solution The concrete dam is very long compared to either its width or height in the xy plane. Thus the dam can be considered as in plane strain. Figure 14.27 shows the division of the dam into 20 elements. The number of nodes, elements, and indicator for plane strain is number of nodes, NP = 18; number of elements, NE = 20; type of analysis, NAN = 2. The nodal coordinates are given in Table 14.4(a), while the element/node connectivity data are shown in Table 14.4(b).

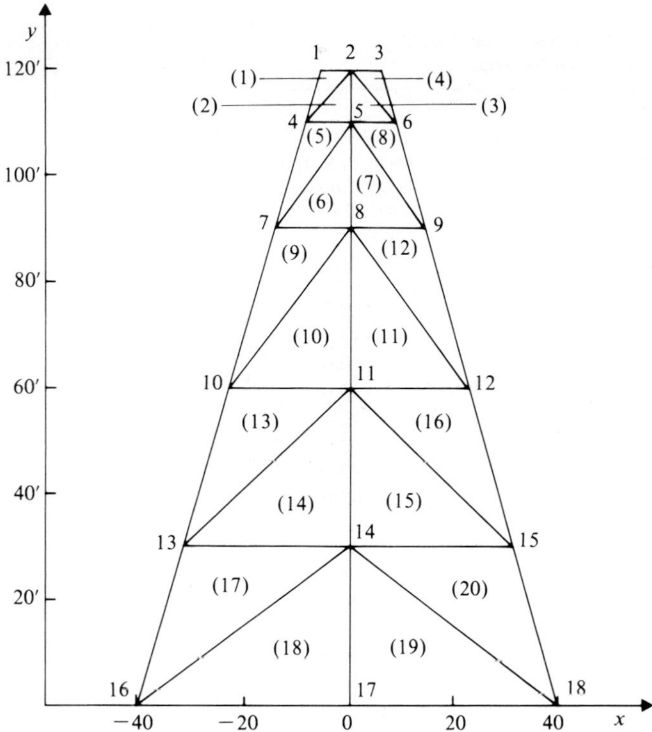

Figure 14.27 Node and element numbering (Example 14.4—Concrete dam with hydrostatic pressure)

Element properties are given in Table 14.4(c). For convenience, the element thickness is taken as unity: $t = 1.0$ (ft). The properties are the same for all elements.

Hydrostatic pressure is proportional to the depth of the water. In this case,

$$p = (110-y)\gamma_{\text{water}} = (110-y)62.4$$

where y is in feet. The pressures at the node points are

Node	Hydrostatic Pressure (lbf/ft²)
6	0.0
9	1,248.0
12	3,120.0
15	4,992.0
18	6,864.0

The element matrix developed in the previous section takes into account distributed loads, but such loads are not included in program STRESS. Therefore, equivalent nodal force components must be calculated.

Table 14.4
Input Data for Program STRESS (Example 14.4: Concrete Dam with Hydrostatic Pressure)

(a) Nodal Coordinates

Global Node Number (INODE)	Horizontal Nodal Coordinate (XNODE) (inches)	Vertical Nodal Coordinate (YNODE) (inches)
1	−5.0	120.0
2	0.0	120.0
3	5.0	120.0
4	−7.9167	110.0
5	0.0	110.0
6	7.9167	110.0
7	−13.750	90.0
8	0.0	90.0
9	13.750	90.0
10	−22.50	60.0
11	0.0	60.0
12	22.50	60.0
13	−31.250	30.0
14	0.0	30.0
15	31.250	30.0
16	−40.0	0.0
17	0.0	0.0
18	40.0	0.0

The force acting on the side of each element is obtained as the product of the average pressure times the length of the side times the element thickness t. For side 1,

Element	Equivalent Hydrostatic Force on Side of Element (lbf)
8	12,480
12	65,520
16	121,680
20	177,840

The nodal forces are then taken as the average of the hydrostatic forces on element sides adjacent to the node.

Node	Nodal Force, F_i (lbf)
6	6,240
9	39,000
12	93,600
15	149,760
18	88,920

Table 14.4 *Continued*

(b) Element/Node Connectivity Data

Element Number (NEL)	Element Type (NETYP)	Global Node Number		
		Node i (NODEI)	Node j (NODEJ)	Node k (NODEK)
1	31	4	2	1
2	31	4	5	2
3	31	5	6	2
4	31	6	3	2
5	31	7	5	4
6	31	7	8	5
7	31	8	9	5
8	31	9	6	5
9	31	10	8	7
10	31	10	11	8
11	31	11	12	8
12	31	12	9	8
13	31	13	11	10
14	31	13	14	11
15	31	14	15	11
16	31	15	12	11
17	31	16	14	13
18	31	16	17	14
19	31	17	18	14
20	31	18	15	14

(c) Element Properties (Only One Element Is Input Since All Have the Same Values)

Element Number (IEL)	Element Properties		
	Young's Modulus (YME) (lbf/ft^2)	Poisson's Ratio (MUE)	Thickness (TE) (ft)
1	4.3×10^9	0.11	1.0

These nodal force values F_i must be resolved into x,y components by the relation $F_{ix} = -F_i\cos\theta$, $F_{iy} = -F_i\sin\theta$, where θ is the angle of the outward normal to the dam relative to the positive x axis. For this case,

$$\theta = \tan^{-1}\frac{35}{120} = 16.26°$$

The results for the force components are given in Table 14.4(d).

Table 14.4 *Continued*

(d) Prescribed Nodal Forces

Node Number (IFORCE)	Prescribed Nodal Forces	
	Direction of Force (NFORCE) ($1=x$, $2=y$)	Component of Force (TFORCE) (lbf)
6	1	−5,990
6	2	−1,750
9	1	−37,440
9	2	−10,920
12	1	−89,860
12	2	−26,210
15	1	−143,770
15	2	−41,930
18	1	−85,360
18	2	−24,900

(e) Prescribed Nodal Displacements

Node Number (IDISP)	Prescribed Nodal Displacements	
	Direction of Force (NDISP)	Magnitude of Displacement (TDISP) (feet)
16	1	0.0
16	2	0.0
17	1	0.0
17	2	0.0
18	1	0.0
18	2	0.0

The nodal displacements are plotted in Figure 14.28 (displacements are greatly exaggerated to show the deformed shape). The top of the dam moves to the left somewhat like a cantilevered beam would. Table 14.5 gives the nodal displacements. Figure 14.29 shows the location of some of the high-stress elements. It is interesting to note that element 1 is actually in tension, although the stresses are very low.

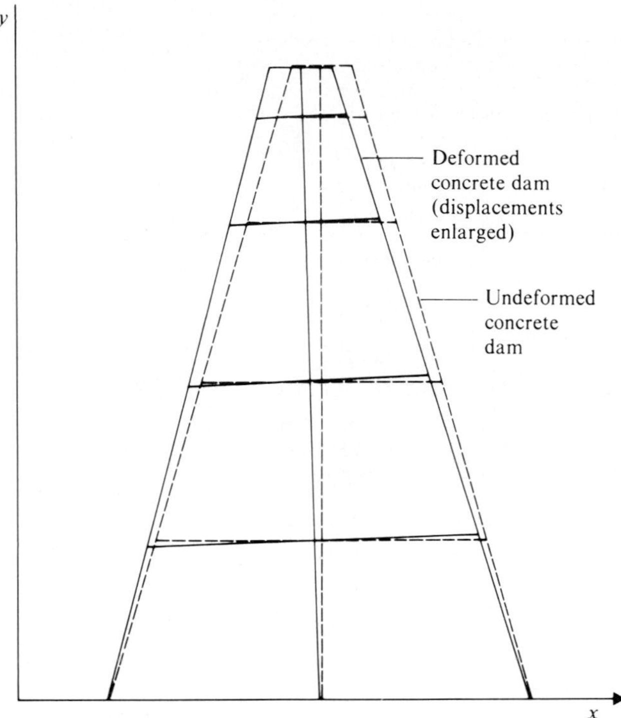

Figure 14.28 Deformed and undeformed dam (Example 14.4—Concrete dam with hydrostatic pressure)

Table 14.5
Nodal Displacements (Example 14.4: Concrete Dam with Hydrostatic Pressure)

Node	*x* Nodal Displacement (inches × 10³)	*y* Nodal Displacement (inches × 10³)
1	−2.654	−0.212
2	−2.653	−0.118
3	−2.653	−0.031
4	−2.470	−0.268
5	−2.472	−0.118
6	−2.478	0.019
7	−2.074	−0.037
8	−2.082	−0.116
9	−2.120	−0.114
10	−1.381	−0.450
11	−1.409	−0.106
12	−1.529	0.022
13	−0.634	−0.342
14	−0.675	−0.082
15	−0.878	0.217
16	0	0
17	0	0
18	0	0

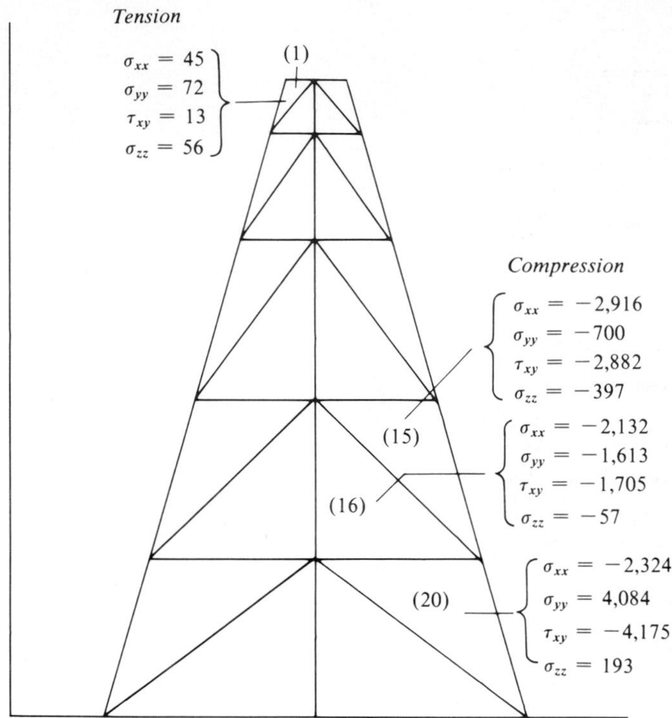

Figure 14.29 Elements with high stresses (Example 14.4—Concrete dam with hydrostatic pressure)

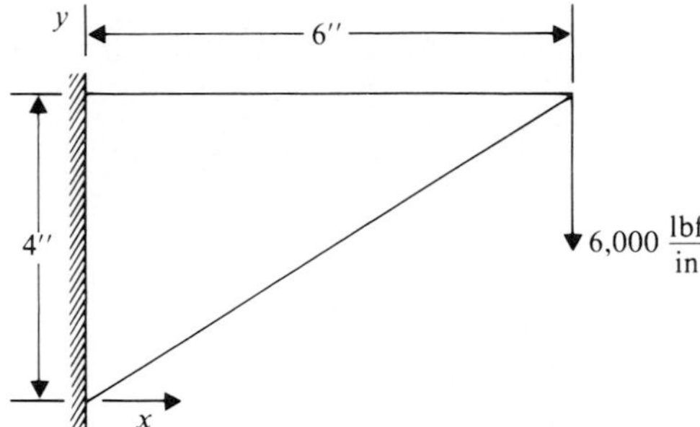

Figure 14.30 Thin aluminum plate for Problem 14.1

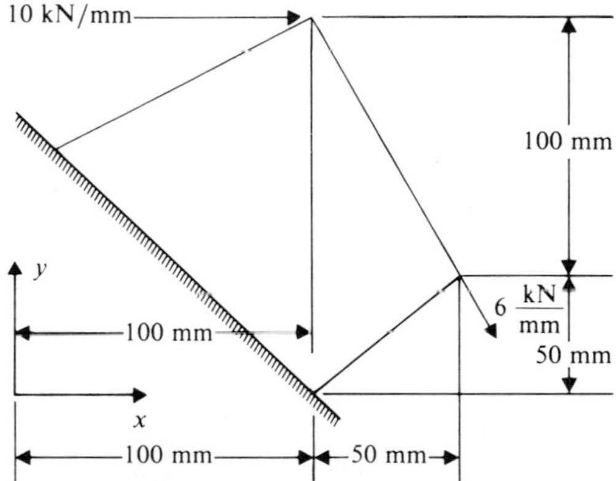

Figure 14.31 Thin steel plate for Problem 14.2

14.7 Problems

14.1 A thin triangular aluminum plate is shown in Figure 14.30. The plate has the properties $E = 1.0 \times 10^7$ lbf/in², $\nu = 0.33$, and $t = 0.1$ in. An applied force of 6,000 lbf/in (into the paper) acts vertically downward, as shown. Use a one-element (simplex) model of the plate to solve for the nodal displacements, element strains, and element stresses. Carry out the solution by hand.

14.2 A thin steel plate is shown in Figure 14.31. It has the properties $E = 207$ kN/mm², $\nu = 0.29$, and $t = 1$ mm, with the dimensions shown. The applied forces are 10 kN/mm (into the paper) on the uppermost point and 6 kN/mm at the rightmost point. Use a two-element (simplex) model to obtain the nodal displacements, element strains, and element stresses. Employ program STRESS.

14.3 A concrete base for a vertical column is 30 feet under water. Figure 14.32(a) shows the geometry. The concrete base is very long (into the paper) and has the properties $E = 4.3 \times 10^9$ lbf/ft² and $\nu = 0.11$. At the top of the base, the column applies a vertical

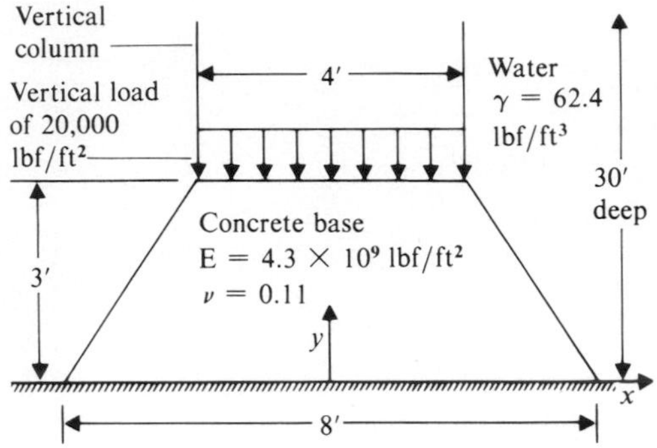

(a) Geometry for concrete base

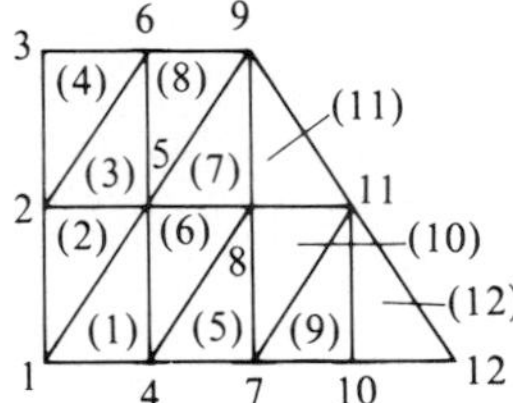

(b) Finite element model

Figure 14.32 Concrete base for Problem 14.3

load of 20,000 lbf/ft². Also, the water applies a hydrostatic force along the sides, which can be calculated for the given depth of water. The base is assumed to be rigidly connected to the ground ($u = 0$, $v = 0$).

Use the 12-element model (taking symmetry into account) to determine the nodal displacements and element stresses. Resolve all applied forces into components acting at the nodes. Use program STRESS. Sketch the resulting displaced shape and indicate the elements with the largest stress components.

14.4 Redo Problem 14.3 with the assumption that the vertical displacements are zero ($v = 0$) along the ground but that the horizontal displacements are not restricted (except for $u = 0$ at the origin, due to symmetry). Sketch the results and compare to the results for Problem 14.3.

14.5 A thin aluminum plate with properties $E = 1.0 \times 10^7$ lbf/in², $\nu = 0.33$, and $t = 1.0$ in is shown in Figure 14.33(a). The rectangular plate of dimensions 40 inches by 20 inches is subject to a uniform shear stress τ_{xy} along its end face with value $\tau_{xy} = -500$ lbf/in². This results in a total downward force on the end of

$$F_y = \tau_{xy} A = -500 \frac{\text{lbf}}{\text{in}^2} \times 20 \text{ in} \times 1.0 \text{ in} = 10{,}000 \text{ lbf}$$

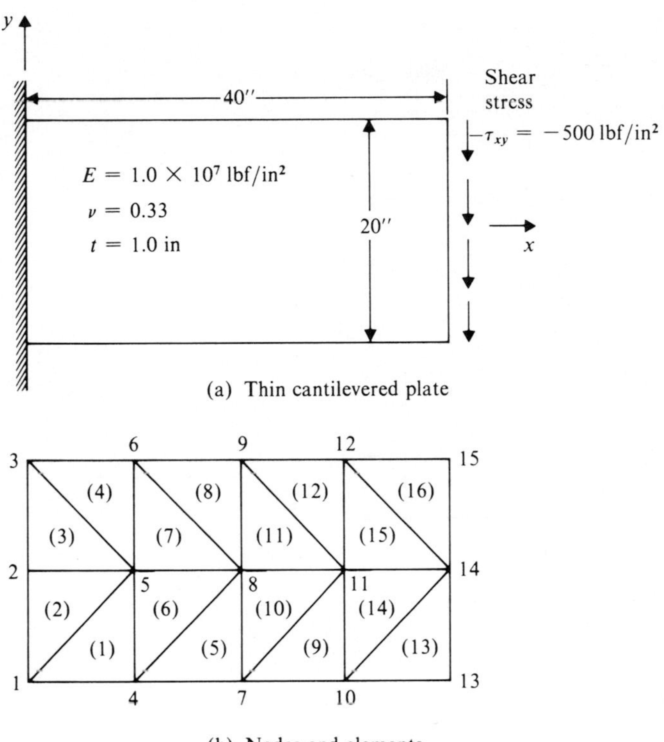

Figure 14.33 Thin cantilevered aluminum plate for Problem 14.5

Use the 16-element (all simplex) model shown in Figure 14.33(b) and program STRESS to determine the displacements at the nodes and element stresses. The deflection of the centerline, $y = 0$, due to a concentrated load F at the end is given by simple Bernoulli–Euler beam theory:

$$v\bigg|_{y=0} = -\frac{FL^3}{3EI}\left[\frac{3}{2}\left(\frac{x}{L}\right)^2 - \frac{1}{2}\left(\frac{x}{L}\right)^3\right], \quad I = \frac{1}{12}bh^3$$

Sketch this curve and the vertical displacements of the plate along the centerline on the same set of axes. Also plot the deformation of the nodes on the upper and lower surfaces of the plate.

14.6 A thin steel plate has the properties $E = 207$ kN/mm², $\nu = 0.29$, and $t = 1$ mm. Figure 14.34(a) shows the geometry. A circular hole of radius 100 mm has been cut in the center. A distributed normal stress of 20 kN/mm² occurs along each end of the plate, as shown.

Solve for the displacements and stresses in the upper right quadrant of the plate (by symmetry) as shown in Figure 14.34(b). Along the x axis $v = 0$, while along the y axis, $u = 0$ due to symmetry. The nodal forces on the right side are easily calculated from the applied 20 kN/mm² stress, and the equivalent nodal forces required to keep the plate

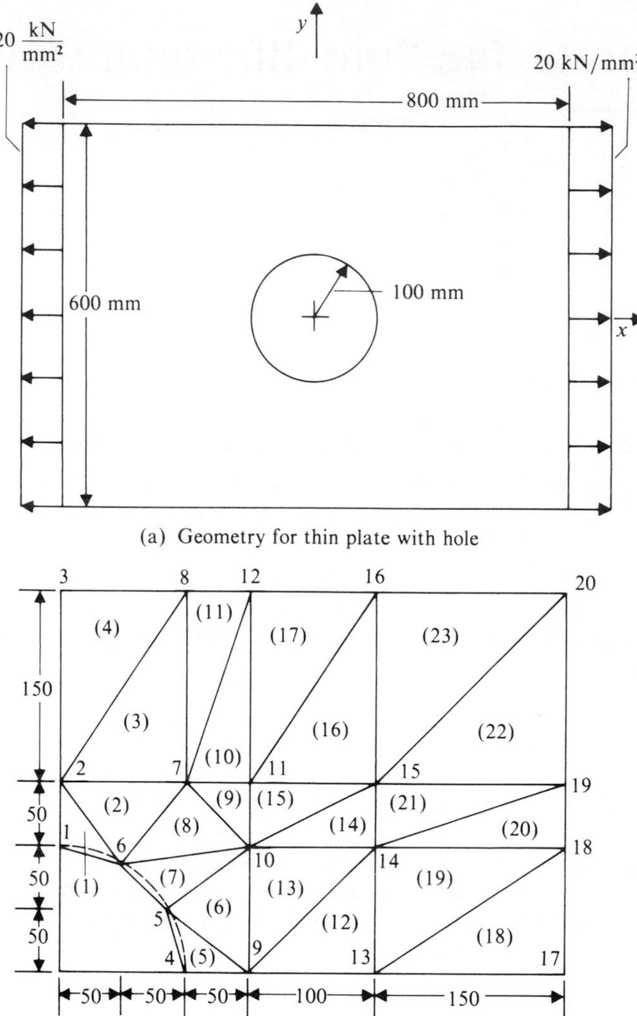

Figure 14.34 Thin plate with hole for Problem 14.6

quadrant in equilibrium ($\Sigma F_x = 0$) can also be calculated. Figure 14.34(b) gives the nodes and elements to be employed in program STRESS. Sketch the deformed plate and indicate the elements in which the largest stress components occur.

14.7 Do Problem 14.6 with a hole of radius 50 mm. If you have done Problem 14.6, note that this requires only minor changes in the input data for program STRESS. Compare the largest element stresses for the two problems.

Chapter 15

Galerkin Methods for Two-Dimensional Field Problems

15.1 Introduction

The Galerkin finite element method for one-dimensional problems was developed in Chapter 8. This chapter extends those results to linear two-dimensional field problems of first and second order. Only the element matrix approach will be used because of its relative simplicity.

As noted in Chapter 8, the Galerkin method has the major advantage over the variational method that it may be extended to many problems for which no functional can be found, even though the differential equation can be written. Differences between the two are emphasized in this chapter.

A. Differential Equation

The most general second-order differential equation may be written in the form

$$\begin{aligned}&\frac{\partial}{\partial x}\left[K_x(x,y)\,\frac{\partial T}{\partial x}\right] + \frac{\partial}{\partial x}\left[\frac{1}{2}\,K_{xy}(x,y)\,\frac{\partial T}{\partial y}\right]\\ &+ \frac{\partial}{\partial y}\left[\frac{1}{2}\,K_{xy}(x,y)\,\frac{\partial T}{\partial x}\right] + \frac{\partial}{\partial y}\left[K_y(x,y)\,\frac{\partial T}{\partial y}\right]\\ &+ M_x(x,y)\,\frac{\partial T}{\partial x} + M_y(x,y)\,\frac{\partial T}{\partial y} + P(x,y)\,T\\ &+ Q(x,y) = 0\end{aligned} \tag{15.1}$$

Along part of the boundary L_1 the value of T is specified as

$$T = T_0(x,y) \text{ on } L_1 \tag{15.2}$$

while on the remainder of the boundary L_2 the general derivative boundary condition is of the form

$$\begin{aligned}&K_x(x,y)\,\frac{\partial T}{\partial x}\,n_x + \frac{1}{2}\,K_{xy}(x,y)\,\frac{\partial T}{\partial y}\,n_x + \frac{1}{2}\,K_{yx}(x,y)\,\frac{\partial T}{\partial x}\,n_y\\ &+ K_y(x,y)\,\frac{\partial T}{\partial y}\,n_y + \alpha(x,y)\,T + \beta(x,y) = 0 \text{ on } L_2\end{aligned} \tag{15.3}$$

The total, L_1 plus L_2, must equal the length of the contour L bounding the two-dimensional area A over which the solution is to be found.

It is easy to see that the above differential equation reduces to the variational form when $K_{xy} = 0$ and $M_x = M_y = 0$. Equation (9.5) results. Also the boundary condition on L_2 becomes the same as the variational method if $K_{xy} = 0$. Then Equation (9.7) results.

Let the differential equation be expressed in operator form:

$$L(x,y,T) = 0 \tag{15.4}$$

where

$$L(x,y,T) = \frac{\partial}{\partial x}\left[K_x \frac{\partial T}{\partial x}\right] + \frac{\partial}{\partial x}\left[\frac{1}{2} K_{xy} \frac{\partial T}{\partial y}\right] + \frac{\partial}{\partial y}\left[\frac{1}{2} K_{xy} \frac{\partial T}{\partial x}\right] + \frac{\partial}{\partial y}\left[K_y \frac{\partial T}{\partial y}\right] + M_x \frac{\partial T}{\partial x} + M_y \frac{\partial T}{\partial y} + PT + Q$$

This is only for convenience in the following derivation.

B. Galerkin's Method

A relatively simple function T^* is chosen as an approximation to T. Ideally, T^* has a shape similar to that of T and satisfies the same boundary conditions. Often T^* is a series of simple functions, such as polynomials. Substituting the approximate solution into the differential equation (expressed in operator form),

$$L(x,y,T^*) = \epsilon \tag{15.5}$$

The approximation produces an error that is a function of x and y: $\epsilon = \epsilon(x,y)$.

A weighting function $W(x,y)$ is chosen and multiplied by the error $\epsilon(x,y)$. The result is then integrated over the entire solution area to form the residual R, which is set equal to zero. The negative sign is explained later.

$$R = -\iint_A W \epsilon \, dA = 0 \tag{15.6}$$

If a series of simple functions are chosen for T^*, usually a series of N weighting functions $W_1(x,y)$, $W_2(x,y)$, $W_3(x,y)$, . . . , $W_N(x,y)$ is also chosen. There will then be N residuals of the form

$$\begin{aligned} R_1 &= -\iint_A W_1 \epsilon \, dA = 0 \\ R_2 &= -\iint_A W_2 \epsilon \, dA = 0 \\ R_3 &= -\iint_A W_3 \epsilon \, dA = 0 \\ &\ldots \end{aligned} \tag{15.7}$$

Continued on next page

. . .

$$R_N = -\iint_A W_N \epsilon \, dA = 0 \tag{15.7 cont.}$$

The Galerkin method consists of choosing the weighting functions to have the same form as the simple approximating functions for T^*. These equations must now be placed in appropriate finite element form.

Within an element, T is approximated by the interpolation functions

$$T \simeq T^* = \sum_{e=1}^{E} T^{(e)}$$

Substituting this into the differential equation (15.1)

$$L\,(x,y,\,T^*) = \sum_{e=1}^{E} \epsilon^{(e)} \tag{15.8}$$

where the element error is a function of x and y, since the interpolation function is $\epsilon^{(e)} = \epsilon^{(e)}\,(x,y)$. Also, each of the weighting functions is separated into element-size weighting functions of the form

$$W_1 = \sum_{e=1}^{E} W_1^{(e)}$$

$$W_2 = \sum_{e=1}^{E} W_2^{(e)}$$

$$W_3 = \sum_{e=1}^{E} W_3^{(e)} \tag{15.9}$$

. . .

$$W_N = \sum_{e=1}^{E} W_N^{(e)}$$

In the general case, these may or may not be composed of element pyramid functions.

For a finite element solution involving N unknown nodal points, the N residual equations have the form

$$R_1 = -\sum_{e=1}^{E} \iint_{A^{(e)}} W_1^{(e)} \epsilon^{(e)}\, dA = 0$$

$$R_2 = -\sum_{e=1}^{E} \iint_{A^{(e)}} W_2^{(e)} \epsilon^{(e)}\, dA = 0$$

$$R_3 = -\sum_{e=1}^{E} \iint_{A^{(e)}} W_3^{(e)} \epsilon^{(e)}\, dA = 0 \tag{15.10}$$

. . .

Continued on next page

. . .

$$R_N = -\sum_{e=1}^{E} \iint_{A^{(e)}} W_N^{(e)} \epsilon^{(e)} \, dA = 0 \tag{15.10 cont.}$$

The residual has the general form

$$\begin{aligned} R_n = -\Bigg[&\iint_{A^{(1)}} W_n^{(1)} \epsilon^{(1)} \, dA + \iint_{A^{(2)}} W_n^{(2)} \epsilon^{(2)} \, dA \\ &+ \iint_{A^{(3)}} W_n^{(3)} \epsilon^{(3)} \, dA + \ldots \\ &+ \iint_{A^{(E)}} W_n^{(E)} \epsilon^{(E)} \, dA \Bigg] = 0, \quad n = 1, 2, 3, \ldots, N \end{aligned} \tag{15.11}$$

The exact form depends upon the order of the equation, number of dimensions, and other factors. Specific cases are shown in what follows.

15.2 First-Order Problems

First-order differential equations in two dimensions are not too commonly found, but they do govern some physical problems. One example is the adiabatic temperature rise in thin fluid films. Viscous heat generation in the film produces temperature gradients along the film. When heat conduction (second-order terms) can be neglected in the energy equation, only first-order derivative terms due to fluid convection are left in the energy equation. In what follows a general first-order partial differential equation is used.

A. Boundary Value Problem

Consider the general form of the first-order partial differential equation of the form

$$M_x(x,y) \frac{\partial T}{\partial x} + M_y(x,y) \frac{\partial T}{\partial y} + P(x,y)T + Q(x,y) = 0 \tag{15.12}$$

where $M_x(x,y)$, $M_y(x,y)$, $P(x,y)$, $Q(x,y)$ are given coefficient functions. Also the boundary condition for prescribed values of T is

$$T = T_0(x,y) \text{ on } L_1 \tag{15.13}$$

where L_1 is only part of the full boundary L.

B. Element Error

Within the element, the dependent variable T is approximated by a linear interpolation function of the form

$$T \simeq T^{(e)} = N_i^{(e)} T_i + N_j^{(e)} T_j + N_k^{(e)} T_k$$

Substituting this into the differential equation gives

$$M_x^{(e)} \frac{\partial T^{(e)}}{\partial x} + M_y^{(e)} \frac{\partial T^{(e)}}{\partial y} + P^{(e)}\, T^{(e)} + Q^{(e)} = \epsilon^{(e)} \tag{15.14}$$

where M_x, M_y, P, Q are coefficient functions, taken constant within element e. Since T is a function of x,y, $\epsilon^{(e)} = \epsilon^{(e)}(x,y)$.

C. Element Residuals

The theory developed in chapter 8 for element coefficient matrices and columns is easily extended to two dimensions. For the element e, the residuals associated with the i, j, k nodal equations are

$$R_i^{(e)} = -\iint_{A^{(e)}} W_i^{(e)}\, \epsilon^{(e)}\, dA$$

$$R_j^{(e)} = -\iint_{A^{(e)}} W_j^{(e)}\, \epsilon^{(e)}\, dA$$

$$R_k^{(e)} = -\iint_{A^{(e)}} W_k^{(e)}\, \epsilon^{(e)}\, dA$$

Here the negative sign is used to correspond to the negative sign in the variational formulation, Equation (9.14). Also

$R_i^{(e)}$ = Contribution of element e to residual in equation for node i

$R_j^{(e)}$ = Contribution of element e to residual in equation for node j

$R_k^{(e)}$ = Contribution of element e to residual in equation for node k

Figure 15.1 shows the triangular element and the pyramid functions. The weighting functions are chosen as the pyramid functions in the Galerkin technique:

$$W_i^{(e)} = N_i^{(e)},\ W_j^{(e)} = N_j^{(e)},\ W_k^{(e)} = N_k^{(e)} \tag{15.15}$$

The complete weighting function W_i is summed over all elements that have node point i as one corner. All other elements (not containing node i) have element weighting functions of zero.

Equation (15.14) for the first residual becomes

$$R_i^{(e)} = -\iint_{A^{(e)}} \left\{ N_i \left[\frac{M_x}{2A}(b_iT_i + b_jT_j + b_kT_k) + \frac{M_y}{2A}(c_iT_i + c_jT_j + c_kT_k) + P(N_iT_i + N_jT_j + N_kT_k) + Q \right] \right\}^{(e)} dA$$

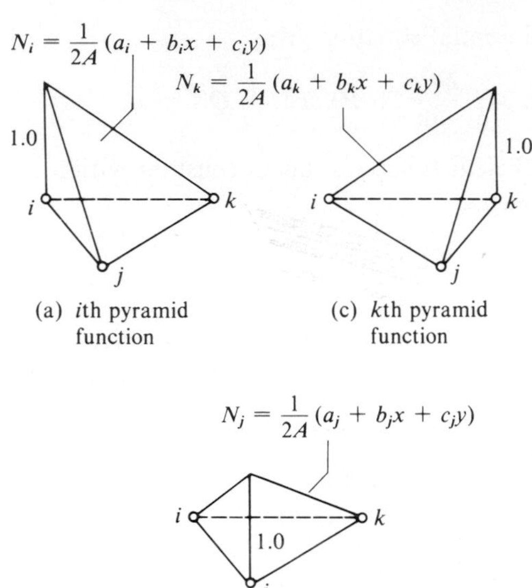

(a) *i*th pyramid function

(c) *k*th pyramid function

(b) *j*th pyramid function

Figure 15.1 Pyramid functions for two-dimensional simplex element

Evaluating the integrals with Equation (3.27) yields

$$R_i^{(e)} = -\left[\frac{M_x}{6}(b_iT_i + b_jT_j + b_kT_k) + \frac{M_y}{6}(c_iT_i + c_jT_j + c_kT_k) + \frac{PA}{12}(2T_i + T_j + T_k) + \frac{1}{3}QA\right]^{(e)}$$

Reorganizing this into nodal form gives

$$\begin{aligned} R_i^{(e)} = {} & \left[-\frac{1}{6}(M_xb_i + M_yc_i) - \frac{1}{6}PA\right]^{(e)} T_i \\ & + \left[-\frac{1}{6}(M_xb_j + M_yc_j) - \frac{1}{12}PA\right]^{(e)} T_j \\ & + \left[-\frac{1}{6}(M_xb_k + M_yc_k) - \frac{1}{12}PA\right]^{(e)} T_k - \left[\frac{1}{3}QA\right]^{(e)} \end{aligned} \tag{15.16}$$

The other two residuals are evaluated in similar fashion.

D. Element Matrices

Writing out the element coefficient matrix gives

$$\begin{Bmatrix} R_i \\ R_j \\ R_k \end{Bmatrix} = [B]^{(e)} \begin{Bmatrix} T_i \\ T_j \\ T_k \end{Bmatrix} - \{C\}^{(e)} \tag{15.17}$$

where

$$[B]^{(e)} = \underbrace{\begin{bmatrix} & i & j & k \\ i & B_{ii} & B_{ij} & B_{ik} \\ j & B_{ji} & B_{jj} & B_{jk} \\ k & B_{ki} & B_{kj} & B_{kk} \end{bmatrix}^{(e)}}_{\text{Element coefficient matrix}}$$

$$\{C\}^{(e)} = \underbrace{\begin{Bmatrix} i & C_i \\ j & C_j \\ k & C_k \end{Bmatrix}^{(e)}}_{\text{Element column}}$$

This is somewhat similar to Equation (9.25) for the variational formulation. Each B component is

$$\begin{aligned}
B_{ii} &= \left[-\frac{1}{6}(M_x b_i + M_y c_i) - \frac{1}{6} PA\right]^{(e)} \\
B_{ij} &= \left[-\frac{1}{6}(M_x b_j + M_y c_j) - \frac{1}{12} PA\right]^{(e)} \\
B_{ik} &= \left[-\frac{1}{6}(M_x b_k + M_y c_k) - \frac{1}{12} PA\right]^{(e)} \\
B_{ji} &= \left[-\frac{1}{6}(M_x b_i + M_y c_i) - \frac{1}{12} PA\right]^{(e)} \\
B_{jj} &= \left[-\frac{1}{6}(M_x b_j + M_y c_j) - \frac{1}{6} PA\right]^{(e)} \\
B_{jk} &= \left[-\frac{1}{6}(M_x b_k + M_y c_k) - \frac{1}{12} PA\right]^{(e)} \\
B_{ki} &= \left[-\frac{1}{6}(M_x b_i + M_y c_i) - \frac{1}{12} PA\right]^{(e)} \\
B_{kj} &= \left[-\frac{1}{6}(M_x b_j + M_y c_j) - \frac{1}{12} PA\right]^{(e)} \\
B_{kk} &= \left[-\frac{1}{6}(M_x b_k + M_y c_k) - \frac{1}{6} PA\right]^{(e)}
\end{aligned} \qquad \textbf{(15.18)}$$

It is easily seen that the element coefficient matrix is not symmetric by considering B_{ij} and B_{ji}. Also the C components are

$$\begin{aligned}
C_i &= \left[\frac{1}{3} QA\right]^{(e)} \\
C_j &= \left[\frac{1}{3} QA\right]^{(e)} \\
C_k &= \left[\frac{1}{3} QA\right]^{(e)}
\end{aligned} \qquad \textbf{(15.19)}$$

15.3 First-Order Example

Only one example is given here for a first-order differential equation in two dimensions. Since the principles for the calculation of element properties and assembly are the same as discussed previously, this should be sufficient.

Example 15.1 *Temperature Distribution in Thin Fluid Film*

Given A thin film of fluid flows between two solid parallel surfaces, the lower one infinite in extent and the upper one shaped like a trapezoid. The upper plate moves over the lower surface with velocity components U and V in the x and y directions. The fluid is assumed to have constant density, viscosity, and specific heat. A diagram is shown in Figure 15.2. The energy differential equation is

$$J\rho c_v\left(\frac{1}{2}U\frac{\partial T}{\partial x} + \frac{1}{2}V\frac{\partial T}{\partial y}\right) - \mu\left[\left(\frac{U}{h}\right)^2 + \left(\frac{V}{h}\right)^2\right] = 0$$

where J = conversion factor = 778 ft-lbf/Btu, ρ = density = 1.7 lbf-sec^2/ft^4, c_v = specific heat at constant volume = 15.13 Btu-ft/lbf-sec^2-°F, h = film thickness = 2.5×10^{-4} ft, U = surface x velocity = 50 ft/sec, V = surface y velocity = 20 ft/sec, μ = fluid viscosity = 4.4×10^{-3} lbf-sec/ft^2, and T_0 = 100°F. Here the first two terms represent the convection of fluid through the space and the second two terms represent heat generation due to viscous shear stresses. Heat conduction is assumed to be small. The boundary condition is simply that the entering fluid is at $T = T_0$ when the upper plate passes over a point in space.

$$5.00 \times 10^5 \frac{\partial T}{\partial x} + 2.00 \times 10^5 \frac{\partial T}{\partial y} - 2.042 \times 10^8 = 0$$

Objective Use two elements to calculate the temperature at the point (0,0).

Solution The upper surface is divided into two elements as shown in Figure 15.2(c). For element 1, the properties are

$$b_1^{(1)} = -0.2 \qquad c_1^{(1)} = 0$$

$$b_3^{(1)} = 0.2 \qquad c_3^{(1)} = -0.3 \qquad A^{(1)} = 0.03$$

$$b_4^{(1)} = 0 \qquad c_4^{(1)} = 0.3$$

$$B_{11}^{(1)} = \left[-\frac{1}{6}(5.00 \times 10^5)(-0.2) - \frac{1}{6}(2.00 \times 10^5)(0) - \frac{1}{6}(0)(0.03)\right] = 1.667 \times 10^4$$

and so on. The element matrices are

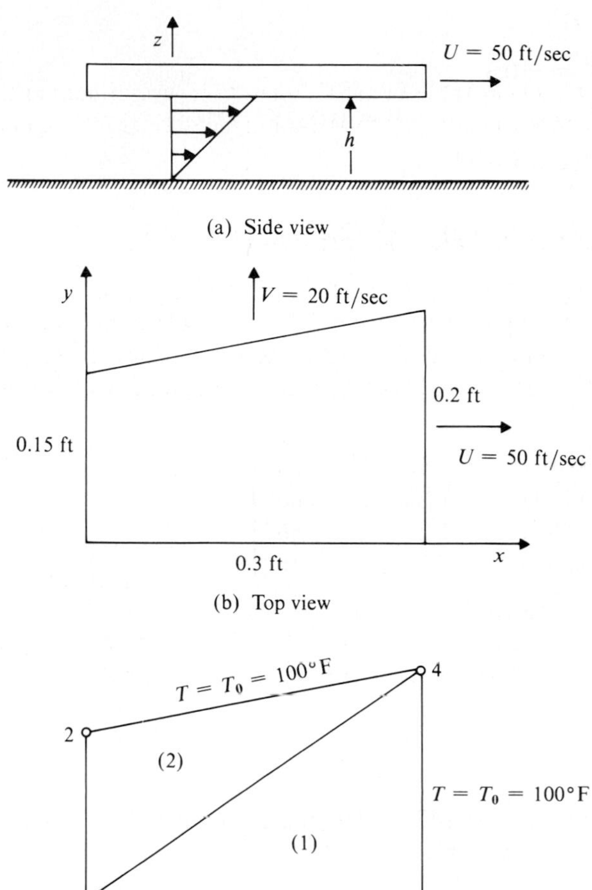

Figure 15.2 Diagram for Example 15.1

$$[B]^{(1)} = 10^4 \begin{matrix} & 1 & 3 & 4 \\ 1 \\ 3 \\ 4 \end{matrix} \begin{bmatrix} 1.667 & -0.667 & -1.000 \\ 1.667 & -0.667 & -1.000 \\ 1.667 & -0.667 & -1.000 \end{bmatrix}^{(1)}$$

$$\{C\}^{(1)} = 10^6 \begin{matrix} 1 \\ 3 \\ 4 \end{matrix} \begin{Bmatrix} -2.042 \\ -2.042 \\ -2.042 \end{Bmatrix}$$

For element 2

$$b_1^{(2)} = 0.05 \qquad c_1^{(2)} = -0.3$$

$$b_4^{(2)} = 0.15 \qquad c_4^{(2)} = 0 \qquad A^{(2)} = 0.0225$$

$$b_2^{(2)} = -0.2 \qquad c_2^{(2)} = 0.3$$

$$B_{11}^{(2)} = \left[-\frac{1}{6}(5.00 \times 10^5)(0.05) - \frac{1}{6}(2.00 \times 10^5)(-0.3) - \frac{1}{6}(0)(0.0225) \right]$$

$$= 0.5833 \times 10^4$$

The element matrices are

$$[B]^{(2)} = 10^4 \begin{matrix} & 1 & 4 & 2 \\ 1 & \\ 4 & \\ 2 & \end{matrix} \begin{bmatrix} 0.5833 & 1.250 & 0.667 \\ 0.5833 & 1.250 & 0.667 \\ 0.5833 & 1.250 & 0.667 \end{bmatrix}$$

$$\{C\}^{(2)} = 10^6 \begin{matrix} 1 \\ 4 \\ 2 \end{matrix} \begin{Bmatrix} -1.532 \\ -1.532 \\ -1.532 \end{Bmatrix}$$

Assembling only equation 1 gives

$$(2.25\ T_1 + 0.667\ T_2 - 0.667\ T_3 + 0.250\ T_4) \times 10^4 = 3.574 \times 10^6$$

Using $T_2 = T_3 = T_4 = 100$, the result is $T_1 = 147°\text{F}$. The full matrix could also have been solved using the normal method described in Chapter 11.

15.4 Second-Order Differential Equations

Second-order partial differential equations with appropriate boundary conditions are certainly widespread in engineering practice. Two-dimensional cases involving field problems will be considered here.

A. Second-Order Terms

Recall the most general second-order differential equation written out as Equation (15.1) with the general boundary conditions (15.2) and (15.3). The general two-dimensional Galerkin method is discussed in Section 15.1. A special treatment of the second order terms

$$\frac{\partial}{\partial x}\left[K_x(x,y)\frac{\partial T}{\partial x}\right] + \frac{\partial}{\partial x}\left[\frac{1}{2}K_{xy}(x,y)\frac{\partial T}{\partial y}\right] + \frac{\partial}{\partial y}\left[\frac{1}{2}K_{xy}(x,y)\frac{\partial T}{\partial x}\right] + \frac{\partial}{\partial y}\left[K_y(x,y)\frac{\partial T}{\partial y}\right]$$

is necessary. Consider the second-order terms in the ith residual (associated with node i):

$$J = -\iint_{A^{(e)}} \left\{ N_i \left[\frac{\partial}{\partial x}\left(K_x \frac{\partial T}{\partial x}\right) + \frac{\partial}{\partial x}\left(\frac{1}{2} K_{xy} \frac{\partial T}{\partial y}\right) + \frac{\partial}{\partial y}\left(\frac{1}{2} K_{xy} \frac{\partial T}{\partial x}\right) + \frac{\partial}{\partial y}\left(K_y \frac{\partial T}{\partial y}\right)\right]\right\}^{(e)} dA$$

where J is defined for convenience.

A method similar to that employed in Chapter 9 to obtain the functional is given here. Start with the identity

$$\begin{aligned}
&\frac{\partial}{\partial x}\left(N_i K_x \frac{\partial T}{\partial x}\right) + \frac{\partial}{\partial x}\left(\frac{1}{2} N_i K_{xy} \frac{\partial T}{\partial y}\right) \\
&+ \frac{\partial}{\partial y}\left(\frac{1}{2} N_i K_{xy} \frac{\partial T}{\partial x}\right) + \frac{\partial}{\partial y}\left(N_i K_y \frac{\partial T}{\partial y}\right) \\
&= N_i \frac{\partial}{\partial x}\left(K_x \frac{\partial T}{\partial x}\right) + N_i \frac{\partial}{\partial x}\left(\frac{1}{2} K_{xy} \frac{\partial T}{\partial y}\right) \\
&+ N_i \frac{\partial}{\partial y}\left(\frac{1}{2} K_{xy} \frac{\partial T}{\partial x}\right) + N_i \frac{\partial}{\partial y}\left(K_y \frac{\partial T}{\partial y}\right) \\
&+ \frac{\partial N_i}{\partial x}\left(K_x \frac{\partial T}{\partial x}\right) + \frac{\partial N_i}{\partial x}\left(\frac{1}{2} K_{xy} \frac{\partial T}{\partial y}\right) \\
&+ \frac{\partial N_i}{\partial y}\left(\frac{1}{2} K_{xy} \frac{\partial T}{\partial x}\right) + \frac{\partial N_i}{\partial y}\left(K_y \frac{\partial T}{\partial y}\right)
\end{aligned}$$

Transferring the last four terms to the other side of the equal sign and substituting the result into J yields

$$\begin{aligned}
J = -\iint_{A^{(e)}} \Bigg\{ &\frac{\partial}{\partial x}\left(N_i K_x \frac{\partial T}{\partial x} + \frac{1}{2} N_i K_{xy} \frac{\partial T}{\partial y}\right) \\
&+ \frac{\partial}{\partial y}\left(\frac{1}{2} N_i K_{xy} \frac{\partial T}{\partial x} + N_i K_y \frac{\partial T}{\partial y}\right) \\
&- \left[\frac{\partial N_i}{\partial x}\left(K_x \frac{\partial T}{\partial x}\right) + \frac{\partial N_i}{\partial x}\left(\frac{1}{2} K_{xy} \frac{\partial T}{\partial y}\right)\right. \\
&\left. + \frac{\partial N_i}{\partial y}\left(\frac{1}{2} K_{xy} \frac{\partial T}{\partial x}\right) + \frac{\partial N_i}{\partial y}\left(K_y \frac{\partial T}{\partial y}\right)\right]\Bigg\}^{(e)} dA
\end{aligned}$$

Now the first half of the terms can be converted from area to line integrals.

Green's theorem has the form

$$\iint_A \left(\frac{\partial f_x}{\partial x} + \frac{\partial f_y}{\partial y}\right) dA = \oint_L (f_x n_x + f_y n_y)\, ds$$

where s is the coordinate along the boundary L. Then

$$J = -\int_{L^{(e)}} \left[\left(N_i K_x \frac{\partial T}{\partial x} + \frac{1}{2} N_i K_{xy} \frac{\partial T}{\partial y} \right) n_x + \left(\frac{1}{2} N_i K_{xy} \frac{\partial T}{\partial x} + N_i K_y \frac{\partial T}{\partial y} \right) n_y \right]^{(e)} ds$$

$$+ \iint_{A^{(e)}} \left[\frac{\partial N_i}{\partial x} \left(K_x \frac{\partial T}{\partial x} \right) + \frac{\partial N_i}{\partial x} \left(\frac{1}{2} K_{xy} \frac{\partial T}{\partial y} \right) + \frac{\partial N_i}{\partial y} \left(\frac{1}{2} K_{xy} \frac{\partial T}{\partial x} \right) + \frac{\partial N_i}{\partial y} \left(K_y \frac{\partial T}{\partial y} \right) \right]^{(e)} dA$$

If the value of T is specified along L on the side of the element, then the weighting function N_i has the value zero there. On the portion of the boundary L_2 where derivative boundary conditions occur,

$$\left(K_x \frac{\partial T}{\partial x} + \frac{1}{2} K_{xy} \frac{\partial T}{\partial y} \right) n_x + \left(\frac{1}{2} K_{xy} \frac{\partial T}{\partial x} + K_y \frac{\partial T}{\partial y} \right) n_y = -\alpha T - \beta \text{ on } L_2$$

from Equation (15.3), so J becomes

$$J = \iint_{A^{(e)}} \left[\frac{\partial N_i}{\partial x} \left(K_x \frac{\partial T}{\partial x} \right) + \frac{1}{2} \frac{\partial N_i}{\partial x} \left(K_{xy} \frac{\partial T}{\partial y} \right) + \frac{1}{2} \frac{\partial N_i}{\partial y} \left(K_{xy} \frac{\partial T}{\partial x} \right) + \frac{\partial N_i}{\partial y} \left(K_y \frac{\partial T}{\partial y} \right) \right]^{(e)} dA + \int_{L_2^{(e)}} N_i (\alpha T + \beta)\, dL_2$$

Substituting the values for N and T yields

$$J = \iint_{A^{(e)}} \left[\frac{K_x}{4A^2} b_i (b_i T_i + b_j T_j + b_k T_k) + \frac{K_{xy}}{8A^2} b_i (c_i T_i + c_j T_j + c_k T_k) + \frac{K_{xy}}{8A^2} c_i (b_i T_i + b_j T_j + b_k T_k) + \frac{K_y}{4A^2} c_i (c_i T_i + c_j T_j + c_k T_k) \right]^{(e)} dA$$

$$+ \int_{L_2^{(e)}} [\alpha N_i (N_i T_i + N_j T_j + N_k T_k) + \beta N_i]^{(e)}\, dL_2$$

Now the integrals over A are evaluated in the normal manner, and the integrals on the boundary are obtained using the methods of Section 9.5. The full element matrices are determined by adding the terms from Equations (15.18) and (15.19).

B. Element Matrices

The complete element matrices are written as

$$B_{ii} = \left\{\frac{1}{4A}\left[K_x b_i^2 + K_{xy} b_i c_i + K_y c_i^2\right] - \frac{1}{6}(M_x b_i + M_y c_i) - \frac{1}{6} PA\right\}^{(e)} + \left(\frac{1}{3}\alpha L_{ij}\right)^{(e)}_{\text{on side } ij} + \left(\frac{1}{3}\alpha L_{ik}\right)^{(e)}_{\text{on side } ik}$$

$$B_{ij} = \left\{\frac{1}{4A}\left[K_x b_i b_j + \frac{1}{2} K_{xy}(b_i c_j + b_j c_i) + K_y c_i c_j\right] - \frac{1}{6}(M_x b_j + M_y c_j) - \frac{1}{12} PA\right\}^{(e)} + \left(\frac{1}{6}\alpha L_{ij}\right)^{(e)}_{\text{on side } ij}$$

$$B_{ik} = \left\{\frac{1}{4A}\left[K_x b_i b_k + \frac{1}{2} K_{xy}(b_i c_k + b_k c_i) + K_y c_i c_k\right] - \frac{1}{6}(M_x b_k + M_y c_k) - \frac{1}{12} PA\right\}^{(e)} + \left(\frac{1}{6}\alpha L_{ik}\right)^{(e)}_{\text{on side } ik}$$

$$B_{ji} = \left\{\frac{1}{4A}\left[K_x b_j b_i + \frac{1}{2} K_{xy}(b_j c_i + b_i c_j) + K_y c_j c_i\right] - \frac{1}{6}(M_x b_i + M_y c_i) - \frac{1}{12} PA\right\}^{(e)} + \left(\frac{1}{6}\alpha L_{ij}\right)^{(e)}_{\text{on side } ij}$$

$$B_{jj} = \left\{\frac{1}{4A}\left[K_x b_j^2 + K_{xy} b_j c_j + K_y c_j^2\right] - \frac{1}{6}(M_x b_j + M_y c_j) - \frac{1}{6} PA\right\}^{(e)} + \left(\frac{1}{3}\alpha L_{ij}\right)^{(e)}_{\text{on side } ij} + \left(\frac{1}{3}\alpha L_{jk}\right)^{(e)}_{\text{on side } jk} \qquad \textbf{(15.20)}$$

$$B_{jk} = \left\{\frac{1}{4A}\left[K_x b_j b_k + \frac{1}{2} K_{xy}(b_j c_k + b_k c_j) + K_y c_j c_k\right] - \frac{1}{6}(M_x b_k + M_y c_k) - \frac{1}{12} PA\right\}^{(e)} + \left(\frac{1}{6}\alpha L_{jk}\right)^{(e)}_{\text{on side } jk}$$

$$B_{ki} = \left\{\frac{1}{4A}\left[K_x b_k b_i + \frac{1}{2} K_{xy}(b_k c_i + b_i c_k) + K_y c_k c_i\right] - \frac{1}{6}(M_x b_i + M_y c_i) - \frac{1}{12} PA\right\}^{(e)} + \left(\frac{1}{6}\alpha L_{ik}\right)^{(e)}_{\text{on side } ik}$$

$$B_{kj} = \left\{\frac{1}{4A}\left[K_x b_k b_j + \frac{1}{2} K_{xy}(b_k c_j + b_j c_k) + K_y c_k c_j\right] - \frac{1}{6}(M_x b_j + M_y c_j) - \frac{1}{12} PA\right\}^{(e)} + \left(\frac{1}{6}\alpha L_{jk}\right)^{(e)}_{\text{on side } jk}$$

Continued on next page

$$B_{kk} = \left\{ \frac{1}{4A} \left[K_x b_k^2 + K_{xy} b_k c_k + K_y c_k^2 \right] - \frac{1}{6} (M_x b_k + M_y c_k) - \frac{1}{6} PA \right\}^{(e)}$$
$$+ \left(\frac{1}{3} \alpha L_{ik} \right)^{(e)}_{\text{on side } ik} + \left(\frac{1}{3} \alpha L_{jk} \right)^{(e)}_{\text{on side } jk} \qquad \textbf{(15.20 cont.)}$$

Equation (9.27) gives the element column. Also L_{ij} represents the length ij, and so on for the other two sides.

Example 15.2 *Solution of Differential Equation*

Given The second-order partial differential equation

$$\frac{\partial^2 T}{\partial x^2} + \frac{\partial^2 T}{\partial y^2} - 20 \frac{\partial T}{\partial x} - 10 \frac{\partial T}{\partial y} + T - 1.25x - 2.5y + 50 = 0 \text{ in } A$$

has both first and second derivative terms that can be solved with the Galerkin method. Figure 15.3 shows the problem geometry.

The boundary conditions are specified values of T along three sides and a derivative boundary condition on the fourth side. Along the bottom (slanted) side, the derivative boundary condition is

$$\frac{\partial T}{\partial x} n_x + \frac{\partial T}{\partial y} n_y + 0.67705 = 0 \text{ along } x = 2y$$

The value of T along the right side is

$$T = \sin\left[\frac{1}{\sqrt{5}} (2 - 2y)\right] + 2.5y + 2.5 \text{ along } x = 2$$

On the top,

$$T = \sin\left[\frac{1}{\sqrt{5}} (x - 8)\right] + 1.25x + 10 \text{ along } y = 4$$

and along the left side,

$$T = \sin\left[\frac{1}{\sqrt{5}} (-2y)\right] + 2.5y \text{ along } x = 0$$

These are evaluated at the node points.

Objective Use computer program GALERK to solve this boundary value problem. Employ a mesh of four quadrilaterals by four quadrilaterals as shown in Figure 15.4. This is to be further subdivided into four simplex elements per quadrilateral. There are 16 quadrilaterals and 64 elements. Also there are 41 node points.

The exact solution to this differential equation and boundary conditions is

$$T = \sin\left[\frac{1}{\sqrt{5}} (x - 2y)\right] + 1.25x + 2.5y$$

Compare this to the finite element solution at a number of node points. Calculate the error.

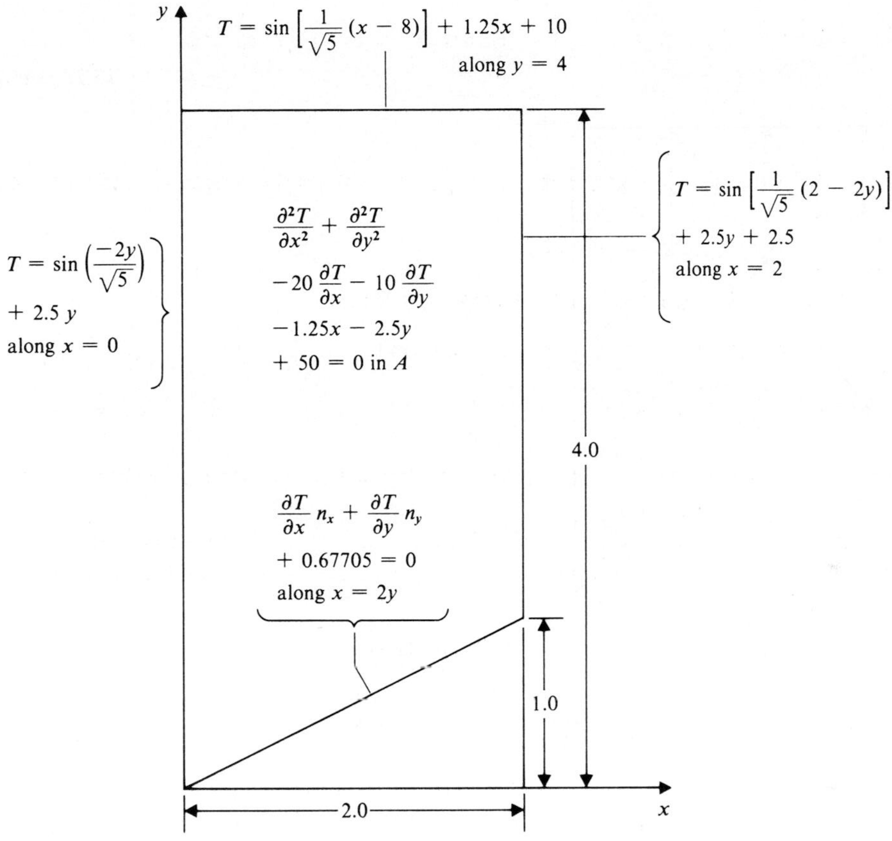

Figure 15.3 Differential equation and boundary conditions for Example 15.2

Solution The nodal coordinates and element/node connectivity are determined as usual. These values are not given here. The coefficient functions are evaluated at the center of each element. Also the boundary conditions are calculated for the appropriate node or element side. None of these values are presented here because of space limitations.

Table 15.1 compares the nodal values for the finite element solution at certain node points and the exact solution at those node points. In all cases the errors are only a few percent.

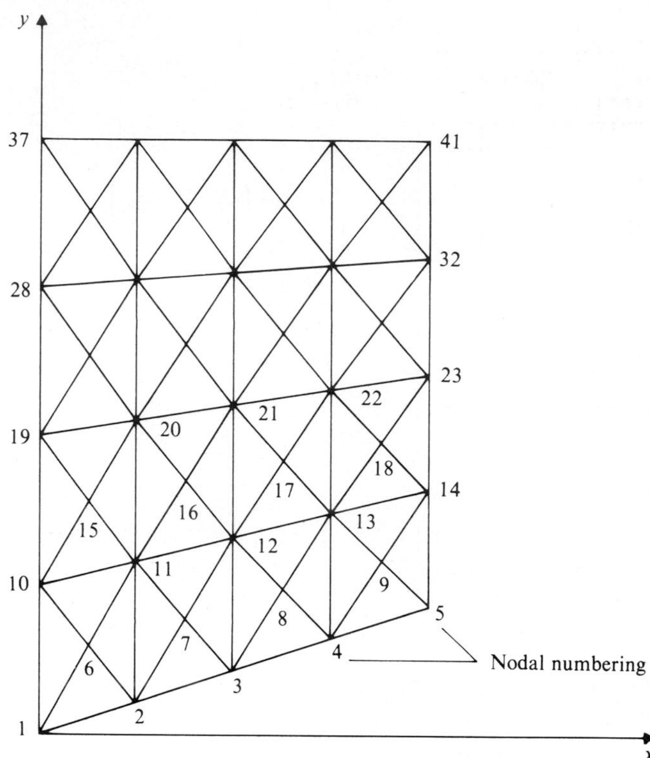

Figure 15.4 Division into elements

Table 15.1
Comparison of Finite Element and Exact Solution at Node Points

Node Number	Finite Element Solution	Exact Solution
1	0.000	0.000
2	1.239	1.250
3	2.484	2.500
4	3.717	3.750
5	5.000	5.000
10	1.720	1.720
11	3.035	3.031
12	4.348	4.342
13	5.668	5.655
14	6.968	6.968
19	4.024	4.024
20	5.282	5.274

Continued on next page

Table 15.1 *Continued*

Node Number	Finite Element Solution	Exact Solution
21	6.542	6.524
22	7.797	7.774
23	9.024	9.024
28	7.058	7.058
29	8.006	8.000
30	8.959	8.945
31	9.928	9.895
32	10.850	10.850

C. Numerical Oscillations

Numerical oscillations, also called numerical instabilities, can occur in the finite element solution of differential equations with first and second derivative terms. These can occur in either one- or two-dimensional problems. Usually they occur when the ratio of first derivative terms to second derivative terms is too large. A full discussion of this problem is beyond the scope of this text, but some discussion is worthwhile.

One case for which numerical oscillations are a real problem is fluid mechanics. The first derivative terms, called convective terms, are often large compared to the second derivative terms, called diffusion terms. A method of reducing or eliminating these problems is called upwinding. It is described in advanced books and journal articles.

The first derivative terms can be approximated in the x direction by

$$M_x \frac{\partial T}{\partial x} \cong M_x \frac{\Delta T}{\Delta x}$$

where Δx = element length in x direction and ΔT = change in T over element. Similarly, for the second-order term,

$$\frac{\partial}{\partial x}\left(K_x \frac{\partial T}{\partial x}\right) \cong K_x \frac{\Delta T}{(\Delta x)^2}$$

Taking the ratio of these terms gives

$$\frac{M_x \dfrac{\partial T}{\partial x}}{\dfrac{\partial}{\partial x}\left(K_x \dfrac{\partial T}{\partial x}\right)} \cong \frac{M_x \dfrac{\Delta T}{\Delta x}}{K_x \dfrac{\Delta T}{(\Delta x)^2}} = \frac{M_x (\Delta x)}{K_x}$$

This ratio is called the Péclet number and written as Pe.

$$\text{Pe} = \frac{M_x (\Delta x)}{K_x}$$

A similar ratio is easily obtained in the y direction.

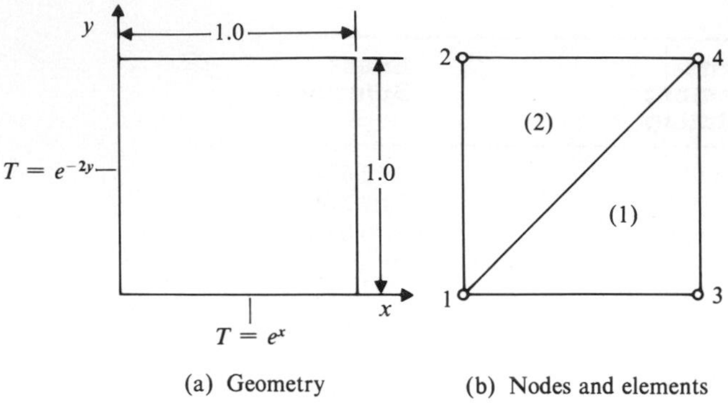

Figure 15.5 Geometry for Problem 15.1

When the Péclet number is less than or equal to 2, the solution does not have numerical oscillations. This limitation

$$\frac{M_x\,(\Delta x)}{K_x} \leq 2$$

can be satisfied by making Δx small. This simply means that the element size must be small enough. If the Péclet number is larger than this value, some numerical oscillations may or may not occur, depending upon the boundary conditions and size of the region. Example 15.2 shows a problem where the Péclet number is 5 but no oscillations are found. This is because the value of T is specified on three sides.

15.5 Problems

15.1 A differential equation has the form

$$M_x \frac{\partial T}{\partial x} + M_y \frac{\partial T}{\partial y} = 0$$

The coefficients have the values $M_x = -20$ and $M_y = -10$. A solution is to be found over the square of size 1 by 1 shown in Figure 15.5(a). Along the left side and bottom of the square, the value of T varies as indicated in the Figure 15.5(a).

Use a two-element model, shown in Figure 15.5(b), to determine the value of T at the upper right-hand corner (node 4) of the square. Employ the element matrices obtained with Galerkin's method in this chapter. Compare the finite element result to the exact solution

$$T = \exp\left[20\left(\frac{x}{20} - \frac{y}{10}\right)\right]$$

at node 4 ($T_4 = 0.3679$).

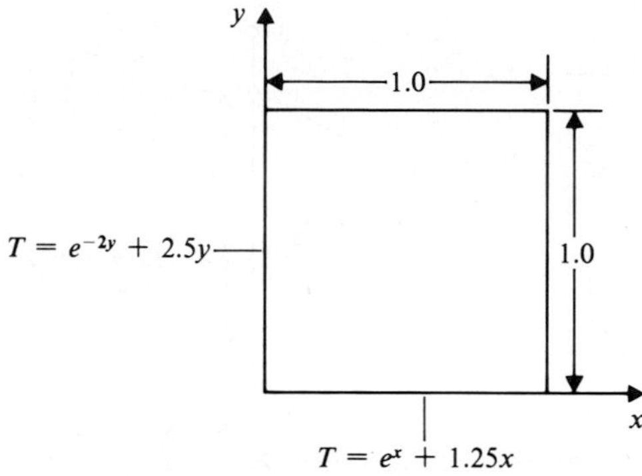

Figure 15.6 Geometry for Problem 15.2

15.2 The differential equation

$$M_x \frac{\partial T}{\partial x} + M_y \frac{\partial T}{\partial y} + Q = 0$$

where $M_x = -20$, $M_y = -10$, and $Q = 50$ represents an extension of the previous problem with the addition of the Q term. The geometry and boundary conditions are shown in Figure 15.6.

Use the same two-element model as employed in Problem 15.1 and shown in Figure 15.5(b). Solve for the value of T at node 4. Compare to the exact solution

$$T = \exp\left[20\left(\frac{x}{20} - \frac{y}{10}\right)\right] + 25\left(\frac{x}{20} + \frac{y}{10}\right)$$

at node 4.

15.3 This problem is very similar to Example 15.1. A thin film of fluid flows between two parallel solid surfaces. The lower plate is infinite in extent and not moving. The upper plate is triangular and moves over the lower surface with velocity components U, V in the x, y directions. Figure 15.7 shows the geometry. The differential (energy) equation for the temperature is

$$J\rho c_v \left(\frac{1}{2} U \frac{\partial T}{\partial x} + \frac{1}{2} V \frac{\partial T}{\partial y}\right) - \mu\left[\left(\frac{U}{h}\right)^2 + \left(\frac{V}{h}\right)^2\right] = 0$$

where $J = 778$ ft-lbf/Btu, $\rho = 1.7$ lbf-sec^2/ft^4, $c_v = 12$ Btu-ft/lbf-sec^2-°F, $h = 1.0 \times 10^{-4}$ ft, $U = 10$ ft/sec, $V = 0$ ft/sec, and $\mu = 4.0 \times 10^{-3}$ lbf-sec/ft^2. The fluid entering the region under the top pad is at temperature T_0 of 120°F. Use one simplex element to determine the temperature at node 1. Carry out the solution by hand to see how the method works.

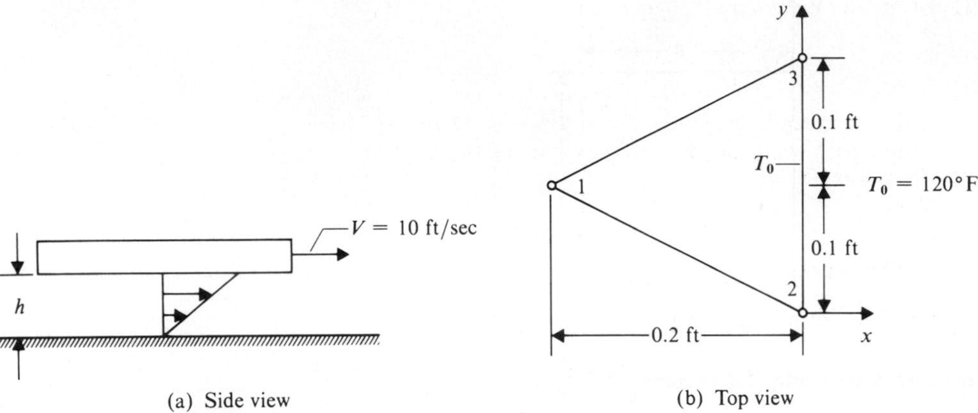

(a) Side view

(b) Top view

Figure 15.7 Geometry for Problem 15.3

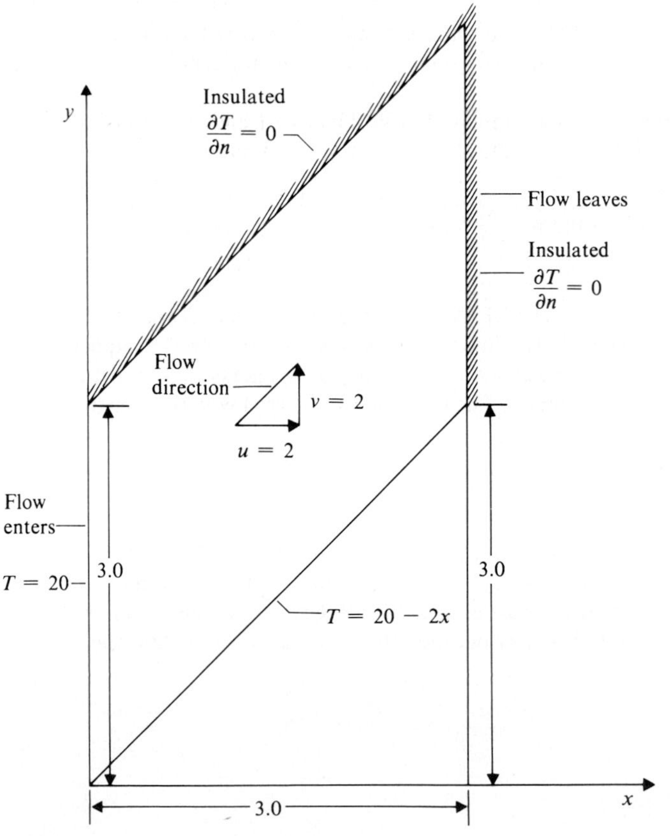

Figure 15.8 Geometry for Problem 15.5

15.4 Consider the differential equation

$$M_x \frac{\partial T}{\partial x} + M_y \frac{\partial T}{\partial y} = 0$$

with the boundary condition $T = T_0$ on L_1. Derive an element matrix for this boundary value problem, using the subdomain weighting function

$$W = \begin{cases} 1.0, \text{ in element} \\ 0.0, \text{ outside element} \end{cases}$$

and a simplex element.

15.5 A two-dimensional energy equation has the form

$$\underbrace{\frac{\partial}{\partial x}\left(k \frac{\partial T}{\partial x}\right) + \frac{\partial}{\partial y}\left(k \frac{\partial T}{\partial y}\right)}_{\text{conduction terms}} - \underbrace{\rho c_p u \frac{\partial T}{\partial x} - \rho c_p v \frac{\partial T}{\partial y}}_{\text{convection terms}} = 0$$

where $k = 1$ (thermal conductivity), $\rho = 1$ (density), $c_p = 1$ (specific heat), $u = 2$ (horizontal velocity), and $v = 2$ (vertical velocity) in some appropriate set of units. Figure 15.8 shows the geometry. Fluid enters the region on the left side at $T = 20$ and leaves on the right through an insulated section where

$$\frac{\partial T}{\partial x} = \frac{\partial T}{\partial n} = 0$$

Along the upper side, the flow is insulated, and along the lower side the temperature is maintained at $T = 20 - 2x$.

Use computer program GALERK to solve for the temperature distribution in the region. Employ at least 20 elements.

Part 4
Time-Transient Problems

Chapter 16

Time-Dependent Problems

16.1 Introduction

Up to this point in the text, all problems considered have been independent of time. This chapter provides an introduction to some time-dependent engineering analyses. Both explicit methods, which solve for nodal equations without simultaneous equations, and weighted implicit methods, which employ simultaneous equations, are discussed.

Only problems with both spatial and time dependence (no purely time-dependent problems) are discussed. First-order purely time-dependent problems can be treated with the first-order Galerkin method presented in Section 8.3. Only partial (not ordinary) differential equations are discussed in this chapter.

A. Approximations in Time and Space

As is shown shortly, time-dependent problems of engineering interest involve both space and time. That is, variations of the temperature, pressure, displacement, or other variables occur in both space and time. There is a fundamental difference in these variations. The time variations start from some initial conditions and then propagate through time. They are not subject to the types of geometric complexities that boundary value problems in space are. Thus, the techniques for treating the time variations are often different from the finite element treatments for spatial variations that have been discussed earlier in the text.

B. Finite Differences in Time

Finite difference methods are commonly employed for the time derivatives, while finite elements are retained for the spatial derivatives. This technique leads directly to quite straightforward solutions. It is assumed in this chapter that the reader has been exposed to a minimal introduction to finite difference techniques elsewhere. Concepts such as forward, backward, and central differences are not developed from basic principles here. These finite difference approximations for the time derivatives in the differential equations are not difficult to apply. Generally, finite element matrices $[B]$ and $\{C\}$ reappear for the spatial derivatives with some minor modifications.

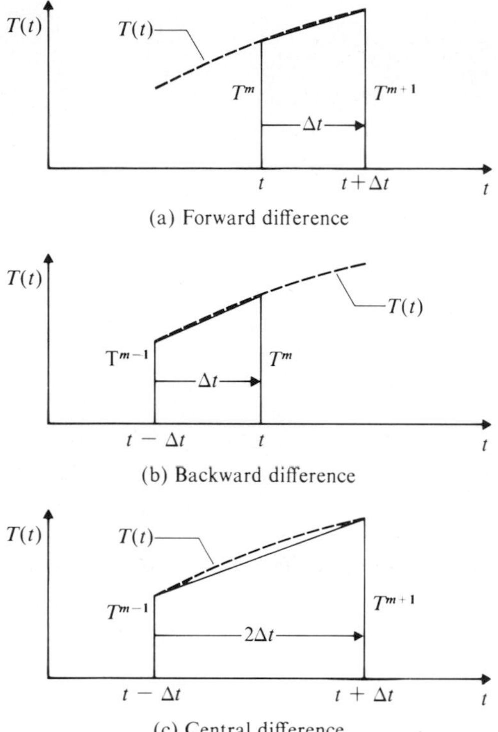

Figure 16.1 Finite difference approximations for $\partial T/\partial t$

Consider some preliminary examples of finite differences applied to a first-order time derivative $\partial T/\partial t$. Here t represents time and T represents the variable to be calculated. Let the current time be t and the time step be denoted Δt, as indicated in Figure 16.1. The forward difference in time is approximated by

$$\frac{\partial T}{\partial t} = \frac{T^{m+1} - T^m}{\Delta t} \tag{16.1}$$

where $T^m = T$ evaluated at time t and $T^{m+1} = T$ evaluated at time $t + \Delta t$. Note that in this section T^m is taken as the value of T at the instant at which the finite element approximation in space is employed.

An alternative approximation is the backward difference

$$\frac{\partial T}{\partial t} = \frac{T^m - T^{m-1}}{\Delta t} \tag{16.2}$$

Here $T^{m-1} = T$ evaluated at time $t - \Delta t$ and $T^m = T$ evaluated at time t. Figure 16.1(b) illustrates this. The forward and backward difference formulas are very similar in format, but the resulting solution algorithms are significantly different. This is explored in some detail in later sections.

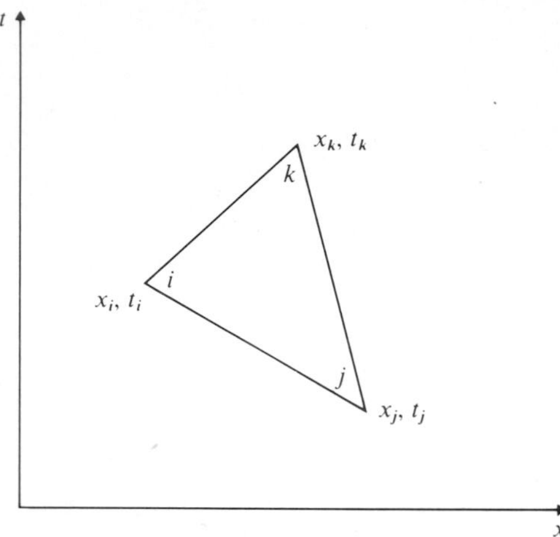

Figure 16.2 Simplex finite element in space and time

Central differences can be thought of as an averaging of the forward and backward differences. Assume Δt to be constant. Then

$$\frac{\partial T}{\partial t} = \underbrace{\frac{1}{2}\left(\frac{T^{m+1} - T^{m}}{\Delta t}\right)}_{\text{Forward difference}} + \underbrace{\frac{1}{2}\left(\frac{T^{m} - T^{m-1}}{\Delta t}\right)}_{\text{Backward difference}}$$

with the result being the central difference approximation

$$\frac{\partial T}{\partial t} = \frac{T^{m+1} - T^{m-1}}{2\,\Delta t} \tag{16.3}$$

Note that the central term at T^m cancels out. Central differences thus involve the average slope over the interval $2\Delta t$, as indicated in Figure 16.1(c).

C. Finite Elements in Time

It is certainly feasible to use finite element approximations in time as well as space. A number of algorithms employing this technique exist. Basically the time dependence is treated as if it were simply another space dimension. Thus a function $T(x,t)$ is analyzed as if t replaces the second space variable y. Figure 16.2 illustrates a simplex element. Interpolation functions have the form

$$T^{(e)} = N_i T_i + N_j T_j + N_k T_k$$

where

$$N_i = N_i(x,t), \quad N_j = N_j(x,t), \quad N_k = N_k(x,t)$$

The Galerkin methods of Chapter 15 are employed if the time derivatives are of first order.

16.2 Explicit Method

A. Explicit and Implicit Methods

An important concept in the numerical analysis of time dependence is that of explicit and implicit methods. Both methods are widely employed, so both are discussed here. Generally, the explicit methods are much easier to apply and should be tried first. If they work, a solution is obtained and implicit methods are not invoked. Implicit methods have often been found successful when explicit methods are not.

As shown in this section, explicit methods evaluate the function $T(x,t)$ at time $t + \Delta t$ solely upon the basis of the solution obtained at time t. They do not solve a boundary value problem at each time step because the value of $T(x,t)$ is already known at time t. Therefore, they are often called marching methods. Starting with the initial time, one marches forward in time to the times $t_0 = t_0$, $t_1 = t_0 + \Delta t$, $t_2 = t_0 + 2\Delta t$, . . . , $t_f = t_0 + m\Delta t$, until the desired final time t_f (after m time steps) is evaluated. The solution for each time step is very fast. Unfortunately the size of the time step Δt is limited by numerical instabilities, which may require that a very large number of time steps be taken for some problems.

Implicit methods, discussed in detail in later sections, determine the solution at time t based upon those values of $T(x,t)$ at both time t and at time $t + \Delta t$. Thus, a boundary value problem with its associated set of simultaneous algebraic global equations in space must be solved for each time step. More computer time—as well as programming work—is involved, but greater accuracy results at each time step. Often, much larger time steps can be taken with implicit methods than with explicit methods. The advantage of one over the other depends upon the problem parameters and the familiarity of the user with one approach or the other.

B. Time-Dependent Parabolic Problem

Consider a general time-dependent parabolic differential equation of the form

$$\frac{\partial}{\partial x}\left(K_x \frac{\partial T}{\partial x}\right) - K_t \frac{\partial T}{\partial t} = 0 \qquad \textbf{(16.4)}$$

Here the coefficient functions are functions of space and time: $K_t = K_t(x,t)$ and $K_x = K_x(x,t)$. The time derivative is of first order and has a different sign from the second-order spatial derivative term. This is typical of parabolic equations. Other terms, such as $PT + Q$, will be added in Section 16.5.

At the initial time $t = t_0$, the initial condition on $T(x,t)$ is

$$T = T_{t0} \qquad \textbf{(16.5)}$$

where T_{t0} is known for $x_0 \leq x \leq x_L$. Figure 16.3 shows the geometry. On the boundaries at x_0 and x_L for all time $t > t_0$, the boundary conditions are

$$T = T_{x0} \text{ or } K_{x0} \frac{dT}{dx} = 0 \text{ at } x = x_0 \qquad \textbf{(16.6a)}$$

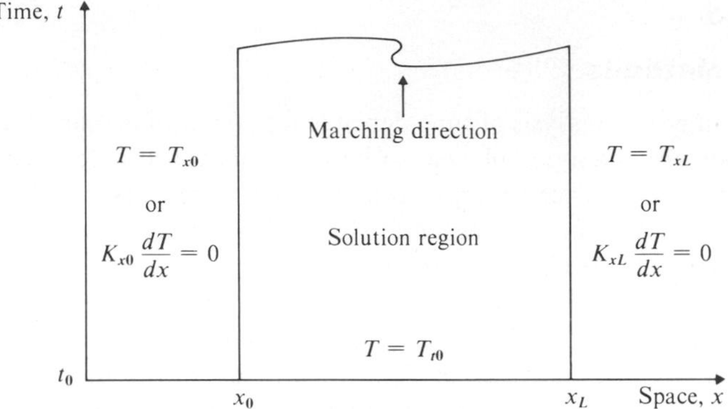

Figure 16.3 Parabolic boundary value problem in space and time

$$T = T_{xL} \text{ or } K_{xL}\frac{dT}{dx} = 0 \text{ at } x = x_L \tag{16.6b}$$

The specified values T_{x0} and T_{xL} are assumed known over time. The derivative boundary conditions would correspond to an insulated boundary in a transient heat-conduction problem.

C. Finite Difference in Time

The explicit method uses a forward finite difference approximation for the time derivative term

$$\frac{\partial T}{\partial t} = \frac{T^{m+1} - T^m}{\Delta t}$$

where $T^m = T$ evaluated at time t and $T^{m+1} = T$ evaluated at time $t + \Delta t$. Backward differences produce the implicit method; this is discussed in Section 16.4. The space derivative term is evaluated at time t (for the explicit method) as

$$\frac{\partial}{\partial x}\left(K_x\frac{\partial T}{\partial x}\right) = \frac{\partial}{\partial x}\left(K_x\frac{\partial T^m}{\partial x}\right)$$

With these approximations, the differential equation (16.4) becomes

$$\frac{\partial}{\partial x}\left(K_x\frac{\partial T^m}{\partial x}\right) - K_t\left(\frac{T^{m+1} - T^m}{\Delta t}\right) = 0 \tag{16.7}$$

In this equation, only the term T^{m+1} is an unknown. All of the other terms involving T^m are known from the previous time step.

D. Approximation in Space

Galerkin's method is employed for the finite element formulation in space. After substituting the approximation $T^*(x,t)$, the error is

$$r(x,t) = \sum_{e=1}^{E} r^{(e)} = \sum_{e=1}^{E} \left[\frac{\partial}{\partial x}\left(K_x \frac{\partial T^m}{\partial x}\right) - K_t \frac{T^{m+1} - T^m}{\Delta t}\right]^{(e)}$$

The element weighting functions are $W_i = N_i$, $W_j = N_j$ as usual.

The element residuals can be written as the summation of the space derivative terms and the time derivative terms. For the ith node in element (e), the residual is

$$R_i^{(e)} = [R_i]_{\text{space}}^{(e)} + [R_i]_{\text{time}}^{(e)} \tag{16.8}$$

where

$$[R_i]_{\text{space}}^{(e)} = \int_{L^{(e)}} \left\{N_i \left[\frac{\partial}{\partial x}\left(K_x \frac{\partial T^m}{\partial x}\right)\right]\right\}^{(e)} dx$$

$$[R_i]_{\text{time}}^{(e)} = -\int_{L^{(e)}} \left\{N_i \left[-K_t \frac{T^{m+1} - T^m}{\Delta t}\right]\right\}^{(e)} dx$$

Also for the jth node in element (e), the residual is

$$R_j^{(e)} = [R_j]_{\text{space}}^{(e)} + [R_j]_{\text{time}}^{(e)} \tag{16.9}$$

where

$$[R_j]_{\text{space}}^{(e)} = -\int_{L^{(e)}} \left\{N_j \left[\frac{\partial}{\partial x}\left(K_x \frac{\partial T^m}{\partial x}\right)\right]\right\}^{(e)} dx$$

$$[R_j]_{\text{time}}^{(e)} = -\int_{L^{(e)}} \left\{N_j \left[-K_t \frac{T^{m+1} - T^m}{\Delta t}\right]\right\}^{(e)} dx$$

Now the space and time terms are treated separately.

The space derivative terms are easily evaluated by either the variational methods (of Chapter 5) or Galerkin's method (in Chapter 8). The result is

$$[R_i]_{\text{space}}^{(e)} = \left[\frac{K}{L} b_i b_i\right]^{(e)} T_i^m + \left[\frac{K}{L} b_i b_j\right]^{(e)} T_j^m$$

$$[R_j]_{\text{space}}^{(e)} = \left[\frac{K}{L} b_i b_j\right]^{(e)} T_i^m + \left[\frac{K}{L} b_j b_j\right]^{(e)} T_j^m$$

These could be put in element matrix form, but they are left written for this section to illustrate the method. Note that the terms T_i and T_j are evaluated at time t (time step m). Thus the space terms are taken as *known* in the explicit method.

E. Approximation in Time—Uncoupled Equations

The simplest approximation for the time derivative terms is to assume that they do not vary with x. This approximation is used in this chapter to introduce the reader to explicit time-dependent problems. (More accurate approximations for the time derivative term are discussed later.) This simple approximation produces a set of uncoupled equations that are very similar to those developed for explicit finite difference methods (in both time and space).

Element residuals for the time derivative are

$$[R_i]^{(e)}_{\text{time}} = -\int_{L^{(e)}} \left\{ N_i \underbrace{\left[-K_t \frac{T^{m+1} - T^m}{\Delta t} \right]}_{\substack{\text{Assumed constant} \\ \text{over element}}} \right\}^{(e)} dx$$

and

$$[R_j]^{(e)}_{\text{time}} = -\int_{L^{(e)}} \left\{ N_j \underbrace{\left[-K_t \frac{T^{m+1} - T^m}{\Delta t} \right]}_{\substack{\text{Assumed constant} \\ \text{over element}}} \right\}^{(e)} dx$$

Evaluating the integrals yields

$$[R_i]^{(e)}_{\text{time}} = \left[\frac{L}{2} \frac{K_t}{\Delta t} \right]^{(e)} T_i^{m+1} - \left[\frac{L}{2} \frac{K_t}{\Delta t} \right]^{(e)} T_i^m$$

$$[R_j]^{(e)}_{\text{time}} = \left[\frac{L}{2} \frac{K_t}{\Delta t} \right]^{(e)} T_j^{m+1} - \left[\frac{L}{2} \frac{K_t}{\Delta t} \right]^{(e)} T_j^m$$

Again an element matrix is not used to better illustrate where each term is employed.

The element residuals are the sum of the space and time terms. Recall that for simplex elements in one dimension $b_i = -1$ and $b_j = 1$. These are used in the next expression to make it simpler without loss of generality.

$$\begin{aligned} R_i^{(e)} = {} & \left[\frac{K_x}{L} \right]^{(e)} T_i^m + \left[-\frac{K_x}{L} \right]^{(e)} T_j^m \\ & + \left[\frac{L}{2} \frac{K_t}{\Delta t} \right]^{(e)} T_i^{m+1} - \left[\frac{L}{2} \frac{K_t}{\Delta t} \right]^{(e)} T_i^m \end{aligned} \tag{16.10}$$

$$\begin{aligned} R_j^{(e)} = {} & \left[-\frac{K_x}{L} \right]^{(e)} T_i^m + \left[\frac{K_x}{L} \right]^{(e)} T_j^m \\ & + \left[\frac{L}{2} \frac{K_t}{\Delta t} \right]^{(e)} T_j^{m+1} - \left[\frac{L}{2} \frac{K_t}{\Delta t} \right]^{(e)} T_j^m \end{aligned} \tag{16.11}$$

In the explicit method, it is not necessary to solve a boundary value problem (a set of simultaneous algebraic equations) in space. Thus it is much more convenient to write the desired equations in nodal form rather than in element form.

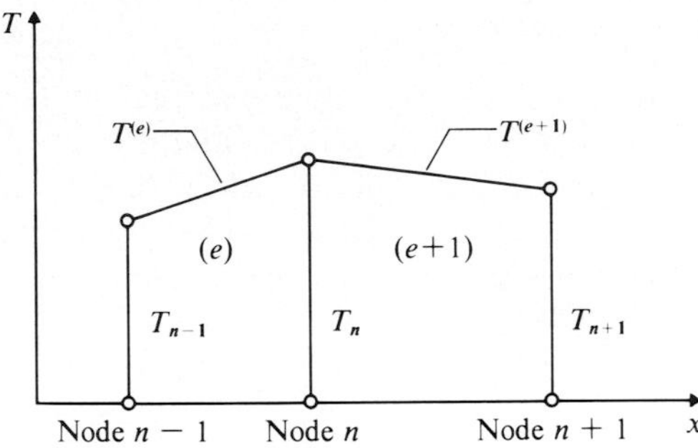

Figure 16.4 Geometry for node n

F. Nodal Equations

Consider two adjacent elements (e) and $(e+1)$ with associated node points $n-1$, n, $n+1$. Figure 16.4 illustrates the elements and nodes. For element (e), residual R_j is used, while with element $(e+1)$, residual R_i is employed. From Equations (16.10) and (16.11) the nodal equation is assembled as

$$\left[-\frac{K_x}{L}\right]^{(e)} T_{n-1}^{m} + \left[\frac{K_x}{L}\right]^{(e)} T_n^m + \left[\frac{L}{2}\frac{K_t}{\Delta t}\right]^{(e)} T_n^{m+1} - \left[\frac{L}{2}\frac{K_t}{\Delta t}\right]^{(e)} T_n^m$$
$$+ \left[\frac{K_x}{L}\right]^{(e+1)} T_n^m + \left[-\frac{K_x}{L}\right]^{(e+1)} T_{n+1}^m + \left[\frac{L}{2}\frac{K_t}{\Delta t}\right]^{(e+1)} T_n^{m+1} - \left[\frac{L}{2}\frac{K_t}{\Delta t}\right]^{(e+1)} T_n^m = 0 \tag{16.12}$$

Because this is now the nodal residual (for node n), the equation is set to zero.

The various nodal values are T_n^{m+1}, which is unknown, and T_{n-1}^m, T_n^m, and T_{n+1}^m, which are known. Clearly Equation 16.12 can be solved for the unknown nodal value at time step $m+1$. The final result is

$$T_n^{m+1} = T_n^m - \frac{\left[-\frac{K_x}{L}\right]^{(e)} T_{n-1}^m + \left[\frac{K_x}{L}\right]^{(e)} T_n^m}{\left[\frac{L}{2}\frac{K_t}{\Delta t}\right]^{(e)} + \left[\frac{L}{2}\frac{K_t}{\Delta t}\right]^{(e+1)}}$$
$$- \frac{\left[\frac{K_x}{L}\right]^{(e+1)} T_n^m + \left[-\frac{K_x}{L}\right]^{(e+1)} T_{n+1}^m}{\left[\frac{L}{2}\frac{K_t}{\Delta t}\right]^{(e)} + \left[\frac{L}{2}\frac{K_t}{\Delta t}\right]^{(e+1)}} \tag{16.13}$$

Note that the solution for T_n at time step $m+1$ is easily solved (or programmed for solution very easily). This is the objective of the explicit method. Further note that this expression is still fully general with regard to variable coefficients K_x, K_t, and length L from element to element in space. Also the size of the time step can be variable over time.

At a given column of node points in space, the spatial derivatives are evaluated at time m, as shown by the small circles in Figure 16.5. The nodes on either side, $n-1$ and

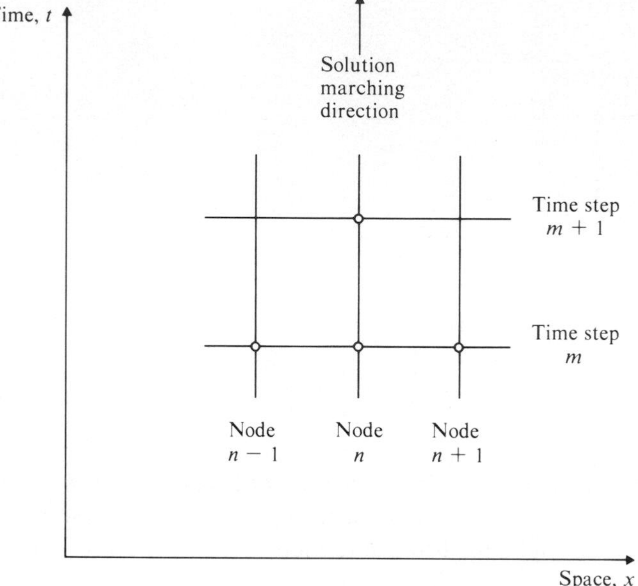

Figure 16.5 Computational nodes for explicit method

$n+1$, are related to the node n by the finite element modeling in space. At time step $m+1$, only the node n is involved. Figure 16.5 shows the computational nodes for the explicit method discussed here. The solution marches outward in time for as many time steps as desired.

G. Constant Space Parameters

As a special case, consider the element length to be constant, that is, $L^{(e)} = L^{(e+1)}$, and the coefficient functions to be constant, that is,

$$K_x^{(e)} = K_x^{(e+1)}$$

$$K_t^{(e)} = K_t^{(e+1)}$$

Equation (16.13) reduces to

$$T_n^{m+1} = T_n^m + \left(\frac{K_x}{L^2}\frac{\Delta t}{K_t}\right)^{(e)}(T_{n-1}^m - 2T_n^m + T_{n+1}^m) \qquad \textbf{(16.14)}$$

This is exactly the same as the result for a forward difference in time and a central difference in space.

Note that a single parameter, denoted by r, is found in the explicit expression for the unknown node.

$$r^{(e)} = \left(\frac{K_x}{L^2}\frac{\Delta t}{K_t}\right)^{(e)} \qquad \textbf{(16.15)}$$

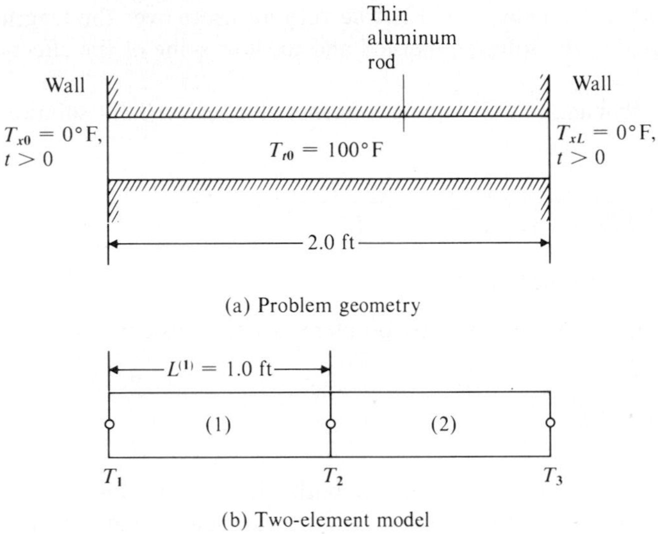

Figure 16.6 Insulated thin aluminum rod (Example 16.1—Explicit transient heat conduction with two elements)

This is a dimensionless number called the Courant number. Then

$$T_n^{m+1} = T_n^m + r^{(e)}(T_{n-1} - 2T_n + T_{n+1}) \tag{16.16}$$

This is the expression used in the first example.

Example 16.1 *Explicit Transient Heat Conduction with Two Elements*

Given Consider a thin aluminum rod surrounded by an insulating material except at the two ends. Figure 16.6(a) illustrates the problem. The temperature distribution is a function of position x and time t. Because the rod is thin, it is assumed that the temperature is constant over its cross section. Properties for the problem are $L = 2$ ft (rod length); $k = 0.0370$ Btu/sec-ft-°F (thermal conductivity); $c = 0.212$ Btu/lbm-°F (specific heat); $\rho = 168$ lbm/ft^3 (density). Initially the rod is at 100°F. At time $t = 0$, the ends of the rod have their temperature suddenly reduced to 0°F.

Objective Use a two-element model, shown in Figure 16.6(b), to solve for the temperature at the center of the rod as a function of time. Consider time step sizes for four cases:

Case 1 $\Delta t = 800$ sec
Case 2 $\Delta t = 400$ sec
Case 3 $\Delta t = 200$ sec
Case 4 $\Delta t = 100$ sec

This example uses only two elements. Thus, it will not be very accurate over the length of the rod, but it serves to illustrate the solution method and to show some of the effects of time-step size.

The exact solution to this problem is well known in terms of a series. That solution is

$$T = \frac{4T_{t0}}{\pi} \sum_{n=0}^{\infty} \frac{1}{(2n+1)} \exp\left[-\frac{K_x}{K_t}\frac{(2n+1)^2\pi^2 t}{L^2}\right] \sin\frac{(2n+1)\pi x}{L}$$

It can be compared with the finite element solution.

Solution The differential equation governing this problem is the transient heat conduction equation

$$\frac{\partial}{\partial x}\left(k\frac{\partial T}{\partial x}\right) - \rho c\frac{\partial T}{\partial t} = 0$$

The rod cross sectional area A could have been included in both terms, but it would cancel out. The initial condition at $t = 0$ is $T = T_{t0} = 100°\text{F}$, and the boundary conditions are $T = T_{x0} = 0°\text{F}$ at $x = 0$ and $T = T_{xL} = 0°\text{F}$ at $x = L$ for $t > 0$.

Comparing the general differential equation (16.4)

$$\frac{\partial}{\partial x}\left(K_x\frac{\partial T}{\partial x}\right) - K_t\frac{\partial T}{\partial t} = 0$$

yields $K_x = k = 0.0370$ Btu/sec-ft-°F and $K_t = \rho c = 35.6$ Btu/ft^3-°F. The boundary conditions are readily apparent.

Equation (16.16) is used here for the middle node. It becomes

$$T_2^{m+1} = T_2^m + r^{(e)}(T_1 - 2T_2 + T_3)$$

where the dimensionless parameter r was defined earlier (16.15) as

$$r^{(e)} = \left(\frac{K_x}{L^2}\frac{\Delta t}{K_t}\right)^{(e)}$$

For this example, the Courant number is

Case	*Time step* Δt	*Courant number* $r^{(e)}$
1	800	0.831
2	400	0.416
3	200	0.208
4	100	0.104

A very simple computer program was written to evaluate this for each time step.

Generally it is not a good idea to set the boundary nodes to zero for the first time step. One half of the change usually gives good solutions. Thus $T_1 = 50°\text{F}$, $T_3 = 50°\text{F}$, for $t = \Delta t$ $(m=1)$. For the remainder of the time steps, the value 0°F is employed: $T_1 = 0°\text{F}$, $T_3 = 0°\text{F}$, for $t = 2\Delta t, 3\Delta t, \ldots$ $(m = 2,3, \ldots)$. Setting the boundary nodes immediately to zero tends to set up undesirable spatial variations in the first few time steps.

Table 16.1
Temperature versus Time at 100-Second Increments
(Example 16.1: Explicit Transient Heat Conduction with Two Elements)

Time (Sec)	Temperature (°F)				
	Case 1 $\Delta t = 800$ Sec $r = 0.831$	Case 2 $\Delta t = 400$ Sec $r = 0.416$	Case 3 $\Delta t = 200$ Sec $r = 0.208$	Case 4 $\Delta t = 100$ Sec $r = 0.104$	Exact Solution
100				89.61	94.35
200			79.22	70.99	75.83
300				56.24	58.96
400		58.45	46.30	44.55	45.65
500				35.30	35.36
600			27.06	27.96	27.35
700				22.15	21.16
800	16.89	9.87	15.82	17.55	16.37
900				13.90	12.67
1000			9.24	11.02	9.81
1100				8.73	7.59
1200		1.67	5.40	6.91	5.87
1300				5.48	4.54
1400			3.16	4.34	3.52
1500				3.44	2.72
1600	−11.18	0.28	1.85	2.72	2.11
1700				2.16	1.63
1800			1.08	1.71	1.26
1900				1.35	0.98
2000		0.05	0.63	1.07	0.76
2100				0.85	
2200			0.37	0.67	
2300				0.53	
2400	7.41	0.01	0.22	0.42	

Case 1 ($\Delta t = 800$ sec)

Here the Courant number is $r = 0.831$. At time $t = 0$ ($m = 1$), the nodal values are $T_1^1 = 50$, $T_2^1 = 100$, $T_3^1 = 50$. The middle node at time $t = \Delta t$ ($m+1=2$) is

$$T_2^2 = T_2^1 + r(T_1^1 - 2T_2^1 + T_3^1)$$

or

$$T_2^2 = 100 + (0.831)[50 - 2(100) + 50]$$

The result is $T_2^2 = 16.89°$F. This is a very large drop in temperature for a single time step. Note that using the boundary condition of 0°F for the first time step gives the predicted temperature of $T_2^2 = -66.2°$F. This value is very far from the exact solution at 800 sec as shown in Table 16.1.

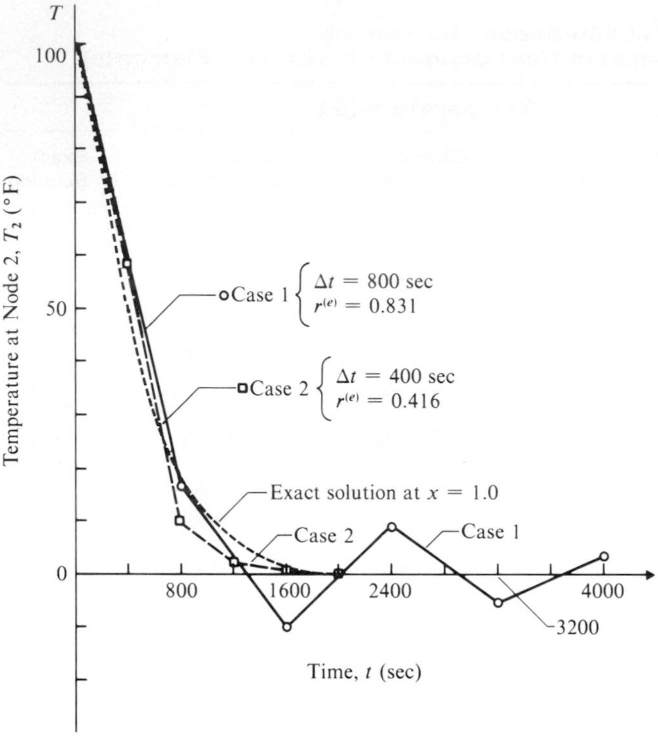

Figure 16.7 Temperature versus time for Cases 1 and 2 (Example 16.1—Explicit transient heat conduction with two elements)

For time step $t = 2\Delta t$ $(m+1=3)$, the equation becomes

$$T_2^3 = T_2^3 + r(T_1^2 - 2T_2^2 + T_3^2)$$

or

$$T_2^3 = 16.89 + (0.831)[0 - 2(16.89) + 0]$$

The actual boundary conditions of 0°F are employed here. The result is $T_2^3 = -11.18$°F. This is below zero—a temperature that cannot occur physically. (No temperature below the steady-state solution $T = 0$°F is realistic.) Thus the time step of $\Delta t = 800$ sec is too large.

Carrying out the method with $t = 800$ sec for more time steps is easily done. Table 16.1 gives the results for several time steps. The calculated temperatures oscillate about the steady-state solution. Figure 16.7 plots the temperature versus time for Case 1. Generally this type of behavior indicates that a smaller time step must be used.

Case 2 $(\Delta t = 400 \text{ sec})$

The value for the Courant number is one-half of the value for Case 1: $r = 0.416$. For the first time step, $t = \Delta t = 400$ sec,

$$T_2^2 = 100 + (0.416)[50 - 2(100) + 50] = 58.45$$

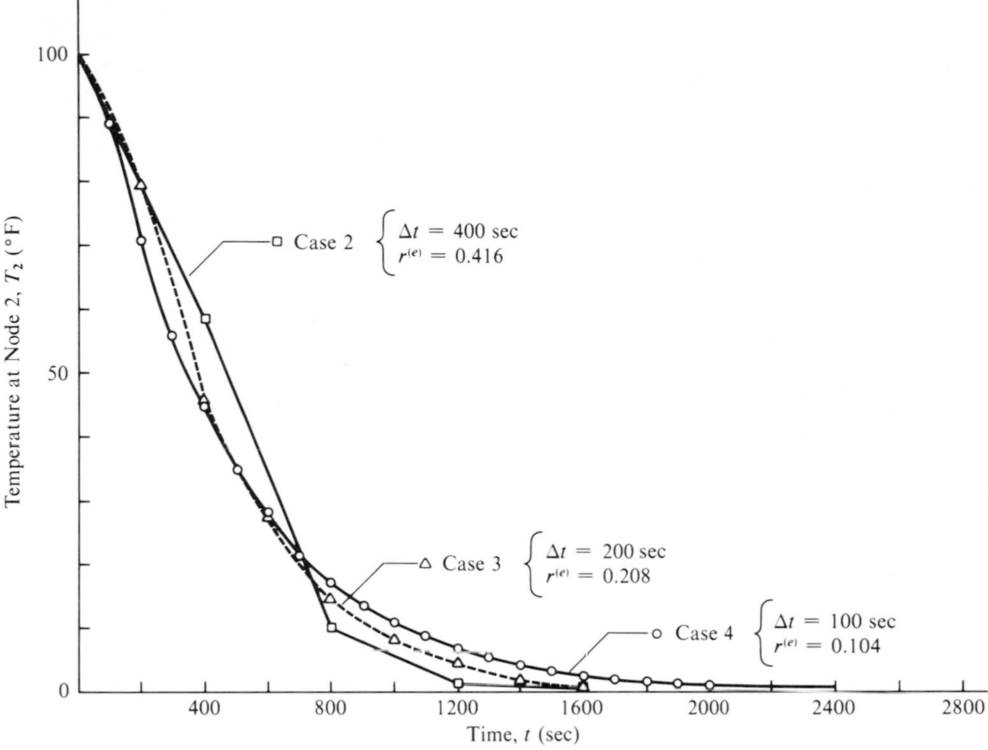

Figure 16.8 Temperature versus time for Cases 2, 3, and 4 (Example 16.1—Explicit transient heat conduction with two elements)

and for the second time step, $t = 2\Delta t = 800$ sec,

$$T_2^3 = 58.45 + (0.416)[0 - 2(58.45) + 0] = 9.87$$

Nodal values for subsequent time steps are given in Table 16.1.

Figure 16.7 plots the nodal temperature versus time for Case 2. It gives a much smoother decreasing function of temperature than Case 1. The oscillations observed in Case 1 are no longer present. Also, the Case 2 temperature never goes below the steady state value of 0°F.

The exact solution is also plotted in Figure 16.7. It is easily seen that the Case 2 finite element solution starts out above the exact solution, then falls below it. The error at time 800 sec is about 45 percent. Again note that if the boundary condition of 0°F is used for the first time step, the result is $T_2^2 = 16.8$°F, which gives a large error.

Case 3 ($\Delta t = 200$ sec)

Again the time step is one-half of the previous value; the Courant number is reduced to $r = 0.208$. The numerical values are calculated from Equation (16.16) and presented in Table 16.1. Figure 16.8 illustrates the results for Cases 2, 3, 4. It is easily seen that

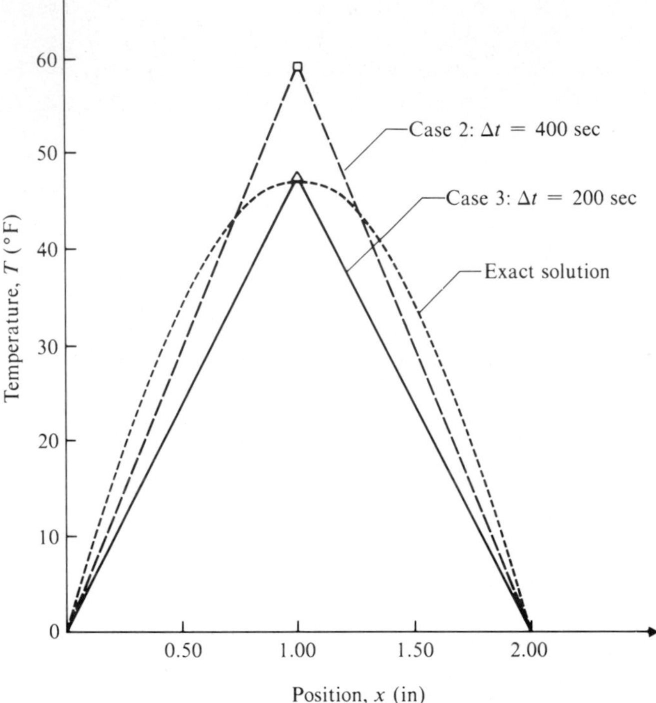

Figure 16.9 Temperature versus position at time $t = 400$ sec (Example 16.1—Explicit transient heat conduction with two elements)

Case 3 yields a smooth curve that fairly accurately predicts the temperature at the center of the rod. Figure 16.9 shows the exact solution versus position at time $t = 400$ sec and compares with the finite element solution for Cases 2 and 3.

Case 4 ($\Delta t = 100$ sec)

This is the final case with a small time step. The Courant number is $r = 0.104$. The nodal temperatures are given in Table 16.1 and plotted in Figure 16.8. The solution is now very smooth and accurate. Of course, more calculation time is required for the small time step solutions. The exact solution is so close to the finite element Case 4 that it is not plotted in Figure 16.8.

16.3 Spatial Oscillations and Stability with Explicit Method

The explicit method is usually the easiest to formulate and the fastest to solve for a given time step. It has two major types of problems, however. In one type, spatial oscillations above and below the exact solution occur, but the solution still converges to the final steady state solution correctly. It is called an *oscillatory stable* solution. The second type oscillates as well but does not converge to the final solution at all. It is called an *unstable* solution. The two cases will be illustrated in the next example.

Usually the stability is related to the numerical value of the dimensionless Courant number $r^{(e)}$, which appears in Equation (16.15)

$$r^{(e)} = \left(\frac{K_x}{L^2}\frac{\Delta t}{K_t}\right)^{(e)} \tag{16.17}$$

which must be considered for all elements. The explicit method can be shown to have the following properties

$0 < r^{(e)} \leq \frac{1}{4}$	No oscillations (in space) and stable
$\frac{1}{4} < r^{(e)} \leq \frac{1}{2}$	Oscillatory (in space) stable
$\frac{1}{2} < r^{(e)}$	Unstable

If a problem has elements in more than one category (say with variable K_x or L), it is unclear what will happen.

Consider that $K_x^{(e)}$, $K_t^{(e)}$ are determined by the differential equation and $L^{(e)}$ by the size of element needed for accuracy in space. Then the nonoscillatory condition $r \leq 1/4$ may be thought of as a limit on the size of time step Δt that can be used. Solving for Δt gives

$$\Delta t \leq \left(\frac{L^2}{4}\frac{K_t}{K_x}\right)^{(e)} \tag{16.18}$$

where the value used on the right side is that of the element with the smallest value of the term in brackets. Often the required time step is quite small, so that many calculations are required. Similarly the condition for a stable solution that converges to the final solution is

$$\Delta t \leq \left(\frac{L^2}{2}\frac{K_t}{K_x}\right)^{(e)} \tag{16.19}$$

Thus, half as many steps are required.

Example 16.2 Explicit Transient Heat Conduction Showing Oscillations and Stability

Given The same thin aluminum rod discussed in Example 16.1, except that the boundary conditions are $T_{x0} = 100°\text{F}$ at $x = 0$ and $T_{xL} = 0°\text{F}$ at $x = L$ for $t > 0$. Thus only the right end is dropped suddenly to 0°F at time zero; the left end is kept at 100°F. Figure 16.10 illustrates the geometry.

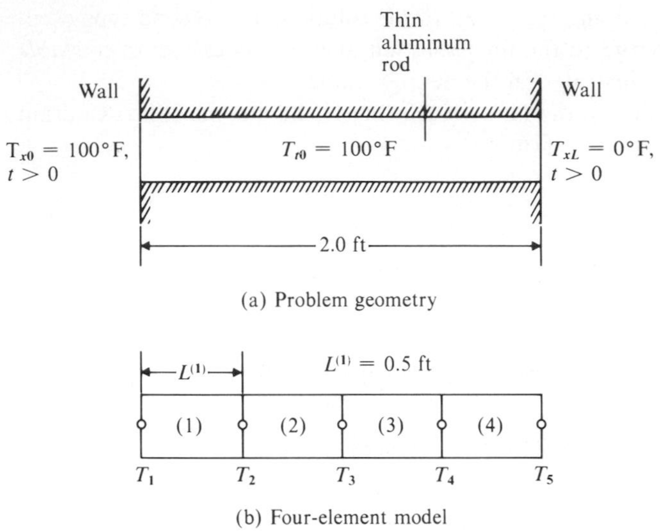

Figure 16.10 Insulated aluminum rod with four elements (Example 16.2—Explicit transient heat conduction showing oscillations and stability)

Objective Use a four-element model to solve for the temperature at the nodes 2, 3, 4 with the explicit method. Again consider four cases

Case 1 $\Delta t = 200$ sec
Case 2 $\Delta t = 100$ sec
Case 3 $\Delta t = 50$ sec
Case 4 $\Delta t = 25$ sec

These produce the same Courant numbers as in Example 16.1.

It is easily seen that the steady-state solution must be linear from $T_1 = 100$ on the left to $T_5 = 0$ on the right. Then after many time steps, the nodal temperatures should be $T_1 = 100°F$, $T_2 = 75°F$, $T_3 = 50°F$, $T_4 = 25°F$, and $T_5 = 0°F$. All numerical solutions should converge to these values as the number of time steps becomes large.

Solution Example 16.1 had only one unknown node point, at the middle of the rod. It did not show an unstable solution because the two boundary conditions (on each side) restrained it. This example will demonstrate an unstable solution.

The element length is now one half (0.5 ft) of the value in the previous example. The Courant numbers for this example

$$r^{(e)} = \left(\frac{K_x}{L^2}\frac{\Delta t}{K_t}\right)^{(e)}$$

are the same for all four cases because Δt has been reduced by a factor of four from the values used in Example 16.1.

The differential equation is the same as in Example 16.1: $K_x = k = 0.0370$ Btu/sec-ft-°F, and $K_t = \rho c = 35.6$ Btu/ft^3-°F. Now the Courant number has these values:

Case	*Time step Δt (sec)*	*Courant number $r^{(e)}$*
1	200	0.831
2	100	0.416
3	50	0.208
4	25	0.104

For the first time step, the right boundary is $T_5 = 50$°F for $t = \Delta t$ ($m=1$) and zero after that.

The basic equation used here is Equation (16.16),

$$T_n^{m+1} = T_n^m + r^{(e)}\,(T_{n-1}^m - 2T_n^m + T_{n+1}^m)$$

It is applied at all three unknown node points

$$T_2^{m+1} = T_2^m + r^{(e)}\,(T_1^m - 2T_2^m + T_3^m)$$

$$T_3^{m+1} = T_3^m + r^{(e)}\,(T_2^m - 2T_3^m + T_4^m)$$

$$T_4^{m+1} = T_4^m + r^{(e)}\,(T_3^m - 2T_4^m + T_5^m)$$

These three equations are used to obtain the solution by marching along in time.

Case 1 ($\Delta t = 200$ sec)

For Case 1, with $\Delta t = 200$, the three nodal equations become

$$T_2^{m+1} = T_2^m + 0.831(T_1^m - 2T_2^m + T_3^m)$$

$$T_3^{m+1} = T_3^m + 0.831(T_2^m - 2T_3^m + T_4^m)$$

$$T_4^{m+1} = T_4^m + 0.831(T_3^m - 2T_4^m + T_5^m)$$

At time $t = 200$, the first time step corresponds to $m=1$ and $m+1=2$. The initial nodal values are $T_2^1 = 100.0$, $T_3^1 = 100.0$, and $T_4^1 = 50.0$, and the nodal equations become

$$T_2^2 = 100.0 + 0.831(100.0 - 200.0 + 100.0)$$

$$T_3^2 = 100.0 + 0.831(100.0 - 200.0 + 100.0)$$

$$T_4^2 = 100.0 + 0.831(100.0 - 200.0 + 50.0)$$

The results are $T_2^2 = 100.0$, $T_3^2 = 100.0$, and $T_4^2 = 58.4$. These are indicated in Table 16.2(a).

At the next time step, $t = 400$, $m=2$, and $m+1=3$. The nodal equations are

$$T_2^3 = 100.0 + 0.831(100.0 - 200.0 + 100.0)$$

$$T_3^3 = 100.0 + 0.831(100.0 - 200.0 + 58.4)$$

$$T_4^3 = 58.4 + 0.831(100.0 - 116.8 + 0)$$

and the results are $T_2^3 = 100.0$, $T_3^3 = 65.5$, and $T_4^3 = 44.4$. Note that the temperature change at the boundary has only propagated to Node 3 so far.

Table 16.2
Temperature versus Time at All Node Points for Cases 1 and 2 (Example 16.2: Explicit Transient Heat Conduction Showing Oscillations and Stability)

(a) Case 1: Δt = 200 Sec

Time (Sec)	Temperatures (°F)		
	Node 2 T_2	Node 3 T_3	Node 4 T_4
200	100.0	100.0	58.4
400	100.0	65.5	44.4
600	71.3	76.7	25.0
800	99.6	29.3	47.2
1000	41.5	102.6	−6.9
1200	140.9	−39.2	89.8

(b) Case 2: Δt = 100 Sec

Time (Sec)	Temperatures (°F)		
	Node 2 T_2	Node 3 T_3	Node 4 T_4
100	100.0	100.0	79.2
200	100.0	91.4	54.9
300	96.4	79.8	47.2
400	91.0	73.2	41.1
500	87.3	67.3	37.4
600	84.3	63.2	34.3
700	82.0	60.0	32.0
800	80.3	57.5	30.3
900	79.0	55.7	29.0
1000	78.0	54.3	28.0
1100	77.3	53.3	27.3
1200	76.7	52.5	26.7

The marching method illustrated for times 200 sec and 400 sec continues in the same simple manner for as many time steps as desired. Table 16.2(a) shows the results for Case 1 at all of the node points. Also Figure 16.11 plots the nodal values versus time. It is easily seen that the solution oscillates strongly after only a few time steps.

In another plot, Figure 16.12 illustrates the calculated temperatures at the center of the aluminum rod (Node 3). Case 1 shows oscillations that grow with time. They do not approach the solution for large time, which is 50° F. Note that this is purely a numerical problem—the actual temperature does not do anything like this.

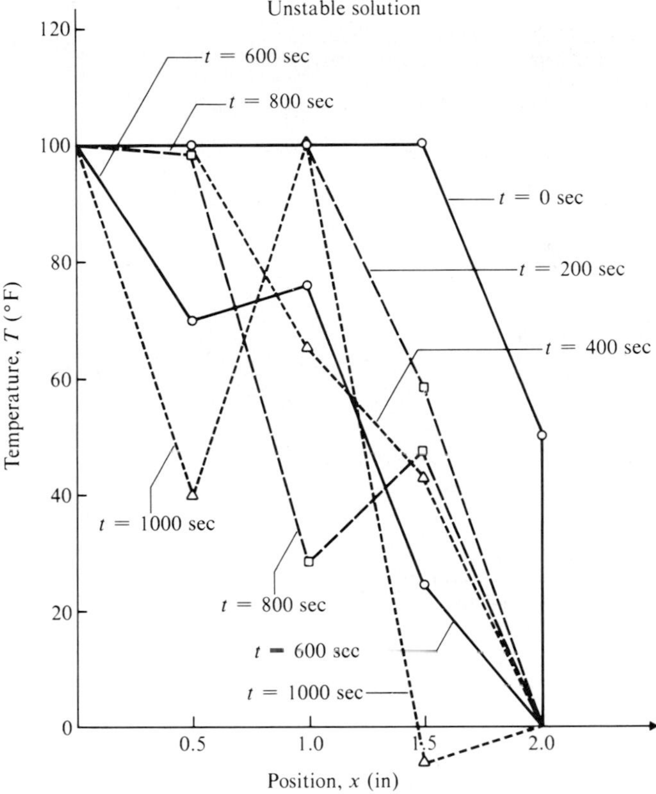

Figure 16.11 Case 1: Δt = 200 sec (Example 16.2—Explicit transient heat conduction showing oscillations and stability)

Case 2 (Δt = 100 sec)

Here the time step is decreased by a factor of two. The calculated results are quite different. Now the three nodal equations are

$$T_2^{m+1} = T_2^m + 0.416(T_1^m - 2T_2^m + T_3^m)$$

$$T_3^{m+1} = T_3^m + 0.416(T_2^m - 2T_3^m + T_4^m)$$

$$T_4^{m+1} = T_4^m + 0.416(T_3^m - 2T_4^m + T_5^m)$$

Only the value of the Courant number r has changed compared to Case 1.

For the first time step, $m=1$ and $m+1=2$. The nodal equations are similar to those of Case 1. These results and those for the rest of the time steps are given in Table 16.2(b).

Figure 16.13 plots the results. The temperature profiles across the rod move in an orderly fashion toward the steady-state solution. Figure 16.12 shows the calculations for Node 3 versus time. It is easily seen that the curve is now smooth rather than oscillatory (unstable) in Case 1.

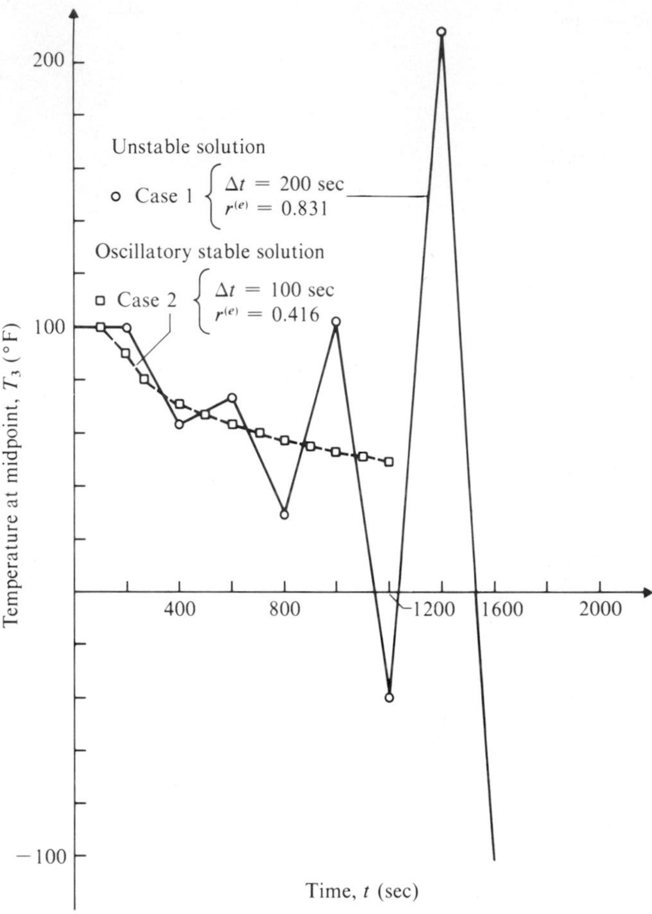

Figure 16.12 Numerical oscillations at middle node for Cases 1 and 2 (Example 16.2—Explicit transient heat conduction showing oscillations and stability)

A small oscillation in the solution for Node 3 can be seen in Figure 16.12 at times 100, 200, 300, 400. The solution starts out a little too high and then goes a little too low. The oscillatory behavior is small for this example, but it can be of significant size in other problems.

Case 3 ($\Delta t = 50$ sec)

The calculation procedure for Case 3 is the same as the first two cases except that $r = 0.208$. Results are given in Table 16.3. The behavior at Node 3 is plotted for Cases 2, 3, 4. It is easily seen that the curve is very smooth over time. No oscillations occur.

Case 4 ($\Delta t = 25$ sec)

Case 4 was calculated as above. Only the results for Node 3 are shown in Figure 16.14. It is somewhat more smooth than the curve for Case 3, but the differences are very small.

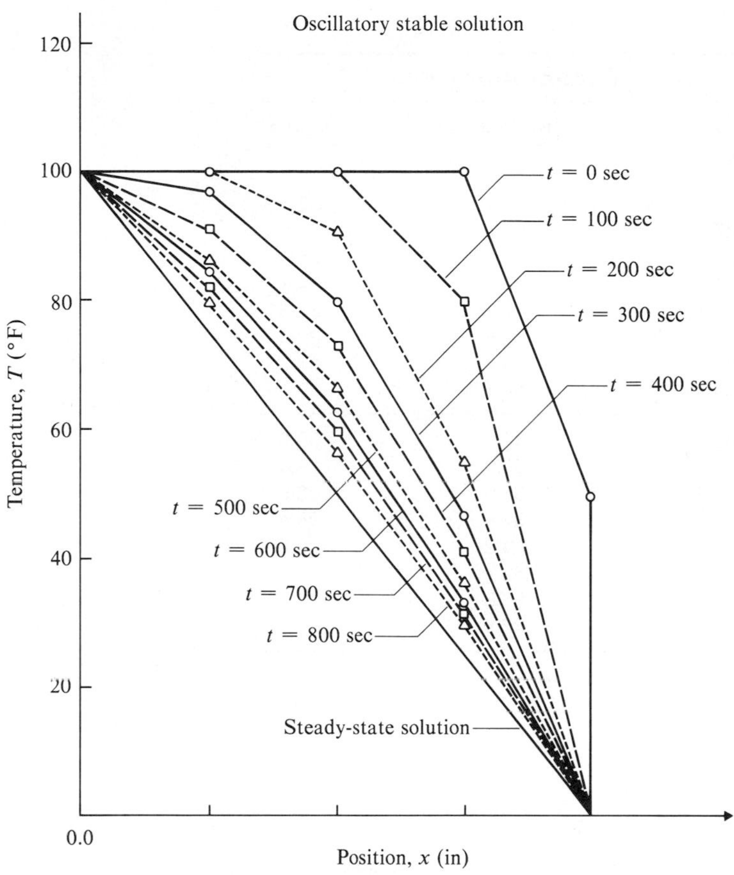

Figure 16.13 Case 2: $\Delta t = 100$ sec (Example 16.2—Explicit transient heat conduction showing oscillations and stability)

Table 16.3
Temperature versus Time at All Node Points for Case 3 (Example 16.2: Explicit Transient Heat Conduction Showing Oscillations and Stability)

Time (Sec)	Temperatures (°F)		
	Node 2 T_2	Node 3 T_3	Node 4 T_4
50	100.0	100.0	89.6
100	100.0	97.8	73.1
150	99.6	93.2	63.1
200	98.3	88.2	56.2
250	96.6	83.7	51.2
300	94.6	79.6	47.3
350	92.6	76.0	44.2
400	90.7	72.8	41.6

Continued on next page

Table 16.3 *Continued*

Time (Sec)	Temperatures (°F)		
	Node 2 T_2	Node 3 T_3	Node 4 T_4
450	88.9	70.1	39.5
500	87.3	67.6	37.6
550	85.9	65.5	36.0
600	84.6	63.6	34.7
650	83.4	61.9	33.5
700	82.4	60.5	32.4
750	81.5	59.2	31.5
800	80.7	58.1	30.7
850	80.0	57.1	30.0
900	79.4	56.2	29.4
950	78.9	55.5	28.9
1000	78.4	54.8	28.4
1050	78.0	54.2	28.0
1100	77.6	53.7	27.6
1150	77.3	53.3	27.3
1200	77.0	52.9	27.0

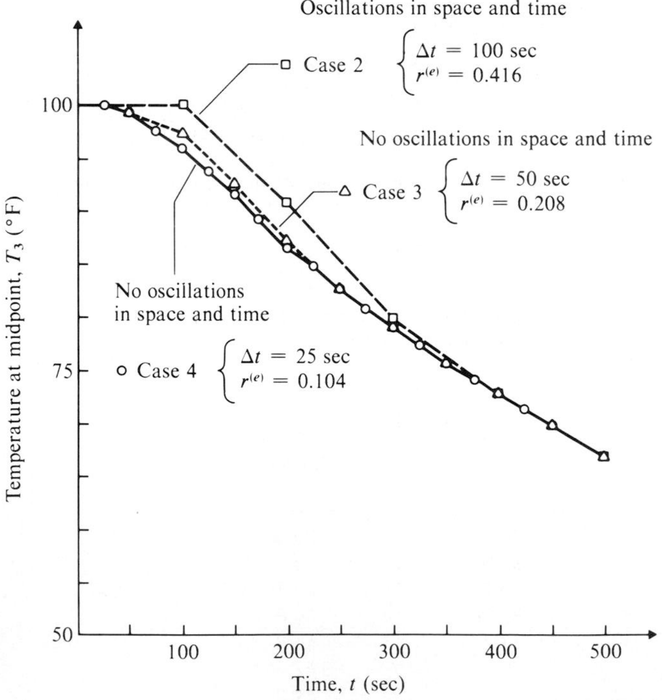

Figure 16.14 Temperature versus time for Cases 2, 3, and 4 (Example 16.2—Explicit transient heat conduction showing oscillations and stability)

16.4 Weighted Explicit and Implicit Method

This section develops a more general time transient method that combines the explicit and implicit methods. A weighted summation of the two methods is presented. It gives the user the option of easily varying weighting. In some cases, one weighting value will give optimum accuracy and stability. In other problems another weighting value is better.

A. Weighting Constant

Let θ be the weighting constant. It has the range $0 \leq \theta \leq 1$. Three numerical values of θ produce the following special cases: $\theta = 0$, fully explicit; $\theta = 1/2$, half explicit–half implicit (Crank-Nicholson); and $\theta = 1$, fully implicit. The first case has already been discussed. As indicated, the half explicit–half implicit method has the name Crank-Nicholson. It has been fairly extensively discussed in the literature. The last case is the fully implicit method.

As noted in Section 16.1, there is a major difference between explicit and implicit methods. The implicit method solves a set of simultaneous algebraic equations for at least some of the spatial nodes at each time step. It is more accurate and more stable than the explicit method but more costly in terms of both computer time and storage at each time step. Usually larger time steps can be used with the implicit method, which is therefore computationally better than the explicit method.

B. Time-Dependent Boundary Value Problem

The differential equation considered in this section is the same as in the previous section

$$\frac{\partial}{\partial x}\left(K_x \frac{\partial T}{\partial x}\right) - K_t \frac{\partial T}{\partial t} = 0 \qquad \textbf{(16.20)}$$

Figure 16.15 shows the solution domain in space and time. The initial condition is

$$T = T_{t0}, \quad t = t_0 \qquad \textbf{(16.21)}$$

The general boundary conditions are

$$T = T_{x0} \text{ or } K_{x0} \frac{\partial T}{\partial x} = 0 \text{ at } x = x_0 \qquad \textbf{(16.22)}$$

$$T = T_{xL} \text{ or } K_{xL} \frac{\partial T}{\partial x} = 0 \text{ at } x = x_L \qquad \textbf{(16.23)}$$

Only one boundary condition applies at each end.

C. Finite Differences

The time derivative term is approximated by

$$\frac{\partial T}{\partial t} = \frac{T^{m+1} - T^m}{\Delta t}$$

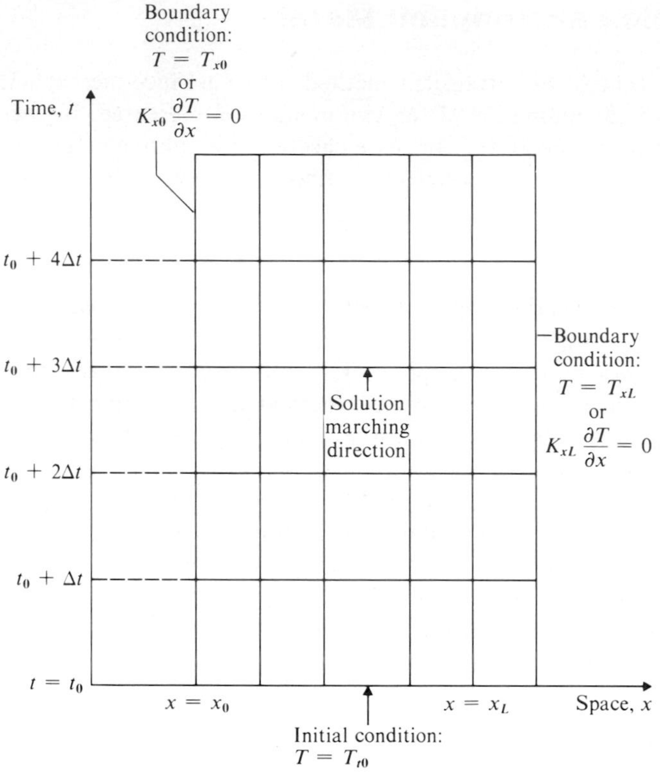

Figure 16.15 Solution region for general boundary value problem

The weighting of the implicit and explicit methods is done with the spatial terms. This becomes

$$\frac{\partial}{\partial x}\left(K_x \frac{\partial T}{\partial x}\right) = \theta \underbrace{\frac{\partial}{\partial x}\left(K_x \frac{\partial T^{m+1}}{\partial x}\right)}_{\substack{\text{Implicit space}\\ \text{term}\\ \text{(evaluated at}\\ \text{time step}\\ m+1)}} + (1-\theta) \underbrace{\frac{\partial}{\partial x}\left(K_x \frac{\partial T^{m}}{\partial x}\right)}_{\substack{\text{Explicit}\\ \text{space term}\\ \text{(evaluated at}\\ \text{time step } m)}} \tag{16.24}$$

The weighting role of θ is obvious here: $\theta = 0$ gives a fully explicit space term while $\theta = 1$ yields the fully implicit.

The differential equation becomes

$$\theta \frac{\partial}{\partial x}\left(K_x \frac{\partial T^{m+1}}{\partial x}\right) + (1-\theta)\frac{\partial}{\partial x}\left(K_x \frac{\partial T^{m}}{\partial x}\right) + K_t \frac{T^{m+1} - T^{m}}{\Delta t} = 0$$

With the weighted explicit-implicit method, there are more terms evaluated at time step $t + \Delta t$ (time step $m+1$) than in the explicit method. These are unknown terms. Because the implicit space terms are evaluated at time step $m+1$, the finite difference expression is equivalent to the backward difference discussed in Section 16.1. The remaining terms are known from time step m. For $\theta = 1/2$, the result is the same as the central difference approximation.

D. Approximation in Space

As in the explicit method, Galerkin's method is employed. For the ith node, the residual is

$$R_i^{(e)} = [R_i]_{\text{space}}^{(e)} + [R_i]_{\text{time}}^{(e)}$$

where

$$[R_i]_{\text{space}}^{(e)} = -\int_{L^{(e)}} \left\{ N_i \left[\theta \frac{\partial}{\partial x} \left(K_x \frac{\partial T^{m+1}}{\partial x} \right) \right] \right\}^{(e)} dx - \int_{L^{(e)}} \left\{ N_i \left[(1-\theta) \frac{\partial}{\partial x} \left(K_x \frac{\partial T^m}{\partial x} \right) \right] \right\}^{(e)} dx$$

$$[R_i]_{\text{time}}^{(e)} = -\int_{L^{(e)}} \left\{ N_i \left[-K_t \frac{T^{m+1} - T^m}{\Delta t} \right] \right\}^{(e)} dx$$

Similarly for the jth node, the residual is

$$R_j^{(e)} = [R_j]_{\text{space}}^{(e)} + [R_j]_{\text{time}}^{(e)}$$

where

$$[R_j]_{\text{space}}^{(e)} = -\int_{L^{(e)}} \left\{ N_j \left[\theta \frac{\partial}{\partial x} \left(K_x \frac{\partial T^{m+1}}{\partial x} \right) \right] \right\}^{(e)} dx - \int_{L^{(e)}} \left\{ N_j \left[(1-\theta) \frac{\partial}{\partial x} \left(K_x \frac{\partial T^m}{\partial x} \right) \right] \right\}^{(e)} dx$$

$$[R_j]_{\text{time}}^{(e)} = -\int_{L^{(e)}} \left\{ N_j \left[-K_t \frac{T^{m+1} - T^m}{\Delta t} \right] \right\}^{(e)} dx$$

All of these are easily evaluated as in previous sections.

The space terms in the residual for Node i are easily evaluated as

$$[R_i]_{\text{space}}^{(e)} = \theta \left[\frac{K_x}{L} \right]^{(e)} T_i^{m+1} + \theta \left[-\frac{K_x}{L} \right]^{(e)} T_j^{m+1} + (1-\theta) \left[\frac{K_x}{L} \right]^{(e)} T_i^m + (1-\theta) \left[-\frac{K_x}{L} \right]^{(e)} T_j^m$$

A set of simultaneous equations must be solved in the weighted explicit-implicit method (if θ is not equal to zero). Thus there is no need to make the simple approximation (uncoupled) for the time derivative term used in the explicit method developed in Section 16.2. The normal finite element interpolation function is employed in the time residual as

$$[R_i]^{(e)}_{\text{time}} = -\int_{L^{(e)}} \left\{ N_i \left[-K_t \frac{T^{m+1} - T^m}{\Delta t} \right] \right\}^{(e)} dx$$

$$= -\int_{L^{(e)}} \left\{ N_i \left[-\frac{K_t}{\Delta t} (N_i T_i^{m+1} + N_j T^{m+1}) + \frac{K_t}{\Delta t} (N_i T_i^m + N_j T^m) \right] \right\}^{(e)} dx$$

$$= \left[\frac{LK_t}{6\Delta t} (2\,T_i^{m+1} + T_j^{m+1}) - \frac{LK_t}{6\Delta t} (2\,T_i^m + T_j^m) \right]^{(e)}$$

Finally this becomes

$$[R_i]^{(e)}_{\text{time}} = \left[\frac{LK_t}{3\Delta t} \right]^{(e)} T_i^{m+1} + \left[\frac{LK_t}{6\Delta t} \right]^{(e)} T_j^{m+1} - \left[\frac{LK_t}{3\Delta t} \right]^{(e)} T_i^m - \left[\frac{LK_t}{6\Delta t} \right]^{(e)} T_j^m$$

Note that this expression involves both the nodal values at Nodes i and j. The simple (uncoupled) approximation used in Section 16.2 involved only the nodal value at Node i for residual i.

The space terms in residual j are

$$[R_j]^{(e)}_{\text{space}} = \theta \left[\frac{K_x}{L} \right]^{(e)} T_i^{m+1} + \theta \left[\frac{K_x}{L} \right] T_j^{m+1} + (1-\theta) \left[-\frac{K_x}{L} \right]^{(e)} T_i^m + (1-\theta) \left[-\frac{K_x}{L} \right] T_j^m$$

In a manner similar to that used for the time terms in residual i, the regular interpolation function gives

$$[R_j]^{(e)}_{\text{time}} = \left[\frac{LK_t}{6\Delta t} \right]^{(e)} T_i^{m+1} + \left[\frac{LK_t}{3\Delta t} \right]^{(e)} T_j^{m+1} - \left[\frac{LK_t}{6\Delta t} \right]^{(e)} T_i^m - \left[\frac{LK_t}{3\Delta t} \right]^{(e)} T_j^m$$

This completes all of the residuals.

E. Element Matrices

It is convenient to organize the residuals in element matrix form because a set of simultaneous nodal equations must be solved at each time step. Then the normal assembly procedure can be used. The terms involving nodal values at time step $m+1$ are unknown and appear in the element matrix. All other terms are known and go in the element column. They have the form

$$\begin{Bmatrix} R_i \\ R_j \end{Bmatrix}^{(e)} = [B]^{(e)} \begin{Bmatrix} T_i^{m+1} \\ T_j^{m+1} \end{Bmatrix} - \{C\}^{(e)}$$

The particular components can now be evaluated.

The element residuals are written as

$$\begin{Bmatrix} R_i \\ R_j \end{Bmatrix}^{(e)} = \begin{Bmatrix} R_i \\ R_j \end{Bmatrix}^{(e)}_{\text{space}} + \begin{Bmatrix} R_i \\ R_j \end{Bmatrix}^{(e)}_{\text{time}}$$

Substituting the previous evaluations for the space residuals yields

$$\begin{aligned} \begin{Bmatrix} R_i \\ R_j \end{Bmatrix}^{(e)}_{\text{space}} &= \theta \left(\frac{K_x}{L}\right)^{(e)} \begin{bmatrix} 1 & -1 \\ -1 & 1 \end{bmatrix} \begin{Bmatrix} T_i^{m+1} \\ T_j^{m+1} \end{Bmatrix} \\ &\quad + (1-\theta)\left(\frac{K_x}{L}\right)^{(e)} \begin{bmatrix} 1 & -1 \\ -1 & 1 \end{bmatrix} \begin{Bmatrix} T_i^{m} \\ T_j^{m} \end{Bmatrix} \end{aligned} \tag{16.25}$$

and the time residuals gives

$$\begin{aligned} \begin{Bmatrix} R_i \\ R_j \end{Bmatrix}^{(e)}_{\text{time}} &= \left(\frac{LK_t}{6\Delta t}\right)^{(e)} \begin{bmatrix} 2 & 1 \\ 1 & 2 \end{bmatrix} \begin{Bmatrix} T_i^{m+1} \\ T_j^{m+1} \end{Bmatrix} \\ &\quad - \left(\frac{LK_t}{6\Delta t}\right)^{(e)} \begin{bmatrix} 2 & 1 \\ 1 & 2 \end{bmatrix} \begin{Bmatrix} T_i^{m} \\ T_j^{m} \end{Bmatrix} \end{aligned} \tag{16.26}$$

Each has a set of terms evaluated at time step $m+1$ and another at time step m.

Combining the terms at time step $m+1$ gives the element matrix

$$[B]^{(e)} = \underbrace{\theta \left(\frac{K_x}{L}\right)^{(e)} \begin{bmatrix} 1 & -1 \\ -1 & 1 \end{bmatrix}}_{\text{Implicit space term}} + \underbrace{\left(\frac{LK_t}{6\Delta t}\right)^{(e)} \begin{bmatrix} 2 & 1 \\ 1 & 2 \end{bmatrix}}_{\text{Time term}} \tag{16.27}$$

Also the element column is

$$\begin{aligned} \{C\}^{(e)} &= \underbrace{-(1-\theta)\left(\frac{K_x}{L}\right)^{(e)} \begin{bmatrix} 1 & -1 \\ -1 & 1 \end{bmatrix} \begin{Bmatrix} T_i^{m} \\ T_j^{m} \end{Bmatrix}}_{\text{Explicit space term}} \\ &\quad + \underbrace{\left(\frac{LK_t}{6\Delta t}\right) \begin{bmatrix} 2 & 1 \\ 1 & 2 \end{bmatrix} \begin{Bmatrix} T_i^{m} \\ T_j^{m} \end{Bmatrix}}_{\text{Time term}} \end{aligned} \tag{16.28}$$

where all terms are known from the previous time step.

It is easy to see that many of the terms in the element matrix (16.27) and column (16.28) repeat themselves. Define a space matrix involving the K_x terms here as

$$[B_x]^{(e)} = \underbrace{\left(\frac{K_x}{L}\right)^{(e)} \begin{bmatrix} 1 & -1 \\ -1 & 1 \end{bmatrix}}_{\text{Element space matrix}} \tag{16.29}$$

and a time matrix with the remaining K_t terms as

$$[B_t]^{(e)} = \underbrace{\left(\frac{K_t}{6\Delta t}\right)^{(e)} \begin{bmatrix} 2 & 1 \\ 1 & 2 \end{bmatrix}}_{\text{Element time matrix}} \tag{16.30}$$

Then the element matrix and column can be written in a more compact form as

$$[B]^{(e)} = \underbrace{\theta[B_x]^{(e)} + [B_t]^{(e)}}_{\text{Element matrix}} \tag{16.31}$$

$$\{C\}^{(e)} = \underbrace{-(1-\theta)[B_x]^{(e)}\{T^m\} + [B_t]^{(e)}\,\{T^m\}}_{\text{Element column}} \tag{16.32}$$

where

$$\{T^m\} = \begin{Bmatrix} T_i^m \\ T_j^m \end{Bmatrix}$$

This is a fairly general expression for the element matrices.

F. Weighting Constant Cases

The three special cases discussed now are $\theta = 0$, fully explicit; $\theta = 1/2$, Crank-Nicholson; and $\theta = 1$, fully implicit. These cases reduce to the following expressions for the element matrices:

$$[B]^{(e)} = \begin{cases} [B_t]^{(e)}, & \text{fully explicit} \\ \dfrac{1}{2}\,[B_x]^{(e)} + [B_t]^{(e)}, & \text{Crank-Nicholson} \\ [B_x]^{(e)} + [B_t]^{(e)}, & \text{fully implicit} \end{cases}$$

Clearly the simplest expression is the fully explicit. The Crank-Nicholson and the fully implicit have about the same level of complexity. Also, the element columns are

$$\{C_x\}^{(e)} = \begin{cases} -[B_x]^{(e)}\{T^m\} + [B_t]^{(e)}\{T^m\}, & \text{fully explicit} \\ -\dfrac{1}{2}\,[B_x]^{(e)}\{T^m\} + [B_t]^{(e)}\{T^m\}, & \text{Crank-Nicholson} \\ [B_t]^{(e)}\{T^m\}, & \text{fully implicit} \end{cases}$$

The element column is somewhat simpler for the implicit method.

Table 16.4
Element Time Matrices

(a) Uncoupled Element Time Matrix

$$[B_t]^{(e)} = \left(\frac{LK_t}{2\Delta t}\right)^{(e)} \begin{bmatrix} 1 & 0 \\ 0 & 1 \end{bmatrix}$$

(b) Coupled Element Time Matrix

$$[B_t]^{(e)} = \left(\frac{LK_t}{6\Delta t}\right)^{(e)} \begin{bmatrix} 2 & 1 \\ 1 & 2 \end{bmatrix}$$

G. Element Time Matrices

Two different element time matrices $[B_t]$ can be considered. Table 16.4 shows the form of the two types. Both are of order 2 by 2.

The first type, called the uncoupled element time matrix, was used with the explicit method developed in Sections 16.2 and 16.3. Table 16.4(a) shows the matrix. It is a purely diagonal matrix. The terms associated with Node i are not directly coupled to those for Node j, as illustrated in some detail in Example 16.2. Thus it is called uncoupled. With it, the explicit method does not require the solution of simultaneous nodal equations at each time step.

The second type is called the coupled element time matrix, and it is given in Table 16.4(b). It has off-diagonal terms coupling Node i to Node j, and vice versa. It is more accurate than the uncoupled matrix but requires simultaneous solutions. With the weighted method, simultaneous solutions in space are required anyway, so the more accurate coupled form should be employed.

16.5 General Formulation for Weighted Method

The methods developed in the previous sections can be fairly easily extended to a general formulation. An element matrix approach is used in a manner consistent with the non-time-dependent boundary value problems discussed in the first fifteen chapters of this text. A computer program called TIME1D has been developed with the method shown in this section.

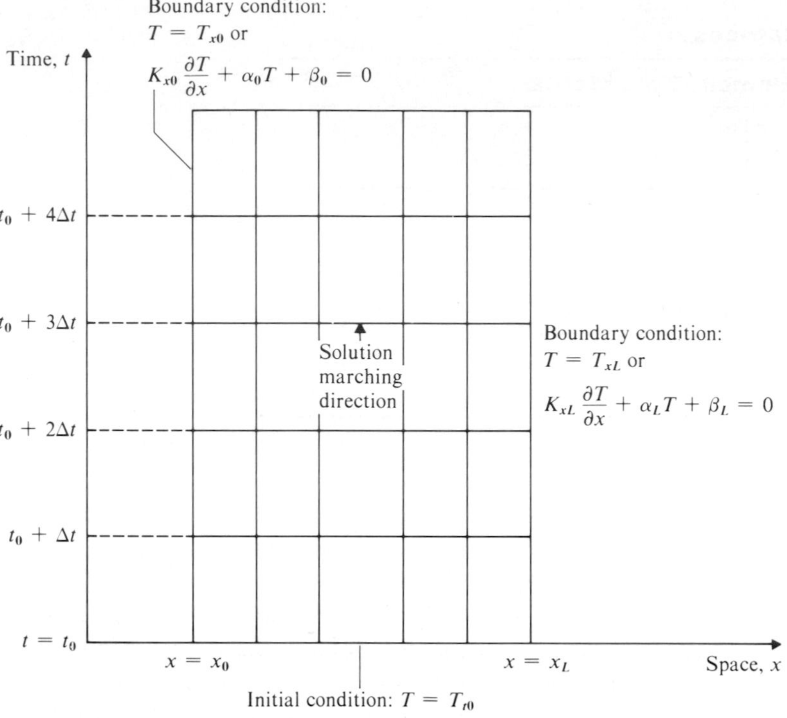

Figure 16.16 Solution region for general boundary value problem

A. Time-Dependent Boundary Value Problem

The general differential equation considered here is

$$\underbrace{\frac{\partial}{\partial x}\left(K_x \frac{\partial T}{\partial x}\right) + M_x \frac{\partial T}{\partial x} + PT + Q}_{\text{Space terms}} - \underbrace{K_t \frac{\partial T}{\partial t}}_{\text{Time term}} = 0 \tag{16.33}$$

Here the coefficients are functions of both space and time. The equation is second-order in the space derivatives and first-order in the time derivative.

Figure 16.16 shows the solution domain in time and space. Various length elements in space are illustrated, but the value of the time step Δt is held constant. It is not necessary to use equal-length time steps.

The initial condition is

$$T = T_{t0} \text{ at } t = t_0 \text{ and } x_0 \leq x \leq x_L \tag{16.34}$$

The general boundary condition at the left end is

$$T = T_{x0} \text{ or } K_{x0}\frac{\partial T}{\partial x} + \alpha_0 T + \beta_0 = 0 \text{ at } x = x_0 \text{ and } t > t_0 \tag{16.35}$$

and at the right end

$$T = T_{xL} \text{ or } K_{xL}\frac{\partial T}{\partial x} + \alpha_L T + \beta_L = 0 \text{ at } x = x_L \text{ and } t > t_0 \qquad \textbf{(16.36)}$$

As in all of the previous work, only one of the boundary conditions applies at each end.

B. Finite Differences

The time derivative term is approximated by

$$\frac{\partial T}{\partial t} = \frac{T^{m+1} - T^m}{\Delta t}$$

The weighting of the implicit and explicit methods on the spatial terms gives

$$\frac{\partial}{\partial x}\left(K_x \frac{\partial T}{\partial x}\right) + M_x \frac{\partial T}{\partial x} + PT + Q$$

$$= \underbrace{\theta\left[\frac{\partial}{\partial x}\left(K_x \frac{\partial T^{m+1}}{\partial x}\right) + M_x \frac{\partial T^{m+1}}{\partial x} + PT^{m+1} + Q\right]}_{\text{Implicit space term}}$$

$$+ \underbrace{(1-\theta)\left[\frac{\partial}{\partial x}\left(K_x \frac{\partial T^m}{\partial x}\right) + M_x \frac{\partial T^m}{\partial x} + PT + Q\right]}_{\text{Explicit space term}}$$

C. Approximation in Space

Following the methods used in Section 16.4, the element residuals for the space derivatives have the form

$$\begin{Bmatrix} R_i \\ R_j \end{Bmatrix}^{(e)}_{\text{space}} = \theta[B_x]^{(e)}\begin{Bmatrix} T_i^{m+1} \\ T_j^{m+1} \end{Bmatrix} - \theta\{C_x\}^{(e)}$$

$$+ (1-\theta)[B_x]^{(e)}\begin{Bmatrix} T_i^m \\ T_j^m \end{Bmatrix} - (1-\theta)\{C_x\}^{(e)}$$

where the general element space matrix $[B_x]$ and element space column $\{C_x\}$ are shown in Table 16.5. These are the same element matrices as developed in Chapter 8 using the Galerkin method. Here the first two terms on the right are the implicit ones and the second two are the explicit terms.

The element residuals for the time derivatives are

$$\begin{Bmatrix} R_i \\ R_j \end{Bmatrix}^{(e)}_{\text{time}} = [B_t]^{(e)}\begin{Bmatrix} T_i^{m+1} \\ T_j^{m+1} \end{Bmatrix}^{(e)} - [B_t]^{(e)}\begin{Bmatrix} T_i^m \\ T_j^m \end{Bmatrix}$$

As discussed in Section 16.4, either the uncoupled element time matrix or the coupled form may be employed. Both are shown in Table 16.6.

Table 16.5
Element Space Matrix and Column

(a) Terms in Element Space Matrix $[B_x]^{(e)}$

$$B_{ii} = \left[\frac{K_x}{L} b_i b_i - \frac{1}{2} M_x b_i - \frac{1}{3} PL - \alpha_0\right]^{(e)}$$

$$B_{ij} = \left[\frac{K_x}{L} b_i b_j - \frac{1}{2} M_x b_j - \frac{1}{6} PL\right]^{(e)}$$

$$B_{ji} = \left[\frac{K_x}{L} b_i b_j - \frac{1}{2} M_x b_i - \frac{1}{6} PL\right]^{(e)}$$

$$B_{jj} = \left[\frac{K_x}{L} b_j b_j - \frac{1}{2} M_x b_j - \frac{1}{3} PL + \alpha_L\right]^{(e)}$$

(b) Terms in Element Space Column $\{C_x\}^{(e)}$

$$C_i = \left[\frac{1}{2} QL + \beta_0\right]^{(e)}$$

$$C_j = \left[\frac{1}{2} QL - \beta_L\right]^{(e)}$$

Table 16.6
Element Time Matrix

(a) Uncoupled Element Time Matrix

$$[B_t]^{(e)} = \left(\frac{L}{2} \frac{K_t}{\Delta t}\right)^{(e)} \begin{bmatrix} 1 & 0 \\ 0 & 1 \end{bmatrix}$$

(b) Coupled Element Time Matrix

$$[B_t]^{(e)} = \left(\frac{L}{6} \frac{K_t}{\Delta t}\right)^{(e)} \begin{bmatrix} 2 & 1 \\ 1 & 2 \end{bmatrix}$$

D. Element Matrices

Summing the element space and time residuals

$$\begin{Bmatrix} R_i \\ R_j \end{Bmatrix}^{(e)} = \begin{Bmatrix} R_i \\ R_j \end{Bmatrix}^{(e)}_{\text{space}} + \begin{Bmatrix} R_i \\ R_j \end{Bmatrix}^{(e)}_{\text{time}}$$

yields

$$\begin{aligned} \begin{Bmatrix} R_i \\ R_j \end{Bmatrix}^{(e)} &= \theta[B_x]^{(e)}\begin{Bmatrix} T_i^{m+1} \\ T_j^{m+1} \end{Bmatrix} - \theta\{C_x\}^{(e)} \\ &+ (1-\theta)[B_x]^{(e)}\begin{Bmatrix} T_i^{m} \\ T_j^{m} \end{Bmatrix} - (1-\theta)\{C_x\}^{(e)} \\ &+ [B_t]^{(e)}\begin{Bmatrix} T_i^{m+1} \\ T_j^{m+1} \end{Bmatrix} - [B_t]^{(e)}\begin{Bmatrix} T_i^{m} \\ T_j^{m} \end{Bmatrix} \end{aligned}$$

The terms involving T at time step $m+1$ form the element matrix

$$\underbrace{[B]^{(e)} = \theta[B_x]^{(e)} + [B_t]^{(e)}}_{\substack{\text{Element matrix for weighted} \\ \text{explicit/implicit method}}} \tag{16.37}$$

All remaining terms form the element column

$$\underbrace{\begin{aligned} \{C\}^{(e)} &= -(1-\theta)[B_x]^{(e)}\begin{Bmatrix} T_i^{m} \\ T_j^{m} \end{Bmatrix} + \{C_x\}^{(e)} \\ &+ [B_t]^{(e)}\begin{Bmatrix} T_i^{m} \\ T_j^{m} \end{Bmatrix} \end{aligned}}_{\substack{\text{Element column for weighted} \\ \text{explicit/implicit method}}} \tag{16.38}$$

These general expressions are employed in Program TIME1D.

Example 16.3 *Element Matrices for Two-Element Problem*

Given Redo the simple thin rod in Example 16.1 using the general formulation for the weighted method. Consider the case of $\Delta t = 200$ sec.

Objective Obtain the space and time matrices for the problem. Develop the element matrix and column for the first time step to illustrate the process and how it compares to the method in Section 16.2.

Solution Comparing the differential equation with the general equation gives the coefficients $K_x = k = 0.0370$ Btu/sec-ft-°F, $P = 0$, $Q = 0$, and $K_t = \rho c = 35.6$ Btu/ft^3-°F. Here $\Delta t = 200$ sec.

Both elements have the same properties so the element matrices are the same. The space matrices are

$$[B_x]^{(e)} = \frac{K_x}{L}\begin{bmatrix} 1 & -1 \\ -1 & 1 \end{bmatrix} = \frac{0.037}{1}\begin{bmatrix} 1 & -1 \\ -1 & 1 \end{bmatrix}$$

$$[B_x]^{(e)} = \begin{bmatrix} 0.037 & -0.037 \\ -0.037 & 0.037 \end{bmatrix}$$

$$\{C_x\}^{(e)} = \begin{Bmatrix} 0 \\ 0 \end{Bmatrix}$$

The uncoupled element time matrix for $\Delta t = 200$ sec is

$$[B_t]^{(e)} = \left(\frac{L}{2}\frac{K_t}{\Delta t}\right)\begin{bmatrix} 1 & 0 \\ 0 & 1 \end{bmatrix} = \frac{1.0 \times 35.6}{2 \times 200}\begin{bmatrix} 1 & 0 \\ 0 & 1 \end{bmatrix}$$

$$[B_t]^{(e)} = \begin{bmatrix} 0.089 & 0 \\ 0 & 0.089 \end{bmatrix}$$

At time step $m = 1$ $(t = 0)$, the temperature columns for the two elements are

$$\begin{Bmatrix} T_1^1 \\ T_2^1 \end{Bmatrix}^{(1)} = \begin{Bmatrix} 50 \\ 100 \end{Bmatrix}, \qquad \begin{Bmatrix} T_2^1 \\ T_3^1 \end{Bmatrix}^{(2)} = \begin{Bmatrix} 100 \\ 50 \end{Bmatrix}$$

The weighting function is $\theta = 0$ for a fully explicit solution. The element matrix is then

$$[B]^{(e)} = \theta[B_x]^{(e)} + [B_t]^{(e)} = [B_t]^{(e)}$$

and

$$[B_t]^{(e)} = \begin{bmatrix} 0.089 & 0 \\ 0 & 0.089 \end{bmatrix}$$

The element columns are given by

$$\{C\}^{(e)} = -(1-\theta)[B_x]^{(e)}\begin{Bmatrix} T_i^m \\ T_j^m \end{Bmatrix} + \{C_x\}^{(e)} + [B_t]^{(e)}\begin{Bmatrix} T_i^m \\ T_j^m \end{Bmatrix}$$

For element (1)

$$\{C\}^{(1)} = -\begin{bmatrix} 0.037 & -0.037 \\ -0.037 & 0.037 \end{bmatrix}\begin{Bmatrix} 50 \\ 100 \end{Bmatrix} + \begin{Bmatrix} 0 \\ 0 \end{Bmatrix} + \begin{bmatrix} 0.089 & -0.089 \\ -0.089 & 0.089 \end{bmatrix}\begin{Bmatrix} 50 \\ 100 \end{Bmatrix}$$

and the element column reduces to

$$\{C\}^{(1)} = \begin{Bmatrix} 6.30 \\ 7.05 \end{Bmatrix}$$

Similarly for element (2)

$$\{C\}^{(2)} = -\begin{bmatrix} 0.37 & -0.37 \\ -0.37 & 0.37 \end{bmatrix} \begin{Bmatrix} 100 \\ 50 \end{Bmatrix} + \begin{Bmatrix} 0 \\ 0 \end{Bmatrix}$$

$$+ \begin{bmatrix} 0.089 & -0.089 \\ -0.089 & 0.089 \end{bmatrix} \begin{Bmatrix} 100 \\ 50 \end{Bmatrix}$$

with the result

$$\{C\}^{(2)} = \begin{Bmatrix} 7.05 \\ 6.30 \end{Bmatrix}$$

Assembling the global matrices gives

$$\begin{bmatrix} 0.089 & -0.089 & 0 \\ -0.089 & 0.178 & -0.089 \\ 0 & -0.089 & 0.089 \end{bmatrix} \begin{Bmatrix} T_1 \\ T_2 \\ T_3 \end{Bmatrix} = \begin{Bmatrix} 6.30 \\ 14.10 \\ 6.30 \end{Bmatrix}$$

The calculated temperature is

$$\begin{Bmatrix} T_1^m \\ T_2^m \\ T_3^m \end{Bmatrix} = \begin{Bmatrix} 0 \\ 79.22 \\ 0 \end{Bmatrix}$$

This is the result as in Example 16.1, as expected. The time columns in $\{C\}$ are updated for each new time step, but the rest of the numbers remain the same.

Example 16.4 *Weighted Explicit/Implicit Heat Conduction with Eight Elements*

Given Again consider the insulated aluminum rod from Examples 16.1 and 16.3. Now use eight elements for more accuracy than in the previous problems. Employ the uncoupled time matrix in the calculations. The problem geometry is shown in Figure 16.17.

Objective Find the transient solution for three cases

Case 1 Explicit method ($\theta = 0$)
Case 2 Crank-Nicholson method ($\theta = 1/2$)
Case 3 Implicit method ($\theta = 1$)

The time step sizes are given in Table 16.7.

Solution Program TIME1D was used to solve the various cases. Because of symmetry, only half of the results are plotted.

Case 1 Explicit method ($\theta = 0$)

The results for the explicit method were calculated and plotted in Figures 16.18 and 16.19 for times $t = 80$ sec and $t = 400$ sec. The cases for $\Delta t = 20$ and $\Delta t = 10$ are fairly close to the exact solution at $t = 80$ sec. At the later time $t = 400$ sec, the large time increment (with Courant number 0.669) is unstable and off the plot. Again both $\Delta t = 10$ and $\Delta t = 20$ are quite accurate.

As seen in the plots, the explicit method is limited to Courant numbers below one half. No worthwhile solution is obtained above this value.

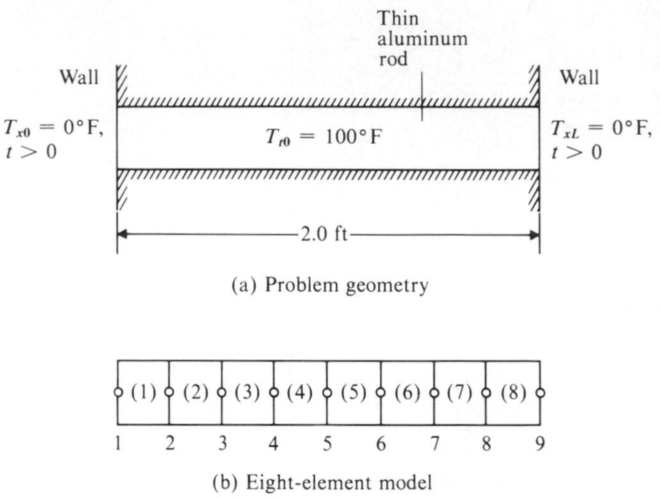

Figure 16.17 Example 16.4—Weighted explicit/implicit heat conduction with eight elements

Table 16.7
Cases for Example 16.4: Weighted Explicit/Implicit Heat Conduction with Eight Elements

Case	Weighting Factor	Time Step Δt (Sec)	Courant Number
1a	0	10	0.167
1b	0	20	0.335
1c	0	40	0.669
2a	1/2	10	0.167
2b	1/2	20	0.335
2c	1/2	40	0.669
2d	1/2	80	1.335
3a	1	10	0.167
3b	1	20	0.335
3c	1	40	0.669
3d	1	80	1.335

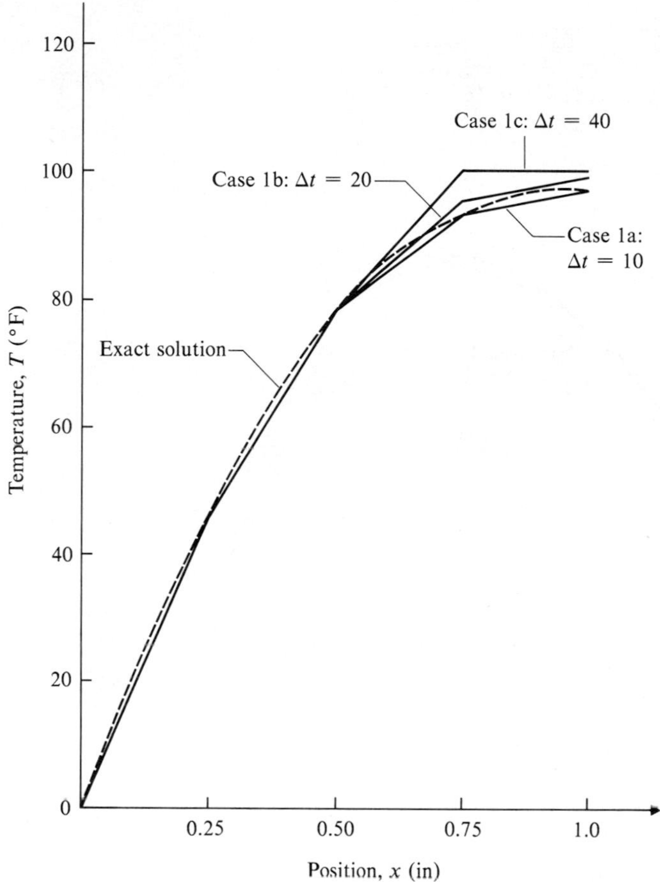

Figure 16.18 Case 1 at time 80 sec (Example 16.4—Weighted explicit/implicit heat conduction with eight elements)

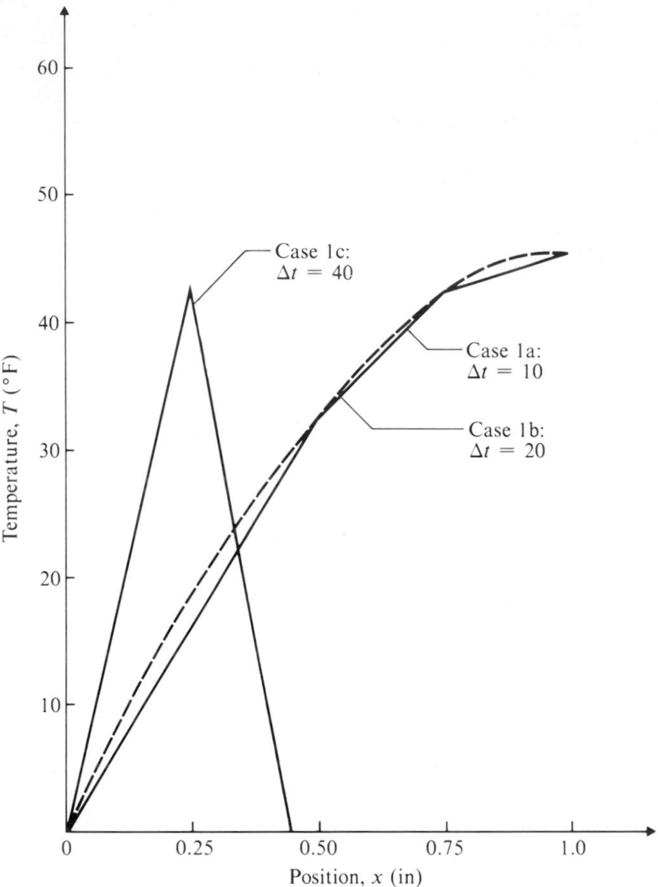

Figure 16.19 Case 1 at time 400 sec (Example 16.4—Weighted explicit/implicit heat conduction with eight elements)

Case 2 Crank-Nicholson method ($\theta = 1/2$)

Two similar plots at times 80 and 400 sec are shown in Figures 16.20 and 16.21 for the Crank-Nicholson case. The cases of $\Delta t = 10$ and 20 are quite accurate. At $\Delta t = 40$, the errors are only a few percent. Even with $\Delta t = 80$, the errors at $x = 0.5$ are only 5 percent.

Case 3 Fully implicit method ($\theta = 1$)

Figures 16.22 and 16.23 show the comparable results for the fully implicit case. The errors are generally larger than for the Crank-Nicholson case. However, both the Crank-Nicholson and fully implicit solutions give answers within about 5 percent, even with a time step of 80 sec. The explicit method requires a time step of about 20 sec and four times as many time steps to obtain a solution.

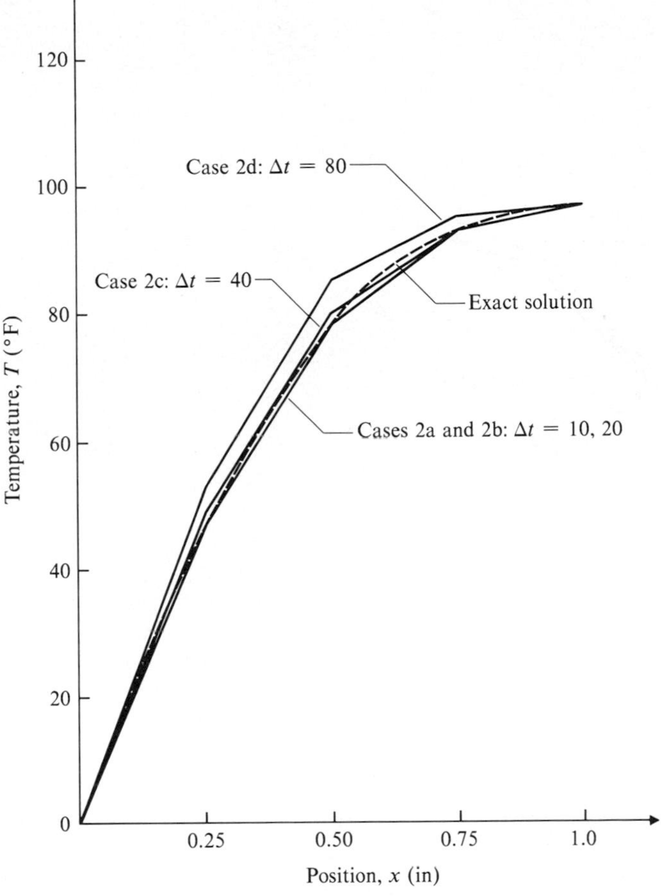

Figure 16.20 Case 2 at time 80 sec (Example 16.4—Weighted explicit/implicit heat conduction with eight elements)

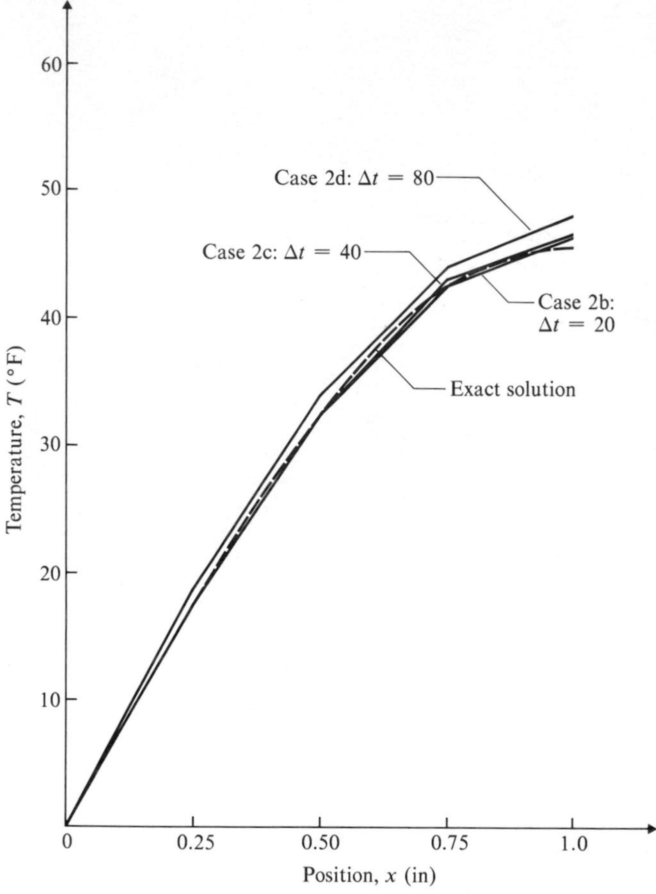

Figure 16.21 Case 2 at time 400 sec (Example 16.4—Weighted explicit/implicit heat conduction with eight elements)

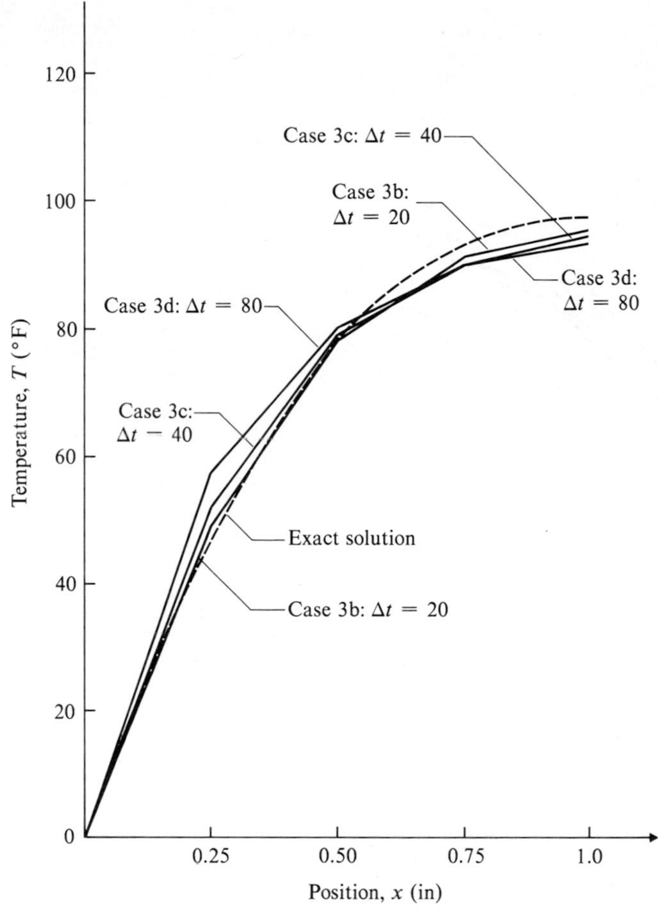

Figure 16.22 Case 3 at time 80 sec (Example 16.4—Weighted explicit/implicit heat conduction with eight elements)

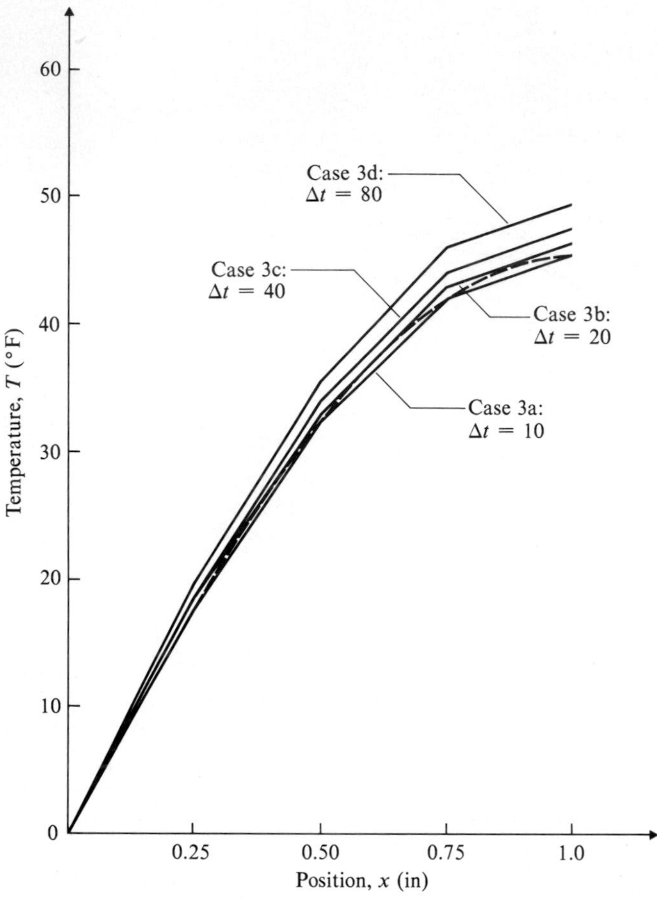

Figure 16.23 Case 3 at time 400 sec (Example 16.4—Weighted explicit/implicit heat conduction with eight elements)

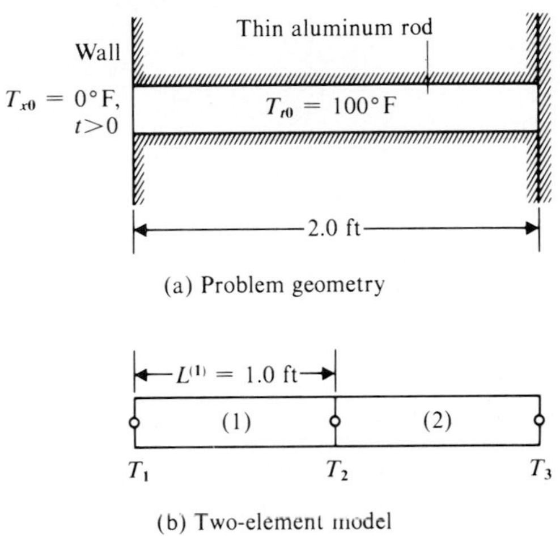

Figure 16.24 Geometry for Problems 16.6, 16.7, and 16.8

16.6 Problems

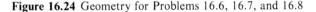

16.1 A copper body is infinitely long in two dimensions, y and z. In the x direction, the body has length 4.0 m and the following other properties: $k = 386$ W/m-°C, $\rho = 8{,}950$ kg/m³, and $c = 0.383$ kJ/Kg-°C. The temperature is initially at 150°C. At time $t = 0$, the ends of the rod are lowered to 50°C. Use a two-element model and a time step with a Courant number r equal to 0.20 to obtain an explicit time transient solution. Carry out the solution by hand for at least ten time steps. Plot the center node temperature versus time.

16.2 Repeat Problem 16.1 but use a time step such that the Courant number is 0.40.

16.3 Repeat Problem 16.2 but use a time step such that the Courant number is 0.80. Only six time steps need to be evaluated.

16.4 Write a computer program to evaluate a four-element (five-node) time transient problem using the explicit method.

16.5 Solve Problem 16.1 using four elements and the computer program developed in Problem 16.4. Evaluate fifty time steps and plot the results of the center node versus time.

16.6 A thin aluminum rod is insulated as shown in Figure 16.24. It has the properties $L = 2$ ft, $k = 0.0370$ Btu/sec-ft-°F, $c = 0.212$ Btu/lbm-°F, and $\rho = 168$ lbm/ft³. The initial rod temperature is 100°F. At time $t = 0$, the temperature of the left end drops to zero while the right end is insulated. Use a two-element model to solve for the time transient temperature at the center and right ends of the rod. Use the explicit method with a time step such that the Courant number is 0.40. Plot these nodal temperatures versus time for at least six time steps. Solve by hand or write a small computer program to carry this out.

16.7 Solve Problem 16.6 with a fully implicit method.

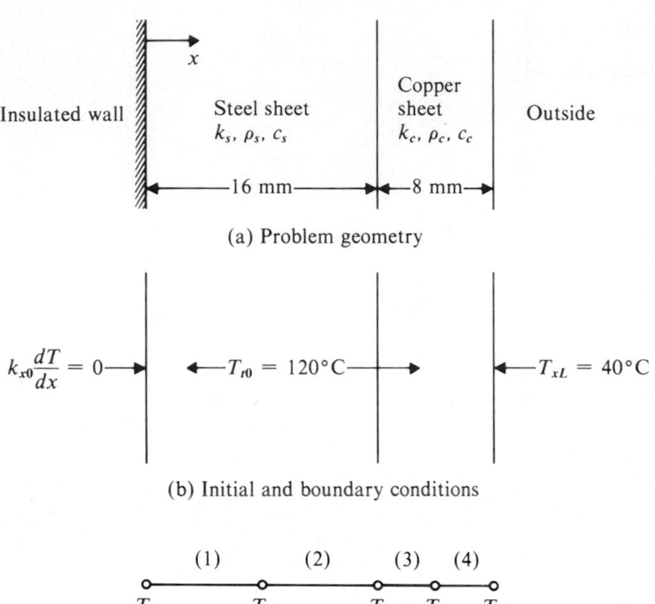

Figure 16.25 Geometry for Problem 16.9

16.8 Write a computer program to solve Problem 16.6 with a weight explicit/implicit method. Solve using weighting factors of 0.5 and 1.0. Let the time step be such that the Courant number has two values (two cases): *Case 1*, $r = 0.50$, *Case 2*, $r = 5.00$. Evaluate enough time steps to determine the errors (assume that Case 1 is the exact solution). Which value of weighting factor gives the most accurate solution for a large time step?

16.9 A sheet of carbon steel 16 mm thick is attached to an insulated wall as shown in Figure 16.25. A thinner sheet of copper 8 mm thick is bonded to the steel. The two materials have the following properties:

Steel: $k_s = 43$ W/m-°C (thermal conductivity), $\rho_s = 7{,}800$ kg/m³ (density), and $c_s = 0.473$ kJ/kg-°C (specific heat).

Copper: $k_c = 386$ W/m-°C (thermal conductivity), $\rho_c = 8{,}950$ kg/m³ (density), and $c_c = 0.383$ kJ/kg-°C (specific heat).

Note that the thermal conductivity k is much higher for copper than for steel. Both sheets are initially at temperature $T_{t0} = 120°C$. The left wall is insulated, so the boundary condition is

$$K_{x0}\frac{dT}{dx} = 0, \qquad x = 0$$

At the outside wall, the temperature drops suddenly to $T_{xL} = 40°C$. This completes the initial and boundary conditions shown in Figure 16.25.

Use a four-element model to estimate the length of time required for the temperature to reach 45°C. Consider three cases: *Case 1*, $\theta = 0$ (fully explicit); *Case 2*, $\theta = 1/2$ (Crank-Nicholson); *Case 3*, $\theta = 1$ (fully implicit). Modify computer program FINONE to solve this problem.

Appendixes

Appendix A

Computer Program FINONE for One-Dimensional Field Problems

A.1 Program Description

A computer program, called FINONE, is presented in this appendix both for instructional purposes and as a useful program to solve engineering problems. It solves a general second-order ordinary differential equation in one dimension of the form

$$\frac{d}{dx}\left[K(x)\frac{dT}{dx}\right] + PT + Q = 0$$

with the boundary conditions

$$T = T_0 \text{ and } K_0\frac{dT}{dx} + \alpha_0 T + \beta_0 = 0 \text{ at } x{=}x_0 \text{ (only one applies)}$$

$$T = T_L \text{ and } K_L\frac{dT}{dx} + \alpha_L T + \beta_L = 0 \text{ at } x{=}x_L \text{ (only one applies)}$$

The theory is developed in Chapter 5, and examples of input and output are given there as well.

FINONE is written as a main program with several subroutines. Table A.1 gives an outline, including the major sections of the main program and longer subroutines. In addition, each major section is broken down further into subsections where appropriate.

Several features have been incorporated into the program to increase readability. A large number of comment lines have been used (approximately half of the total lines are comment lines). At the beginning, the required input data and format are given. Major sections of the program are labeled 1, 2, 3, etc., while statement numbers ending in zero and statements within DO loops are all indented two spaces. Nested DO loops are indented two spaces per loop. Very long DO loops have statement numbers 1000, 2000, 3000, etc.

Table A.1
Outline of Computer Program FINONE

Main Program—FINONE

1. Input and output of problem data.
2. Calculation of element properties and assembly of global matrices.
3. Inclusion of prescribed nodal values into global coefficient matrix and column.
4. Calculation of nodal T values, element derivatives, and integrals.

Subroutine ELEMT1

1. Calculation of element coefficient matrices and columns.
2. Output of element matrices and columns.

Subroutine HEAD

1. Writing of titles at top of each page.

Subroutine PROP1

1. Calculation of pyramid function terms and area for simplex one-dimensional element.

Subroutine GAUSS1

1. Construction of upper banded triangular form.
2. Backward substitution to obtain unknowns.

Subroutine WRITEM

1. Writing out of global matrices and columns.

A.2 Listing of Program FINONE

The following pages give a complete listing of computer program FINONE with an example output from Example 5.9.

```
      PROGRAM FINONE (INPUT,OUTPUT,TAPE60=INPUT,TAPE61=OUTPUT,TAPE99)
C*************************************************************
C
C                     PROGRAM FINONE
C         VERSION 1.2                    OCTOBER, 1982
C
C                     DR PAUL ALLAIRE
C                     PROFESSOR
C                     CHIP QUEITZSCH
C                     GRADUATE RESEARCH ASSISTANT
C                     DEPARTMENT OF MECHANICAL AND
C                      AEROSPACE ENGINEERING
C                     UNIVERSITY OF VIRGINIA
C                     CHARLOTTESVILLE, VIRGINIA 22903
C
C
C*************************************************************
C         PROGRAM DESCRIPTION
C*************************************************************
C         THIS COMPUTER PROGRAM SOLVES A SECOND ORDER ORDINARY
C           DIFFERENTIAL EQUATION IN ONE DIMENSION USING
C           THE FINITE ELEMENT METHOD.  THE VARIATIONAL METHOD
C           WAS USED TO FORMULATE THE APPROPRIATE ELEMENT
C           COEFFICIENT MATRICES AND COLUMNS.
C
C         IT SOLVES A DIFFERENTIAL EQUATION OF THE FORM
C           D(KX*DT/DX)/DX+P*T+Q=0
C         WITH THE BOUNDARY CONDITIONS
C           T=TO, ON L1
C           KX*DT/DX+ALPHA*T+BETA=0, ON L2
C
C         THE COEFFICIENT FUNCTIONS KX,P,Q,ALPHA,BETA
C           ARE KNOWN FUNCTIONS OF X.
C
C*************************************************************
C         UNITS
C*************************************************************
C*************************************************************
C     1.  REQUIRED INPUT DATA AND FORMAT
C*************************************************************
C         INPUT DATA IS ORGANIZED INTO SECTIONS LABELED
C           1,2,3,ETC.  THE NUMBER OF DATA CARDS WITHIN
C           EACH SECTION WILL VARY ACCORDING TO THE NUMBER
C           NODE POINTS, ELEMENTS, PRESCRIBED NODAL VALUES,
C           AND DERIVATIVE BOUNDARY CONDITIONS.
C
C+++++++++++++
C     1.1.  TITLE AND GLOBAL PARAMETERS
C+++++++++++++
C         CARD  1.  TITLE1    FORMAT (20A4)
C           TITLE1=ANY ALPHA NUMERIC STRING FOR TITLE
C
C         CARD  2.  TITLE2    FORMAT (20A4)
C           TITLE2=ANY ALPHA NUMERIC STRING FOR TITLE
C
C         CARD  3.  NP,NE    FORMAT (2I5)
C           NP=NUMBER OF NODES
C           NE=NUMBER OF ELEMENTS
C
```

```
C             THIS PROGRAM IS CURRENTLY LIMITED TO NP=100
C               AND NE=99 DUE TO ARRAY DIMENSIONING.  THE ARRAY
C               SIZES MAY BE INCREASED BY CHANGING THE APPROPRIATE
C                DIMENSION AND COMMON STATEMENTS.
C
C+++++++++++++
C       1.2.  NODAL COORDINATES
C+++++++++++++
C           CARDS  1 THRU NP.  INODE,XNODE
C                                   FORMAT (I5,5X,G10.1)
C             INODE=GLOBAL NODE NUMBER
C             XNODE=X COORDINATE OF NODE
C
C+++++++++++++
C       1.3.  ELEMENT/NODE CONNECTIVITY DATA
C+++++++++++++
C           CARDS  1 THRU NE.  NEL,NODEI,NODEJ
C                                   FORMAT (3I5)
C             NEL  =ELEMENT NUMBER
C             NODEI=GLOBAL NUMBER OF FIRST NODE IN ELEMENT
C             NODEJ=GLOBAL NUMBER OF SECOND NODE IN ELEMENT
C
C+++++++++++++
C        1.4  CALCULATE BANDWIDTH OF GLOBAL COEFFICIENT MATRIX
C+++++++++++++
C               (NO DATA INPUT REQUIRED)
C
C               BAND WIDTH IS CURRENTLY LIMITED TO NBW=2 WHICH
C               SHOULD BE SUFFICIENT FOR FOR THE DIFFERENTIAL
C               EQUATION INDICATED ABOVE.
C
C+++++++++++++
C       1.5.  COEFFICIENT FUNCTIONS
C+++++++++++++
C           CARD  1.  ICO    FORMAT (I5)
C             ICO=COEFFICIENT PARAMETER
C               IF ICO=1, KX,P,Q HAVE THE SAME VALUE FOR
C                         ALL ELEMENTS SO ONLY ONE DATA CARD
C                         IS NEEDED.
C               IF ICO=2, VALUES FOR KX,P,Q MUST BE INPUT
C                         FOR EACH ELEMENT ON A DATA CARD.
C
C           CARD  2 (IF ICO=1).  IEL,KXE,PE,QE
C                                   FORMAT (I5,5X,3G10.1)
C             IEL=ELEMENT NUMBER
C             KXE=VALUE OF KX IN ELEMENT
C             PE =VALUE OF P IN ELEMENT
C             QE =VALUE OF Q IN ELEMENT
C
C           NEXT NE CARDS (IF ICO=2).  IEL,KXE,PE,QE
C                                   FORMAT (I5,5X,3G10.1)
C
C+++++++++++++
C       1.6.  PRESCRIBED NODAL VALUES
C+++++++++++++
C           CARD  1.  NBOUND    FORMAT (I5)
C             NBOUND=NUMBER OF PRESCRIBED NODAL VLAUES
C
C           NEXT NBOUND CARDS.  IBOUND,TBOUND
C                                   FORMAT (I5,5X,G10.1)
```

```
C           IBOUND=GLOBAL NODE NUMBER
C           TBOUND=PRESCRIBED NODAL VALUES
C
C+++++++++++++
C     1.7.  DERIVATIVE BOUNDARY VALUES
C+++++++++++++
C         CARD  1.  NDERIV    FORMAT (I5)
C           NDERIV=TOTAL NUMBER OF NODES FOR WHICH
C                  DERIVATIVE BOUNDARY VALUES ARE GIVEN
C
C         NEXT NDERIV CARDS.  INODIR,ALPHAE,BETAE
C                                 FORMAT (I5,2G10.1)
C           INODIR=NODE NUMBER OF DERIVATIVE
C                    BOUNDARY VALUE
C           ALPHAE=VALUE OF ALPHA ON BOUNDARY NODE
C           BETAE =VALUE OF BETA ON BOUNDARY NODE
C
C+++++++++++++
C     1.8. OUTPUT PARAMETERS
C+++++++++++++
C         CARD 1. NOUT     FORMAT(I5)
C           NOUT=OUTPUT PARAMETER
C             IF NOUT=0, ALL PROGRAM OUTPUT IS PRINTED.
C             IF NOUT=1, OUTPUT OF ELEMENT PROPERTIES IS OMITTED.
C             IF NOUT=2, OUTPUT OF ELEMENT PROPERTIES AND GLOBAL
C                        MATRICES IS OMITTED.
C
C************************************************************
C     2.  ELEMENT OUTPUT DATA
C************************************************************
C         OUTPUT DATA FOR EACH ELEMENT INCLUDES CONNECTIVITY,
C           NODAL COORDINATES, INTERPOLATION FUNCTION
C           PROPERTIES, DERIVATIVE BOUNDARY CONDITIONS,
C           AND ELEMENT COEFFICIENT MATRICES.
C
C************************************************************
C     3.  OUTPUT OF GLOBAL MATRICES
C************************************************************
C         CONDENSED GLOBAL MATRICES AND COLUMNS ARE OUTPUT
C           JUST AFTER ASSEMBLY OF ELEMENT MATRICES, AFTER
C           INCLUSION OF PRESCRIBED NODAL VALUES, AND AFTER
C           FORMATION OF UPPER TRIANGLUAR FORM.
C
C************************************************************
C     4.  OUTPUT OF NODAL VALUES
C************************************************************
C         NODAL VALUES, ELEMENT DERIVATIVES, AND INTEGRALS
C           ARE OUTPUT.
C
C************************************************************
C         PROGRAM READABLILITY FEATURES
C************************************************************
C         PROGRAM SECTIONS ARE LABELED WITH COMMENT CARDS.
C           MAJOR SECTIONS ARE LABELED WITH 1,2,3,ETC WHILE
C           SUBSECTIONS ARE LABELED WITH 1.1,1.2,1.3,ETC
C           WITHIN EACH MAJOR SECTION.
C
C         ALL DO LOOPS END WITH A STATEMENT NUMBER ENDING
C           IN ZERO AND STATEMENTS WITHIN THE DO LOOP ARE
C           INDENTED TWO SPACES.  VERY LONG DO LOOPS HAVE
```

```
C           STATEMENT NUMBERS 1000,2000,ETC.
C
C************************************************************
C
C
      DIMENSION INODE(100),XNODE(100)
      DIMENSION NEL(99),NODEI(99),NODEJ(99)
      DIMENSION KXE(99),PE(99),QE(99)
      DIMENSION IBOUND (2),TBOUND(2)
      DIMENSION INODIR(2),ALPHAE(2),BETAE(2)
      DIMENSION X(2),NN(2),ALPHA(2),BETA(2)
      DIMENSION B(2)
      DIMENSION EBM(2,2),EC(2)
      REAL KX,KXE,KXA,LENG
      COMMON /ALP10/ TITLE1(20),TITLE2(20)
      COMMON /INT10/ NP,NE,NBW,NBC,NOUT
      COMMON /REAL10/ ACOND(100,2),F(100),T(100)
      DATA IN,IO/60,61/
C************************************************************
C     1.  INPUT AND OUTPUT OF PROBLEM DATA
C************************************************************
C
C+++++++++++++
C     1.1.  INPUT AND OUTPUT OF TITLE AND GLOBAL PARAMETERS
C+++++++++++++
      READ (IN,11) TITLE1
      READ (IN,11) TITLE2
   11 FORMAT (20A4)
      READ (IN,12) NP,NE
   12 FORMAT (2I5)
      CALL HEAD
      WRITE (IO,15)
   15 FORMAT (//10X,36H1.  INPUT AND OUTPUT OF PROBLEM DATA)
      WRITE (IO,16)
   16 FORMAT (////10X,23H1.1.  GLOBAL PARAMETERS)
      WRITE (IO,17) NP,NE
   17 FORMAT ( /10X,15HNUMBER OF NODES,I5,/
     1    10X,18HNUMBER OF ELEMENTS,I5)
C+++++++++++++
C     1.2.  INPUT AND OUTPUT OF NODAL COORDINATES
C+++++++++++++
      WRITE (IO,21)
   21 FORMAT (////10X,23H1.2.  NODAL COORDINATES,
     1    //10X,5HINODE,10X,1HX)
      DO 20 I=1,NP
        READ (IN,22) INODE(I),XNODE(I)
   22   FORMAT (I5,5X,G10.1)
        WRITE (IO,23) INODE(I),XNODE(I)
   23   FORMAT (10X,I5,5X,G12.5)
   20 CONTINUE
C+++++++++++++
C     1.3.  INPUT AND OUTPUT OF ELEMENT/NODE CONNECTIVITY DATA
C+++++++++++++
      WRITE (IO,31)
   31 FORMAT (////10X,36H1.3.  ELEMENT/NODE CONNECTIVITY DATA,
     1    //10X,7HELEMENT,15X,5HNODES)
      DO 30 I=1,NE
        READ (IN,32) NEL(I),NODEI(I),NODEJ(I)
   32   FORMAT (3I5)
        WRITE (IO,33) NEL(I),NODEI(I),NODEJ(I)
```

```
 33     FORMAT (10X,I5,10X,2I5)
 30  CONTINUE
C+++++++++++++
C      1.4.  CALCULATE BANDWIDTH OF GLOBAL COEFFICIENT MATRIX
C            NBW=BANDWIDTH
C            FOR A 1-D PROGRAM, THE BANDWIDTH IS ALWAYS TWO.
C+++++++++++++
       NBW=2
       WRITE (IO,41) NBW
 41    FORMAT (////10X,27H1.4.  CALCULATED BANDWIDTH=,I5)
C++++++++++++++
C      1.5.  INPUT AND OUTPUT OF COEFFICIENT FUNCTIONS
C++++++++++++++
       WRITE (IO,51)
 51    FORMAT (////10X,27H1.5.  COEFFICIENT FUNCTIONS,/
     1     /10X,7HELEMENT,10X,2HKX,10X,1HP,11X,1HQ)
       READ (IN,52) ICO
 52    FORMAT(I5)
       IF (ICO .EQ. 1) READ (IN,54) IEL,KXA,PA,QA
 54    FORMAT (I5,5X,3G10.1)
       DO 50 I=1,NE
         IEL=I
         IF (ICO .EQ. 1) GO TO 55
         READ (IN,54) IEL,KXA,PA,QA
 55      CONTINUE
         KXE(IEL)=KXA
         PE(IEL)=PA
         QE(IEL)=QA
         WRITE (IO,56) IEL,KXE(IEL),PE(IEL),QE(IEL)
 56      FORMAT (10X,I5,5X,3G12.5)
 50    CONTINUE
       IF (ICO .EQ. 1) WRITE (IO,57) ICO
 57    FORMAT (//10X,34HKX,P,Q VALUES ARE THE SAME FOR ALL,
     1     9H ELEMENTS,/10X,4HICO=,I5)
C+++++++++++++
C      1.6.  INPUT AND OUTPUT OF PRESCRIBED NODAL VALUES
C+++++++++++++
       WRITE (IO,61)
 61    FORMAT (////10X,29H1.6.  PRESCRIBED NODAL VALUES)
       READ (IN,62) NBOUND
 62    FORMAT (I5)
       IF (NBOUND .LE. 0) GO TO 66
       WRITE (IO,63) NBOUND
 63    FORMAT (/ 10X,34HNUMBER OF PRESCRIBED NODAL VALUES=,I5,
     1     /10X,4HNODE,10X,11HNODAL VALUE)
       DO 60 I=1,NBOUND
         READ (IN,64) IBOUND(I),TBOUND(I)
 64      FORMAT (I5,5X,G10.1)
         WRITE (IO,65) IBOUND(I),TBOUND(I)
 65      FORMAT (10X,I5,10X,G12.5)
 60    CONTINUE
 66    IF (NBOUND .LE. 0) WRITE (IO,67) NBOUND
 67    FORMAT (42HNO PRESCRIBED NODAL VALUES HAVE BEEN GIVEN,
     1     /10X,7HNBOUND=,I5)
C+++++++++++++
C      1.7.  INPUT AND OUTPUT OF ELEMENT DERIVATIVE BOUNDARY
C            VALUES
C+++++++++++++
       WRITE (IO,71)
 71    FORMAT (////10X,32H1.7.  DERIVATIVE BOUNDARY VALUES)
```

```
      READ (IN,72) NDERIV
   72 FORMAT (I5)
      IF (NDERIV .LE. 0) GO TO 76
      WRITE (IO,73) NDERIV
   73 FORMAT (/ 10X,37HNUMBER OF DERIVATIVE BOUNDARY VALUES=,
     1      I5, /10X,4HNODE,10X,5HALPHA,
     2      5X,4HBETA)
      DO 70 I=1,NDERIV
        READ (IN,74) INODIR(I),ALPHAE(I),BETAE(I)
   74   FORMAT (I5,2G10.1)
        WRITE (IO,75) INODIR(I),ALPHAE(I),BETAE(I)
   75   FORMAT (10X,I5,10X,2G12.5)
   70 CONTINUE
   76 IF (NDERIV .LE. 0) WRITE (IO,77) NDERIV
   77 FORMAT (////10X,32HNO DERIVATIVE BOUNDARY CONDITONS,
     1      6H GIVEN,/10X,7HNDERIV=,I5)
C+++++++++++++
C     1.8.  OUTPUT PARAMETER
C
C+++++++++++++
      READ(IN,81) NOUT
      WRITE(IO,82) NOUT
   81 FORMAT(I5)
   82 FORMAT(////10X,22H1.8.  OUTPUT PARAMETER/12X,5HNOUT=,I5)
C*************************************************************
C     2.  CALCULATION OF ELEMENT PROPERTIES AND ASSEMBLY
C           OF GLOBAL MATRICES
C*************************************************************
C
C+++++++++++++
C     2.1.  WRITE TITLE AND ZERO GLOBAL MATRICES
C+++++++++++++
      CALL HEAD
      WRITE (IO,101)
  101 FORMAT (////10X,37H2.  CALCULATION OF ELEMENT PROPERTIES)
      DO 120 I=1,NP
        DO 130 J=1,NBW
          ACOND(I,J)=0.0
  130   CONTINUE
        F(I)=0.0
  120 CONTINUE
C+++++++++++++
C     2.2.  FORM LARGE DO LOOP OVER ALL ELEMENTS
C+++++++++++++
      DO 1000 I=1,NE
C+++++++++++++
C     2.3.  ASSIGN LOCAL NODAL COORDINATES
C+++++++++++++
        IEL=NEL(I)
        NN(1)=NODEI(IEL)
        NN(2)=NODEJ(IEL)
        X(1)=XNODE(NN(1))
        X(2)=XNODE(NN(2))
C+++++++++++++
C     2.4.  ASSIGN LOCAL ELEMENT COEFFICIENT FUNCTIONS
C+++++++++++++
        KX=KXE(IEL)
        P=PE(IEL)
        Q=QE(IEL)
C+++++++++++++
```

```
C     2.5.  ASSIGN LOCAL ELEMENT DERIVATIVE BOUNDARY CONDTIONS
C++++++++++++
      DO 140 J=1,2
        ALPHA(J)=0.0
        BETA(J)=0.0
 140  CONTINUE
      ISUM=0
      DO 150 K=1,NDERIV
        IF (INODIR(K) .NE. NN(1)) GO TO 160
        ISUM=1
        ALPHA(1)=ALPHAE(K)
        BETA(1)=BETAE(K)
 160    CONTINUE
        IF(INODIR(K) .NE. NN(2)) GO TO 150
        ISUM=1
        ALPHA(2)=ALPHAE(K)
        BETA(2)=BETAE(K)
 150    CONTINUE
C+++++++++++++
C     2.6.  CALL SUBROUTINE ELEMT1
C           ALL ELEMENT PROPERTIES ARE OUTPUT FROM ELEMT1
C+++++++++++++
      CALL ELEMT1 (IEL,X,NN,KX,
     1      P,Q,ALPHA,BETA,EBM,EC)
C+++++++++++++
C     2.7.  ASSEMBLY OF ELEMENT MATRICES INTO GLOBAL MATRICES
C           A    =GLOBAL COEFFICIENT MATRIX (WHICH IS TOO
C                   LARGE TO STORE SO IT IS NOT CALCULATED)
C           ACOND=CONDENSED GLOBAL COEFFICIENT MATRIX
C                   (FOR BANDED SYMMETRIC A MATRIX ONLY THE
C                   UPPER TRIAGONAL NONZERO TERMS ARE STORED)
C           F    =GLOBAL COLUMN
C           IROW =ROW OF GLOBAL A MATRIX
C           ICOL =COLUMN OF GLOBAL A MATRIX
C           JCOL =COLUMN OF CONDENSED GLOBAL ACOND MATRIX
C+++++++++++++
      DO 170 J=1,2
        DO 180 K=1,2
          IROW=NN(J)
          ICOL=NN(K)
          JCOL=ICOL-IROW+1
          IF (JCOL .LT. 1) GO TO 181
          ACOND(IROW,JCOL)=ACOND(IROW,JCOL)+EBM(J,K)
 181      CONTINUE
 180    CONTINUE
        F(IROW)=F(IROW)+EC(J)
 170  CONTINUE
1000  CONTINUE
C+++++++++++++
C     2.8.  OUTPUT OF CONDENSED GLOBAL MATRIX AND COLUMN
C+++++++++++++
      CALL WRITEM
      WRITE (IO,183)
 183  FORMAT (//10X,37H(GLOBAL MATRICES JUST AFTER ASSEMBLY))
C*************************************************************
C     3.  INCLUSION OF PRESCRIBED NODAL VALUES INTO GLOBAL
C           COEFFICIENT MATRIX AND COLUMN
C*************************************************************
C
C+++++++++++++
```

```
C     3.1.  MODIFY ROW AND COLUMN OF ACOND MATRIX AND MODIFY
C             ROW OF F COLUMN
C             IROW  =ROW OF PRESCRIBED NODAL VALUE
C             JROW  =ROW OF TERM TO BE MOVED TO RIGHT SIDE
C                    OF EQUAL SIGN
C             JCOL  =COLUMN OF ACOND TERM TO BE SET TO ZERO
C+++++++++++++
      DO 200 I=1,NBOUND
        IROW=IBOUND(I)
C
C         TRANSFER TERMS IN UPPER MATRIX TO RIGHT SIDE
        DO 210 J=2,NBW
          JROW=IROW-NBW+J-1
          JCOL=NBW-J+2
          IF (JROW .LT. 1) GO TO 210
          F(JROW)=F(JROW)-ACOND(JROW,JCOL)*TBOUND(I)
          ACOND(JROW,JCOL)=0.0
 210    CONTINUE
C
C         TRANSFER TERMS IN LOWER MATRIX TO RIGHT SIDE
        DO 220 J=2,NBW
          JROW=IROW+J-1
          JCOL=J
          IF (JROW .GT. NP) GO TO 220
          F(JROW)=F(JROW)-ACOND(IROW,JCOL)*TBOUND(I)
 220    CONTINUE
C
C         SET ACOND TERMS TO ZERO IN IROW AND MODIFY RIGHT
C           HAND SIDE
        DO 230 JCOL=2,NBW
          ACOND(IROW,JCOL)=0.0
 230    CONTINUE
        F(IROW)=ACOND(IROW,1)*TBOUND(I)
 200  CONTINUE
C+++++++++++++
C     3.2.  OUTPUT OF CONDENSED GLOBAL MATRIX AND COLUMN
C+++++++++++++
      CALL WRITEM
      WRITE (IO,233)
 233  FORMAT (//10X,36H(GLOBAL MATRICES AFTER INCLUSION OF
     1    ,24HPRESCRIBED NODAL VALUES))
C***********************************************************
C     4.  CALCULATION OF NODAL T VALUES, ELEMENT DERIVATIVES
C           AND INTEGRALS
C***********************************************************
C
C+++++++++++++
C     4.1.  CALCULATION OF NODAL VALUES
C+++++++++++++
C
C         SOLUTION OF BANDED LINEAR SIMULTANEOUS EQUATIONS
C           FOR THE NODAL T VALUES
      CALL GAUSS1
      CALL WRITEM
      WRITE (IO,404)
 404  FORMAT (//10X,36H(GLOBAL MATRICES AFTER FORMATION OF
     1    ,22HUPPER TRIANGULAR FORM))
 408  FORMAT(////10X,28H3.  GLOBAL MATRIX AND COLUMN///10X,
     1       49HCONDENSATION AND UPPER TRIANGULAR FORM COMPLETED.
     2       /12X,5HNOUT=,I5)
```

```
C
C           OUTPUT OF NODAL VALUES
        CALL HEAD
        WRITE (IO,403)
  403   FORMAT (////10X,37H4.  NODAL VALUES, ELEMENT DERIVATIVES
       1    ,14H, AND INTEGRAL)
        WRITE (IO,405)
  405   FORMAT (////10X,18H4.1.  NODAL VALUES)
        WRITE (IO,402)
  402   FORMAT (   /10X,4HNODE, 8X,12HNODAL VALUES)
        DO 400 I=1,NP
          WRITE (IO,401) INODE(I),T(INODE(I))
  401     FORMAT (10X,I5,5X, G15.8)
  400   CONTINUE
C+++++++++++++
C       4.2.  ELEMENT DERIVATIVES AND INTEGRALS
C+++++++++++++
        WRITE (IO,543)
  543   FORMAT (////10X,30H4.2.  ELEMENT DERIVATIVES AND ,
       1     , 8HINTEGRAL )
        WRITE (IO,542)
  542   FORMAT (////10X,7HELEMENT,10X,
       1     ,11HDERIVATIVES,8X,8HINTEGRAL)
        DO 500 I=1,NE
          IEL=NEL(I)
          NN(1)=NODEI(IEL)
          NN(2)=NODEJ(IEL)
          X(1)=XNODE(NN(1))
          X(2)=XNODE(NN(2))
          CALL PROP1 (X,B,LENG)
          DTDX=0.0
          DO 520 J=1,2
            DTDX=DTDX+B(J)*T(NN(J))
  520     CONTINUE
          DTDX=DTDX/LENG
          TINT=(T(NN(1))+T(NN(2)))*LENG/2.0
          WRITE (IO,541) NEL(I),DTDX,TINT
  541     FORMAT (10X,I5,5X,G15.8,5X,G15.8)
  500   CONTINUE
        END
C
C
        SUBROUTINE ELEMT1 (IEL,X,NN,KX,P,Q,ALPHA,
       1    BETA,EBM,EC)
C**********************************************************
C           SUBROUTINE ELEMT1 CALCULATES THE ELEMENT
C             COEFFICIENT MATRIX AND COLUMN FOR A LINEAR
C             INTERPOLATION FUNCTION IN A ONE DIMENSIONAL
C             ELEMENT. (THIS ROUTINE USES SIMPLEX ELEMENTS.)
C
C           THE DIFFERENTIAL EQUATION HAS THE FORM
C             D(KX*DT/DX)/DX+P*T+Q=0
C           WITH THE BOUNDARY CONDITIONS
C             T=T0, ON L1
C             KX*DT/DX*NX+ALPHA*T+BETA=0, ON L2
C**********************************************************
        DIMENSION X(2),NN(2),
       1    ALPHA(2),BETA(2),EBM(2,2),EC(2)
        DIMENSION B(2)
        REAL KX,LENG
```

```
      COMMON /INT10/ NP,NE,NBW,NBC,NOUT
      DATA IN,IO/60,61/
C***************************************************************
C     1.  CALCULATION OF ELEMENT COEFFICIENT MATRICES
C         AND COLUMNS
C***************************************************************
C
C++++++++++++
C     1.1.  CALCULATION OF INTERPOLATION FUNCTION PROPERTIES
C           B    =X GRADIENT TERMS
C           LENG=ELEMENT LENGTH
C++++++++++++
      CALL PROP1 (X,B,LENG)
C++++++++++++
C     1.2.  ELEMENT COEFFICIENT MATRIX AND COLUMN
C           KX     =ELEMENT VALUE OF KX COEFFICIENT FUNCTION
C           P      =ELEMENT VALUE OF P  COEFFICIENT FUNCTION
C           Q      =ELEMENT VALUE OF Q  COEFFICIENT FUNCTION
C           EBM    =ELEMENT B MATRIX
C           EC     =ELEMENT C COLUMN
C++++++++++++
      COEFKX=KX
      DO 10 I=1,2
        DO 20 J=1,2
          EBM(I,J)=COEFKX*B(I)*B(J)/LENG
          IF (I .EQ. J) EBM(I,J)=EBM(I,J)-P*LENG/3.0
          IF (I .NE. J) EBM(I,J)=EBM(I,J)-P*LENG/6.0
   20   CONTINUE
        EC(I)=Q*LENG/2.0
   10 CONTINUE
C+++++++++++++
C     1.3.  DERIVATIVE BOUNDARY CONDITIONS
C+++++++++++++
        EBM(1,1)=EBM(1,1)-ALPHA(1)
        EBM(1,2)=EBM(1,2)
        EBM(2,1)=EBM(1,2)
        EBM(2,2)=EBM(2,2)+ALPHA(2)
        EC(1)=EC(1)+BETA(1)
        EC(2)=EC(2)-BETA(2)
C***************************************************************
C     2.  OUTPUT OF ELEMENT MATRICES AND COLUMNS
C***************************************************************
C     CHECK OUTPUT PARAMETER
      IF (NOUT .GT. 0) GO TO 999
C++++++++++++
C     2.1.  WRITE ELEMENT CONNECTIVITY DATA AND NODAL COORD
C++++++++++++
      WRITE (IO,111) IEL
  111 FORMAT (////10X,30HELEMENT PROPERTIES FOR ELEMENT,I5)
      WRITE (IO,114)
  114 FORMAT (/10X,37H2.1.  ELEMENT CONNECTIVITY AND NODAL
     1     ,11HCOORDINATES)
      WRITE (IO,112)
  112 FORMAT (  10X,7HELEMENT, 5X,4HNODE, 6X,1HX)
      DO 110 I=1,2
        WRITE (IO,113) IEL,NN(I),X(I)
  113   FORMAT (10X,I5, 5X,I5,F10.4)
  110 CONTINUE
C+++++++++++++
C     2.2.  WRITE ELEMENT COEFFICIENT FUNCTIONS
```

```
C+++++++++++++
      WRITE (IO,123)
  123 FORMAT (/10X,37H2.2.  ELEMENT INTERPOLATION FUNCTIONS)
      WRITE (IO,121)
  121 FORMAT( 12X,2HKX,10X,1HP,10X,1HQ)
        WRITE (IO,122) KX,P,Q
  122   FORMAT (10X,3G12.5)
  120 CONTINUE
C++++++++++++
C     2.3.  WRITE ELEMENT INTERPOLATION FUNCTIONS
C++++++++++++
      WRITE (IO,134)
  134 FORMAT (/10X,37H2.3.  ELEMENT INTERPOLATION FUNCTION
     1     ,10HPROPERTIES)
      WRITE (IO,132)
  132 FORMAT (10X,4HNODE,4X,1HB)
      DO 130 I=1,2
        WRITE (IO,133) NN(I),B(I)
  133   FORMAT (10X,I5,G12.5)
  130 CONTINUE
      WRITE (IO,131) LENG
  131 FORMAT (   10X,7HLENGTH=,G12.5)
C+++++++++++++
C     2.4.  WRITE ELEMENT DERIVATIVE BOUNDARY CONDITIONS
C+++++++++++++
      IF (ISUM .EQ. 0) WRITE (IO,141)
  141 FORMAT ( /10X,41H2.4.  NO DERIVATIVE BOUNDARY CONDITIONS
     1     ,10HIN ELEMENT)
      IF (ISUM .NE. 0) WRITE (IO,145)
  145 FORMAT (/10X,34H2.4.  ELEMENT DERIVATIVE BOUNDARY
     1     ,9HCONDITION)
      IF (ISUM .NE. 0) WRITE (IO,142)
  142 FORMAT ( 10X,4HNODE,10X,5HALPHA, 7X,4HBETA)
      DO 140 I=1,2
        WRITE (IO,144) NN(I),ALPHA(I),BETA(I)
  144   FORMAT (10X,I5,5X,2G12.5)
  140 CONTINUE
C+++++++++++++
C     2.5.  WRITE ELEMENT COEFFICIENT MATRIX AND COLUMN
C+++++++++++++
      WRITE (IO,153)
  153 FORMAT (/10X,37H2.5.  ELEMENT COEFFICIENT MATRIX AND
     1     , 6HCOLUMN)
      WRITE (IO,151)
  151   FORMAT ( 10X,4HNODE,13X,18HCOEFFICIENT MATRIX,15X,
     1     6HCOLUMN)
      DO 150 I=1,2
        WRITE (IO,152) NN(I),(EBM(I,J),J=1,2),EC(I)
  152 FORMAT (10X,I5,10X,2G12.5,10X,G12.5)
  150 CONTINUE
  999 CONTINUE
      RETURN
      END
C
C
      SUBROUTINE HEAD
C*************************************************************
C           WRITE TITLES ON TOP OF PAGE
C*************************************************************
      COMMON /ALP10/ TITLE1(20),TITLE2(20)
```

```
      DATA IN,IO/60,61/
      WRITE (IO,13) TITLE1
   13 FORMAT (1H1//10X,20A4)
      WRITE (IO,14) TITLE2
   14 FORMAT (10X,20A4)
      RETURN
      END
C
C
      SUBROUTINE PROP1 (X,B,LENG)
C*************************************************************
C           SUBROUTINE PROP1 CALCULATES THE X GRADIENT
C             TERMS AND LENGTH FOR A LINEAR INTERPOLATION
C             FUNCTION IN A LINE ELEMENT.
C*************************************************************
      DIMENSION X(2),B(2)
      REAL LENG
      B(1)=1.
      B(2)=-1.
      LENG=X(2)-X(1)
      RETURN
      END
C
C
      SUBROUTINE GAUSS1
C*************************************************************
C           SUBROUTINE GAUSS1 SOLVES A BANDED SYMMETRIC SET OF
C             LINEAR ALGEBRAIC EQUATIONS USING A BANDED GAUSS
C             ELIMINATION TECHNIQUE.
C
C           THE MATRIX IS STORED IN CONDENSED FORM USING ONLY
C             THE UPPER TRIANGULAR BANDWITH.  THUS ALL ROWS
C             OF THE SQUARE MATRIX ARE SHIFTED TO THE LEFT SO
C             THAT THE MAIN DIAGONAL TERMS ARE NOW IN COLUMN
C             ONE OF THE RECTANGULAR MATRIX ACOND.  ACOND HAS
C             DIMENSIONS NP BY NBW.
C*************************************************************
      COMMON /INT10/ NP,NE,NBW,NBC,NOUT
      COMMON /REAL10/ ACOND(100,2),F(100),T(100)
      DATA IN,IO/60,61/
C+++++++++++++
C     1.  CONSTRUCTION OF UPPER BANDED TRIANGULAR FORM
C+++++++++++++
      NPM=NP-1
      DO 100 I=1,NPM
        IP=I+1
        DO 110 J=2,NBW
          IF (ACOND(I,J) .EQ. 0.0) GO TO 110
          C=ACOND(I,J)/ACOND(I,1)
          L=I+J-1
          M=0
          DO 120 K=J,NBW
            M=M+1
            ACOND(L,M)=ACOND(L,M)-C*ACOND(I,K)
  120     CONTINUE
          F(L)=F(L)-C*F(I)
  110   CONTINUE
  100 CONTINUE
C+++++++++++++
C     2.  BACKWARD SUBSTITUTION TO OBTAIN UNKNOWNS
```

```
C+++++++++++++
      DO 200 I=1,NP
        L=NP-I+1
        T(L)=F(L)
        K=L
        DO 210 J=2,NBW
          K=K+1
          IF (K .GT. L+NBW-1) GO TO 210
          IF (K .GT. NP) GO TO 210
          T(L)=T(L)-ACOND(L,J)*T(K)
  210   CONTINUE
        T(L)=T(L)/ACOND(L,1)
  200 CONTINUE
      RETURN
      END
C
C
      SUBROUTINE WRITEM
C*************************************************************
C          SUBROUTINE WRITEM WRITES OUT THE ACOND MATRIX AND
C            F COLUMN.
C*************************************************************
      COMMON /ALP10/ TITLE1(20),TITLE2(20)
      COMMON /INT10/ NP,NE,NBW,NBC,NOUT
      COMMON /REAL10/ ACOND(100,2),F(100),T(100)
      DATA IN,IO/60,61/
      IF (NOUT .GT. 1) GO TO 999
      CALL HEAD
      WRITE (IO,501)
  501 FORMAT (////10X,28H3.  GLOBAL MATRIX AND COLUMN)
      WRITE (IO,533)
  533 FORMAT (////10X,35H3.1.  CONDENSED GLOBAL COEFFICIENT
     1    ,14H(ACOND) MATRIX)
      WRITE (IO,535)
  535 FORMAT ( /10X,4HNODE,20X, 5HACOND)
      DO 510 I=1,NP
      WRITE (IO,511) I,(ACOND(I,J),J=1,NBW)
  511 FORMAT (10X,I5,10X,6G12.5)
  510 CONTINUE
      WRITE (IO,534)
  534 FORMAT (////10X,21H3.2.  GLOBAL F COLUMN)
      WRITE (IO,537)
  537 FORMAT ( /10X,4HNODE,12X,1HF)
      DO 520 I=1,NP
        WRITE (IO,521) I,F(I)
  521   FORMAT ( 10X,I5,5X,G12.5)
  520 CONTINUE
  999 CONTINUE
      RETURN
      END
```

```
  SAMPLE INPUT FOR PROGRAM FINONE
  EXAMPLE 5.9 - HEAT CONDUCTION AND CONVECTION IN THIN FIN

1.  INPUT AND OUTPUT OF PROBLEM DATA

1.1.  GLOBAL PARAMETERS

NUMBER OF NODES     4
NUMBER OF ELEMENTS     3

1.2.  NODAL COORDINATES

INODE          X
    1       0.
    2        40.000
    3        80.000
    4        120.00

1.3.  ELEMENT/NODE CONNECTIVITY DATA

ELEMENT                  NODES
    1               1    2
    2               2    3
    3               3    4


1.4.  CALCULATED BANDWIDTH=     2

1.5.  COEFFICIENT FUNCTIONS

ELEMENT            KX            P             Q
    1        40.000      -.64000E-01   1.9200
    2        40.000      -.64000E-01   1.9200
    3        40.000      -.64000E-01   1.9200


KX,P,Q VALUES ARE THE SAME FOR ALL ELEMENTS
ICO=    1

1.6.  PRESCRIBED NODAL VALUES

NUMBER OF PRESCRIBED NODAL VALUES=     2
NODE           NODAL VALUE
    1              330.00
    4              30.000


1.7.  DERIVATIVE BOUNDARY VALUES

NO DERIVATIVE BOUNDARY CONDITONS GIVEN
NDERIV=    0

1.8.  OUTPUT PARAMETER
  NOUT=    0
```

```
  SAMPLE INPUT FOR PROGRAM FINONE
  EXAMPLE 5.9 - HEAT CONDUCTION AND CONVECTION IN THIN FIN

2.  CALCULATION OF ELEMENT PROPERTIES

ELEMENT PROPERTIES FOR ELEMENT    1

2.1.  ELEMENT CONNECTIVITY AND NODAL COORDINATES
ELEMENT      NODE       X
    1          1     0.0000
    1          2    40.0000

2.2.  ELEMENT INTERPOLATION FUNCTIONS
  KX           P           Q
 40.000      -.64000E-01  1.9200

2.3.  ELEMENT INTERPOLATION FUNCTION PROPERTIES
NODE     B
    1  1.0000
    2 -1.0000
LENGTH=  40.000

2.4.  ELEMENT DERIVATIVE BOUNDARY CONDITION
NODE            ALPHA        BETA
    1        0.            0.
    2        0.            0.

2.5.  ELEMENT COEFFICIENT MATRIX AND COLUMN
NODE               COEFFICIENT MATRIX                 COLUMN
    1               1.8533      -.57333                38.400
    2              -.57333       1.8533                38.400

ELEMENT PROPERTIES FOR ELEMENT    2

2.1.  ELEMENT CONNECTIVITY AND NODAL COORDINATES
ELEMENT      NODE       X
    2          2    40.0000
    2          3    80.0000

2.2.  ELEMENT INTERPOLATION FUNCTIONS
  KX           P           Q
 40.000      -.64000E-01  1.9200

2.3.  ELEMENT INTERPOLATION FUNCTION PROPERTIES
NODE     B
    2  1.0000
    3 -1.0000
LENGTH=  40.000

2.4.  ELEMENT DERIVATIVE BOUNDARY CONDITION
NODE            ALPHA        BETA
    2        0.            0.
    3        0.            0.

2.5.  ELEMENT COEFFICIENT MATRIX AND COLUMN
NODE               COEFFICIENT MATRIX                 COLUMN
    2               1.8533      -.57333                38.400
    3              -.57333       1.8533                38.400
```

```
ELEMENT PROPERTIES FOR ELEMENT    3

2.1.  ELEMENT CONNECTIVITY AND NODAL COORDINATES
ELEMENT     NODE      X
    3         3    80.0000
    3         4   120.0000

2.2.  ELEMENT INTERPOLATION FUNCTIONS
  KX          P            Q
 40.000      -.64000E-01   1.9200

2.3.  ELEMENT INTERPOLATION FUNCTION PROPERTIES
NODE    B
    3  1.0000
    4 -1.0000
LENGTH=  40.000

2.4.  ELEMENT DERIVATIVE BOUNDARY CONDITION
NODE          ALPHA        BETA
    3      0.           0.
    4      0.           0.

2.5.  ELEMENT COEFFICIENT MATRIX AND COLUMN
NODE             COEFFICIENT MATRIX                COLUMN
    3             1.8533      -.57333               38.400
    4            -.57333       1.8533               38.400

  SAMPLE INPUT FOR PROGRAM FINONE
  EXAMPLE 5.9 - HEAT CONDUCTION AND CONVECTION IN THIN FIN

3.  GLOBAL MATRIX AND COLUMN

3.1.  CONDENSED GLOBAL COEFFICIENT (ACOND) MATRIX

NODE                    ACOND
    1             1.8533      -.57333
    2             3.7067      -.57333
    3             3.7067      -.57333
    4             1.8533      0.

3.2.  GLOBAL F COLUMN

NODE           F
    1       38.400
    2       76.800
    3       76.800
    4       38.400


(GLOBAL MATRICES JUST AFTER ASSEMBLY)
```

SAMPLE INPUT FOR PROGRAM FINONE
EXAMPLE 5.9 - HEAT CONDUCTION AND CONVECTION IN THIN FIN

3. GLOBAL MATRIX AND COLUMN

3.1. CONDENSED GLOBAL COEFFICIENT (ACOND) MATRIX

NODE	ACOND	
1	1.8533	0.
2	3.7067	-.57333
3	3.7067	0.
4	1.8533	0.

3.2. GLOBAL F COLUMN

NODE	F
1	611.60
2	266.00
3	94.000
4	55.600

(GLOBAL MATRICES AFTER INCLUSION OF PRESCRIBED NODAL VALUES)

SAMPLE INPUT FOR PROGRAM FINONE
EXAMPLE 5.9 - HEAT CONDUCTION AND CONVECTION IN THIN FIN

3. GLOBAL MATRIX AND COLUMN

3.1 CONDENSED GLOBAL COEFFICIENTS (ACOND) MATRIX

NODE	ACOND	
1	1.8533	0.
2	3.7067	-.57333
3	3.6180	0.
4	1.8533	0.

3.2. GLOBAL F COLUMN

NODE	F
1	611.60
2	266.00
3	135.14
4	55.600

(GLOBAL MATRICES AFTER FORMATION OF UPPER TRIANGULAR FORM)

SAMPLE INPUT FOR PROGRAM FINONE
EXAMPLE 5.9 - HEAT CONDUCTION AND CONVECTION IN THIN FIN

4. NODAL VALUES, ELEMENT DERIVATIVES, AND INTEGRAL

4.1. NODAL VALUES

NODE	NODAL VALUES
1	330.00000
2	77.540266
3	37.353351
4	30.000000

4.2. ELEMENT DERIVATIVES AND INTEGRAL

ELEMENT	DERIVATIVES	INTEGRAL
1	6.3114933	8150.8053
2	1.0046729	2297.8723
3	.18383376	1347.0670

Appendix B

Computer Program FINTWO for Two-Dimensional Field Problems

B.1 Program Description

An example computer program, called FINTWO, is presented in this appendix both for instructional purposes and as a useful program to solve engineering problems. It solves a second-order partial differential equation in two dimensions of the form

$$\frac{\partial}{\partial x}\left[K_x\,(x,y)\,\frac{\partial T}{\partial x}\right] + \frac{\partial}{\partial y}\left[K_y\,(x,y)\,\frac{\partial T}{\partial y}\right] + P(x,y)T + Q(x,y) = 0$$

with the boundary conditions $T = T_0$ on L_1 and

$$K_x\,(x,y)\,\frac{\partial T}{\partial x}\,n_x + K_y\,(x,y)\,\frac{\partial T}{\partial y}\,n_y + \alpha(x,y)T + \beta(x,y) = 0, \text{ on } L_2$$

The finite element theory is developed in Chapter 9. Also, all the examples given in Chapter 10, were carried out with this program. The input and output is discussed in Chapter 11.

The program can be used to solve many types of problems, such as heat conduction and convection, ideal fluid flow, viscous flow in lubrication problems, torsion of noncircular sections, and electrical field problems. A general form of input and output has been employed so that the most flexibility has been retained. An efficient storage scheme for the global matrices and an efficient banded Gauss elimination method for the algebraic equations have been incorporated into the program. The theory is covered in Chapter 11. These techniques make the program efficient enough for industrial use. The user should label the nodes to minimize the bandwidth.

FINTWO is written as a main program with several subroutines. Table B.1 gives an outline including the major sections of the main program and longer subroutines. In addition, each major section is broken down further into subsections where appropriate.

Several features have been incorporated into the program to increase readability. A large number of comment lines have been used (approximately half of the lines are comment lines). At the beginning, the required input data and format are given. Major sections of the program are labeled 1, 2, 3, etc., while subsections are labeled 1.1, 1.2, 1.3, etc. All DO loops end with a statement number ending in zero, and statements within a DO loop are all indented two spaces. Nested DO loops are indented two spaces per loop. Very long DO loops have statement numbers 1000, 2000, 3000, etc.

Table B.1
Outline of Computer Program FINTWO

Main Program—FINTWO

1. Input and output of problem data.
2. Calculation of element properties and assembly of global matrices.
3. Inclusion of prescribed nodal values into global coefficient matrix and column.
4. Calculation of nodal *T* values, element derivatives, and integrals.

Subroutine ELEMNT

1. Calculation of element coefficient matrices and columns.
2. Output of element matrices and columns.

Subroutine HEAD

1. Writing of titles at top of each page.

Subroutine PROP1

1. Calculation of pyramid function terms and area for simplex triangular element.

Subroutine GAUSS1

1. Construction of upper banded triangular form.
2. Backward substitution to obtain unknowns.

Subroutine WRITEM

1. Writing of global matrices and columns.

B.2 Listing of Program FINTWO

The following pages give a complete listing of computer program FINTWO with an example output from Example 10.2

```
      PROGRAM FINTWO (INPUT,OUTPUT,TAPE60=INPUT,TAPE61=OUTPUT)
C*******************************************************************
C
C                    PROGRAM FINTWO
C
C                    VERSION 1.4
C                    DECEMBER, 1983
C
C                    DR PAUL ALLAIRE
C                    PROFESSOR
C                    MECHANICAL AND AEROSPACE
C                      ENGINEERING DEPARTMENT
C                    UNIVERSITY OF VIRGINIA
C                    CHARLOTTESVILLE, VIRGINIA 22903
C
C*******************************************************************
C          PROGRAM DESCRIPTION
C*******************************************************************
C          THIS COMPUTER PROGRAM SOLVES A SECOND ORDER PARTIAL
C            DIFFERENTIAL EQUATION IN TWO DIMENSIONS USING
C            THE FINITE ELEMENT METHOD.  THE VARIATIONAL METHOD
C            WAS USED TO FORMULATE THE APPROPRIATE ELEMENT
C            COEFFICIENT MATRICES AND COLUMNS.
C
C          IT SOLVES A DIFFERENTIAL EQUATION OF THE FORM
C            D(KX*DT/DX)/DX+D(KY*DT/DY)/DY+P*T+Q=0
C          WITH THE BOUNDARY CONDITIONS
C            T=TO, ON L1
C            KX*DT/DX*NX+KY*DT/DY*NY+ALPHA*T+BETA=0, ON L2
C
C          THE COEFFICIENT FUNCTIONS KX,KY,P,Q,ALPHA,BETA
C            ARE KNOWN FUNCTIONS OF X AND Y.
C
C*******************************************************************
C      1.  REQUIRED INPUT DATA AND FORMAT
C*******************************************************************
C          INPUT DATA IS ORGANIZED INTO SECTIONS LABELED
C            1,2,3,ETC.  THE NUMBER OF DATA LINES WITHIN
C            EACH SECTION WILL VARY ACCORDING TO THE NUMBER
C            NODE POINTS, ELEMENTS, PRESCRIBED NODAL VALUES,
C            AND DERIVATIVE BOUNDARY CONDITIONS.
C
C+++++++++++++
C      1.1.  TITLE AND GLOBAL PARAMETERS
C+++++++++++++
C          LINE  1.  TITLE1    FORMAT (20A4)
C            TITLE1=ANY ALPHA NUMERIC STRING FOR TITLE
C
C          LINE  2.  TITLE2    FORMAT (20A4)
C            TITLE2=ANY ALPHA NUMERIC STRING FOR TITLE
C
C          LINE  3.  NP,NE    FORMAT (2I5)
C            NP=NUMBER OF NODES
C              -NP MUST BE LESS THAN OR EQUAL TO 150.
C            NE=NUMBER OF ELEMENTS
C              -NE MUST BE LESS THAN OR EQUAL TO 200.
C
C            THIS PROGRAM IS CURRENTLY LIMITED TO NP=150
C              AND NE=200 DUE TO ARRAY DIMENSIONING. THE
```

```
C               ARRAY SIZES MAY BE INCREASED BY CHANGING THE
C               APPROPRIATE DIMENSION AND COMMON STATEMENTS.
C
C+++++++++++++
C      1.2.  NODAL COORDINATES
C+++++++++++++
C          LINES  1 THRU NP.  INODE,XNODE,YNODE
C                                  FORMAT (I5,5X,2G10.1)
C            INODE=GLOBAL NODE NUMBER
C            XNODE=X COORDINATE OF NODE
C            YNODE=Y COORDINATE OF NODE
C
C+++++++++++++
C      1.3.  ELEMENT/NODE CONNECTIVITY DATA
C+++++++++++++
C          LINES  1 THRU NE.  NEL,NODEI,NODEJ,NODEK
C                                  FORMAT (4I5)
C            NEL  =ELEMENT NUMBER
C            NODEI=GLOBAL NUMBER OF FIRST NODE IN ELEMENT
C            NODEJ=GLOBAL NUMBER OF SECOND NODE IN ELEMENT
C            NODEK=GLOBAL NUMBER OF THIRD NODE IN ELEMENT
C
C            PLEASE REMEMBER TO NUMBER ELEMENTS IN COUNTER
C              CLOCKWISE FASHION.
C
C+++++++++++++
C      1.4.  BANDWIDTH
C++++++++++++
C
C            NO INPUT IS NEEDED.
C
C            BANDWIDTH IS CURRENTLY LIMITED TO NBW=20 DUE TO
C              ARRAY DIMENSIONING.  IT MAY BE INCREASED BY
C              CHANGING THE COMMON /REAL10/ STATEMENTS.
C
C+++++++++++++
C      1.5.  COEFFICIENT FUNCTIONS
C+++++++++++++
C          LINE  1.  ICO    FORMAT (I5)
C            ICO=COEFFICIENT PARAMETER
C              IF ICO=1, KX,KY,P,Q HAVE THE SAME VALUE FOR
C                        ALL ELEMENTS SO ONLY ONE DATA LINE
C                        IS NEEDED.
C              IF ICO=2, VALUES FOR KX,KY,P,Q MUST BE INPUT
C                        FOR EACH ELEMENT ON A DATA LINE.
C
C          LINE  2 (IF ICO=1).  IEL,KXE,KYE,PE,QE
C                                  FORMAT (I5,5X,4G10.1)
C            IEL=ELEMENT NUMBER
C            KXE=VALUE OF KX IN ELEMENT
C            KYE=VALUE OF KY IN ELEMENT
C            PE =VALUE OF P IN ELEMENT
C            QE =VALUE OF Q IN ELEMENT
C
C          NEXT NE LINES (IF ICO=2).  IEL,KXE,KYE,PE,QE
C                                  FORMAT (I5,5X,4G10.1)
C
C+++++++++++++
C      1.6.  PRESCRIBED NODAL VALUES
C+++++++++++++
```

```
C         LINE  1.  NBOUND    FORMAT (I5)
C           NBOUND=NUMBER OF PRESCRIBED NODAL VLAUES
C                 -NBOUND MUST BE LESS THAN OR EQUAL TO 100.
C
C         NEXT NBOUND LINES.  IBOUND,TBOUND
C                                 FORMAT (I5,5X,G10.1)
C           IBOUND=GLOBAL NODE NUMBER
C           TBOUND=PRESCRIBED NODAL VALUES
C
C+++++++++++++
C     1.7.  DERIVATIVE BOUNDARY VALUES
C+++++++++++++
C         LINE  1.  NDERIV    FORMAT (I5)
C             NDERIV=TOTAL NUMBER OF SIDES ON WHICH
C                    DERIVATIVE BOUNDARY VALUES ARE GIVEN
C                   -NDERIV MUST BE LESS THAN OR EQUAL TO 100.
C
C         NEXT NDERIV LINES.  IELDIR,NSIDE,ALPHAE,BETAE
C                                 FORMAT (2I5,2G10.1)
C           IELDIR=ELEMENT NUMBER OF DERIVATIVE
C                    BOUNDARY VALUE
C           NSIDE =LOCAL NUMBER OF SIDE ON BOUNDARY
C           ALPHAE=VALUE OF ALPHA ON BOUNDARY SIDE
C           BETAE =VALUE OF BETA ON BOUNDARY SIDE
C
C           SIDES ARE NUMBERED 1 THRU 3 COUNTER CLOCKWISE STARTING
C             IMMEDIATELY FOLLOWING THE FIRST NODE (NODEI) ENTERED
C             IN SECTION 1.3 FOR THAT ELEMENT.
C
C+++++++++++++
C     1.8.  OUTPUT PARAMETERS
C+++++++++++++
C         LINE  1.  NOUT    FORMAT (I5)
C           NOUT  =OUTPUT PARAMETER
C             IF NOUT=0, ALL PROGRAM OUTPUT IS PRINTED.
C             IF NOUT=1, OUTPUT OF ELEMENT PROPERTIES IS OMITTED.
C             IF NOUT=2, OUTPUT OF ELEMENT PROPERTIES AND GLOBAL
C                        MATRICES IS OMITTED.
C
C*************************************************************
C     2.  ELEMENT OUTPUT DATA
C*************************************************************
C         OUTPUT DATA FOR EACH ELEMENT INCLUDES CONNECTIVITY,
C           NODAL COORDINATES, INTERPOLATION FUNCTION
C           PROPERTIES, DERIVATIVE BOUNDARY CONDITIONS,
C           AND ELEMENT COEFFICIENT MATRICES.
C
C*************************************************************
C     3.  OUTPUT OF GLOBAL MATRICES
C*************************************************************
C         CONDENSED GLOBAL MATRICES AND COLUMNS ARE OUTPUT
C           JUST AFTER ASSEMBLY OF ELEMENT MATRICES, AFTER
C           INCLUSION OF PRESCRIBED NODAL VALUES, AND AFTER
C           FORMATION OF UPPER TRIANGLUAR FORM.
C
C*************************************************************
C     4.  OUTPUT OF NODAL VALUES
C*************************************************************
C         NODAL VALUES, ELEMENT DERIVATIVES, AND INTEGRALS
C           ARE OUTPUT.
```

```
C
C*************************************************************
C          PROGRAM READABLILITY FEATURES
C*************************************************************
C          PROGRAM SECTIONS ARE LABELED WITH COMMENT CARDS.
C            MAJOR SECTIONS ARE LABELED WITH 1,2,3,ETC WHILE
C            SUBSECTIONS ARE LABELED WITH 1.1,1.2,1.3,ETC
C            WITHIN EACH MAJOR SECTION.
C
C          ALL DO LOOPS END WITH A STATEMENT NUMBER ENDING
C            IN ZERO AND STATEMENTS WITHIN THE DO LOOP ARE
C            INDENTED TWO SPACES.  VERY LONG DO LOOPS HAVE
C            STATEMENT NUMBERS 1000,2000,ETC.
C
C*************************************************************
C
C
      DIMENSION INODE(150),XNODE(150),YNODE(150)
      DIMENSION NEL(200),NODEI(200),NODEJ(200),NODEK(200)
      DIMENSION KXE(200),KYE(200),PE(200),QE(200)
      DIMENSION IBOUND (100),TBOUND(100)
      DIMENSION IELDIR(100),NSIDE(100),ALPHAE(100),BETAE(100)
      DIMENSION X(3),Y(3),NN(3),ISIDE(3),ALPHA(3),BETA(3)
      DIMENSION B(3),C(3)
      DIMENSION EBM(3,3),EC(3)
      REAL KX,KY,KXE,KYE,KXA,KYA
      COMMON /ALP10/ TITLE1(20),TITLE2(20)
      COMMON /INT10/ NP,NE,NBW,NOUT
      COMMON /REAL10/ ACOND(150,20),F(150),T(150)
      DATA IN,IO/60,61/
C*************************************************************
C      1.  INPUT AND OUTPUT OF PROBLEM DATA
C*************************************************************
C
C+++++++++++++
C      1.1.  INPUT AND OUTPUT OF TITLE AND GLOBAL PARAMETERS
C+++++++++++++
      READ (IN,11) TITLE1
      READ (IN,11) TITLE2
   11 FORMAT (20A4)
      READ (IN,12) NP,NE
   12 FORMAT (2I5)
      CALL HEAD
      WRITE (IO,15)
   15 FORMAT (//10X,36H1.  INPUT AND OUTPUT OF PROBLEM DATA)
      WRITE (IO,16)
   16 FORMAT (////10X,23H1.1.  GLOBAL PARAMETERS)
      WRITE (IO,17) NP,NE
   17 FORMAT ( /10X,15HNUMBER OF NODES,I5,/
     1     10X,18HNUMBER OF ELEMENTS,I5)
C+++++++++++++
C      1.2.  INPUT AND OUTPUT OF NODAL COORDINATES
C+++++++++++++
      WRITE (IO,21)
   21 FORMAT (////10X,23H1.2.  NODAL COORDINATES,
     1     //10X,5HINODE,10X,1HX,10X,1HY)
      DO 20 I=1,NP
        READ (IN,22) INODE(I),XNODE(I),YNODE(I)
   22   FORMAT (I5,5X,2G10.1)
        WRITE (IO,23) INODE(I),XNODE(I),YNODE(I)
```

```
 23     FORMAT (10X,I5,5X,2G12.5)
 20   CONTINUE
C+++++++++++++
C       1.3.  INPUT AND OUTPUT OF ELEMENT/NODE CONNECTIVITY DATA
C+++++++++++++
      WRITE (IO,31)
 31   FORMAT (////10X,36H1.3.  ELEMENT/NODE CONNECTIVITY DATA,
     1     //10X,7HELEMENT,15X,5HNODES)
      DO 30 I=1,NE
        READ (IN,32) NEL(I),NODEI(I),NODEJ(I),NODEK(I)
 32     FORMAT (4I5)
        WRITE (IO,33) NEL(I),NODEI(I),NODEJ(I),NODEK(I)
 33     FORMAT (10X,I5,10X,3I5)
 30   CONTINUE
C+++++++++++++
C       1.4.  CALCULATE BANDWIDTH OF GLOBAL COEFFICIENT MATRIX
C             NBW=BANDWIDTH
C+++++++++++++
      NDMAX=0
      DO 40 I=1,NE
        NDIF1=IABS(NODEI(I)-NODEJ(I))
        NDIF2=IABS(NODEJ(I)-NODEK(I))
        NDIF3=IABS(NODEK(I)-NODEI(I))
        IF (NDMAX .LT. NDIF1) NDMAX=NDIF1
        IF (NDMAX .LT. NDIF2) NDMAX=NDIF2
        IF (NDMAX .LT. NDIF3) NDMAX=NDIF3
 40   CONTINUE
      NBW=NDMAX+1
      WRITE (IO,41) NBW
 41   FORMAT (//10X,21HCALCULATED BANDWIDTH=,I5)
C++++++++++++++
C       1.5.  INPUT AND OUTPUT OF COEFFICIENT FUNCTIONS
C++++++++++++++
      WRITE (IO,51)
 51   FORMAT (////10X,27H1.4.  COEFFICIENT FUNCTIONS,/
     1     /10X,7HELEMENT,10X,2HKX,10X,2HKY,10X,1HP,11X,1HQ)
      READ (IN,52) ICO
 52   FORMAT(I5)
      IF (ICO .EQ. 1) READ (IN,54) IEL,KXA,KYA,PA,QA
 54   FORMAT (I5,5X,4G10.1)
      DO 50 I=1,NE
        IEL=I
        IF (ICO .EQ. 1) GO TO 55
        READ (IN,54) IEL,KXA,KYA,PA,QA
 55     CONTINUE
        KXE(IEL)=KXA
        KYE(IEL)=KYA
        PE(IEL)=PA
        QE(IEL)=QA
        WRITE (IO,56) IEL,KXE(IEL),KYE(IEL),PE(IEL),QE(IEL)
 56     FORMAT (10X,I5,5X,4G12.5)
 50   CONTINUE
      IF (ICO .EQ. 1) WRITE (IO,57) ICO
 57   FORMAT (//10X,37HKX,KY,P,Q VALUES ARE THE SAME FOR ALL,
     1     9H ELEMENTS,/10X,4HICO=,I5)
C+++++++++++++
C       1.6.  INPUT AND OUTPUT OF PRESCRIBED NODAL VALUES
C+++++++++++++
      WRITE (IO,61)
 61   FORMAT (////10X,29H1.5.  PRESCRIBED NODAL VALUES)
```

```
      READ (IN,62) NBOUND
   62 FORMAT (I5)
      IF (NBOUND .LE. 0) GO TO 66
      WRITE (IO,63) NBOUND
   63 FORMAT (/ 10X,34HNUMBER OF PRESCRIBED NODAL VALUES=,I5,
     1    /10X,4HNODE,10X,11HNODAL VALUE)
      DO 60 I=1,NBOUND
        READ (IN,64) IBOUND(I),TBOUND(I)
   64   FORMAT (I5,5X,G10.1)
        WRITE (IO,65) IBOUND(I),TBOUND(I)
   65   FORMAT (10X,I5,10X,G12.5)
   60 CONTINUE
   66 IF (NBOUND .LE. 0) WRITE (IO,67) NBOUND
   67 FORMAT (42HNO PRESCRIBED NODAL VALUES HAVE BEEN GIVEN,
     1    /10X,7HNBOUND=,I5)
C++++++++++++
C     1.7.  INPUT AND OUTPUT OF ELEMENT DERIVATIVE BOUNDARY
C           VALUES
C++++++++++++
      WRITE (IO,71)
   71 FORMAT (////10X,32H1.6.  DERIVATIVE BOUNDARY VALUES)
      READ (IN,72) NDERIV
   72 FORMAT (I5)
      IF (NDERIV .LE. 0) GO TO 76
      WRITE (IO,73) NDERIV
   73 FORMAT (/ 10X,37HNUMBER OF DERIVATIVE BOUNDARY VALUES=,
     1    I5, /10X,7HELEMENT, 4X,11HSIDE NUMBER, 9X,5HALPHA,
     2    5X,4HBETA)
      DO 70 I=1,NDERIV
        READ (IN,74) IELDIR(I),NSIDE(I),ALPHAE(I),BETAE(I)
   74   FORMAT (2I5,2G10.1)
        WRITE (IO,75) IELDIR(I),NSIDE(I),ALPHAE(I),BETAE(I)
   75   FORMAT (10X,I5,5X,I5,10X,2G12.5)
   70 CONTINUE
   76 IF (NDERIV .LE. 0) WRITE (IO,77) NDERIV
   77 FORMAT (////10X,32HNO DERIVATIVE BOUNDARY CONDITONS,
     1    6H GIVEN,/10X,7HNDERIV=,I5)
      IF (NBOUND .LE. 0 .AND. NDERIV .LE. 0) GO TO 999
C+++++++++++++
C     1.8.  OUTPUT PARAMETERS
C+++++++++++++
      WRITE (IO,91)
   91 FORMAT (////10X,23H1.8.  OUTPUT PARAMETERS)
      READ (IN,92) NOUT
   92 FORMAT (I5)
      WRITE (IO,93) NOUT
   93 FORMAT( /10X,5HNOUT=,I5)
C*************************************************************
C     2.  CALCULATION OF ELEMENT PROPERTIES AND ASSEMBLY
C           OF GLOBAL MATRICES
C*************************************************************
C
C+++++++++++++
C     2.1.  WRITE TITLE AND ZERO GLOBAL MATRICES
C+++++++++++++
      CALL HEAD
      WRITE (IO,101)
  101 FORMAT (////10X,37H2.  CALCULATION OF ELEMENT PROPERTIES)
      DO 120 I=1,NP
        DO 130 J=1,NBW
```

```
             ACOND(I,J)=0.0
 130      CONTINUE
          F(I)=0.0
 120   CONTINUE
C++++++++++++
C     2.2.  FORM LARGE DO LOOP OVER ALL ELEMENTS
C++++++++++++
       DO 1000 I=1,NE
C+++++++++++++
C     2.3.  ASSIGN LOCAL NODAL COORDINATES
C+++++++++++++
          IEL=NEL(I)
          NN(1)=NODEI(IEL)
          NN(2)=NODEJ(IEL)
          NN(3)=NODEK(IEL)
          X(1)=XNODE(NN(1))
          X(2)=XNODE(NN(2))
          X(3)=XNODE(NN(3))
          Y(1)=YNODE(NN(1))
          Y(2)=YNODE(NN(2))
          Y(3)=YNODE(NN(3))
C++++++++++++
C     2.4.  ASSIGN LOCAL ELEMENT COEFFICIENT FUNCTIONS
C++++++++++++
          KX=KXE(IEL)
          KY=KYE(IEL)
          P=PE(IEL)
          Q=QE(IEL)
C++++++++++++
C     2.5.  ASSIGN LOCAL ELEMENT DERIVATIVE BOUNDARY CONDTIONS
C++++++++++++
          DO 140 J=1,3
            ISIDE(J)=0
            ALPHA(J)=0.0
            BETA(J)=0.0
 140      CONTINUE
          DO 150 K=1,NDERIV
            IF (IELDIR(K) .NE. IEL) GO TO 150
            DO 160 J=1,3
              JSIDE=NSIDE(K)
              IF (J .NE. JSIDE) GO TO 160
              ISIDE(JSIDE)=JSIDE
              ALPHA(JSIDE)=ALPHAE(K)
              BETA(JSIDE)=BETAE(K)
 160        CONTINUE
 150      CONTINUE
C+++++++++++++
C     2.6.  CALL SUBROUTINE ELEMNT
C           ALL ELEMENT PROPERTIES ARE OUTPUT FROM ELEMNT
C+++++++++++++
       CALL ELEMNT (IEL,X,Y,NN,KX,KY,
     1       P,Q,ISIDE,ALPHA,BETA,EBM,EC)
C++++++++++++
C     2.7.  ASSEMBLY OF ELEMENT MATRICES INTO GLOBAL MATRICES
C           A     =GLOBAL COEFFICIENT MATRIX (WHICH IS TOO
C                   LARGE TO STORE SO IT IS NOT CALCULATED)
C           ACOND=CONDENSED GLOBAL COEFFICIENT MATRIX
C                  (FOR BANDED SYMMETRIC A MATRIX ONLY THE
C                   UPPER TRIAGONAL NONZERO TERMS ARE STORED)
C           F     =GLOBAL COLUMN
```

```
C               IROW =ROW OF GLOBAL A MATRIX
C               ICOL =COLUMN OF GLOBAL A MATRIX
C               JCOL =COLUMN OF CONDENSED GLOBAL ACOND MATRIX
C++++++++++++
           DO 170 J=1,3
             DO 180 K=1,3
               IROW=NN(J)
               ICOL=NN(K)
               JCOL=ICOL-IROW+1
               IF (JCOL .LT. 1) GO TO 181
               ACOND(IROW,JCOL)=ACOND(IROW,JCOL)+EBM(J,K)
  181          CONTINUE
  180        CONTINUE
             F(IROW)=F(IROW)+EC(J)
  170      CONTINUE
 1000    CONTINUE
C+++++++++++++
C        2.8.  OUTPUT OF CONDENSED GLOBAL MATRIX AND COLUMN
C+++++++++++++
         CALL WRITEM
         WRITE (IO,183)
  183    FORMAT (//10X,37H(GLOBAL MATRICES JUST AFTER ASSEMBLY))
C*************************************************************
C        3.  INCLUSION OF PRESCRIBED NODAL VALUES INTO GLOBAL
C               COEFFICIENT MATRIX AND COLUMN
C*************************************************************
C
C+++++++++++++
C        3.1.  MODIFY ROW AND COLUMN OF ACOND MATRIX AND MODIFY
C               ROW OF F COLUMN
C               IROW  =ROW OF PRESCRIBED NODAL VOLAUE
C               JROW  =ROW OF TERM TO BE MOVED TO RIGHT SIDE
C                        OF EQUAL SIGN
C               JCOL  =COLUMN OF ACOND TERM TO BE SET TO ZERO
C+++++++++++++
         IF (NBOUND .LE. 0) GO TO 299
         DO 200 I=1,NBOUND
           IROW=IBOUND(I)
C
C            TRANSFER TERMS IN UPPER MATRIX TO RIGHT SIDE
           DO 210 J=2,NBW
             JROW=IROW-NBW+J-1
             JCOL=NBW-J+2
             IF (JROW .LT. 1) GO TO 210
             F(JROW)=F(JROW)-ACOND(JROW,JCOL)*TBOUND(I)
             ACOND(JROW,JCOL)=0.0
  210      CONTINUE
C
C            TRANSFER TERMS IN LOWER MATRIX TO RIGHT SIDE
           DO 220 J=2,NBW
             JROW=IROW+J-1
             JCOL=J
             IF (JROW .GT. NP) GO TO 220
             F(JROW)=F(JROW)-ACOND(IROW,JCOL)*TBOUND(I)
  220      CONTINUE
C
C            SET ACOND TERMS TO ZERO IN IROW AND MODIFY RIGHT
C               HAND SIDE
           DO 230 JCOL=2,NBW
             ACOND(IROW,JCOL)=0.0
```

```
 230    CONTINUE
        F(IROW)=ACOND(IROW,1)*TBOUND(I)
 200  CONTINUE
 299  CONTINUE
C+++++++++++++
C     3.2.  OUTPUT OF CONDENSED GLOBAL MATRIX AND COLUMN
C+++++++++++++
      CALL WRITEM
      WRITE (IO,233)
 233  FORMAT (//10X,36H(GLOBAL MATRICES AFTER INCLUSION OF
     1     ,24HPRESCRIBED NODAL VALUES))
C****************************************************************
C     4.  CALCULATION OF NODAL T VALUES, ELEMENT DERIVATIVES
C            AND INTEGRALS
C****************************************************************
C
C+++++++++++++
C     4.1.  CALCULATION OF NODAL VALUES
C+++++++++++++
C
C         SOLUTION OF BANDED LINEAR SIMULTANEOUS EQUATIONS
C            FOR THE NODAL T VALUES
      CALL GAUSS1
      CALL WRITEM
      WRITE (IO,404)
 404  FORMAT (//10X,36H(GLOBAL MATRICES AFTER FORMATION OF
     1     ,22HUPPER TRIANGULAR FORM))
C
C         OUTPUT OF NODAL VALUES
      CALL HEAD
      WRITE (IO,403)
 403  FORMAT (////10X,37H4.  NODAL VALUES, ELEMENT DERIVATIVES
     1     ,14H, AND INTEGRAL)
      WRITE (IO,405)
 405  FORMAT (////10X,18H4.1.  NODAL VALUES)
      WRITE (IO,402)
 402  FORMAT (   /10X,4HNODE, 8X,12HNODAL VALUES)
      DO 400 I=1,NP
        WRITE (IO,401) INODE(I),T(INODE(I))
 401    FORMAT (10X,I5,5X, G15.8)
 400  CONTINUE
C+++++++++++++
C     4.2.  ELEMENT DERIVATIVES AND INTEGRALS
C+++++++++++++
      WRITE (IO,543)
 543  FORMAT (////10X,29H4.2.  ELEMENT DERIVATIVES AND ,
     1     , 8HINTEGRAL )
      WRITE (IO,542)
 542  FORMAT (////10X,7HELEMENT,10X,
     1     ,11HDERIVATIVES,20X,8HINTEGRAL)
      DO 500 I=1,NE
        IEL=NEL(I)
        NN(1)=NODEI(IEL)
        NN(2)=NODEJ(IEL)
        NN(3)=NODEK(IEL)
        X(1)=XNODE(NN(1))
        X(2)=XNODE(NN(2))
        X(3)=XNODE(NN(3))
        Y(1)=YNODE(NN(1))
        Y(2)=YNODE(NN(2))
```

```
            Y(3)=YNODE(NN(3))
            CALL PROP1 (X,Y,B,C,AREA)
            DTDX=0.0
            DTDY=0.0
            DO 520 J=1,3
              DTDX=DTDX+B(J)*T(NN(J))
              DTDY=DTDY+C(J)*T(NN(J))
  520       CONTINUE
            DTDX=DTDX/(2.0*AREA)
            DTDY=DTDY/(2.0*AREA)
            TINT=(T(NN(1))+T(NN(2))+T(NN(3)))*AREA/3.0
            WRITE (IO,541) NEL(I),DTDX,DTDY,TINT
  541       FORMAT (10X,I5,5X,2G15.8,5X,G15.8)
  500     CONTINUE
  999     CONTINUE
          END
C
C
          SUBROUTINE ELEMNT (IEL,X,Y,NN,KX,KY,P,Q,ISIDE,ALPHA,
     1        BETA,EBM,EC)
C*************************************************************
C             SUBROUTINE ELEMNT CALCULATES THE ELEMENT
C               COEFFICIENT MATRIX AND COLUMN FOR A LINEAR
C               INTERPOLATION FUNCTION IN A TRIAGULAR ELEMENT
C               IN TWO DIMENSIONS.
C
C             THE DIFFERENTIAL EQUATION HAS THE FORM
C               D(KX*DT/DX)/DX+D(KY*DT/DY)/DY+P*T+Q=0
C             WITH THE BOUNDARY CONDITIONS
C               T=T0, ON C1
C               KX*DT/DX*NX+KY*DT/DY*NY+ALPHA*T+BETA=0, ON C2
C*************************************************************
          DIMENSION X(3),Y(3),NN(3),
     1        ISIDE(3),ALPHA(3),BETA(3),EBM(3,3),EC(3)
          DIMENSION B(3),C(3)
          REAL KX,KY
          COMMON /INTIO/ NP,NE,NBW,NOUT
          DATA IN,IO/60,61/
C*************************************************************
C         1.  CALCULATION OF ELEMENT COEFFICIENT MATRICES
C             AND COLUMNS
C*************************************************************
C
C++++++++++++
C         1.1.  CALCULATION OF INTERPOLATION FUNCTION PROPERTIES
C               B    =X GRADIENT TERMS
C               C    =Y GRADIENT TERMS
C               AREA=ELEMENT AREA
C++++++++++++
          CALL PROP1 (X,Y,B,C,AREA)
C++++++++++++
C         1.2.  ELEMENT COEFFICIENT MATRIX AND COLUMN
C               KX     =ELEMENT VALUE OF KX COEFFICIENT FUNCTION
C               KY     =ELEMENT VALUE OF KY COEFFICIENT FUNCTION
C               P      =ELEMENT VALUE OF P  COEFFICIENT FUNCTION
C               Q      =ELEMENT VALUE OF Q  COEFFICIENT FUNCTION
C               EBM    =ELEMENT B MATRIX
C               EC     =ELEMENT C COLUMN
C++++++++++++
          COEFKX=KX/(AREA*4.0)
```

```
      COEFKY=KY/(AREA*4.0)
      DO 10 I=1,3
        DO 20 J=1,3
          EBM(I,J)=COEFKX*B(I)*B(J)+COEFKY*C(I)*C(J)
          IF (I .EQ. J) EBM(I,J)=EBM(I,J)-P*AREA/6.0
          IF (I .NE. J) EBM(I,J)=EBM(I,J)-P*AREA/12.0
   20   CONTINUE
        EC(I)=Q*AREA/3.0
   10 CONTINUE
C++++++++++++++
C     1.3.  DERIVATIVE BOUNDARY CONDITIONS
C++++++++++++++
      DO 40 I=1,3
        IF (ISIDE(I) .LE. 0) GO TO 40
        IF (ISIDE(I) .EQ. 1) GO TO 41
        IF (ISIDE(I) .EQ. 2) GO TO 42
        IF (ISIDE(I) .EQ. 3) GO TO 43
        GO TO 40
C         DERIVATIVE BOUNDARY CONDITIONS ON SIDE 1
   41   CONTINUE
        SL=SQRT((X(1)-X(2))**2+(Y(1)-Y(2))**2)
        EBM(1,1)=EBM(1,1)+2.0*ALPHA(1)*SL/6.0
        EBM(1,2)=EBM(1,2)+1.0*ALPHA(I)*SL/6.0
        EBM(2,1)=EBM(1,2)
        EBM(2,2)=EBM(2,2)+2.0*ALPHA(I)*SL/6.0
        EC(1)=EC(1)-BETA(I)*SL/2.0
        EC(2)=EC(2)-BETA(I)*SL/2.0
        GO TO 40
C         DERIVATIVE BOUNDARY CONDITIONS ON SIDE 2
   42   CONTINUE
        SL=SQRT((X(2)-X(3))**2+(Y(2)-Y(3))**2)
        EBM(2,2)=EBM(2,2)+2.0*ALPHA(I)*SL/6.0
        EBM(2,3)=EBM(2,3)+1.0*ALPHA(I)*SL/6.0
        EBM(3,2)=EBM(2,3)
        EBM(3,3)=EBM(3,3)+2.0*ALPHA(I)*SL/6.0
        EC(2)=EC(2)-BETA(I)*SL/2.0
        EC(3)=EC(3)-BETA(I)*SL/2.0
        GO TO 40
C         DERIVATIVE BOUNDARY CONDITIONS ON SIDE 3
   43   CONTINUE
        SL=SQRT((X(3)-X(1))**2+(Y(3)-Y(1))**2)
        EBM(3,3)=EBM(3,3)+2.0*ALPHA(I)*SL/6.0
        EBM(3,1)=EBM(3,1)+1.0*ALPHA(I)*SL/6.0
        EBM(1,3)=EBM(3,1)
        EBM(1,1)=EBM(1,1)+2.0*ALPHA(I)*SL/6.0
        EC(3)=EC(3)-BETA(I)*SL/2.0
        EC(1)=EC(1)-BETA(I)*SL/2.0
   40 CONTINUE
C**************************************************************
C     2.  OUTPUT OF ELEMENT MATRICES AND COLUMNS
C**************************************************************
      IF (NOUT .GT. 0) GO TO 999
C+++++++++++++
C     2.1.  WRITE ELEMENT CONNECTIVITY DATA AND NODAL COORD
C+++++++++++++
      WRITE (IO,111) IEL
  111 FORMAT (////10X,30HELEMENT PROPERTIES FOR ELEMENT,I5)
      WRITE (IO,114)
  114 FORMAT (/10X,37H2.1.  ELEMENT CONNECTIVITY AND NODAL
     1    ,11HCOORDINATES)
```

```
      WRITE (IO,112)
112   FORMAT (  10X,7HELEMENT, 5X,4HNODE, 6X,1HX, 9X,1HY)
      DO 110 I=1,3
        WRITE (IO,113) IEL,NN(I),X(I),Y(I)
113     FORMAT (10X,I5, 5X,I5,2F10.4)
110   CONTINUE
C+++++++++++++
C     2.2.  WRITE ELEMENT COEFFICIENT FUNCTIONS
C+++++++++++++
      WRITE (IO,123)
123   FORMAT (/10X,37H2.2.  ELEMENT COEFFICIENT FUNCTIONS  )
      WRITE (IO,121)
121   FORMAT( 10X,4HNODE,12X,2HKX,10X,2HKY,10X,1HP,10X,1HQ)
      DO 120 I=1,3
        WRITE (IO,122) NN(I),KX,KY,P,Q
122     FORMAT (10X,I5,5X,4G12.5)
120   CONTINUE
C+++++++++++++
C     2.3.  WRITE ELEMENT INTERPOLATION FUNCTIONS
C+++++++++++++
      WRITE (IO,134)
134   FORMAT (/10X,37H2.3.  ELEMENT INTERPOLATION FUNCTION
     1     ,10HPROPERTIES)
      WRITE (IO,132)
132   FORMAT (10X,4HNODE,10X,1HB,11X,1HC)
      DO 130 I=1,3
        WRITE (IO,133) NN(I),B(I),C(I)
133     FORMAT (10X,I5,5X,2G12.5)
130   CONTINUE
      WRITE (IO,131) AREA
131   FORMAT (  10X,5HAREA=,G12.5)
      IF (AREA .LT. 0.0) WRITE (IO,136)
136   FORMAT (/10X,42HCALCULATED AREA LESS THAN ZERO SO ELEMENT/,
     1     53HNODE CONNECTIVITY DATA IS PROBABLY NUMBERED CLOCKWISE)
C+++++++++++++
C     2.4.  WRITE ELEMENT DERIVATIVE BOUNDARY CONDITIONS
C+++++++++++++
      ISUM=ISIDE(1)+ISIDE(2)+ISIDE(3)
      IF (ISUM .LE. 0) WRITE (IO,141)
141   FORMAT ( /10X,41H2.4.  NO DERIVATIVE BOUNDARY CONDITIONS
     1     ,10HIN ELEMENT)
      IF (ISUM .GE. 1) WRITE (IO,145)
145   FORMAT (/10X,34H2.4.  ELEMENT DERIVATIVE BOUNDARY
     1     ,10HCONDITIONS)
      IF (ISUM .GE. 1) WRITE (IO,142)
142   FORMAT ( 10X,4HSIDE,10X,5HALPHA, 7X,4HBETA)
      DO 140 I=1,3
        IF (ISIDE(I) .LE. 0) GO TO 143
        WRITE (IO,144) ISIDE(I),ALPHA(I),BETA(I)
144     FORMAT (10X,I5,5X,2G12.5)
143     CONTINUE
140   CONTINUE
C+++++++++++++
C     2.5.  WRITE ELEMENT COEFFICIENT MATRIX AND COLUMN
C+++++++++++++
      WRITE (IO,153)
153   FORMAT (/10X,37H2.5.  ELEMENT COEFFICIENT MATRIX AND
     1     , 6HCOLUMN)
      WRITE (IO,151)
151     FORMAT ( 10X,4HNODE,16X,18HCOEFFICIENT MATRIX,18X,
```

```
     1    6HCOLUMN)
      DO 150 I=1,3
        WRITE (IO,152) NN(I),(EBM(I,J),J=1,3),EC(I)
152   FORMAT (10X,I5,5X,3G12.5,5X,G12.5)
150   CONTINUE
999   CONTINUE
      RETURN
      END
C
C
      SUBROUTINE HEAD
C*****************************************************************
C          WRITE TITLES ON TOP OF PAGE
C*****************************************************************
      COMMON /ALP10/ TITLE1(20),TITLE2(20)
      DATA IN,IO/60,61/
      WRITE (IO,13) TITLE1
 13   FORMAT (1H1//10X,20A4)
      WRITE (IO,14) TITLE2
 14   FORMAT (10X,20A4)
      RETURN
      END
C
C
      SUBROUTINE PROP1 (X,Y,B,C,AREA)
C*****************************************************************
C          SUBROUTINE PROP1 CALCULATES THE X AND Y GRADIENT
C            TERMS AND AREA FOR A LINEAR INTERPOLATION
C            FUNCTION IN A TRIAGULAR ELEMENT.
C*****************************************************************
      DIMENSION X(3),Y(3),B(3),C(3)
      B(1)=Y(2)-Y(3)
      B(2)=Y(3)-Y(1)
      B(3)=Y(1)-Y(2)
      C(1)=X(3)-X(2)
      C(2)=X(1)-X(3)
      C(3)=X(2)-X(1)
      AREA=(X(2)*Y(3)+X(3)*Y(1)+X(1)*Y(2)-X(2)*Y(1)
     1    -X(3)*Y(2)-X(1)*Y(3))/2.0
      RETURN
      END
C
C
      SUBROUTINE GAUSS1
C*****************************************************************
C          SUBROUTINE GAUSS1 SOLVES A BANDED SYMMETRIC SET OF
C            LINEAR ALGEBRAIC EQUATIONS USING A BANDED GAUSS
C            ELIMINATION TECHNIQUE.
C
C          THE MATRIX IS STORED IN CONDENSED FORM USING ONLY
C            THE UPPER TRIANGULAR BANDWITH.  THUS ALL ROWS
C            OF THE SQUARE MATRIX ARE SHIFTED TO THE LEFT SO
C            THAT THE MAIN DIAGONAL TERMS ARE NOW IN COLUMN
C            ONE OF THE RECTANGULAR MATRIX ACOND.  ACOND HAS
C            DIMENSIONS NP BY NBW.
C*****************************************************************
      COMMON /INT10/ NP,NE,NBW,NOUT
      COMMON /REAL10/ ACOND(150,20),F(150),T(150)
      DATA IN,IO/60,61/
C+++++++++++++
```

```
C      1.  CONSTRUCTION OF UPPER BANDED TRIANGULAR FORM
C+++++++++++++
       NPM=NP-1
       DO 100 I=1,NPM
         IP=I+1
         DO 110 J=2,NBW
           IF (ACOND(I,J) .EQ. 0.0) GO TO 110
           C=ACOND(I,J)/ACOND(I,1)
           L=I+J-1
           M=0
           DO 120 K=J,NBW
             M=M+1
             ACOND(L,M)=ACOND(L,M)-C*ACOND(I,K)
  120      CONTINUE
           F(L)=F(L)-C*F(I)
  110    CONTINUE
  100  CONTINUE
C+++++++++++++
C      2.  BACKWARD SUBSTITUTION TO OBTAIN UNKNOWNS
C+++++++++++++
       DO 200 I=1,NP
         L=NP-I+1
         T(L)=F(L)
         K=L
         DO 210 J=2,NBW
           K=K+1
           IF (K .GT. L+NBW-1) GO TO 210
           IF (K .GT. NP) GO TO 210
           T(L)=T(L)-ACOND(L,J)*T(K)
  210    CONTINUE
         T(L)=T(L)/ACOND(L,1)
  200  CONTINUE
       RETURN
       END
C
C
       SUBROUTINE WRITEM
C*************************************************************
C            SUBROUTINE WRITEM WRITES OUT THE ACOND MATRIX AND
C              F COLUMN.
C*************************************************************
       COMMON /ALP10/ TITLE1(20),TITLE2(20)
       COMMON /INT10/ NP,NE,NBW,NOUT
       COMMON /REAL10/ ACOND(150,20),F(150),T(150)
       DATA IN,IO/60,61/
       IF (NOUT .GT. 1) GO TO 999
       CALL HEAD
       WRITE (IO,501)
  501  FORMAT (////10X,28H3.  GLOBAL MATRIX AND COLUMN)
       WRITE (IO,533)
  533  FORMAT (////10X,35H3.1.  CONDENSED GLOBAL COEFFICIENT
      1     ,14H(ACOND) MATRIX)
       WRITE (IO,535)
  535  FORMAT ( /10X,4HNODE,20X, 5HACOND)
       DO 510 I=1,NP
       WRITE (IO,511) I,(ACOND(I,J),J=1,NBW)
  511  FORMAT (10X,I5,10G12.5)
  510  CONTINUE
       WRITE (IO,534)
  534  FORMAT (////10X,21H3.2.  GLOBAL F COLUMN)
```

```
      WRITE (IO,537)
537   FORMAT ( /10X,4HNODE,12X,1HF)
      DO 520 I=1,NP
        WRITE (IO,521) I,F(I)
521     FORMAT ( 10X,I5,5X,G12.5)
520   CONTINUE
999   CONTINUE
      RETURN
      END
```

SAMPLE INPUT DATA FOR PROGRAM FINTWO
DATA TAKEN FROM EXAMPLE 10.2

1. INPUT AND OUTPUT OF PROBLEM DATA

1.1. GLOBAL PARAMETERS

NUMBER OF NODES 9
NUMBER OF ELEMENTS 8

1.2. NODAL COORDINATES

INODE	X	Y
1	0.	0.
2	0.	3.0000
3	0.	5.0000
4	2.0000	0.
5	2.0000	3.0000
6	2.2667	5.0000
7	4.0000	2.0000
8	4.0000	3.0000
9	4.0000	4.0000

1.3. ELEMENT/NODE CONNECTIVITY DATA

ELEMENT		NODES	
1	1	4	2
2	5	2	4
3	2	5	3
4	6	3	5
5	4	7	5
6	8	5	7
7	5	8	6
8	9	6	8

CALCULATED BANDWIDTH= 4

1.4. COEFFICIENT FUNCTIONS

ELEMENT	KX	KY	P	Q
1	.25000	.25000	0.	0.
2	.25000	.25000	0.	0.
3	.25000	.25000	0.	0.
4	.25000	.25000	0.	0.
5	.25000	.25000	0.	0.
6	.25000	.25000	0.	0.
7	.25000	.25000	0.	0.
8	.25000	.25000	0.	0.

KX,KY,P,Q VALUES ARE THE SAME FOR ALL ELEMENTS
ICO= 1

1.5. PRESCRIBED NODAL VALUES

NUMBER OF PRESCRIBED NODAL VALUES= 6

NODE	NODAL VALUE
1	400.00
2	400.00
3	400.00
4	200.00
6	400.00
7	100.00

1.6. DERIVATIVE BOUNDARY VALUES

NUMBER OF DERIVATIVE BOUNDARY VALUES= 4

ELEMENT	SIDE NUMBER	ALPHA	BETA
1	1	0.	0.
6	3	.10000E-02	-.50000E-01
8	1	.10000E-02	-.50000E-01
8	3	.10000E-02	-.50000E-01

1.8. OUTPUT PARAMETERS

NOUT= 0

SAMPLE INPUT DATA FOR PROGRAM FINTWO
DATA TAKEN FROM EXAMPLE 10.2

2. CALCULATION OF ELEMENT PROPERTIES

ELEMENT PROPERTIES FOR ELEMENT 1

2.1. ELEMENT CONNECTIVITY AND NODAL COORDINATES

ELEMENT	NODE	X	Y
1	1	0.0000	0.0000
1	4	2.0000	0.0000
1	2	0.0000	3.0000

2.2. ELEMENT COEFFICIENT FUNCTIONS

NODE	KX	KY	P	Q
1	.25000	.25000	0.	0.
4	.25000	.25000	0.	0.
2	.25000	.25000	0.	0.

2.3. ELEMENT INTERPOLATION FUNCTION PROPERTIES

NODE	B	C
1	-3.0000	-2.0000
4	3.0000	0.
2	0.	2.0000

AREA= 3.0000

2.4. ELEMENT DERIVATIVE BOUNDARY CONDITIONS

SIDE	ALPHA	BETA
1	0.	0.

```
2.5.  ELEMENT COEFFICIENT MATRIX AND COLUMN
NODE                COEFFICIENT MATRIX                      COLUMN
   1       .27083      -.18750     -.83333E-01      0.
   4      -.18750       .18750     0.               0.
   2      -.83333E-01 0.            .83333E-01      0.


ELEMENT PROPERTIES FOR ELEMENT    2

2.1.  ELEMENT CONNECTIVITY AND NODAL COORDINATES
ELEMENT      NODE      X         Y
   2           5    2.0000    3.0000
   2           2    0.0000    3.0000
   2           4    2.0000    0.0000

2.2.  ELEMENT COEFFICIENT FUNCTIONS
NODE            KX          KY            P           Q
   5       .25000      .25000      0.          0.
   2       .25000      .25000      0.          0.
   4       .25000      .25000      0.          0.

2.3.  ELEMENT INTERPOLATION FUNCTION PROPERTIES
NODE            B           C
   5       3.0000      2.0000
   2      -3.0000     0.
   4      0.          -2.0000
AREA=  3.0000

2.4.  NO DERIVATIVE BOUNDARY CONDITIONS  IN ELEMENT

2.5.  ELEMENT COEFFICIENT MATRIX AND COLUMN
NODE                COEFFICIENT MATRIX                      COLUMN
   5       .27083      -.18750     -.83333E-01      0.
   2      -.18750       .18750     0.               0.
   4      -.83333E-01 0.            .83333E-01      0.


ELEMENT PROPERTIES FOR ELEMENT    3
2.1.  ELEMENT CONNECTIVITY AND NODAL COORDINATES
ELEMENT      NODE      X         Y
   3           2    0.0000    3.0000
   3           5    2.0000    3.0000
   3           3    0.0000    5.0000

2.2.  ELEMENT COEFFICIENT FUNCTIONS
NODE            KX          KY            P           Q
   2       .25000      .25000      0.          0.
   5       .25000      .25000      0.          0.
   3       .25000      .25000      0.          0.

2.3.  ELEMENT INTERPOLATION FUNCTION PROPERTIES
NODE            B           C
   2      -2.0000     -2.0000
   5       2.0000     0.
   3      0.           2.0000
AREA=  2.0000
```

```
2.4.  NO DERIVATIVE BOUNDARY CONDITIONS  IN ELEMENT

2.5.  ELEMENT COEFFICIENT MATRIX AND COLUMN
NODE                COEFFICIENT MATRIX                     COLUMN
    2        .25000      -.12500     -.12500          0.
    5      -.12500        .12500     0.               0.
    3      -.12500      0.             .12500         0.


ELEMENT PROPERTIES FOR ELEMENT    4

2.1.  ELEMENT CONNECTIVITY AND NODAL COORDINATES
ELEMENT     NODE       X          Y
    4          6    2.2667     5.0000
    4          3    0.0000     5.0000
    4          5    2.0000     3.0000

2.2.  ELEMENT COEFFICIENT FUNCTIONS
NODE            KX            KY            P            Q
    6        .25000        .25000      0.           0.
    3        .25000        .25000      0.           0.
    5        .25000        .25000      0.           0.

2.3.  ELEMENT INTERPOLATION FUNCTION PROPERTIES
NODE             B             C
    6        2.0000        2.0000
    3       -2.0000        .26670
    5       0.            -2.2667
AREA=  2.2667

2.4.  NO DERIVATIVE BOUNDARY CONDITIONS  IN ELEMENT

2.5.  ELEMENT COEFFICIENT MATRIX AND COLUMN
NODE                COEFFICIENT MATRIX                     COLUMN
    6      .22058       -.95585E-01 -.12500          0.
    3    -.95585E-01  .11225      -.16669E-01        0.
    5     -.12500      -.16669E-01  .14167 0.
ELEMENT PROPERTIES FOR ELEMENT    5

2.1.  ELEMENT CONNECTIVITY AND NODAL COORDINATES
ELEMENT     NODE       X          Y
    5          4    2.0000     0.0000
    5          7    4.0000     2.0000
    5          5    2.0000     3.0000

2.2.  ELEMENT COEFFICIENT FUNCTIONS
NODE            KX            KY            P            Q
    4        .25000        .25000      0.           0.
    7        .25000        .25000      0.           0.
    5        .25000        .25000      0.           0.

2.3.  ELEMENT INTERPOLATION FUNCTION PROPERTIES
NODE             B             C
    4      -1.0000       -2.0000
    7       3.0000       0.
    5      -2.0000        2.0000
AREA=  3.0000
```

2.4. NO DERIVATIVE BOUNDARY CONDITIONS IN ELEMENT

2.5. ELEMENT COEFFICIENT MATRIX AND COLUMN

NODE	COEFFICIENT MATRIX			COLUMN
4	.10417	-.62500E-01	-.41667E-01	0.
7	-.62500E-01	.18750	-.12500	0.
5	-.41667E-01	-.12500	.16667	0.

ELEMENT PROPERTIES FOR ELEMENT 6

2.1. ELEMENT CONNECTIVITY AND NODAL COORDINATES

ELEMENT	NODE	X	Y
6	8	4.0000	3.0000
6	5	2.0000	3.0000
6	7	4.0000	2.0000

2.2. ELEMENT COEFFICIENT FUNCTIONS

NODE	KX	KY	P	Q
8	.25000	.25000	0.	0.
5	.25000	.25000	0.	0.
7	.25000	.25000	0.	0.

2.3. ELEMENT INTERPOLATION FUNCTION PROPERTIES

NODE	B	C
8	1.0000	2.0000
5	-1.0000	0.
7	0.	-2.0000

AREA= 1.0000

2.4. ELEMENT DERIVATIVE BOUNDARY CONDITIONS

SIDE	ALPHA	BETA
3	.10000E-02	-.50000E-01

2.5. ELEMENT COEFFICIENT MATRIX AND COLUMN

NODE	COEFFICIENT MATRIX			COLUMN
8	.31283	-.62500E-01	-.24983	.25000E-01
5	-.62500E-01	.62500E-01	0.	0.
7	-.24983	0.	.25033	.25000E-01

ELEMENT PROPERTIES FOR ELEMENT 7

2.1. ELEMENT CONNECTIVITY AND NODAL COORDINATES

ELEMENT	NODE	X	Y
7	5	2.0000	3.0000
7	8	4.0000	3.0000
7	6	2.2667	5.0000

2.2. ELEMENT COEFFICIENT FUNCTIONS

NODE	KX	KY	P	Q
5	.25000	.25000	0.	0.
8	.25000	.25000	0.	0.
6	.25000	.25000	0.	0.

2.3. ELEMENT INTERPOLATION FUNCTION PROPERTIES

NODE	B	C
5	-2.0000	-1.7333
8	2.0000	-.26670
6	0.	2.0000

AREA= 2.0000

2.4. NO DERIVATIVE BOUNDARY CONDITIONS IN ELEMENT

2.5. ELEMENT COEFFICIENT MATRIX AND COLUMN

NODE	COEFFICIENT MATRIX			COLUMN
5	.21889	-.11055	-.10833	0.
8	-.11055	.12722	-.16669E-01	0.
6	-.10833	-.16669E-01	.12500	0.

ELEMENT PROPERTIES FOR ELEMENT 8

2.1. ELEMENT CONNECTIVITY AND NODAL COORDINATES

ELEMENT	NODE	X	Y
8	9	4.0000	4.0000
8	6	2.2667	5.0000
8	8	4.0000	3.0000

2.2. ELEMENT COEFFICIENT FUNCTIONS

NODE	KX	KY	P	Q
9	.25000	.25000	0.	0.
6	.25000	.25000	0.	0.
8	.25000	.25000	0.	0.

2.3. ELEMENT INTERPOLATION FUNCTION PROPERTIES

NODE	B	C
9	2.0000	1.7333
6	-1.0000	0.
8	-1.0000	-1.7333

AREA= .86665

2.4. ELEMENT DERIVATIVE BOUNDARY CONDITIONS SIDE ALPHA BETA 1

.10000E-02 -.50000E-01 3 .10000E-02 -.50000E-01

2.5. ELEMENT COEFFICIENT MATRIX AND COLUMN NODE COEFFICIENT MATRIX COLUMN 9

.50613 -.14390 -.36073 .75027E-01 6 -.14390 .72784E-01 .721
.50027E-01 8 -.36073 .72117E-01 .28911 .25000E-01

```
SAMPLE INPUT DATA FOR PROGRAM FINTWO
DATA TAKEN FROM EXAMPLE 10.2

3.  GLOBAL MATRIX AND COLUMN

3.1.  CONDENSED GLOBAL COEFFICIENT (ACOND) MATRIX

NODE                   ACOND
   1  .27083     -.83333E-01  0.           -.18750
   2  .52083     -.12500      0.           -.31250
   3  .23725     0.           -.16669E-01  -.95585E-01
   4  .37500     -.12500      0.           -.62500E-01
   5  .98555     -.23333      -.12500      -.17305
   6  .41837     0.            .55448E-01  -.14390
   7  .43783     -.24983      0.           0.
   8  .72917     -.36073      0.           0.
   9  .50613     0.           0.           0.

3.2.  GLOBAL F COLUMN

NODE         F
   1      0.
   2      0.
   3      0.
   4      0.
   5      0.
   6        .50027E-01
   7        .25000E-01
   8        .50000E-01
   9        .75027E-01

(GLOBAL MATRICES JUST AFTER ASSEMBLY)

SAMPLE INPUT DATA FOR PROGRAM FINTWO
DATA TAKEN FROM EXAMPLE 10.2

3.  GLOBAL MATRIX AND COLUMN

3.1.  CONDENSED GLOBAL COEFFICIENT (ACOND) MATRIX

NODE                   ACOND
   1  .27083     0.           0.           0.
   2  .52083     0.           0.           0.
   3  .23725     0.           0.           0.
   4  .37500     0.           0.           0.
   5  .98555     0.           0.           -.17305
   6  .41837     0.           0.           0.
   7  .43783     0.           0.           0.
   8  .72917     -.36073      0.           0.
   9  .50613     0.           0.           0.
```

3.2. GLOBAL F COLUMN

NODE	F
1	108.33
2	208.33
3	94.901
4	75.000
5	262.50
6	167.35
7	43.783
8	2.8541
9	57.635

(GLOBAL MATRICES AFTER INCLUSION OF PRESCRIBED NODAL VALUES)

SAMPLE INPUT DATA FOR PROGRAM FINTWO
DATA TAKEN FROM EXAMPLE 10.2

3. GLOBAL MATRIX AND COLUMN

3.1. CONDENSED GLOBAL COEFFICIENT (ACOND) MATRIX

NODE	ACOND			
1	.27083	0.	0.	0.
2	.52083	0.	0.	0.
3	.23725	0.	0.	0.
4	.37500	0.	0.	0.
5	.98555	0.	0.	-.17305
6	.41837	0.	0.	0.
7	.43783	0.	0.	0.
8	.69878	-.36073	0.	0.
9	.31991	0.	0.	0.

3.2. GLOBAL F COLUMN

NODE	F
1	108.33
2	208.33
3	94.901
4	75.000
5	262.50
6	167.35
7	43.783
8	48.947
9	82.903

(GLOBAL MATRICES AFTER FORMATION OF UPPER TRIANGULAR FORM)

SAMPLE INPUT DATA FOR PROGRAM FINTWO
DATA TAKEN FROM EXAMPLE 10.2

4. NODAL VALUES, ELEMENT DERIVATIVES, AND INTEGRAL

4.1. NODAL VALUES

NODE	NODAL VALUES
1	400.00000
2	400.00000
3	400.00000
4	200.00000
5	302.13676
6	400.00000
7	100.00000
8	203.82133
9	259.14171

4.2. ELEMENT DERIVATIVES ANDINTEGRAL

ELEMENT	DERIVATIVES		INTEGRAL
1	-100.00000	0.	1000.0000
2	-48.931620	34.045587	902.13676
3	-48.931620	0.	734.75784
4	0.	48.931620	832.73780
5	-84.045587	34.045587	602.13676
6	-49.157717	103.82133	201.98603
7	-49.157717	55.486801	603.97206
8	-49.349744	55.320381	249.29564

Appendix C

Computer Program STRESS for Plane Stress, Plane Strain, and Trusses

C.1 Program Description

Program STRESS is a general plane stress/plane strain/truss program using simplex bar, triangular, and rectangular elements. The basic program structure is similar to program FINTWO (described in Appendix B). The two programs utilize similar input/output formats and the same global matrix assembly and Gauss elimination solution routines. The theory behind this program is developed in Chapters 4 and 14, where examples also appear. The program may be used to solve any problem for which the region in question may be approximated as a combination of straight bars or flat triangles and rectangles. Examples in the text include the stresses, strains, and displacements in trusses, thin plates, and dams.

Symmetry of the element matrices is taken into account in the storage of the global system matrix. The banded nature of the global system of equations is utilized in an efficient banded Gauss elimination procedure for problem solution. The user is responsible for choosing a nodal numbering scheme to minimize bandwidth.

STRESS is written as a main program with a number of subroutines. Table C.1 presents an outline of the program and major subroutines. To improve program readability, program information is listed at the beginning in comment form. In addition, the program is liberally documented with comment lines. Major sections in the program are labeled 1, 2, 3, etc., while subsections are labeled 1.1, 1.2, 1.3, etc.

Table C.1
Outline of Computer Program STRESS

Main Program—STRESS

1. Input and output of problem data.
2. Calculation of element properties.
3. Inclusion of prescribed nodal values into global coefficient matrix and column.
4. Output of nodal displacements and element stresses and strains.

Subroutine ELEM21

1. Calculation of element coefficient matrix for a two-node simplex bar element
2. Output of element coefficient matrix.

Subroutine RESS20

1. Calculation of bar element stresses and strains.

Subroutine ELEM31

1. Calculation of element coefficient matrix for a triangular simplex element.
2. Output of element coefficient matrix.

Subroutine RESS30

1. Calculation of triangular element stresses and strains.

Subroutine ELEM41

1. Calculation of element coefficient matrix for a quadrilateral simplex element.
2. Output of element coefficient matrix.

Subroutine RESS40

1. Calculation of quadrilateral element stresses and strains.

Subroutine HEAD

1. Writing of titles at the top of each page.

Subroutine PROP31

1. Calculation of x and y gradient terms and the area for a linear interpolation function in a simplex triangular element.

Subroutine PROP41

1. Calculation of element length, width, area, and coordinate transformation angle for a linear rectangular element.

Subroutine GAUSS1

1. Construction of upper triangular form of global stiffness matrix.
2. Backward substitution to obtain unknowns.

Subroutine WRITEM

1. Writing of global matrices and columns.

```
      PROGRAM STRESS (INPUT,OUTPUT,TAPE60=INPUT,TAPE61=OUTPUT,TAPE99)
C**************************************************************
C
C                    PROGRAM STRESS
C        VERSION 1.2                    SEPTEMBER, 1982
C
C                    DR PAUL ALLAIRE
C                    PROFESSOR
C                    CHIP QUEITZSCH
C                    GRADUATE RESEARCH ASSISTANT
C                    MECHANICAL AND AEROSPACE
C                      ENGINEERING DEPARTMENT
C                    UNIVERSITY OF VIRGINIA
C                    CHARLOTTESVILLE, VIRGINIA 22903
C
C**************************************************************
C        PROGRAM DESCRIPTION
C**************************************************************
C        THIS COMPUTER PROGRAM SOLVES TWO DIMENSIONAL
C          PROBLEMS FOR
C            1.  PLANE STRESS
C            2.  PLANE STRAIN
C            3.  PLANE TRUSSES
C          USING THE FINITE ELEMENT METHOD.  THREE DIFFERENT
C          ELEMENTS ARE INCLUDED IN THE ELEMENT LIBRARY
C            1.  TWO NODE SIMPLEX BAR ELEMENTS
C            2.  THREE NODE SIMPLEX TRIANGULAR ELEMENT
C            3.  FOUR NODE BILINEAR RECTANGULAR ELEMENTS
C
C**************************************************************
C        UNITS
C**************************************************************
C        ANY CONSISTENT SET OF UNITS MAY BE USED WITH THIS
C          PROGRAM.  ONLY TWO UNITS ARE EMPLOYED - LENGTH
C          AND FORCE.  SOME EXAMPLES ARE GIVEN BELOW IN UNITS
C          OF MILLIMETERS (MM) AND KILONEWTONS (KN).
C            LENGTH=MM
C            DISPLACEMENT=MM
C            THICKNESS=MM
C            AREA=MM**2
C            STRAIN=MM/MM
C            FORCE=KN
C            YOUNG S MODULUS=KN/MM**2
C            STIFFNESS=KN/MM
C            STRESS=KN/MM**2
C          BRITISH UNITS OF INCHES (IN) AND POUND FORCE (LBF)
C          COULD ALSO BE USED.
C
C**************************************************************
C    1.  REQUIRED INPUT DATA AND FORMAT
C**************************************************************
C        INPUT DATA IS ORGANIZED INTO SECTIONS LABELED
C          1,2,3,ETC.  THE NUMBER OF DATA CARDS WITHIN
C          EACH SECTION WILL VARY ACCORDING TO THE NUMBER
C          NODE POINTS, ELEMENTS, AND PRESCRIBED NODAL VALUES.
C
C++++++++++++++
C    1.1.  TITLE AND GLOBAL PARAMETERS
C++++++++++++++
```

```
C          CARD  1.  TITLE1    FORMAT (20A4)
C            TITLE1=ANY ALPHA NUMERIC STRING FOR TITLE
C
C          CARD  2.  TITLE2    FORMAT (20A4)
C            TITLE2=ANY ALPHA NUMERIC STRING FOR TITLE
C
C          CARD  3.  NP,NE,NAN    FORMAT (3I5)
C            NP=NUMBER OF NODES
C            NE=NUMBER OF ELEMENTS
C            NAN=TYPE OF ANALYSIS - PLANE STRESS, PLANE STRAIN,
C                        OR PLANE TRUSS
C              IF NAN=1, PLANE STRESS IS ANALYSED.
C              IF NAN=2, PLANE STRESS IS ANALYSED.
C              IF NAN=3, A PLANE TRUSS IS ANALYSED.
C
C            THIS PROGRAM IS CURRENTLY LIMITED TO NP=100
C              AND NE=200 DUE TO ARRAY DIMENSIONING.  THE
C              ARRAY SIZES MAY BE INCREASED BY CHANGING THE
C              APPROPRIATE DIMENSION AND COMMON STATEMENTS.
C
C+++++++++++++
C      1.2.  NODAL COORDINATES
C+++++++++++++
C          CARDS  1 THRU NP.  INODE,XNODE,YNODE
C                                  FORMAT (I5,5X,2G10.1)
C            INODE=GLOBAL NODE NUMBER
C            XNODE=X COORDINATE OF NODE
C            YNODE=Y COORDINATE OF NODE
C
C+++++++++++++
C      1.3.  ELEMENT/NODE CONNECTIVITY DATA
C+++++++++++++
C          CARDS  1 THRU NE.  NEL,NETYP,NODEI,NODEJ,NODEK,NODEL
C                                  FORMAT (6I5)
C            NEL  =ELEMENT NUMBER
C            NETYP=ELEMENT TYPE
C              IF NETYP=21, IT IS A TWO NODE BAR ELEMENT.
C              IF NETYP=31, IT IS A THREE NODE TRIANGULAR ELEMENT.
C              IF NETYP=41, IT IS A FOUR NODE RECTANGULAR ELEMENT.
C            NODEI=GLOBAL NUMBER OF FIRST NODE IN ELEMENT
C            NODEJ=GLOBAL NUMBER OF SECOND NODE IN ELEMENT
C            NODEK=GLOBAL NUMBER OF THIRD NODE IN ELEMENT
C              USE NODEK ONLY FOR TRIANGULAR AND RECTANGULAR
C              ELEMENTS.
C            NODEL=GLOBAL NUMBER OF FOURTH NODE IN ELEMENT
C              USE NODEL ONLY FOR RECTANGULAR ELEMENTS.
C
C            PLEASE REMEMBER TO NUMBER NODES IN THE
C              COUNTER CLOCKWISE DIRECTION FOR TRIANGULAR AND
C              RECTANGULAR ELEMENTS.
C
C+++++++++++++
C      1.4.  CALCULATE BANDWIDTH
C+++++++++++++
C            NO INPUT REQUIRED
C
C            BANDWIDTH IS CURRENTLY LIMITED TO NBW=40 IN THE
C              ACOND MATRIX.  COMMON REAL10 MUST BE MODIFIED
C              TO INCREASE IT IF THE USER DESIRES A LARGER
C              BANWIDTH CAPABILITY.
```

```
C
C+++++++++++++
C     1.5.  ELEMENT PROPERTIES
C+++++++++++++
C           CARD  1.  ICO    FORMAT (I5)
C             ICO=COEFFICIENT PARAMETER
C               IF ICO=1, YOUNGS MODULUS,POISSONS RATIO,THICKNESS HAVE
C                         THE SAME VALUE FOR
C                         ALL ELEMENTS SO ONLY ONE DATA CARD
C                         IS NEEDED.
C               IF ICO=2, VALUES FOR YOUNGS MODULUS,POISSONS RATIO,
C                         THICKNESS MUST BE INPUT
C                         FOR EACH ELEMENT ON A DATA CARD.
C
C           CARD  2 (IF ICO=1).  IEL,YME,MUE,TE
C                                   FORMAT (I5,5X,3G10.1)
C             IEL=ELEMENT NUMBER
C             YME=VALUE OF YOUNGS MODULUS IN ELEMENT
C             MUE=VALUE OF POISSONS RATIO IN ELEMENT
C             TE =VALUE OF THICKNESS IN RECTANGULAR AND TRIANGULAR
C                 ELEMENTS AND AREA IN BAR ELEMENTS
C
C           NEXT NE CARDS (IF ICO=2).  IEL,YME,MUE,TE
C                                   FORMAT (I5,5X,3G10.1)
C
C+++++++++++++
C     1.6.  PRESCRIBED NODAL FORCES
C+++++++++++++
C           CARD  1.  NPF    FORMAT (I5)
C             NPF=NUMBER OF PRESCRIBED NODAL FORCES
C
C           NEXT NPF CARDS.  IFORCE,NFORCE,TFORCE
C                                   FORMAT (I5,I5,G10.1)
C             IFORCE=GLOBAL NODE NUMBER
C             NFORCE=GLOBAL DIRECTION OF FORCE (1 OR 2)
C               IF NFORCE=1, THE X COMPONENT IS CHOSEN.
C               IF NFORCE=2, THE Y COMPONENT IS CHOSEN.
C             TFORCE=COMPONENT OF PRESCRIBED NODAL FORCE IN GLOBAL
C                    DIRECTION
C
C+++++++++++++
C     1.7.  PRESCRIBED NODAL DISPLACEMENTS
C+++++++++++++
C           CARD 1. NPD     FORMAT (I5)
C             NPD=NUMBER OF PRESCRIBED NODAL DISPLACEMENTS
C
C           NEXT NPD CARDS.  IDISP,NDISP,TDISP
C                                   FORMAT (I5,I5,G10.1)
C             IDISP=GLOBAL NODE NUMBER
C             NDISP=GLOBAL DIRECTION OF DISPLACEMENT (1 OR 2)
C               IF NDISP=1, THE X COMPONENT IS CHOSEN.
C               IF NDISP=2, THE Y COMPONENT IS CHOSEN.
C             TDISP=COMPONENT OF PRESCRIBED NODAL DISPLACEMENT IN
C                   GLOBAL DIRECTION
C
C+++++++++++++
C     1.8.  OUTPUT PARAMETERS
C+++++++++++++
C           CARD  1.  NOUT    FORMAT (I5)
C             NOUT=OUTPUT PARAMETER
```

```
C              IF NOUT=0, ALL PROGRAM OUTPUT IS PRINTED.
C              IF NOUT=1, OUTPUT OF ELEMENT PROPERTIES IS OMITTED.
C              IF NOUT=2, OUTPUT OF ELEMENT PROPERTIES AND GLOBAL
C                         MATRICES IS OMITTED.
C
C***********************************************************************
C      2.  ELEMENT OUTPUT DATA
C***********************************************************************
C          OUTPUT DATA FOR EACH ELEMENT INCLUDES CONNECTIVITY,
C            NODAL COORDINATES, INTERPOLATION FUNCTION
C            PROPERTIES, AND ELEMENT COEFFICIENT MATRICES.
C
C***********************************************************************
C      3.  OUTPUT OF GLOBAL MATRICES
C***********************************************************************
C          CONDENSED GLOBAL MATRICES AND COLUMNS ARE OUTPUT
C            JUST AFTER ASSEMBLY OF ELEMENT MATRICES, AFTER
C            INCLUSION OF PRESCRIBED NODAL VALUES, AND AFTER
C            FORMATION OF UPPER TRIANGLUAR FORM.
C
C***********************************************************************
C      4.  OUTPUT OF NODAL AND ELEMENT VALUES
C***********************************************************************
C          NODAL VALUES AND ELEMENT STRESSES AND STRAINS
C            ARE OUTPUT.
C
C***********************************************************************
C          PROGRAM READABLILITY FEATURES
C***********************************************************************
C          PROGRAM SECTIONS ARE LABELED WITH COMMENT CARDS.
C            MAJOR SECTIONS ARE LABELED WITH 1,2,3,ETC WHILE
C            SUBSECTIONS ARE LABELED WITH 1.1,1.2,1.3,ETC
C            WITHIN EACH MAJOR SECTION.
C
C          ALL DO LOOPS END WITH A STATEMENT NUMBER ENDING
C            IN ZERO AND STATEMENTS WITHIN THE DO LOOP ARE
C            INDENTED TWO SPACES.  VERY LONG DO LOOPS HAVE
C            STATEMENT NUMBERS 1000,2000,ETC.
C
C***********************************************************************
      DIMENSION XPOINT(7),YPOINT(7)
      DIMENSION INODE(200),XNODE(200),YNODE(200)
      DIMENSION NEL(200),NODEI(200),NODEJ(200),NODEK(200)
      DIMENSION NODEL(200),NETYP(200)
      DIMENSION YME(200),MUE(200),TE(200)
      DIMENSION IBOUND (200),NBOUND(200)
      DIMENSION IFORCE(200),TFORCE(200),NFORCE(200)
      DIMENSION IDISP(200),TDISP(200),NDISP(200)
      DIMENSION STRNXX(200),STRNYY(200),STRNZZ(200),STRNXY(200)
      DIMENSION STRSXX(200),STRSYY(200),STRSZZ(200),STRSXY(200)
      DIMENSION AFORCE(200)
      DIMENSION X(4),Y(4),NN(4),U(4),V(4)
      DIMENSION B(3),C(3)
      DIMENSION EBM(8,8),ETRAN(8,8),ROT(8,8),ROTT(8,8),REBM(8,8)
      DIMENSION AA(8,8),BB(8,8),CC(8,8)
      REAL MUE,MUA,MU
      COMMON /ALP10/ TITLE1(20),TITLE2(20)
      COMMON /INT10/ NP,NE,NBW,NOUT
      COMMON /REAL10/ ACOND(200,40),F(200),T(200)
      DATA IN,IO/60,61/
```

```
C***************************************************************
C     1.  INPUT AND OUTPUT OF PROBLEM DATA
C***************************************************************
C
C++++++++++++
C     1.1.  INPUT AND OUTPUT OF TITLE AND GLOBAL PARAMETERS
C++++++++++++
      READ (IN,11) TITLE1
      READ (IN,11) TITLE2
   11 FORMAT (20A4)
      READ (IN,12) NP,NE,NAN
   12 FORMAT (3I5)
      CALL HEAD
      IF (NAN .LT. 1 .OR. NAN .GT. 3) WRITE (IO,14) NAN
      IF (NAN.EQ.1) WRITE(IO,18) NAN
      IF (NAN.EQ.2) WRITE(IO,19) NAN
      IF (NAN.EQ.3) WRITE(IO,13) NAN
   14 FORMAT(10X,17HINPUT ERROR, NAN=,I5)
   18 FORMAT(//10X,21HPLANE STRESS ANALYSIS/12X,4HNAN=,I5)
   13 FORMAT(//10X,20HPLANE TRUSS ANALYSIS /12X,4HNAN=,I5)
   19 FORMAT(//10X,21HPLANE STRAIN ANALYSIS/12X,4HNAN=,I5)
      WRITE (IO,15)
   15 FORMAT (//10X,36H1.  INPUT AND OUTPUT OF PROBLEM DATA)
      WRITE (IO,16)
   16 FORMAT (////10X,23H1.1.  GLOBAL PARAMETERS)
      WRITE (IO,17) NP,NE
   17 FORMAT ( /10X,15HNUMBER OF NODES,I5,/
     1    10X,18HNUMBER OF ELEMENTS,I5)
C++++++++++++
C     1.2.  INPUT AND OUTPUT OF NODAL COORDINATES
C++++++++++++
      WRITE (IO,21)
   21 FORMAT (////10X,23H1.2.  NODAL COORDINATES,
     1    //10X,5H NODE,10X,1HX,10X,1HY)
      DO 20 I=1,NP
        READ (IN,22) INODE(I),XNODE(I),YNODE(I)
   22   FORMAT (I5,5X,2G10.1)
        WRITE (IO,23) INODE(I),XNODE(I),YNODE(I)
   23   FORMAT (10X,I5,5X,2G12.5)
   20 CONTINUE
C++++++++++++
C     1.3.  INPUT AND OUTPUT OF ELEMENT/NODE CONNECTIVITY DATA
C++++++++++++
      WRITE (IO,31)
   31 FORMAT (////10X,36H1.3.  ELEMENT/NODE CONNECTIVITY DATA,
     1    //10X,8HELEM NO.,3X,9HELEM TYPE,5X,5HNODES)
      DO 30 I=1,NE
        READ (IN,32) IEL,NTYP,NODI,NODJ,NODK,NODL
   32   FORMAT (6I5)
        NEL(IEL)=IEL
        NETYP(IEL)=NTYP
        NODEI(IEL)=NODI
        NODEJ(IEL)=NODJ
        NODEK(IEL)=NODK
        NODEL(IEL)=NODL
        IF(NETYP(IEL).LT.40) GOTO 34
        WRITE(IO,33) NEL(I),NETYP(I),NODEI(I),NODEJ(I),
     1               NODEK(I),NODEL(I)
   33   FORMAT (10X,I5,5X,I5,5X,4I5)
        GOTO 30
```

```
 34     IF(NETYP(IEL).LT.30) GOTO 36
        WRITE(IO,35) NEL(I),NETYP(I),NODEI(I),NODEJ(I),
     1           NODEK(I)
 35     FORMAT(10X,I5,5X,I5,5X,3I5)
        GOTO 30
 36     WRITE(IO,37) NEL(I),NETYP(I),NODEI(I),NODEJ(I)
 37     FORMAT(10X,I5,5X,I5,5X,2I5)
 30   CONTINUE
C++++++++++++
C
C     1.4.  CALCULATE BANDWIDTH OF GLOBAL COEFFICIENT MATRIX
C           NBW=BANDWIDTH
C++++++++++++
      NDMAX=0
      DO 40 I=1,NE
        NDIF1=IABS(NODEI(I)-NODEJ(I))
      IF(NETYP(I).LT.30) GOTO 43
        NDIF2=IABS(NODEJ(I)-NODEK(I))
        NDIF3=IABS(NODEK(I)-NODEI(I))
        IF(NETYP(I).LT.40) GOTO 42
        NDIF4=IABS(NODEI(I)-NODEL(I))
        NDIF5=IABS(NODEJ(I)-NODEL(I))
        NDIF6=IABS(NODEK(I)-NODEL(I))
        IF(NDMAX.LT.NDIF4) NDMAX=NDIF4
        IF(NDMAX.LT.NDIF5) NDMAX=NDIF5
        IF(NDMAX.LT.NDIF6) NDMAX=NDIF6
 42     IF (NDMAX .LT. NDIF3) NDMAX=NDIF3
        IF (NDMAX .LT. NDIF2) NDMAX=NDIF2
 43     IF (NDMAX .LT. NDIF1) NDMAX=NDIF1
 40   CONTINUE
      NBW=(NDMAX+1)*2
      WRITE (IO,41) NBW
 41   FORMAT (////10X,27H1.4.  CALCULATED BANDWIDTH=,I5)
C+++++++++++++
C     1.5.  INPUT AND OUTPUT OF ELEMENT PROPERTIES
C+++++++++++++
      WRITE (IO,51)
 51   FORMAT (////10X,24H1.5.  ELEMENT PROPERTIES,/
     1     /10X,7HELEMENT,10X,2HYM,10X,2HMU,10X,1HT)
      READ (IN,52) ICO
 52   FORMAT(I5)
      IF (ICO .EQ. 1) READ (IN,54) IEL,YMA,MUA,TA
 54   FORMAT (I5,5X,3G10.1)
      DO 50 I=1,NE
        IEL=I
        IF (ICO .EQ. 1) GO TO 55
        READ (IN,54) IEL,YMA,MUA,TA
 55     CONTINUE
        YME(IEL)=YMA
        MUE(IEL)=MUA
        TE(IEL)=TA
        WRITE (IO,56) IEL,YME(IEL),MUE(IEL),TE(IEL)
 56     FORMAT (10X,I5,5X,3G12.5)
 50   CONTINUE
      IF (ICO .EQ. 1) WRITE (IO,57) ICO
 57   FORMAT (//10X,35HYM,MU,T VALUES ARE THE SAME FOR ALL,
     1     9H ELEMENTS,/10X,4HICO=,I5)
C++++++++++++
C     1.6.  INPUT AND OUTPUT OF PRESCRIBED NODAL FORCES
C++++++++++++
```

```
      WRITE (IO,71)
   71 FORMAT (////10X,25H1.6.  APPLIED NODAL FORCE)
      READ (IN,72) NPF
   72 FORMAT (I5)
      IF (NPF .LE. 0) GO TO 76
      WRITE (IO,73) NPF
   73 FORMAT (/ 10X,31HNUMBER OF APPLIED NODAL FORCES=,
     1    I5, /10X,4HNODE, 4X,9HDIRECTION,4X,5HFORCE)
      DO 70 I=1,NPF
        READ (IN,74) IFORCE(I),NFORCE(I),TFORCE(I)
   74   FORMAT (I5,I5,G10.1)
        WRITE (IO,75) IFORCE(I),NFORCE(I),TFORCE(I)
   75   FORMAT (10X,I5,5X,I5,5X,G12.5)
   70 CONTINUE
   76 IF (NPF .LE. 0) WRITE (IO,77) NPF
   77 FORMAT (///10X,23HNO APPLIED NODAL FORCES,
     1    6H GIVEN,/10X,4HNPF=,I5)
C+++++++++++++
C     1.7.  PRESCRIBED NODAL DISPLACEMENTS
C+++++++++++++
      READ(IN,82) NPD
      IF(NPD.LE.0) GOTO 86
      WRITE(IO,81)
   81 FORMAT(///10X,36H1.7.  PRESCRIBED NODAL DISPLACEMENTS)
   82 FORMAT(I5)
      WRITE(IO,83) NPD
   83 FORMAT(/ 10X,38HNUMBER OF APPLIED NODAL DISPLACEMENTS=,
     1       I5, /10X,4HNODE,4X,9HDIRECTION,8X,12HDISPLACEMENT)
      DO 80 I=1,NPD
        READ(IN,84) IDISP(I),NDISP(I),TDISP(I)
   84   FORMAT(I5,I5,G10.1)
        WRITE(IO,85) IDISP(I),NDISP(I),TDISP(I)
   85   FORMAT(10X,I5,5X,I5,5X,G12.5)
   80 CONTINUE
   86 IF(NPD.LE.0) WRITE(IO,87) NPD
   87 FORMAT(///10X,34HNO PRESCRIBED NODAL DISPLACEMENTS,
     1       6H GIVEN,/10X,4HNPD=,I5)
C+++++++++++++
C     1.8.  OUTPUT PARAMETERS
C+++++++++++++
C
      WRITE(IO,91)
   91 FORMAT (////10X,23H1.8.  OUTPUT PARAMETERS)
      READ (IN,92) NOUT
   92 FORMAT (I5)
      WRITE (IO,93) NOUT
   93 FORMAT (/ 10X,5HNOUT=,I5)
C**************************************************************
C     2.  CALCULATION OF ELEMENT PROPERTIES AND ASSEMBLY
C           OF GLOBAL MATRICES
C**************************************************************
C
C+++++++++++++
C     2.1.  WRITE TITLE AND ZERO GLOBAL MATRICES
C+++++++++++++
      CALL HEAD
      WRITE (IO,101)
  101 FORMAT (////10X,37H2.  CALCULATION OF ELEMENT PROPERTIES)
      NP2=NP*2
      DO 120 I=1,NP2
```

```
        DO 130 J=1,NBW
          ACOND(I,J)=0.0
 130    CONTINUE
        F(I)=0.0
 120  CONTINUE
C++++++++++++
C     2.2.  FORM LARGE DO LOOP OVER ALL ELEMENTS
C++++++++++++
      DO 1000 I=1,NE
C+++++++++++++
C     2.3.  ASSIGN LOCAL NODAL COORDINATES
C+++++++++++++
      NNOD=2
        IEL=NEL(I)
        NN(1)=NODEI(IEL)
        NN(2)=NODEJ(IEL)
        X(1)=XNODE(NN(1))
        X(2)=XNODE(NN(2))
        Y(1)=YNODE(NN(1))
        Y(2)=YNODE(NN(2))
        IF(NETYP(I).LT.30) GOTO 121
        NN(3)=NODEK(IEL)
        X(3)=XNODE(NN(3))
        Y(3)=YNODE(NN(3))
        NNOD=3
        IF(NETYP(I).LT.40) GOTO 121
        NN(4)=NODEL(IEL)
        X(4)=XNODE(NN(4))
        Y(4)=YNODE(NN(4))
        NNOD=4
  121 CONTINUE
C++++++++++++
C     2.4.  ASSIGN LOCAL ELEMENT COEFFICIENT FUNCTIONS
C++++++++++++
        YM=YME(IEL)
        MU=MUE(IEL)
        TH=TE(IEL)
C+++++++++++++
C     2.6.  CALL SUBROUTINE ELEMNT
C           ALL ELEMENT PROPERTIES ARE OUTPUT FROM ELEMNT
C+++++++++++++
      IF(NETYP(I).EQ.21) CALL ELEM21 (IEL,X,Y,NN,YM,TH,EBM)
      IF(NETYP(I).EQ.31) CALL ELEM31 (IEL,X,Y,NN,YM,MU,TH,EBM,NAN)
      IF(NETYP(I).EQ.41) CALL ELEM41 (IEL,X,Y,NN,YM,MU,TH,EBM,NAN)
C++++++++++++
C     2.7.  ASSEMBLY OF ELEMENT MATRICES INTO GLOBAL MATRICES
C           A    =GLOBAL COEFFICIENT MATRIX (WHICH IS TOO
C                   LARGE TO STORE SO IT IS NOT CALCULATED)
C           ACOND=CONDENSED GLOBAL COEFFICIENT MATRIX
C                   (FOR BANDED SYMMETRIC A MATRIX ONLY THE
C                   UPPER TRIAGONAL NONZERO TERMS ARE STORED)
C           F    =GLOBAL COLUMN
C           IROW =ROW OF GLOBAL A MATRIX
C           ICOL =COLUMN OF GLOBAL A MATRIX
C           JCOL =COLUMN OF CONDENSED GLOBAL ACOND MATRIX
C++++++++++++
        DO 170 J=1,NNOD
          J2=J*2
          J1=J2-1
          DO 180 K=1,NNOD
```

```
                  K2=K*2
                  K1=K2-1
                  IROW2=NN(J)*2
                  IROW1=IROW2-1
                  ICOL2=NN(K)*2
                  ICOL1=ICOL2-1
                  JCOL1=ICOL1-IROW1+1
                  JCOL2=JCOL1+1
                  JCOL3=JCOL1-1
                  IF (JCOL1 .LT. 1) GO TO 179
                  ACOND(IROW2,JCOL1)=ACOND(IROW2,JCOL1)+EBM(J2,K2)
                  ACOND(IROW1,JCOL1)=ACOND(IROW1,JCOL1)+EBM(J1,K1)
  179             IF(JCOL2.LT.1) GOTO 181
                  ACOND(IROW1,JCOL2)=ACOND(IROW1,JCOL2)+EBM(J1,K2)
                  IF(JCOL3.LT.1) GOTO 181
                  ACOND(IROW2,JCOL3)=ACOND(IROW2,JCOL3)+EBM(J2,K1)
  181             CONTINUE
  180           CONTINUE
  170         CONTINUE
 1000       CONTINUE
            IF(NPF.LE.0) GOTO 176
            DO 175 J=1,NPF
              JJ=IFORCE(J)*2+NFORCE(J)-2
              F(JJ)=TFORCE(J)
  175       CONTINUE
  176       CONTINUE
C+++++++++++++
C           2.8.  OUTPUT OF CONDENSED GLOBAL MATRIX AND COLUMN
C+++++++++++++
            CALL WRITEM
            WRITE (IO,183)
  183       FORMAT (//10X,37H(GLOBAL MATRICES JUST AFTER ASSEMBLY))
C*************************************************************
C           3.  INCLUSION OF PRESCRIBED NODAL VALUES INTO GLOBAL
C                 COEFFICIENT MATRIX AND COLUMN
C*************************************************************
C
C+++++++++++++
C           3.1.  MODIFY ROW AND COLUMN OF ACOND MATRIX AND MODIFY
C                 ROW OF F COLUMN
C                 IROW  =ROW OF PRESCRIBED NODAL VOLAUE
C                 JROW  =ROW OF TERM TO BE MOVED TO RIGHT SIDE
C                          OF EQUAL SIGN
C                 JCOL  =COLUMN OF ACOND TERM TO BE SET TO ZERO
C+++++++++++++
             IF(NPD.LE.0) GOTO 260
            DO 200 I=1,NPD
              IROW=IDISP(I)*2+NDISP(I)-2
C
C               TRANSFER TERMS IN UPPER MATRIX TO RIGHT SIDE
              DO 210 J=2,NBW
                JROW=IROW-NBW+J-1
                JCOL=NBW-J+2
                IF (JROW .LT. 1) GO TO 210
                F(JROW)=F(JROW)-ACOND(JROW,JCOL)*TDISP(I)
                ACOND(JROW,JCOL)=0.0
  210         CONTINUE
C
C               TRANSFER TERMS IN LOWER MATRIX TO RIGHT SIDE
              DO 220 J=2,NBW
```

```
          JROW=IROW+J-1
          JCOL=J
          IF (JROW .GT. NP) GO TO 220
          F(JROW)=F(JROW)-ACOND(IROW,JCOL)*TDISP(I)
 220    CONTINUE
C
C         SET ACOND TERMS TO ZERO IN IROW AND MODIFY RIGHT
C           HAND SIDE
        DO 230 JCOL=2,NBW
          ACOND(IROW,JCOL)=0.0
 230    CONTINUE
        F(IROW)=ACOND(IROW,1)*TDISP(I)
 200  CONTINUE
 260  CONTINUE
C+++++++++++++
C     3.2.  OUTPUT OF CONDENSED GLOBAL MATRIX AND COLUMN
C+++++++++++++
      CALL WRITEM
      WRITE (IO,233)
 233  FORMAT (//10X,36H(GLOBAL MATRICES AFTER INCLUSION OF
     1    ,49HPRESCRIBED NODAL VALUES AND BOUNDARY CONSTRAINTS))
C*****************************************************************
C     4.  CALCULATION OF NODAL T VALUES, ELEMENT STRESSES
C           AND STRAINS
C*****************************************************************
C
C+++++++++++++
C     4.1.  CALCULATION OF NODAL VALUES
C+++++++++++++
C
C         SOLUTION OF BANDED LINEAR SIMULTANEOUS EQUATIONS
C           FOR THE NODAL T VALUES
      CALL GAUSS1
      CALL WRITEM
      WRITE (IO,404)
 404  FORMAT (//10X,36H(GLOBAL MATRICES AFTER FORMATION OF
     1    ,22HUPPER TRIANGULAR FORM))
C
C         OUTPUT OF NODAL VALUES
      CALL HEAD
      WRITE (IO,403)
 403  FORMAT (////10X,35H4.  NODAL DISPLACEMENTS AND ELEMENT,
     1     21H STRESSES AND STRAINS)
      WRITE (IO,405)
 405  FORMAT (////10X,25H4.1.  NODAL DISPLACEMENTS)
      WRITE (IO,402)
 402  FORMAT (   /10X,4HNODE, 8X,1HX,12X,1HY)
      DO 400 I=1,NP
        IN2=INODE(I)*2
        IN1=IN2-1
        WRITE (IO,401) INODE(I),T(IN1),T(IN2)
 401    FORMAT (10X,I5,5X,2G15.8)
 400  CONTINUE
C+++++++++++++
C     4.2.  ELEMENT STRESSES AND STRAINS
C+++++++++++++
      WRITE (IO,609)
 609  FORMAT (////10X,34H4.2.  ELEMENT STRESSES AND STRAINS)
      DO 600 I=1,NE
        IEL=NEL(I)
```

```
         NN(1)=NODEI(IEL)
         NN(2)=NODEJ(IEL)
         X(1)=XNODE(NN(1))
         X(2)=XNODE(NN(2))
         Y(1)=YNODE(NN(1))
         Y(2)=YNODE(NN(2))
         IF(NETYP(I).LT.30) GOTO 610
         NN(3)=NODEK(IEL)
         X(3)=XNODE(NN(3))
         Y(3)=YNODE(NN(3))
         IF(NETYP(I).LT.40) GOTO 610
         NN(4)=NODEL(IEL)
         X(4)=XNODE(NN(4))
         Y(4)=YNODE(NN(4))
 610     CONTINUE
         IF(NETYP(I).EQ.21) CALL RESS20 (STRNXX,STRSXX,AFORCE,X,Y,
     1     IEL,YME,TE,NN)
         IF(NETYP(I).EQ.31) CALL RESS30(STRNXX,STRNYY,STRNZZ,STRNXY,
     1     X,Y,B,C,STRSXX,STRSYY,STRSZZ,STRSXY,IEL,NAN,YME,MUE,NN)
         IF(NETYP(I).GE.40) CALL RESS40(STRNXX,STRNYY,STRNZZ,STRNXY,
     1       X,Y,STRSXX,STRSYY,STRSZZ,STRSXY,IEL,NAN,YME,MUE,NN)
 600   CONTINUE
C
C        PRINT ELEMENT STRESSES AND STRAINS
       IF (NAN .EQ. 1) GO TO 601
       IF (NAN .EQ. 2) GO TO 602
       IF (NAN .EQ. 3) GO TO 603
C
C        RESULTS FOR PLANE STRESS ANALYSIS
 601   CONTINUE
 631   FORMAT(1H1)
       WRITE (IO,604)
 604   FORMAT(/10X,21HPLANE STRESS ANALYSIS//10X,15HELEMENT STRAINS,
     1        /9X,7HELEMENT,8X,6HSTRNXX,10X,6HSTRNYY,10X,6HSTRNZZ,
     2        10X,6HSTRNXY)
       DO 620 I=1,NE
         IEL=I
         IF(NETYP(IEL).GT.30) WRITE(IO,621) IEL,STRNXX(IEL),STRNYY(IEL),
     1                         STRNZZ(IEL),STRNXY(IEL)
 621     FORMAT(10X,I5,5X,4(2X,G14.6))
         IF(NETYP(IEL).LT.30) WRITE(IO,623) IEL,STRNXX(IEL)
 623     FORMAT (10X,I5,12X,G14.6,5X,13H(BAR ELEMENT))
 620   CONTINUE
       WRITE(IO,608)
  608 FORMAT(//10X,39HRESULTS GIVEN IN GLOBAL REFERENCE FRAME,
     1      23H EXCEPT FOR BAR RESULTS)
       WRITE(IO,632)
 632   FORMAT(////10X,21HPLANE STRESS ANALYSIS//10X,8HELEMENT ,
     1      8HSTRESSES/9X,7HELEMENT,8X,6HSTRSXX,8X,6HSTRSYY,
     2      8X,6HSTRSXY)
       DO 640 I=1,NE
         IEL=NEL(I)
         IF(NETYP(IEL).GT.30) WRITE(IO,641) IEL,STRSXX(IEL),
     1                        STRSYY(IEL),STRSXY(IEL)
 641     FORMAT(10X,I5,5X,3G14.6)
         IF(NETYP(IEL).LT.30) WRITE(IO,623) IEL,STRSXX(IEL)
 640   CONTINUE
       WRITE(IO,608)
       GO TO 699
C
```

```
C        RESULTS FOR PLANE STRAIN ANALYSIS
  602  CONTINUE
       WRITE(IO,605)
  605  FORMAT(/10X,21HPLANE STRAIN ANALYSIS//10X,8HELEMENT ,
     1        7HSTRAINS/9X,7HELEMENT,8X,6HSTRNXX,18X,
     2        6HSTRNYY,18X,6HSTRNXY)
       DO 660 I=1,NE
         IEL=NEL(I)
         IF(NETYP(IEL).GT.30) WRITE(IO,661)  IEL,STRNXX(IEL),
     1                         STRNYY(IEL),STRNXY(IEL)
  661    FORMAT(10X,I5,5X,3G14.6)
         IF(NETYP(IEL).LT.30) WRITE(IO,623) IEL,STRNXX(IEL)
  660  CONTINUE
       WRITE(IO,608)
       WRITE(IO,665)
  665  FORMAT(////10X,21HPLANE STRAIN ANALYSIS//10X,8HELEMENT ,
     1        8HSTRESSES/9X,7HELEMENT,8X,6HSTRSXX,8X,6HSTRSYY,
     2        8X,6HSTRSZZ,8X,6HSTRSXY)
       DO 670 I=1,NE
         IEL=NEL(I)
         IF(NETYP(IEL).GT.30) WRITE(IO,671) IEL,STRSXX(IEL),
     1                  STRSYY(IEL),STRSZZ(IEL),STRSXY(IEL)
  671    FORMAT(10X,I5,5X,4G14.6)
         IF(NETYP(IEL).LT.30) WRITE(IO,623) IEL,STRSXX(IEL)
  670  CONTINUE
       WRITE(IO,608)
       GO TO 699
C
C        RESULTS FOR PLANE TRUSS PROBLEMS
  603  CONTINUE
       WRITE (IO,681)
  681  FORMAT (/10X,20HPLANE TRUSS ANALYSIS,
     1     /10X,7HELEMENT,8X,12HAXIAL STRAIN,4X,12HAXIAL STRESS,
     1     4X,11HAXIAL FORCE)
       DO 680 I=1,NE
         IEL=I
         WRITE (IO,682) IEL,STRNXX(IEL),STRSXX(IEL),AFORCE(IEL)
  682  FORMAT (10X,I5,5X,3(2X,G14.6))
  680  CONTINUE
  699  CONTINUE
       END
C
C
       SUBROUTINE ELEM21(IEL,X,Y,NN,YM,AREA,EBM)
C*****************************************************************
C         SUBROUTINE ELEM21 CALCULATES THE ELEMENT
C           COEFFICIENT MATRIX FOR A SIMPLEX
C           BAR ELEMENT IN TWO DIMENSIONS.
C*****************************************************************
       DIMENSION X(4,4),Y(4,4),NN(4,4),EBM(8,8)
       REAL LENGTH
       REAL KX,KXY,KY
       COMMON /INTIO/ NP,NE,NBW,NOUT
       DATA IN,IO/60,61/
C*****************************************************************
C      1. CALCULATION OF ELEMENT COEFFICIENT MATRIX
C*****************************************************************
C
C+++++++++++++
C      1.1.  CALCULATE ELEMENT PROPERTIES
```

```
C++++++++++++++
      DX=X(2)-X(1)
      DY=Y(2)-Y(1)
      THETA=ATAN2(DY,DX)
      LENGTH=SQRT(DX*DX+DY*DY)
      CT=COS(THETA)
      ST=SIN(THETA)
      KX=AREA*YM*CT*CT/LENGTH
      KY=AREA*YM*ST*ST/LENGTH
      KXY=AREA*YM*CT*ST/LENGTH
C++++++++++++++
C     1.2.  EVALUATE ELEMENT MATRIX
C++++++++++++++
      EBM(1,1)=KX
      EBM(1,2)=KXY
      EBM(1,3)=-KX
      EBM(1,4)=-KXY
      EBM(2,1)=KXY
      EBM(2,2)=KY
      EBM(2,3)=-KXY
      EBM(2,4)=-KY
      EBM(3,1)=-KX
      EBM(3,2)=-KXY
      EBM(3,3)=KX
      EBM(3,4)=KXY
      EBM(4,1)=-KXY
      EBM(4,2)=-KY
      EBM(4,3)=KXY
      EBM(4,4)=KY
C*****************************************************************
C     2.  OUTPUT OF ELEMENT MATRICES
C*****************************************************************
      IF (NOUT.GE.1) GOTO 900
C++++++++++++++
C     2.1.  WRITE ELEMENT CONNECTIVITY DATA AND NODAL COORD
C++++++++++++++
      WRITE(IO,111) IEL
 111  FORMAT (////10X,30HELEMENT PROPERTIES FOR ELEMENT,I5)
      WRITE (IO,114)
 114  FORMAT (/10X,37H2.1.  ELEMENT CONNECTIVITY AND NODAL
     1      ,11HCOORDINATES)
      WRITE (IO,112)
 112  FORMAT (  10X,7HELEMENT,5X,4HNODE,6X,1HX,9X,1HY)
      DO 110 I=1,2
        WRITE(IO,113)IEL,NN(I),X(I),Y(I)
 113  FORMAT (10X,I5,5X,I5,2F10.4)
 110  CONTINUE
      WRITE (IO,115) LENGTH
 115  FORMAT (/10X, 7HLENGTH=,2X,F10.4)
C++++++++++++++
C     2.2.  WRITE ELEMENT MATERIAL PROPERTIES
C++++++++++++++
      WRITE (IO,123)
 123  FORMAT(/10X,33H2.2.  ELEMENT MATERIAL PROPERTIES)
      WRITE (IO,121)
 121  FORMAT(  26X2HYM,8X,4HAREA)
      WRITE(IO,122) YM,AREA
 122  FORMAT (20X,2G12.5)
      AK=YM*AREA/LENGTH
      WRITE (IO,155) AK
```

```
 155   FORMAT (/10X,10HSTIFFNESS=,2X,E14.5)
C++++++++++++++
C      2.3  WRITE ELEMENT COEFFICIENT MATRIX
C++++++++++++++
       WRITE (IO,153)
 153   FORMAT (/10X,32H2.3.  ELEMENT COEFFICIENT MATRIX)
       WRITE (IO,151)
 151   FORMAT ( 10X,4HNODE,6X,9HDIRECTION,11X,18HCOEFFICIENT MATRIX)
       DO 150 I=1,2
         II2=2*I
         II1=II2-1
         WRITE(IO,154) NN(I),(EBM(II1,J),J=1,4)
 154     FORMAT (10X,I5,9X,1H1,4X,4G12.5)
         WRITE(IO,152) NN(I),(EBM(II2,J),J=1,4)
 152     FORMAT (10X,I5,9X,1H2,4X,4G12.5)
 150   CONTINUE
 900   CONTINUE
       RETURN
       END
C
C
       SUBROUTINE RESS20 (STRNXX,STRSXX,AFORCE,X,Y,IEL,YME,TE,NN)
C*************************************************************
C           SUBROUTINE RESS20 CALCULATES THE STRESS, STRAIN
C            AND AXIAL FORCE IN A SIMPLEX BAR ELEMENT.
C*************************************************************
       DIMENSION STRNXX(200),STRSXX(200),AFORCE(200),X(4),Y(4)
       DIMENSION YME(200),NN(4),U(4),TE(200)
       REAL LENGTH
       COMMON /REAL10/ ACOND(200,40),F(200),T(200)
       DATA IN,IO/60,61/
       DX=X(2)-X(1)
       DY=Y(2)-Y(1)
       LENGTH=SQRT(DX*DX+DY*DY)
       THETA=ATAN2(DY,DX)
       DO 100 I=1,2
         IY=NN(I)*2
         IX=IY-1
         U(I)=T(IX)*COS(THETA)+T(IY)*SIN(THETA)
 100   CONTINUE
       STRNXX(IEL)=(U(2)-U(1))/LENGTH
       STRSXX(IEL)=YME(IEL)*STRNXX(IEL)
       AFORCE(IEL)=STRSXX(IEL)*TE(IEL)
       RETURN
       END
C
C
       SUBROUTINE ELEM31 (IEL,X,Y,NN,YM,MU,TH,EBM,NAN)
C*************************************************************
C          SUBROUTINE ELEM31 CALCULATES THE ELEMENT
C            COEFFICIENT MATRIX FOR A SIMPLEX TRIANGULAR ELEMENT
C            IN TWO DIMENSIONAL PLANE STRESS OR PLANE STRAIN.
C*************************************************************
       DIMENSION X(4),Y(4),NN(4),EBM(8,8)
       DIMENSION B(3),C(3)
       COMMON /INT10/ NP,NE,NBW,NOUT
       REAL MU
       DATA IN,IO/60,61/
C*************************************************************
C      1.  CALCULATION OF ELEMENT COEFFICIENT MATRICES
```

```
C**************************************************************
C
C+++++++++++++
C     1.1.  CALCULATION OF INTERPOLATION FUNCTION PROPERTIES
C           B    =X GRADIENT TERMS
C           C    =Y GRADIENT TERMS
C           AREA=ELEMENT AREA
C+++++++++++++
      CALL PROP31 (X,Y,B,C,AREA)
C+++++++++++++
C     1.2.  ELEMENT COEFFICIENT MATRIX
C           YM=ELEMENT VALUE OF YOUNGS MODULUS
C           MU=ELEMENT VALUE OF POISSONS RATIO
C           TH=ELEMENT THICKNESS
C           EBM   =ELEMENT B MATRIX
C+++++++++++++
      IF(NAN-1) 21,22,23
   21 WRITE(IO,24) NAN
   24 FORMAT(//10X,10HERROR,NAN=,I5)
   22 COEFA=YM*TH/(4.*AREA*(1.-MU*MU))
   25 GOTO 26
   23 COEFA=YM*TH*(1.-MU)/(4.*AREA*(1.+MU)*(1.-2.*MU))
   26 COEFB=YM*TH/(8.*AREA*(1.+MU))
      DO 10 I=1,3
        II2=2*I
        II1=II2-1
        DO 20 J=1,3
          JJ2=J*2
          JJ1=JJ2-1
          EBM(II1,JJ1)=COEFA*B(I)*B(J)+COEFB*C(I)*C(J)
          EBM(II2,JJ2)=COEFA*C(I)*C(J)+COEFB*B(I)*B(J)
          EBM(II2,JJ1)=COEFA*MU*C(I)*B(J)+COEFB*B(I)*C(J)
          EBM(II1,JJ2)=EBM(II2,JJ1)
   20   CONTINUE
   10 CONTINUE
C**************************************************************
C     2.  OUTPUT OF ELEMENT MATRICES
C**************************************************************
      IF (NOUT.GE.1) GOTO 900
C+++++++++++++
C     2.1.  WRITE ELEMENT CONNECTIVITY DATA AND NODAL COORD
C+++++++++++++
      WRITE (IO,111) IEL
  111 FORMAT (////10X,30HELEMENT PROPERTIES FOR ELEMENT,I5)
      WRITE (IO,114)
  114 FORMAT (/10X,37H2.1.  ELEMENT CONNECTIVITY AND NODAL
     1     ,11HCOORDINATES)
      WRITE (IO,112)
  112 FORMAT (  10X,7HELEMENT, 5X,4HNODE, 6X,1HX, 9X,1HY)
      DO 110 I=1,3
        WRITE (IO,113) IEL,NN(I),X(I),Y(I)
  113   FORMAT (10X,I5, 5X,I5,2F10.4)
  110 CONTINUE
C++++++++++++++
C     2.2.  WRITE ELEMENT MATERIAL PROPERTIES
C++++++++++++++
      WRITE (IO,123)
  123 FORMAT (/10X,33H2.2.  ELEMENT MATERIAL PROPERTIES)
      WRITE (IO,121)
  121 FORMAT( 26X,2HYM,10X,2HMU,10X,1HT)
```

```
          WRITE (IO,122) YM,MU,TH
  122     FORMAT (20X,3G12.5)
C++++++++++++
C     2.3.  WRITE ELEMENT INTERPOLATION FUNCTIONS
C++++++++++++
        WRITE (IO,134)
  134   FORMAT (/10X,37H2.3.  ELEMENT INTERPOLATION FUNCTION
     1     ,10HPROPERTIES)
        WRITE (IO,132)
  132   FORMAT (10X,4HNODE,10X,1HB,11X,1HC)
        DO 130 I=1,3
          WRITE (IO,133) NN(I),B(I),C(I)
  133     FORMAT (10X,I5,5X,2G12.5)
  130   CONTINUE
        WRITE (IO,131) AREA
  131   FORMAT (  10X,5HAREA=,G12.5)
C+++++++++++++
C     2.4.  WRITE ELEMENT COEFFICIENT MATRIX
C+++++++++++++
        WRITE (IO,153)
  153   FORMAT (/10X,32H2.4.  ELEMENT COEFFICIENT MATRIX)
        WRITE (IO,151)
  151     FORMAT ( 10X,4HNODE,6X,9HDIRECTION,11X,18HCOEFFICIENT MATRIX)
        DO 150 I=1,3
          II2=2*I
          II1=II2-1
          WRITE(IO,154) NN(I),(EBM(II1,J),J=1,6)
  154     FORMAT(10X,I5,9X,1H1,4X,6G12.5)
          WRITE (IO,152) NN(I),(EBM(II2,J),J=1,6)
  152   FORMAT (10X,I5,9X,1H2,4X,6G12.5)
  150   CONTINUE
  900   CONTINUE
        RETURN
        END
C
C
        SUBROUTINE RESS30 (STRNXX,STRNYY,STRNZZ,STRNXY,X,Y,B,C,
     1       STRSXX,STRSYY,STRSZZ,STRSXY,IEL,NAN,YME,MUE,NN)
C*************************************************************
C          SUBROUTINE RESS30 CALCULATES THE ELEMENT STRAINS
C            AND STRESSES FOR A SIMPLEX TRIANGULAR ELEMENT IN
C            PLANE STRESS OR PLANE STRAIN.
C*************************************************************
        DIMENSION STRNXX(200),STRNYY(200),STRNZZ(200),STRNXY(200)
        DIMENSION STRSXX(200),STRSYY(200),STRSZZ(200),STRSXY(200)
        DIMENSION X(4),Y(4),B(4),C(4),NEL(200)
        DIMENSION YME(200),MUE(200),NN(4)
        COMMON /REAL10/ ACOND(200,40),F(200),T(200)
        REAL MU,MUE
        DATA IN,IO/60,61/
        CALL PROP31(X,Y,B,C,AREA)
C+++++++++++++
C     1.1.  CALCULATE ELEMENT STRAINS
C+++++++++++++
        A2=AREA*2.
        STRNXX(IEL)=0.0
        STRNYY(IEL)=0.0
        STRNXY(IEL)=0.0
        DO 100 J=1,3
          IY=NN(J)*2
```

```
         IX=IY-1
         DISPX=T(IX)
         DISPY=T(IY)
         STRNXX(IEL)=B(J)*DISPX/A2+STRNXX(IEL)
         STRNYY(IEL)=C(J)*DISPY/A2+STRNYY(IEL)
         STRNXY(IEL)=C(J)*DISPX/A2+B(J)*DISPY/A2+STRNXY(IEL)
  100  CONTINUE
       YM=YME(IEL)
       MU=MUE(IEL)
       IF(NAN-1) 110,120,130
  110  WRITE(IO,112)
  112  FORMAT(10X,11HERROR, NAN=,I5)
C+++++++++++++
C      1.2.  CALCULATE ELEMENT STRESSES FOR PLANE STRESS
C+++++++++++++
  120  STRSXX(IEL)=YM*(STRNXX(IEL)+STRNYY(IEL)*MU)/(1.-MU*MU)
       STRSYY(IEL)=YM*(STRNYY(IEL)+STRNXX(IEL)*MU)/(1.-MU*MU)
       STRSXY(IEL)=YM*STRNXY(IEL)/(2.+2.*MU)
       STRNZZ(IEL)=-MU*(STRSXX(IEL)+STRSYY(IEL))/YM
       GOTO 125
C+++++++++++++
C      1.3.  CALCULATE ELEMENT STRESSES FOR PLANE STRAIN
C+++++++++++++
  130  CONTINUE
       CON1=(1.+MU)*(1.-2.*MU)
       CON2=1.-MU
       STRSXX(IEL)=YM*(CON2*STRNXX(IEL)+MU*STRNYY(IEL))/CON1
       STRSYY(IEL)=YM*(CON2*STRNYY(IEL)+MU*STRNXX(IEL))/CON1
       STRSZZ(IEL)=YM*MU*(STRNXX(IEL)+STRNYY(IEL))/CON1
       STRSXY(IEL)=YM*STRNXY(IEL)/(2.+2.*MU)
  125  CONTINUE
       RETURN
       END
C
C
       SUBROUTINE ELEM41 (IEL,X,Y,NN,YM,MU,TH,EBM,NAN)
C**************************************************************
C          SUBROUTINE ELEM41 CALCULATES THE ELEMENT MATRIX
C            FOR A BILINEAR RECTANGULAR ELEMENT IN TWO
C            DIMENSIONAL PLANE STRESS OR PLANE STRAIN.
C**************************************************************
       DIMENSION X(4),Y(4),NN(4),EBM(8,8),ETRAN(8,8)
       DIMENSION ROT(8,8),ROTT(8,8)
       COMMON /INTIO/ NP,NE,NBW,NOUT
       REAL MU,LENGTH
       DATA IN,IO/60,61/
C**************************************************************
C      1.  CALCULATION OF ELEMENT COEFFICIENT MATRICES
C**************************************************************
C
C++++++++++++
C      1.1.  CALCULATION OF INTERPOLATION FUNCTION PROPERTIES
C            YM=ELEMENT VALUE OF YOUNGS MODULUS
C            MU=ELEMENT VALUE OF POISSONS RATIO
C            TH=ELEMENT THICKNESS
C            A=ELEMENT HEIGHT
C            B=ELEMENT WIDTH
C++++++++++++
       CALL PROP41(X,Y,A,B,AREA,TRAN,ITRAN)
C+++++++++++++
```

```
C      1.2.  PROPERTIES FOR PLANE STRESS OR PLANE STRAIN
C++++++++++++++
       DO 50 I=1,8
         DO 60 J=1,8
           ROT(I,J)=0.
           ROTT(I,J)=0.
   60    CONTINUE
   50  CONTINUE
       IF(NAN-1) 21,22,23
   21  WRITE(IO,24) NAN
   24  FORMAT(//10X,10HERROR,NAN=,I5)
   22  G1=YM/(1.-MU*MU)
       G2=YM*MU/(1.-MU*MU)
       G3=YM/(2.*(1+MU))
   25  GOTO 26
   23  G1=YM*(1.-MU)/((1.+MU)*(1.-2.*MU))
       G2=YM*MU/((1.+MU)*(1.-2.*MU))
       G3=YM/(2.*(1.+MU))
   26  CONTINUE
       A2=A/2.
       B2=B/2.
       C1=TH*A2*A2*G1/(3.*AB4)
       AB4=A2*B2
       C1=TH*(A2*A2*G1+B2*B2*G3)/(3.*AB4)
       C2=TH*(B2*B2*G1+A2*A2*G3)/(3.*AB4)
C++++++++++++++
C      1.3.  ELEMENT COEFFICIENT MATRIX
C            EBM   =ELEMENT B MATRIX
C++++++++++++++
       DO 10 I=1,4
         IY=2*I
         IX=IY-1
         EBM(IX,IX)=C1
         EBM(IY,IY)=C2
   10  CONTINUE
       EBM(5,1)=-C1/2.
       EBM(7,3)=-C1/2.
       EBM(6,2)=-C2/2.
       EBM(8,4)=-C2/2.
       C2=TH*B2*B2*G3/(3.*AB4)
       EBM(3,1)=-C1+C2/2.
       EBM(7,1)=C1/2.-C2
       EBM(5,3)=C1/2.-C2
       EBM(7,5)=-C1+C2/2.
       C1=TH*B2*B2*G1/(3.*AB4)
       C2=TH*A2*A2*G3/(3.*AB4)
       EBM(4,2)=C1/2.-C2
       EBM(8,2)=-C1+C2/2.
       EBM(6,4)=-C1+C2/2.
       EBM(8,6)=C1/2.-C2
       C1=TH*(G2+G3)/4.
       C2=TH*(G2-G3)/4.
       EBM(2,1)=C1
       EBM(3,2)=-C2
       EBM(5,2)=-C1
       EBM(7,2)=C2
       EBM(4,1)=C2
       EBM(4,3)=-C1
       EBM(5,4)=-C2
       EBM(7,4)=C1
```

```
      EBM(6,1)=-C1
      EBM(6,3)=C2
      EBM(6,5)=C1
      EBM(7,6)=-C2
      EBM(8,1)=-C2
      EBM(8,3)=C1
      EBM(8,5)=C2
      EBM(8,7)=-C1
      DO 20 I=1,7
        I2=I+1
        DO 30 J=I2,8
          EBM(I,J)=EBM(J,I)
   30   CONTINUE
   20 CONTINUE
C***************************************************************
C     2.  OUTPUT OF ELEMENT MATRICES
C***************************************************************
C
      IF (NOUT.GE.1) GOTO 900
C+++++++++++++
C     2.1.  WRITE ELEMENT CONNECTIVITY DATA AND NODAL COORD
C+++++++++++++
      WRITE (IO,111) IEL
  111 FORMAT (////10X,30HELEMENT PROPERTIES FOR ELEMENT,I5)
      WRITE (IO,114)
  114 FORMAT (/10X,37H2.1.  ELEMENT CONNECTIVITY AND NODAL
     1    ,11HCOORDINATES)
      WRITE (IO,112)
  112 FORMAT (  10X,7HELEMENT, 5X,4HNODE, 6X,1HX, 9X,1HY)
      DO 110 I=1,4
        WRITE (IO,113) IEL,NN(I),X(I),Y(I)
  113   FORMAT (10X,I5, 5X,I5,2F10.4)
  110 CONTINUE
      DTRAN=TRAN*180./3.14159
      WRITE(IO,115) DTRAN
  115 FORMAT(/10X,21HTRANSFORMATION ANGLE=,F8.3,8H DEGREES)
C++++++++++++++
C     2.2.  WRITE ELEMENT MATERIAL PROPERTIES
C++++++++++++++
      WRITE (IO,123)
  123 FORMAT (/10X,33H2.2.  ELEMENT MATERIAL PROPERTIES)
      WRITE (IO,121)
  121 FORMAT( 26X,2HYM,10X,2HMU,10X,1HT)
        WRITE (IO,122) YM,MU,TH
  122   FORMAT (20X,3G12.5)
C+++++++++++++
C     2.3. WRITE ELEMENT PHYSICAL PROPERTIES
C+++++++++++++
      WRITE(IO,134)
  134 FORMAT(/10X,33H2.3.  ELEMENT PHYSICAL PROPERTIES)
      WIDTH=B
      HEIGHT=A
      WRITE(IO,132) HEIGHT,WIDTH,AREA
  132 FORMAT(/10X,7HHEIGHT=,G10.3/10X,6HWIDTH=,G10.3/
     1         10X,5HAREA=,G10.3)
C+++++++++++++
C     2.4.  WRITE ELEMENT COEFFICIENT MATRIX
C++++++++++++++
      WRITE (IO,153)
  153 FORMAT (/10X,32H2.4.  ELEMENT COEFFICIENT MATRIX/
```

```
      1         15X,34H(BEFORE COORDINATE TRANSFORMATION))
155  WRITE (IO,151)
151    FORMAT ( 10X,4HNODE,6X,9HDIRECTION,11X,18HCOEFFICIENT MATRIX)
     DO 150 I=1,4
       II2=2*I
       II1=II2-1
       WRITE(IO,154) NN(I),(EBM(II1,J),J=1,8)
154    FORMAT(10X,I5,9X,1H1,4X,8G12.5)
       WRITE (IO,152) NN(I),(EBM(II2,J),J=1,8)
152  FORMAT (10X,I5,9X,1H2,4X,8G12.5)
150  CONTINUE
     IF(ITRAN.EQ.0) GOTO 200
     WRITE(IO,156)
156  FORMAT(/10X,43HELEMENT COEFFICIENT MATRIX AFTER COORDINATE,
    1     15H TRANSFORMATION)
     ITRAN=0
     CO=COS(TRAN)
     SI=SIN(TRAN)
     DO 160 I=1,4
       I2=I*2
       I1=I2-1
       ROT(I1,I1)=CO
       ROT(I2,I2)=CO
       ROT(I1,I2)=SI
       ROT(I2,I1)=-SI
       ROTT(I1,I1)=CO
       ROTT(I2,I2)=CO
       ROTT(I1,I2)=-SI
       ROTT(I2,I1)=SI
160  CONTINUE
     CALL MATMUL (ROTT,EBM,ETRAN,8,8,8)
     CALL MATMUL (ETRAN,ROT,EBM,8,8,8)
200  CONTINUE
900  CONTINUE
     RETURN
     END
C
C
     SUBROUTINE RESS40 (STRNXX,STRNYY,STRNZZ,STRNXY,X,Y,
    1       STRSXX,STRSYY,STRSZZ,STRSXY,IEL,NAN,YME,MUE,NN)
C************************************************************
C          SUBROUTINE RESS40 CALCULATES THE ELEMENT STRAINS
C            AND STRESS FOR A BILINEAR RECTANGULAR ELEMENT IN
C            PLANE STRESS OR PLANE STRAIN.
C************************************************************
     DIMENSION STRNXX(200),STRNYY(200),STRNZZ(200),STRNXY(200)
     DIMENSION STRSXX(200),STRSYY(200),STRSZZ(200),STRSXY(200)
     DIMENSION X(4),Y(4),NEL(200)
     DIMENSION YME(200),MUE(200),NN(4)
     DIMENSION U(4),V(4)
     COMMON /REAL10/ ACOND(200,40),F(200),T(200)
     REAL MU,MUE
     DATA IN,IO/60,61/
     CALL PROP41(X,Y,A,B,AREA,TRAN,ITRAN)
C++++++++++++++
C     1.1.  CALCULATE ELEMENT STRAINS
C++++++++++++++
     DO 100 I=1,4
       IY=NN(I)*2
       IX=IY-1
```

```
        U(I)=T(IX)*COS(TRAN)+T(IY)*SIN(TRAN)
        V(I)=-T(IX)*SIN(TRAN)+T(IY)*COS(TRAN)
 100  CONTINUE
      STNXX=(-U(1)+U(2)+U(3)-U(4))/(2.*B)
      STNYY=(-V(1)-V(2)+V(3)+V(4))/(2.*A)
      STNXY=(-U(1)-U(2)+U(3)+U(4))/(2.*A)
     1     +(-V(1)+V(2)+V(3)-V(4))/(2.*B)
      YM=YME(IEL)
      MU=MUE(IEL)
      IF(NAN-1) 110,120,130
 110  WRITE(IO,112)
 112  FORMAT(10X,11HERROR, NAN=,I5)
C+++++++++++++
C     1.2.  CALCULATE ELEMENT STRESSES FOR PLANE STRESS
C+++++++++++++
 120  STSXX=YM*(STNXX+STNYY*MU)/(1-MU*MU)
      STSYY=YM*(STNYY+STNXX*MU)/(1-MU*MU)
      STSXY=YM*STNXY/(2+2*MU)
      STRNZZ(IEL)=-MU*(STSXX+STSYY)/YM
      GOTO 125
C+++++++++++++
C     1.3.  CALCULATE ELEMENT STRESSES FOR PLANE STRAIN
C+++++++++++++
 130  CONTINUE
      CON1=(1+MU)*(1-2*MU)
      CON2=1-MU
      STSXX=YM*(CON2*STNXX+MU*STNYY)/CON1
      STSYY=YM*(CON2*STNYY+MU*STNXX)/CON1
      STRSZZ(IEL)=YM*MU*(STNXX+STNYY)/CON1
      STSXY=YM*STNXY/(2+2*MU)
 125  CONTINUE
C+++++++++++++
C     1.4.  APPLY COORDINATE TRANSFORMATION TO ROTATE ELEMENT
C           STRAINS AND STRESSES TO GLOBAL COORDINATE SYSTEM.
C+++++++++++++
      CO=COS(TRAN)
      SI=SIN(TRAN)
      STRNXX(IEL)=STNXX*CO*CO+STNYY*SI*SI+2.*STNXY*CO*SI
      STRNYY(IEL)=STNXX*SI*SI+STNYY*CO*CO-2.*STNXY*CO*SI
      STRNXY(IEL)=(STNYY-STNXX)*CO*SI+STNXY*(CO*CO-SI*SI)
      STRSXX(IEL)=STSXX*CO*CO+STSYY*SI*SI+2.*STSXY*CO*SI
      STRSYY(IEL)=STSXX*SI*SI+STSYY*CO*CO-2.*STSXY*CO*SI
      STRSXY(IEL)=(STSYY-STSXX)*CO*SI+STSXY*(CO*CO-SI*SI)
      RETURN
      END
      SUBROUTINE MATMUL(AA,BB,CC,II,JJ,KK)
C*************************************************************
C         SUBROUTINE MATMUL MULTIPLIES MATRIX A TIMES MATRIX B
C           AND STORES THE RESULT IN MATRIX C.
C*************************************************************
      DIMENSION AA(II,JJ),BB(JJ,KK),CC(II,KK)
      DO 10 I=1,II
        DO 10 K=1,KK
         SUM=0.
          DO 20 J=1,JJ
  20      SUM=SUM+AA(I,J)*BB(J,K)
  10  CC(I,K)=SUM
      RETURN
      END
C
```

```
C
      SUBROUTINE HEAD
C*****************************************************************
C           WRITE TITLES ON TOP OF PAGE
C*****************************************************************
      COMMON /ALP10/ TITLE1(20),TITLE2(20)
      DATA IN,IO/60,61/
      WRITE (IO,13) TITLE1
   13 FORMAT (1H1//10X,20A4)
      WRITE (IO,14) TITLE2
   14 FORMAT (10X,20A4)
      RETURN
      END
C
C
      SUBROUTINE PROP31 (X,Y,B,C,AREA)
C*****************************************************************
C           SUBROUTINE PROP31 CALCULATES THE X AND Y GRADIENT
C             TERMS AND AREA FOR A LINEAR INTERPOLATION
C             FUNCTION IN A TRIAGULAR ELEMENT.
C*****************************************************************
      DIMENSION X(4),Y(4),B(3),C(3)
      B(1)=Y(2)-Y(3)
      B(2)=Y(3)-Y(1)
      B(3)=Y(1)-Y(2)
      C(1)=X(3)-X(2)
      C(2)=X(1)-X(3)
      C(3)=X(2)-X(1)
      AREA=(X(2)*Y(3)+X(3)*Y(1)+X(1)*Y(2)-X(2)*Y(1)
     1   -X(3)*Y(2)-X(1)*Y(3))/2.0
      RETURN
      END
C
C
      SUBROUTINE PROP41(X,Y,A,B,AREA,TRAN,ITRAN)
C*********************************************************
C           SUBROUTINE PROP41 CALCULATES THE ELEMENT LENGTH,
C             WIDTH,AREA, AND COORDINATE TRANSFORMATION ANGLE
C             FOR A LINEAR RECTANGULAR ELEMENT.
C*****************************************************************
      DIMENSION X(4),Y(4)
      B=SQRT((X(2)-X(1))*(X(2)-X(1))+(Y(2)-Y(1))*(Y(2)-Y(1)))
      A=SQRT((X(3)-X(2))*(X(3)-X(2))+(Y(3)-Y(2))*(Y(3)-Y(2)))
      AREA=A*B
      IF(Y(2)-Y(1)) 20,10,20
   10 TRAN=0.
      ITRAN=0
      GOTO 30
C      CALCULATE COORDINATE TRANSFORMATION ANGLE
   20 TRAN=ATAN2((Y(2)-Y(1)),(X(2)-X(1)))
      ITRAN=1
   30 CONTINUE
      RETURN
      END
C
C
      SUBROUTINE GAUSS1
C*****************************************************************
C           SUBROUTINE GAUSS1 SOLVES A BANDED SYMMETRIC SET OF
C             LINEAR ALGEBRAIC EQUATIONS USING A BANDED GAUSS
```

```
C            ELIMINATION TECHNIQUE.
C
C          THE MATRIX IS STORED IN CONDENSED FORM USING ONLY
C            THE UPPER TRIANGULAR BANDWITH.  THUS ALL ROWS
C            OF THE SQUARE MATRIX ARE SHIFTED TO THE LEFT SO
C            THAT THE MAIN DIAGONAL TERMS ARE NOW IN COLUMN
C            ONE OF THE RECTANGULAR MATRIX ACOND.  ACOND HAS
C            DIMENSIONS NP BY NBW.
C************************************************************
      COMMON /INT10/ NP,NE,NBW,NOUT
      COMMON /REAL10/ ACOND(200,40),F(200),T(200)
      DATA IN,IO/60,61/
C++++++++++++++
C     1.  CONSTRUCTION OF UPPER BANDED TRIANGULAR FORM
C++++++++++++++
      NPM=NP*2-1
      DO 100 I=1,NPM
        IP=I+1
        DO 110 J=2,NBW
          IF (ACOND(I,J) .EQ. 0.0) GO TO 110
          C=ACOND(I,J)/ACOND(I,1)
          L=I+J-1
          M=0
          DO 120 K=J,NBW
            M=M+1
            ACOND(L,M)=ACOND(L,M)-C*ACOND(I,K)
 120      CONTINUE
          F(L)=F(L)-C*F(I)
 110    CONTINUE
 100  CONTINUE
C++++++++++++++
C     2.  BACKWARD SUBSTITUTION TO OBTAIN UNKNOWNS
C++++++++++++++
      NP2=NP*2
      DO 200 I=1,NP2
        L=NP2-I+1
        T(L)=F(L)
        K=L
        DO 210 J=2,NBW
          K=K+1
          IF (K .GT. L+NBW-1) GO TO 210
          IF (K .GT. NP2) GO TO 210
          T(L)=T(L)-ACOND(L,J)*T(K)
 210    CONTINUE
        T(L)=T(L)/ACOND(L,1)
 200  CONTINUE
      RETURN
      END
C
C
      SUBROUTINE WRITEM
C************************************************************
C          SUBROUTINE WRITEM WRITES OUT THE ACOND MATRIX AND
C            F COLUMN.
C************************************************************
      COMMON /ALP10/ TITLE1(20),TITLE2(20)
      COMMON /INT10/ NP,NE,NBW,NOUT
      COMMON /REAL10/ ACOND(200,40),F(200),T(200)
      DATA IN,IO/60,61/
      CALL HEAD
```

```
      WRITE (IO,501)
  501 FORMAT (////10X,28H3.  GLOBAL MATRIX AND COLUMN)
      WRITE (IO,533)
  533 FORMAT (////10X,35H3.1.  CONDENSED GLOBAL COEFFICIENT
     1     ,14H(ACOND) MATRIX)
      WRITE (IO,535)
  535 FORMAT ( /10X,3HDOF,20X, 5HACOND)
      IF(NBW.GT.20) GOTO 540
      NP2=NP*2
      NPN=NBW
      IF(NBW.GT.10) NPN=10
      DO 510 I=1,NP2
      WRITE (IO,511) I,(ACOND(I,J),J=1,NPN)
      IF(NBW.GT.10) WRITE(IO,511) I,(ACOND(I,J),J=11,NBW)
  511 FORMAT (7X,I5,3X,10G12.4)
  510 CONTINUE
      WRITE (IO,534)
  534 FORMAT (////10X,21H3.2.  GLOBAL F COLUMN)
      WRITE (IO,537)
  537 FORMAT ( /10X,3HDOF,5X,4HNODE,4X,9HDIRECTION,5X,1HF)
      DO 520 I=1,NP
        NP2=I*2
        NP1=NP2-1
        WRITE (IO,521) NP1,I,F(NP1)
  521   FORMAT ( 10X,I5,5X,I5,5X,1HX,5X,G12.5)
        WRITE(IO,522) NP2,I,F(NP2)
  522   FORMAT( 10X,I5,5X,I5,5X,1HY,5X,G12.5)
  520 CONTINUE
      GOTO 550
  540 WRITE(IO,542)
  542 FORMAT (//10X,19HBANDWIDTH TOO LARGE/10X,
     1           33HGLOBAL MATRIX WILL NOT BE PRINTED)
  550 CONTINUE
      RETURN
      END
```

```
  SAMPLE INPUT FOR PROGRAM STRESS
  EXAMPLE 4.8 - BRIDGE TRUSS

PLANE TRUSS ANALYSIS
  NAN=    3

1.  INPUT AND OUTPUT OF PROBLEM DATA

1.1.  GLOBAL PARAMETERS

NUMBER OF NODES    6
NUMBER OF ELEMENTS    9

1.2.  NODAL COORDINATES

 NODE          X            Y
    1      0.           0.
    2        96.000     0.
    3       192.00      0.
    4       48.000       36.000
    5       144.00       36.000
    6        96.000      72.000

1.3.  ELEMENT/NODE CONNECTIVITY DATA

ELEM NO.    ELEM TYPE      NODES
    1          21          1    2
    2          21          2    3
    3          21          3    5
    4          21          5    6
    5          21          6    4
    6          21          4    1
    7          21          4    2
    8          21          5    2
    9          21          6    2

1.4.  CALCULATED BANDWIDTH=   10

1.5.  ELEMENT PROPERTIES

ELEMENT             YM           MU           T
    1        .10000E+08 0.                4.0000
    2        .10000E+08 0.                4.0000
    3        .10000E+08 0.                4.0000
    4        .10000E+08 0.                4.0000
    5        .10000E+08 0.                4.0000
    6        .10000E+08 0.                4.0000
    7        .10000E+08 0.                2.0000
    8        .10000E+08 0.                2.0000
    9        .10000E+08 0.                2.0000

1.6.  APPLIED NODAL FORCE

NUMBER OF APPLIED NODAL FORCES=    2
NODE    DIRECTION     FORCE
    5         2       -10000.
    6         2       -15000.
```

```
1.7.  PRESCRIBED NODAL DISPLACEMENTS

NUMBER OF APPLIED NODAL DISPLACEMENTS=    3
NODE    DIRECTION         DISPLACEMENT
    1          1      0.
    1          2      0.
    3          2      0.

1.8.  OUTPUT PARAMETERS

NOUT=    0

  SAMPLE INPUT FOR PROGRAM STRESS
  EXAMPLE 4.8 - BRIDGE TRUSS

2.  CALCULATION OF ELEMENT PROPERTIES

ELEMENT PROPERTIES FOR ELEMENT    1

2.1.  ELEMENT CONNECTIVITY AND NODAL COORDINATES
ELEMENT     NODE       X ??        Y
    1          1     0.0000     0.0000
    1          2    96.0000     0.0000

LENGTH=     96.0000

2.2.  ELEMENT MATERIAL PROPERTIES
                YM          AREA
            .10000E+08  4.0000

STIFFNESS=       .41667E+06

2.3.  ELEMENT COEFFICIENT MATRIX
NODE      DIRECTION              COEFFICIENT MATRIX
    1          1       .41667E+06 0.          -.41667E+06 0.
    1          2      0.          0.          0.          0.
    2          1      -.41667E+06 0.           .41667E+06 0.
    2          2      0.          0.          0.          0.

ELEMENT PROPERTIES FOR ELEMENT    2

2.1.  ELEMENT CONNECTIVITY AND NODAL COORDINATES
ELEMENT     NODE        X          Y
    2          2    96.0000     0.0000
    2          3   192.0000     0.0000

LENGTH=     96.0000
2.2.  ELEMENT MATERIAL PROPERTIES
                YM          AREA
            .10000E+08  4.0000

STIFFNESS=       .41667E+06
```

```
2.3.  ELEMENT COEFFICIENT MATRIX
NODE      DIRECTION              COEFFICIENT MATRIX
   2         1       .41667E+06  0.          -.41667E+06  0.
   2         2      0.           0.          0.           0.
   3         1      -.41667E+06  0.           .41667E+06  0.
   3         2      0.           0.          0.           0.

ELEMENT PROPERTIES FOR ELEMENT    3

2.1.  ELEMENT CONNECTIVITY AND NODAL COORDINATES
ELEMENT      NODE      X          Y
   3           3  192.0000    0.0000
   3           5  144.0000   36.0000

LENGTH=     60.0000

2.2.  ELEMENT MATERIAL PROPERTIES
                 YM         AREA
          .10000E+08  4.0000

STIFFNESS=       .66667E+06

2.3.  ELEMENT COEFFICIENT MATRIX
NODE      DIRECTION              COEFFICIENT MATRIX
   3         1       .42667E+06 -.32000E+06 -.42667E+06  .32000E+06
   3         2      -.32000E+06  .24000E+06  .32000E+06 -.24000E+06
   5         1      -.42667E+06  .32000E+06  .42667E+06 -.32000E+06
   5         2       .32000E+06 -.24000E+06 -.32000E+06  .24000E+06

ELEMENT PROPERTIES FOR ELEMENT    4

2.1.  ELEMENT CONNECTIVITY AND NODAL COORDINATES
ELEMENT      NODE      X          Y
   4           5  144.0000   36.0000
   4           6   96.0000   72.0000

LENGTH=     60.0000

2.2.  ELEMENT MATERIAL PROPERTIES
                 YM         AREA
          .10000E+08  4.0000

STIFFNESS=       .66667E+06

2.3.  ELEMENT COEFFICIENT MATRIX
NODE      DIRECTION              COEFFICIENT MATRIX
   5         1       .42667E+06 -.32000E+06 -.42667E+06  .32000E+06
   5         2      -.32000E+06  .24000E+06  .32000E+06 -.24000E+06
   6         1      -.42667E+06  .32000E+06  .42667E+06 -.32000E+06
   6         2       .32000E+06 -.24000E+06 -.32000E+06  .24000E+06

ELEMENT PROPERTIES FOR ELEMENT    5

2.1.  ELEMENT CONNECTIVITY AND NODAL COORDINATES
ELEMENT      NODE      X          Y
   5           6   96.0000   72.0000
   5           4   48.0000   36.0000
```

```
LENGTH=     60.0000

2.2.  ELEMENT MATERIAL PROPERTIES
               YM         AREA
          .10000E+08  4.0000

STIFFNESS=       .66667E+06

2.3.  ELEMENT COEFFICIENT MATRIX
NODE      DIRECTION               COEFFICIENT MATRIX
    6         1        .42667E+06  .32000E+06 -.42667E+06 -.32000E+06
    6         2        .32000E+06  .24000E+06 -.32000E+06 -.24000E+06
    4         1       -.42667E+06 -.32000E+06  .42667E+06  .32000E+06
    4         2       -.32000E+06 -.24000E+06  .32000E+06  .24000E+06

ELEMENT PROPERTIES FOR ELEMENT    6

2.1.  ELEMENT CONNECTIVITY AND NODAL COORDINATES
ELEMENT     NODE      X          Y
    6         4   48.0000    36.0000
    6         1    0.0000     0.0000

LENGTH=     60.0000

2.2.  ELEMENT MATERIAL PROPERTIES
               YM         AREA
          .10000E+08  4.0000

STIFFNESS=       .66667E+06

2.3.  ELEMENT COEFFICIENT MATRIX
NODE      DIRECTION               COEFFICIENT MATRIX
    4         1        .42667E+06  .32000E+06 -.42667E+06 -.32000E+06
    4         2        .32000E+06  .24000E+06 -.32000E+06 -.24000E+06
    1         1       -.42667E+06 -.32000E+06  .42667E+06  .32000E+06
    1         2       -.32000E+06 -.24000E+06  .32000E+06  .24000E+06

ELEMENT PROPERTIES FOR ELEMENT    7

2.1.  ELEMENT CONNECTIVITY AND NODAL COORDINATES
ELEMENT     NODE      X          Y
    7         4   48.0000    36.0000
    7         2   96.0000     0.0000

LENGTH=     60.0000

2.2.  ELEMENT MATERIAL PROPERTIES
               YM         AREA
          .10000E+08  2.0000
STIFFNESS=       .33333E+06

2.3.  ELEMENT COEFFICIENT MATRIX
NODE      DIRECTION               COEFFICIENT MATRIX
    4         1        .21333E+06 -.16000E+06 -.21333E+06  .16000E+06
    4         2       -.16000E+06  .12000E+06  .16000E+06 -.12000E+06
    2         1       -.21333E+06  .16000E+06  .21333E+06 -.16000E+06
    2         2        .16000E+06 -.12000E+06 -.16000E+06  .12000E+06
```

```
ELEMENT PROPERTIES FOR ELEMENT    8

2.1.  ELEMENT CONNECTIVITY AND NODAL COORDINATES
ELEMENT      NODE       X          Y
    8          5  144.0000   36.0000
    8          2   96.0000    0.0000

LENGTH=     60.0000

2.2.  ELEMENT MATERIAL PROPERTIES
                    YM          AREA
            .10000E+08  2.0000

STIFFNESS=       .33333E+06

2.3.  ELEMENT COEFFICIENT MATRIX
NODE       DIRECTION            COEFFICIENT MATRIX
    5          1        .21333E+06   .16000E+06  -.21333E+06  -.16000E+06
    5          2        .16000E+06   .12000E+06  -.16000E+06  -.12000E+06
    2          1       -.21333E+06  -.16000E+06   .21333E+06   .16000E+06
    2          2       -.16000E+06  -.12000E+06   .16000E+06   .12000E+06

ELEMENT PROPERTIES FOR ELEMENT    9

2.1.  ELEMENT CONNECTIVITY AND NODAL COORDINATES
ELEMENT      NODE       X          Y
    9          6   96.0000   72.0000
    9          2   96.0000    0.0000

LENGTH=     72.0000

2.2.  ELEMENT MATERIAL PROPERTIES
                    YM          AREA
            .10000E+08  2.0000

STIFFNESS=       .27778E+06

2.3.  ELEMENT COEFFICIENT MATRIX
NODE       DIRECTION            COEFFICIENT MATRIX
    6          1        .81713E-24   .47642E-09  -.81713E-24  -.47642E-09
    6          2        .47642E-09   .27778E+06  -.47642E-09  -.27778E+06
    2          1       -.81713E-24  -.47642E-09   .81713E-24   .47642E-09
    2          2       -.47642E-09  -.27778E+06   .47642E-09   .27778E+06

  SAMPLE INPUT FOR PROGRAM STRESS
  EXAMPLE 4.8 - BRIDGE TRUSS

3.  GLOBAL MATRIX AND COLUMN
```

3.1. CONDENSED GLOBAL COEFFICIENT (ACOND) MATRIX

DOF	ACOND						
1	.8433E+06	.3200E+06	-.4167E+06	0.	0.	0.	-.4267E+06
2	.2400E+06	0.	0.	0.	0.	-.3200E+06	-.2400E+06
3	.1260E+07	.4764E-09	-.4167E+06	0.	-.2133E+06	.1600E+06	-.2133E+06
4	.5178E+06	0.	0.	.1600E+06	-.1200E+06	-.1600E+06	-.1200E+06
5	.8433E+06	-.3200E+06	0.	0.	-.4267E+06	.3200E+06	0.
6	.2400E+06	0.	0.	.3200E+06	-.2400E+06	0.	0.
7	.1067E+07	.4800E+06	0.	0.	-.4267E+06	-.3200E+06	0.
8	.6000E+06	0.	0.	-.3200E+06	-.2400E+06	0.	0.
9	.1067E+07	-.4800E+06	-.4267E+06	.3200E+06	0.	0.	0.
10	.6000E+06	.3200E+06	-.2400E+06	0.	0.	0.	0.
11	.8533E+06	.4764E-09	0.	0.	0.	0.	0.
12	.7578E+06	0.	0.	0.	0.	0.	0.

3.2. GLOBAL F COLUMN

DOF	NODE	DIRECTION	F
1	1	X	0.
2	1	Y	0.
3	2	X	0.
4	2	Y	0.
5	3	X	0.
6	3	Y	0.
7	4	X	0.
8	4	Y	0.
9	5	X	0.
10	5	Y	-10000.
11	6	X	0.
12	6	Y	-15000.

(GLOBAL MATRICES JUST AFTER ASSEMBLY)

SAMPLE INPUT FOR PROGRAM STRESS
EXAMPLE 4.8 - BRIDGE TRUSS

3. GLOBAL MATRIX AND COLUMN

3.1. CONDENSED GLOBAL COEFFICIENT (ACOND) MATRIX

DOF	ACOND						
1	.8433E+06	0.	0.	0.	0.	0.	0.
2	.2400E+06	0.	0.	0.	0.	0.	0.
3	.1260E+07	.4764E-09	-.4167E+06	0.	-.2133E+06	.1600E+06	-.2133E+06
4	.5178E+06	0.	0.	.1600E+06	-.1200E+06	-.1600E+06	-.1200E+06
5	.8433E+06	0.	0.	0.	-.4267E+06	.3200E+06	0.
6	.2400E+06	0.	0.	0.	0.	0.	0.
7	.1067E+07	.4800E+06	0.	0.	-.4267E+06	-.3200E+06	0.
8	.6000E+06	0.	0.	-.3200E+06	-.2400E+06	0.	0.
9	.1067E+07	-.4800E+06	-.4267E+06	.3200E+06	0.	0.	0.
10	.6000E+06	.3200E+06	-.2400E+06	0.	0.	0.	0.
11	.8533E+06	.4764E-09	0.	0.	0.	0.	0.
12	.7578E+06	0.	0.	0.	0.	0.	0.

```
-.3200E+06  0.          0.
0.          0.          0.
-.1600E+06  -.8171E-24  -.4764E-09
-.4764E-09  -.2778E+06  0.
0.          0.          0.
0.          0.          0.
0.          0.          0.
0.          0.          0.
0.          0.          0.
0.          0.          0.
0.          0.          0.
0.          0.          0.
```

```
0.          0.          0.
0.          0.          0.
-.1600E+06  -.8171E-24  -.4764E-09
-.4764E-09  -.2778E+06  0.
0.          0.          0.
0.          0.          0.
0.          0.          0.
0.          0.          0.
0.          0.          0.
0.          0.          0.
0.          0.          0.
0.          0.          0.
```

3.2. GLOBAL F COLUMN

DOF	NODE	DIRECTION	F
1	1	X	0.
2	1	Y	0.
3	2	X	0.
4	2	Y	0.
5	3	X	0.
6	3	Y	0.
7	4	X	0.
8	4	Y	0.
9	5	X	0.
10	5	Y	-10000.
11	6	X	0.
12	6	Y	-15000.

(GLOBAL MATRICES AFTER INCLUSION OF PRESCRIBED NODAL VALUES AND BOUNDARY CONSTRAINTS)

SAMPLE INPUT FOR PROGRAM STRESS
EXAMPLE 4.8 - BRIDGE TRUSS

3. GLOBAL MATRIX AND COLUMN

3.1. CONDENSED GLOBAL COEFFICIENT (ACOND) MATRIX

DOF	ACOND						
1	.8433E+06	0.	0.	0.	0.	0.	0.
2	.2400E+06	0.	0.	0.	0.	0.	0.
3	.1260E+07	.4764E-09	-.4167E+06	0.	-.2133E+06	.1600E+06	-.2133E+06
4	.5178E+06	.1575E-09	0.	.1600E+06	-.1200E+06	-.1600E+06	-.1200E+06
5	.7055E+06	0.	-.7055E+05	.5291E+05	-.4972E+06	.2671E+06	-.1252E-24
6	.2400E+06	0.	0.	0.	0.	0.	0.
7	.9741E+06	.5495E+06	-.3639E+05	.3670E+05	-.4267E+06	-.2342E+06	0.
8	.2380E+06	.4782E+05	-.4822E+05	-.7932E+05	-.1723E+06	0.	0.
9	.6197E+06	-.3449E+06	-.4267E+06	.2600E+06	0.	0.	0.
10	.2477E+06	.8256E+05	-.1858E+06	0.	0.	0.	0.
11	.3187E+06	.8095E+05	0.	0.	0.	0.	0.
12	.1587E+06	0.	0.	0.	0.	0.	0.

3.2. GLOBAL F COLUMN

DOF	NODE	DIRECTION	F
1	1	X	0.
2	1	Y	0.
3	2	X	0.
4	2	Y	0.
5	3	X	0.
6	3	Y	0.
7	4	X	0.
8	4	Y	0.
9	5	X	0.
10	5	Y	-10000.
11	6	X	3333.3
12	6	Y	-233.47.

(GLOBAL MATRICES AFTER FORMATION OF UPPER TRIANGULAR FORM)

0.	0.	0.
0.	0.	0.
-.1600E+06	-.8171E-24	-.4764E-09
-.4764E-09	-.2778E+06	0.
-.7303E-10	0.	0.
0.	0.	0.
0.	0.	0.
0.	0.	0.
0.	0.	0.
0.	0.	0.
0.	0.	0.
0.	0.	0.

D

E

F

G

U

V

W

Y